# Nucleic Acids in Chemistry and Biology

3rd Edition

# Nucleic Acids in Chemistry and Biology

## 3rd Edition

**Edited by**

G. Michael Blackburn
*Centre for Chemical Biology, Department of Chemistry, University of Sheffield, Sheffield, UK*

Michael J. Gait
*Medical Research Council, Laboratory of Molecular Biology, Cambridge, UK*

David Loakes
*Medical Research Council, Laboratory of Molecular Biology, Cambridge, UK*

David M. Williams
*Centre for Chemical Biology, Department of Chemistry, University of Sheffield, Sheffield, UK*

RSC Publishing

ISBN-10: 0-85404-654-2
ISBN-13: 978-0-85404-654-6

A catalogue record for this book is available from the British Library



Published by The Royal Society of Chemistry,
Thomas Graham House, Science Park, Milton Road,
Cambridge CB4 0WF, UK

Registered Charity Number 207890

For further information see our web site at www.rsc.org

Typeset by Macmillan India Ltd, Bangalore, India
Printed by Henry Ling Ltd, Dorchester, Dorset, UK

# Foreword

It was just 62 years ago that we finally learned that DNA was the genetic material – the master blueprint of life. Since then, the nucleic acids DNA and RNA have been studied in exquisite detail and both their chemical and biochemical properties are firmly established. Indeed, the double helical structure of DNA has become an icon of our time appearing widely not only in the scientific literature, but also in the popular press and most recently as jewelry. A thorough knowledge of nucleic acids and their properties is now a key ingredient in the education of both biologists and chemists. Ten years ago the second edition of "Blackburn & Gait" was published and seemed sufficiently comprehensive that only small additions would be needed if it were ever to be rewritten. Its popularity is attested to by its now being out of print – it has also inevitably become out of date. Much has changed in the last 10 years and a new edition is now both necessary and most welcome.

One major discovery within the biological arena has been the phenomenon of RNA interference, which was not even mentioned in the last edition, and yet at this time several companies have been formed to capitalize on it and at least one product is heading into clinical trials. We also now know that short microRNAs play key roles in development and are probably of ubiquitous importance in controlling gene expression. These and other small RNAs are likely to play a much more critical and subtle role in the lives of cells than we might ever have imagined. I find this personally very satisfying, since, when we discovered split genes and RNA splicing in 1977, the introns were almost immediately labeled "junk". It now seems that at least some of these intronic sequences play positive roles in controlling gene expression and their involvement in other processes may still await discovery. Studies of small RNAs in eukaryotes are proceeding quickly and I eagerly await the results from similar studies in bacteria and archaea. It seems likely that great discoveries lie ahead although new methods may be required to make them. The development of such methods will be greatly facilitated by a thorough knowledge of the chemistry and biology of nucleic acids – the subject of this book.

Among the great technical achievements of the last 10 years have been several breakthroughs in the scale of DNA sequencing. First came the complete sequence of a simple bacterium, *Haemophilus influenzae*, quickly followed by that of the first archaea, *Methanocaldococcus jannaschii.* A key feature of these projects was the use whole-genome shotgun sequencing pioneered by Craig Venter. These "small" genomes were soon followed by draft sequences for a number of eukaryotic genomes including, of course, the draft human genome sequence announced in 2003 and coinciding with the 50th anniversary of the determination of the structure of DNA by Jim Watson and Francis Crick. With more recent advances in sequencing technologies that use highly parallel methodology, one machine can now generate enough data for a small bacterial genome in a few hours, at a quite reasonable price. We can anticipate an even more massive influx of new data in the next few years. The accumulation of sequence data far exceeds our experimental capacity to probe it. Fortunately, bioinformatics stands ready to help and with appropriate experimental input, should allow us to make sense of the terabases ($10^{12}$) of DNA sequence data that will soon be present in GenBank. In parallel with these improvements in DNA sequence determination, techniques for DNA synthesis have progressed rapidly. It has now become so simple and inexpensive that many laboratories find it more expedient to have the genes of interest synthesized rather than to clone them. Among other things, this allows the introduction of desirable codons tailored to the expression system to be used.

All of this new work serves to highlight the intertwining of chemistry and biology that has taken place over the last 50 years. Those wishing to understand this interrelationship and appreciate the excitement currently present in the field can do no better than browse the many excellent chapters in this third edition of *Nucleic Acids in Chemistry and Biology*.

*Richard J. Roberts*

# Preface

The first edition of *Nucleic Acids in Chemistry and Biology* in 1990 met the pressing need for a single volume that integrated the chemistry and biology of nucleic acids in an introductory yet authoritative text. That book was so very well received that in 1996 we produced the second edition, which was completely revised and rewritten by very much the same team of international experts.

Ten years on we have responded to the still growing need for this book with a fully revised and updated third edition. Two irresistible pressures have driven this activity. First, the expansion in the chemistry and biology of nucleic acids continues unabated. The human and numerous lesser genomes have been fully sequenced since we presented the second edition and there has been a veritable explosion in the chemistry and biology of RNA. Many exciting crystal and NMR structures of nucleic acids and their protein complexes, including the ribosome, have been published. Changes of such magnitude have inevitably made significant parts of the 1996 text out of date. We have addressed these issues by expansion of the appropriate sections of the book and also by new authorship. Second, the second edition sold out several years ago. Indeed second-hand copies are occasionally available on the web at a handsome premium!

In planning this third edition, we first expanded the team of editors to include two younger colleagues, David Loakes and David Williams. We then changed publishing house to move under the roof of the Royal Society of Chemistry. For a variety of reasons it has been necessary to make changes to the team of principal authors and we thank especially Stephanie Allen, Martin Egli, Julie Fisher, Andy Flavell, Ihtshamul Haq, Charles Laughton, Ben Luisi, Anna Marie Pyle, Elliott Stollar, and Nick Williams for their essential and scholarly contributions. With the active support of the Royal Society of Chemistry and its commissioning and production teams, we have made significant changes in the style of presentation of this new edition. It now has a bibliography of primary and secondary sources that are referenced throughout the text. While we have maintained a number of multi-colour illustrations in addition to our standard two-colour format, we have abandoned the use of stereo-pair illustrations and the end-of-section summaries. These changes have created space for some expansion – but not enough for our needs: the third edition has grown substantially compared to its predecessor! This has enabled the authors to introduce a great deal of new material. In doing so, we have retained the essential core of chemistry and biology that has made this book so effective as a teaching resource at every level of study and an initiation into the molecular basics of nucleic acids. A selection of figures that may have value for course teachers are available electronically at the following website: http://www.rsc.org/books/nucleicacids

Above all, we have endeavoured to maintain the quality of the earlier editions, both of which have been widely appreciated for their easy readability, simplicity of exposition, clarity of illustration, and uniformity of style. That has underpinned our efforts to deliver a new edition that once again fulfils the needs of students and new research workers, primarily those having a chemical and biochemical background who seek to understand this great subject at a molecular level. Indeed, we know that *Nucleic Acids in Chemistry and*

*Biology* has become the course-book of choice in universities across three continents. At the same time, from many favourable comments on editions 1 and 2 we know that this book has also reached out to more senior scientists across many disciplines.

*G Michael Blackburn*
*Michael J Gait*
*David Loakes*
*David M Williams*

# Acknowledgements

Two Mikes and two Davids are extremely grateful for the efforts of all who have supported the production of this book. Our unqualified thanks are given above all to our 10 expert and understanding co-authors, whose contributions have made possible this third edition. We express our sincere appreciation of their patience, tolerance, and enthusiastic diligence during the numerous revision processes required for the production of the completed work.

We are also very grateful to very many colleagues and fellow scientists who have provided us with valuable comments on the first two editions of the text and especially those who have read portions of the new edition. They include Jason Betley, Chris Calladine, Rick Cosstick, Steve Fodor, Dan Gewirth, Alec Jeffreys, David Lilley, Kiyoshi Nagai, Barbara Nawrot, Frank Seela, Jean Thomas, Andrew Travers, and David Wilson. We are once again indebted to Joachim Engels for updating and expanding the glossary and to Rich Roberts for writing the forward for this edition.

The technical production of this book has been enabled by many skilled individuals. We particularly appreciate the efforts of all the staff involved in the Royal Society of Chemistry for their enthusiasm, patience, and highly professional production of the completed work. We and our co-authors gratefully acknowledge the efforts of Fred Anston, John Brazier, Pat Mellor, Sabuj Pattanayek, and Wenke Zhang who have facilitated the completion of figures and text in various chapters. In particular we are greatly indebted to Annette Lenton who has redrawn, recoloured, or reworked very many of the figures in order to achieve a homogeneous standard and style. We thank Venki Ramakrishnan for providing original graphics for the cover of the third edition of the book and Richard Dickerson, Stephen Lippard, and Dinshaw Patel for illustrative figures.

Last but not least, we have enjoyed receiving a large number of positive and helpful comments from readers of the first two editions. We have endeavoured to incorporate all constructive criticisms into the new edition. In particular we thank colleagues around the world whose strong support provided much motivation for the creation of this third edition.

Despite all our careful work, it is inevitable that there will be some errors and we accept full responsibility for them. Finally, we look forward to your advice, comments, and suggestions for future revisions.

# Contents

# Contributors

**Stephanie Allen**, *School of Pharmacy, University of Nottingham, University Park, Nottingham NG7 2RD, UK.*

**G. Michael Blackburn**, *Department of Chemistry, University of Sheffield, Brook Hill, Sheffield S3 7HF, UK.*

**Martin Egli**, *Department of Biochemistry, Vanderbilt University, School of Medicine, Nashville, TN 37232, USA.*

**Julie Fisher**, *School of Chemistry, University of Leeds, Woodhouse Lane, Leeds LS2 9JT, UK.*

**Andrew J. Flavell**, *Plant Research Unit, University of Dundee at SCRI, Invergowrie, Dundee DD2 5DA, UK.*

**Michael J. Gait**, *MRC Laboratory of Molecular Biology, Hills Road, Cambridge CB2 2QH, UK.*

**Ihtshamul Haq**, *Department of Chemistry, University of Sheffield, Brook Hill, Sheffield S3 7HF, UK.*

**Charles Laughton**, *School of Pharmacy, University of Nottingham, University Park, Nottingham NG7 2RD, UK.*

**David Loakes**, *MRC Laboratory of Molecular Biology, Hills Road, Cambridge CB2 2QH, UK.*

**Ben Luisi**, *Department of Biochemistry, University of Cambridge, 80 Tennis Court Road, Cambridge CB2 1GA, UK.*

**Anna Marie Pyle**, *Department of Molecular Biophysics and Biochemistry, Yale University, 266 Whitney Avenue, P.O. Box 208114, New Haven, CT 06520-8114, USA.*

**Elliott Stollar**, *The Hospital for Sick Children Research Institute, 555 University Avenue, Toronto, Ont., Canada M5G 1X8.*

**David M. Williams**, *Department of Chemistry, University of Sheffield, Brook Hill, Sheffield S3 7HF, UK.*

**Nicholas H. Williams**, *Department of Chemistry, University of Sheffield, Brook Hill, Sheffield S3 7HF, UK.*

# Glossary

AGAROSE: A polysaccharide isolated from seaweed used as a matrix in gel electrophoresis.

ALLELE: One of two alternate forms of a gene occupying a given locus on the chromosome.

ALLOSTERIC CONTROL: The ability of an interaction at one site of a protein to influence (positively or negatively) the activity at another site.

ALU FAMILY: A set of short (*ca.* 300 bp) related sequences dispersed throughout the human genome. Refers to the property of these sequences to be cleaved once by the restriction enzyme *AluI*. Genomes of other mammals contain similar families. Their role is unknown.

AMPLIFICATION: The production of extra copies of a chromosomal sequence found either as intra- or extra-chromosomal DNA. With respect to plasmids it refers to the increase in the number of plasmid copies per cell induced by certain treatments of transformed cells.

ANNEAL (RE-ANNEAL): The (re)establishment of base pairing between complementary strands of DNA or a DNA and an RNA strand.

ANOMERIZATION: The interconversion of stereoisomers of a sugar that differ only in the stereochemistry at the carbonyl carbon in their cyclic (furanose or pyranose) form. For D-ribofuranose and D-2-deoxyribofuranose this relates to the $\alpha$- and $\beta$-forms at C-1.

ANTIBODY: A protein that is produced in response to and specifically recognizes and binds to an antigen.

ANTICODON: A triplet of nucleotides in a constant position in the structure of tRNA that is complementary to the triplet codon(s) in mRNA to which the tRNA responds.

ANTIGEN: Any molecule which, upon entry into the organism, causes the production of antibodies (immunoglobulins).

ANTISENSE: A strand of DNA or RNA that has the sequence complementary to mRNA (also non-coding strand).

APOPTOSIS: The programmed death of a cell within a multi-cellular organism, which follows an ordered process.

APTAMER: DNA or RNA molecules that have been selected from random pools based on their ability to bind other molecules.

ARRAY: A spatial arrangement of *e.g.* oligonucleotides or peptides, which can be at high density ($\geq$10,000 individual sequences).

AUTORADIOGRAPHY: The detection of radioactively labelled molecules present for example in a gel or on a filter by exposing an X-ray film to it.

AUXOTROPHY: The inability of microorganisms to live on minimal medium without supplemented (auxiliary) nutrients.

BACK MUTATION: Reverses the effect of a mutation that had inactivated a gene.

BACTERIOPHAGE: A virus that infects bacteria; often abbreviated as phage.

BASE PAIR (BP): A duplex of A with T or of C with G in a DNA or RNA double helix; other pairs are possible in RNA under some circumstances.

BLOTTING: Transfer of DNA, RNA, or protein from a gel to nitrocellulose or other "paper".

CAP: The structure at the 5′-end of eukaryotic mRNA introduced after transcription by linking the 5′-end of a guanine nucleotide to the terminal base of the mRNA and methylating at least the additional G; the structure is $7Me_G{}^{5'}ppp^{5'}Np$.

CATENANE: A molecule in which two or more closed rings are interlocked thus holding the structure together without any covalent bond between the separate rings. A DNA catenane is a topoisomer of its components, *i.e.* it is a distinct topological structure that can be acted on by topoisomerase.

cDNA: A single-stranded DNA complementary to the RNA synthesized from it by *in vitro* reverse transcription.

CENTROMERE: The most condensed and constricted region of a chromosome; point of attachment of the spindle fiber during mitosis.

CHAIN TERMINATION SEQUENCING: See Sanger–Coulson sequencing.

CHROMATIN: Basic organizational unit of eukaryotic chromosomes; consists of DNA and associated proteins assembled into fibers of average diameter 30 nm that are produced by the compaction of 10-nm nucleosome fibers.

CHROMOSOME: A discrete unit of the genome carrying many genes, consisting of a very long molecule of DNA, complexed with a large number of different proteins (mostly histones). Chromosomes are visible as a morphological entity only during the act of cell division.

*cis*-ACTING: The ability of a DNA (or RNA) sequence to effect its influence only on the molecule from which it forms a part. Usually implies that the sequence does not code for a protein. When applied to a protein it means that the protein acts only on the DNA (or RNA) molecule from which it was expressed.

CISTRON: The genetic unit defined by the *cis/trans* test; equivalent to gene in comprising a unit of DNA representing a protein.

CLONE: A large number of cells or molecules genetically identical with a single ancestral cell or molecule.

CODON: A triplet of nucleotides that corresponds to an amino acid or a termination signal.

COMPETENT: A culture of bacteria or yeast cells treated in such a way that their ability to take up DNA molecules without transduction or conjugation has been enhanced.

COMPLEMENTARY SEQUENCE: Nucleic acid sequence of bases that can form a double-stranded structure by virtue of Watson–Crick base pairing e.g. A-T, C-G.

COMPLEMENTATION: The ability of independent (non-allelic) genes to provide diffusible products that produce wild phenotype when two mutants are tested in *trans*-configuration in a heterozygote.

CONJUGATION: Directional transfer of DNA between two bacteria.

CONSENSUS SEQUENCE: An idealized sequence in which each position represents the base most often found when many actual sequences are compared.

COPY NUMBER: The average number of copies of a particular (recombinant) plasmid present in a single host cell. Also used for individual genes.

COSMIDS: Plasmids into which phage lambda cos sites have been inserted; as a result, the plasmid DNA can be packaged *in vitro* into the phage coat.

CO-TRANSFORMATION: Introduction of two or more genes carried on separate DNA molecules into a cell.

CROSS-LINKING: Introduction of covalent intra- or intermolecular bonds between groups that are normally not covalently linked. Used to detect proximity of (parts of) (macro) molecules.

CUT: A double-strand scission in the duplex polynucleotide in distinction to the single-strand "nick".

DELETION: The removal of a sequence of DNA, the regions on either side being joined together.

DENATURATION (OF PROTEIN): Conversion from the native conformation into some other (inactive) conformation.

DIFFERENTIAL LYSIS: A method to enrich for sperm DNA in a mixture of sperm and epithelial cells by preferentially lysing the latter using detergent and protease, so that sperm nuclei can be recovered by centrifugation.

DIRECT REPEATS: Identical (or closely related) sequences present in two or more copies in the same orientation on the same DNA (or RNA) molecule; they are not necessarily adjacent.

DNA FINGERPRINTING: Generation of a pattern of bands, by Southern blotting and hybridization with a multi-locus probe, which is highly individual-specific.

DNAZYME: A short catalytic single-stranded DNA molecule.

DOMAIN (OF A CHROMOSOME): Ether a discrete structural entity defined as a region within which supercoiling is independent of other domains, or an extensive region including an expressed gene that has heightened sensitivity to degradation by the enzyme DNase I.

DOMAIN (OF A PROTEIN): A discrete continuous part of the amino acid sequence that can be equated with a particular function or a particular substructure of the tertiary structure.

DOMINANT (ALLELE): Determines the phenotype displayed in a heterozygote with another (recessive) allele.

DOWNSTREAM: Sequences that proceed further in the direction of expression; for example, the coding region is downstream from the initation codon.

ELECTROPHEROGRAM: The graphical output of electrophoresis devices in STR (see short tandem repeat) and sequencing analysis, showing fluorescence intensity as a function of molecular weight. The peak at a particular wavelength (colour) corresponds to a specifically labelled molecule of a particular size.

END LABELLING: The addition of a radioactively labelled group to one end (5′ or 3′) of a DNA or RNA strand.

ENDONUCLEASE: An enzyme that cleaves bonds within a nucleic acid chain. It may be specific for RNA or for single-stranded or double-stranded DNA.

ENHANCER ELEMENT: A DNA sequence that increases the utilization of (some) eukaryotic promoters in *cis*-configuration, but can function in any location, upstream or downstream, relative to the promoter.

EPITOPE: Any part of a molecule that acts as an antigenic determinant. A macromolecule can have many different epitopes each stimulating the production of a different specific antibody.

EUKARYOTIC: Any organism that contains a nucleus.

EXCISION-REPAIR: A repair system that removes a single-stranded sequence of DNA containing damaged or mispaired bases and replaces it in the duplex by synthesis of a sequence complementary to the remaining strand.

EXON: Any segment of an interrupted gene that is represented in the mature RNA product.

EXONUCLEASE: An enzyme that cleaves nucleotides one at a time from the end of a polynucleotide chain. Such enzymes may be specific for either the 5′- or 3′-end of DNA or RNA.

EXPRESSION VECTOR: A cloning vector designed in such a way that a foreign gene inserted into the vector will be expressed in the host organism.

FINGERPRINT: The characteristic array of oligopeptides or oligonucleotides obtained upon two-dimensional electrophoresis of a protein digested with a specific endopeptidase or an RNA digested with a specific endonuclease.

FOOTPRINTING: A technique for identification of the site of DNA or RNA bound by some protein by virtue of the protection of bonds in this region against attack by nucleases or by chemicals.

FORENSIC GENETICS: The application of genetics for the resolution of disputes at law.

FUSION GENE: A recombinant gene constructed from parts of two different genes.

FUSION PROTEIN: The protein expressed by a fusion gene containing parts of the coding sequence of two different genes.

GAPMER: An antisense oligonucleotide where the central section is either unmodified or contains modifications, such as phosphorothioate, that permit recognition by RNase H, and where the 5′- and 3′-flanking regions contain other chemical modifications.

GEL ELECTROPHORESIS: Electrophoresis performed in a gel matrix (usually agarose or polyacrylamide) that allows separation of molecules of similar electric charge density on the basis of their difference in molecular weight.

GENE: A DNA sequence involved in the production of an RNA or protein molecule as the final product. Includes both the transcribed region and any sequences upstream and/or downstream responsible for its correct and regulated expression (*e.g.* promotor and operator sequences).

GENETIC CODE: The complete set of codons specifying the various amino acids, including the nonsense codons. The code is usually written in the form in which it occurs in mRNA. (It can be different in mitochondrial DNA.)

GENOME: The entire genetic material of a cell.

G-TETRAD: A structure that involves four oligonucleotide strands in which there is participation from one guanine base in each strand.

HAIRPIN: The double-stranded region formed by base pairing of adjacent complementary sequences in the same DNA or RNA strand.

HAPTEN: A small molecule that acts as an antigen when it is conjugated to a large (carrier) molecule.

HETERODUPLEX (HYBRID) DNA: DNA that is generated by base pairing between partly non-complementary single strands derived from the different parental duplex molecules. It occurs during genetic recombination.

HOLLIDAY JUNCTION: A structure that occurs during homologous recombination between two chromosomes; with the two chromosomes side-by-side, one strand of DNA on each chromosome is broken and then attached to the broken strand of DNA on the alternate chromosome. The crossover point is called the Holliday junction.

HOLOENZYME: The complete enzyme including all its subunits. Often used in reference to RNA and DNA polymerases.

HOMOLOGY: The degree of identity existing between the nucleotide sequences of two related but not complementary DNA or RNA molecules. 70% homology means that on average 70 out of every 100 nucleotides are identical. The same term is used in comparing the amino acid sequences of related proteins.

HYBRIDIZATION: The pairing of complementary RNA and DNA strands to give an RNA–DNA hybrid. It is also used to describe the pairing of two single-stranded DNA molecules.

HYBRIDOMA: The cell line produced by fusion of a myeloma cell with a lymphocyte. It continues indefinitely to express the immunoglobulins of both parents.

HYPERCHROMICITY: The increase of optical density that occurs when DNA is denatured.

i-MOTIF: A structure composed of two parallel-stranded duplexes held together in an antiparallel orientation. The structure is stabilised by hemiprotonated $C:C^+$ base pairs.

INCOMPATIBILITY: The inability of certain bacterial plasmids to coexist in the same cell.

INDUCER: A small molecule that triggers gene transcription by binding to a regulator protein.

INITATION CODON: AUG (sometimes GUG), three bases that code for the first amino acid in a protein sequence (*N*-formylmethionine in prokaryotes). This fMet is often removed post-translationally.

*IN SITU* HYBRIDIZATION: A technique in which the DNA of cells is denatured by squashing on a microscope slide so that reaction is possible with an added single-stranded RNA or DNA. The added preparation is radioactively labelled and its hybridization is followed by autoradiography.

INTASOME: A protein–DNA complex between the phage lambda integrase (Int) and the phage lambda attachment site (*attP*).

INTRON: A segment of DNA that is transcribed, but is removed from within the transcript by splicing together the sequences (exons) on either side of it. The occurrence of introns is almost exclusively limited to eukaryotic cells.

*IN VITRO*: (lit. "in glass"): Any experimental (biological) process that occurs outside the living cell.

*IN VIVO*: Any biological process that occurs within the living cell or organism.

IPTG: Isopropyl β-D-thiogalactoside; an artificial inducer of the *lac* operon (physiological inducer: allolactose).

kb: Abbreviation for 1000 base pairs of DNA or 1000 bases of RNA.

KINASE: An enzyme that catalyzes the transfer of a phosphate group from ATP or GTP to an acceptor, usually a protein or a nucleotide.

KLENOW FRAGMENT: An N-terminal truncation of DNA Polymerase I that retains polymerase activity, but has lost the $5'\rightarrow3'$ exonuclease activity.

*LAC* OPERON: An inducible operon in *Escherichia coli* that codes for three genes involved in the metabolism of lactose.

LEADER SEQUENCE: The sequence at the $5'$-end of an mRNA that is not translated into protein. It contains the coded information that the ribosome and special proteins read to tell it where to begin the synthesis of the polypeptide.

LIBRARY: A set of cloned fragments together representing the entire genome.

LIGASE: (DNA LIGASE): An enzyme that catalyzes the formation of a phosphodiester bond at the site of a single-strand break in duplex DNA. Some DNA ligases can also ligate blunt-end DNA molecules. RNA ligase covalently links separate RNA molecules.

LIGATION: The formation of a phosphate diester linkage between two adjacent nucleosides separated by a nick in one strand of a double helix of DNA. (The term can also be applied to blunt-end ligation and to joining of RNA.)

LINKER (FRAGMENT): A short synthetic duplex oligonucleotide containing the target site for some restriction enzyme. A linker may be added to the end of a DNA fragment prepared by cleavage with some other enzyme during reconstruction of recombinant DNA.

LTR: An abbreviation for long-terminal repeat, a sequence directly repeated at both ends of a retroviral DNA.

LYSIS: The death of bacteria at the end of a phage infective cycle when they burst open to release the progeny of an infecting phage.

M13: An *E. coli* phage containing single-stranded circular DNA that forms the basis for a series of cloning vectors.

MATCH PROBABILITY: The chance of two unrelated people sharing a DNA profile.

MAXAM–GILBERT SEQUENCING: A DNA sequencing technique based on specific chemical modification of each of the four bases.

MELTING TEMPERATURE ($T_m$): The temperature where hyperchromicity is half-maximal.

MINIMAL MEDIUM: A chemically fully defined medium containing only inorganic sources of the essential elements as well as an organic carbon source.

MINISATELLITES: Loci made up of a number (~10–1000) of tandemly repeated sequences, each typically 10–100 bp in length, which are usually GC-rich and often hypervariable.

MODIFIED BASES: All those except the usual five from which DNA and RNA (A, C, G, T, and U) are synthesized. They result from post-synthetic changes in the nucleic acid or chemical synthesis.

MONOCLONAL ANTIBODY: The unique immunoglobulin molecule (1° protein sequence) produced by a clone of cells derived from the fusion of a B lymphocyte with a myeloma cell. The antibody is directed against a single epitope of the antigen used to raise the antibody.

MULTICOPY PLASMIDS: Present in bacteria at amounts greater than one per chromosome.

MULTIPLE DISPLACEMENT AMPLIFICATION: A method for whole-genome amplification using a highly processive polymerase from bacteriophage $\phi$29 and random primers to synthesize long molecules from the template.

MUTAGENS: Molecules that increase the rate of mutation by causing changes in DNA.

MUTATION: Any change in the sequence of genomic DNA.

NICK TRANSLATION: The ability of *E. coli* DNA polymerase I to use a nick as a starting point from which one strand of a duplex DNA can be degraded and replaced by resynthesis of new material; is used to introduce radioactively labelled nucleotides into DNA *in vitro*.

NONSENSE CODON: Any one of three triplets (UAG, UAA, UGA) that cause termination of protein synthesis (UAG is known as *amber*, UAA as *ochre*, UGA as *opal*).

NORTHERN BLOTTING: A technique for transferring RNA from an agarose gel to a nitrocellulose filter on which it can be hybridized to a complementary DNA.

NUCLEOLUS: The region in the nucleus where rRNA synthesis takes place.

NUCLEOSOME: The fundamental repeating unit of a eukaryotic cell and which consists of DNA and histones.

OKAZAKI FRAGMENTS: Separate, contiguous DNA sequences of 1000–2000 bases produced during discontinuous replication; they are later joined together to give an intact strand.

OLIGOMER: Term often used in place of oligonucleotide.

OLIGONUCLEOTIDE: Polymer comprising of nucleotide units (usually less than 50) joined typically by $5'\rightarrow3'$ phosphate diester linkages. Those comprised of DNA and RNA can be distinguished where necessary by using 'oligodeoxyribonucleotide' and 'oligoribonucleotide' respectively.

ONCOGENE: A retroviral gene that causes transformation of the mammalian infected cell. Oncogenes are slightly changed equivalents of normal cellular genes called proto-oncogenes. The viral version is designated by the prefix v, the cellular version by the prefix c.

OPEN READING FRAME (ORF): A series of triplets coding for amino acids terminated by a termination codon; sequence is (potentially) translatable into protein.

OPERATOR: The site on DNA at which a repressor protein binds to prevent transcription from initiating at the adjacent promoter.

OPERON: A complete unit of bacterial gene expression and regulation, including structural genes, regulator gene(s), and control elements in DNA recognized by regulator gene product(s).

ORIGIN (ORI): A sequence of DNA at which replication is initiated.

PALINDROME: A sequence of double-stranded DNA that is the same when one strand is read left to right or its complement is read right to left; consists of adjacent inverted repeats.

PATERNITY TESTING: The determination of whether or not a particular man is the father of a child, using genetic analysis. This generally uses similar autosomal markers to individual identification work.

pBR322: One of the standard plasmid cloning vectors.

PCR: Polymerase chain reaction, an *in vitro* amplification of DNA based on primer, template, and a thermostable DNA polymerase.

PCR STUTTER: A PCR artefact in which, as well as a band of the expected size, an additional band is seen that is typically one repeat unit smaller, resulting from slippage synthesis errors by the PCR polymerase.

PHAGE (BACTERIOPHAGE): A bacterial virus.

PLASMID: An autonomous self-replicating extrachromosomal circular DNA.

PLASTID: A family of membrane-bound organelles unique to plant cells; only one type is found in each cell while all types derive from a common precursor organelle called a proplastid.

POLYADENYLATION: The post-transcriptional attachment of up to 200 AMP residues to the $3'$-terminus of most eukaryotic mRNAs.

POLYLINKER: A synthetic double-stranded DNA oligonucleotide containing a number of different restriction sites.

POLYMERASE: An enzyme that catalyzes the assembly of nucleotides into RNA or of deoxynucleotides into DNA; usually the enzyme requires single-stranded DNA (sometimes RNA) as a template.

POLYMORPHISM: The simultaneous occurrence in the population of genomes showing allelic variations (as seen either on alleles producing different phenotypes or, for example, in changes in DNA affecting the restriction pattern).

PHOSPHATASE: A class of enzymes that hydrolyses (terminal) phosphoryl groups from nucleotides as well as from proteins.

PRIMER: A short sequence (of DNA or RNA) that is paired with one strand of DNA and provides a free 3′-OH end at which a DNA polymerase starts synthesis of a deoxyribonucleotide chain.

PROBE (HYBRIDIZATION): A labelled DNA or RNA molecule used to detect a complementary sequence by molecular hybridization.

PROKARYOTIC: Any organism that lacks a membrane-enclosed nucleus.

PROMOTER: (IN BACTERIA): The region of the gene involved in binding of the RNA polymerase. (In eukaryotes) usually all regions of the gene required for maximum expression (excluding enhancer sequences).

PROTEIN A: A protein from *Staphylococcus aureus* that binds specifically to immunoglobulin G molecules. Used in detection of proteins by immunological techniques.

PROTEINASE K: A protease used to remove contaminating protein from preparations of nucleic acids. The enzyme also degrades itself.

PROTEIN KINASE: A class of enzymes that phosphorylates a protein with the help of ATP, the phosphorylation takes place preferentially at tyrosines.

PROTOPLAST: A cell without cell wall but with intact cell membrane; gram-positive bacterium after removal of the cell wall.

PSEUDOKNOT: An RNA secondary structure that is minimally composed of two helical segments connected by single-stranded regions or loops.

QUADRUPLEX: A four-stranded box-like structure, with a central cavity, composed of successive stacking of two or more G-tetrads.

RECOMBINANT DNA: Any DNA molecule created by ligating pieces of DNA that normally are not contiguous.

RECOMBINATION: A genetic rearrangement occurring during sperm and egg cell formation.

RENATURATION (OF DNA OR RNA): The re-establishment of the DNA duplex or intrastrand hairpin structures in an RNA molecule after denaturation. (Of a protein); the conversion from an inactive into a biologically active conformation.

REPLICON: The regulatory unit of an origin and proteins necessary for initiation of replication (specific for this origin).

REPRESSION: The blocking of the synthesis of certain enzymes when their products are present; more generally, refers to inhibition of transcription (or translation) by binding of repressor protein to specific site on DNA (or mRNA).

RESTRICTION ENZYME: An enzyme that recognizes specific short sequences of (usually) unmethylated DNA and cleaves the respective DNA molecule (sometimes at target site, sometimes elsewhere (in trans), depending on type).

RESTRICTION FRAGMENT: A duplex DNA fragment obtained by cutting a larger fragment with either a single or two different restriction enzymes.

RETROTRANSPOSON: The major class of eukaryotic transposable elements, which are able to transpose into other genomic DNA sites *via* an RNA intermediate by use of retrotransposon-encoded reverse transcriptase.

RETROVIRUS: A virus containing a single-stranded RNA genome that propagates *via* conversion into double-stranded DNA by reverse transcription.

REVERSE TRANSCRIPTASE: RNA-dependent DNA polymerase. Originally detected in retroviruses. It is, however, also present in normal eukaryotic cells and even in *E. coli.*

REVERSION (OF MUTATION): A change in DNA that either reverses the original alteration (true reversion) or compensates for it (second site reversion in the same gene).

RIBOSOMES: Subcellular particles consisting of several RNA and numerous protein molecules. Involved in translating the genetic code in mRNA into the amino acid sequence of the corresponding protein.

RIBOSWITCH: A part of an mRNA molecule that can directly bind a small target molecule, where the binding of the target affects the activity of the RNA.

RIBOZYME: A naturally occurring folded RNA structure that cuts cognate RNA through an intramolecular *trans*-esterification reaction. Can also refer to any single-stranded catalytic RNA molecule.

RNA EDITING: A series of consecutive "cut and paste" reactions carried out by complex cell machinery; results in a change of sequence of RNA following transcription.

SIRNA: Short interfering RNA; an intermediate in the RNAi process in which the long double-stranded RNA has been cut up into short (~21 nucleotides) double-stranded RNA. The SIRNA stimulates the cellular machinery to cut up other single-stranded RNA having the same sequence as the SIRNA.

SANGER–COULSON SEQUENCING: DNA sequencing technique based on transcription of single-stranded DNA by a polymerase in the presence of dideoxynucleotides. The same technique can also be used for sequencing of RNA.

SATELLITE DNA: The many tandem repeats (identical or related) of a short basic repeating unit.

SDS (SODIUM DODECYLSULFATE): A detergent.

SDS GEL ELECTROPHORESIS: Gel electrophoresis of proteins in polyacrylamide gels in the presence of SDS. Molecules of SDS associate with the protein molecules giving them all a similar electric charge density and thus allowing separation on the basis of differences in molecular weight.

SELECTION: The use of particular conditions to allow survival only of cells with a particular phenotype.

SELEX: A technique that allows the simultaneous screening of highly diverse pools of different RNA or DNA molecules in order to obtain a particular feature.

SEQUENCING GEL: A very thin (0.1–1 mm) high-resolution polyacrylamide gel.

SHINE–DALGARNO SEQUENCE: Part or all of the polypurine sequence AGGAGG located on bacterial mRNA just prior to an AUG initiation condon; is complementary to the sequence at the 3′-end of 16S rRNA; involved in binding of ribosome to mRNA.

SHORT TANDEM REPEAT (STR): A DNA sequence containing a variable number (typically =50) of tandemly repeated short (2–6 bp) sequences, such as $(GATA)_n$. Forensic STRs are usually tetranucleotide repeats, which show little PCR stutter.

SHUTTLE VECTOR: A vector which is able to replicate in different host organisms *e.g. E. coli*, COS cells.

SIGMA FACTOR: A subunit of bacterial RNA polymerase needed for initiation; is the major influence on selection of binding sites (promoters).

SIGNAL HYPOTHESIS: The process by which proteins synthesized in the cytoplasm are exported either out of the cell or into one of the cellular organelles. The signal peptide of the protein plays an important role in this process.

SIGNAL PEPTIDE: The region (usually N-terminal) of a protein that ensures its export out of the cell or its import into one of the cellular organelles (s. leader).

SIGNAL TRANSDUCTION: Molecular mechanism of transferring the information from the outside of a cell, a receptor, to the nucleus. The stimulus may be, *e.g.* a hormone or cytokine, the transferring molecules are second messengers, protein kinases, and phosphatases and finally transcription factors.

SIMPLE STRS: Short tandem repeat loci composed of uninterrupted runs of a single repeat type.

SINGLE NUCLEOTIDE POLYMORPHISM (SNP): A common DNA sequence variation among individuals of the same species.

SITE-DIRECTED MUTAGENESIS: Introduction in the test tube of a specific mutation(s) into a DNA molecule at a predetermined site.

SOUTHERN BLOTTING: A procedure for transferring denatured DNA from an agarose gel to a nitrocellulose filter where it can be hybridized with a complementary nucleic acid.

SPLICEOSOME: A complex of several RNAs and proteins responsible for removing the non-coding parts of RNA (introns) from unprocessed mRNA.

SPLICING: Describes the removal of introns and joining of exons in RNA; thus introns are spliced out, while exons are spliced together.

STEM: The base-paired segment of a hairpin.

STOP CODON: Same as termination codon.

STRUCTURAL GENE: Gene coding for any RNA or protein product other than a regulator.

STUTTER: See PCR Stutter.

SUBCLONING: The cloning of fragments of an already cloned DNA sequence.

SUPERCOIL: A closed circular double-stranded DNA molecule that is twisted on itself. Typically a conformation of a circular double-stranded nucleic acid in which strain derived from an excess or deficit of turns of the double-stranded helix is relieved by a counter-helical winding of the circular nucleic acid (imaged as in a skein of wool).

TAC-PROMOTOR: A chimeric bacterial promotor of high strength constructed from parts of the Trp and lac promotors of *E. coli.*

TATA (HOGENESS) BOX: A conserved A-T-rich heptamer found about 25 bp before the start-point of each eukaryotic RNA polymerase II transcription unit; involved in positioning the enzyme for correct initiation.

TELOMERE: A region of highly repetitive DNA at the end of a chromosome.

TEMPLATE: Portion of single-stranded DNA or RNA used to direct the synthesis of a complementary polynucleotide.

TERMINATION CODON: One of three triplet sequences, UAG (*amber*), UAA (*ochre*), or UGA (*opal*), that cause termination of protein synthesis; they are also called nonsense codons.

TOLL-LIKE RECEPTOR: In vertebrates, receptor molecules that are able to stimulate activation of the adaptive immune system, linking innate and acquired immune responses.

TOPOISOMERASES: Enzymes that act on the topology of DNA; needed to unravel DNA strands that are topologically linked or knotted; they catalyze and guide the unknotting of DNA.

TRANS-ACTING: Referring to mutations of, for example, a repressor gene, that act through a diffusable protein product and can therefore act at a distance not simply on the DNA molecule in which they occur.

TRANSCRIPTION: Usually the synthesis of RNA on a DNA template. Also used to describe the synthesis of DNA on an RNA template by reverse transcriptase, the copying of a (primed) single-stranded DNA by DNA polymerase and the copying of RNA by (viral) RNA polymerase.

TRANSDUCTION: The transfer of a bacterial gene from one bacterium to another by a phage; phage carrying host as well as its own genes is called transducing phage.

TRANSFECTION: The acquisition of native protein-free DNA of a phage by bacteria.

TRANSFORMATION: The acquisition by a cell of new genetic markers by incorporation of added DNA. In eukaryotic cells it also refers to conversion to a state of unrestrained growth in culture resembling or identical to the tumorigenic condition.

TRANSITION: A mutation in which a purine is replaced by another purine (*e.g.* G to A) or a pyrimidine by another pyrimidine (*e.g.* T to C).

TRANSPOSABLE ELEMENT: A heterogeneous class of genetic element that can insert into a new location within chromosomes.

TRANSVERSION: A mutation in which a purine is replaced by a pyrimidine or *vice versa.*

TRIPLET: A sequence of three nucleotides in DNA or RNA. Usually means the same as codon.

TWO-DIMENSIONAL GEL ELECTROPHORESIS: A technique in which a second electrophoretic separation is carried out perpendicular to the first. The two separations are based on different criteria (*e.g.* electric charge and molecular weight).

UPSTREAM: Sequences that proceed in the opposite direction from expression. For example, the bacterial promoter is upstream from the transcription unit, the initiation codon is upstream from the coding region.

WATSON–CRICK RULES: The base-pairing rules that underlie gene structure and expression. G pairs with C; A pairs with T (A pairs with U in RNA).

WESTERN BLOTTING: Transfer of proteins from a gel to a nitrocellulose filter on which they can subsequently be detected by immunological screening.

WILD-TYPE: The genotype or phenotype that is found in nature or in the standard laboratory stock for a given organism; the phenotype of a particular organism when first seen in nature.

WOBBLE HYPOTHESIS: The ability of a tRNA to recognize more than one codon by non-Watson–Crick (non-G-C, A-T) pairing with the third base of a codon.

CHAPTER 1

# Introduction and Overview

## CONTENTS



## 1.1 THE BIOLOGICAL IMPORTANCE OF DNA

From the beginning, the study of nucleic acids has drawn together, as though by a powerful unseen force, a galaxy of scientists of the highest ability.[1,2] Striving to tease apart its secrets, these talented individuals have brought with them a broad range of skills from other disciplines while many of the problems they have encountered have proved to be soluble only by new inventions. Looking at their work, one is constantly made aware that scientists in this field appear to have enjoyed a greater sense of excitement in their work than is given to most. Why?

For over 60 years, such men and women have been fascinated and stimulated by their awareness that the study of nucleic acids is central to the knowledge of life. Let us start by looking at Fred Griffith, who was employed as a scientific civil servant in the British Ministry of Health investigating the nature of epidemics. In 1923, he was able to identify the difference between a virulent, *S*, and a non-virulent, *R*, form of the pneumonia bacterium. Griffith went on to show that this bacterium could be made to undergo a permanent, hereditable change from non-virulent to virulent type. This discovery was a bombshell in bacterial genetics.

Oswald Avery and his group at the Rockefeller Institute in New York set out to identify the molecular mechanism responsible for the change Griffith had discovered, now technically called **bacterial transformation**. They achieved a breakthrough in 1940 when they found that non-virulent *R* pneumococci could be transformed *irreversibly* into a virulent species by treatment with a pure sample of high molecular weight DNA.[3] Avery had purified this DNA from heat-killed bacteria of a virulent strain and showed that it was active at a dilution of 1 part in $10^9$.

Avery concluded that '*DNA is responsible for the transforming activity*' and published that analysis in 1944, just 3 years after Griffith had died in a London air-raid. The staggering implications of Avery's work turned a searchlight on the molecular nature of nucleic acids and it soon became evident that ideas on the chemistry of nucleic acid structure at that time were wholly inadequate to explain such a momentous discovery. As a result, a new wave of scientists directed their attention to DNA and discovered that large parts of the accepted tenets of nucleic acid chemistry had to be set aside before real progress was possible. We need to examine some of the earliest features of that chemistry to fully appreciate the significance of later progress.

## 1.2 THE ORIGINS OF NUCLEIC ACIDS RESEARCH

Friedrich Miescher started his research career in Tübingen by looking into the physiology of human lymph cells. In 1868, seeking a more readily available material, he began to study human pus cells, which he obtained in abundant supply from the bandages discarded from the local hospital. After defatting the cells with alcohol, he incubated them with a crude preparation of pepsin from pig stomach and so obtained a grey precipitate of pure cell nuclei. Treatment of this with alkali followed by acid gave Miescher a precipitate of a phosphorus-containing substance, which he named **nuclein**. He later found this material to be a common constituent of yeast, kidney, liver, testicular and nucleated red blood cells.[4]

After Miescher moved to Basel in 1872, he found the sperm of Rhine salmon to be a more plentiful source of nuclein. The pure nuclein was a strongly acidic substance, which existed in a salt-like combination with a nitrogenous base that Miescher crystallized and called protamine. In fact, his nuclein was really a nucleoprotein and it fell subsequently to Richard Altman in 1889 to obtain the first protein-free material, to which he gave the name **nucleic acid**.

Following William Perkin's invention of mauveine in 1856, the development of aniline dyes had stimulated a systematic study of the colour-staining of biological specimens. Cell nuclei were characteristically stained by basic dyes, and around 1880, Walter Flemming applied that property in his study of the rod-like segments of chromatin (called so because of their colour-staining characteristic), which became visible within the cell nucleus only at certain stages of cell division. Flemming's speculation that the chemical composition of these **chromosomes** was identical to that of Miescher's nuclein was confirmed in 1900 by E.B. Wilson who wrote

> Now chromatin is known to be closely similar to, if not identical with, a substance known as nuclein which analysis shows to be a tolerably definite chemical compound of nucleic acid and albumin. And thus we reach the remarkable conclusion that inheritance may, perhaps, be affected by the physical transmission of a particular compound from parent to offspring.

While this insight was later to be realized in Griffith's 1928 experiments, all of this work was really far ahead of its time. We have to recognize that, at the turn of the century, tests for the purity and identity of substances were relatively primitive. Emil Fischer's classic studies on the chemistry of high molecular weight, polymeric organic molecules were in question until well into the twentieth century. Even in 1920, it was possible to argue that there were only two species of nucleic acids in nature: animal cells were believed to provide **thymus nucleic acid** (DNA), while nuclei of plant cells were thought to give **pentose nucleic acid** (RNA).

## 1.3 EARLY STRUCTURAL STUDIES ON NUCLEIC ACIDS

Accurate molecular studies on nucleic acids essentially date back to 1909 when Levene and Jacobs began a reinvestigation of the structure of **nucleotides** at the Rockefeller Institute. Inosinic acid, which Liebig had isolated from beef muscle in 1847, proved to be hypoxanthine-riboside 5′-phosphate. Guanylic acid, isolated from the nucleoprotein of pancreas glands, was identified as guanine-riboside 5′-phosphate (Figure 1.1). Each of these nucleotides was cleaved by alkaline hydrolysis to give phosphate and the corresponding **nucleosides**, inosine and guanosine, respectively. Since then, all nucleosides are characterized as the condensation products of a pentose and a nitrogenous base while nucleotides are the phosphate esters of one of the hydroxyl groups of the pentose.

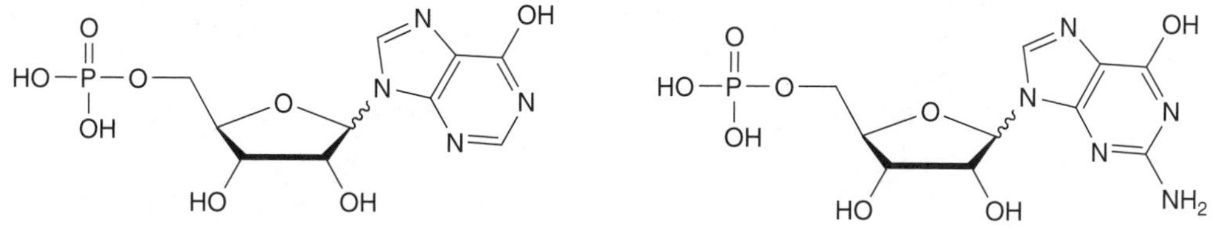

inosinic acid (as enolic tautomer) guanylic acid (as enolic tautomer)

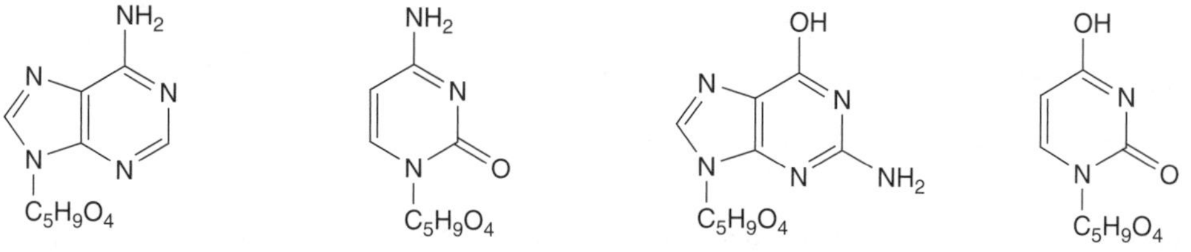

adenosine cytidine guanosine (as enolic tautomer) uridine (as enolic tautomer)

**Figure 1.1** *Early nucleosides and nucleotide structures (using the enolic tautomers originally employed). Wavy lines denote unknown stereochemistry at C-1′*

Thymus nucleic acid, which was readily available from calf tissue, was found to be resistant to alkaline hydrolysis. It was only successfully degraded into deoxynucleosides in 1929 when Levene adopted enzymes to hydrolyse the deoxyribonucleic acid followed by mild acidic hydrolysis of the deoxynucleotides. He identified its pentose as the hitherto unknown 2-deoxy-D-ribose. These deoxynucleosides involved the four heterocyclic bases, adenine, cytosine, guanine and thymine, with the latter corresponding to uracil in ribonucleic acid.

Up to 1940, most groups of workers were convinced that hydrolysis of nucleic acids gave the appropriate four bases in **equal relative proportions**. This erroneous conclusion probably resulted from the use of impure nucleic acid or from the use of analytical methods of inadequate accuracy and reliability. It led, naturally enough, to the general acceptance of a **tetranucleotide hypothesis** for the structure of both thymus and yeast nucleic acids, which materially retarded further progress on the molecular structure of nucleic acids.

Several of these tetranucleotide structures were proposed. They all had four nucleosides (one for each of the bases) with an arbitrary location of the two purines and two pyrimidines. They were joined together by four phosphate residues in a variety of ways, among which there was a strong preference for phosphodiester linkages. In 1932, Takahashi showed that yeast nucleic acid contained neither pyrophosphate nor phosphomonoester functions and so disposed of earlier proposals in preference for a neat, cyclic structure which joined the pentoses exclusively using phosphodiester units (Figure 1.2). It was generally accepted that these bonded 5′- to 3′-positions of adjacent deoxyribonucleosides, but the linkage positions in ribonucleic acid were not known.

One property stuck out like a sore thumb from this picture: the molecular mass of nucleic acids was greatly in excess of that calculated for a tetranucleotide. The best DNA samples were produced by Einar Hammarsten in Stockholm and one of his students, Torjbörn Caspersson, who showed that this material was greater in size than protein molecules. Hammarsten's DNA was examined by Rudolf Signer in Bern whose flow-birefringence studies revealed rod-like molecules with a molecular mass of $0.5–1.0 \times 10^6$ Da. The same material provided Astbury in Leeds with X-ray fibre diffraction measurements that supported Signer's conclusion. Finally, Levene estimated the molecular mass of native DNA to be between 200,000 and $1 \times 10^6$ Da, based on ultracentrifugation studies.

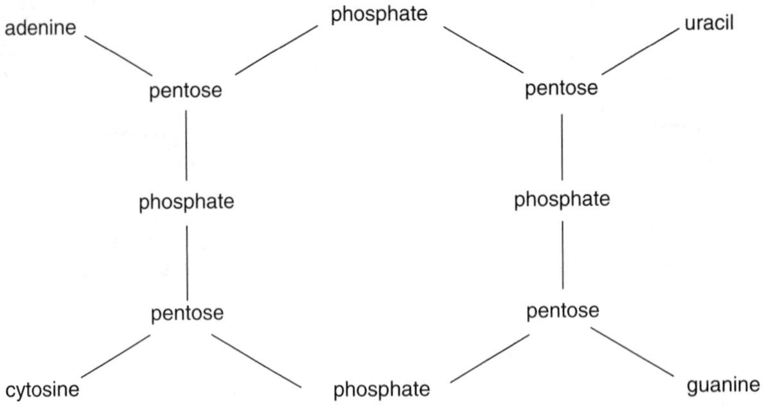


**Figure 1.2** *The tetranucleotide structure proposed for nucleic acids by Takahashi (1932)*

The scientists compromised. In his Tilden Lecture of 1943, Masson Gulland suggested that the concept of nucleic acid structures of polymerized, uniform tetranucleotides was limited, but he allowed that they could 'form a practical working hypothesis'.

This then was the position in 1944 when Avery published his great work on the transforming activity of bacterial DNA. One can sympathize with Avery's hesitance to press home his case. Levene, in the same Institute, and others were strongly persuaded that the tetranucleotide hypothesis imposed an invariance on the structure of nucleic acids, which denied them any role in biological diversity. In contrast, Avery's work showed that DNA was responsible for completely transforming the behaviour of bacteria. It demanded a fresh look at the structure of nucleic acids.

## 1.4 THE DISCOVERY OF THE STRUCTURE OF DNA

From the outset, it was evident that DNA exhibited greater resistance to selective chemical hydrolysis than did RNA. So, the discovery in 1935 that DNA could be cut into **mononucleotides** by an enzyme doped with arsenate was invaluable. Using this procedure, Klein and Thannhauser obtained the four crystalline deoxyribonucleotides, whose structures (Figure 1.3) were later put beyond doubt by total chemical synthesis by Alexander Todd[5] and the Cambridge school he founded in 1944. Todd established the D-configuration and the glycosylic linkage for ribonucleosides in 1951, but found the chemical synthesis of the 2′-deoxyribonucleosides more taxing. The key to success for the Cambridge group was the development of methods of phosphorylation, for example for the preparation of the 3′- and 5′-phosphates of deoxyadenosine[6] (Figure 1.4).

All the facts were now available to establish the primary structure of DNA as a **linear polynucleotide** in which each deoxyribonucleoside is linked to the next by means of a 3′- to 5′-phosphate diester (see Figure 2.15). The presence of only diester linkages was essential to explain the stability of DNA to chemical hydrolysis, since phosphate triesters and monoesters, not to mention pyrophosphates, are more labile. The measured molecular masses for DNA of about $1 \times 10^6$ Da meant that a single strand of DNA would have some 3000 nucleotides. Such a size was much greater than that of enzyme molecules, but entirely compatible with Staudinger's established ideas on macromolecular structure for synthetic and natural polymers. But by the mid-twentieth century, chemists could advance no further with the primary structure of DNA. Neither of the key requirements for sequence determination was to hand: there were no methods for obtaining pure samples of DNA with homogeneous base sequence nor were methods available for the cleavage of DNA strands at a specific base residue. Consequently, all attention came to focus on the secondary structure of DNA.

Two independent experiments in biophysics showed that DNA possesses an ordered secondary structure. Using a sample of DNA obtained from Hammarsten in 1938, Astbury obtained an X-ray diffraction pattern

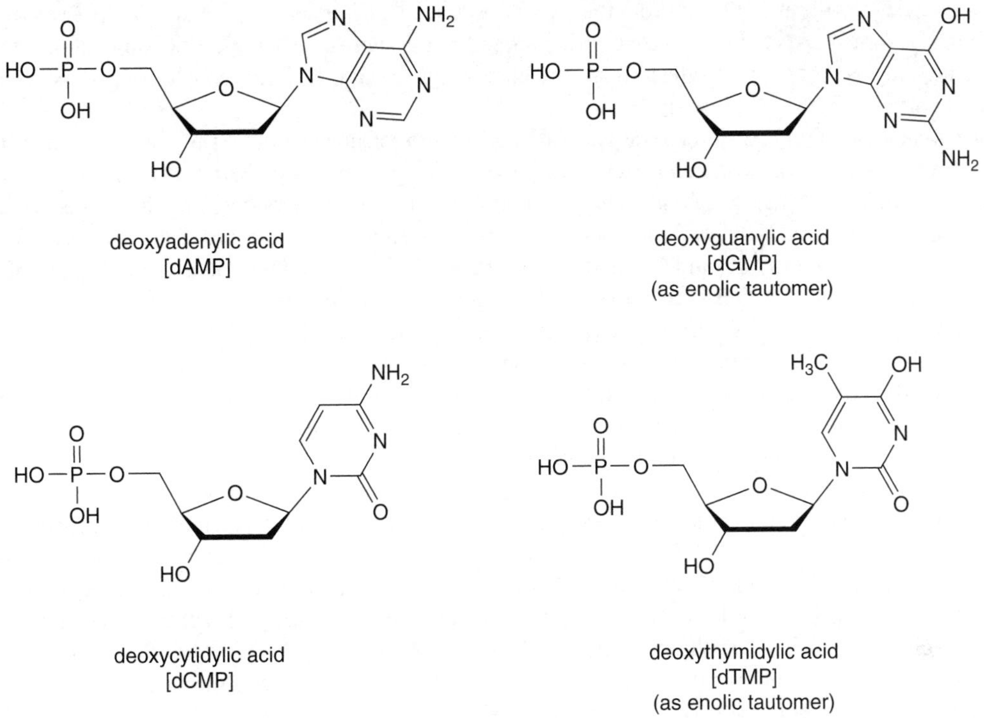


**Figure 1.3** *Structures of 5′-deoxyribonucleotides (original tautomers for dGMP and dTMP)*

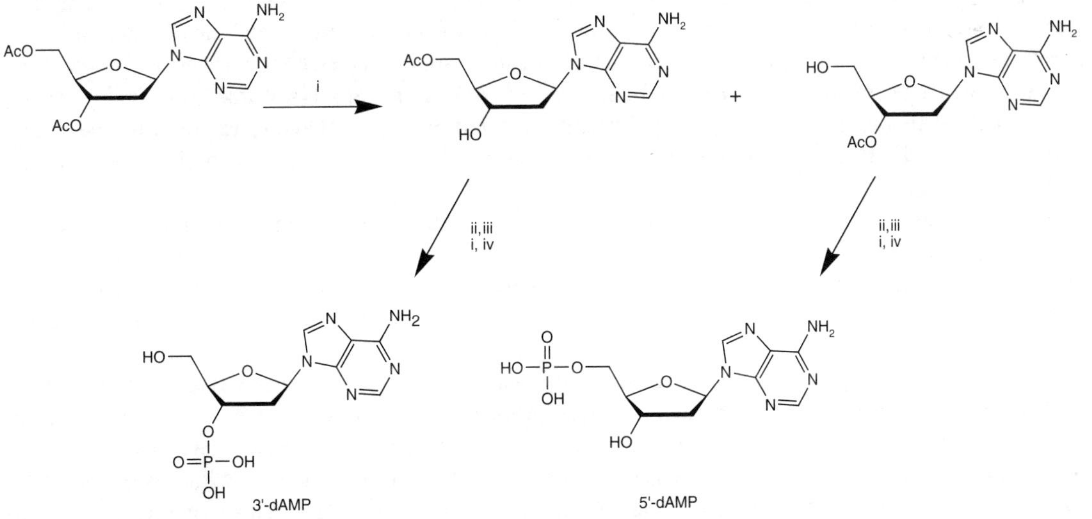


**Figure 1.4** *Todd's synthesis of deoxyadenosine 3′- and 5′-phosphates Reagents: (i) MeOH, $NH_3$ (ii) $(PhO)_2P(O)OP(H)(O)OCH_2Ph$ (iii) N-chlorosuccinimide (iv) $H_2/PdC$*
(D.H. Hayes, A.M. Michelson and A.R. Todd, *J. Chem. Soc.*, 1955, 808–815)

from stretched, dry fibres of DNA. From the rather obscure data he deduced '... A spacing of 3.34 Å along the fibre axis corresponds to that of a close succession of flat or flattish nucleotides standing out perpendicularly to the long axis of the molecule to form a relatively rigid structure.' These conclusions roundly contradicted the tetranucleotide hypothesis.

Some years later, Gulland studied the viscosity and flow-birefringence of calf thymus DNA and thence postulated the presence of hydrogen bonds linking the purine–pyrimidine **hydroxyl** groups and some of the amino groups. He suggested that these hydrogen bonds could involve nucleotides either in adjacent chains or within a single chain, but he somewhat hedged his bets between these alternatives. Sadly, Astbury returned to the investigation of proteins and Gulland died prematurely in a train derailment in 1947. Both of them left work that was vital for their successors to follow, but each contribution contained a misconception that was to prove a stumbling block for the next half-a-dozen years. Thus, Linus Pauling's attempt to create a helical model for DNA located the pentose-phosphate backbone in its core and the **bases pointing outwards** – as Astbury had decided. Gulland had subscribed to the wrong tautomeric forms for the heterocyclic bases thymine and guanine, believing them to be **enolic** and having hydroxyl groups. The importance of the true **keto forms** was only appreciated in 1952.

Erwin Chargaff began to investigate a very different type of order in DNA structure. He studied the base composition of DNA from a variety of sources using the new technique of paper chromatography to separate the products of hydrolysis of DNA and employing one of the first commercial ultraviolet spectrophotometers to quantify their relative abundance.[7] His data showed that there is a variation in base composition of DNA between species that is overridden by a universal 1:1 ratio of adenine with thymine and guanine with cytosine. This meant that the proportion of purines, (A + G), is always equal to the proportion of pyrimidines, (C + T). Although the ratio (G + C)/(A + T) varies from species to species, different tissues from a single species give DNA of the same composition. Chargaff's results finally discredited the tetranucleotide hypothesis, because it called for equal proportions of all four bases in DNA.

In 1951, Francis Crick and Jim Watson joined forces in the Cavendish Laboratory in Cambridge to tackle the problem of DNA structure. Both of them were persuaded that the model-building approach that had led Pauling and Corey to the α-helix structure for peptides should work just as well for DNA. Almost incredibly, they attempted no other line of direct experimentation but drew on the published and unpublished results of other research teams in order to construct a variety of models, each to be discarded in favour of the next until they created one which satisfied all the facts.[8,9]

The best X-ray diffraction results were to be found in King's College, London. There, Maurice Wilkins had observed the importance of keeping DNA fibres in a moist state and Rosalind Franklin had found that the X-ray diffraction pattern obtained from such fibres showed the existence of an A-form of DNA at low humidity, which changed into a B-form at high humidity. Both forms of DNA were highly crystalline and clearly helical in structure. Consequently, Franklin decided that this behaviour required the phosphate groups to be exposed to water on the **outside** of the helix, with the corollary that the bases were on the **inside** of the helix.

Watson decided that the number of nucleotides in the unit crystallographic cell favoured a double-stranded helix. Crick's physics-trained mind recognized the symmetry implications of the space-group of the A-form diffraction pattern, monoclinic *C*2. There had to be local twofold symmetry axes normal to the helix, a feature, which called for a double-stranded helix, whose two chains must run in opposite directions.

Crick and Watson thus needed merely to solve the final problem: how to construct the core of the helix by packing the bases together in a regular structure. Watson knew about Gulland's conclusions regarding hydrogen bonds joining the DNA bases. This convinced him that the crux of the matter had to be a rule governing hydrogen bonding between bases. Accordingly, Watson experimented with models using the **enolic** tautomeric forms of the bases (Figure 1.3) and pairing like with like. This structure was quickly rejected by Crick because it had the wrong symmetry for B-DNA. **Self-pairing** had to be rejected because it could not explain Chargaff's 1:1 base ratios, which Crick had perceived were bound to result if you had **complementary base pairing**.

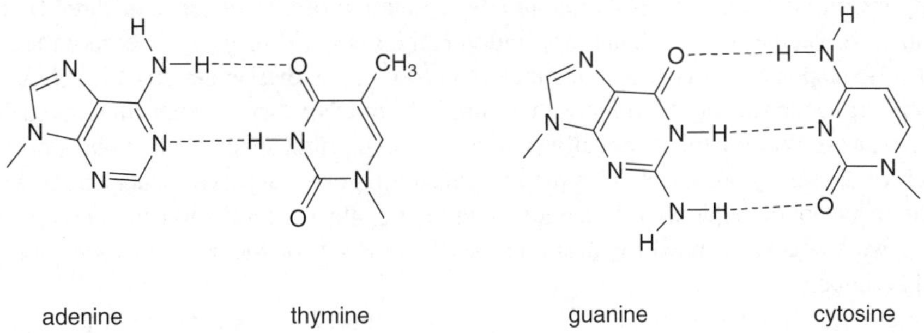


**Figure 1.5** *Complementary hydrogen-bonded base-pairs as proposed by Watson and Crick (thymine and guanine in the revised keto forms). The G···C structure was later altered to include three hydrogen bonds*

On the basis of the advice from Jerry Donohue in the Cavendish Laboratory, Watson turned to manipulating models of the bases in their **keto forms** and paired adenine with thymine and guanine with cytosine. Almost at once, he found a compellingly simple relationship involving two hydrogen bonds for an A···T pair and two or three hydrogen bonds for a G···C pair. The special feature of this base-pairing scheme is that the relative geometry of the bonds joining the bases to the pentoses is virtually identical for the A···T and G···C pairs (Figure 1.5). It follows that if a purine always pairs with a pyrimidine then an irregular sequence of bases in a single strand of DNA could nevertheless be paired regularly in the centre of a double helix and without loss of symmetry.[10]

Chargaff's 'rules' were straightaway revealed as an obligatory consequence of a double-helical structure for DNA. Above all, since the base sequence of one chain automatically determines that of its partner, Crick and Watson could easily visualize how one single chain might be the template for creation of a second chain of complementary base sequence.

The structure of the core of DNA had been solved and the whole enterprise fittingly received the ultimate accolade of the scientific establishment when Crick, Watson and Wilkins shared the Nobel prize for chemistry in 1962, just 4 years after Rosalind Franklin's early death.

## 1.5 THE ADVENT OF MOLECULAR BIOLOGY

It is common to describe the publication of Watson and Crick's paper in *Nature* in April 1953 as the end of the 'classical' period in the study of nucleic acids, up to which time basic discoveries were made by a few gifted academics in an otherwise relatively unexplored field. The excitement aroused by the model of the double helix drew the attention of a much wider scientific audience to the importance of nucleic acids, particularly because of the biological implications of the model rather than because of the structure itself. It was immediately apparent that locked into the irregular sequence of nucleotide bases in the DNA of a cell was all the information required to specify the diversity of biological molecules needed to carry out the functions of that cell. The important question now was what was the key, the **genetic code**, through which the sequence of DNA could be translated into protein?[11]

The solution to the coding problem is often attributed to the laboratories in the USA of Marshall Nirenberg and of Severo Ochoa who devised an elegant cell-free system for translating enzymatically synthesized polynucleotides into polypeptides and who by the mid-1960s had established the genetic code for a number of amino acids.[12,13] In reality, the story of the elucidation of the code involves numerous strands of knowledge obtained from a variety of workers in different laboratories. An essential contribution came from Alexander Dounce in Rochester, New York, who in the early 1950s postulated that RNA, and not DNA, served as a template to direct the synthesis of cellular proteins and that a sequence of three nucleotides might specify a single amino acid. Sydney Brenner and Leslie Barnett in Cambridge, later (1961) confirmed the code to be both triplet and non-overlapping. From Robert Holley in Cornell University, New York, and

Hans Zachau in Cologne, came the isolation and determination of the sequence of three **transfer RNAs** (tRNA) 'adapter' molecules that each carry an individual amino acid ready for incorporation into protein and which are also responsible for recognizing the triplet code on the **messenger RNA** (mRNA). The mRNA species contain the sequences of individual genes copied from DNA (see Chapter 6). Gobind Khorana and his group in Madison, Wisconsin, chemically synthesized all 64 ribotrinucleoside diphosphates and, using a combination of chemistry and enzymology, synthesized a number of polyribonucleotides with repeating di-, tri-, and tetranucleotide sequences.[14] These were used as synthetic mRNA to help identify each triplet in the code. This work was recognized by awarding the Nobel Prize for Medicine in 1968 jointly to Holley, Khorana and Nirenberg.

Nucleic acid research in the 1950s and 1960s was preoccupied by the solution to the coding problem and the establishment of the biological roles of tRNA and mRNA. This was not surprising bearing in mind that at that time the smaller size and attainable homogeneity made isolation and purification of RNA a much easier task than it was for DNA. It was clear that in order to approach the fundamental question of what constituted a **gene** – a single hereditable element of DNA that up to then could be defined genetically but not chemically – it was going to be necessary to break down DNA into smaller, more tractable pieces in a specific and predictable way.

The breakthrough came in 1968 when Meselson and Yuan reported the isolation of a **restriction enzyme** from the bacterium *Escherichia coli*. Here at last was an enzyme, a nuclease, which could recognize a defined sequence in a DNA and cut it specifically (see Section 5.3.1). The bacterium used this activity to break down and hence inactive invading (*e.g.* phage) DNA. It was soon realized that this was a general property of bacteria, and the isolation of other restriction enzymes with different specificities soon followed. But it was not until 1973 that the importance of these enzymes became apparent. At this time, Chang and Cohen at Stanford and Helling and Boyer at the University of California were able to construct in a test tube, a biologically functional DNA that combined genetic information from two different sources. This **chimera** was created by cleaving DNA from one source with a restriction enzyme to give a fragment that could then be joined to a carrier DNA, a **plasmid**. The resultant **recombinant DNA** was shown to be able to replicate and express itself in *E. coli*.[15]

This remarkable demonstration of genetic manipulation was to revolutionize biology. It soon became possible to dissect out an individual gene from its source DNA, to amplify it in a bacterium or other organism (**cloning**, see Section 5.2), and to study its expression by the synthesis first of RNA and then of protein (see Chapters 6 and 7). This single advance by the groups of Cohen and Boyer truly marked the dawn of modern molecular biology.

## 1.6 THE PARTNERSHIP OF CHEMISTRY AND BIOLOGY

In the 1940s and 1950s, the disciplines of chemistry and biology were so separate that it was a rare occurrence for an individual to embrace both. Two young scientists who were just setting out on their careers at that time were exceptional in recognizing the potential of chemistry in the solution of biological problems and both, in their different ways, were to have a substantial and lasting effect in the field of nucleic acids.

One was Frederick Sanger, a product of the Cambridge Biochemistry School, who in the early 1940s set out to determine the sequence of a protein, insulin. This feat had been thought unattainable, since it was widely supposed that proteins were not discrete species with defined primary sequence. Even more remarkably, he went on to develop methods for sequence determination first of RNA and then of DNA (see Section 5.1). These methods involved a subtle blend of enzymology and chemistry that few would have thought possible to combine.[16] The results of his efforts transformed **DNA sequencing** in only a few years into a routine procedure. In the late 1980s, the procedure was adapted for use in automated sequencing machines and the 1990s saw worldwide efforts to sequence whole organism genomes. In 2003, exactly 50 years after the discovery of the structure of the DNA double helix, it was announced that the human genome sequence had been completed. The award of two Nobel prizes to Sanger (1958 and 1980) hardly seems recognition enough!

The other scientist has already been mentioned in connection with the elucidation of the genetic code. Not long after his post-doctoral studies under George Kenner and Alexander Todd in Cambridge, Gobind Khorana was convinced that chemical synthesis of polynucleotides could make an important contribution to the study of the fundamental process of information flow from DNA to RNA to protein. Having completed the work on the genetic code in the mid-1960s and aware of Holley's recently determined (1965) sequence for an alanine tRNA, he then established a new goal of total synthesis of the corresponding DNA duplex, the gene specifying the tRNA. Like Sanger, he ingeniously devised a combination of nucleic acid chemistry and enzymology to form a general strategy of **gene synthesis**, which in principle remains unaltered to this day (see Section 5.4).[17] Knowledge became available by the early 1970s about the signals required for gene expression and the newly emerging recombinant DNA methods of Cohen and Boyer allowed a second synthetic gene, this time specifying the precursor of a tyrosine suppressor tRNA (Figure 1.6) to be cloned and shown to be fully functional.

It is ironic that even up to the early 1970s many biologists thought Khorana's gene syntheses unlikely to have practical value. This view changed dramatically in 1977 with the demonstration by the groups of Itakura (a chemist) and Boyer (a biologist) of the expression in a bacterium of the hormone somatostatin (and later insulin A and B chains) from a chemically synthesised gene.[18] This work spawned the biotechnology

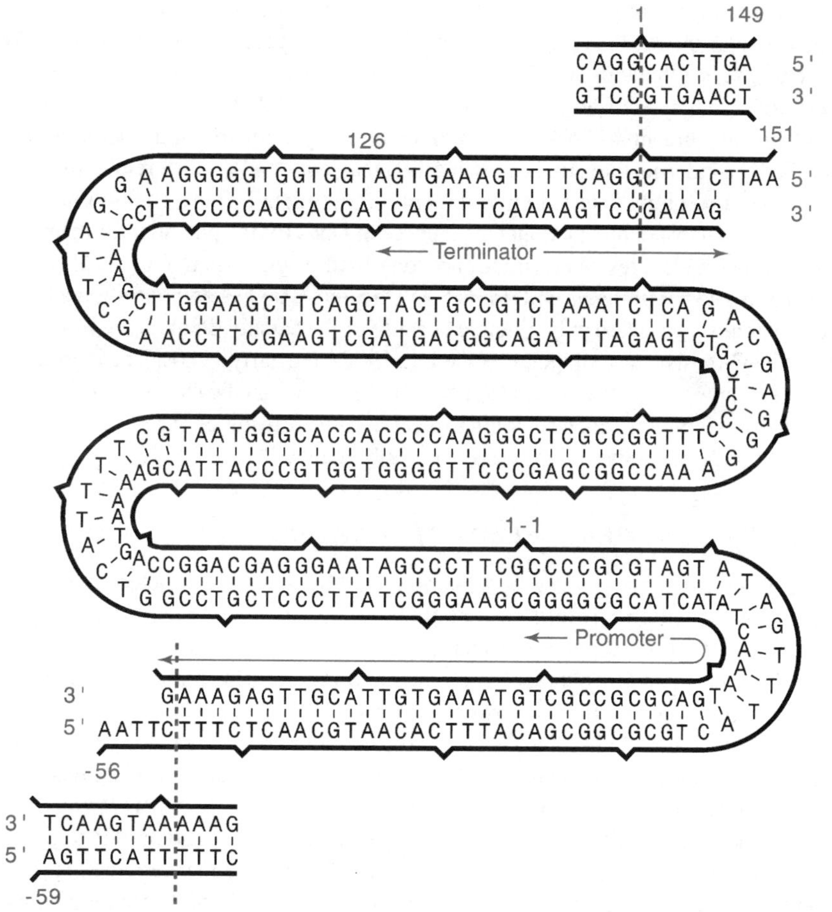


**Figure 1.6** *Khorana's totally synthetic DNA corresponding to the tyrosine suppressor transfer RNA gene* (reprinted from Belagaje, R. *et al.*, in *Chemistry and Biology of Nucleosides and Nucleotides*, R.E. Harmon, R.K. Robins and L.B. Townsend (eds), Academic Press, New York. © (1978), with permission from Elsevier)

industry and synthetic genes became routinely used in the production of proteins. Further, **oligodeoxyribonucleotides**, the short pieces of single-stranded DNA for which Khorana developed the first chemical syntheses, became invaluable general tools in the manipulation of DNA, for example, as primers in DNA sequencing, as probes in gene detection and isolation, and as mutagenic agents to alter the sequence of DNA. From the late 1980s, research accelerated into synthetic **oligonucleotide analogues** as **antisense** modulators of gene expression in cells, as therapeutic agents (see Section 5.7) and for the construction of microarray chips for gene expression analysis.

The availability of synthetic DNA also provided new impetus in the study of DNA structure. In the early 1970s, new X-ray crystallographic techniques had been developed and applied to solve the structure of the dinucleoside phosphate, ApU, by Rich and co-workers in Cambridge, USA. This was followed by the complete structure of yeast phenylalanine tRNA, determined independently by Rich and by Klug and colleagues in Cambridge, England. For the first time, the complementary base pairing between two strands could be seen in greater detail than was previously possible from studies of DNA and RNA fibres. ApU formed a double helix by end-to-end packing of molecules, with Watson–Crick pairing clearly in evidence between each strand. The tRNA showed not only Watson–Crick pairs, but also a variety of alternative base pairs and base triples, many of which were entirely novel (see Sections 2.3.3 and 7.1.2).

Then in 1978, the structure of synthetic d(pATAT) was solved by Kennard and her group in Cambridge. This tetramer also formed an extended double helix, but excitingly revealed that there was a substantial sequence-dependence in its conformation. The angles between neighbouring dA and dT residues were quite different between the A–T sequence and the T–A sequence elements. Soon after, Wang and colleagues discovered that synthetic d(CGCGCG) adopted a totally unpredicted, left-handed Z-conformation. This was soon followed by the demonstration of both a **B-DNA** helix in a synthetic dodecamer by Dickerson in California and an A-DNA helix in an octamer by Kennard, and finally put paid to the concept that DNA had a rigid, rod-like structure. Clearly, DNA could adopt different conformations dependent on sequence and also on its external environment (see Section 2.3). More importantly, an immediate inference could be drawn that conformational differences in DNA (or the potential for their formation) might be recognized by other molecules. Thus, it was not long before synthetic DNA was also being used in the study of DNA binding to carcinogens and drugs (see Chapters 8 and 9) and to proteins (see Chapter 10).

These spectacular advances were only possible because of the equally dramatic improvements in methods of oligonucleotide synthesis that took place in the late 1970s and early 1980s. The laborious manual work of the early gene synthesis days was replaced by reliable automated DNA synthesis machines, which, within hours, could assemble sequences well in excess of 100 residues (see Section 4.1.4). Khorana's vision of the importance of synthetic DNA has been fully realized.

## 1.7 FRONTIERS IN NUCLEIC ACIDS RESEARCH

The last decade of the twentieth century was characterized by the quest to determine the complete DNA sequence of the **human genome**. Efforts by a publicly funded international consortium gathered considerable pace in the late 1990s in response to a challenge from a private company and the resultant concerns over the availability of sequencing data to the research community. The completion of the human genome sequence was duly announced by the consortium in April 2003, 50 years after papers on the discovery of the structure of the DNA double helix had been published and only 25 years since the first simple bacteriophage genome sequences were obtained. Genome sequences of many other organisms have also been completed, for example, mouse, nematode, zebrafish, yeast and parasites such as *Plasmodium falciparium* (see Section 6.5). The vast quantity of DNA sequence information generated has led to the founding of the new discipline of **Bioinformatics** in order to analyse and compare sequence data. One big surprise was that the human genome contains far fewer genes than expected, only about 24,500. We now know that production of the considerably larger number of human proteins and their regulation during cell division and biological development involves control of gene expression at many different stages (*e.g.* transcription, alternative splicing, RNA editing, translation, see Chapters 6 and 7), a full understanding of which is likely to occupy biologists well into the twenty-first century. A recent technical advance here is the development of **microarrays** of synthetic oligonucleotides or **cDNAs** as hybridisation probes of DNA or RNA sequences both for mutational and gene expression analysis (see Section 5.5.4). This has led to the science of '-omics', such as **genomics** and *ribonomics*, where DNA sequence variations can be studied and global effects of particular pathological states or external stimuli can be gauged on a whole genome basis.

A number of other advances have also been made in nucleic acids chemistry. First, a strong revival in the synthesis of nucleoside analogues has led to a number of therapeutic agents being approved for clinical use in treatment of AIDS and HIV infection as well as herpes and hepatitis viruses (see Section 3.7.2). Further, synthetic oligonucleotide analogues have become clinical agents for the treatment of viral infections and some cancers, although few have passed full regulatory approval as yet. The exploitation of the 'antisense' technology as a principle of therapeutic **gene modulation** has led to the investigation of a large number of nucleic acid analogues to enhance activity (see Section 5.7.1). As the twenty-first century arrived, gene modulation technology was finding increasing use to validate gene targets in cell lines and animals. At the same time, there was increasing recognition that other mechanisms of action can contribute to therapeutic effects of oligonucleotides in humans, such as stimulation of the immune system by 'CpG' domains (see Section 5.7.1), which may be harnessed perhaps for use as vaccine adjuvants.

The provision of synthetic RNA has also become routine (see Section 4.2) resulting in major advances in our understanding of catalytic RNA (**ribozymes**, see Sections 5.7.3 and 7.6.2) and protein-RNA interactions (see Section 10.9). New techniques of *in vitro* selection of RNA sequences have extended the potential of ribozymes and **aptamers** to carry out artificial reactions or bind unusual substrates, for example to act as 'riboswitches' responsive to certain analytes (see Section 5.7.3). A considerable upsurge of research in RNA biology has paralleled the availability of synthetic RNA. New ways have been elucidated for specific RNA sequences and structures to play important roles in gene regulation (*e.g.* **microRNA**, see Section 5.7.2). The exciting discovery of '**RNA interference**' as a natural cell mechanism has led to the development of short synthetic RNA duplexes (siRNA and shRNA) as new gene control reagents that now rival, and may well surpass, antisense oligonucleotides for therapeutic and diagnostic use (see Section 5.7.2).

Dramatic advances have also been made in high-resolution structural determination of DNA and RNA sequences and their complexes with proteins (see Chapter 10), which are providing useful insights into molecular recognition and suggesting new approaches for drug design. In addition, the study of DNA recognition by small molecules in the minor groove has taken a major leap forward with the development of hairpin polyamides as a novel class of DNA-specific reagents with potential as drugs (see Section 9.7.4). Targeting of unusual DNA telomeric G-tetraplex structures is also an active area of current drug design (see Section 9.10).

The heady days of the discovery of the double helix and the elucidation of the genetic code are long gone, but in their place have come even more exciting times when many more of us now have the opportunity to answer fundamental questions about genetic structure and function and can utilise the insights and tools now available in the nucleic acids. '*You ain't heard nothin' yet folks*' (Al Jolson, *The Jazz Singer*; July 1927).

## REFERENCES

1. J.S. Fruton, *Molecules and life*. Wiley Interscience, New York, 1972, 180–224.
2. F.H. Portugal and J.S. Cohen, *A century of DNA*. MIT Press, Cambridge, MA, 1977.
3. O.T. Avery, C.M. MacLeod and M. McCarty, Studies on the chemical nature of the substance inducing transformation of pneumococcal types. *J. Exp. Med.*, 1944, **79**, 137–158.
4. F. Miescher, *Die histochemischen und physiologischen arbeiten*. Vogel, Leipzig, 1897.
5. J.G. Buchanan and Lord Todd. *Adv. Carbohydr. Chem.*, 2000, **55**, 1–13.
6. D.H. Hayes, A.M. Michelson and A.R. Todd, Mononucleotides derived from deoxyadenosine and deoxyguanosine. *J. Chem. Soc.*, 1955, 808–815.
7. E. Chargaff, Chemical specificity of nucleic acids and mechanism of their enzymatic degradation. *Experientia*, 1950, **6**, 201–209.
8. J.D. Watson, *The Double Helix*. Athenaeum Press, New York, 1968.

9. R. Olby, *The Path to the Double Helix*. Macmillan, London, 1973.
10. J.D. Watson and F.H.C. Crick, A structure for deoxyribose nucleic acid. *Nature*, 1953, **171**, 737–738.
11. H.F. Judson, *The Eighth Day of Creation*. Jonathan Cape, London, 1979.
12. M.W. Nirenberg, J.H. Matthei, O.W. Jones, R.G. Martin and S.H. Barondes, Approximation of genetic code via cell-free protein synthesis directed by template RNA. *Fed. Proc.*, 1963, **22**, 55–61.
13. S. Ochoa, Synthetic polynucleotides and the genetic code. *Fed. Proc.*, 1963, **22**, 62–74.
14. H.G. Khorana, Polynucleotide synthesis and the genetic code. *Fed. Proc.*, 1965, **24**, 1473–1487.
15. S.N. Cohen, The manipulation of genes. *Sci. Am.*, 1975, **233**, 24–33.
16. F. Sanger, Sequences, sequences and sequences. *Ann. Rev. Biochem.*, 1988, **57**, 1–28.
17. H.G. Khorana, Total synthesis of a gene. *Science*, 1979, **203**, 614–625.
18. K. Itakura, T. Hirose, R. Crea, A.D. Riggs, H.L. Heyneker, F. Bolivar and H.W. Boyer, Expression in *Escherichia coli* of a chemically synthesized gene for the hormone somatostatin. *Science*, 1977, **198**, 1056–1063.

CHAPTER 2

# DNA and RNA Structure

## CONTENTS

## 2.1 STRUCTURES OF COMPONENTS

Nucleic acids are very long, thread-like polymers, made up of a linear array of monomers called **nucleotides**. Different nucleic acids can have from around 80 nucleotides, as in tRNA, to over $10^8$ nucleotide pairs in a single eukaryotic chromosome. The unit of size of a nucleic acid is the base pair (for double-stranded species) or base (for single-stranded species). The abbreviation* bp is generally used, as are the larger units Mbp (million base pairs) and kbp (thousand base pairs). The chromosome in *Escherichia coli* has $4 \times 10^6$ base pairs, 4 Mbp, which gives it a molecular mass of $3 \times 10^9$ Da and a length of 1.5 mm. The size of the fruit fly genome (haploid) is 180 Mbp which, shared between four chromosomes, gives a total length of 56 mm. The genomic DNA of a single human cell has 3900 Mbp and is 990 mm long. How are these extraordinarily long molecules constructed?

### 2.1.1 Nucleosides and Nucleotides

Nucleotides are the phosphate esters of nucleosides and these are components of both ribonucleic acid (RNA) and deoxyribonucleic acid (DNA). RNA is made up of ribonucleotides whereas the monomers of DNA are 2′-deoxyribonucleotides.

All nucleotides are constructed from three components: a nitrogen heterocyclic **base**, a pentose **sugar** and a **phosphate** residue. The major bases are monocyclic **pyrimidines** or bicyclic **purines**, some species of tRNA have tricyclic minor bases such as the Wye (Figure 3.17). The major purines are **adenine (A)** and **guanine (G)** and are found in both DNA and RNA. The major pyrimidines are **cytosine (C)**, **thymine (T)** and **uracil (U)** (Figure 2.1).

In **nucleosides**, the purine or pyrimidine base is joined from a ring nitrogen to carbon-1 of a pentose sugar. In RNA, the pentose is **D-ribose** which is locked into a five-membered **furanose** ring by the bond from C-1 of the sugar to N-1 of C or U or to N-9 of A or G. This bond is on the same side of the sugar ring as the C-5 hydroxymethyl group and is defined as a β-glycosylic linkage (Figure 2.2).

In DNA, the pentose is 2-deoxy-D-ribose and the four nucleosides are **deoxyadenosine**, **deoxyguanosine**, **deoxycytidine** and **deoxythymidine** (Figure 2.3). In DNA, the methylated pyrimidine base thymine takes the place of uracil in RNA, and its nucleoside with deoxyribose is still commonly called thymidine. However, since the discovery of **ribothymidine** as a regular component of tRNA species, it has been preferable to use the name deoxythymidine rather than thymidine. Unless indicated otherwise, it is assumed that nucleosides, nucleotides and oligonucleotides are derived from D-pentofuranose sugars.

The **phosphate** esters of nucleosides are **nucleotides**, and the simplest of them have one of the hydroxyl groups of the pentose esterified by a single phosphate monoester function. Adenosine 5′-phosphate is a **5′-ribonucleotide**, also called adenylic acid and abbreviated to AMP (Figure 2.4). Similarly, deoxycytidine 3′-phosphate is a **3′-deoxyribonucleotide**, identified as 3′-dCMP. Nucleotides containing two phosphate

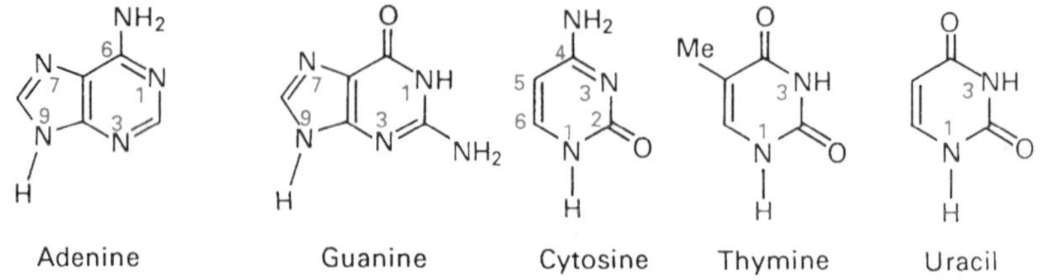


**Figure 2.1** *Structures of the five major purine and pyrimidine bases of nucleic acids in their dominant tautomeric forms and with the IUPAC numbering systems for purines and pyrimidines*

*A useful source for IUPAC nomenclature of nucleic acids can be found at http://www.chem.qmul.ac.uk/iupac/misc/naabb.html and for polynucleotide conformation at http://www.chem.qmul.ac.uk/iupac/misc/pnuc2.html#300.

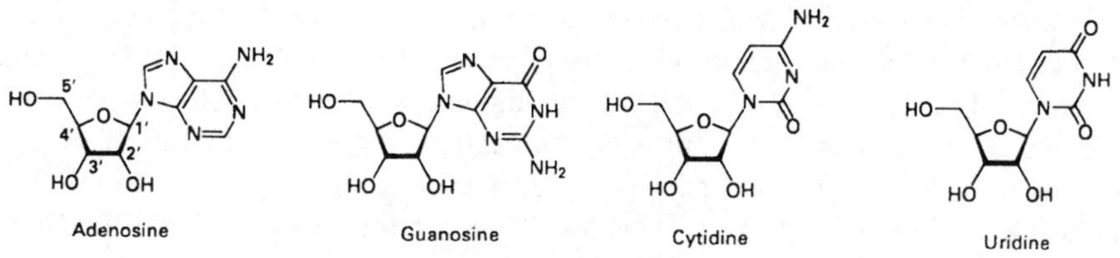


**Figure 2.2** *Structures of the four ribonucleosides. The bases retain the same numbering system and the pentose carbons are numbered 1′ through 5′. By convention, the furanose ring is drawn with its ring oxygen at the back and C-2′ and C-3′ at the front. Hydrogen atoms are usually omitted for clarity*

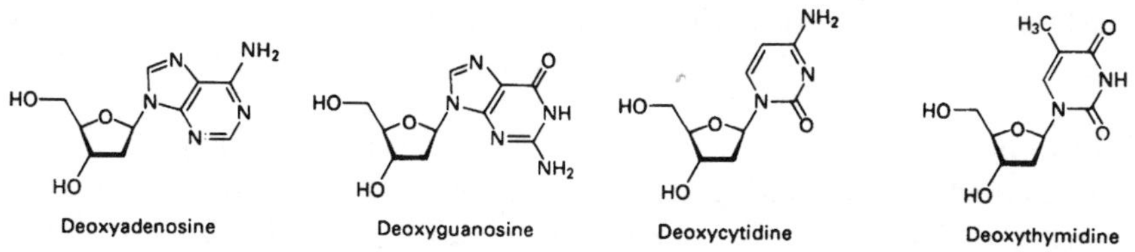


**Figure 2.3** *Structures of the four major deoxyribonucleosides. By convention, only hydrogens bonded to oxygen or nitrogen are depicted*

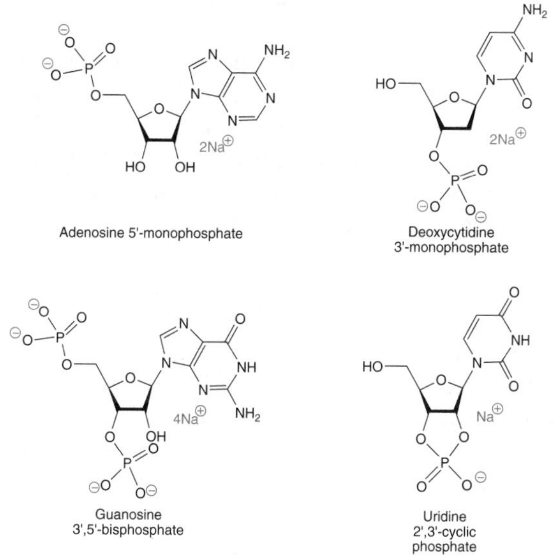


**Figure 2.4** *Structures of some common nucleotides. All are presented as their sodium salts in the state of ionization observed at neutral pH*

monoesters on the same sugar are called **nucleoside bisphosphates** whereas nucleoside monoesters of pyrophosphoric acid are **nucleoside diphosphates**. By extension, nucleoside esters of tripolyphosphoric acid are **nucleoside triphosphates** of which the classic example is adenosine 5′-triphosphate (ATP) (Section 3.3.2). Finally, cyclic nucleotides are nucleosides which have two neighbouring hydroxyl groups on the same pentose esterified by a single phosphate as a diester. The most important of these is adenosine 3′,5′-cyclic phosphate (cAMP).

In the most abbreviated nomenclature currently employed, **pN** stands for 5′-nucleotide, **Np** for a 3′-nucleotide and **dNp** for a 3′-deoxynucleotide (to be precise, a 2′-deoxyribonucleoside 3′-phosphate). This shorthand notation is based on the convention that an oligonucleotide chain is drawn horizontally with its 5′-hydroxyl group at the left- and its 3′-hydroxyl group at the right-hand end. Thus, pppGpp is the shorthand representation of the 'magic spot' nucleotide, guanosine 3′-diphosphate 5′-triphosphate, whereas ApG is short for adenylyl-(3′→5′)-guanosine, whose 3′→5′ internucleotide linkage runs from the nucleoside on the left to that on the right of the phosphate.

### 2.1.2 Physical Properties of Nucleosides and Nucleotides

Owing to their polyionic character, nucleic acids are soluble in water up to about 1% w/v according to size and are precipitated by the addition of alcohol. Their solutions are quite viscous, and the long nucleic acid molecules are easily sheared by stirring or by passage through a fine nozzle such as a hypodermic needle or a fine pipette.

*2.1.2.1 Ionisation.* The acid–base behaviour of a nucleotide is its most important physical characteristic. It determines its charge, its tautomeric structure, and thus its ability to donate and accept hydrogen bonds, which is the key feature of the base:base recognition. The $pK_a$ values for the five bases in the major nucleosides and nucleotides are listed in Table 2.1.

It is clear that all of the bases are uncharged in the physiological range $5 < pH < 9$. The same is true for the pentoses, where the ribose 2′,3′-diol only loses a proton above pH 12 while isolated hydroxyl groups ionise only above pH 15. The nucleotide phosphates lose one proton at pH 1 and a second proton (in the case of monoesters) at pH 7. This pattern of proton equilibria is shown for AMP across the whole pH range (Figure 2.5).

The three amino bases, A, C and G, each becomes protonated on one of the ring nitrogens rather than on the exocyclic amino group since this does not interfere with de-localisation of the $NH_2$ electron lone pair into the aromatic system. The C—$NH_2$ bonds of A, C and G are about 1.34 Å long, which means that they have 40–50% double bond order, while the C=O bonds of C, G, T and U have some 85–90% double bond order. It is also noteworthy that the proximity of negative charge of the phosphate residues has a secondary effect, making the ring nitrogens more basic ($\Delta pK_a \approx +0.4$) and the amine protons less acidic ($\Delta pK_a \approx +0.6$).

*2.1.2.2 Tautomerism.* A tautomeric equilibrium involves alternative structures that differ only in the location of hydrogen atoms. The choices available to nucleic acid bases are illustrated by the **keto–enol** equilibrium between 2-pyridone and 2-hydroxypyridine and the **amine–imine** equilibrium for 2-aminopyridine (Figure 2.6). Ultraviolet, NMR and IR spectroscopies have established that the five major bases exist overwhelmingly (>99.99%) in the **amino**- and **keto**-tautomeric forms at physiological pH (Figure 2.1) and not in the benzene-like **enol** tautomers, in common use before 1950 (Figure 1.3).

**Table 2.1** *$pK_a$ values for bases in nucleosides and nucleotides*

| *Bases (site of protonation)* | *Nucleoside* | *3′-Nucleotide* | *5′-Nucleotide* |
|---|---|---|---|
| Adenine (N-1) | 3.63 | 3.74 | 3.74 |
| Cytosine (N-3) | 4.11 | 4.30 | 4.56 |
| Guanine (N-7) | 2.20 | 2.30 | 2.40 |
| Guanine (N-1) | 9.50 | 9.36 | 9.40 |
| Thymine (N-3) | 9.80 | — | 10.00 |
| Uracil (N-3) | 9.25 | 9.43 | 9.50 |

*Note*: These data approximate to 20°C and zero salt concentration. They correspond to *loss* of a proton for $pK_a > 9$ and *capture* of a proton for $pK_a > 5$.

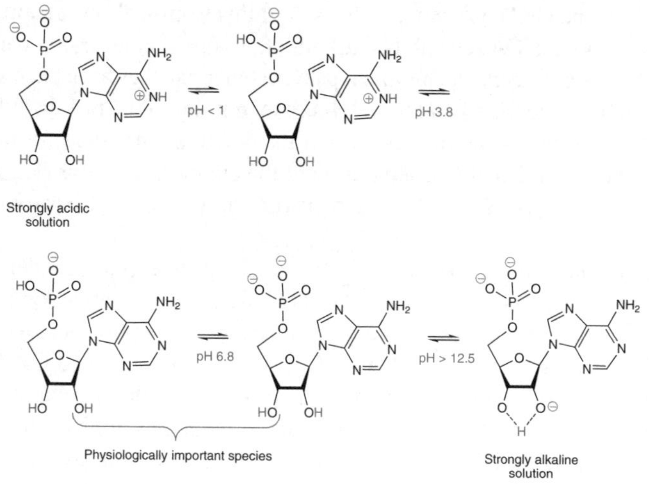


**Figure 2.5** *States of protonation of adenosine 5′-phosphate (AMP) from strongly acidic solution (left) to strongly alkaline solution (right)*

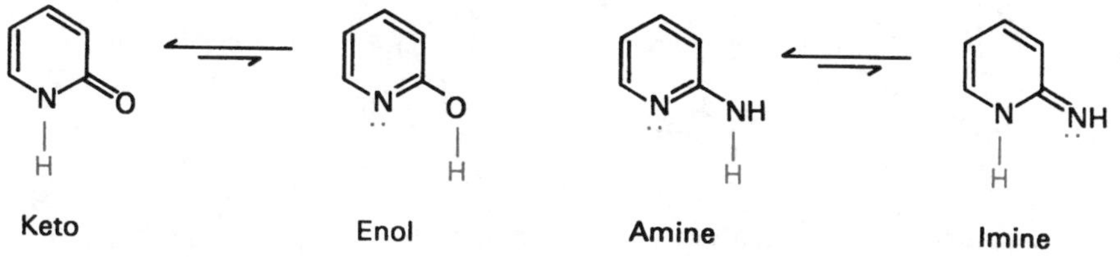


**Figure 2.6** *Keto–enol tautomers for 2-pyridone:2-hydroxypyridine (left) and amine–imine tautomerism for 2-aminopyridine (right)*

*2.1.2.3 Hydrogen Bonding.* The mutual recognition of A by T and of C by G uses hydrogen bonds to establish the fidelity of DNA transcription and translation. The NH groups of the bases are good hydrogen bond donors (**d**), while the $sp^2$-hybridised electron pairs on the oxygens of the base C=O groups and on the ring nitrogens are much better hydrogen bond acceptors (**a**) than are the oxygens of either the phosphate or the pentose. The **a·d** hydrogen bonds so formed are largely electrostatic in character, with a charge of about $+0.2e$ on the hydrogens and about $-0.2e$ on the oxygens and nitrogens, and they seem to have an average strength of 6–10 kJ $mol^{-1}$.

The predominant amino–keto tautomer for cytosine has a pattern of hydrogen bond acceptor and donor sites for which O-2·N-3·N-4 can be expressed as **a·a·d** (Figure 2.7). Its minor tautomer has a very different pattern: **a·d·a**. In the same way, we can establish that the corresponding pattern for the dominant tautomer of dT is **a·d·a** whereas the pattern for N-2·N-1·O-6 of dG is **d·d·a** (Figure 2.7) and that for dA is **(–)·a·d**.

When Jim Watson was engaged in DNA model-building studies in 1952 (Section 1.4), he recognised that the hydrogen bonding capability of an A·T base pair uses complementarity of **(–)·a·d** to **a·d·a** whereas a C·G pair uses the complementarity of **a·a·d** to **d·d·a**. This base-pairing pattern rapidly became known as **Watson–Crick pairing** (Figure 2.8). There are two hydrogen bonds in an A·T pair and three in a C·G pair. The geometry of the pairs has been fully analysed in many structures from dinucleoside phosphates through oligonucleotides to tRNA species, both by the use of X-ray crystallography and, more recently, by NMR spectroscopy.

In planar base pairs, the hydrogen bonds join nitrogen and oxygen atoms that are 2.8–2.95 Å apart. This geometry gives a C-1′···C-1′ distance of 10.60 ± 0.15 Å with an angle of 68 ± 2° between the two glycosylic bonds for both the A·T and the C·G base pairs. As a result of this **isomorphous geometry**, the four base pair combinations A·T, T·A, C·G and G·C can all be built into the same regular framework of the DNA duplex.

While Watson–Crick base pairing is the dominant pattern, other pairings have been suggested of which the most significant to have been identified so far are **Hoogsteen pairs** and Crick **'wobble' pairs**. Hoogsteen pairs, illustrated for A·T, are not isomorphous with Watson–Crick pairs because they have an 80° angle between the glycosylic bonds and an 8.6 Å separation of the anomeric carbons (Figure 2.8). In the case of reverse Hoogsteen pairs and reverse Watson–Crick pairs (not shown), one base is rotated through 180° relative to the other.

Francis Crick proposed the existence of 'wobble' base pairings to explain the degeneracy of the genetic code (Section 7.3.1). This phenomenon calls for a single base in the 5′-anticodon position of tRNA to be able to recognise either of the pyrimidines or, alternatively, either of the purines as its 3′-codon base partner. Thus a G·U 'wobble' pair has two hydrogen bonds, G—N1—H···O2—U and G—O6···H—N3—U, and this requires a sideways shift of one base relative to its positions in the regular Watson–Crick geometry (Figure 2.9). The resulting loss of a hydrogen bond leads to reduced stability which can be offset in part by the improved **base stacking** (Section 2.3.1) that results from such sideways base displacement.

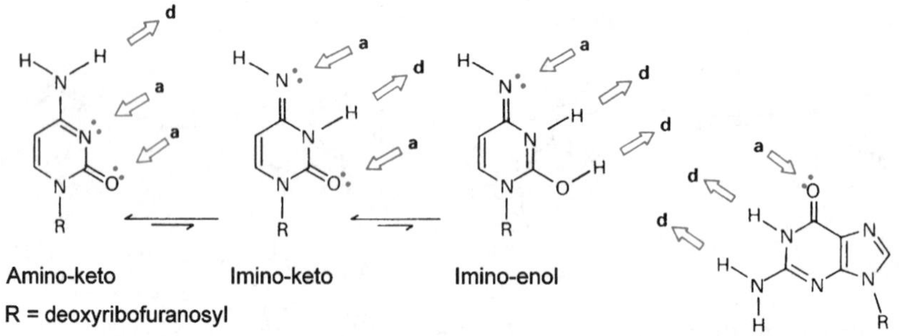


**Figure 2.7** *Tautomeric equilibria for deoxycytidine showing hydrogen-bond acceptor* ***a*** *and donor* ***d*** *sites as used in nucleic acid base pairing. The major tautomer for deoxyguanosine is drawn to show its characteristic* ***d·d·a*** *hydrogen-bond donor–acceptor capacity*

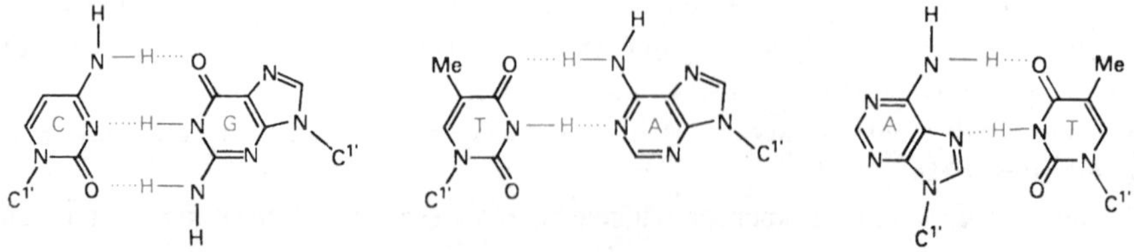

**Figure 2.8** *Watson–Crick base pairing for C·G (left) and T·A (centre). Hoogsteen base pairing for A·T (right)*

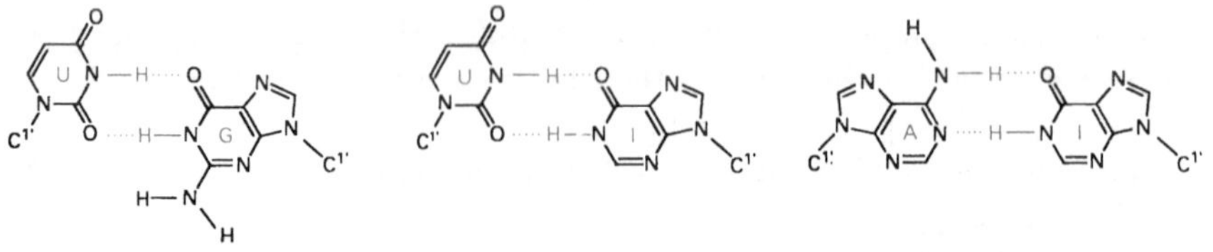

**Figure 2.9** *'Wobble' pairings for U·G (left), U·I (centre) and A·I (right)*

Base pairings of these and other non-Watson–Crick patterns is significant in three structural situations. First, the compact structures of RNAs maximise both base pairing and base stacking wherever possible. This has led to the identification of a considerable variety of reverse Hoogsteen and 'wobble' base pairs as well as of tertiary base pairs (or base-triplets) (Section 7.1.2). Second, where there are triple-stranded helices for DNA and RNA, such as (poly(dA)·2poly(dT)) and (poly(rG)·2poly(rC)), the second pyrimidine chain binds to the purine in the major groove by Hoogsteen hydrogen bonds and runs parallel to the purine chain (Sections 2.3.6 and 2.4.5). Third, mismatched base pairs are necessarily identified with anomalous hydrogen bonding and many such patterns have been revealed by X-ray studies on synthetic oligodeoxyribonucleotides (Section 2.3.2). They are also targets for some DNA repair enzymes (Section 8.11).

### 2.1.3 Spectroscopic Properties of Nucleosides and Nucleotides

Neither the pentose nor the phosphate components of nucleotides show any significant UV absorption above 230 nm. This means that both nucleosides and nucleotides have UV absorption profiles rather similar to those of their constituent bases and absorb strongly with $\lambda_{max}$ values close to 260 nm and molar extinction coefficients of around $10^4$ (Table 2.2).

The light absorptions of isolated nucleoside bases given above are measured in solution in high dilution. They undergo marked changes when they are in close proximity to neighbouring bases, as usually shown in ordered secondary structures of **oligo**- and poly-nucleotides. In such ordered structures, the bases can stack face-to-face and thus share $\pi$–$\pi$ electron interactions that profoundly affect the transition dipoles of the bases. Typically such changes are manifest in a marked reduction in the intensity of UV absorption (by up to 30%), which is known as **hypochromicity** (Section 5.5.1). This phenomenon is reversed on unstacking of the bases.

There are two important applications of this phenomenon. First, it is used in the determination of temperature-dependent and pH-dependent changes in base-stacking. Second, it permits the monitoring of changes in the asymmetric environment of the bases by circular dichroism (CD), or by optical rotatory dispersion (ORD) effects. Both of these techniques are especially valuable for studying helix-coil transitions (Section 11.1.3).

Infrared analysis of nucleic acid components has been less widely used, but the availability of laser Raman and Fourier transform IR methods is making a growing contribution (Section 11.1.4).

Nuclear magnetic resonance has had a dramatic effect on studies of oligonucleotides largely as a result of a variety of complex spin techniques such as NOESY and COSY for proton spectra, the use of $^{17}O$, $^{18}O$ and sulfur substituent effects in $^{31}P$ NMR, and the analysis of nuclear Overhauser effects (nOe) (Section 11.2). These provide a useful measure of inter-nuclear distances and with computational analysis can provide solution conformations of oligonucleotides (Section 2.2). Nucleosides, nucleotides and their analogues have relatively simple $^1H$ NMR spectra. The aromatic protons of the pyrimidines and purines resonate at low field

**Table 2.2** *Some light absorption characteristics for nucleotides*

| | | *pH 1–2* | | *pH > 11* | |
|---|---|---|---|---|---|
| *Compound* | *[$\alpha_D$]** | *$\lambda_{MAX}$ ($nm^{-1}$)* | *$10^{-4} \times \varepsilon$* | *$\lambda_{MAX}$ ($nm^{-1}$)* | *$10^{-4} \times \varepsilon$* |
| Ado 5′-P | −26° | 257 | 1.5 | 259 | 1.54 |
| Guo 3′-P | −57° | 257 | 1.22 | 257 | 1.13 |
| Cyd 3′-P | +27° | 279 | 1.3 | 272 | 0.89 |
| Urd 2′-P | +22° | 262 | 0.99 | 261 | 0.73 |
| Thd 5′-P | +7.3° | 267 | 1.0 | 267[a] | 1.0 |
| 3′,5′-cAMP | −51.3° | 256 | 1.45 | 260[b] | 1.5 |

[a] pH 7.0.
[b] pH 6.0.
* Specific molar rotation.

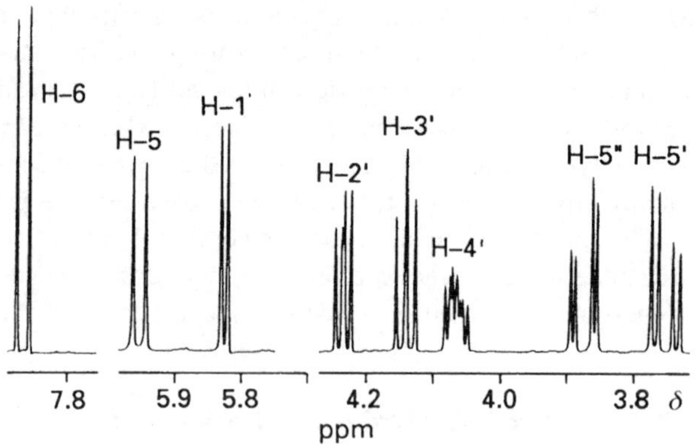


**Figure 2.10** *Proton NMR spectrum for cytidine (run in $D_2O$ at 400 MHz)*

(δ7.6 to δ8.3 with C5–H close to δ5.9). The anomeric hydrogen is a doublet for ribonucleosides and a double-doublet for 2′-deoxynucleosides at δ5.8–6.4. The pentoses provide a multi-spin system that generally moves from low to high field in the series: H-2′, H-3′, H-4′, H-5′ and H-5″ in the region δ4.3 to 3.7. Lastly, 2′-deoxynucleosides have H-2′ and H-2′ as an ABMX system near δ2.5. The 400 MHz spectrum of a simple nucleoside, cytidine (Figure 2.10), shows why two-dimensional (2D) spin techniques are required for the complete analysis of the spectrum in a large oligomer, which may be equivalent to a dozen such monomer spectra superimposed.

### 2.1.4 Shapes of Nucleotides

Nucleotides have rather compact shapes with several interactions between non-bonded atoms. Their molecular geometry is so closely related to that of the corresponding nucleotide units in oligomers and nucleic acid helices that it was once argued that helix structure is a consequence of the conformational preferences of individual nucleotides. However, the current view is that sugar–phosphate backbone appears to act as no more than a constraint on the range of conformational space accessible to the base pairs and that **π–π interactions** between the base pairs provide the driving force for the different conformations of DNA (Section 2.3.1).

The details of conformational structure are accurately defined by the torsion angles α, β, γ, δ, ε, and ζ in the phosphate backbone, $\theta_0$–$\theta_4$ in the furanose ring, and χ for the glycosylic bond (Figure 2.11). Because many of these torsional angles are inter-dependent, we can more simply describe the shapes of nucleotides in terms of four parameters: the sugar pucker, the *syn–anti* conformation of the glycosylic bond, the orientation of C4′–C5′ and the shape of the phosphate ester bonds.

*2.1.4.1 Sugar Pucker.* The furanose rings are twisted out of plane to minimise non-bonded interactions between their substituents. This 'puckering' is described by identifying the major displacement of carbons-2′ and -3′ from the median plane of C1′–O4′–C4′. Thus, if the *endo* displacement of C-2′ is greater than the *exo* displacement of C-3′, the conformation is called C2′-*endo* and so on (Figure 2.11). The *endo* face of the furanose is on the same side as C-5′ and the base; the *exo* face is on the opposite face to the base. These sugar puckers are located in the north (***N***) and south (***S***) domains of the **pseudorotation cycle** of the furanose ring and so spectroscopists frequently use ***N*** and ***S*** designations, which also fortuitously reflect the relative shapes of the C—C—C—C bonds in the C2′-*endo* and -*exo* forms, respectively.[1]

In solution, the *N* and *S* conformations are in rapid equilibrium and are separated by an energy barrier of less than 20 kJ $mol^{-1}$. The average position of the equilibrium can be estimated from the magnitudes of the $^3J$ NMR coupling constants linking H1′—H2′ and H3′—H4′. This is influenced by (1) the preference of electronegative substituents at C-2′ and C-3′ for axial orientation, (2) the orientation of the base (*syn* goes

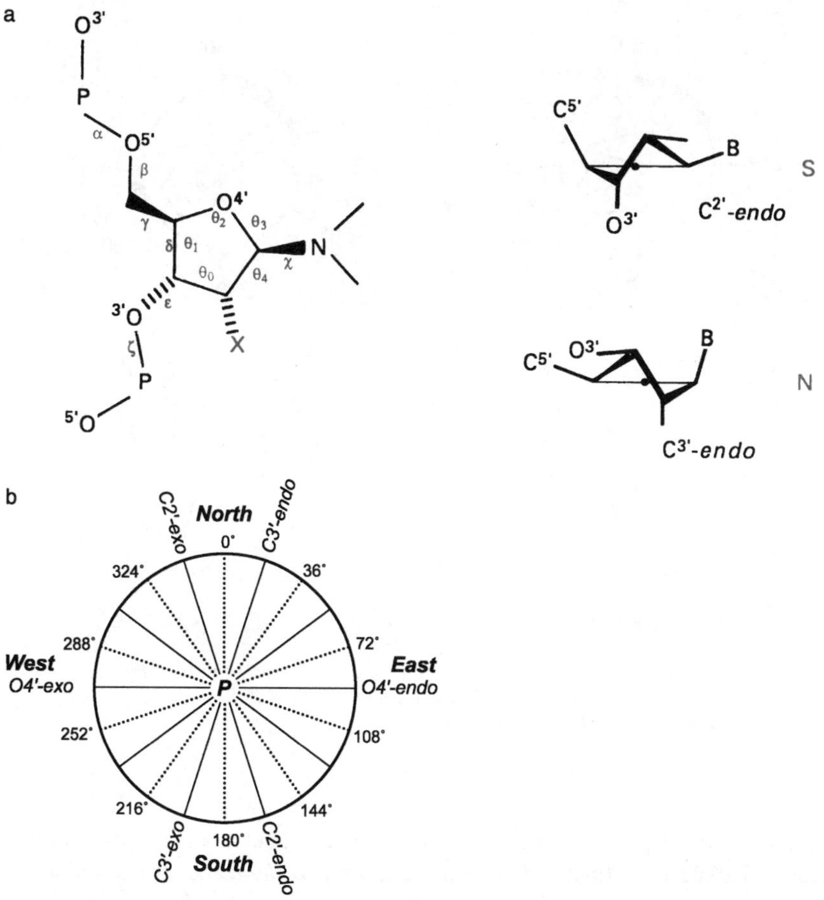


**Figure 2.11** *(a) Torsion angle notation (IUPAC) for poly-nucleotide chains and structures for the C2′-endo(S) and C3′-endo(N) preferred sugar puckers. (b) Schematic of the pseudorotation phase angle (P) cycle with the angle ranges of selected pucker types indicated*

with C2′-*endo*), and (3) the formation of an intra-strand hydrogen bond from O-2′ in one RNA residue to O-4′ in the next which favours C3′-*endo* pucker. However, in RNA helical regions, this latter hydrogen bond is not often observed and an axial C—H···O interaction between the C2′—H2′ ($n$) group and the O-4′ ($n + 1$) atom appears to make a more important contribution to the stability of RNA helices.

*2.1.4.2 Syn–Anti Conformation.* The plane of the bases is almost perpendicular to that of the sugars and approximately bisects the O4′—C1′—C2′ angle. This allows the bases to occupy either of two principal orientations. The *anti* conformer has the smaller H-6 (pyrimidine) or H-8 (purine) atom above the sugar ring, whereas the *syn* conformer has the larger O-2 (pyrimidine) or N-3 (purine) in that position. Pyrimidines occupy a narrow range of *anti* conformations (Figure 2.12) whereas purines are found in a wider range of *anti* conformations that can even extend into the high-*anti* range for 8-azapurine nucleosides such as formycin.

One inevitable consequence of this *anti* conformation for the glycosylic bonds is that the backbone chains for A- and B-forms DNA run downwards on the right of the minor groove and run upwards on the left of the minor groove, depicted as (↑↓).

There is one important exception to the general preference for *anti* forms. Nuclear magnetic resonance, CD and X-ray analyses all show that guanine prefers the *syn* glycoside in mono-nucleotides, in alternating

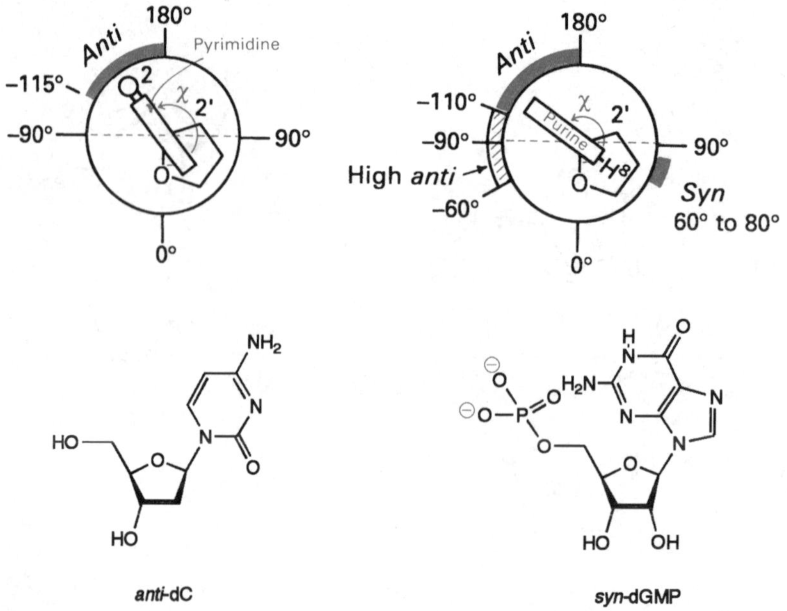


**Figure 2.12** *Anti and syn conformational ranges for glycosylic bonds in pyrimidine (left) and purine (right) nucleosides, and drawings of the anti conformation for deoxycytidine (lower left) and the syn conformation for deoxyguanosine 5′-phosphate (lower right)*

oligomers such as d(CpGpCpG) and in Z-DNA. Theoretical calculations suggest that this effect comes from a favourable electrostatic attraction between the phosphate anion and the C2-amino group in guanine nucleotides. It results from polarisation of one of the nitrogen non-bonding electrons towards the ring. Most unusually, this *syn* conformation can only be built into left-handed helices.

*2.1.4.3 C4′—C5′ Orientation.* The conformation of the exocyclic C4′—C5′ bond determines the position of the 5′-phosphate relative to the sugar ring. The three favoured conformers for this bond are the classical **synclinal (sc)** and **antiperiplanar (ap)** rotamers. For pyrimidine nucleosides, **+sc** is preferred whereas for purine nucleosides **+sc** and **ap** are equally populated. However, in the nucleotides, the 5′-phosphate reduces the conformational freedom and the dominant conformer for this γ-bond is **+sc** (Figure 2.13). Once again, the demands of Z-DNA have a major effect and the **ap** conformer is found for the *syn* guanine deoxynucleotides.

*2.1.4.4 C—O and P—O Ester Bonds.* Phosphate diesters are tetrahedral at phosphorus and show antiperiplanar conformations for the C5′—O5′ bond. Similarly, the C3′—O3′ bond lies in the **antiperiplanar** to **anticlinal** sector. This conformational uniformity has led to the use of the **virtual bond concept** in which the chains P5′—O5′—C5′—C4′ and P3′—O3′—C3′—C4′ can be analysed as rigid, planar units linked at phosphorus and at C-4′. Such a simplification has been used to speed up initial calculations of some complex polymeric structures.

Our knowledge of P—O bond conformations comes largely from X-ray structures of tRNA and DNA oligomers. In general, H4′—C4′—C5′—O5′—P adopts an extended W-conformation in these structures. A skewed conformation for the C—O—P—O—C system has been observed in structures of simple phosphate diesters such as dimethyl phosphate and also for polynucleotides. This has been described as an **anomeric** effect and attributed to the favourable interactions of a non-bonding electron pair on O-5′ with the P—O3′ bond, and vice versa for the P—O5′ bond (Figure 2.14). This may arise from interaction of the

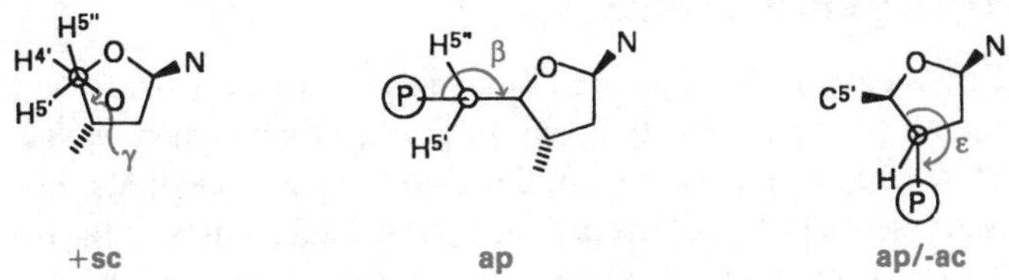


**Figure 2.13** *Preferred nucleotide conformations: +**sc** for C-4′–C-5′ (left); **ap** for C-5′–O-5′ (centre); and **ap/–ac** for C-3′–O-3′ (right)*

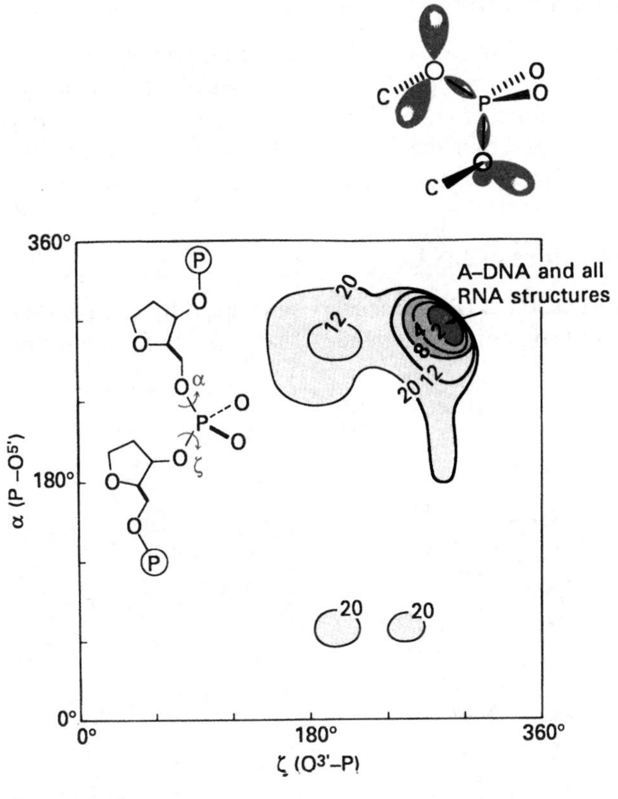


**Figure 2.14** *(Upper) Gauche conformation for phosphate diesters showing the antiperiplanar alignment of an occupied non-bonding oxygen orbital with the adjacent P—O bond. (Lower) Contour map for P—O bond rotations calculated for diribose triphosphate (energies in kJ mol$^{-1}$)*
(Adapted from G. Govil, Biopolymers, 1976, **15**, 2303–2307. © (1976), with permission from John Wiley and Sons, Inc.)

electron lone pair with either phosphorus d orbitals or, more likely, with the P—O anti-bonding σ orbital. The interaction has been calculated at 30 kJ mol$^{-1}$ more favourable than the extended W-conformation for the C—O—P—O—C system. Other non-bonded interactions dictate that α and ζ both have values close to +300° in helical structures though values of +60° are seen in some dinucleoside phosphate structures.

Other P—O conformations have been observed in non-helical nucleotides while left-handed helices also require changed P—O conformations. These changes take place largely in the rotamers for α. In Z-DNA, these are +sc for guanines but broadly **antiperiplanar** for the cytosines whereas ζ is +sc for cytosines but broadly **synperiplanar** for guanines (Section 2.2.2).

## 2.2 STANDARD DNA STRUCTURES

Structural studies on DNA began with the nature of the primary structure of DNA. The classical analysis, completed in mid twentieth century, is easily taken for granted today when we have machines for DNA oligomer synthesis that pre-suppose the integrity of the 3′-to-5′ phosphate diester linkage. Nonetheless, the classical analysis was the essential key that opened the door to later studies on the regular secondary structure of double-stranded DNA and thereby primed the modern revolution known as molecular biology. **Standard structures** for DNA have generally been determined on heterogeneous duplex material and are thus independent of sequence and apply only to Watson–Crick base-pairing.

### 2.2.1 Primary Structure of DNA

Klein and Thannhauser's work (Section 1.4) established that the primary structure of DNA has each nucleoside joined by a phosphate diester from its 5′-hydroxyl group to the 3′-hydroxyl group of one neighbour and by a second phosphate diester from its 3′-hydroxyl group to the 5′-hydroxyl of its other neighbour. There are no 5′-5′ or 3′-3′ linkages in the regular DNA primary structure (Figure 2.15). This means that the uniqueness of a given DNA primary structure resides solely in the sequence of its bases.

### 2.2.2 Secondary Structure of DNA

In the first phase of investigation of DNA secondary structure, diffraction studies on heterogeneous DNA fibres identified two distinct conformations for the DNA double helix.[2] At **low humidity** (and high salt) the

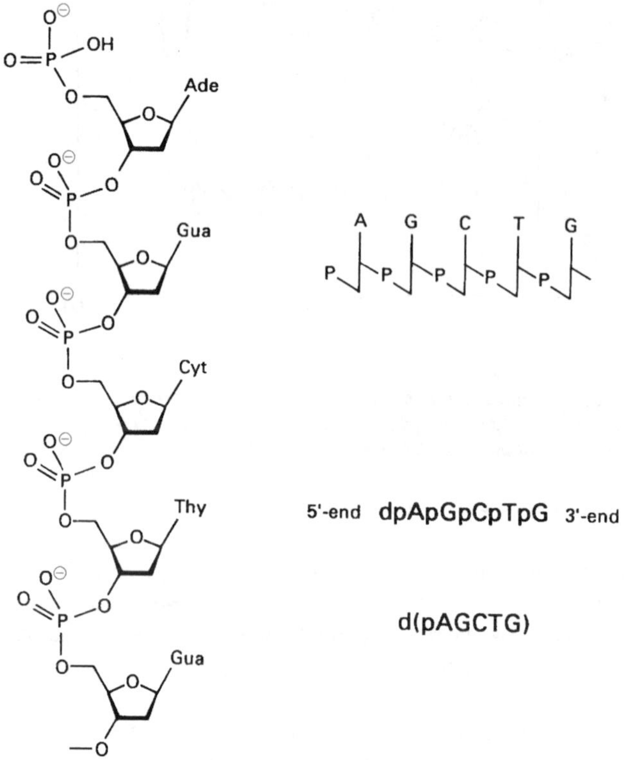


**Figure 2.15** *The primary structure of DNA (left) and three of the common shorthand notations: 'Fischer' (upper right), linear alphabetic (centre right) and condensed alphabetic (lower right)*

favoured form is the highly crystalline **A-DNA** whereas at **high humidity** (and low salt) the dominant structure is **B-DNA**. We now recognise that there is a wide variety of right-handed double helical DNA conformations and this **structural polymorphism** is denoted by the use of the letters A to T as illustrated by A, A′, B, α–B′, β–B′, C, C′, D, E and T forms of DNA. In broad terms, all of these can be classified in two generically different DNA families: A and B. These are associated with the sugar pucker C3′-*endo* for the A-family and C2′-*endo* (or the equivalent C3′-*exo*) for the B-family. However, as we shall see later it is the energetics of base-stacking which determines the conformation of the helix and sugar pucker is largely consequential. We shall also see that in B-form DNA the base pairs sit directly on the helix axis and are nearly perpendicular to it. In A-form DNA the base pairs are displaced off-axis towards the minor groove and are inclined.

The unexpected discovery by Wang, Rich and co-workers in 1979 that the hexamer d(CGCGCG) adopts a left-handed helical structure, now named **Z-DNA**, was one of the first dramatic results to stem from the synthesis of oligonucleotides in sufficient quantity for crystallisation and X-ray diffraction analysis.[3] Since then, over 100 different oligodeoxynucleotide structures have been solved and these have provided the details on which standard DNA structures are now based.[4,5] The main features of A-, B- and Z-DNA are shown in Figures 2.16–2.19 and structural parameters are provided for a range of standard helices in Tables 2.3 and 2.4.

As more highly resolved structures have become available, the idea that these three families of DNA conformations are restricted to standard structures has been whittled away.[6,7] We now accept that there are local, sequence-dependent modulations of structures that are primarily associated with the changes in the orientation of bases. Such changes seek to minimise non-bonded interactions between adjacent bases and

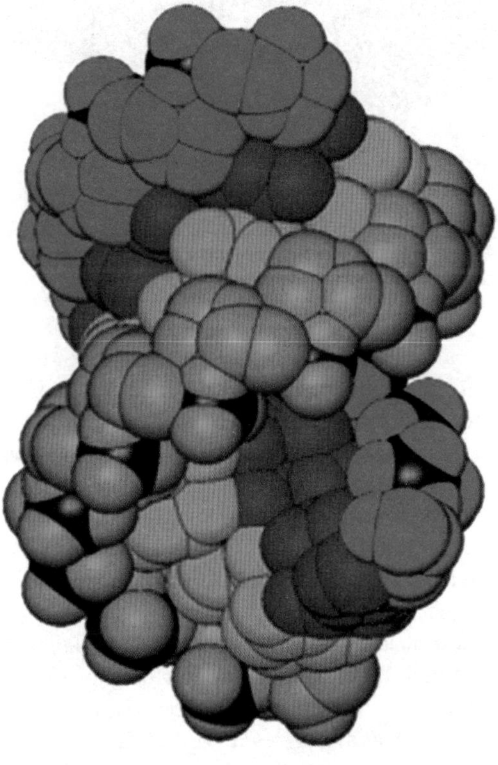

**Figure 2.16** *Van der Waals representation of 10 bp of A-form DNA. The view is across the major (bottom) and minor grooves (top). Atoms of the sugar–phosphate backbones of strands are coloured in red and green, respectively, and the corresponding nucleoside bases are coloured in pink and blue, respectively. Phosphorus atoms are highlighted in black*

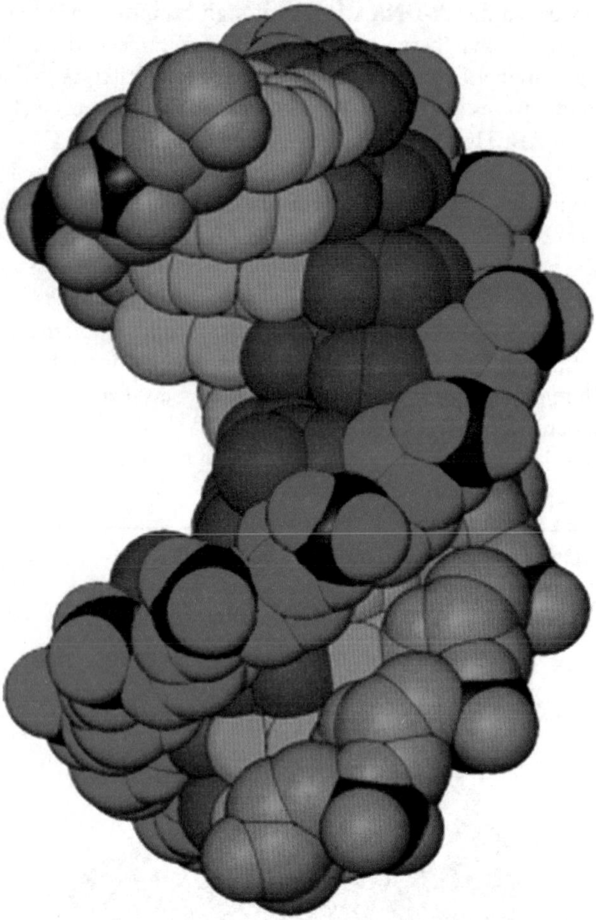

**Figure 2.17** *Van der Waals representation of 10 bp of B-form DNA. The view is across the major (top) and minor grooves (bottom). The colour code is identical to that in Figure 2.16*

maximise base-stacking. They are generally tolerated by the relatively flexible sugar–phosphate backbone. Other studies have explored perturbations in regular helices, which result from deliberate mismatching of base pairs and of lesions caused by chemical modification of bases, such as **base methylation** and **thymine photodimers** (Section 8.8.1). In all of these areas, the results derived from X-ray crystallography have been carried into solution phase by high-resolution NMR analysis, and rationalised by molecular modelling.

Finally, our knowledge of higher order structures, which began with Vinograd's work on DNA **supercoiling** in 1965, has been extended to studies on DNA cruciform structures to 'bent' DNA and to other unusual features of DNA structures.

Regular DNA structures are described by a range of characteristic features.[8,9] The global parameters of **average rise** (**$D_z$**) and **helix rotation** (**$\Omega$**) per base pair define the pitch of the helix. Sideways tilting of the base pairs through a **tilt angle $\tau$** permits the separation of the bases along the **helix axis** $D_z$ to be smaller than the van der Waals distance, 3.4 Å and so gives a shorter, fatter cylindrical envelope for DNA. The angle $\tau$ is positive for A-DNA (positive means a clock-wise rotation of the base pair when viewed end-on and towards the helix axis) but is smaller and negative for B-DNA helices. At the same time, the base pairs are displaced laterally from the helix axis by a distance $D_a$. This parameter together with the groove width defines the depth of the **major groove** and the **minor groove** (Table 2.3).

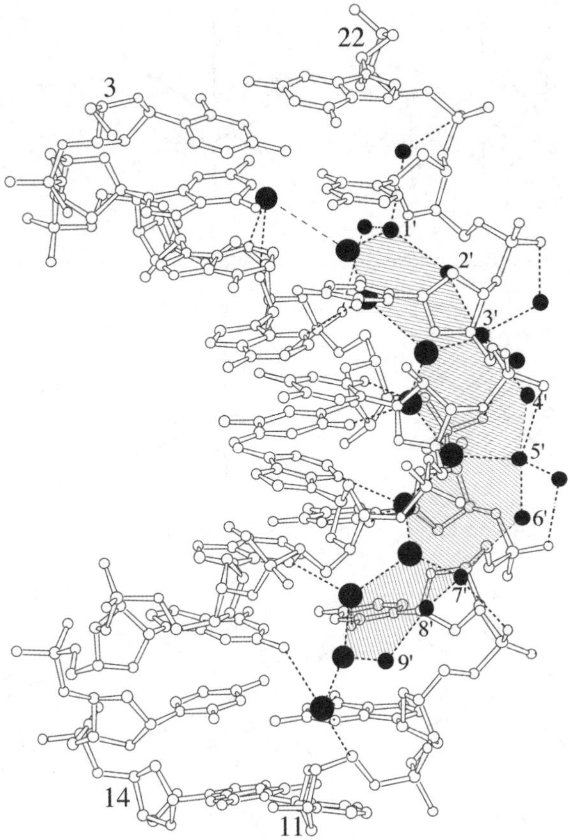


**Figure 2.18** *The minor groove hydration 'ribbon' in the dodecamer d(CGCGAATTCGCG). The inner and outer water (1′–9′) spines define four fused hexagons that dissect the minor groove. Only 10 bp are shown and terminal residues are numbered*
(Adapted from V. Tereshko *et al.*, *J. Am. Chem. Soc.*, 1999, **121**, 3590–3595. © (1999), with permission from the American Chemical Society)

### 2.2.3 A-DNA

Among the first synthetic oligonucleotides to be crystallised in the late 1970s were d(GGTATACC), an iodinated-d(CCGG) and d(GGCCGGCC). They all proved to have A-type DNA structures, similar to the classical A-DNA deduced from fibre analysis at low resolution. Several other oligomers, mostly octamers, also form crystals of the A-structure, but NMR studies suggest that some of these may have the B-form in solution. It is conceivable that crystal packing might especially favour A-DNA for octanucleotides.

The general anatomy of A-DNA follows the Watson–Crick model with anti-parallel, right-handed double helices. The sugar rings are parallel to the helix axis and the phosphate backbone is on the outside of a cylinder of about 24 Å diameter (Figure 2.16).

X-ray diffraction at atomic resolution shows that the bases are displaced 4.5 Å away from the helix axis and this creates a hollow core down the axis around 3 Å in diameter. There are 11 bases in each turn of 28 Å, which gives a vertical rise of 2.56 Å per base pair. To maintain the normal van der Waals separation of 3.4 Å, the stacked bases are tilted sideways through 20°. The sugar backbone has skewed phosphate ester bonds, and anti-periplanar conformations for the adjacent C—O ester bonds. Finally, the furanose ring has a C3′-*endo* pucker and the glycosylic bond is in the *anti* conformation (Table 2.3). As a result of these features, the major groove of A-DNA is cavernously deep and the minor groove is extremely shallow, as

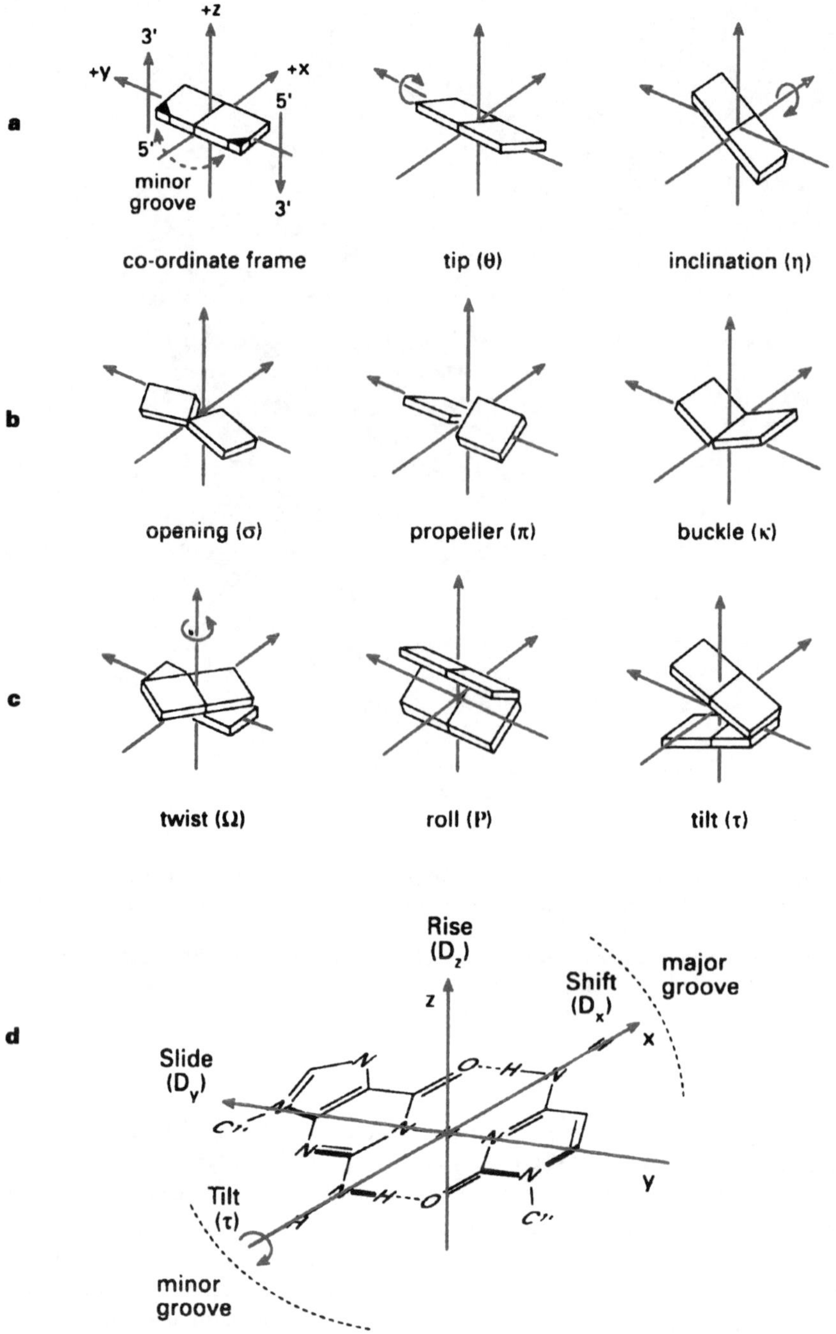


**Figure 2.19** *Diagrams illustrating movements of bases in sequence-dependent structures. Rows (a) to (c) show local rotational helix parameters from the Cambridge DNA nomenclature accord*
(See R.E. Dickerson *et al.*, *Nucleic Acids Res.*, 1989, **17**, 1797–1803)
*Note:* Within each of the three vertical columns, rotations are around the x, y and z axes, from right to left, respectively. (a) Bases of a pair moving in concert. (b) Bases of a pair moving in opposition. (c) Steps between two base pairs. (d) Trans-locational movements of base pairs relative to the helix axis and to the major and minor grooves

**Table 2.3** *Average helix parameters for the major DNA conformations*

| *Structure type* | *Helix sense* | *Residues per turn* | *Twist per bp Ω°* | *Displacement bp D/Å* | *Rise per bp/Å* | *Base tilt (τ°)* | *Sugar pucker* | *Groove (minor)* | *Width/Å (major)* | *Groove (minor)* | *Depth/Å (major)* |
|---|---|---|---|---|---|---|---|---|---|---|---|
| A-DNA | R | 11 | 32.7 | 4.5 | 2.56 | 20 | C3′-*endo* | 11.0 | 2.7 | 2.8 | 13.5 |
| DGGCCGGCC | R | 11 | 32.6 | 3.6 | 3.03 | 12 | C3′-*endo* | 9.6 | 7.9 | — | — |
| B-DNA | R | 10 | 36 | −0.2 to −1.8 | 3.3−3.4 | −6 | C2′-*endo* | 5.7 | 11.7 | 7.5 | 8.8 |
| dCGCGAATTCGCG | R | 9.7 | 37.1 | — | 3.34 | −1.2 | C2′-*endo* | 3.8 | 11.7 | — | — |
| C-DNA | R | 9.33 | 38.5 | −1.0 | 3.31 | −8 | C3′-*exo* | 4.8 | 10.5 | 7.9 | 7.5 |
| D-DNA | R | 8 | 45 | −1.8 | 3.03 | −16 | C3′-*exo* | 1.3 | 8.9 | 6.7 | 5.8 |
| T-DNA | R | 8 | 45 | −1.43 | 3.4 | −6 | C2′-*endo* | Narrow | Wide | Deep | Shallow |
| Z-DNA | L | 12 | −9, −51 | −2 to −3 | 3.7 | −7 | C3′-*endo (syn)* | 2.0 | 8.8 | 13.8 | 3.7 |
| A-RNA | R | 11 | 32.7 | 4.4 | 2.8 | 16–19 | C3′-*endo* | | | | |
| A′-RNA | R | 12 | 30 | 4.4 | 3.0 | 10 | C3′-*endo* | | | | |

**Table 2.4** *Comparison of helix parameters for A-DNA and B-DNA crystal structures and for a model Z-DNA helix*

*1. Base step parameters*

| *Helix* | *Step* | *Roll* | *Tilt* | *Cup* | *Slide* | *Twist* | *Rise* | $D_{xy}$ | $Rad_p$ |
|---|---|---|---|---|---|---|---|---|---|
| B | All | 0.6° | 0.0° | 10.0° | 0.4 Å | 36.1° | 3.36 Å | 3.5 Å | 9.4 Å |
| A | All | 6.3° | — | — | −1.6 Å | 31.1° | 2.6 Å | — | 9.5 Å |
| Z | C–G | −5.8° | 0.0° | 12.5° | 5.4 Å | −9.4° | 3.92 Å | 5.0 Å | 6.3 Å |
| Z | C–G | 5.8° | 0.0° | −12.5° | −1.1 Å | −50.6° | 3.51 Å | 6.0 Å | 7.3 Å |

*2. Base pair parameters*

| | *Base* | *Tip* | *Inclination* | *Propeller* | *Buckle* | *Shift* | *Slide* | *P–P*[a] |
|---|---|---|---|---|---|---|---|---|
| B | All | 0.0° | 2.4° | −11.1° | −0.2° | 0.8 Å | 0.1 Å | 8.8–14 Å |
| A | All | 11.0° | 12.0° | −8.3° | −2.4° | −4.1 Å | — | 11.5–11.9 Å |
| Z | C | 2.9° | −6.2° | −1.3° | −6.2° | 3.0 Å | −2.3 Å | 13.7 Å |
| Z | G | −2.9° | −6.2° | −1.3° | 6.2° | 3.0 Å | 2.3 Å | 7.7 Å |

[a] P–P is the shortest inter-strand distance across the minor groove.

can be appreciated from the 3D picture of the helix (Figure 2.16). This is further characterised by an approximate 5.4 Å P···P separation between adjacent intra-strand phosphorus atoms.

### 2.2.4 The B-DNA Family

The general features of the B-type structure, obtained from DNA fibres at high relative humidity (95% RH) were first put into sharper focus by X-ray studies on the dodecamer d(CGCGAATTCGCG) and its C-5 bromo-derivative at cytosine-9. The structure of the so-called Dickerson–Drew dodecamer has now been revealed at atomic resolution.[10] The B-conformation has been observed in crystals of numerous oligomers and initial standard parameters were averaged from structures of ten isomorphous oligodeoxynucleotides (Figure 2.17).

In B-form DNA, the base pairs sit directly on the helix axis so that the major and minor grooves are of similar depth (Table 2.3). Its bases are stacked predominantly above their neighbours in the same strand and are perpendicular to the helix axis (Table 2.4). The sugars have the C2′-*endo* pucker (with some displaying puckers in the neighbouring ranges of the pseudorotation phase cycle, such as C1′-*exo* or O4′-*endo*), all the glycosides have the *anti* conformation, and most of the other rotamers have normal populations (Table 2.5). Adjacent phosphates in the same chain are further apart, P···P = 6.7 Å, than in A-DNA (Table 2.4).

The interaction of water molecules around a DNA double helix can be very important in stabilising helix structure,[11] to the extent that **hydration** has sometimes been described as the ‘fourth component’ of DNA structure, after bases, sugars and phosphates. Just how many water molecules per base pair can be seen in an X-ray structure depends on the quality of structure resolution. In the best structures, up to 14 unique waters per base pair have been resolved. For B-DNA, whose stability is closely linked to high humidity (Section 2.3.1), highly ordered water molecules can be seen in both major and minor grooves. The broad major groove is ‘coated’ by a uni-molecular layer of water molecules that interact with exposed C═O, N and NH functions and also extensively solvate the phosphate backbone. The narrow minor groove contains an inner and an outer zig-zag chain of water molecules that form four regular planar hexagons in the central A·T region of the Dickerson–Drew dodecamer (Figure 2.18).[12] The inner **spine of hydration** consists of alternating water molecules that are buried at the floor of the groove, directly contacting the bases, and located in the second-shell, above and between first-shell water molecules and closer to the periphery of the groove, respectively.

To a first approximation, the differences between the A, B and other polymorphs of DNA can be described in terms of just two coordinates: slide ($D_y$) and roll ($\sigma$). Clearly, A-DNA has high roll and negative slide

**Table 2.5** *Average torsion angles (°) for DNA helices*

| *Structure type* | $\alpha$ | $\beta$ | $\gamma$ | $\delta$ | $\varepsilon$ | $\zeta$ | $\chi$ |
|---|---|---|---|---|---|---|---|
| A-DNA[a] | −50 | 172 | 41 | 79 | −146 | −78 | −154 |
| GGCCGGCC | −75 | 185 | 56 | 91 | −166 | −75 | −149 |
| B-DNA[a] | −41 | 136 | 38 | 139 | −133 | −157 | −102 |
| CGCGAATTCGCG | −63 | 171 | 54 | 123 | −169 | −108 | −117 |
| Z-DNA (C residues) | −137 | −139 | 56 | 138 | −95 | 80 | −159 |
| Z-DNA (G residues) | 47 | 179 | −169 | 99 | −104 | −69 | 68 |
| DNA–RNA decamer | −69 | 175 | 55 | 82 | −151 | −75 | −162 |
| A-RNA | −68 | 178 | 54 | 82 | −153 | −71 | −158 |

[a] Fibres.

whereas B-DNA has little roll and small positive slide. These and other movements of base pairs are illustrated in Figure 2.19(a) and values of the parameters given in Table 2.4. This results in a greater hydrophobic surface area of the bases being exposed in A-DNA per base pair. From this, it has been argued that B-DNA will have the lesser energy of solvation, explaining its greater stability at high humidity (95%) and that this hydrophobic effect may well tip the balance between the A- and B-form helices.

Other B-DNA structures have much lower significance. C-DNA is obtained from the lithium salt of natural DNA at rather low humidity.[2] It has 28 bases and three full turns of the helix. D-DNA is observed for alternating A·T regions of DNA and has an overwound helix compared to B-DNA with 8 bp per turn. In phage T2 DNA, where cytosine bases have been replaced by glucosylated 5-hydroxymethylcytosines, the B-conformation observed at high humidity changes into a T-DNA form at low humidity (<60% RH), which also has eightfold symmetry around the helix (see Table 2.3).

## 2.2.5 Z-DNA

Two of the earliest crystalline oligodeoxyribonucleotides, d(CGCGCG) and d(CGCG), provided structures of a new type of DNA conformer, the left-handed Z-DNA, which has also been found for d(CGCATGCG). Initially it was thought that left-handed DNA had a strict requirement for alternating purine–pyrimidine sequences. We now know that this condition is neither necessary nor sufficient since left-handed structures have been found for crystals of d(CGATCG) in which cytosines have been modified by C-5 bromination or methylation and have been identified for GTTTG and GACTG sequences by supercoil relaxation studies (Section 2.3.4).

The Z-helix is also an anti-parallel duplex but is a radical departure from the A- and B-forms of DNA. It is best typified by an alternating $(dG\text{–}dC)_n$ polymer. Its two backbone strands run downwards at the left of the minor groove and upwards at the right (↓↑), and this is the opposite from those of A- and B-DNA (↑↓) (NB: the forward direction is defined as the sequence O3′→P→O5′). In an idealised left-handed duplex, such reversed chain directions would require all the nucleosides to have the *syn* conformation for their glycosylic bonds. However, this is not possible for the pyrimidines because of the clash between O-2 of the pyrimidine and the sugar furanose ring (Section 2.1.4). So the cytosines take the *anti* conformation and the guanines the *syn* conformation. The name Z-DNA results from this *anti–syn* feature of the glycosylic bonds that **alternates regularly along the backbone** (Figure 2.20). It causes a local chain reversal that generates a **zig-zag backbone** path and produces a helical repeat consisting of two successive bases (purine-*plus*-pyrimidine) and with an overall chain sense that is the opposite of that of A- and B-DNA. The *syn* conformation of Z-DNA guanines is represented by glycosylic angles $\chi$ close to 60° while the sugar pucker is C2′-*endo* at dC and C3′-*endo* at dG residues (Table 2.3).[13]

The switch from B- to Z-DNA conformation appears to be driven by the energetics of **π–π base-stacking**. In Z-DNA the GpC step is characterised by helical twist of –50.6° and a base pair slide of –1.1 Å. However, for the CpG steps the twist is –9° and the slide is 5.4 Å (Table 2.4; see Figure 2.19 for an explanation of these terms). These preferences occupy the two extremes of the slide axis and thus appear to be incompatible

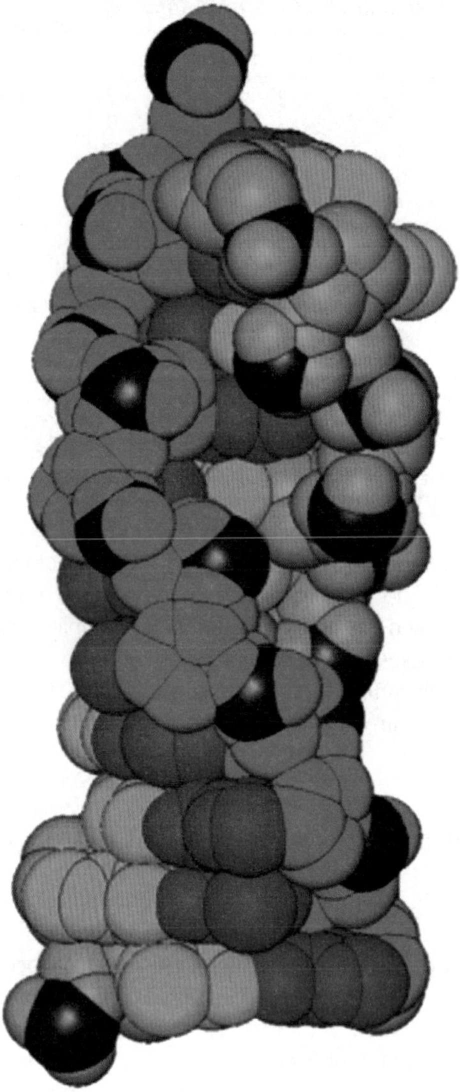

**Figure 2.20** *Van der Waals representation of 10 bp of Z-form DNA. The drawing illustrates the narrow minor groove, visible in the centre of the top half of the duplex, and the lack of an effective major groove that takes on the shape of a convex surface instead, visible on the left-hand side of the bottom half of the duplex. The colour code is identical to that in Figure 2.16*

with a standard right-handed helix, for which helix twist is 36° and base pair slide is 0.4 Å in B-DNA. However, these extremes taken together can be accommodated by a left-handed, Z-type helix. A similar analysis also explains the preference for a Z-helix in the polymer $(dG–dT)_n \cdot (dA–dC)_n$.

The net result of these changes is that the minor groove of Z-DNA is so deep that it actually contains the helix axis whilst the major groove of Z-DNA has become a convex surface on which cytosine-C-5 and guanine-N-7 and-C-8 are exposed (Figure 2.20 and Table 2.3).

Solution studies on poly(dG–dC) have shown a salt-dependent transition between conformers that can be monitored by CD or by $^{31}$P NMR (Section 11.2). In particular, there is a near inversion in the CD spectrum above 4 M NaCl, which has been identified as a change from B- to Z-DNA. It appears that a high salt concentration stabilises the Z-conformation because it has a much smaller separation between the phosphate anions in opposing strands than is the case for B-DNA, 8 Å as opposed to 11.7 Å. A detailed stereochemical examination of this conformational change shows that it calls for an elaborate mechanism and this has posed a problem known as the **chain–sense paradox**: 'How does one reverse the sense of direction of the chains in a B-helix (↑↓) to its opposite in a Z-helix (↓↑) without unpairing the bases?' Further consideration to this problem will be given later (Section 2.5.4).

The scanning tunnelling microscope has the power to resolve the structure of biological molecules with atomic detail (Section 11.5.2). Much progress has been made with dried samples of duplex DNA, in recording images of DNA in wet state, and in revealing details of single-stranded poly(dA). Such STM microscopy has provided images of poly(dG-me$^5$dC)·poly(dG-me$^5$dC) in the Z-form. Both the general appearance of the fibres and measurements of helical parameters are in good agreement with models derived from X-ray diffraction data.

## 2.3 REAL DNA STRUCTURES

### 2.3.1 Sequence-Dependent Modulation of DNA Structure

So far we have emphasised the importance of hydrogen bonds in base-pairing and DNA structure and have said little about base stacking. We shall see later that both these two features are important for the energetics and dynamics of DNA helices (Section 2.5), but it is now time to look at the major part played by base stacking in real DNA structures. Two particular hallmarks of B-DNA, in contrast to the A- and Z-forms, are its flexibility and its capacity to make small adjustments in local helix structure in response to particular base sequences.[14]

Different base sequences have their own characteristic signature: they influence groove width, helical twist, curvature, mechanical rigidity and resistance to bending. It seems probable that these features help proteins to read and recognise one base sequence in preference to another (Chapter 10), possibly only through changes in the positions of the phosphates in the backbone. What do we know about these sequence-dependent structural features?

One surprise to emerge from single-crystal structure analyses of synthetic DNA oligomers has been the breadth of variation of local helix parameters relative to the mean values broadly derived from fibre diffraction analysis and used for the standard A- and B-form DNA structures described earlier. Dickerson has compared eight dodecamer and three decamer B-DNA structures.[15] The mean value of the **helical twist** angle between neighbouring base pairs is 36.1° but the standard deviation (SD) is 5.9° and the range is from 24° to 51°. Likewise, the mean **helical rise** per base pair is 3.36 Å with a SD of 0.46 Å but with a range from 2.5 to 4.4 Å. (NB: because rise is a parameter measured between the C-1′ atoms of adjacent base-pairs, it can be smaller than the thickness of a base pair if the ends of the two base pairs bow towards each other. Such **bowing** is also defined as 'positive cup'.) Roll angles between successive base pairs average +0.6° but with a SD of 6.0° and a range from –18° to +16°. These variations in twist and roll have the effect of substantially re-orienting the potential hydrogen bond acceptors and donors at the edges of the bases along the floor of the DNA grooves, so they may well be a significant component of the sequence-recognition process used by drugs and proteins (Chapters 9 and 10). These and other modes of local changes in the geometry of base pairs are illustrated in Figure 2.19.

The major irregularities in the positions of the bases in real DNA structures contrast with only secondary, small conformational changes in their sugar–phosphate backbones. The main characteristic of these sequence-dependent modulations is **propeller twist**. This results when the bases rotate by some 5° to 25° relative to their hydrogen-bonded partner around the long axis through C-8 of the purine and C-6 of the pyrimidine (Figure 2.19b, centre). Sections of oligonucleotides with consecutive A residues, as in d(CGCAAAAAAGCG)·d(CGCTTTTTTGCG) have unusually high propeller twist (approximately 25°) and these permit the formation of a three-centred hydrogen bonding network in the major groove between adenine-N-6 and two thymine-O-4 residues, the first being the Watson–Crick base pair partner and the second being its 3′-neighbour, both in the opposing strand. This network of hydrogen bonds gives added rigidity to the duplex and may explain why long runs of adenines are not found in the more sharply curved tracts of chromosomes (Section 2.6.2), yet are found at the end of nucleosomal DNA with decreased supercoiling.

Why should the bases twist in this way?[16] The advantage of propeller twist is that it gives improved face-to-face contact between adjacent bases in the same strand and this leads to increased stacking stability in the double-helix. However, there is a penalty! The larger purine bases occupy the centre of the helix so that in alternating purine–pyrimidine sequences they overlap with neighbouring purines in the opposite strand. Consequently, propeller twist causes a clash between such pairs in adjacent purines in opposite strands. For pyrimidine-(3→5′)–purine steps, these purine–purine clashes take place in the minor groove where they involve guanine-N-3 and-N-2 and adenine-N-3 atoms. For purine-(3′→5′)–pyrimidine steps, they take place in the major groove between guanine-O-6 and adenine-N-6 atoms (Figure 2.21). There are no such clashes for purine–purine and pyrimidine–pyrimidine sequences.

One of the consequences of these effects is that bends may occur at junctions between polyA tracts and mixed-sequence DNA as a result of propeller twist, base pair inclination and base-stacking differences on two sides of the junction (see below).

*2.3.1.1 Electrostatic Interactions between Bases.* There are two principal types of base–base interaction that drive the local variations in helix parameters described above and in Figure 2.19a–c. First, there are repulsive steric interactions between proximate bases and sugars. They are associated with steric interactions between thymine methyl groups, the guanine amino group and the configuration of the step pyrimidine–purine (described as YR), purine–pyrimidine (described as RY) and RR/YY. Second, there are $\pi$–$\pi$ stacking interactions that are determined by the distribution of $\pi$-electron density above and below the planar bases.

Chris Hunter has identified four principal contributions to the energy of $\pi$–$\pi$ interactions between DNA base–pairs:[17]

(1) van der Waals interactions (designated *vdW* and vary as $r^{-6}$)
(2) Electrostatic interactions between partial atomic charges (designated *atom–atom* and vary as $r^{-1}$)
(3) Electrostatic interactions between the charge distributions associated with the $\pi$-electron density above and below the plane of the bases (designated $\pi\sigma$–$\pi\sigma$ and vary approximately as $r^{-5}$)
(4) Electrostatic interactions between the charge distributions associated with the $\pi$-electron density and the partial atomic charges (designated as atom–$\pi\sigma$, this is the cross-term of (2) and (3) and varies as $r^{-4}$)

He has used these components to calculate the $\pi$–$\pi$ interaction energies between pairs of stacked bases and applied the results to interpret the source of slide, roll and helical twist, of propeller twist, and of a range of other conformational preferences that are sequence-dependent. In addition, his calculations correlate

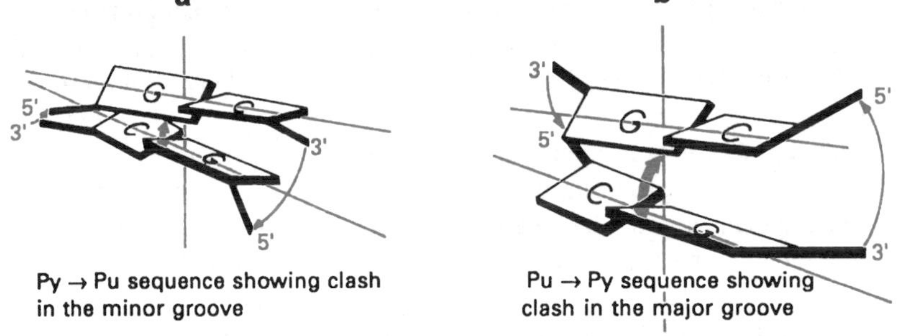


**Figure 2.21** *Diagrams illustrating (a) clockwise propeller twist for a C-(3′→5′)-G clash between guanines in the* ***minor*** *groove and (b) clockwise propeller twist for a G-(3′→5′)-C sequence showing purines clashing in the* ***major*** *groove*

very well with experimental observations on polymorphic forms of DNA. The main conclusions can broadly be summarised as follows:

- vdW–steric interactions are seen cross-strand at pyrimidine–purine (YR) and CX/XG steps and can be diminished by reducing propeller twist, reducing helical twist, or by positive slide or positive roll. They are seen as same-strand clashes between the thymine methyl group and the neighbouring 5′-sugar in AX/XT steps which are avoided by introducing negative propeller twist, reducing helical twist, or generating negative slide coupled with negative roll.
- Electrostatic interactions cause positive or negative slide with the sole exception of AA/TT. These slide effects are opposed by the hydrophobic effect, which tends to force maximum base overlap and favours a zero-slide B-type conformation.
- Atom–atom interactions are most important for C·G base-pairs, where there are large regions of charge and lead to strong conformational preferences for positive slide in CG steps and negative slide in GC steps (see Table 2.4). This leads poly(dCG) to adopt the Z-form left-handed duplex.
- Atom–$\pi\sigma$ interactions lead to sequence-dependent effects, which are repulsive in AX/XT, TX/XA and CX/XG steps where they can be reduced by negative propeller twist, by positive or negative slide, or by introducing buckle.
- $\pi\sigma$–$\pi\sigma$ electrostatic interactions tend to be swamped by other effects and play a relatively minor role in sequence-dependent conformations.

In sequence-dependent structures, propeller twist is most marked for purines on opposing strands in successive base pairs. The 'purine–purine' clash is much more pronounced for YR steps, where the clash is in the minor groove (Figure 2.21a), than for RY steps, where the clash is seen in the major groove (Figure 2.21b). Although its origin was at first thought to result solely from van der Waals interactions, it seems now to be better explained by the total electrostatic interaction picture (see above).

Taken together, these sequence-dependent features suggest that DNA should most easily be unwound and/or unpaired in A-T rich sequences, which have only two hydrogen bonds per base pair, and in pyrimidine–purine steps. It is noteworthy that the dinucleotide TpA satisfies both of these requirements and has been identified as the base step that serves as a nucleus for DNA unwinding in many enzymatic reactions requiring strand separation.

*2.3.1.2 Calladine's Rules.* Notwithstanding the apparent success of the above calculations, the evidence from analyses of X-ray structures suggests that base step conformations are influenced by the nature of neighbouring steps. It follows that a better sequence–structure correlation is likely to emerge from examining each step in the context of its flankers: three successive base steps, or a tetrad of four successive base pairs. However, until a majority of the 136 possible triads has been sampled by analysis of real structures, a set of empirical rules enunciated by Chris Calladine in 1982 will remain useful.[9]

Calladine observed that B-DNA structures respond to minimise the problems of sequence-dependent base clashes in four ways, which he articulated as follows:

- Flatten the propeller twist locally for either or both base-pairs
- Roll the base pairs away from their clashing edges
- Slide one or both of the base pairs along their axis to push the purine away from the helix axis
- Unwind the helix axis locally to diminish inter-strand purine–purine overlap.

The relative motions required to achieve these effects are described by six parameters, of which the most significant are $\rho$ for roll, $D_y$ for slide and $\Omega$ for helix twist. These motions are illustrated for neighbouring G·C base pairs (Figure 2.19).

In practice, the structures of crystalline oligomers have exhibited the following six types of conformational modulation which are sequence-dependent and which support these rules:

1. The B-DNA helix axis need not be straight but can curve with a radius of 112 Å.
2. The twist angle, $\Omega$ is not constant at 36° but can vary from 28° to 43°.

3. Propeller twist averages –11° for C·G pairs and +17° for A·T pairs.
4. Base pairs 'roll' along their long axes to reduce clashing.
5. Sugar pucker varies from C3′-*exo* to O4′-*endo* to C2′-*endo*.
6. There can be local improved overlap of bases by slide, as in d(TCG) where C-2 moves towards the helix axis to increase stacking with G-3.

The Calladine model is incomplete because it ignores such important factors as electrostatic interactions, hydrogen bonding and hydration. For example, a major stabilising influence proposed for the high propeller twist in sequences with consecutive adenines is the existence of cross-strand hydrogen bonding between adenine N-6 in one strand with thymine O-4 of the next base pair in the opposite strand (see above).

Modulations of B-DNA structure, which have been observed in the solid state, have been mirrored to some extent by the results of solution studies for d(GCATGC) and d(CTGGATCCAG) obtained by a combination of NMR analysis and restrained molecular dynamics calculations. These oligomers have B-type structures, which show clear, sequence-dependent variations in torsion angles and helix parameters. There is a strong curvature to the helix axis of the hexamer, which results from large positive roll angles at the pyrimidine–purine steps. The decamer has a straight central core but there are bends in the helix axis at the second (TpG) and eighth (CpA) steps, which result from positive roll angles and large slide values.

Taken together, these X-ray and NMR analyses give good support for the general conclusion that minor groove clashes at pyrimidine–purine steps are twice as severe as major groove clashes at purine–pyrimidine steps. As a result, it is possible to calculate the behaviour of the helix twist angle, $\Omega$, using sequence data only.

*2.3.1.3 The Continuum of Right-Handed DNA Conformations.* The simple concept that the standard conformations for right-handed DNA represent discontinuous states, only stable in very different environments, has undergone marked revision. In addition to the range of conformations seen in crystal structures, CD and NMR analyses of solution structures have also undermined that naïve picture. In particular, CD studies have shown that there is a continuum of helix conformations in solution that is sequence-dependent while both CD analysis of the complete TFIIIA binding site of 54 bp and the crystal structure of a nonameric fragment from it have identified a conformer that is intermediate between the canonical A- and B-DNA forms. Crystallographic analysis of complexes between TATA-box binding protein (TBP) and DNA fragments containing TATA boxes has revealed a DNA structure that shares features of A- and B-DNA. In addition, A- and B-DNA polymorphs can co-exist, as seen in the crystal structure of d(GGBrUABrUACC), and stable intermediates between the A- and B-DNA forms have been trapped in crystal structures.[18] By use of 13 separate structures of the hexamer duplex $[d(GGCGCC)]_2$ in different crystallographic environments, P. Shing Ho and collaborators were able to map the transition from B-DNA to A-DNA.[19] Their analysis demonstrated that little correlation exists between helix type and base pair inclination and that the single parameter with which to follow the B→A transition appears to be $x$-displacement (Figure 2.19).

*2.3.1.4 Bending at Helix Junctions.* Bent DNA was first identified as a result of modelling the junction between an A- and a B-type helix. The best solution to this problem requires a bend of 26° in the helix axis to maintain full stacking of the bases. Bent DNA has gained support not only from NMR and CD studies on a DNA·RNA hybrid [poly(dG)·$(rC)_{11}$– $(dC)_{16}$], but also from studies on regular homo-polymers which contain $(dA)_5$·$(dT)_5$ sections occurring in phase in each turn of a 10- or 11-fold helix. Moreover, bent DNA containing such dA·dT repeats has been investigated from a variety of natural sources.

It appears that bending of this sort happens at junctions between the stiff [dA·dT] helix and the regular B-helix (see above). In situations where such junctions occur every five bases and in an alternating sense, the net result is a progression of bends, which is equivalent to a continuous curve in the DNA.

### 2.3.2 Mismatched Base–Pairs

The fidelity of transmission of the genetic code rests on the specific pairings of A·T and C·G bases. Consequently, if changes in shape result from base mismatches, such as A·G, they must be recognised and be repaired by enzymes with high efficiency (Section 8.11.6).

X-ray analysis of DNA fragments with potential mispairs cannot give any information about the transient occurrence of rare tautomeric forms at the instant of replication. However, it can define the structure of a DNA duplex, which incorporates mismatched base pairs and provides details of the hydrogen bonding scheme, the response of the duplex to the mismatch, the influence of neighbouring sequence on the structure and stability of the mismatch, and the effect of global conformation.[20] All these are intended to provide clues about the ways in which mismatches might be recognised by the proteins that constitute repair systems. High-resolution NMR studies have extended the picture to solution conformations. The different types of base pair mismatch can be grouped into **transition mismatches**, which pair a purine with the wrong pyrimidine, and **transversion mismatches**, which pair either two purines or two pyrimidines.

*2.3.2.1 Transition Mismatches.* The G·T base pair has been observed in crystal structures for A-, B- and Z-conformations of oligonucleotides. In every case it has been found to be a typical 'wobble' pair having *anti–anti* glycosylic bonds. The structure of the dodecamer, d(CGCGAATTTGCG), which has two G·T-9 mismatches, can be superimposed on that of the regular dodecamer and shows excellent correspondence of backbone atomic positions.

The A·C pair has been examined in the dodecamer d(CGCAAATTCGCG) and once again the two A-4·C mismatches are typical 'wobble' pairs, achieved by the protonation of adenine-N-1 (Figure 2.22). It is notable that there is no significant worsening of base-stacking and little perturbation of the helix conformation. However, it appears that no water molecules are bonded to these bases in the minor groove.

*2.3.2.2 Transversion Mismatches.* The G·A mismatch is thoroughly studied in the solution and solid states and two different patterns have been found. Crystals of the dodecamer d(CGCGAATTAGCG) have an (*anti*)G·A(*syn*) mismatch with hydrogen bonds from Ade-N-7 to Gua-N-1 and from Ade-N-6 to Gua-O-6 (Figure 2.23). A similar (*anti*)I·A(*syn*) mismatch has been identified in a related dodecamer structure. Calculations on both of these mismatches suggest that they can be accommodated into a regular B-helix with minimal perturbation.

This work contrasts with both NMR and X-ray studies on d(CGAA**GA**TTGG) and NMR work on d(CG**A**GAATTC**G**CG), which have identified (*anti*)G·A(*anti*) pairings with two hydrogen bonds. The X-ray analysis of the dodecamer shows a typical B-helix with a broader minor groove and a changed

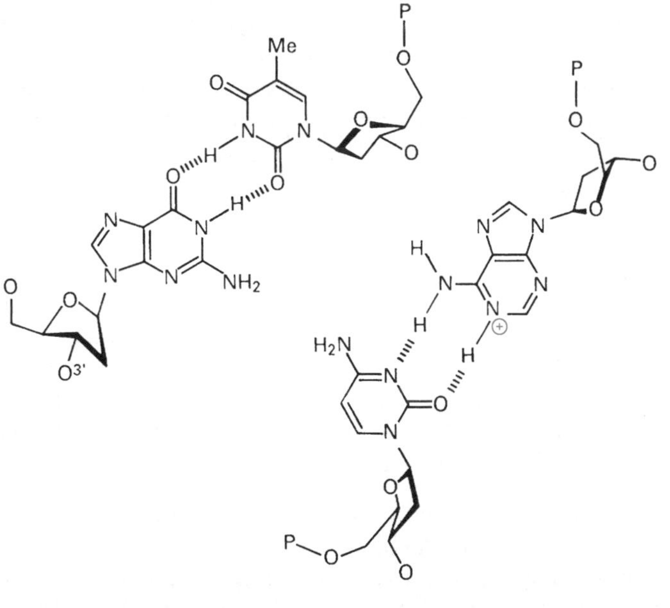

**Figure 2.22** *'Wobble' pairs for transition mismatches G·T (left) and A·C (right)*

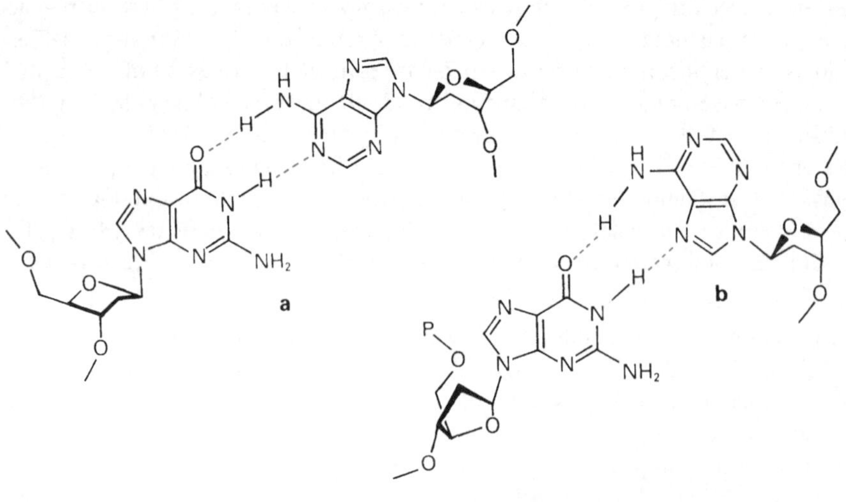


**Figure 2.23** *Mismatched G·A pairings (a) for the decamer d(CCAAGATTGG) with (anti)G·A(anti) and (b) for the dodecamer d(CGAGAATTCGCG) with (anti)G·A(syn) conformation*

pattern of hydration. This arises in part because the two mismatched G·A pairs are 2.0 Å wider (from C-1′ to C-1′) than a conventional Watson–Crick pair.

*2.3.2.3 Insertion–Deletion Mispairs.* When one DNA strand has one nucleotide more than the other, the extra residue can either be accommodated in an intra-strand position or be forced into an extra-strand location. Tridecanucleotides containing an extra A, C or T residue have been examined in the crystalline solid and solution states. In one case, an extra A has been accommodated into the helix stack while in others a C or A is seen to be extruded into an extrahelical, unstacked location.

In addition to such work on mismatched base pairs, related investigations have made good progress into structural changes caused by covalent modification of DNA. On the one hand, crystal structures of DNA adducts with cisplatin have characterised its monofunctional linking to guanine sites in a B-DNA helix (Section 8.5.4). On the other, NMR studies of $O^4$-methylthymine residues and of thymine photodimers and psoralen:DNA photoproducts are advancing our understanding of the modifications to DNA structure that result from such lesions (Section 8.8.2). It seems likely that the range of patterns of recognition of structural abnormalities may be as wide as the range of enzymes available to repair them!

## 2.3.3 Unusual DNA Structures

Since 1980, there has been a rapid expansion in our awareness of the heterogeneity of DNA structures which has resulted from a widening use of new analytical techniques notably structure-dependent nuclease action, structure-dependent chemical modification and physical analysis.[21] Unusual structures are generally sequence-specific, as we have already described for the A–B helix junction (Section 2.3.1). Some of them are also dependent on DNA supercoiling, which provides the necessary driving energy for their formation due to the release of torsional strain, as is particularly well defined for cruciform DNA. Consequently, much use has been made of synthetic DNA both in short oligonucleotides and cloned into circular DNA plasmids where the effect of DNA supercoiling can be explored.

*2.3.3.1 Curved DNA.* The axial flexibility of DNA is one of the significant factors in DNA–protein interactions (Chapter 10).[22,23] DNA duplexes up to 150 bp long behave in solution as stiff, although not necessarily straight, rods. By contrast, many large DNA–protein complexes have DNA that is tightly bent. One of the best examples is the bending of DNA in the eukaryotic chromosome where 146 bp of DNA are wrapped around a protein core of histones (Section 10.6.1) to form nearly two complete turns on a left-handed superhelix with a radius of curvature of 43 Å. To achieve this, the major and minor grooves are compressed on the inside of the curve and stretched on its outside. At the same time, the helix axis must change direction.

DNA curvature has also been examined in kinetoplast DNA from trypanosomatids. It provides a source of *open* DNA mini-circles whose curvature is sequence-dependent rather than being enforced by covalent closure of the circles. Such circles can be examined by electron microscopy and have 360° curvature for about 200 bp. Such kinetoplast DNA has short adenine tracts spaced at 10 bp intervals by general sequence. This fact led to solution studies on synthetic oligomers with repeated sets of four $CA_{5-6}T$ sequences spaced by 2–3 bp. These behave as though they have a 20°–25° bend for each repeat, which led to the simple idea that DNA bending is an inherent property of poly(dA) tracts (Figure 2.24a). In conflict with this idea, poly(dA) tracts in the crystal structures of several oligonucleotides are seen to be straight. What then is the real origin of DNA curvature?[24,25]

Richard Dickerson has examined helix bending in a range of B-form crystal structures of oligonucleotides containing poly(dA) tracts and has concluded that poly(dA) tracts are straight and not bent and that regions of A·T base pairs exhibit a narrow minor groove, large propeller twist and a spine of hydration in the minor groove. He argues that DNA curvature results from the direct combination of two general features of DNA structure.

1. General sequence DNA writhes
2. Poly(dA) tract are straight.

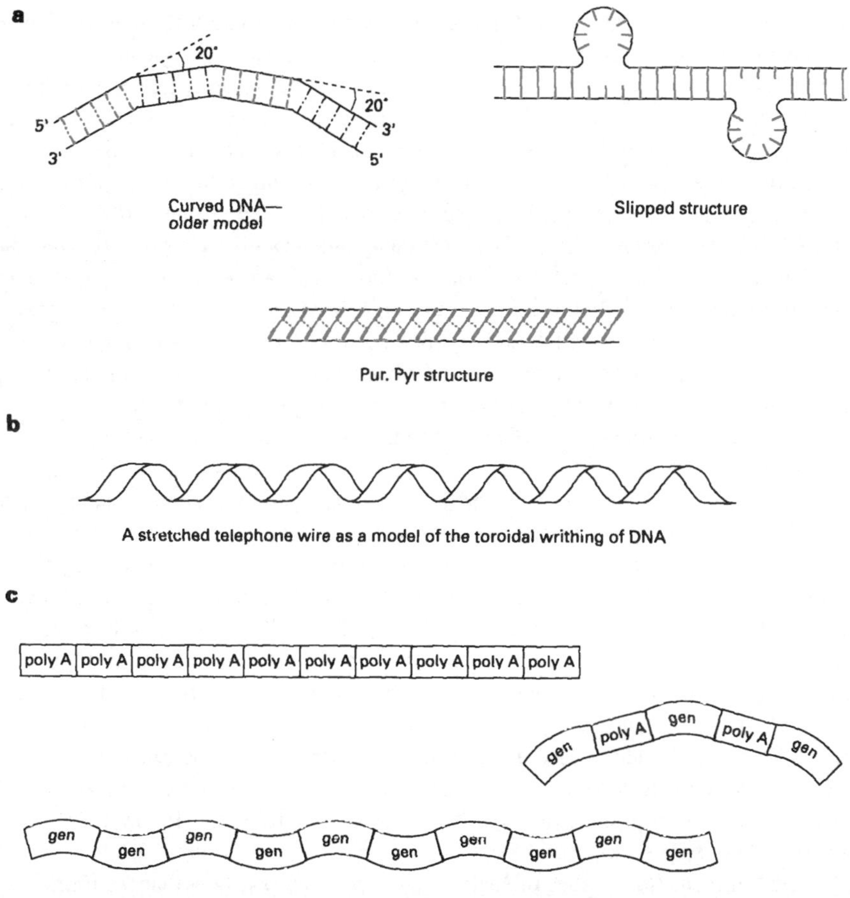


**Figure 2.24** *Curved DNA illustrated by straight poly(dA) tracts (upper), consecutive tracts of writhing DNA of general sequence (lower), and curved DNA (right) of alternating segments of linear poly(dA) and curved tracts of general sequence*

Studies on the hydrodynamic properties of DNA show that general-sequence DNA migrates through gels more slowly than expected which is because the DNA helix occupies a cylindrical volume that has a larger diameter than that of a simple B-helix (Section 11.4.3). This phenomenon is a result of **DNA writhing**, which involves a continuously curved distortion of the helix axis to generate a spiral form and is nicely illustrated by the extension of a coiled telephone wire (Figure 2.24b). It follows that the repeated alternation of straight A-tracts with short sections of general sequence, each having half of a writhing turn, will generate curved DNA (Figure 2.24c). A detailed structural analysis of this explanation says that curvature of B-DNA involves rolling of base pairs, compresses the major groove (which corresponds to positive roll), has a sequence-determined continuum in the bending behaviour, and shows anisotropy of flexible bending.

*2.3.3.2 DNA Bending.* Such intrinsic, sequence-dependent curvature must be distinguished from the bending of DNA, which results from the application of an external force. Dickerson has also examined the bending of the DNA helix that occurs in many crystal structures of the B-form. It is associated with the step from a G·C to an A·T base pair and results from rolling one base pair over the next along their long axes in a direction that compresses the major groove (Figure 2.19c). He suggests that this junction is a flexible hinge that is capable of bending or not bending. Such 'facultative bending' responds to the influence of local forces, typically interactions with other macromolecules, for example control proteins or a nucleosome core. By contrast, poly(dA) tracts are known to resist bending in nucleosome reconstitution experiments. It can thus be seen that sequence-dependent variation in **DNA bendability** is an important factor in DNA recognition by proteins.

*One important conclusion emerges*: DNA has evolved conformationally to interact with other macromolecules. A free, linear DNA helix in solution may, in fact, be the least biologically relevant state of all.[24]

**Slipped structures** have been postulated to occur at direct repeat sequences, and they have been found up-stream of important regulatory sites. The structures described (Figure 2.24a) are consistent with the pattern of cleavage by single-strand nucleases but otherwise are not well characterised.

**Purine–pyrimidine tracts** manifest an unusual structure at low temperature with a long-range, sequence-dependent single base shift in base-pairing in the major groove. For the dodecamer d(ACCGGCGCCACA)·d(TGTGGCGCCGGT), the bases in the d(CA)$_n$ tract have high propeller twist (–32°) and are so strongly tilted in the 3′-direction that there is disruption of Watson–Crick pairing in the major groove and formation of interactions with the 5′-neighbour of the complementary base. This alteration propagates along the B-form helix for at least half a turn with a domino-like motion. As a result, the DNA structure is normal when viewed from the minor groove and mismatched when seen from the major groove. Since (CA)$_n$ tracts are involved both in recombination and in transcription, this new recognition pattern has to be considered in the analysis of the various processes involved with reading of genetic information.

**Anisomorphic DNA** is the description given to DNA conformations associated with direct repair, DR2, sequences at 'joint regions' in viral DNA, which are known to have unusual chemical and physical properties. The two complementary strands have different structures and this leads to structural aberrations at the centre of the tandem sequences that can be seen under conditions of torsional stress induced by negative supercoiling.

**Hairpin loops** are formed by oligonucleotide single strands which have a segment of inverted complementary sequence. For example, the 16-mer d(CGCGCGTTTTCGCGCG) has a hexamer repeat and its crystal structure shows a hairpin with a loop of four Ts and a Z-DNA hexamer stem (Figure 2.25a). When such inverted sequences are located in a DNA duplex, the conditions exist for formation of a cruciform.

**Cruciforms** involve intra-strand base-pairing and generate two stems and two hairpin loops from a single unwound duplex region.[26] The inverted sequence repeats are known as **palindromes**, which have a given DNA duplex sequence followed after a short break by the same duplex sequence in the opposite direction. This is illustrated for a segment of the bacterial plasmid pBR322 (Figure 2.25b), where a palindrome of two undecamer sequences exists.

X-ray, NMR and sedimentation studies of such stem–loop structures show that the four arms are aligned in pairs to give an oblique X structure with continuity of base-stacking and helical axes across the junctions (Section 6.8.1). Also, the loops have an optimum size of from four to six bases. Residues in the loops

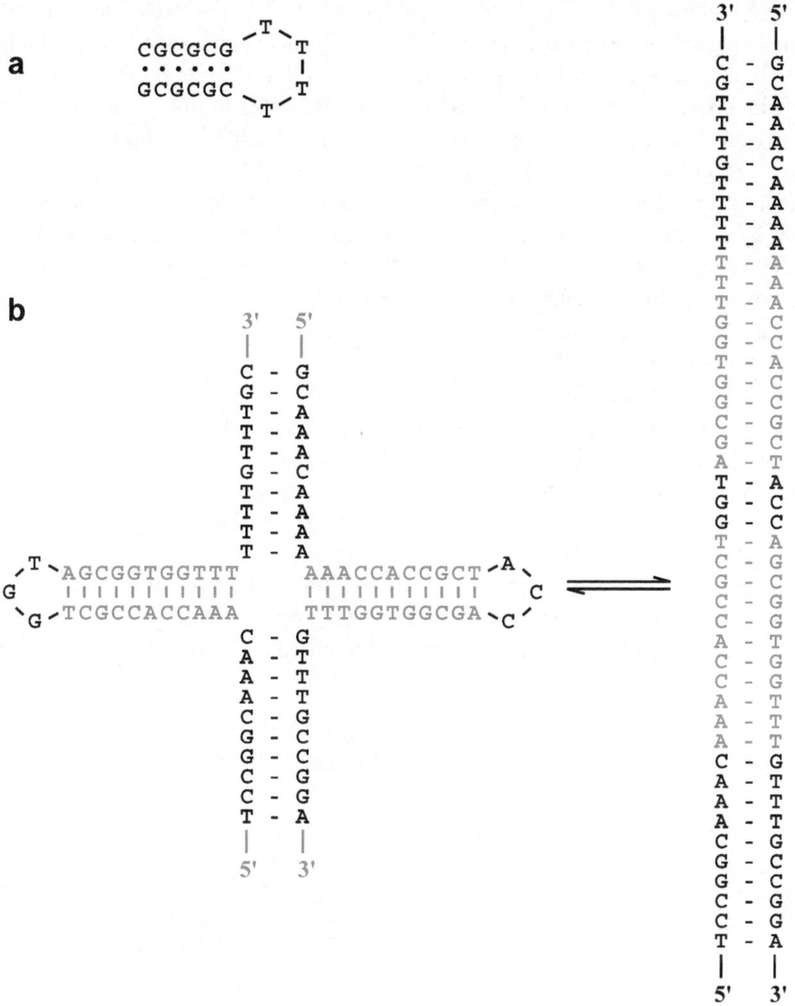


**Figure 2.25** *(a) The hairpin loop formed by d(CGCGCGTTTTCGCGCG). (b) Formation of a cruciform from an inverted repeat sequence of the bacterial plasmid pBR322. The inverted palindromic regions, each of 11 bp, are shown in colour*

are sensitive to single-strand nucleases, such as S1 and P1, and especially to chemical reagents such as bromoacetaldehyde, osmium tetroxide, bisulfite and glyoxal (Chapter 8). In addition, the junctions are cleavage sites for yeast resolvase and for T4 endonuclease VII.

David Lilley has shown that the formation of two such loops requires the unpairing and unstacking of three or more base pairs and so will be thermodynamically unstable compared to the corresponding single helix.[27] While there can be some stacking of bases in the loops, the adverse energy of formation of a single cruciform has been calculated to be some 75 kJ $mol^{-1}$. In experiments on cruciforms using closed circular superhelical DNA, this energy can be provided by the release of strain energy in the form of negative supercoiling (see the following section) and is directly related to the length of the arms of the cruciform: the formation of an arm of 10.5 bp unwinds the supercoil by a single turn.

There is also a kinetic barrier to cruciform formation and Lilley has suggested two mechanisms that have clearly distinct physical parameters and may be sequence-dependent. The faster process for cruciform formation, the S-pathway, has $\Delta G^{\ddagger}$ of about 100 kJ $mol^{-1}$ with a small positive entropy of activation.

This more common pathway is typified by the behaviour of plasmid pIRbke8. Following the formation of a relatively small unpaired region, a proto-cruciform intermediate is produced, which then grows to equilibrium size by branch migration through the four-way junction (Figure 2.26). The slower mechanism, the C-pathway, involves the formation of a large bubble followed by its condensation to give the fully developed cruciform. This behaviour explains the data for the pColl315 plasmid whose cruciform kinetics show $\Delta G^{\ddagger}$ about 180 kJ $mol^{-1}$ with a large entropy of activation.

Such extrusion of cruciforms provides the most complete example of the characterisation of unusual DNA structures by combined chemical, enzymatic, kinetic and spectroscopic techniques. However, it is not clear whether cruciforms have any role *in vivo*. One reason may simply be that intracellular superhelical densities may be too low to cause extrusion of inverted repeat sequences. Equally, the kinetics of the process may also be too slow to be of physiological significance. However, cruciforms are formally equivalent to Holliday junctions and these four-way junctions involve two DNA duplexes that are formed during homologous recombination (Section 6.8).

Several X-ray crystallographic studies have provided a detailed picture of the 3D structure of the Holliday junction.[28] Interestingly, DNA decamers with sequences CCGGG<u>ACC</u>GG, CCGGT<u>ACC</u>GG and TCGGT<u>ACC</u>GA fold into four-way junctions instead of adopting the expected B-form double helical geometry (Figure 2.27). The tri-nucleotide ACC (underlined) forms the core of the junction and its 3′-C·G base pairs helps to stabilise the arrangement by engaging in direct and water-mediated hydrogen bonds to phosphate groups at the strand crossover. The four strands exhibit a stacked-X conformation whereby the two inter-connected duplexes form coaxially stacked arms that cross at an angle of *ca.* 40°. Stable Holliday junctions were also observed with DNA decamers that featured an AC($Me^5C$) tri-nucleotide core and the tri-nucleotide AGC when covalently intercalated by psoralen.

*2.3.3.3 Role of Metal Ions.* NMR in solution, X-ray crystallography and computational simulations (molecular dynamics, MD) have all shed light on the locations of metal cations surrounding nucleic acid

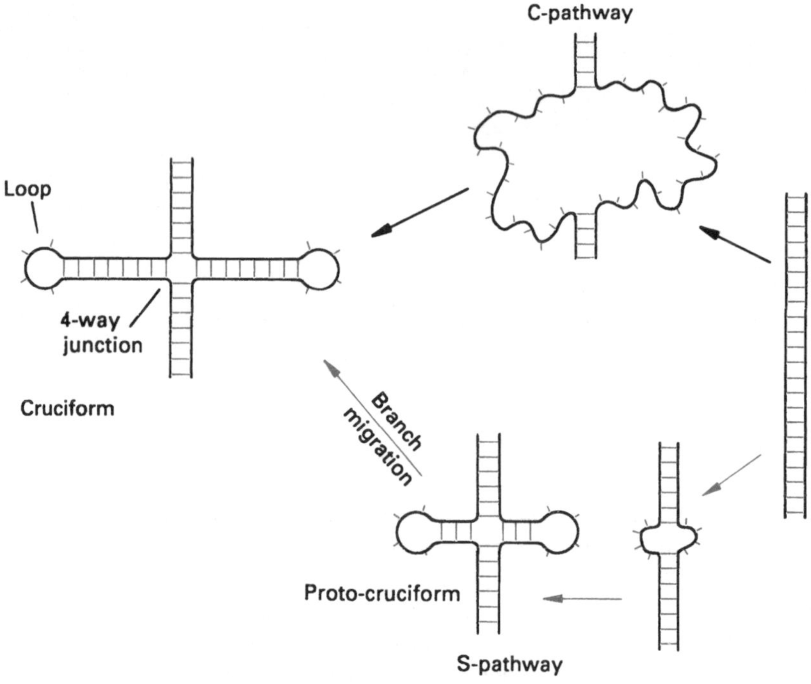


**Figure 2.26** *Structures of a cruciform and alternative pathways for its formation (base-paired sections are helical throughout)*

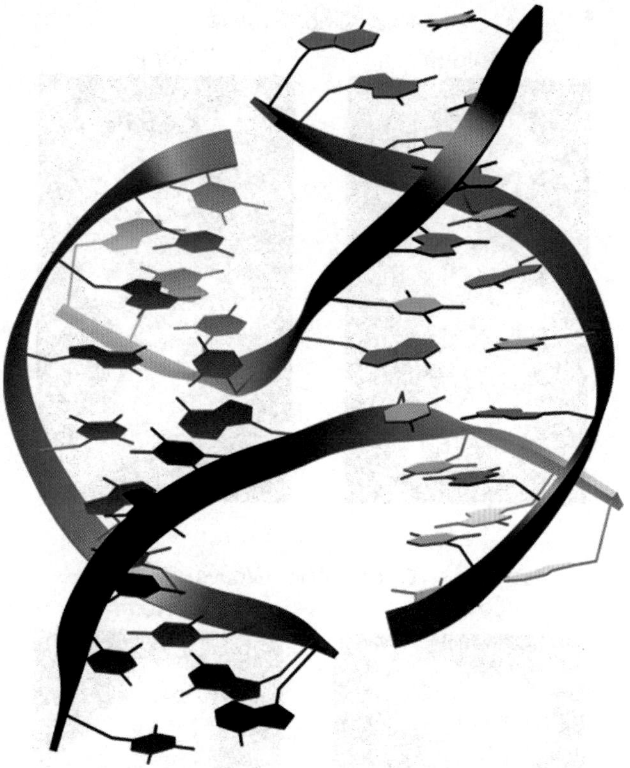

**Figure 2.27** *The Holliday junction adopted by four DNA decamers with sequence TCGGTACCGA (PDB: 1M6G). The view illustrates the side-by-side arrangement of the two duplex portions, with the A nucleotides (red) C nucleotides (pink) of two decamers that form the core of the junction visible near the centre. The positions of phosphate groups in the backbones of individual oligonucleotides are traced with ribbons*

molecules.[29] But while ions are often visible in structures of nucleic acids, it is not straightforward to determine how they affect the structure. The question of whether cations can assume specific roles in the control of DNA duplex conformation has stirred up controversy in recent years. Some in the field, notably Nicholas Hud and Loren Williams, believe that cation localisation within the grooves of DNA represents a significant factor in sequence-specific helical structure. By contrast, probably a majority of those studying the structure of DNA is of the opinion that the specific sequence dictates local DNA conformation and thus binding of metal cations. According to this second view, metal ions can bind to DNA in a sequence-specific manner and in turn modulate the local structure, but ions should not be considered the single most significant driving force of a number of DNA conformational phenomena.

To study the possible effects of metal ions on DNA conformation, all sequences can be divided into three principal groups: **A-tracts**, **G-tracts** and generic DNA, the latter representing the vast majority of DNA sequences.[30] A-tracts have an unusually narrow minor groove, are straight and have high base pair propeller twist (see also Figures 2.18 and 2.19). G-tracts have a propensity to undergo the B-form→A-form transition at increased ionic strength. The proponents of the 'ions are dominant' model believe that the DNA grooves are flexible ionophores and that DNA duplex structure is modulated by a tug of war between the two grooves for cation localisation. They argue that the duplex geometry adopted by A-tracts (referred to as B*-DNA, Figure 2.28a) is due to ion localisation in the minor groove as a result of the highly negative electrostatic potentials there. Conversely, G-tract DNA exhibits a highly negative electrostatic potential in the major groove (Figure 2.28b), leading to preferred localisation of cations there and consequently a collapse of the DNA around the ions. Generic DNA on the other hand would have a more balanced occupation of

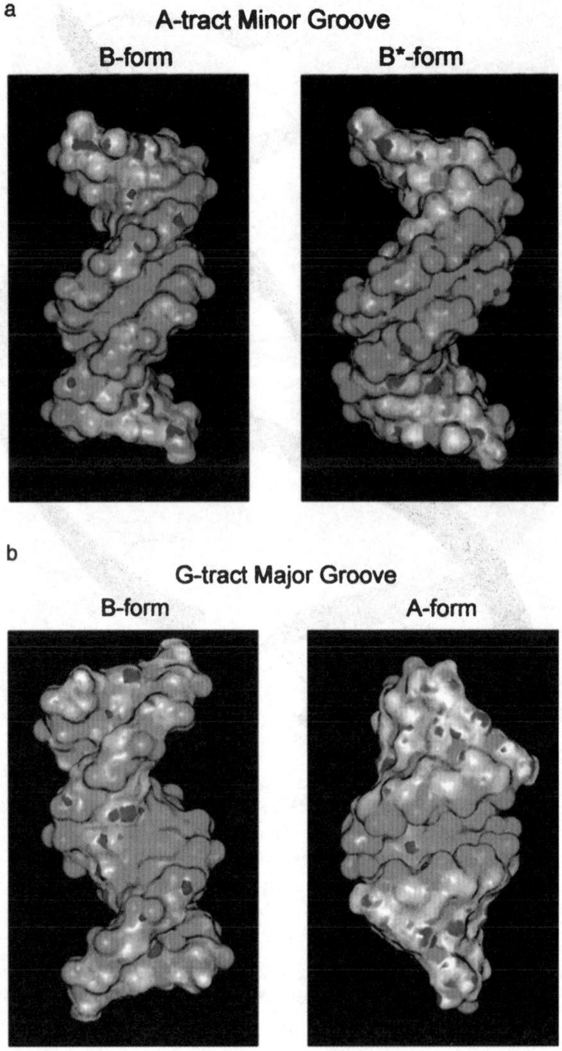


**Figure 2.28** *Graphical representations of electrostatic surface potentials (ESPs) calculated at the solvent accessible surfaces of model (a) A-tract and (b) G-tract DNA duplexes, each in two helical forms. The model for the A-tract is the duplex $(dA)_{12}{\cdot}(dT)_{12}$ and the model for the G-tract is the duplex $(dG)_{12}{\cdot}(dC)_{12}$. Colours of the DNA surfaces range from red, –8 kT/e, to blue, +3 kT/e, with increasing electrostatic potential* (Adapted from N.V. Hud and J. Plavec, *Biopolymers,* 2003, **69**, 144–159. © (2003), with permission from John Wiley and Sons, Inc.)

its major and minor grooves by cations, consistent with a more or less canonical B-form geometry. By contrast, those who emphasise the dominating role of sequence in the control of DNA conformation argue that it is the sequence that shapes the DNA in the first place and that the narrow minor groove of A-tract or B*-DNA is narrow even before ions settle in the groove.

Therefore, it is difficult to settle the issue of the relative importance of sequence and ions in governing DNA duplex conformation, and no single experimental or, certainly, theoretical method alone will provide a definitive answer. Although the 'ions first' hypothesis has a number of attractive features – *i.e.* it provides a link between sequence-specific cation localisation and sequence-directed curvature of DNA (Section 2.3.3) – it cannot be overlooked that high-resolution crystal structures of oligodeoxynucleotides containing A-tracts have shown no variation of groove width as a consequence of different types and concentrations of alkali metal ions present in the crystallisations. Moreover, MD simulations of A-tract DNA in the presence of different classes and varying localisations of metal cations have not provided a picture that is consistent with a crucial role for metal ions with regard to the structure of duplex DNA. Thus, there will undoubtedly be more studies directed at a refined understanding of the relative importance of sequence and cation co-ordination in governing the structure of double helical DNA.

### 2.3.4 B–Z Junctions and B–Z Transitions

Segments of left-handed Z-DNA can exist in a single duplex in continuity with segments of right-handed B-DNA. This phenomenon has been observed both *in vitro* and *in vivo*. Because the backbone chains of these polymorphs run in opposite directions (↓↑and ↑↓) respectively (Section 2.2.5), there has to be a transitional region between two such segments, and this boundary is known as a **B–Z junction**. Such structures are polymorphic and sequence-specific and six features have been described:

1. B–Z junction can be as small as 3 bp.
2. At least one base pair has neither the B- nor the Z-conformation.
3. Hydrogen bonds between the base pairs are intact below 50°C.
4. Chemical reagents specific for single-stranded DNA (chloroacetaldehyde, bromoacetaldehyde and glyoxal; Section 8.5.3) show high reactivity with the junction bases.
5. Junctions are sites for enhanced intercalation for psoralens (Section 8.8.2).
6. Junctions are neither strongly bent nor particularly flexible.

This conformational B–Z transition between the right- and left-handed helices has a high energy of activation (about 90 kJ $mol^{-1}$) but is practically independent of temperature ($\Delta G°$ about 0 kJ $mol^{-1}$) (Section 2.5.3). Thus, the B–Z transition is co-operative and propagates readily along the helix chains.

In the absence of structural data at high resolution, two different models had been suggested to explain the conformational switch that has to occur as a B–Z junction migrates, rather like a bubble, along a double helix. In the first model, the bases unpair, guanine flips into the *syn* conformation, the entire deoxycytidine undergoes a conformational switch, and the base pairs reform their hydrogen bonds. This model appears to be at variance with NMR studies that suggest the bases remain paired because their imino-protons do not become free to exchange with solvent water. In the second model, the backbone is stretched until one base pair has sufficient room to rotate 180° about its glycosyl bonds (**tip**, as shown in Figure 2.19a), and the bases re-stack. However, one might expect this 'expand–rotate–collapse' process to be impeded by linking bulky molecules to the edge of the base pairs. Yet, bonding *N*-acetoxy-*N*-acetyl-2-aminofluorene to guanine actually facilitates the B–Z transition. Thus, the dynamics of the B–Z transition poses a major conformational problem, and this has sometimes been called the chain–sense paradox.

In addition, Ansevin and Wang suggested an alternative zig-zag model for the left-handed double-helical form of DNA that avoids this paradox and is accessible from B-DNA by simple untwisting. Their **W-DNA** has a Watson–Crick chain sense (↑↓) like B-DNA but similar glycosyl geometry to that of Z-DNA. It has reversed sugar puckers, C3′-*endo* at cytosine and C2′-*endo* at guanine, while in both W- and Z-DNA the minor groove is deep and the major groove broad and very shallow. In addition, this W-model explains (1) the incompatibility of poly(dA–dT)·poly(dA–dT) with a left-handed state, (2) the very slow rate of exchange of hydrogens in the 2-$NH_2$ group of guanine in left-handed DNA and (3) the incompatibility of left-handed helix with replacement of $O_R$ oxygens by a methyl group. They argue that Z-DNA has a lower energy than W-DNA and so is adopted in crystals of short oligonucleotides but it may be conformationally inaccessible to longer stretches of DNA in solution.

Finally, more than 25 years after the discovery of Z-DNA, a crystal structure of a B–Z junction has solved the mystery of how DNA switches from the right-handed low energy to the left-handed high energy form.[31] The junction was trapped in the crystal by stabilising the Z-DNA portion at one end of a 15-bp segment with a Z-DNA binding protein, with the rest of the DNA assuming the B-form geometry. Continuous stacking of bases between B-DNA and Z-DNA is found with the breaking of one base pair at the junction and the two bases extruded from the helix on either side. A sharp turn accommodates the reversal in the backbone direction and at the junction the DNA is bent by *ca.* 10° and the helical axes of the B- and Z-form duplexes are displaced from each other by *ca.* 5 Å.

### 2.3.5 Circular DNA and Supercoiling

The replicative form of bacteriophage ϕX174 DNA was found to be a double-stranded closed circle. It was later shown that bacterial DNA exists as closed circular duplexes, that DNA viruses have either single- or double-helical circular DNA, and that RNA viroids have circular single-stranded RNA as their genomic material. Plasmid DNAs also exist as small, closed circular duplexes.

Topologically unconstrained dsDNA in its linear, relaxed state is either biologically inactive or displays reduced activity in key processes such as recombination, replication or transcription. It follows that topological changes associated with the constraints of circularisation of dsDNA have a profound biological significance.[32] Although such circularisation can be achieved directly by covalent closure, the same effect can be achieved for eukaryotic DNA as a result of holding DNA loops together by means of a protein scaffold.

The molecular topology of closed circular DNA was described by Vinograd in 1965 and is especially associated with the phenomenon of superhelical DNA, which is also called supercoiled or supertwisted DNA. Vinograd's basic observation was that when a planar, relaxed circle of DNA is strained by changing the pitch of its helical turns, it relieves this torsional strain by winding around itself to form a superhelix whose axis is a diameter of the original circle.

This behaviour is most directly observed by following the sedimentation of negatively supercoiled DNA as the pitch of its helix is changed by intercalating a drug, typically ethidium bromide (Section 9.6). Intercalation is the process of slotting the planar drug molecules between adjacent base pairs in the helix. For each ethidium molecule intercalated into the helix there is an increase of about 3.4 Å and a linked decrease of about 36° in twist (Figure 2.19c,d). The DNA helix responds first by reducing the number of negative, right-handed supercoils until it is fully relaxed and then by increasing the number of positive, left-handed supercoils. As this happens, the sedimentation coefficient of the DNA first decreases, reaching a minimum when fully relaxed, and then increases as it becomes positively supercoiled. As a control process, the same circular DNA can be nicked in one strand to make it fully relaxed. The result is that it now shows a low sedimentation coefficient at all concentrations of the intercalator species (Figure 2.29) (Section 11.4.1).

Vinograd showed that the topological state of these covalently closed circles can be defined by three parameters and that the fundamental topological property is **linkage**. The **topological winding number**, $T_w$, is the number of right-handed helical turns in the relaxed, planar DNA circle and the **writhing number**, $W_r$, gives the number of left-handed crossovers in the supercoil. The sum of these two is the **linking number**, $L_k$, which is the number of times one strand of the helix winds around the other (clock-wise is positive) when the circle is constrained to lie in a plane. The simple equation is $L_k = T_w + W_r$.

Such behaviour can be illustrated simply (Figure 2.30) for a relaxed closed circle with 20 helical turns $T_w = 20$, $L_k = 20$, $W_r = 0$. One strand is now cut, unwound two turns, and resealed to give $L_k = T_w = 18$. This circle is thus under-wound by two turns. To restore fully the normal B-DNA base-pairing and base-stacking, the circle needs to gain two right-handed helical turns, $\Delta T_w = +2$ to give $T_w = 20$. Since the DNA circles have remained closed and the linking number stays at 18, the formation of the right-handed helical turns is balanced by the creation of one right-handed supercoil, making $W_r = -2$.

The behaviour of a supercoil can be modelled using a length of rubber tubing. The ends are first held together to form a relaxed closed circle. If the end in your right hand is given one turn clock-wise (right-handed twist) and the other end is given one turn in the opposite sense, the tube will relieve this strain by forming one left-handed supercoil. This is equivalent to unwinding the DNA helix by two turns, which generates one positive supercoil (four turns generate two supercoils, and so on). This model shows the relationship: two turns equals one supercoil.

In practice, it is sometimes useful to describe the degree of supercoiling using the **super-helical density**, $\sigma = W_r/T_w$, which is close to the number of superhelical turns per 10 bp and is typically around 0.06 for

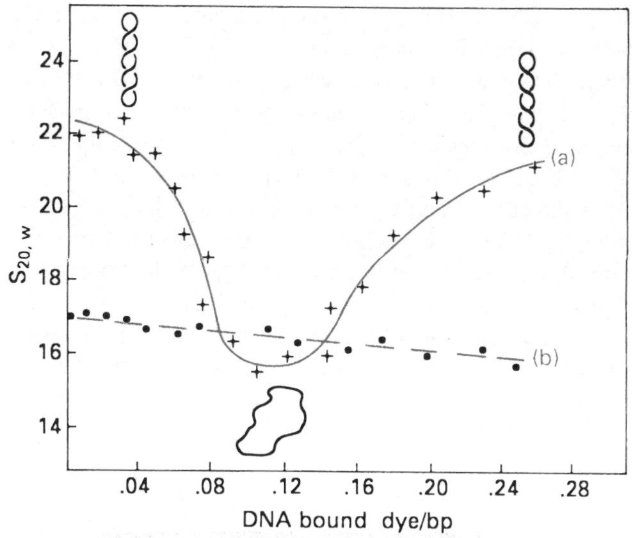


**Figure 2.29** *Sedimentation velocity for SV40 DNA as a function of bound ethidium (a) for closed circular DNA (– + – + – + –) and (b) for nicked circular DNA (–·–·–·– ) showing the transition from a negative supercoil (left) through a relaxed circle (centre) to a positive supercoil (right)*
(Adapted from W. Bauer and J. Vinograd, *J. Mol. Biol.*, 1968, **33**, 141. © (1968), with permission from Elsevier)

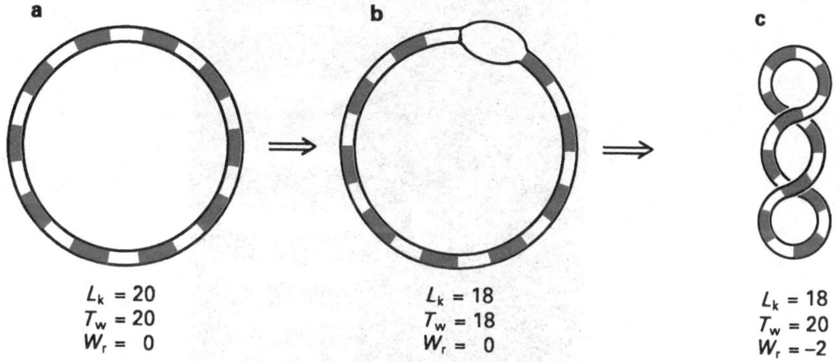


**Figure 2.30** *Supercoil formation in closed circular DNA. (a) Closed circle of 20 duplex turns (alternate turns in colour). (b) Circle nicked, under-wound two turns, and re-sealed. (c) Base pairing and stacking forces result in the formation of B-helix with two new right-handed helix turns and one compensating right-handed supercoil*

superhelical DNA from cells and virions. The **energy of supercoiling** is a quadratic function of the density of supercoils as described by the equation

$$\Delta G_s = 1050\frac{RT}{N}\Delta L_k^2 \text{ kJ mol}^{-1}$$

where $R$ is the gas constant, $T$ the absolute temperature and $N$ the number of base pairs.

**B–Z transitions** are especially important for supercoiling since the conversion of one right-handed B-turn into a left-handed Z-turn causes a change in $T_w$ of –2. This must be complemented by $\Delta W_r + 2$ through the formation of one left-handed superturn.

*2.3.5.1 Enzymology of DNA Supercoiling.* DNA topoisomers are circular molecules, which have identical sequences and differ only in their linking number. A group of enzymes, discovered by Jim Wang, can change that linking number.[33] They fall into two classes: Class I topoisomerases effect integral changes in the linking number, $\Delta L_k = n$, whereas Class II enzymes inter-convert topoisomers with a step rise of $\Delta L_k = \pm 2n$. **Topoisomerase** I enzymes use a 'nick-swivel-close' mechanism to operate on supercoiled DNA. They break a phosphate diester linkage, hold its ends, and reseal them after allowing exothermic (*i.e.* passive) free rotation of the other strand. Such enzymes from eukaryotes can operate on either left- or right-handed supercoils while prokaryotic enzymes only work on negative supercoils. The products of topoisomerase I action on plasmid DNA can be observed by gel electrophoresis, and show a ladder of bands, each corresponding to unit change in $W_r$ as the supercoils are unwound, half at a time (Figure 2.31).

By contrast, Class II topoisomerases use a 'double-strand passage' mechanism to effect unit change in the number of supercoils, $\Delta W_r = \pm 2$, and such prokaryotic enzymes can drive the endothermic supercoiling of DNA by coupling the reaction to hydrolysis of ATP. These topoisomerases cleave two phosphate

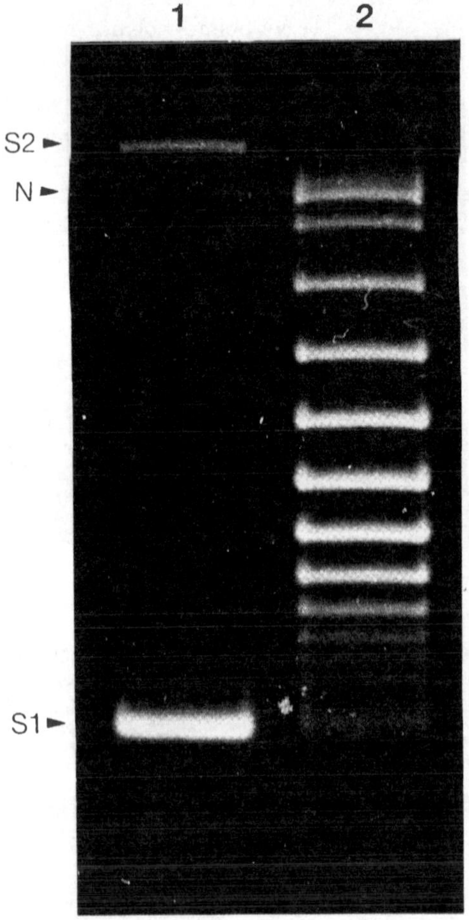


**Figure 2.31** *Topoisomers of plasmid pAT153 after incubation with topoisomerase I to produce partial relaxation. Electrophoresis in a 1% agarose gel: Track 1 shows native supercoiled pAT153 (S1), supercoiled dimer (S2) and nicked circular DNA (N); Track 2 shows products of topoisomerase I where $\Delta L_k$ up to 14 can be seen clearly*
(Adapted from D.M.J. Lilley, *Symp. Soc. Gen. Microbiol.*, 1986, **39**, 105–117. © (1986), with permission from the Society of General Microbiology)

esters to produce an enzyme-bridged gap in both strands. The other DNA duplex is passed through the gap (using energy provided by hydrolysis of ATP), and the gap is resealed.[34] DNA gyrase from *E. coli* is a special example of the Class II enzyme. It is an $A_2B_2$ tetramer with the energy-free topoisomerase activity of the A subunit being inhibited by quinolone antibiotics such as nalidixic acid. The energy-transducing activity of the B subunit can be inhibited by novobiocin and other coumarin antibiotics. We should point out that such topiosomerases also operate on linear DNA that is torsionally stressed by other processes, most notably at the replication fork in eukaryotic DNA.

Supercoiling is important for a growing range of enzymes as illustrated by two examples. RNA polymerase *in vitro* appears to work ten times faster on supercoiled DNA $\sigma = 0.06$, than on relaxed DNA, and this phenomenon appears to be related to the enhanced binding of the polymerase to the promoter sequence. Second, the tyrT promoter in *E. coli* is expressed *in vitro* at least 100 times stronger for supercoiled than for relaxed DNA, and this behaviour seems linked to 'pre-activation' of the DNA promoter region by negative supercoiling.[35]

*2.3.5.2 Catenated and Knotted DNA Circles.* Although Class II topoisomerases usually only effect passage of a duplex from the same molecule through the separated double strands, they can also manipulate a duplex from a second molecule. As a result, two different DNA circles can be inter-linked with the formation of a **catenane** (Figure 2.32). Such catenanes have been identified by electron microscopy and can be artificially generated in high yield from mammalian mitochondria. **Knotted DNA** circles are another unusual topoisomer species, which are also formed by intra-molecular double-strand passage from an incompletely unwound duplex (Figure 2.32).

### 2.3.6 Triple-Stranded DNA

Triple helices were first observed for oligoribonucleotides in 1957. A decade later, the same phenomenon was observed for poly(dCT) binding to poly(dGA)·poly(dCT) and for poly(dG) binding to poly(dG)·poly(dC). Oligonucleotides can bind in the major groove of B-form DNA by forming Hoogsteen or reversed

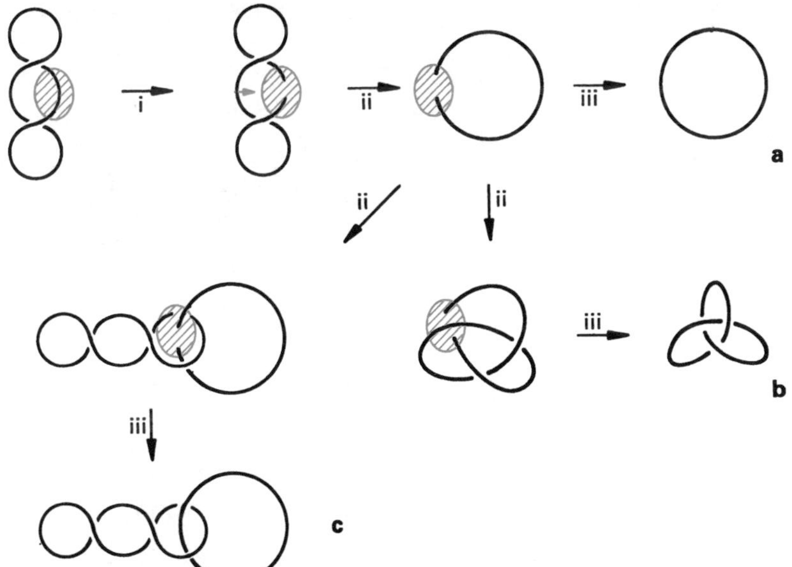


**Figure 2.32** *Action of topoisomerase II (red) on singly supercoiled DNA: (i) double-strand opening; (ii) double-strand passage; (iii) resealing to give (a) relaxed circle, (b) knotted circle and (c) catenated DNA circles*

Hoogsteen hydrogen bonds using N-7 of the purine bases of the Watson–Crick base pairs (Figures 2.33 and 2.34).[36] The resulting base-triplets form the core of a triple helix. In theory, G can form a base-triple with a C·G pair and A with a T·A pair (Figure 2.33b), but the only combinations that have isomorphous location of their C-1′ atoms are the two triplets TxA·T and C·GxC$^+$ (Figure 2.33a), where C$^+$ is the N-3 protonated form of cytosine. This means that the three strands of triple-helical DNA are normally two

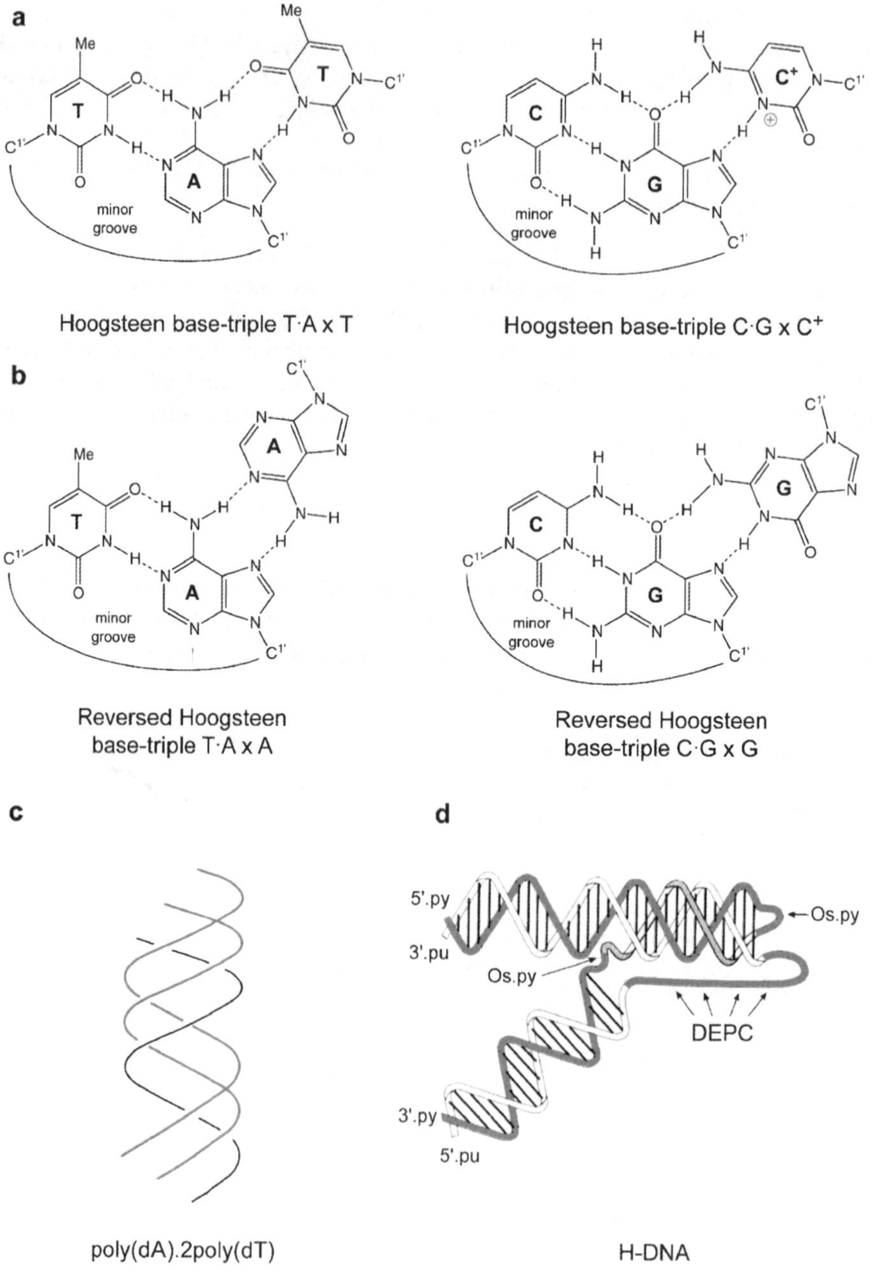


**Figure 2.33** *Base triples formed by (a) Hoogsteen bonding for T·AxT and C·GxC$^+$ and (b) reversed Hoogsteen binding for T·AxA and C·GxG. (c) Model of triple-helical DNA based on fibre diffraction of poly(dA)·2poly(dT) (kindly provided by the late Professor Claude Hélène). (d) A schematic representation of H-DNA showing loci for attack by single-strand specific reagents*

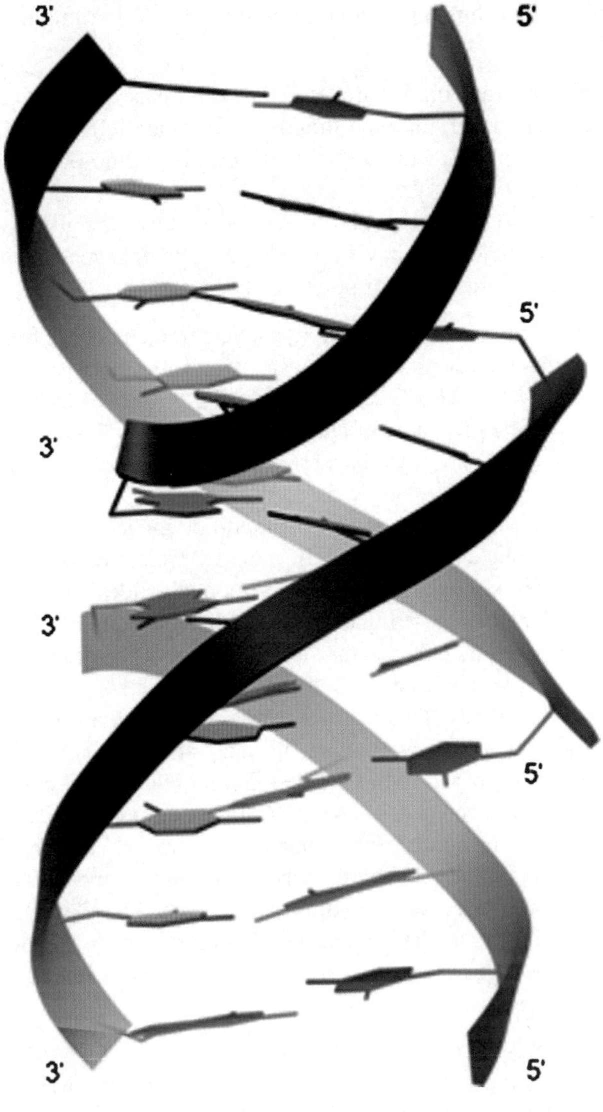


**Figure 2.34** *The triple helical structure formed in the crystal structure of the 1:1 complex between the DNA 12 mer d(CTCCTCCGCGCC) and the 9 mer d(CGCGCGGAG) (PDB: 1D3R; available at: http://www.rcsb.org).[39] The triplex segment consists of two 5′-halves of the 12 mer (underlined, top and bottom left) and two 3′-terminal trimers GAG (underlined; top and bottom right) from two 9 mers that are stacked tail-to-tail in the crystal: 5′-····-GAG-3′\3′-GAG-····-5′ (visible near the centre, left of drawing). DNA bases are coloured red, grey, pink and black for G, A, C and T, respectively, and the directions of the sugar–phosphate backbones are traced by ribbons*

homo-pyrimidines and one homo-purine (Figure 2.33c). However, despite the backbone distortion that must result from the hetero-morphism of other base triplet combinations, oligonucleotides containing G and T, G and A, or G, T and C have been shown to form helices.

Intermolecular triple helices are now well characterised for short oligonucleotides binding in the groove of a longer DNA duplex,[37,38] H-DNA provides an example of an intramolecular triple helix because it has a mirror-repeat sequence relating homo-purine and homo-pyrimidine tracts in a circular double-stranded DNA molecule, and triplex formation is driven by supercoiling (Figure 2.33d).

Several studies on third-strand binding to a homo-purine · homo-pyrimidine duplex have established the following features:

- A third homo-pyrimidine strand binds parallel to the homo-purine strand using Hoogsteen hydrogen bonds (*i.e.* the homo-pyrimidine strands are anti-parallel).
- A third homo-purine strand binds anti-parallel to the original homo-purine strand using reversed Hoogsteen hydrogen bonds.
- The bases in the third strand have a regular *anti* conformation of the glycosylic bond.
- Synthetic oligodeoxyribonucleotides having an α-glycosylic linkage also bind as a third strand, parallel for poly(d-α-T) and anti-parallel for poly(d-α-TC).

Triple helices are less stable than duplexes. Thermodynamic parameters have been obtained from melting curves, from kinetics, and from the use of differential scanning calorimetry (DSC) (Section 11.4.4). Using this last technique, values of $\Delta H^\circ = 22 \pm 2\,\text{kJ mol}^{-1}$ and $\Delta S^\circ = 70 \pm 7\,\text{J mol}^{-1}\,\text{K}^{-1}$ have been found for $d(C^+TTC^+C^+TC^+C^+TC^+T)$. The p$K_a$ value of cytidine in isolation is 4.3 but it is higher in oligonucleotides because of their polyanionic phosphate backbone. So it is to be expected that the stability of triple helices is seen to decrease as the pH rises above 5.

Triple helix stability can be enhanced by the use of modified nucleotides. 5-Methylcytosine increases stability at neutral pH, probably by a hydrophobic effect, and 5-bromouracil can usefully replace thymine. Oligoribonucleotides bind more strongly than do deoxyribonucleotides and 2′-*O*-methylribonucleotides bind even better. Finally, Hélène has shown that at the attachment of an intercalating agent to the 5′- (or 3′-end) of the third strand can greatly enhance the stability of the triple-stranded helix.

The major application of triple helices relates to the specificity of the interaction between the single strand and a much larger DNA duplex. This is because homo-pyrimidines have been identified as potential vehicles for the sequence-specific delivery of agents that can modify DNA and thereby control genes. The DNA of the bacterium *E. coli* has 4.5 Mbp, so the minimum number of base pairs needed to define a unique sequence in its genome is 11bp (*i.e.* $4^{11} = 4{,}194{,}304$ assuming a statistically random distribution of the four bases). The corresponding number for the human genome is about 17 bp. Thus, a synthetic 17-mer could be expected to identify and bind to a unique human DNA target and thus deliver a lethal agent to a specific sequence of DNA. In practice, the energetics of mismatched base triples is complex and depends on nearest neighbours, metal ions and other parameters. However, a value of about $1.5\,\text{kJ mol}^{-1}$ per mismatch seems to fit much of the data and suggests that the specificity of triple helices is at least as good as that of double-helical complexes.

*2.3.6.1 H-DNA.* A new polymorph of DNA was discovered in 1985 within a sequence of $d(A\text{-}G)_{16}$ in the polypurine strand of a recombinant plasmid pEJ4. Its requirement for protons led to the name H-DNA (half of its C residues are protonated, so the transition depends on acid pH as well as on a degree of negative supercoiling). Probes for single-stranded regions of DNA (especially osmium tetroxide:pyridine (Section 8.3) and nuclease P1 cleavage) were used to identify specific sites and provide experimental support for the model advanced earlier for a triple helical H-DNA (Figure 2.33d). This has a Watson–Crick duplex which extends to the centre of the $(dT\text{–}dC)_n\cdot(dG\text{–}dA)_n$ tract and the second half of the homo-pyrimidine tract then folds back on itself, anti-parallel to the first half and winding down the major groove of the helix. The second half of the poly-purine tract also folds back, probably in an unstructured single-stranded form. The energetics of nucleation of H-DNA suggests that it requires at least 15 bp for stability and the consequent loss of twist makes H-DNA favoured by negative supercoiling.

Although antibodies have been raised to detect triple-stranded structures, no evidence has yet been found for their natural existence in cells *in vivo*.

### 2.3.7 Other Non-Canonical DNA Structures

*2.3.7.1 Four-Stranded Motifs.* Both G- and C-rich DNA sequences have been found to adopt four-stranded motifs, also called **tetraplexes** or **quadruplexes** (see also Section 9.10.2). Sequences containing

G- and C-rich strands are found at the telomeric ends of chromosomes (see Section 6.4.5), and such sequences are of fundamental importance in protecting the cell from re-combination and degradation. It is known that when DNA containing a palindromic sequence of bases is subjected to supercoiling stress (Section 2.3.5), a cruciform can be extruded (Figure 2.26). If the tips of the cruciform extrusions contain C residues in one limb and G residues in the other, G- and C-rich quadruplexes can be formed by combining two cruciforms of this type. So it is possible that formation of four-stranded G- and C-rich motifs provides the physical basis for identical DNA sequences to bind together, *i.e.* during meiosis when identical chromosomes line up with each other. Proteins that bind to G-rich quadruplexes have been identified and it is unlikely that the C-rich motif is stable at neutral pH without also binding a protein factor, since the motif is held together by hemiprotonated $C{\cdot}C^+$ base pairs.

The structures of G- and C-rich quadruplexes are fundamentally different. In the case of four-stranded G-rich motifs, guanines join together via cyclic hydrogen bonding that involves four guanines at each level (often called a **G-tetrad** or **G-quartet**) (Figure 2.35). Each G base is engaged in four hydrogen bonds via its Hoogsteen and Watson–Crick faces, such that guanines are related by a four-fold rotation axis and are nearly co-planar. In this way, each guanine directs its O-6 carbonyl oxygen into the central core of the tetrad. Although it had been found as early as 1910 that concentrated solutions of guanylic acid were unusually viscous and formed a clear gel upon cooling (see Section 1.3) and Gellert and Davies had described a four-stranded helix for guanylic acid based on fibre diffraction experiments more than 40 years ago, detailed 3D

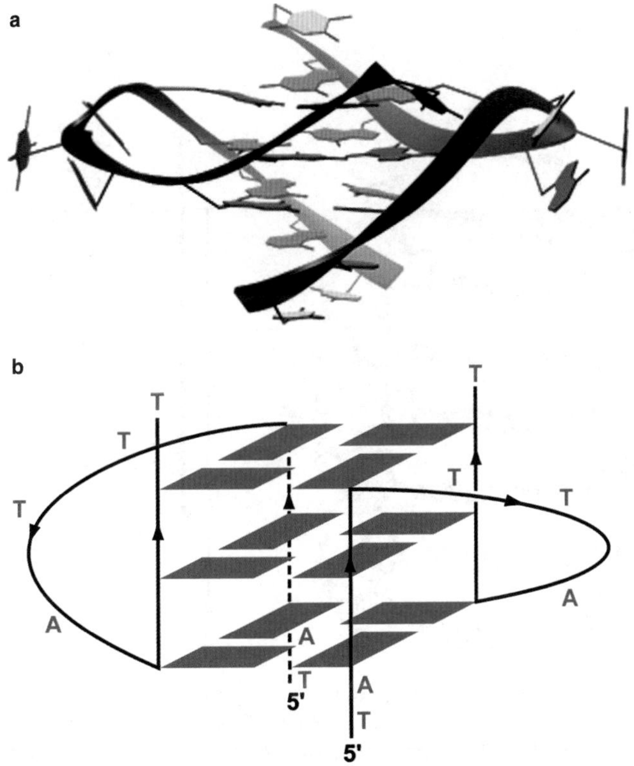


**Figure 2.35** *(a) The parallel-stranded G-tetrad motif formed by two molecules d(TAGGGTTAGGGT) (PDB: 1K8P). The directions of the sugar–phosphate backbones are traced by ribbons. The view is from the side, illustrating the disc-like shape with the three G-tetrad layers on the inside and the TTA 'propellers' on the outside. The 5′-termini of both strands are pointing downwards. (b) Schematic line diagram illustrating the relative orientation of the two strands and the formation of three layers of G-quartets*

structures for G-tetrads have only emerged recently. Dinshaw Patel and co-workers determined the structure of the tetraplex adopted by the human telomeric repeat d($AG_3(T_2AG_3)_3$) in the presence of sodium ions by solution NMR. Under these conditions, the 22 mer adopts a four-stranded motif with three $G_4$ layers and lateral and diagonal loops. The four strands alternate between parallel and anti-parallel orientations and G residues in adjacent layers alternate between the *syn* and *anti* conformations. A similar arrangement had also been found for an intra-molecular quadruplex formed by the $G_4T_4$-repeat sequence from *Oxytricha nova*.

Moreover, by use of X-ray crystallography it was found that the DNA sequences $TG_4T$ (4×; $Na^+$ form), TAGGGTTAGGGT (2×; $K^+$ form) and the above 22 mer $AG_3(T_2AG_3)_3$ (intramolecular; $K^+$ form, Figure 2.35) all adopted quadruplexes with parallel orientations of strands.[40] Most of the deoxyguanosine sugars exhibit the C2′-*endo* pucker. However, in contrast to the above parallel/anti-parallel-type quadruplex, the glycosylic bonds in the four-stranded motifs with parallel orientation of strands adopt exclusively the *anti* conformation. The local G-quartet rise is about 3.13 Å and the four strands writhe in a right-handed fashion with an average twist of around 30° between adjacent layers. Therefore, the G-rich quadruplex is extensively stabilised by π–stacking interactions between layers of guanines. Potassium ions are trapped in the core between stacked G-quartets, spaced at *ca.* 2.7 Å from each of a total of eight O-6 carbonyl groups (Figure 9.2).

A major difference between the four-stranded motifs formed by G- and C-rich sequences is that the former are stable at neutral pH, whereas the latter require protonation of half the cytosine residues and hence are stable only at lower values of pH. Maurice Guéron's laboratory provided the initial structure of the C-rich motif a decade ago, determined by solution NMR methods.[41] They termed it intercalation or **i-motif** for the peculiar four-stranded arrangement involving two parallel intercalating duplexes, each held together by $C{\cdot}C^+$ base pairs (Figure 2.36). The two duplexes are intercalated with opposite polarity, and a

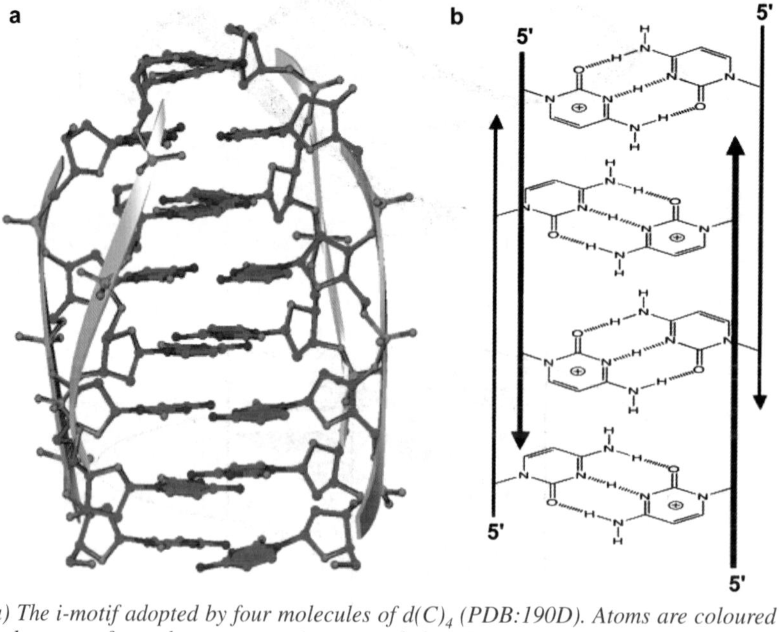


**Figure 2.36** *(a) The i-motif adopted by four molecules of d(C)$_4$ (PDB:190D). Atoms are coloured green, red, blue and magenta for carbon, oxygen, nitrogen and phosphorus, respectively. Positions of phosphorus atoms from the two intercalated $C{\cdot}C^+$-paired duplexes are traced by orange (5′-ends at the top and 3′-ends at the bottom) and grey (5′-ends at the bottom and 3′-ends at the top) ribbons, respectively, to highlight their anti-parallel relative orientation. The absence of significant overlap between cytosine planes from adjacent hemiprotonated base pairs and two wide grooves and two narrow grooves with van der Waals contacts between anti-parallel strands from two duplexes across the latter grooves are hallmarks of the i-motif. (b) Schematic line diagram of the i-motif, illustrating the formation of intercalated parallel-stranded $C{\cdot}C^+$-paired duplexes with opposite polarities*

gentle right-handed helical twist (12°–20°, the rise is 6.2 Å) between covalently linked residues gives the C-rich quadruplex a quasi 2D form. The structure has two broad and two narrow grooves. In the latter, the anti-parallel backbone pairs within intercalated duplexes are in van der Waals contact.

Several crystal structures of C-rich sequences, such as d(CCCC), d(CCCT), d(TCCCCC), d(TCC), d(CCCAAT) and d(TAACCC), have provided details of the conformation, stabilisation and hydration of the i-motif.[42] One surprising finding is the absence of effective stacking between the cytosine rings of adjacent hemi-protonated C·$C^+$ base pairs from intercalated duplexes, an obvious difference to the above G-rich quadruplex. However, a systematic base-on-deoxyribose stacking pattern as well as intracytidine C–H···O hydrogen bonds may partially compensate for the lack of effective base–base-stacking between layers of cytosines. The most unusual feature of the i-motif is a systematic, potentially stabilising C–H···O hydrogen bonding network between the C2′-*endo* puckered deoxyribose sugars of anti-parallel backbones.

*2.3.7.2 The Hoogsteen Duplex.* Virtually all nucleic acid duplexes studied in the last 20 years contain either all G·C base pairs or A·T base pairs flanked by G·C base pairs. Very little 3D-structural work has been carried out on AT-rich sequences, although the functional relevance of such sequences is well known. For example, the promoters of many eukaryotic structural genes contain stretches composed exclusively of A·T base pairs. Further, coding sequences in the yeast genome tend to be clustered with AT-rich sequences separating them, and AT-rich sequences are common in transposable elements (see Section 6.8.3). Crystal structures of TATA boxes bound to the TBP revealed highly distorted B-form conformations of the DNA. Fibre diffraction studies of AT-rich sequences provided indications for considerable structural polymorphism. By contrast, the mostly canonical geometries observed for A·T paired regions in the structures of oligonucleotide duplexes may have resulted from the constraints exerted by the G·C base pair clamps at both ends.

A new crystal structure determined by Juan Subirana and co-workers of the alternating hexamer d(ATATAT) raises interesting questions with regard to the existence of double-stranded DNA species that lack base-pairing of the Watson–Crick type and the possible biological relevance of such alternative DNA conformations.[43] In the crystal structure, but apparently not in solution, two hexametric fragments adopt anti-parallel orientation with Hoogsteen pairing between adenine and thymine (Figure 2.37). The Hoogsteen duplex features an average of 10.6 bp per turn, similar to B-DNA, and all sugars adopt the C2′-*endo* pucker. The diameter of the Hoogsteen duplex is also similar to B-form DNA and the minor groove widths of the two duplexes differ only marginally. A unique characteristic of the Hoogsteen duplex is the *syn* conformation of purine nucleosides. This arrangement generates a pattern of hydrogen bond donors and acceptors in the major and minor grooves that differs between B-DNA and Hoogsteen DNA. It also confers on the latter – a less electro-negative environment in the minor groove – that may lead to preferred interactions with relatively hydrophobic groups at that site. Hoogsteen DNA also differs clearly from triple-stranded arrangements (Section 2.3.6). Thus, in the TxA·T triplex, adenine is always in the *anti* conformation. Moreover, T and A form the Hoogsteen pair in a parallel orientation while in the case of an anti-parallel orientation (Figure 2.33a), base-pairing between A and T is of the reverse Hoogsteen type (Figure 2.33b). Furthermore, in the Ax(A·T) triplex, third-strand adenines always base pair with adenine of the Watson–Crick base pair (Figure 2.33b). Therefore, the anti-parallel Hoogsteen DNA duplex found for d(ATATAT) is not simply a component of the triple helical motifs.

## 2.4 STRUCTURES OF RNA SPECIES

As with DNA, studies on RNA structure began with its primary structure. This quest was pursued in parallel with that of DNA, but had to deal with the extra complexity of the 2′-hydroxyl group in ribonucleosides. Today, we recognise also that RNA has greater structural versatility than DNA in the variety of its species, in its diversity of conformations and in its chemical reactivity. Different natural RNAs can either form long, double-stranded structures or adopt a globular shape composed of short duplex domains connected by single-stranded segments. Watson–Crick base-pairing seems to be the norm, though tRNA structures have provided a rich source of unusual base pairs and base-triplets (Section 7.1.2). In general, it

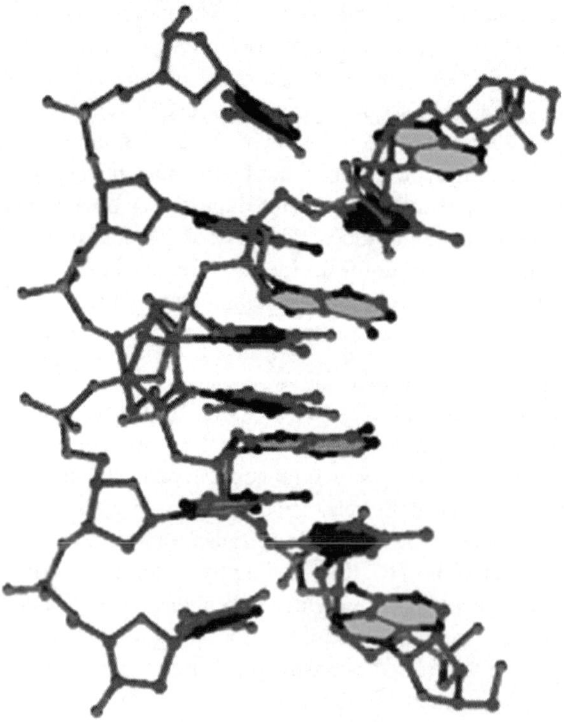

**Figure 2.37** *Structure of the anti-parallel duplex observed in the crystal structure of d(ATATAT) which displays Hoogsteen pairing between adenine and thymine bases (PDB: 1GQU). The view is across the major (right) and minor (left) grooves. Atoms are coloured green, red, blue and magenta for carbon, oxygen, nitrogen and phosphorus, respectively, and base planes of A and T are filled yellow and blue, respectively*

is now possible to predict double-helical sections by computer analysis of primary sequence data, and this technique has been used extensively to identify secondary structural components of ribosomal RNA and viral RNA species. In this section, we shall focus attention mainly on regular RNA secondary structure.

### 2.4.1 Primary Structure of RNA

The first degradation studies of RNA using mild alkaline hydrolysis gave a mixture of mono-nucleotides, originally thought to have only four components – one for each base, A, C, G and U. However, Waldo Cohn used ion-exchange chromatography to separate each of these four into pairs of isomers, which were identified as the ribonucleoside 2′- and 3′-phosphates. This duplicity was overcome by Dan Brown's use of a phosphate diesterase isolated from spleen tissue which digests RNA from its 5′-end to give the four 3′-phosphates Ap, Cp, Gp and Up, while an internal diesterase (snake venom phosphate diesterase was used later) cleaved RNA to the four 5′-phosphates, pA, pC, pG and pU. It follows that RNA chains are made up of nucleotides that have 3′→5′-phosphate diester linkages just like DNA (Figure 2.38).

The 3′→5′ linkage in RNA is, in fact, thermodynamically less stable than the 'unnatural' 2′→5′ linkage, which might therefore have had an evolutionary role. A rare example of such a polymer is produced in vertebrate cells in response to viral infection. Such cells make a glycoprotein called **interferon**, which stimulates the production of an oligonucleotide synthetase. This polymerises ATP to give oligoadenylates with 2′→5′ phosphate diester linkages and from 3 to 8 nucleotides long. Such (2′→5′) $(A)_n$ (Figure 2.39) then activates an interferon-induced ribonuclease, RNase L, whose function seems to be to break down the viral messenger RNA (Note also the 2′→5′ ester linkage is a key feature of self-splicing RNA (Section 7.2.2)).

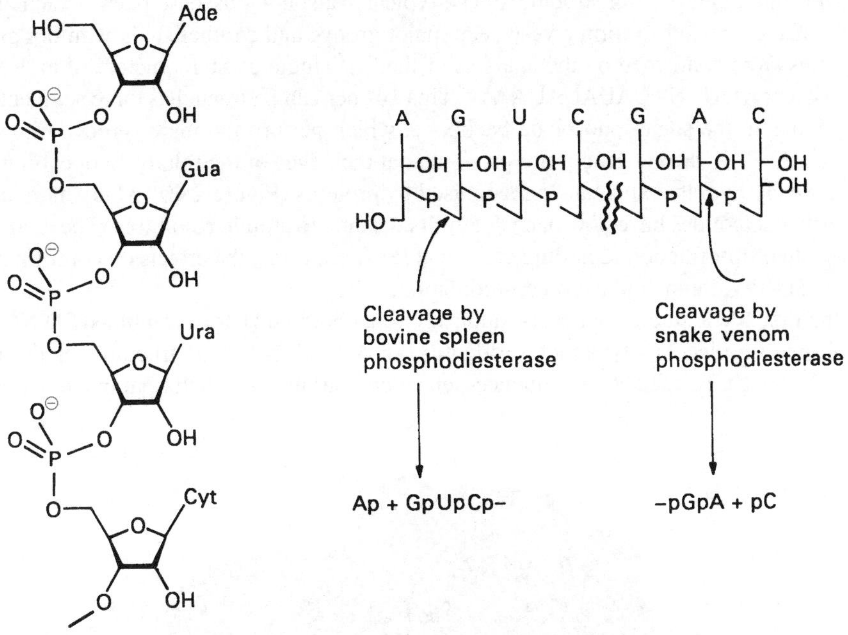


**Figure 2.38** *The primary structure of RNA (left) and cleavage patterns with spleen (centre) and snake venom (right) phosphate diesterases*

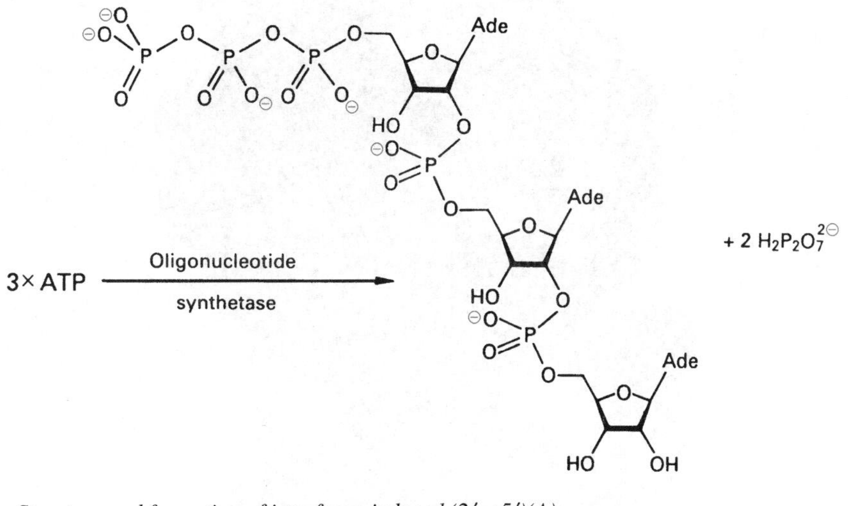


**Figure 2.39** *Structure and formation of interferon-induced (2′→5′)(A)$_n$*

## 2.4.2 Secondary Structure of RNA: A-RNA and A′-RNA

Two varieties of A-type helices have been observed for fibres of RNA species such as poly(rA)·poly(rU). At low ionic strength, **A-RNA** has 11 bp per turn in a right-handed, anti-parallel double-helix. The sugars adopt a C3′-*endo* pucker and the other geometric parameters are all very similar to those for A-DNA (see Tables 2.3 and 2.4). If the salt concentration is raised above 20%, an A′-RNA form is observed which has

12 bp per turn of the duplex. Both structures have typical Watson–Crick base pairs, which are displaced 4.4 Å from the helix axis and so form a very deep major groove and a rather shallow minor groove.

These features were confirmed by the analysis of the first single crystal structure of an RNA oligonucleotide, the 14 mer r(UUAUAUAUAUAUAA).[44] This 14-mer can be treated as three segments of A-helix separated by kinks in the sugar–phosphate backbone, which perturb the major groove dimensions. It is noteworthy that the 2′-hydroxyl groups are prominent at the edges of the relatively open minor groove.[45] They are extensively hydrated and can be recognised by proteins (Figure 2.40). Many more crystal structures of oligoribonucleotides have now been determined, some to atomic resolution. These structures have provided a wealth of information regarding canonical RNA duplexes, the effects on conformation by mismatched base pairs, hydration and cation co-ordination.

In one of the first NMR studies of a RNA duplex, Gronenborn and Clore combined 2D NOE analysis[46] (Section 11.2) with molecular dynamics to identify an A-RNA solution structure for the hexaribonucleotide, 5′-r(GCAUGC)$_2$.[47] It shows sequence-dependent variations in helix parameters, particularly in

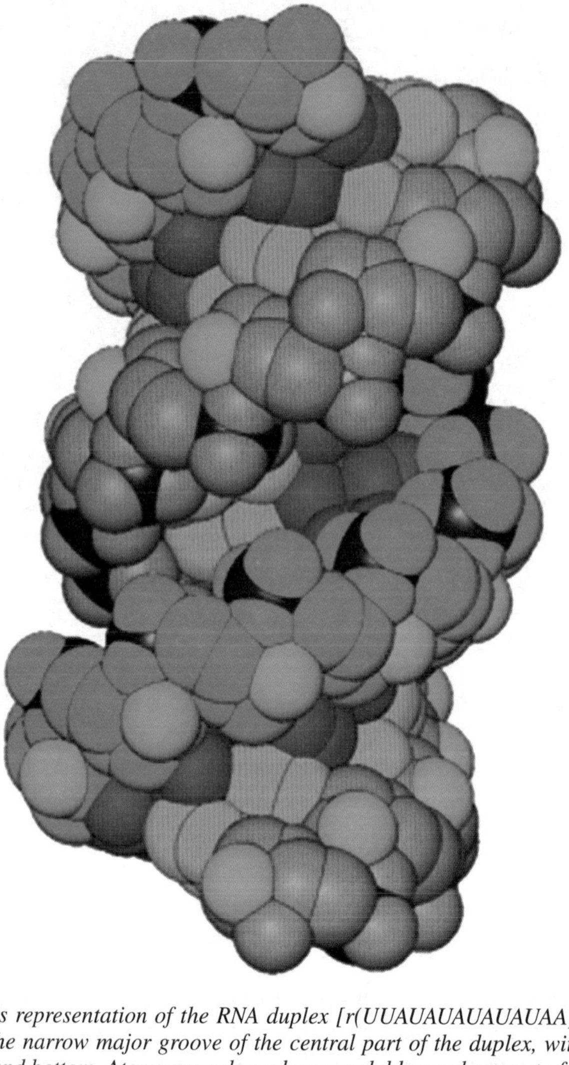

**Figure 2.40** *Van der Waals representation of the RNA duplex [r(UUAUAUAUAUAUAA)]$_2$ (PDB code 1RNA). The view is into the narrow major groove of the central part of the duplex, with the minor groove visible near the top and bottom. Atoms are coloured grey, red, blue and magenta for carbon, oxygen, nitrogen and phosphorus, respectively. 2′-Oxygen atoms lining the minor groove are highlighted in cyan*

helix twist and in base pair roll, slide and propeller twist (Figure 2.19). The extent of variation from base to base is much less than for the corresponding DNA hexanucleotide and seems to be dominated by the need of the structure to achieve very nearly optimal base stacking. This picture supports experimental studies that indicate that base stacking and hydrogen bonding are equally important as determinants of RNA helix stability.

**Antisense RNA** is defined as a short RNA transcript that lacks coding capacity, but has a high degree of complementarity to another RNA, which enables the two to hybridise.[48] The consequence is that such anti-sense, or complementary, RNA can act as a repressor of the normal function or expression of the targeted RNA. Such species have been detected in prokaryotic cells with suggested functions concerning RNA-primed replication of plasmid DNA, transcription of bacterial genes, and messenger translation in bacteria and bacteriophages. Quite clearly, such regulation of gene expression depends on the integrity of RNA duplexes.

A crucial cellular 'security' machinery that also depends on double-stranded RNA is **RNA interference** or **RNAi** (see Section 5.7.2).[49–51] This mechanism has evolved to protect cells from hostile genes as well as to regulate the activity of normal genes during growth and development. Tiny RNAs that are termed short interfering RNAs (**siRNAs**) or micro RNAs (**miRNAs**), depending on their origin, are capable of down-regulating gene expression by binding to complementary mRNAs, resulting either in mRNA elimination or arrest of translation. Although only discovered some 13 years ago in plants, RNA interference has now been found to be ubiquitous in all eukaryotes. The extraordinary specificity of RNAi and the simplicity of administering double-stranded RNAs to organisms with fully sequenced genomes (*i.e. C. elegans*, *D. melanogaster* and *X. lavis*) render RNAi a method of choice for functional genomics. As with potential applications of the anti-sense strategy for therapeutic purposes, the success of RNAi as a drug will depend on breakthroughs in cellular uptake and delivery.

### 2.4.3 RNA·DNA Duplexes

Helices that have one strand of RNA and one of DNA are very important species in biology.

- They are formed when reverse transcriptase makes a DNA complement to the viral RNA.
- They occur when RNA polymerase transcribes DNA into complementary messenger RNA.
- They are a feature in DNA replication of the short primer sequences in Okazaki fragments (Section 6.6.4).
- Anti-sense DNA is a single-stranded oligodeoxynucleotide designed to bind to a short complementary segment of a target nucleic acid (RNA or single-stranded DNA) with the potential for regulation of gene expression (Section 5.7.1).

Such hybrids are formed *in vitro* by annealing together two strands with complementary sequences, such as poly(rA)·poly(dT) and poly(rI)·poly(dC). These two hetero-duplexes adopt the A-conformation common to RNA and DNA, the former giving an 11-fold helix typical of A-RNA and the latter a 12-fold helix characteristic of A′-RNA.

A self-complementary decamer r(GCG)d(TATACGC) also generates a hybrid duplex with Watson–Crick base pairs. It has a helix rotation of 330° with a step-rise of 2.6 Å and C-3′-*endo* sugar pucker typical of A-DNA and A-RNA (see Table 2.1). The thermodynamic stability of RNA·DNA hybrids relative to the corresponding DNA·DNA duplexes is a function of the deoxypyrimidine content in the DNA strands of the former.[52] Hybrids with DNA strands containing 70–100% pyrimidines are more stable to thermal de-naturation than their DNA·DNA counterparts, whereas those with less than 30% deoxypyrimidines are less stable than the DNAs. A pyrimidine content of *ca.* 50% is the 'break-even' point.

The greater stability in some cases of RNA·DNA hetero-duplexes over DNA·DNA homo-duplexes is the basis of the construction of **antisense DNA oligomers**.[53–56] These are intended to enter the cell where they can pair with, and so inactivate, complementary mRNA sequences. Additional desirable features such as membrane permeability and resistance to enzymatic degradation have focused attention on oligonucleotides with chemical modifications in the phosphate, sugar or base moieties (Section 4.4). In some cases, the resulting hetero-duplexes have proved to have higher association constants than the natural DNA·RNA duplexes and the oligonucleotide analogues exhibit increased resistance to phosphate diesterase action (Table 2.6).

The subtle differences in conformation between an RNA·DNA hybrid duplex and either DNA·DNA or RNA·RNA duplexes have significance for enzyme action and also for anti-sense therapy. The therapeutic objective of antisense oligodeoxynucleotides very much depends on their ability to create a duplex with the target RNA and thus make it a substrate for ribonuclease H (Table 2.6). Because RNase H cleaves DNA·RNA hybrids but does not cleave the corresponding RNA·RNA duplexes, it can be induced to degrade an endogeneous mRNA species through hybridisation with a synthetic antisense oligodeoxyribonucleotide.

X-ray structures of crystals of duplexes having DNA and RNA residues in both strands showed them to have pure A-form geometry (see above). Duplexes between RNA and DNA also adopt A-form geometry in the solid state and in two cases it has been shown that self-complementary DNA decamers with a single incorporated ribonucleotide are in the A-form in the crystal although the all-DNA sequences prefer the B-form in the crystal and in solution. It is possible that crystal lattice forces and crystallisation kinetics play a role in the preference of the A-form geometry observed for all crystal structures of DNA·RNA duplexes.

By contrast, the hybrid duplexes d(GTCACATG)·r(CAUGUGAC) and d(GTGAACCTT)·r(AAGUUCAC) have been analysed by 2D NOE NMR in solution and shown to have neither pure A-form nor pure B-form structure.[57] The sugars of the RNA strands have the regular C3′-*endo* conformation but those in the DNA strand have a novel, intermediate C4′-*endo* conformation. Glycosylic torsion angles in the DNA chain are typical of B-form (near –120°) but those in the RNA chain are typically A-form values (near –140°). Overall the global structure is that of an A-form helix in which the base pairs have the small rise and positive inclination typical of an A-form duplex (Figure 2.19a,d). However, the width of the minor groove appears to be intermediate between A- and B-form duplexes and such structures have been modelled into the active site of RNase H. The results suggest that additional interactions of the protein with the DNA strand are possible only for this intermediate hybrid duplex conformation but not for an RNA·RNA duplex. So, it seems possible that these subtle changes in nucleotide conformation may explain the selectivity of RNase H for hybrid DNA·RNA duplexes.[58] Indeed, crystal structures for complexes between a bacterial

**Table 2.6** *Properties of antisense oligonucleotides and 1st and 2nd generation analogues*

| *Oligonucleotide type* | *Duplex stability*[a] | *Nuclease resistance*[b] | *RNase H activation*[c] |
|---|---|---|---|
| Oligodeoxyribonucleotide ($PO_2$) | Par[a] | — | Yes |
| Oligodeoxyribonucleotide phosphorothioate | — | ++ | Yes |
| Oligodeoxyribonucleotide methylphosphonate | — | +++ | No |
| Oligodeoxyribonucleotide phosphoramidate | + | ++ | No |
| Oligoribonucleotide ($PO_2$) | + | — | No |
| Oligo (2′-*O*-Me)ribonucleotide ($PO_2$) | ++ | + | No |
| Oligo (2′-*O*-(2-methoxyethyl)ribonucleotide[d] ($PO_2$) | ++ | + | No |
| Oligo (2′-*O*-(3-aminopropyl)ribonucleotide[e] ($PO_2$) | + | +++ | No |
| Oligo (2′-*O*-(*N*,*N*-dimethylaminooxyethyl)[f] ($PO_2$) | ++ | +++ | No |
| Oligo (2′,4′-methylene ...)[g] ($PO_2$) | +++ | + | No |
| Oligo (2′-fluoroarabinonucleotide)[h] ($PO_2$) | + | + | Yes |
| Peptide nucleic acids[i] | + | +++ | No |
| Oligodeoxy(5-propyne-cytidine) ($PO_2$) | ++ | — | Yes |

[a] Compared to DNA–RNA stability under physiological conditions.
[b] Compared to DNA (phosphate diesterase digestion).
[c] Activation of RNase H by the duplex formed between the oligonucleotide and RNA.
[d] 2′-*O*-MOE.
[e] 2′-*O*-AP.
[f] 2′-*O*-DMAEOE.
[g] Locked nucleic acid (LNA).
[h] FANA.
[i] PNA.

RNase H[59] and the RNase H domain from HIV-1 reverse transcriptase,[60] and hybrid duplexes have revealed that the RNA adopts a standard A-form geometry whereas the DNA exhibits B-form sugar puckers. However, the DNA·RNA hybrid at the active site of the reverse transcriptase domain assumes a canonical A-form[60] and it is important to note that DNA·RNA hybrids at the active sites of enzymes can assume a range of conformations.

### 2.4.4 RNA Bulges, Hairpins and Loops

The functional diversity of RNA species is reflected in the diversity of their 3D structures. Several structural elements have been identified that make up folded RNA and their thermodynamic stabilities relative to the un-folded single strand have been evaluated.[61] The folded conformations are largely stabilised by anti-parallel double-stranded helical regions, in which **intra-strand** and **inter-strand** base-stacking and hydrogen bonding provide most of the stabilisation. Base-paired regions are separated by regions of unpaired bases, either as various types of loops or as single strands, as illustrated for a 55-nucleotide fragment from R17 virus (Figure 2.41). Recent years have brought a flurry of new crystal structures of ever larger RNA molecules, featuring many different non-canonical secondary and tertiary structural motifs and culminating in the high-resolution crystal structure analyses of the large and small ribosomal sub-units (Section 7.3.3).

**Hairpin loops** were first identified as components of tRNA structures (Section 7.1.4) where they contain many bases. In the secondary structure deduced for 16S ribosomal RNA, most of the loops have four unpaired bases and these are known as **tetra-loops** (Section 7.1.4). Smaller tri-loops of three bases can also be formed.

Nuclear magnetic resonance and crystallographic studies on such stable tetra-loop hairpins show that their stems have A-form geometry while the loops have additional, unusual hydrogen bonding and base pair interactions (Section 7.1.2). For example, the GAAA loop has the unusual G·A base pair and UUYG loops have a reverse wobble U · G base pair. As a result, simple models appear to be inadequate to describe RNA hairpin stem-loop structures. The nonanucleotide r(CGCUUUGCG) forms a stable tri-loop hairpin whose thermodynamic stability has been determined by analysis of $T_m$ curves to be $-101\,\text{kJ mol}^{-1}$ for $\Delta H°$ and is close to the calculated value ($-90\,\text{kJ mol}^{-1}$) for this RNA helix. Nuclear magnetic resonance analysis shows that the loop has an A-form stem and the chain reversal appears between residues U5 and U6. The three uridine residues on the tri-loop have the C2′-*endo* conformation and show partial base-stacking, notably involving the first U on the 5′-side of the loop. These very high-resolution NMR results give a structure different from those structures computed by restrained molecular dynamics (Section 11.7.2), indicating that further refinement of the computational model is needed.

The hairpin loop is not only an important and stable component of secondary structure but also a key functional element in a number of well-characterised RNA systems. For example, it is required in the RNA TAR region of HIV (human immunodeficiency virus) for *trans*-activation by the Tat protein, and several viral coat proteins bind to specific hairpin loop structures.

**Bulges** are formed when there is an excess of residues on one side of a duplex. For single base bulges, the extra base can either stack into the duplex, as in the case of an adenine bulge in the coat protein-binding site of R17 phage, or be looped out, as shown by NMR studies in uracil bulges in duplexes. Such bulges can provide high-affinity sites for intercalators such as ethidium bromide (Section 9.6). In general, it appears that a bulge of one or two nucleotides has four effects on structure: (1) it distorts the stacking of bases in the duplex, (2) induces a bend in RNA, (3) reduces the stability of the helix, and (4) increases the major groove accessibility at base pairs flanking the bulge.

**Internal loops** occur where there are non-Watson–Crick mismatched bases. They can involve either one or two base pairs with pyrimidine–pyrimidine opposition (as in Figure 2.41) or mismatched purine–purine or pyrimidine–pyrimidine pairs, of which G·A pairs can form a mismatched base pair compatible with an A-form helix. There are also many examples of larger internal loops. Some of those that are rich in purine residues have been implicated as protein recognition sites. Many of these larger loops show marked resistance to chemical reagents specific for single-strand residues and this, in combination with structural data,

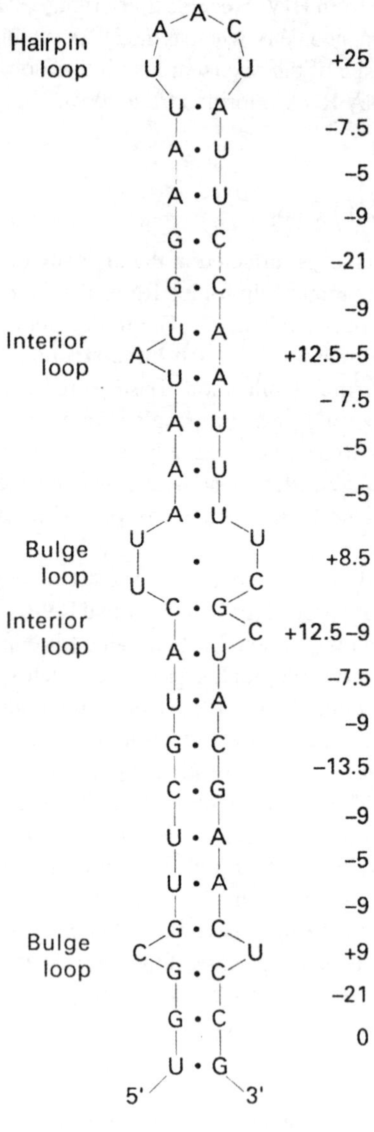


**Figure 2.41** *A possible secondary structure for a 55-nucleotide fragment from R17 virus which illustrates hairpin loop, interior loop and bulge structures. The free energy of this structure has been calculated to have a net $\Delta G°$ of $-90\,kJ\,mol^{-1}$ using appropriate values for base pairs (Table 2.6) and for loops and bulges*

suggests that there is probably a high level of order in such loops, notably of base-stacking and base-triples. One general opinion is that the major differences between loop and stem regions are dynamic rather than structural.

**Junctions** are regions that connect three or more stems (the connecting region for two stems is an internal loop) and are a common feature of computer-generated secondary structures for large RNAs. A prime example is the four-stem junction in the cloverleaf structure of tRNAs in which stacking continuity between the acceptor and the T stems and between the anti-codon and D stems is maintained (Section 7.1.4).[62] A junction of three stems forms the hammerhead structure of self-cleaving RNA (Section 7.6.2) and junctions of up to five stems have been observed for 16S RNA.

*2.4.4.1 Thermodynamics of Secondary Structure Elements.* The free energy of an RNA conformation has to take into account the contributions of interactions between bases, sugars, phosphates, ions and solvent. The most reliable parameters are those derived experimentally from the $T_m$ profiles (Section 2.5.1) of double-helical regions of RNA and data for each of the 10 nearest-neighbour sequences are given in Table 2.7. They are accurate enough to predict the expected thermodynamic behaviour of any RNA duplex to within about 10% of its experimental value.[63]

Other structural features are less easy to predict. It is clear that stacking interactions are more important than base-pairings so that an odd purine nucleotide 'dangling' at the 3′-end of a stem can contribute some $-4\,\mathrm{kJ\,mol^{-1}}$ to the stability of the adjacent duplex. The energies for mispairs or loops are rather less accurate, but always destabilising and change with the size of the loop (Table 2.8). Energies of these irregular secondary structures also depend on base composition, for example a single base bulge for uridine costs about $+8\,\mathrm{kJ\,mol^{-1}}$ and for guanosine about $+14\,\mathrm{kJ\,mol^{-1}}$.

By use of such data, the prediction of secondary structure is a conceptually simple task that can be handled by a modest computer while the more advanced programmes search sub-optimal structures as well as that of lowest free energy.

Interactions between separate regions of secondary structure are defined as tertiary interactions. One example is that of **pseudoknots**, which involve base-pairing between one strand of an internal loop and a distant single-strand region (Section 7.6.3, Figure 7.41). Pseudoknots can also involve base-pairing between components of two separate hairpin loops and examples with 3–8 bp have been described as a

**Table 2.7** *Thermodynamic parameters for RNA helix initiation and propagation in 1 M NaCl*

| *Propagation sequence* | *ΔH (kJ mol⁻¹)* | *ΔS (J K⁻¹ mol⁻¹)* | *ΔG (kJ mol⁻¹)* | *Propagation sequence* | *ΔH (KJ mol⁻¹)* | *ΔS (J K⁻¹ mol⁻¹)* | *ΔG (kJ mol⁻¹)* |
|---|---|---|---|---|---|---|---|
| ↑ A·U / A·U ↓ | −27.7 | −77.3 | −3.8 | ↑ A·U / G·C ↓ | −55.8 | −149 | −9.6 |
| ↑ U·A / A·U ↓ | −23.9 | −65.1 | −3.8 | ↑ U·A / G·C ↓ | −42.8 | −110 | −8.8 |
| ↑ A·U / U·A ↓ | −34.1 | −94.9 | −4.6 | ↑ G·C / C·G ↓ | −33.6 | −81.5 | −8.4 |
| ↑ A·U / C·G ↓ | −44.1 | −117 | −7.6 | ↑ C·G / G·C ↓ | −59.6 | −147 | −14.7 |
| ↑ U·A / C·G ↓ | −31.9 | −80.6 | −7.1 | ↑ G·C / C·G ↓ | −51.2 | −125 | −12.4 |
| Initiation | (0) | −45.4 | 14.6 | | | | |
| Symmetry correction (self-complementary) | 0 | −5.9 | 1.7 | Symmetry correction (non-self-complementary) | 0 | 0 | 0 |

Arrows point in a 5′-3′ direction to designate the stacking of adjacent base pairs.
The enthalpy change for helix initiation is assumed to be zero.

**Table 2.8** *Free energy increments for loops (kJ mol⁻¹ in 1 M NaCl, 37°C)*

| *Loop size* | *Internal loop* | *Bulge loop* | *Hairpin loop* |
|---|---|---|---|
| 1 | — | +14 | — |
| 2 | +4 | +22 | — |
| 3 | +5.4 | +25 | +31 |
| 4 | +7.1 | +28 | +25 |
| 5 | +8.8 | +31 | +18.5 |
| 6 | +10.5 | +34.5 | +18 |

result of both NMR and X-ray analysis. However, the computer prediction of tertiary interactions and base-triples appears to be still beyond the scope of present methodology.

### 2.4.5 Triple-Stranded RNAs

The first triple-stranded nucleic acid was described in 1957 when poly(rU)·poly(rA) was found to form a stable 2:1 complex in the presence of magnesium chloride. The extra poly(rU) strand is parallel to the poly(rA) strand and forms Hoogsteen base-triples in the major groove of an A-form Watson–Crick helix. Triplexes of 2poly(rA)·poly(rU) can also be formed while poly (rC) can form a triplex with poly(rG) at pH 6 which has two cytidines per guanine, one of them being protonated to give the $C^{+}$xG·C base-triple also seen for triple-helical DNA (Figure 2.33). Base triples are also a very common feature of tRNA structure (Section 7.1.4).[62]

The importance of added cations to overcome the repulsion between the anionic chains of the Watson–Crick duplex and the poly-pyrimidine third strand is an essential feature of triple-helix formation. $Co^{3+}(NH_3)_6$ and spermine are also effective counter-ions as well as the more usual $Mg^{2+}$.

Poly(rG) as well as guanosine and GMP can form structures with four equivalent hydrogen-bonded bases in a plane, with all four strands parallel. It is not clear whether this structure has any relevance to RNA folding.

## 2.5 DYNAMICS OF NUCLEIC ACID STRUCTURES

Any over-emphasis on the stable structures of nucleic acids runs the risk of playing down the dynamic activity of nucleic acids that is intrinsic to their function. Pairing and unpairing, breathing and winding are integral features of the behaviour of these species.[64]

Established studies on structural transitions of nucleic acids have for a long time used classical physical methods, which include light absorption, NMR spectroscopy, ultra-centrifugation, viscometry and X-ray diffraction (Chapter 11). More recently, these techniques have been augmented by a range of powerful computational methods (Section 11.7). In each case, the choice of experiment is linked to the time-scale and amplitude of the molecular motion under investigation.

### 2.5.1 Helix-Coil Transitions of Duplexes

Double helices have a lower molecular absorptivity for UV light than would be predicted from the sum of their constituent bases. This **hypochromicity** is usually measured at 256 nm while C·G base pairs can also be monitored at 280 nm. It results from coupling of the transition dipoles between neighbouring stacked bases and is larger in amplitude for A·U and A·T pairs than for C·G pairs. As a result, the UV absorption of a DNA duplex *increases* typically by 20–30% when it is denatured. This transition from a helix to an unstacked, strand-separated coil has a strong entropic component and so is temperature dependent. The mid-point of this thermal transition is known as the **melting temperature** ($T_m$).

Such dissociation of nucleic acid helices in solution to give single-stranded DNA is a function of base composition, sequence and chain length as well as of temperature, salt concentration and pH of the solvent. In particular, early observations of the relationship between $T_m$ and base composition for different DNAs showed that A·T pairs are less stable than C·G pairs, a fact which is now expressed in a linear correlation between $T_m$ and the gross composition of a DNA polymer by the equation:

$$T_m = X + 0.41[\% \text{ (C + G)}] \text{ (°C)}$$

The constant $X$ is dependent on salt concentration and pH and has a value of 69.3°C for 0.3 M sodium ions at pH 7 (Figure 2.42).

A second consequence is that the steepness of the transition also depends on base sequence. Thus, melting curves for homo-polymers have much sharper transitions than those for random-sequence polymers. This is because A·T rich regions melt first to give unpaired regions, which then extend gradually with rising temperature until, finally, even the pure C·G regions have melted (Figures 2.42 and 2.43). In some

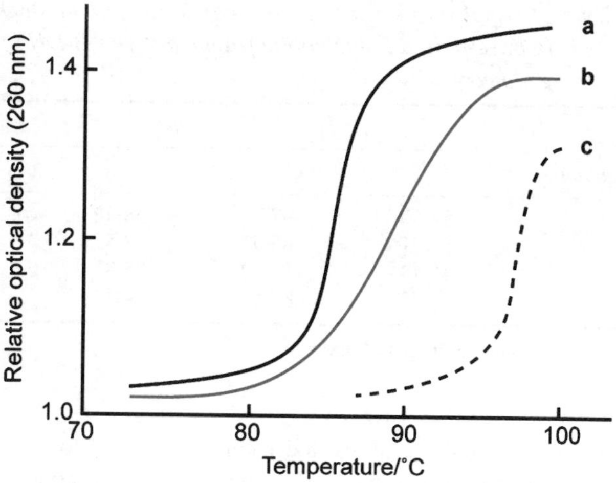


**Figure 2.42** *Thermal denaturation of DNAs as a function of base composition (per cent G·C) for three species of bacteria: (a) Pneumococcus (38% G·C). (b) E. coli (52% G·C). (c) M. phlei (66% G·C)* (Adapted from J. Marmur and P. Doty, *Nature*, 1959, **183**, 1427–1429. © (1959), with permission from Macmillan publishers Ltd)

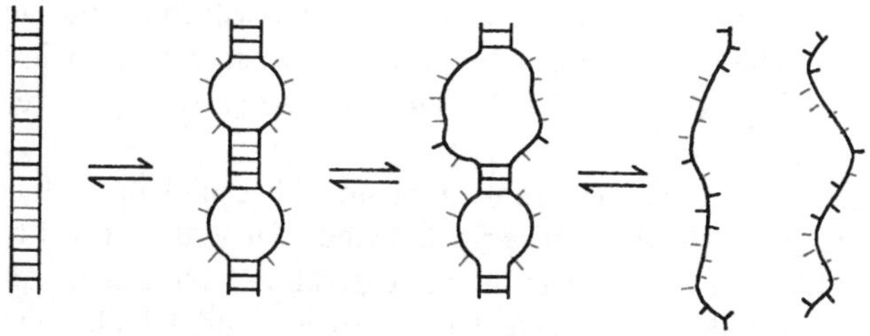

**Figure 2.43** *Scheme illustrating the melting of A·T-rich regions (colour) followed by mixed regions, then by C·G-rich regions (black) with rise in temperature (left→right)*

cases, the shape of the melting curve can be analysed to identify several components of defined composition melting in series.

Because of end-effects, short homo-oligomers melt at lower temperatures and with broader transitions than longer homo-polymers. For example for poly(rA)$_n$·poly(rU)$_m$, the octamer melts at 9°C, the undecamer at 20°C, and long oligomers at 49°C in the same sodium cacodylate buffer at pH 6.9. Consequently, in the design of synthetic, self-complementary duplexes for crystallisation and X-ray structure determination, C·G pairs are often places at the ends of hexamers and octamers to stop them 'fraying'. Lastly, the marked dependence of $T_m$ on salt concentration is seen for DNA from *Diplococcus pneumoniae* whose $T_m$ rises from 70°C at 0.01 M KCl to 87°C for 0.1 M KCl and to 98°C at 1.0 M KCl.

Data from many melting profiles have been analysed to give a **stability matrix** for nearest neighbour stacking (Table 2.9). This can be used to predict $T_m$ for a B-DNA polymer of known sequence with a general accuracy of 2–3°C.[65]

The converse of melting is the **renaturation** of two separated complementary strands to form a correctly paired duplex. In practice, the melting curve for denaturation of DNA is reversible only for relatively short oligomers, where the rate-determining process is the formation of a nucleation site of about 3 bp followed by rapid zipping-up of the strands and where there is no competition from other impeding processes.

When solutions of unpaired, complementary large nucleic acids are incubated at 10–20°C below their $T_m$, renaturation takes place over a period of time. For short DNAs of up to several hundred base pairs,

**Table 2.9** *Thermal stability matrix for nearest-neighbour stacking in base-paired dinucleotide fragments with B-DNA geometry*

| | *3′-Neighbour* | | | |
|---|---|---|---|---|
| *5′-Neighbour* | *A* | *C* | *G* | *T* |
| A | 54.50 | 97.73 | 58.42 | 57.02 |
| C | 54.71 | 85.97 | 72.55 | 58.42 |
| G | 86.44 | 136.12 | 85.97 | 97.73 |
| T | 36.73 | 86.44 | 54.71 | 54.50 |

Numbers give $T_m$ values in °C at 19.5 mM $Na^+$.

nucleation is rate-limiting at low concentrations and each duplex zips to completion almost instantly (>1000 bp $s^{-1}$). The nucleation process is bi-molecular, so renaturation is concentration dependent with a rate constant around $10^6\ M^{-1}\ s^{-1}$.[66] It is also dependent on the complexity of the single strands. Thus, for the simplest cases of homo-polymers and of short heterogeneous oligonucleotides, nucleation sites will usually be fully extended by rapid zipping-up. This gives us an 'all-or-none' model for duplex formation. By contrast, for bacterial DNA each nucleation sequence is present only in very low concentration and the process of finding its correct complement will be slow. Lastly, in the case of eukaryotic DNA the existence of repeated sequences means that locally viable nucleation sites will form and can be propagated to give relatively stable structures. These will not usually have the two strands in their correct overall register. Because such pairings become more stable as the temperature falls, complete renaturation may take an infinitely long time.

Longer nucleic acid strands are able to generate intra-strand hairpin loops, which optimally have about six bases in the loop and paired sections of variable length. They are formed by rapid, uni-molecular processes which can be 100 times faster than the corresponding bimolecular pairing process. Although such hairpins are thermodynamically less stable than a correctly paired duplex, their existence retards the rate of renaturation, so that propagation of the duplex is now the rate-limiting process (Figure 2.44). One notable manifestation of this phenomenon is seen when a hot solution of melted DNA is quickly quenched to +4°C to give stable denatured DNA.

With longer DNA species, Britten and Kohne have shown that the rate of recombination, which is monitored by UV hypochromicity, can be used to estimate the size of DNA in a homogenous sample. The time $t$ for renaturation at a given temperature for DNA of single-strand concentration $C$ and total concentration $C_0$ is related to the rate constant $k$ for the process by an equation which in its simplest form is:

$$C/C_0 = (1 + kC_0t)^{-1}$$

In practice where $C/C_0$ is 0.5 the value of $C_0t$ is closely related to the complexity of the DNA under investigation.

This **annealing** of two complementary strands has found many applications. For DNA oligomers, it provided a key component of Khorana's chemical synthesis of a gene (Section 5.4.1). It is now an integral feature of the insertion of chemically synthesised DNA into vectors. For RNA·DNA duplexes, it has provided a tool of fundamental importance for gene identification (Section 5.5) and is being explored in the applications of antisense DNA (Section 5.7.1).

### 2.5.2 DNA Breathing

Complete separation of two nucleic acid strands in the melting process is a relatively slow, long-range process that is not easily reversible. By contrast, the hydrogen bonds between base pairs can be disrupted

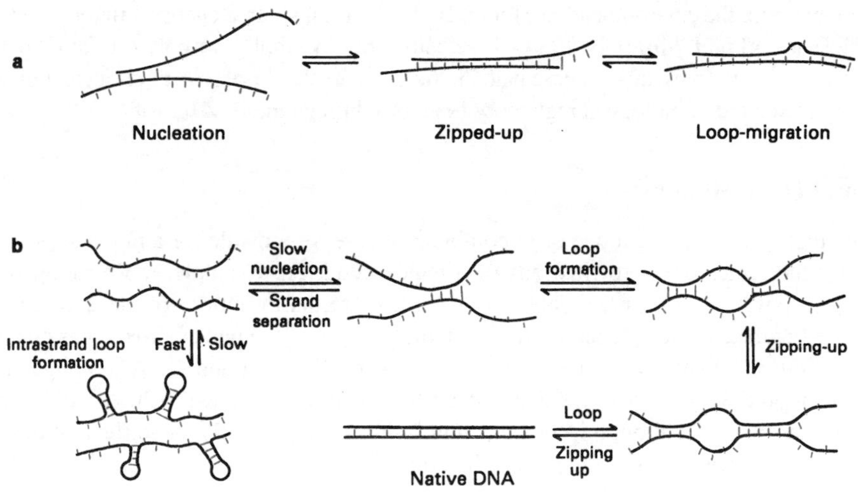


**Figure 2.44** *Renaturation processes (a) for short oligonucleotide and longer homo-polymers and (b) for natural DNA strands*

at temperatures well below the melting temperature to give local, short-range separation of the strands. This readily reversible process is known as **breathing**.

The evidence for such dynamic motion comes from chemical reactions, which take place at atoms that are completely blocked by normal base-pairings. Those used include tritium exchange studies in hydrogen-bonded protons in base pairs, the reactivity of formaldehyde with base NH groups, and NMR studies of imino–proton exchange with solvent water. This last technique can be used on a time scale from minutes down to 10 ms. It shows that in linear DNA the base pairs open singly and transiently with a life time around 10 ms at 15°C.

Because NMR can distinguish between imino- and amino–proton exchange, it can also be used to identify breathing in specific sequences.[67] Some of the most detailed work of this sort has come from studies on tRNA molecules, which show that, with increasing temperature, base-triplets (Figure 2.33) are destabilised first followed by the ribothymidine helix and then the dihydrouridine helix. Finally, the acceptor helix 'melts' after the anti-codon helix (Section 7.1.4).

Another possible motion that might be important for the creation of intercalation sites is known as 'soliton excitation'. The concept here is of a stretching vibration of the DNA chain, which travels like a wave along the helix axis until, given sufficient energy, it leads to local unstacking of adjacent bases with associated deformation of sugar pucker and other bond conformations.

Such pre-melting behaviour may well relate to the process of drug intercalation, to the association of single-strand specific DNA binding proteins (Section 10.3.8), and to the reaction of small electrophilic reagents with imino and amino groups such as cytosine-N-3 (Section 8.5).

### 2.5.3 Energetics of the B–Z Transition

The isomerisation equilibrium between the right-handed B-form and the left-handed Z-form of DNA is determined by three factors:

1. Chemical structure of the polynucleotide (sequence, modified bases)
2. Environmental conditions (solvent, pH, temperature, *etc.*)
3. Degree of topological stress (supercoiling, cruciform formation).

Many quantitative data have been obtained from spectroscopic, hydrodynamic and calorimetric studies and linked to theoretical calculations. Although these have not yet defined the kinetics or complex mechanisms of the B–Z transition, it is evident that the small transition enthalpies involved lie within the range of the thermal energies available from the environment. So, for example, the intrinsic free energy difference between Z- and B-forms is close to 2 kJ $mol^{-1}$ for poly-d(G·C) base pairs, only 1 kJ $mol^{-1}$ for poly-d(G·$m^5$C) base pairs, and greater than 5 kJ $mol^{-1}$ for poly-d(A·T) base pairs. It thus appears that local structural fluctuations may be key elements in the mediation of biological regulatory functions through the B–Z transition.[68,69]

### 2.5.4 Rapid DNA Motions

Rotations of single bonds, either alone or in combination, are responsible for a range of very rapid DNA motions with time scales down to fractions of a nanosecond. For example, the twisting of base pairs around the helix axis has a life time around $10^{-8}$ s while crankshaft rotations of the β, α, ζ and ε C-O-P-O-C bonds (see Figure 2.11) lead to an oscillation in the position of the phosphorus atoms on a millisecond time scale. Various calculations on the inter-conversion of C3′-*endo* and C2′-*endo* sugar pucker have given low activation energy barriers for their inter-conversion, in the range 3–20 kJ $mol^{-1}$, showing that the conformers are in rapid, although weighted, equilibrium at 37°C. Lastly, rapid fluctuations in propeller twist can result from oscillations of the glycosylic bond.

## 2.6 HIGHER-ORDER DNA STRUCTURES

The way in which eukaryotic DNA is packaged in the cell nucleus is one of the wonders of the macromolecular structure. In general, higher organisms have more DNA than lower ones (Table 2.10) and this calls for correspondingly greater condensation of the double helix. Human cells contain a total of $7.8 \times 10^9$ bp, which corresponds to an extended length of about 2 m. The DNA is packed into 46 cylindrical chromosomes of total length 200 μm, which gives a net packaging ratio of about $10^4$ for such metaphase human chromosomes (see also Section 6.4). The overall process has been broken down into two stages: the formation of nucleosomes and the condensation of nucleosomes into chromatin.[70]

### 2.6.1 Nucleosome Structure

The first stage in the condensation of DNA is the **nucleosome**, whose core has been crystallised by Aaron Klug and John Finch and analysed using X-ray diffraction. The DNA duplex is wrapped around a block of eight histone proteins to give 1.75 turns of a left-handed superhelix (Section 10.6, Figure 10.15).[71,72] This process achieves a packing ratio of 7. The number of base pairs involved in nucleosome structures varies from species to species, being 165 bp for yeast, 183 bp for HeLa cells, 196 bp for rat liver and 241 bp for sea urchin sperm. Such nucleosomes are joined by linker DNA whose length ranges from 0 bp in neurons to 80 bp in sea urchin sperm but usually averages 30–40 bp. The details of packaging the histone proteins are discussed later (Section 10.6.1).[73]

**Table 2.10** *Cellular DNA content of various species*

| *Organism* | *Numbers of base pairs* | *DNA length (mm)* | *Number of chromosomes* |
|---|---|---|---|
| *Escherichia coli* | $4 \times 10^6$ | 1.4 | 1 |
| Yeast (*Saccharomyces cerevisiae*) | $1.4 \times 10^6$ | 4.6 | 16 |
| Fruit fly (*Drosophila melanogaster*) | $1.7 \times 10^7$ | 56.0 | 4 |
| Humans (*Homo sapiens*) | $3.9 \times 10^9$ | 990.0 | 23 |

*Note*: Values are provided for haploid genomes.

As the DNA winds around the nucleosome core, the major and minor grooves are compressed on the inside with complementary widening of the grooves on the outside of the curved duplex. Runs of A·T base pairs, which have an intrinsically narrow minor groove should be most favourably placed on the inside of the curved segment while runs of G·C base pairs should be more favourably aligned with minor grooves facing outwards, where they are more accessible to enzyme cleavage. In practice, Drew and Travers measured the periodicities of A·T and G·C base pairs by cleavage with DNase I and found them to be exactly out of phase and having a periodicity of 10.17 ± 0. 5 bp.[74,75] This result was later confirmed by hydroxyl radical cleavage, which avoids the steric constraints of DNase I.

### 2.6.2 Chromatin Structure

**Chromatin** is too large and heterogeneous to yield its secrets to X-ray analysis, so electron microscopy is the chosen experimental probe (Section 11.5.1). At intermediate salt concentration (~1 mM NaCl) the nucleosomes are revealed as 'beads on a string'. Spherical nucleosomes can be seen with a diameter of 7–10 nm joined by variable-length filaments, often about 14 nm long. If the salt concentration is increased to 0.1 M NaCl, the spacing filaments get shorter and a zig-zag arrangement of nucleosomes is seen in a fibre 10–11 nm wide (Section 10.6, Figure 10.15). At even higher salt concentration and in the presence of magnesium, these condense into a 30 nm diameter fibre, called a **solenoid**, which is thought to be either a right-handed or a left-handed helix made up of close-packed nucleosomes with a packing ratio of around 40.

For the further stages in DNA condensation, one of the models proposed suggests that loops of these 30 nm fibres, each containing about 50 solenoid turns and possibly wound in a supercoil, are attached to a central protein core from which they radiate outwards.[76] Organisation of these loops around a cylindrical scaffold could give rise to the observed mini-band structure of chromosomes, which is some 0.84 μm in diameter and 30 nm in thickness. A continuous helix of loops would then constitute the **chromosome**. These ideas are illustrated in a possible scheme (Figure 2.45).

It is clear from all of the relevant biological experiments that the single DNA duplex has to be continuously accessible despite all this condensed structure in order for replication to take place. Some of the most exciting electron micrographs of DNA have been obtained from samples where the histones have been digested away leaving only the DNA as a tangled network of inter-wound superhelices radiating from a central nuclear region where the scaffold proteins remain intact (Figure 2.46).

Even then, the most condensed packing of nucleic acid is found in the sperm cell. Here a series of arginine-rich proteins called protamines bind to DNA, probably with their α-helices in the major groove of the DNA where they neutralise the phosphate charge, and so enable very tight packing of DNA duplexes.

Bacterial DNA is also condensed into a highly organised state (Section 10.6.2). In *E. coli* the genome has 4400 kbp in a closed circle, which is negatively supercoiled. It is condensed around histone-like proteins, HU and HI, to form a **nucleoid** and achieve a compaction of 1000-fold, which is followed by further condensation into supercoil domains. Unlike chromosomal DNA in eukaryotes, there is some additional negative supercoiling in prokaryotes that is not accounted for by protein binding.[77] This is probably a consequence of the activity of the bacterial DNA gyrase, which is capable of actively introducing further negative supercoiling, driven by hydrolysis of ATP. This whole process differs in several respects from assembly of chromatin in eukaryotes.

- There is no apparent regular repeating structure equivalent to the eukaryotic nucleosome although short DNA segments of 60–129 bp are organised by means of their interaction with abundant DNA-binding proteins.
- There is no prokaryotic equivalent to the solenoid structure.
- Bacterial DNA seems to be torsionally strained *in vivo* and organised into independently supercoiled domains of about 100 kbp.

The establishment of DNA architecture in the bacterial chromosome has progressed through the analysis of two types of structure. First, the interaction of a dimer of the HU protein from *B. stearothermophilus*

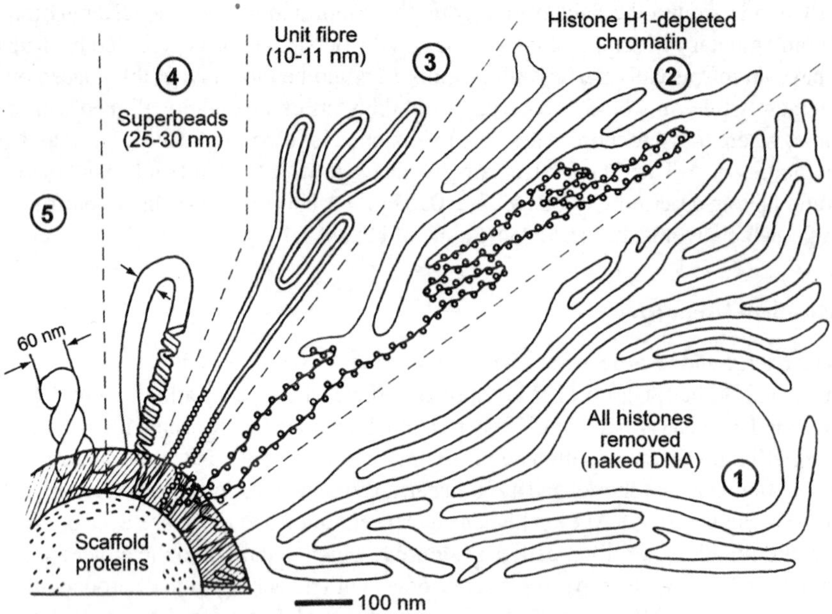


**Figure 2.45** *Schematic drawing to illustrate the gradual organisation of DNA into highly condensed chromatin. (1) DNA fixed to the protein scaffold; (2) DNA complexed with all histones except H1; (3) aggregation into a 100 Å fibre; (4) formation of 'superbeads' and (5) contraction into a 600 Å knob*
(Adapted from K.-P. Rindt and L. Nover, *Biol. Zentralblat.*, 1980, **99**, 641–673. © (1980), with permission from Elsevier)

shows loops of anti-parallel β-sheets inserted into the DNA minor groove (Section 10.6.2, Figure 10.16b) in non-sequence specific binding. Second, a large nucleoprotein complex in bacteria is involved in integration of phage DNA into the host chromosome and is called an **intasome**. This has the phage DNA wrapped as a left-handed supercoil around a complex of proteins including several copies of two DNA-binding proteins, the phase-coded integrase and the IHP protein (integration host factor). The IHP binds to a specific DNA sequence. These developments suggest that structural analysis of the bacterial chromosome may well overtake that of eukaryotic systems.

Tremendous progress has been made in the characterisation of nucleic acid structure during the past three decades. The ability to chemically synthesise oligodeoxynucleotides paved the way to a characterisation of DNA structure in atomic detail. Following the focus of early studies on (a) the conformations of the double helical families, (b) the sequence dependence of their structures, and (c) interactions between DNA and small molecule drugs, attention has progressively shifted to new tertiary structural motifs, such as junctions and four-stranded motifs, some of which were discussed in this chapter, and the interactions between DNA and proteins. Considerable numbers of DNA structures are now deposited in public data bases every year, many of them revealing surprises and shedding light on familiar but hitherto relatively poorly characterised phenomena such as conformational transitions.[78] 'Is there anything then that we still do not know about the structure of DNA?' the reader may ask. Of course, new exploitations of DNA's chemical and conformational versatility warrant structural characterisations. Supramolecular assemblies and nanostructures constructed from DNA are one example.[79,80] Many questions regarding the interactions between proteins and DNA and the important role that DNA plays in them remain to be answered. Suffice it to mention replication and the need for understanding the nature of nucleotide incorporation by high-fidelity and *trans*-lesion polymerases opposite native and lesioned DNA templates.[81] Also studies directed at a chemical etiology of nucleic acid structure based on the creation and characterisation of dozens of artificial pairing systems have created a further need for structural data.[82] With regard to RNA, the last decade has witnessed an explosion in the analysis of its structure and function. In 1994, the only RNA molecule whose

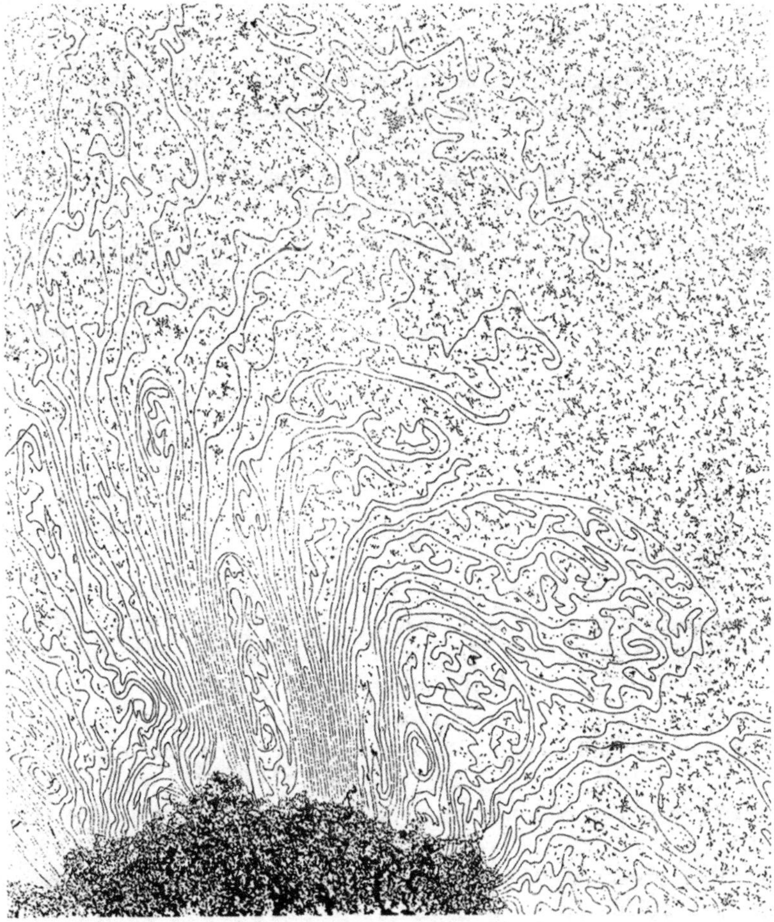

**Figure 2.46** *Electron micrograph of a histone-depleted chromosome showing that the DNA is attached to the scaffold in loops*
(Adapted from J.K. Paulson and U.K. Laemmli *Cell,* 1997, **12** 817–828. © (1977), with permission from Elsevier)

relatively complex tertiary structure had been revealed was transfer RNA. Then came the structures of ribozymes and those of numerous oligonucleotide-sized fragments, offering a glimpse at the repertoire of RNA's tertiary structural motifs, and a flurry of protein–RNA complexes, and – at last – atomic resolution structures of ribosomal subunits and whole ribosomes. Much remains to be discovered in terms of the mechanism of translation,[83,84] but the availability of ribosome structures and oligonucleotide fragments mimicking portions thereof has reinvigorated the interest in RNA as a drug target.[85,86] On the functional side, *in vitro* selection and the emergence of a flurry of the so-called aptamers, RNA molecules with the capacity to recognise and tightly bind small and large molecules have given a boost to the RNA-world hypothesis, and have subsequently led to the identification of natural control elements in messenger RNAs, 'ribsoswitches' (Section 5.7.2) that regulate gene expression.[87,88] And as if this were not enough, the advent of RNA interference (RNAi) has further underscored the importance of RNA in the mediation and control of biological information transfer. Perhaps it does not come as a surprise then that well over one third of the human genes appear to be conserved miRNA targets.[89] New functions come with their structural underpinnings and the structural biology of RNA-mediated gene silencing has already yielded first insights into novel RNA–protein interactions.[90] A little over 50 years after Watson and Crick's model of the DNA double helix there is no end in sight in the quest for the structural analysis of DNA and RNA.

## REFERENCES


1. C. Altona and M. Sundaralingam, Conformational analysis of the sugar ring in nucleosides and nucleotides. A new description using the concept of pseudorotation. *J. Am. Chem. Soc.*, 1972, **94**, 8205–8212.
2. R. Chandrasekaran and S. Arnott, *The structures of DNA and RNA in oriented fibres*. Springer, Berlin.
3. A. Rich and S. Zhang, Z-DNA: the long road to biological function *Nat. Rev. Genet.*, 2003, **4**, 566–572.
4. O. Kennard and W.N. Hunter, Single-crystal X-ray diffraction studies of oligonucleotides and oligonucleotide-drug complexes. *Angew. Chem. Intl. Ed. Engl.*, 1991, **30**, 1254–1277.
5. R.E. Dickerson, Nucleic acids, in *Crystallography of Biological Macromolecules*, Vol. F, M.G. Rossmann and E. Arnold (eds), International Tables of Crystallography, Kluwer Academic Publishers, Dordrecht, 2001, 588–622.
6. F.A. Jurmak and A. McPherson (eds), *Biological Macromolecules and Assemblies*, Vol. 1, Wiley, New York, 1984.
7. B. Hartmann and R. Lavery, DNA structural forms. *Quart. Rev. Biophys.*, 1996, **29**, 309–368.
8. W. Saenger, *Principles of Nucleic Acid Structure*. Springer, New York, 1984.
9. C.R. Calladine and H.R. Drew, *Understanding DNA. The Molecule and How it Works*. Academic Press Ltd, London, 1997.
10. G. Minasov, V. Tereshko and M. Egli, Atomic-resolution crystal structures of B-DNA reveal specific influences of divalent metal ions on conformation and packing. *J. Mol. Biol.*, 1999, **291**, 83–99.
11. M. Egli, V. Tereshko, M. Teplova, G. Minasov, A. Joachimiak, R. Sanishvili, C.M. Weeks, R. Miller, M.A. Maier, H. An, P.D. Cook and M. Manoharan, X-ray crystallographic analysis of the hydration of A- and B-form DNA at atomic resolution. *Biopolymers (Nucl. Acid Sci.)*, 2000, **48**, 234–252.
12. V. Tereshko, G. Minasov and M. Egli, A "hydration" spine in a B-DNA minor groove. *J. Am. Chem. Soc.*, 1999, **121**, 3590–3595.
13. A. Rich, The nucleic acids. A backward glance. In DNA: The double helix, *Ann. NY Acad. Sci.*, 1995, **758**, 97–142.
14. A.A. Gorin, V.B. Zhurkin and W.K. Olson, B-DNA twisting correlates with base-pair morphology. *J. Mol. Biol.*, 1995, **247**, 34–48.
15. K. Yanagi, G.G. Privé and R.E. Dickerson, Analysis of the local helix geometry in three B-DNA decamers and eight dodecamers. *J. Mol. Biol.*, 1991, **217**, 201–214.
16. M.A. El Hassan and C.R. Calladine, Propeller-twisting of base-pairs and the conformational mobility of dinucleotide steps in DNA. *J. Mol. Biol.*, 1996, **259**, 95–103.
17. C.A. Hunter, Sequence-dependent DNA structure, the role of base-stacking interactions. *J. Mol. Biol.*, 1993, **230**, 1025–1054.
18. H.-L. Ng, M.L. Kopka and R.E. Dickerson, The structure of a stable intermediate in the A ⇔ B DNA helix transition. *Proc. Natl. Acad. Sci. USA*, 2000, **97**, 2035–2039.
19. J.M. Vargason, K. Henderson and P.S. Ho, A crystallographic map of the transition from B-DNA to A-DNA. *Proc. Natl. Acad. Sci. USA*, 2001, **98**, 7265–7270.
20. T. Brown and O. Kennard, Structural basis of DNA mutagenesis**.** *Curr. Opin. Struct. Biol.*, 1992, **2**, 354–360.
21. R.D. Wells, Unusual DNA structures. *J. Biol. Chem.*, 1988, **263**, 1095–1098.
22. C.R. Calladine, H.R. Drew and M.J. McCall, The intrinsic curvature of DNA in solution. *J. Mol. Biol.*, 1988, **210**, 127–137.
23. D.M. Crothers, M.R. Gartenberg and T.R. Shrader, DNA bending in protein–DNA complexes. *Prog. Nucl. Acids Mol. Biol.*, 1992, **208**, 118–145.
24. R.E. Dickerson, D.S. Goodsell and S.A. Neidle, "...The tyranny of the lattice...," *Proc. Natl. Acad. Sci. USA*, 1994, **91**, 3579–3583.
25. R.E. Dickerson, DNA bending: the prevalence of kinkiness and the virtues of normality. *Nucleic Acids Res.*, 1998, **26**, 1906–1926.

26. Y. Timsit and D. Moras, Cruciform structures and functions. *Quart. Rev. Biophys.*, 1996, **29**, 279–307.
27. D.M. Lilley, K.M. Sullivan, A.I.H. Murchie and J. Furlong, Cruciform extrusion in supercoiled DNA-mechanisms and contextual influence, in *Unusual DNA structures*, R.D. Wells and S.C. Harvey (eds), Springer, Heidelberg, 1988, 55–72.
28. M. Ortiz-Lombardía, A. González, R. Eritja, J. Aymamí, F. Azorín and M. Coll, Crystal structure of a DNA Holliday junction. *Nat. Struct. Biol.*, 1999, **6**, 913–917.
29. M. Egli, DNA–cation interactions: *quo vadis*? *Chem. Biol.*, 2002, **9**, 277–286.
30. N.V. Hud and J. Plavec, A unified model for the origin of DNA sequence-directed curvature. *Biopolymers*, 2003, **69**, 144–159.
31. S.C. Ha, K. Lowenhaupt, A. Rich, Y.-G. Kim and K.K. Kim, Crystal structure of a junction between B-DNA and Z-DNA reveals two extruded bases. *Nature*, 2005, **437**, 1183–1186.
32. P. Palacek, Local supercoil-stabilized DNA structures. *Crit. Rev. Biochem. Mol. Biol.*, 1991, **26**, 151–226.
33. J.C. Wang, DNA topoisomerases. *Ann. Rev. Biochem.*, 1985, **54**, 665–697.
34. L.F. Liv, DNA topoisomerase poisons as antitumour drugs. *Ann. Rev. Biochem.*, 1989, **58**, 351–375.
35. H.R. Drew and A.A. Travers, DNA structural variations in the *E. coli tyrT* promoter. *Cell*, 1984, **37**, 491–502.
36. V.N. Soyfer and V.N. Potaman, *Triple-Helical Nucleic Acids*, Springer, New York, 1996.
37. H.E. Moser and P.B. Dervan, Sequence-specific cleavage of double helical DNA by triple-helix formation. *Science*, 1987, **238**, 645–650.
38. N.T. Thuong and C. Hélène, Sequence-specific recognition and modification of double-helical DNA by oligonucleotides. *Angew. Chem. Int. Ed. Engl.*, 1993, **32**, 666–690.
39. S. Rhee, Z.-J. Han, K. Liu, T. Miles and D.R. Davies, Structure of a triplex helical DNA with a triplex-duplex junction. *Biochemistry*, 1999, **38**, 16810–16815.
40. G.N. Parkinson, M.P.H. Lee and S. Neidle, Crystal structure of parallel quadruplexes from human telomeric DNA. *Nature*, 2002, **417**, 876–880.
41. K. Gehring, J.-L. Leroy and M. Guéron, A tetrameric DNA structure with protonated cytosine–cytosine base pairs. *Nature*, 1993, **363**, 561–565.
42. L. Chen, L. Cai, X. Zhang and A. Rich, Crystal structure of a four-stranded intercalated DNA: d($C_4$). *Biochemistry*, 1994, **33**, 13540–13546.
43. N.G.A. Abrescia, A. Thompson, T. Hyunh-Dinh and J.A. Subirana, Crystal structure of an antiparallel DNA fragment with Hoogsteen base-pairing. *Proc. Natl. Acad. Sci. USA*, 2002, **99**, 2806–2811.
44. A.C. Dock-Bregeon, B. Chevrier, A. Podjarny, D. Moras, J.S. deBear, G.R. Gough, P.T. Gilham and J.E. Johnson, High resolution structure of the RNA duplex $[U(U\text{-}A)_6A]_2$. *Nature*, 1988, **335**, 375–378.
45. M. Egli, S. Portmann and N. Usman, RNA hydration: a detailed look. *Biochemistry*, 1996, **35**, 8489–8494.
46. D. Neuhaus and M.P. Williamson, *The Nuclear Overhauser Effect in Structural and Conformation Analysis*, Chapters 5 and 12, VCH, Weinheim, 1989.
47. S.C. Happ, E. Happ, M. Nilges, A.M. Gronenborn and G.M. Clore, Refinement of the solution structure of the ribonucleotide $5'r(GCAUGC)_2$. *Biochemistry*, 1988, **27**, 1735–1743.
48. N. Houba-He'rin and M. Inouye, Antisense RNA in *Nucleic Acids and Molecular Biology*, F. Eckstein and D.M.J. Lilley (eds), Vol. 1, Springer, Berlin, 210–221.
49. G.J. Hannon, RNA interference. *Nature*, 2002, **418**, 244–251.
50. J.C. Carrington and V. Ambros, Role of microRNAs in plant and animal development. *Science*, 2003, **301**, 336–338.
51. N.C. Lau and D.P. Bartel, Censors of the genome. *Sci. Am.* 2003, **239**, 34–41.
52. E.A. Lesnik and S.M. Freier, Relative thermodynamic stability of DNA, RNA, and DNA:RNA hybrid duplexes: relationship with base composition and structure. *Biochemistry*, 1995, **34**, 10807–10815.
53. J.S. Cohen (ed), *Oligodeoxynucleotides, Antisense Inhibitors of Gene Expression*. Macmillan, London, 1989.
54. E. Uhlmann and A. Peyman, Antisense oligonucleotides, a new therapeutic principle. *Chem. Rev.*, 1990, **90**, 543–584.

55. Y.S. Sanghvi and P.D. Cook (eds), *Carbohydrate modifications in antisense research. ACS Symp. Ser.*, Vol. 580, American Chemical Society, Washington, DC, 1994.
56. P.E. Nielsen (ed), *Oligonucleotide antisense. Biochim. Biophys. Acta*, 1999, **1489**, 1–206.
57. A.M. Lane, S. Ebel and T. Brown, NMR assignments and solution conformation of the DNA:RNA hybrid d(GCGAACTT)·r(AAGUUCAC). *Eur. J. Biochem.*, 1993, **213**, 297–306.
58. O.Y. Fedoroff, M. Salazar and B.R. Reid, Structure of a DNA:RNA hybrid duplex, Why RNase H does not cleave pure RNA. *J. Mol. Biol.*, 1993, **233**, 509–523.
59. M. Nowotny, S.A. Gaidamakov, R.J. Crouch and W. Yang, Crystal structures of RNase H bound to an RNA/DNA hybrid: substrate specificity and metal-dependent catalysis. *Cell*, 2005, **121**, 1005–1016.
60. S.G. Sarafianos, K. Das, C. Tantillo, A.D. Clark Jr., J. Ding, J.M. Whitcomb, P.L. Boyer, S.H. Hughes and E. Arnold, Crystal structure of HIV-1 reverse transcriptase in complex with a polypurine tract RNA: DNA, *EMBO J.*, 2001, **20**, 1449–1461.
61. J.A. Jaeger, J. SantaLucia and I. Tinoco, Determination of RNA structure and thermodynamics. *Ann. Rev. Biochem.*, 1993, **62**, 255–287.
62. P.R. Schimmel, D. Söll and J.N. Abelson, *Transfer RNA: Structure and dynamics of RNA*. NATO ASI Series. Plenum, New York, 1979.
63. S.M. Ereler, R. Kierzek, J.A. Jaeger, N. Sugimoto, M.H. Caruthers, T. Neilson and D.H. Turner, Improved free energy parameters for predictions of RNA duplex stability. *Proc. Natl. Acad. Sci. USA*, 1986, **83**, 9373–9377.
64. J.A. McCammon and S.C. Harvey, *Dynamics of Proteins and Nucleic Acids*. Cambridge University Press, Cambridge, 1987.
65. K.J. Breslauer, R. Frank, H. Blöcker and L.A. Marky, Predicting DNA duplex stability from base sequence. *Proc. Natl. Acad. Sci. USA*, 1986, **83**, 3746–3750.
66. J.G. Wetmur, Hybridization and renaturation kinetics of nucleic acids. *Ann. Rev. Biophys. Bioeng.*, 1976, **5**, 337–361.
67. T.L. James, Relaxation behaviour of nucleic acids, in *Phosphorus-31 NMR*, D.G. Gorenstein (ed), Academic Press, New York, 349–400.
68. D.M. Soumpasis and T.M. Jovin, Energetics of the B–Z transition, in *Nucleic Acids and Molecular Biology*, Vol. 1, F. Eckstein and D.M.J. Lilley (eds), Springer, Heidelberg, 85–111.
69. M. Guéron and J.-P. Demaret, A simple explanation of the electrostatics of the B-to-Z transition of DNA. *Proc. Natl. Acad. Sci. USA*, 1992, **89**, 5740–5743.
70. Cold Spring Harbor Symposia, Chromatin. *Cold Spring Harbor Symp. Quant. Biol.*, 1978, **42**, 1–1353.
71. K. Luger, A.W. Mäder, R.K. Richmond, D.F. Sargent and T.J. Richmond, Crystal structure of the nucleosome core particle at 2.8 Å resolution. *Nature*, 1997, **389**, 251–260.
72. T.J. Richmond and C.A. Davey, The structure of DNA in the nucleosome core. *Nature*, 2003, **423**, 145–150.
73. D.S. Pederson, F. Thorma and R.T. Simpson, Core particles, fibre and transcriptionally active chromatin structure. *Ann. Rev. Cell Biol.*, 1986, **2**, 117–147.
74. A.A. Travers and A. Klug, The bending of DNA in nucleosomes and its wider implications. *Phil. Trans. Roy. Soc. Lond. B*, 1987, **317**, 537–561.
75. A.A. Travers, DNA conformation and protein binding. *Ann. Rev. Biochem.*, 1989, **58**, 427–452.
76. E.U. Selker, DNA methylation and chromatin structure: a view from below. *Trends Biol. Sci.*, 1990, **15**, 103–107.
77. M.B. Schmid, Structure and function of the bacterial chromosome. *Trends Biol. Sci.*, 1988, **13**, 131–135.
78. M. Egli, Nucleic acid crystallography: current progress. *Curr. Opin. Chem. Biol.*, 2004, **8**, 580–591.
79. P.J. Paukstelis, J. Nowakowski, J.J. Birkoft and N.C. Seeman, Crystal structure of a continuous three-dimensional DNA lattice. *Chem. Biol.*, 2004, **11**, 1119–1126.
80. M. Egli, "Deoxyribo nanonucleic acid": antiparallel, parallel and unparalleled. *Chem. Biol.*, 2004, **11**, 1027–1029.

81. D.T. Nair, R.E. Johnson, S. Prakash, L. Prakash and A.K. Aggarwal, Replication by human DNA polymerase occurs by Hoogsteen base-pairing. *Nature*, 2004, **430**, 377–380.
82. A. Eschenmoser, Chemical etiology of nucleic acid structure. *Science*, 1999, **284**, 2118–2124.
83. L. Ferbitz, T. Maier, H. Pratzelt, B. Bukau, E. Deuerling and N. Ban, Trigger factor in complex with the ribosome forms a molecular cradle for nascent proteins. *Nature*, 2004, **431**, 590–596.
84. S. Takyar, R.P. Hickerson and H.F. Noller, mRNA helicase activity of the ribosome. *Cell*, 2005, **120**, 49–58.
85. A.P. Carter, W.M. Clemons, D.E. Broderson, R.J. Morgan-Warren, B.T. Wimberly and V. Ramakrishnan, Functional insights from the structure of the 30S ribosomal subunit and its interactions with antibiotics. *Nature*, 2000, **407**, 340–348.
86. D. Vourloumis, G.C. Winters, K.B. Simonsen, M. Takahashi, B.K. Ayida, S. Shandrick, Q. Zhao, Q. Han and T. Hermann, Aminoglycoside-hybrid ligands targeting the ribosomal decoding site. *ChemBioChem.*, 2005, **6**, 58–65.
87. W. Winkler, A. Nahvi and R.R. Breaker, Thiamine derivatives bind messenger RNAs directly to regulate bacterial gene expression. *Nature*, 2002, **419**, 952–956.
88. R.T. Batey, S.D. Gilbert and R.K. Montange, Structure of a natural guanine-responsive riboswitch complexed with the metabolite hypoxanthine. *Nature*, 2004, **432**, 411–415.
89. B.P. Lewis, C.B. Burge and D.P. Bartel, Conserved seed pairing, often flanked by adenosines, indicates that thousands of human genes are microRNA targets. *Cell*, 2005, **120**, 15–20.
90. J.-B. Ma, K. Ye and D.J. Patel, Structural basis for overhang-specific small interfering RNA recognition by the PAZ domain. *Nature*, 2004, **429**, 318–322.

CHAPTER 3

# Nucleosides and Nucleotides

## CONTENTS



## 3.1 CHEMICAL SYNTHESIS OF NUCLEOSIDES

The first nucleoside syntheses were planned to prove the structures of adenosine and the other ribo- and deoxyribonucleosides. Modern syntheses have been aimed at producing nucleoside analogues for using them as inhibitors of nucleic acid metabolism (Section 3.7) and for incorporation into synthetic oligonucleotides (Section 4.4.1). These have a variety of uses such as therapeutic applications using antigene or antisense technologies (Section 5.7.1),[1] studying RNA and DNA structure,[2,3] DNA–protein interactions (Chapter 10) and nucleic acid catalysis (Section 5.7.3). In spite of advances in stereospecific synthesis, it remains more economical to produce the major nucleosides by degrading nucleic acids than by total synthesis.

Modified nucleosides are widely distributed naturally. For example, all species of tRNA contain unusual minor bases and many bacteria and fungi provide rich sources of nucleosides modified in the base, in the sugar or in both base and sugar residues. Since some of these have been found to show a wide and useful range of biological activity, thousands of nucleoside analogues have been synthesised in pharmaceutical laboratories across the world. In recent times, industrial targets for this work have been anti-viral and anti-cancer agents. For instance, the arabinose analogues of adenosine and cytidine, *ara*A and *ara*C, are useful as anti-viral and anti-leukaemia drugs, while 5-iodouridine is valuable for treating Herpes simplex infections of the eye (Figure 3.1).

D-Ribose and other pentoses are relatively inexpensive starting materials, which are especially useful in stereochemically controlled synthesis of modified sugars. Three principal strategies for the synthesis of modified nucleosides have been developed. These are illustrated by retrosynthetic analysis (Figure 3.2). First, disconnection A identifies formation of the glycosylic bond by joining the sugar onto a preformed base. In practice, this uses the easy displacement of a leaving group from C-1 of an aldose derivative by a nucleophilic nitrogen (or carbon) atom of the heterocyclic base. Second, the double disconnection B identifies the process of building a heterocyclic base onto a preformed nitrogen or carbon substituent at C-1 of the

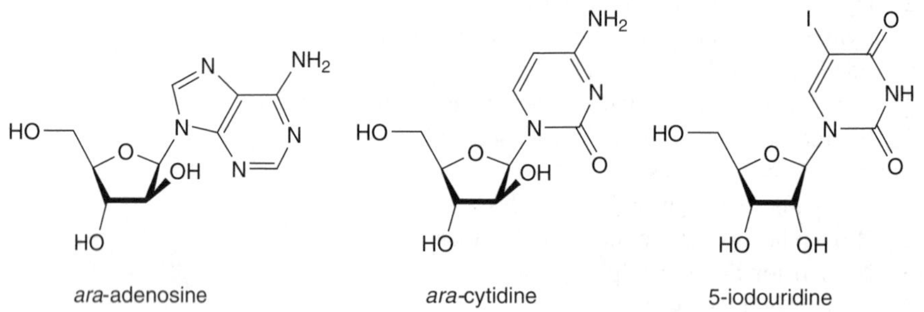


**Figure 3.1** *Modified nucleosides of biological importance*

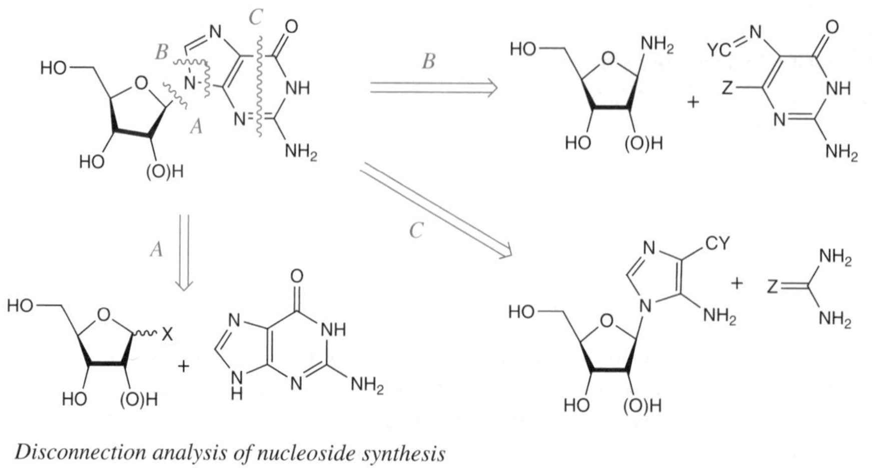

**Figure 3.2** *Disconnection analysis of nucleoside synthesis*

sugar moiety. Third, a double disconnection C shows the formation of a purine base onto a preformed imidazole ribonucleoside. We shall now explore each of these three routes in turn.

### 3.1.1 Formation of the Glycosylic Bond

The synthesis of nucleosides through glycosyl bond formation should ideally address both stereoselectivity (formation of nucleosides with the natural β-configuration at C-1) and regioselectivity (glycosylation of pyrimidines at N-1 and purines at N-9). There are essentially three methods that are used: (a) metal salt procedures, (b) silyl base procedures, and (c) fusion reactions together with various modifications of these. The first two methods are generally more widely applicable and used most frequently. While the following sections describe methods that may be used for the preparation of nucleosides, these mainly refer to ribonucleosides. Although such methods may be extended to the syntheses of 2′-deoxyribonucleosides, they often give poor stereoselectivity during glycosyl bond formation. Modifications to these methods that are more suited to the preparation of 2′-deoxyribonucleosides have been developed and are dealt with later in the chapter.

*3.1.1.1 Heavy Metal Salts of Bases.* Fischer and Helferich,[4] and Koenigs and Knorr introduced the use of a heavy metal salt (initially silver(I)) of a purine to catalyse the nucleophilic displacement of a halogen substituent from C-1 of a protected sugar. In the late 1940s, Todd's group adapted this chemistry to achieve a synthesis of adenosine[5] and guanosine[6] following an initial glycosylation between a protected 1-bromo-ribofuranose derivative and 2,8-dichloroadenine. In a later modification, Davoll and Lowy used mercury(II) salts to improve the yields of products.[7] Typically, chloromercuri-6-benzamidopurine reacts with 2,3,5-tri-*O*-acetyl-D-ribofuranosyl chloride or bromide to give a protected nucleoside from which adenosine is obtained by removal of the protecting groups (Figure 3.3). These syntheses almost invariably gave the desired stereoselectivity, predominantly providing the **β-anomer** at C-1 of ribose owing to the formation of an intermediate **acyloxonium ion** by the sugar component (see Section 3.1.1.7). The chloromercuri salts of a range of purines can be used, provided the nucleophilic substituents are protected. Thus, amino groups have to be protected by acylation, as shown in a synthesis of guanine nucleosides using 2-acetamido-6-chloropurine followed by appropriate hydrolysis (Figure 3.3).

The chloromercuri derivatives of suitable pyrimidines can be used in much the same way as illustrated by a synthesis of cytidine from 4-ethoxypyrimidine-2-one (Figure 3.4).[8] While this type of glycosylation gives

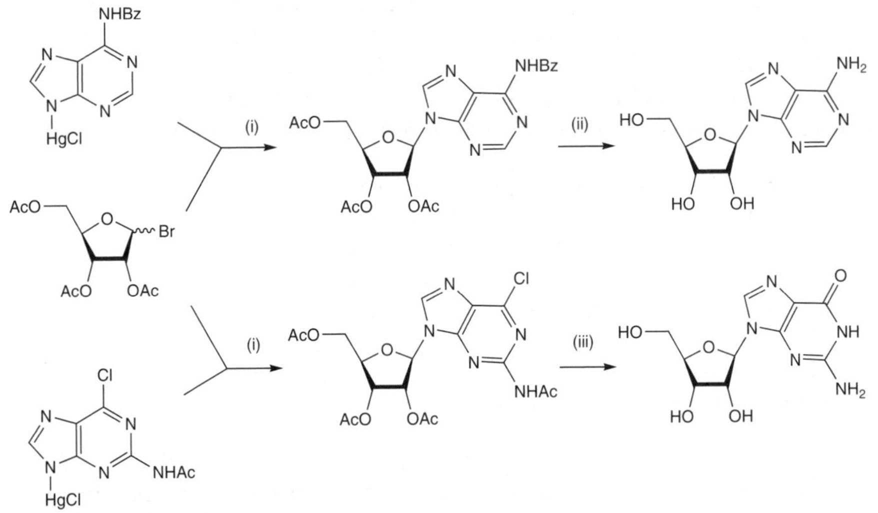


**Figure 3.3** *Chloromercuri route for synthesis of purine nucleosides. Reagents: (i) xylene, 120°C; (ii) $NH_3$, MeOH; and (iii) NaOH aq*

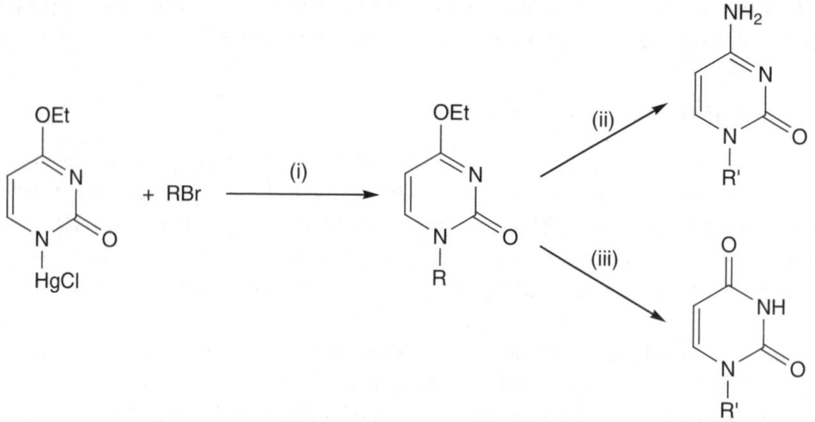


**Figure 3.4** *Chloromercuri route for synthesis of pyrimidine nucleosides. Reagents: (i) xylene, 120°C; (ii) $NH_3$, MeOH; and (iii) NaOH aq. R = protected ribofuranosyl; R′ =* 1-β-D-*ribofuranosyl*

the desired thermodynamic products at N-9 for purines and N-1 for pyrimidines, the condensation reactions are often mechanistically much more complex.[9] Thus, there is considerable evidence for pyrimidines that reaction initially gives an *O*-glycoside or even an $O^2$,$O^4$-diglycoside that is then transformed into the desired *N*-glycoside. For purines, condensation initially takes place on N-3 for adenine and its derivatives or alternatively at N-7, particularly for bases with a 6-keto substituent (*e.g.* $N^2$-acetylguanine and hypoxanthine). The general mechanism for N-3→N-9 glycosylation is shown in Figure 3.5 and proceeds stereoselectively owing to the formation of an acyloxonium ion intermediate (see Section 3.1.1.7).

*3.1.1.2 Fusion Synthesis of Nucleosides.* Two disadvantages of the above methods are the poor solubility of the mercury derivatives and the instability of the halogeno-sugar derivative. Furthermore, the biological activity of a number of nucleosides synthesised in this way has often been wrongly assigned owing to the presence of trace amounts of mercury in samples. One early improvement was the combination of 1-acetoxy sugars with Lewis acids such as $TiCl_4$ or $SnCl_4$ as a means of generating the reactive halogeno-sugar *in situ*. That led to the fusion process, in which a melt of the 1-acetoxy sugar and a suitable base *in vacuo*, often with a trace of an acid catalyst, can give acceptable yields of nucleosides.[10] Thus, 1,2,3,5-tetra-*O*-acetyl-D-ribofuranose fused with 2,6-dichloropurine[11] or 3-bromo-5-nitro-1,2,4-triazole[12] gives useful yields of the corresponding acylated nucleosides (Figure 3.6). This method works best for purines that contain electron-withdrawing groups and have low melting points. Recent examples include the syntheses of 2′-deoxyribonucleosides of purines, but such methods result in anomeric mixtures of nucleosides.

*3.1.1.3 The Quaternization Procedure: Hilbert Johnson Reaction.* Hilbert and Johnson noticed that substituted pyrimidines are sufficiently nucleophilic to react directly with halogeno-sugars without any need for electrophilic catalysis. The method, which bears their name, involves the alkylation of a 2-alkoxypyrimidine with a halogeno-sugar[13,14] and has been reviewed.[15] The initial product is a quaternary salt, which at higher temperatures eliminates an alkyl halide to give an intermediate condensation product. Further chemical modification of substituents on the pyrimidine ring can lead to a range of natural and artificial bases (Figure 3.7). Such condensations frequently give mixtures of α- and β-anomers although the use of $HgBr_2$ increases the proportion of the β-anomer.

*3.1.1.4 Silyl Base Procedure.* A major improvement came from the utilisation of silylated bases (silyl-Hilbert–Johnson method), developed independently by Nishimura,[16] Birkofer[17] and Wittenberg.[18] Silylated bases have three advantages: (1) they are easily prepared, (2) they react smoothly with sugars in homogeneous solution due to their increased solubilities and greater nucleophilicities, and (3) they give intermediate products that can be easily converted into modified bases. The early use of mercuric oxide as

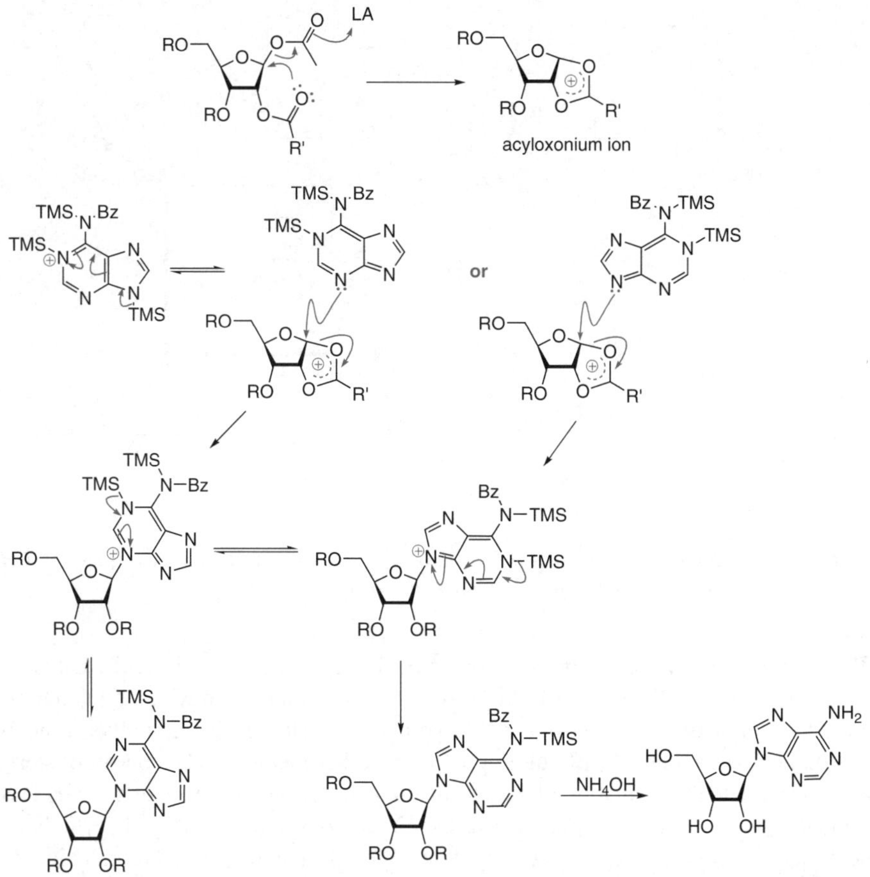


**Figure 3.5** *Rearrangement and formation of thermodynamic product $N^9$-ribosylated purine (LA = mercury salt or Lewis acid e.g. TMSOTf)*

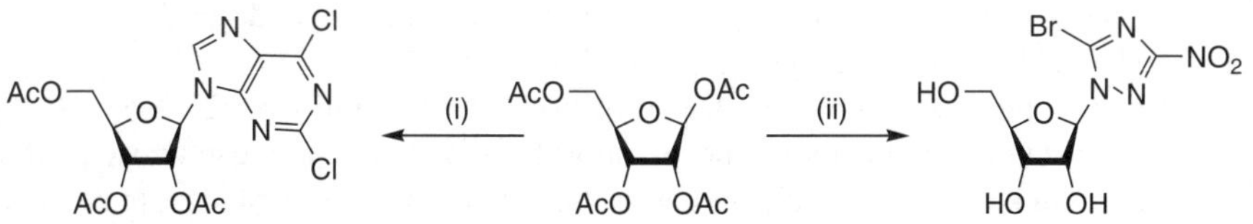


**Figure 3.6** *The fusion method of nucleoside synthesis. Reagents: (i) 2,6-dichloropurine, acetic acid, melt at 150°C; and (ii) 3-bromo-5-nitro-1,2,4-triazole, acetic acid, melt at 150°C*

a catalyst gave way to Lewis acid catalysts[19] (*e.g.* $SnCl_4$ or $Hg(OAc)_2$) and they, in turn, have been superseded by the use of silyl esters of strong acids, notably trimethylsilyl triflate,[20] trimethylsilyl nonaflate or trimethylsilyl perchlorate. Some examples are shown in Figure 3.8. The silylated base is usually generated immediately prior to the glycosylation by heating under reflux with a mixture of hexamethyldisilazane (HMDS) and trimethylsilyl chloride (TMSCl).

Although, one-pot reactions have been described and are more convenient than handling moisture-sensitive silylated bases, they generally result in lower overall yields of product. In earlier methods *bis*(trimethylsilyl)acetamide (BSA) was used, but the mixture of HMDS and TMSCl is generally preferred since the by-product of the reaction (ammonium chloride) does not generally interfere with the subsequent glycosylation reaction.

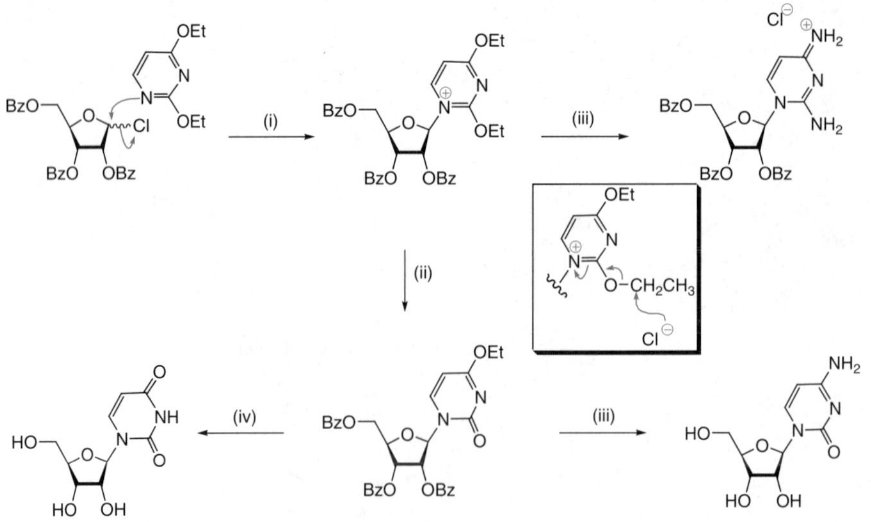


**Figure 3.7** *The quaternization (Hilbert–Johnson) method of nucleoside synthesis. Reagents: (i) $CH_3CN$, 10°C; (ii) $CH_3CN$, reflux; (iii) $NH_3$, MeOH; and (iv) NaOH aq*

The synthetic methodology for the preparation of nucleosides using Lewis acid catalysts, most commonly, trimethylsilyl triflate (TMSOTf, $Me_3Si\text{-}O\text{-}SO_2CF_3$), in combination with silylated bases has come to be known as the **Vorbrüggen procedure.**[21,22] It works very well for a large number of nucleoside analogues with modified bases that are difficult to prepare by other methods. The control of stereochemistry in the ribo-series is due to the formation of an intermediate acyloxonium ion as mentioned earlier (Figure 3.3). Consequently, when the sugar component lacks a 2′-acyloxy substituent, glycosylic bond formation shows reduced stereoselectivity. Regioselectivity depends on the capture of the intermediate oxonium ion by the most electronegative nitrogen on the base and consequently, a mixture of regioisomers can result. Under appropriate conditions, the thermodynamically favoured N-9-alkylated purines and N-1-alkylated pyrimidines can be isolated in good yields.[23] Trimethylsilyl triflate is a weaker Lewis acid than $SnCl_4$ and allows generation of the acyloxonium ion of the sugar without the formation of $\sigma$-complexes with the silylated base.[24] These latter species can dramatically increase the amounts of undesired regioisomers such as N-3-monoalkylated and N-1, N-3 *bis*-alkylated pyrimidines.

The usual glycosyl component employed in the Vorbrüggen preparation of 2′-deoxyribonucleosides, namely 2-deoxy-3,5-di-*O*-(4-toluoyl)ribofuranosyl chloride[25] (**chlorosugar**), can be isolated as the pure, crystalline $\alpha$-anomer; but it undergoes rapid anomerisation at elevated temperatures, in polar solvents and in the presence of Lewis acids. However, the reaction of certain silylated pyrimidine bases with the chlorosugar in chloroform provides a good compromise between the rate and yield of glycosylation on the one hand with minimal **anomerisation** of the sugar component that would otherwise lead to the $\alpha$-nucleoside[26] (Figure 3.8). The addition of CuI can also increase the stereoselectivity by increasing the rate of the nucleophilic substitution.[27] In contrast, pure $\alpha$-nucleoside may be isolated by adding the silylated nucleobase to a solution of the chlorosugar that has been allowed to anomerise by standing in acetonitrile.[26] While the silyl base procedure is still widely used for the synthesis of 2′-deoxyribonucleosides of pyrimidines, the reaction of a purinyl anion with the chlorosugar is generally the method of choice for preparing 2′-deoxyribonucleosides of purines (see Section 3.1.1.8).

*3.1.1.5 Transglycosylation.* It is often relatively easy to convert a natural nucleoside, typically 2′-deoxythymidine, into a nucleoside with a modified sugar residue: for instance the drug 3′-azido-2′,3′-dideoxythymidine (**AZT**). However, it can be difficult to achieve the same chemical transformation of 2′-deoxyadenosine into 3′-azido-2′,3′-dideoxyadenosine. In such cases the sugar moiety can be transferred

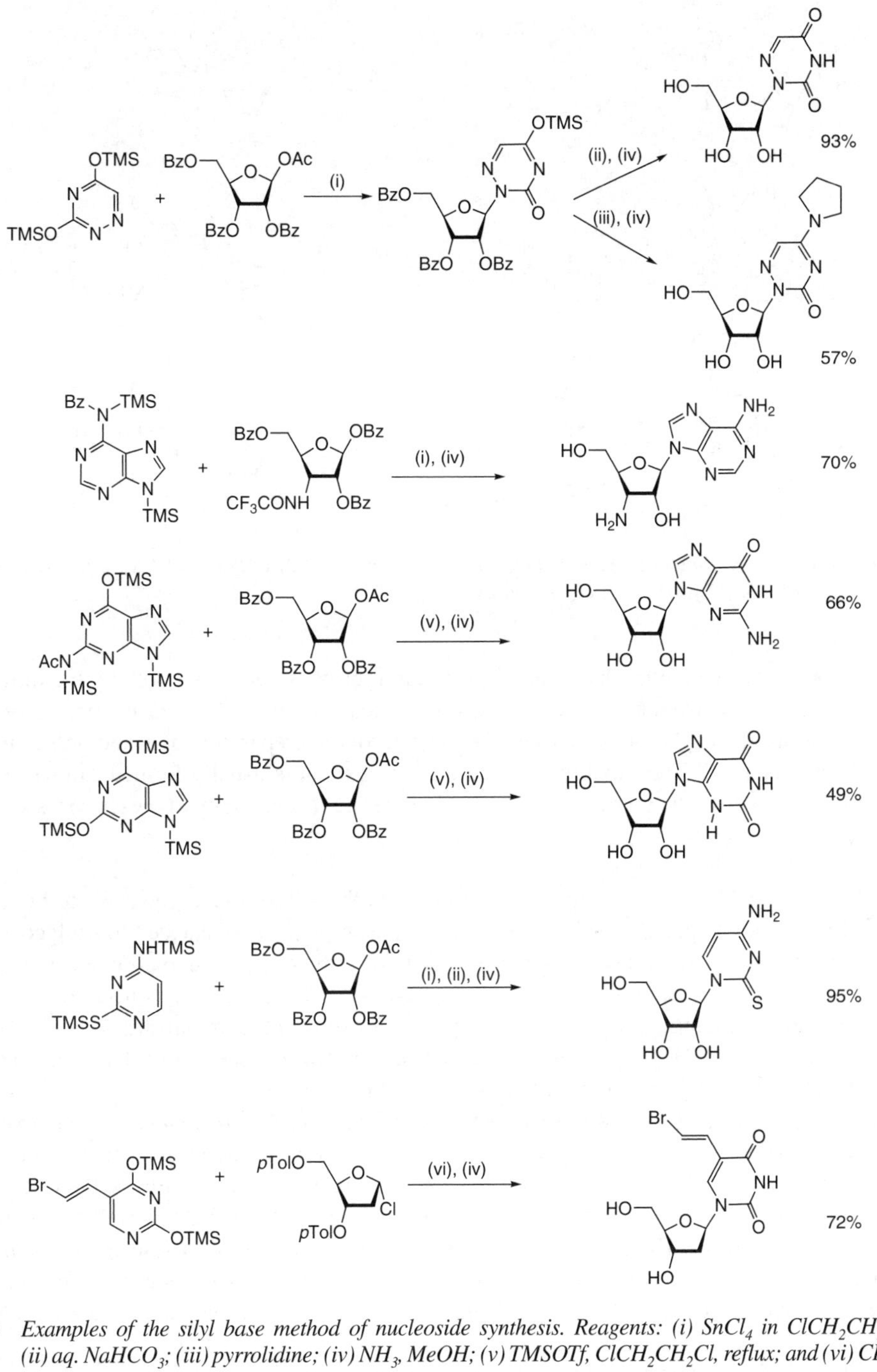


**Figure 3.8** *Examples of the silyl base method of nucleoside synthesis. Reagents: (i) $SnCl_4$ in $ClCH_2CH_2Cl$, 20°C; (ii) aq. $NaHCO_3$; (iii) pyrrolidine; (iv) $NH_3$, MeOH; (v) TMSOTf, $ClCH_2CH_2Cl$, reflux; and (vi) $CHCl_3$, 20°C*

from one base to another by a process known as **transglycosylation**.[28–30] This procedure makes use of the fact that nucleoside formation described in the sections above, in the presence of Lewis acids, is a reversible process. The reaction is particularly effective for transferring sugars from pyrimidines (which are π-deficient heterocycles) to the more basic purines (π-excessive heterocycles). Some examples[31,32] are shown in Figure 3.9.

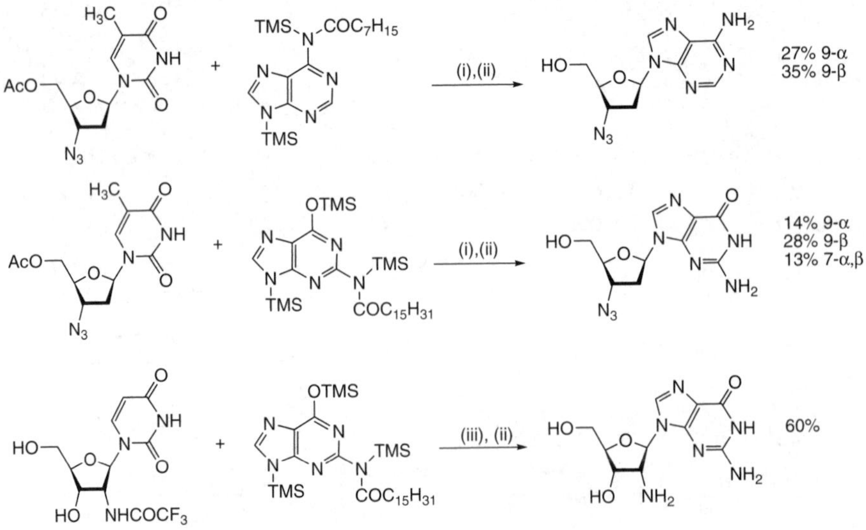


**Figure 3.9** *Transglycosylation synthesis of nucleosides. Reagents: (i) TMSOTf in $CH_3CN$, reflux; (ii) $NH_3$, MeOH; and (iii) TMSOTf, BSA in $CH_3CN$, reflux*

This reaction has all the hallmarks of an $S_N1$ ionization process, as shown both by the intramolecular transfer of a sugar residue from N-7 to N-9 of 6-chloro-1-deazapurine and by the anomerisation of β- into α-nucleosides. Transglycosylation is also a useful method for the preparation of α-anomers of nucleosides from their natural isomers. The mixture of α- and β-species that is usually formed can be separated by chromatography. However, the thermodynamically favoured regioselectivity of these processes is not easily predictable.

*3.1.1.6 Enzymatic Methods.* Hóly, Hutchinson and others have made good use of biotransformations of readily available nucleosides into novel derivatives by enzyme-catalysed transglycosylation.[33,34] Uridine phosphorylase and thymidine phosphorylase degrade uridine, 1-β-D-arabinofuranosyluracil (*ara*-U) and thymidine into the corresponding pentose-1-α-phosphates. These may be converted into the corresponding nucleosides containing purines, modified purines or substituted imidazoles *in situ* in the presence of the new nucleobase and the enzyme **purine nucleoside phosphorylase** (PNP). The method also works for some 3-deazapurines.[35] Some examples are shown in Figure 3.10.

A number of other enzymes have also been used, such as PNP from *Enterobacter aerogenes* that can transform inosine into virazole in the presence of 1,2,4-triazole-3-carboxamide[36] and the enzyme nucleoside 2′-deoxyribosyltransferase from *Lactobacillus leichmannii* that has been used for the large scale transformation of thymidine or 2′-deoxycytidine into corresponding purine nucleosides,[37] as well as into a number of 1-deazapurine nucleosides.[38] The transfer of the sugars 2,3-dideoxy-D-ribofuranose and D-arabinofuranose are also practicable propositions. In general, enzymatic transglycosylations are relatively efficient and highly stereospecific as only β-glycosides are formed, and they can often be employed on a gram scale.

*3.1.1.7 Control of Anomeric Stereochemistry.* Condensation of sugars having a 2-acyloxy substituent with a base invariably gives *N*-glycoside products that have the 1,2-*trans*-configuration. This control of anomeric stereochemistry led Baker to suggest that neighbouring group participation by the acyloxy moiety at the 2-position is responsible.[39] In the case of ribonucleosides, ionization of the leaving group at C-1 of the sugar generates a carbocation that is then captured by the carbonyl group of the adjacent acyl group to form an acyloxonium ion on the lower face of the sugar (Figure 3.3). This is independent of the initial configuration of the sugar halide and is followed by nucleophilic displacement of the base from the opposite

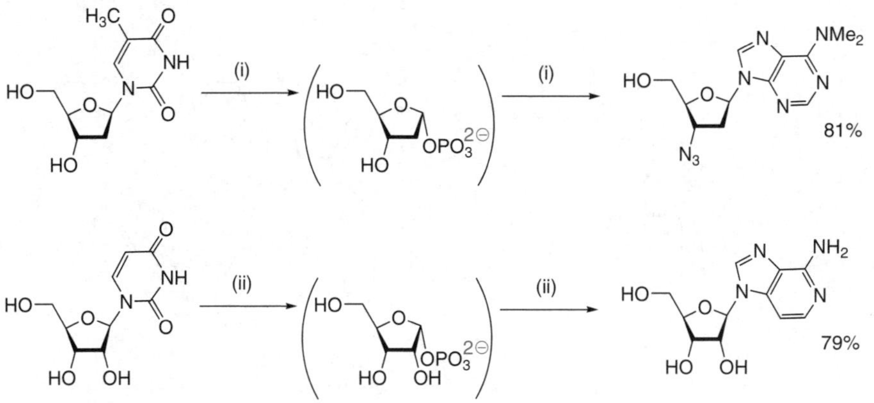


**Figure 3.10** *Enzymatic transglycosylation synthesis of nucleosides. Reagents: (i) Thymidine phosphorylase, purine nucleoside phosphorylase, $N^6$-dimethylamino purine; and (ii) uridine phosphorylase, purine nucleoside phosphorylase, 4-amino-1H-imidazo[4,5-c]pyridine*

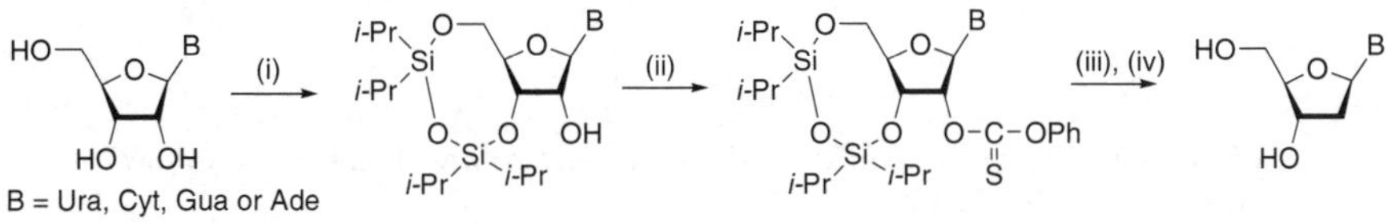


**Figure 3.11** *Conversion of ribonucleosides into 2′-deoxyribonucleosides. Reagents: (i) $(i\text{-}Pr_2SiCl)_2O$, DMF, imidazole; (ii) PhOC(S)Cl, DMAP, $Et_3N$; (iii) $Bu_3SnH$, AIBN; and (iv) TBAF in THF*

face to give the natural β-anomer (**Baker's 1,2-*trans*-rule**). While the formation of the acyloxonium intermediate ensures good stereocontrol in the syntheses of ribonucleosides, since the halides of 2-deoxyribosugars cannot form an acyloxonium ion, mixtures of α- and β-nucleosides result.

This method gives good β-stereochemical control for ribo- and xylo-nucleosides by using peracylated ribose and xylose derivatives; while arabinose and lyxose sugars with a 2-acyloxy substituent will give α-anomers. In cases where a hydroxyl group at C-2 is protected as a benzyl ether or by an isopropylidene or carbonate group cyclized onto the adjacent 3-hydroxyl group, the neighbouring group participation is not possible and mixtures of anomers are formed. Similarly for nucleosides with modified sugars such as 2-deoxy-2-fluoro- or 2-deoxy-2-azido-D-ribofuranose there is no anomeric control. The ability to control the anomeric stereochemistry in the syntheses of ribonucleosides bearing modified bases has led to the development of a number of methods for the preparation of 2′-deoxyribonucleosides by **2′-deoxygenation**, subsequent to the glycosylation step. The most widely used method involves Barton reduction of a 2′-thiocarbonate[40,41] as shown in Figure 3.11. This scheme also illustrates the use of the bifunctional silylating agent 1,3-dichloro-1,1,3,3-tetraisopropyldisiloxane, the **Markiewicz reagent**, which can be used for simultaneous protection of both the 3′- and 5′-hydroxyl groups of ribonucleosides. The direct synthesis of 2′-deoxyribonucleosides with good stereocontrol generally involves reaction of a purine anion or silylated pyrimidine base with an α-chlorosugar under carefully chosen conditions (see Sections 3.1.1.4 and 3.1.1.8).

*3.1.1.8 Nucleobase Anions.* The reaction of the anion of a purine with 2-deoxy-3,5-di-*O*-(4-toluoyl)-D-ribofuranosyl chloride ('chlorosugar') proceeds rapidly in acetonitrile *via* an $S_N2$ process. Useful reviews of this subject have been published.[42–44] In procedures developed by Seela, the potassium salt of the nucleobase is used in acetonitrile.[45] The methodology is applicable to the glycosylation of purines and related deazapurine and azapurine derivatives. In a complementary procedure developed by Robins, sodium hydride is used to generate the nucleophilic purinyl anion that reacts with the chlorosugar in acetonitrile to afford

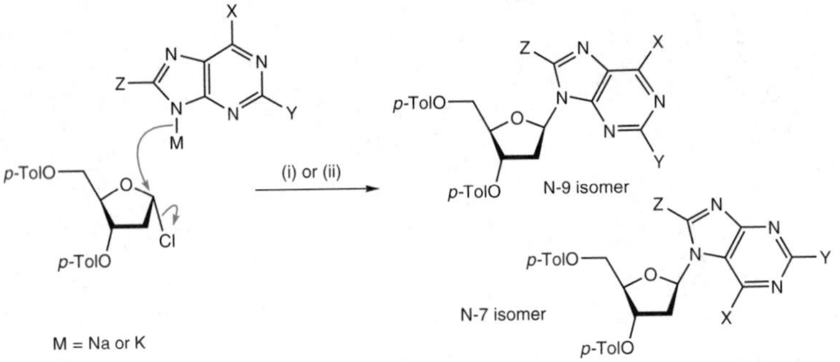


| Conditions | X | Y | Z | Yield (%) N-9 | Yield (%) N-7 |
|---|---|---|---|---|---|
| (i) | OMe | H | H | 44 | 28 |
| (i) | SMe | H | H | 43 | 9 |
| (i) | Cl | H | H | 59 | 13 |
| (i) | OMe | $NH_2$ | H | 48 | 24 |
| (ii) | Cl | Cl | Cl | 56 | - |
| (ii) | Cl | Cl | H | 59 | 13 |
| (ii) | Cl | $NH_2$ | H | 57 | - |
| (ii) | Br | Br | H | 57 | 7 |

**Figure 3.12** *Nucleobase anion route for synthesis of purine nucleosides. Reagents: (i) powdered KOH, TDA-1, $CH_3CN$; and (ii) NaH, $CH_3CN$*

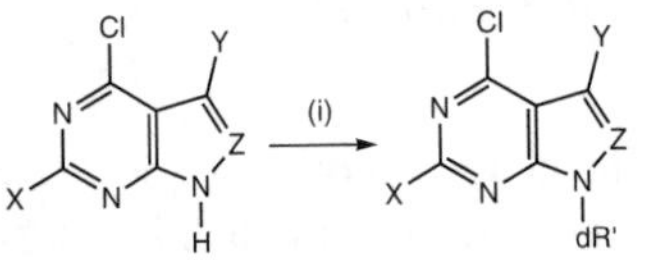


| X | Y | Z | yield (%) |
|---|---|---|---|
| Cl | H | H | 60 |
| H | H | H | 71 |
| MeS | H | H | 66 |
| Cl | I | H | 71 |
| H | H | N | 32 |

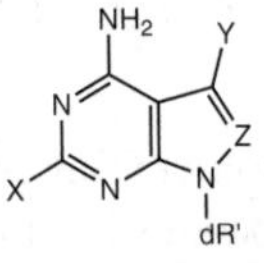


| X | Y | Z | yield (%) |
|---|---|---|---|
| Cl | H | H | 61 |
| H | H | H | 85 |
| MeS | H | H | 72 |
| Cl | I | H | 88 |
| H | H | N | 85 |

**Figure 3.13** *Aminopurine nucleoside analogues prepared by the nucleobase anion route. Reagents: (i) NaH, $CH_3CN$ then 3,5-di-O-p-toluoyl-β-D-ribofuranosyl chloride; and (ii) $NH_3$/MeOH, heat*

good yields of the corresponding 2′-deoxyribonucleoside with the natural β-configuration.[42] The useful stereocontrol achieved in these reactions arises since the anomerisation of the chlorosugar in acetonitrile is much slower than the nucleophilic displacement of chloride by the purinyl anion. The regioselectivity of glycosylation (N-7/N-9) is variable, depending on the purine derivative, but such isomers can usually be separated by chromatography. Some examples are shown in Figure 3.12. Treatment of the nucleosidic products with ammonia in methanol removes the sugar protecting groups, while heating with the same reagent provides a useful route to amino-substituted purine nucleosides (Figure 3.13).

The **nucleobase anion glycosylation** procedure has also been used to synthesise a wide variety of 2′-deoxyribonucleosides of deazapurines (Figure 3.14).[42–44]

*3.1.1.9 C-nucleosides.* A few C-nucleosides have been made by carbanion displacement reactions at C-1 of a suitably protected sugar, although the high basicity of the carbanion can lead to an unwanted 1,2-elimination. A classic example is Brown's synthesis of pseudouridine,[46] a common component of tRNA species, by the reaction of 2,4-*bis*-(*t*-butoxy)-5-lithiopyrimidine with 2,4;3,5-*bis*-*O*-benzylidene-D-ribose. This gave more of the β-pseudouridine (18%) than the α-anomer (8%) (Figure 3.15). Grignard reagents have also been used in carbanion condensations at C-1 of 2-deoxyribose precursors; for example, in the synthesis of fluorinated nucleobase analogues by Kool[47] (Figure 3.15). The use of palladium chemistry has also been exploited.[48]

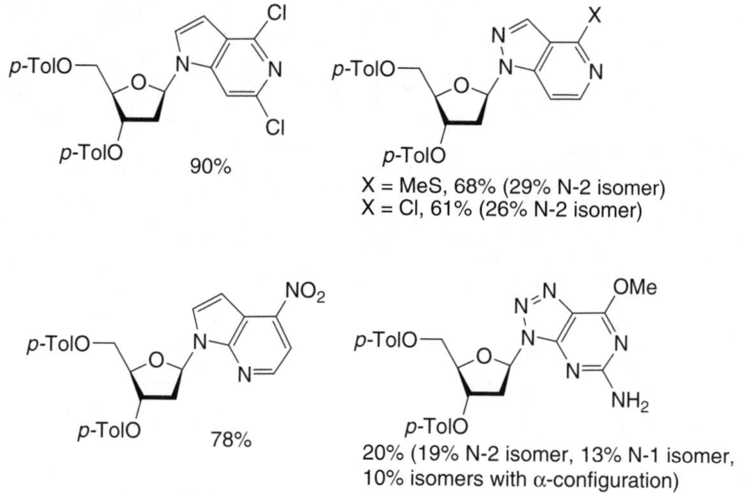


**Figure 3.14** *Purine nucleoside analogues prepared by the nucleobase anion route. Reagents: Purine analogue, KOH, TDA-1, $CH_3CN$ then 3,5-di-O-p-toluoyl-β-D-ribofuranosyl chloride*

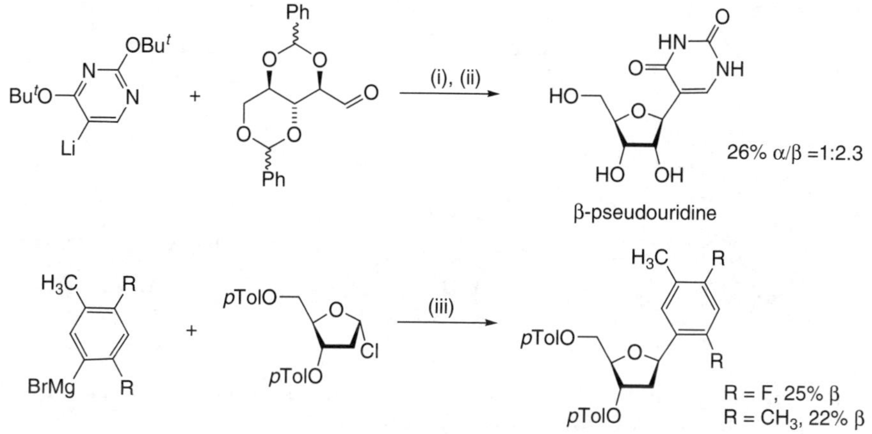


**Figure 3.15** *Syntheses of C-nucleosides via carbanion condensations at C-1 of pentose derivatives. Reagents: (i) THF, −78°C; (ii) mild acid hydrolysis; and (iii) THF, +40°C*

### 3.1.2 Building the Base onto a C-1 Substituent of the Sugar

This approach[49] to nucleoside synthesis has three important features. Historically it was used in Todd's group for a regiospecific synthesis of adenosine (Figure 3.16). Later, it became the preferred route for the synthesis of C-nucleosides and some unusual N-nucleosides. Most recently, it has emerged as the most flexible pathway for the synthesis of nucleosides with highly modified sugars linked to normal or to modified bases.

*3.1.2.1 Nucleosides with Modified Bases.* A good example of the use of this route is the synthesis of the fluorescent base Wyosine, which is found in the anticodon loop of some species of tRNA[50] (Sections 7.2.4 and 7.3.2). In this case, the isocyanate function is the foundation for construction of the tricyclic imidazopurine base. The same isocyanate precursor has been used in a synthesis of 5-azacytidine (Figure 3.17). This nucleoside is elaborated by a *Streptomyces* species and has been used in the treatment of certain leukaemias.

Syntheses of these types based on 1-amino-1-deoxy-β-D-ribofuranose have the general advantage that the place of attachment of the sugar onto the heterocyclic base is unambiguous and is not determined by

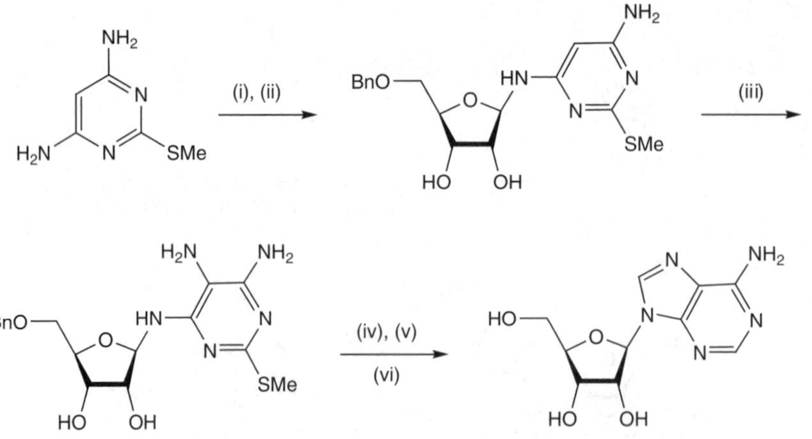


**Figure 3.16** *Todd's synthesis of adenosine. Reagents: (i) 5-O-benzyl-2,3,4-tri-O-acetyl-D-ribose; (ii)* $NH_3$ *(iii) diazonium coupling or nitrosation followed by reduction; (iv) thiourea; (v) Raney nickel desulfurisation; and (vi)* $H_2$*/Pd-C debenzylation*

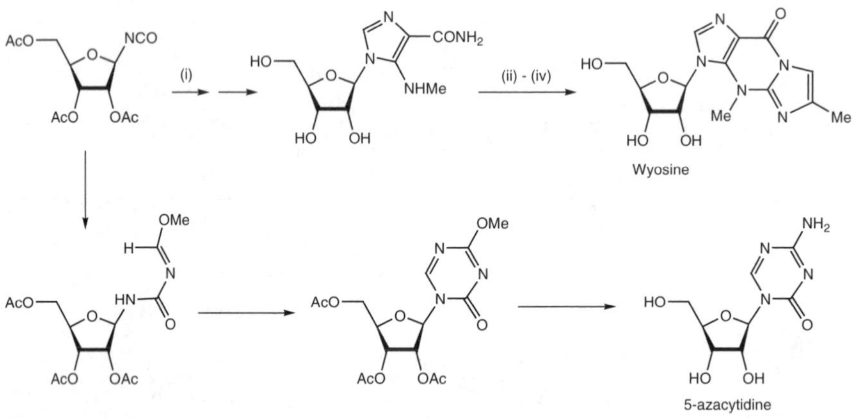


**Figure 3.17** *Building the Wye base and 5-azacytosine onto a C-1 isocyanate. Reagents: (i) three carbon fragment; (ii) CNBr; (iii) NaOEt, EtOH; and (iv)* $BrCH_2COCH_3$

the most nucleophilic heteroatom on the base component. Such syntheses have, therefore, been widely employed for the preparation of the imidazole nucleosides involved in the *de novo* biosynthesis of purine nucleosides (Section 3.4.1), and of modified pyrimidine and purine nucleosides. A typical example of the work of Gordon Shaw in this area is the synthesis of 2-thioribothymidine (Figure 3.18).[51]

In a similar way, a cyanomethyl group at C-1 of D-ribose supports the syntheses of 9-deazainosine, antibiotic oxazinomycin and pseudouridine (Figure 3.18).[52,53]

*3.1.2.2 C-nucleosides.* With the growing availability of chemical reactions having a high degree of stereochemical selectivity, the synthesis of *C*-nucleosides by this route has moved away from sugars as starting materials. **Showdomycin** is a product of *Streptomyces showdowensis* and has useful cytotoxic and enzyme inhibitory properties. A route starting from a tricyclic precursor can branch to give either showdomycin or **psuedouridine** in a stereospecific fashion (Figure 3.19).[54]

The formal replacement of the 4′-oxygen in the sugar by a methylene group gives a carbocyclic nucleoside. Much of the activity in the synthesis of **carbocyclic nucleosides** has been carried out in the search for potential anti-tumour and anti-viral agents, especially provoked by the search for agents effective against human immunodeficiency virus (HIV) (Section 3.7.2). One of the particular values of carbocyclic nucleosides is

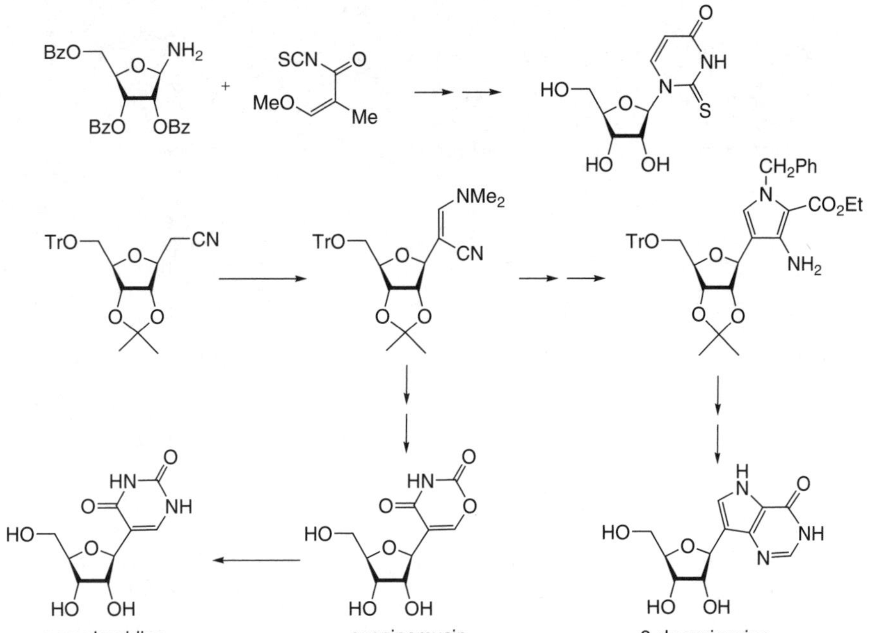


**Figure 3.18** *Syntheses of N- and C-nucleosides by building the base onto the sugar*

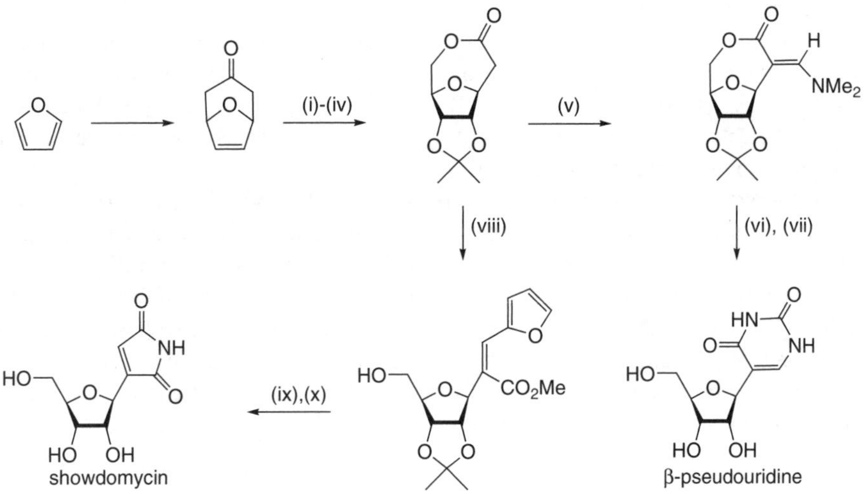


**Figure 3.19** *Synthesis of showdomycin and pseudouridine from furan. Reagents: (i) $OsO_4$, $H_2O_2$; (ii) acetone, $H^+$; (iii) $CF_3CO_3H$; (iv) resolution; (v) dimethylformamide; (vi) urea; (vii) $H_3O^+$; (viii) furan-1-carbaldehyde, NaOMe; (ix) ozone; and (x) $Ph_3P = CHCONH_2$*

their great metabolic stability to the phosphorylase enzymes, which cleaves the glycosylic bond of normal nucleosides. The carbocyclic analogues of adenosine (**aristeromycin**) and **neplanocin** A (Figure 3.20) are both naturally occurring and display anti-tumour and antibiotic activity respectively. Carbovir (Figure 3.20), a carbocyclic 2′,3′-dideoxy-2′,3′-didehydro analogue of guanosine, is a potent inhibitor of HIV replication, as indeed are several corresponding 2′,3′-dideoxy and 2′,3′-dideoxy-2′,3′-didehydro nucleoside analogues of inosine, guanosine and adenosine.

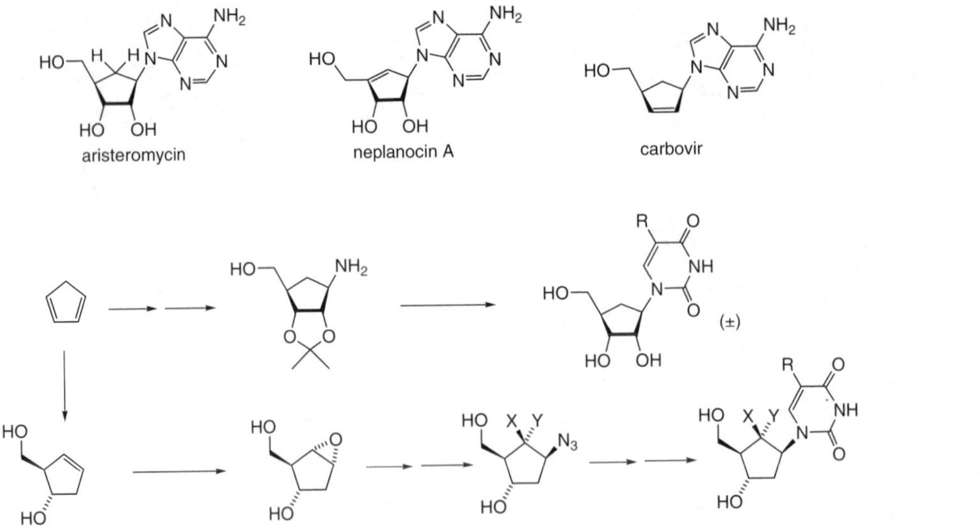


**Figure 3.20** *Structures of aristeromycin and neplanocin A (upper). Synthesis of carbocyclic analogues of deoxy- and ribo-uridine (R = H) and thymidine (R = Me) where X and/or Y are H or F (lower)*

Aristeromycin was first prepared in racemic form by Shealy and Clayson in 1966 and its laevorotatory enantiomer was discovered 2 years later as a metabolite of *Streptomyces citricolour*, now named aristeromycin. New concepts of carbocyclic nucleosides emerged in 1981 with the isolation of neplanocin A from *Ampullariella regularis* (Figure 3.20).

Many syntheses use the key 'carbocyclic ribofuranosylamine' which is made from cyclopentadiene in five steps and then built into pyrimidine or purine carbocyclic nucleosides by standard methods.[55] The adaptation of this route for the introduction of a fluorine atom into the 6-position (which may mimic an oxygen lone pair of electrons in binding to a receptor) presents a good example of the development of such syntheses to highly modified sugars (Figure 3.20).

The use of Pd(0)-catalysed allylic substitution chemistry developed by Tsuji and Trost[48] using activated cyclopentenes has been widely employed in the syntheses of carbocyclic nucleosides (Figure 3.21). The resolution of the two enantiomers of the readily accessible lactone, as shown in Figure 3.22, allows an efficient route to carbocyclic nucleosides. The reactions proceed through the formation of a cationic $\eta^3$-allylpalladium(II) complex that undergoes nucleophilic attack by the nucleobase at the least hindered site. The formation of regioisomers in these reactions, particularly with purines, is common.

The Mitsunobu reaction has also been used for the synthesis of several carbocyclic nucleosides. An example of its use in a synthesis of neplanocin A[56] is shown in Figure 3.23.

*3.1.2.3 Dioxolane and Oxathiolane Nucleosides.* A recent development has been the introduction of a second heteroatom into the 'sugar' ring. For example, 2,3-dioxolane nucleosides have been made and found to have useful anti-HIV activity. The preparation of 2′,3′-dideoxy-3′-oxacytidine by Chu is a good example of stereospecific control in such syntheses (Figure 3.24). Liotta[57] has synthesised the racemic 1,3-oxathiolane analogue of 5-fluorodeoxycytidine and separated the enantiomers by the action of pig liver esterase on their 5′-butyroyl derivatives. Unexpectedly, he found that it is the unnatural L-(−)-isomer, which has both higher anti-viral activity and lower toxicity than the D-(+)-enantiomer (Figure 3.24).

### 3.1.3 Synthesis of Acyclonucleosides

The success of **acyclovir** for the treatment of genital herpes infections has stimulated much work in this area. In these acyclonucleosides (or seco-nucleosides) the base is usually adenine, guanine or a related

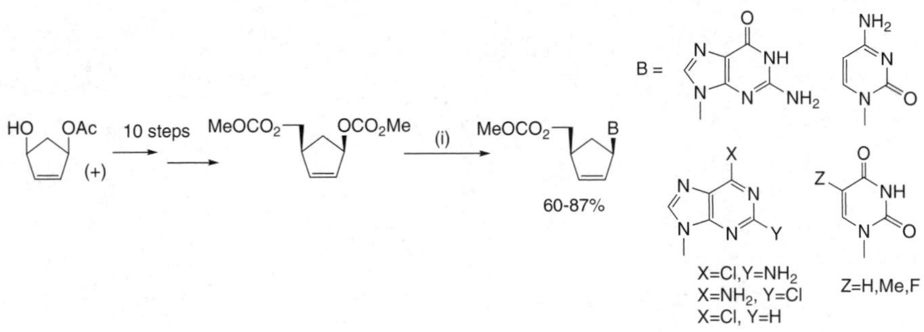


**Figure 3.21** *Syntheses of carbocyclic nucleosides using Pd(0)-catalysed allylic coupling. Reagents: (i) Base (BH), $Pd(PPh_3)_4$, DMF*

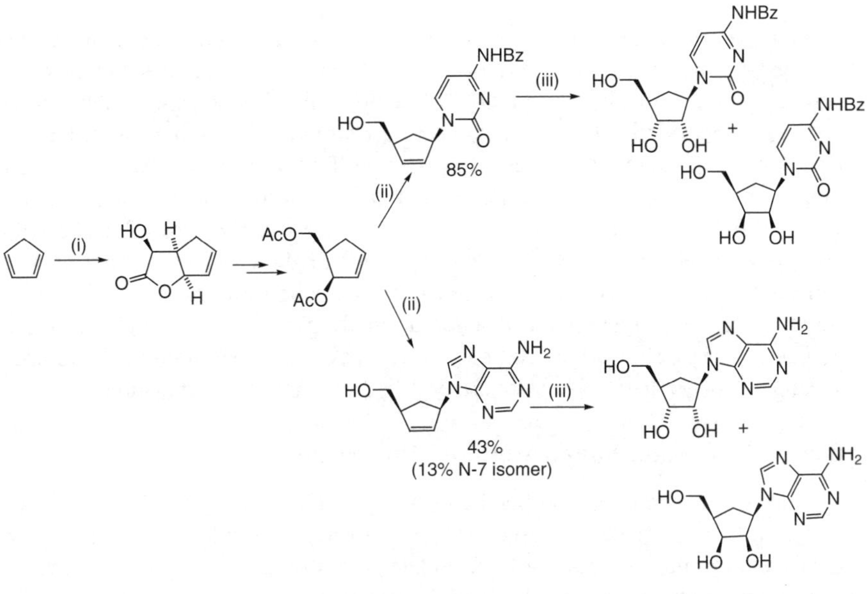


**Figure 3.22** *Syntheses of carbocyclic nucleosides using Pd(0)-catalysed allylic coupling. Reagents: (i) $OCH.CO_2H$; (ii) diastereomeric resolution and deprotection; (iii) Cs salt of base, $Pd(PPh_3)_4$, 55°C, DMF, then $NH_3$/MeOH; and (iv) $OsO_4$, trimethylamine-N-oxide*

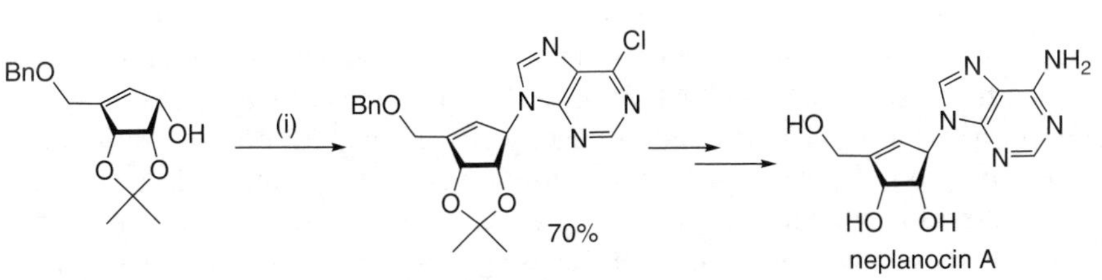


**Figure 3.23** *Use of the Mitsunobu reaction in the synthesis of neplanocin A. Reagents: (i) $Ph_3P$, $EtO_2CN{=}NCO_2Et$, THF, 6-chloropurine*

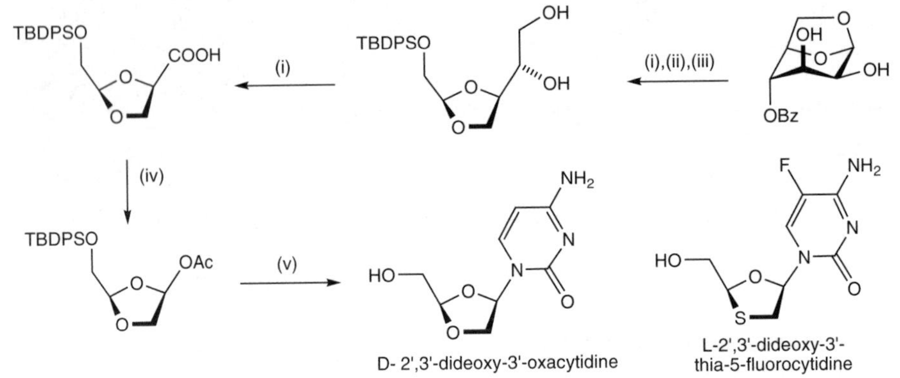


**Figure 3.24** *Synthesis of* D*-2′,3′-dideoxy-3′-oxacytidine from 4-O-benzoyl-1,6-anhydro-*D*-mannose (upper) and structure of* L*-2′,3′-dideoxy-3′-thia-5-fluorocytidine (lower right). Reagents: (i) $NaIO_4$; (ii) $NaBH_4$; (iii) TBDPSCl, pyridine; (iv) $Pb(OAc)_4$; and (v) silylated $N^4$-acetylcytosine, TMS triflate, 1,2-dichloroethane*

purine, which can be converted into adenine or guanine as a result of metabolic deamination or hydroxylation (*i.e.* prodrugs, Section 3.7). The mode of action of acyclovir in Herpes Simplex Virus (**HSV**)-infected cells involves its specific phosphorylation by the thymidine kinase expressed by the virus. This is followed by further phosphorylation by cellular kinases to afford the triphosphate, which is a selective and potent inhibitor of the HSV DNA polymerase. In principle, four sections of the sugar ring can be 'cut away' and promising biological results have been found in three of these areas. Formally, one can excise (1) C-2′, (2) C-3′, (3) C-2′+C-3′, or (4) O-4′+C-4′+C-5′ as shown in Figure 3.25. The syntheses of all of these types of acyclonucleoside are invariably based on N-9 alkylation of a chloropurine precursor, with subsequent amination and manipulation of the necessary protecting groups. Alkylation of the silylated chloropurine in the presence of mercury(II) cyanide normally gives excellent yields of the desired N-9 regioisomer[58] (Figure 3.25). Seco-carbocyclic nucleosides have also been found to have useful antiviral activity. One example is penciclovir, N-9-(4-hydroxy-3-hydroxymethylbutyl)guanine.

### 3.1.4 Syntheses of Base and Sugar-Modified Nucleosides

A vast number of nucleosides bearing modified bases or sugars have been made. Many display significant biological activity, while others have been used for applications in molecular biology, such as nucleic acid sequencing and labelling and investigations of nucleic-acid structure and protein–nucleic acid interactions. However, the majority of these involve modification to the heterocyclic base or modification to the C-2′ and/or C-3′ positions of the sugar. The following is a selective overview of the chemical syntheses of pentofuranosyl nucleosides either modified at C-2′ or C-3′ or on the heterocyclic base. By contrast, modifications at the 5′-position of the nucleoside do not involve stereochemical control. Many transformations typical for chemical modification of primary hydroxyl groups have been used on nucleosides, *e.g.*, displacement of a tosylate, halogenation in the presence of triphenylphosphine or the Mitsunobu reaction.

*3.1.4.1 Modified Bases.* The **halogenation** of pyrimidine nucleosides at C-5[59] and purine nucleosides at C-8[60] is known for all four halogens, although the 5-iodopyrimidine and 8-bromopurine analogues have been most widely used for subsequent functionalization at these positions of the nucleobases. The nucleoside bases 5-iodocytosine and 5-iodouracil can be readily transformed into other 5-substituted analogues by use of palladium chemistry; while nucleophilic displacement of bromine or the palladium-catalysed modification at the 8-position of purine nucleosides gives a variety of 8-substituted analogues.[48]

Uridine and cytidine and their analogues can be halogenated using bromine water or iodine in aqueous acid/chloroform. These reactions appear to involve a 5-halogeno-6-hydroxy-5,6-dihydropyrimidine adduct (Figure 3.26), which is subsequently dehydrated to give the 5-substituted nucleoside. For 2′,3′-isopropylidine-protected ribonucleosides, there is some evidence that an analogous intermediate is formed

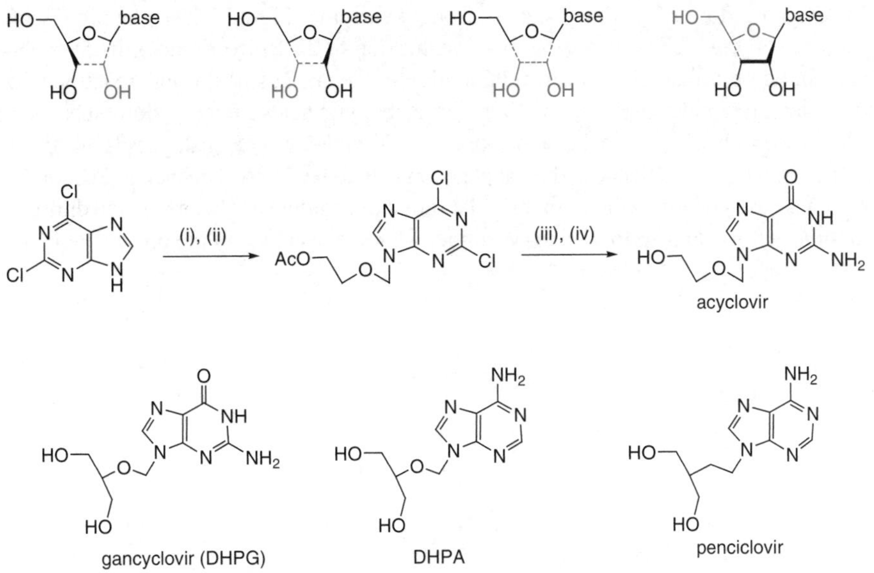


**Figure 3.25** *Relationship of various acyclonucleosides to natural prototypes, with exciseable parts in red (top), synthesis of acyclovir and structures of several acyclonucleosides (bottom). Reagents: (i) HMDS, $(NH_4)_2SO_4$; (ii) 2-(bromomethoxy)ethyl acetate, $Hg(CN)_2$, TMSCl; (iii) NaOH; and (iv) $NH_3$/MeOH*

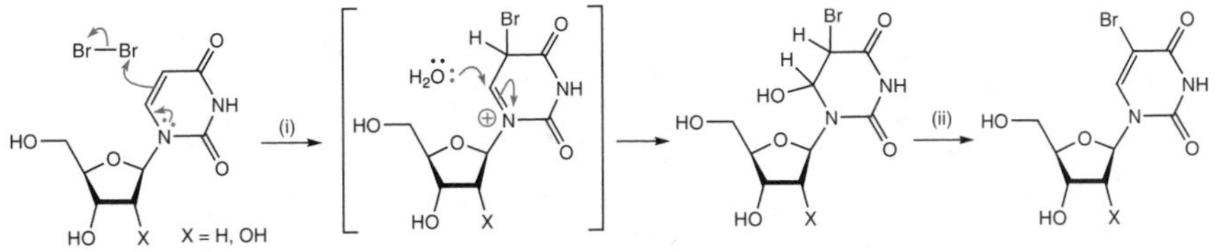


**Figure 3.26** *Mechanism of halogenation at C-5 of pyrimidine nucleosides. Reagents: (i) $Br_2$, $H_2O$; and (ii) EtOH, heat*

through nucleophilic addition of the 5′-hydroxyl group to C-6. The lower reactivity of iodine and the requirement for acidic conditions that might result in some cleavage of the glycosylic bond has led to the use of alternative iodinating agents such as ICl or *N*-iodosuccinimide/dibutylsulfide in DMSO. When performed under anhydrous conditions, halogenation is presumed to proceed by normal electrophilic aromatic substitution. In a variation of this chemistry, the use of ICl and sodium azide in acetonitrile provides excellent yields of 5-iodouridine and **5-iodo-2′-deoxyuridine**. Fluorination of uridine or its analogues can be achieved in high yield by reaction with trichlorofluoromethane in methanol followed by elimination in the presence of triethylamine. The mechanism is analogous to that shown in Figure 3.26.

A number of 5-halopyrimidine nucleosides are known to display biological activity (Sections 3.7.1 and 3.7.2). 5-Iodo-2′-deoxyuridine shows anti-viral activity, while the most notable of these analogues is **5-fluorouracil** and its corresponding 2′-deoxyribonucleoside. The active species *in vivo*, 5-fluoro-2′-deoxyuridine 5′-monophosphate, is a potent inhibitor of thymidylate synthase and displays anti-tumour activity. $^{19}F$- and $^{18}F$-containing species have been used for NMR studies[61,62] involving nucleic acids and for radioimaging of tumours,[63] respectively.

The application of palladium-catalysed chemistry to pyrimidine nucleosides[48] has made 5-iodo-2′-deoxyuridine an important precursor for the preparation of C-5 modified 2′-deoxyuridine analogues. Substitution at C-5 produces analogues that can still form Watson–Crick base pairs, while many C-5-substituted dUTPs are good substrates for DNA polymerase enzymes. The **Sonogashira reaction** allows coupling of terminal alkynes to 5-iodo- or 5-triflate esters of 2′-deoxyuridine (Figure 3.27). Coupling with allylamine or propargylamine (Figure 3.27c) allows the functionalization of 2′-deoxyuridine or its 5′-triphosphate with nucleophilic amino groups that allow further elaboration of the nucleoside through reaction with *N*-hydroxysuccinimidyl esters. Examples include coupling with carboxylic acids of compounds such as imidazole for use in SELEX[64,65] (Section 5.7.3), or in the preparation of 2′,3′-dideoxy-UTP analogues labelled with, for example, biotin or fluorescein,[2,66,67] the latter finding application in dideoxy DNA sequencing (Section 5.1).

In a number of cases, a minor, fluorescent bicyclic furanopyrimidine has been isolated during Sonogashira coupling reactions involving 5-iodo-2′-deoxyuridine, which takes place in the presence of CuI. This latter

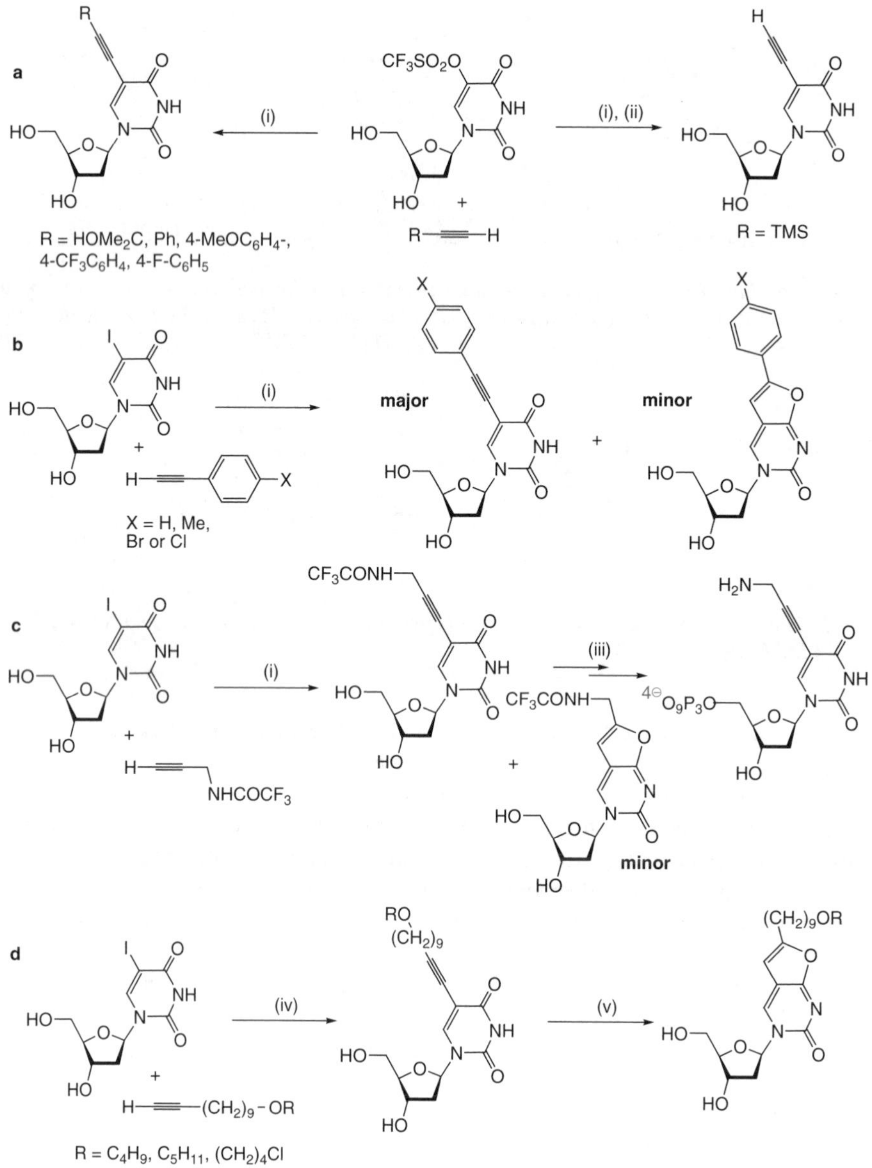


**Figure 3.27** *Palladium-catalysed coupling reactions as routes to C-5-substituted pyrimidine nucleosides. Reagents: (i) Pd(PPh$_3$)$_4$, CuI, Et$_3$N, DMF; (ii) TBAF, MeOH; (iii) synthesis of triphosphate then NH$_4$OH; (iv) Pd(PPh$_3$)$_4$, CuI, i-Pr$_2$EtN, DMF; and (v) CuI, Et$_3$N, MeOH, heat*

transformation has allowed the syntheses of several pyranopyrimidine nucleosides (Figure 3.27d) that have been shown to display important anti-viral activity, especially against Varicella Zoster Virus (VZV).[68]

The related palladium-catalysed **Heck reaction** has also been used to prepare C-5 alkenyl analogues of pyrimidine nucleosides.[48] Initial chemistry developed by Bergstrom[69] used **C-5 chloromercuri**-derivatised pyrimidine nucleosides (Figure 3.28). While these derivatives are still employed, the commercially available and less-toxic C-5-iodopyrimidine nucleosides have been increasingly used (Figure 3.28). In each case the *(E)-, trans*-alkene is the major product.

Halogenation at C-8 of purine nucleosides can be achieved by use of acidified bromine water or *N*-bromosuccinimide in water, while the other halogens are normally incorporated by use of *N*-iodosuccinimide, chlorine or fluorine, respectively.[60] Nucleophilic displacement of the bromide provides convenient access to 8-substituted purine nucleosides, whilst palladium-catalysed substitution of bromine furnishes a wide variety of C-8 alkynyl-, alkenyl- and alkyl-substituted analogues (Figure 3.29).[44,60]

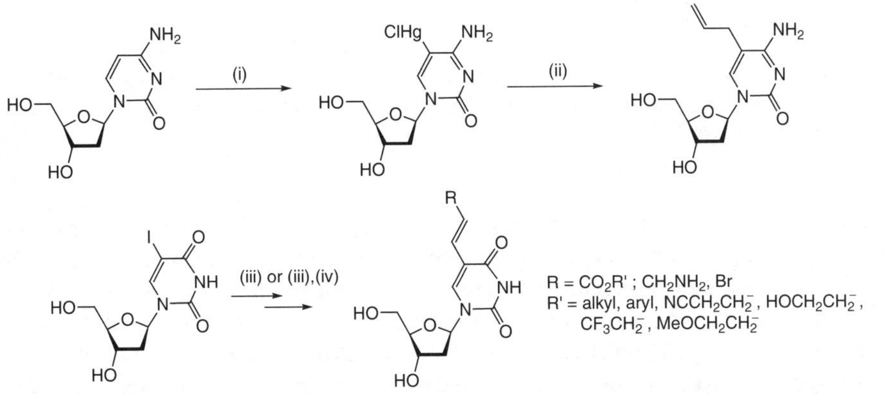


**Figure 3.28** *Palladium-catalysed coupling reactions as routes to C-5 alkenyl-substituted pyrimidine nucleosides. Reagents:(i) $Hg(OAc)_2$ then NaCl; (ii) $Li_2PdCl_4$, $ClCH_2CH{=}CH_2$ in MeOH; (iii) $Pd(OAc)_2$, $Ph_3P$, $Et_3N$, $H_2C{=}CH{-}R$ (for $R = CH_2NH_2$, $CF_3CO$-protected derivative used in coupling) DMF or dioxan; and (iv) $NH_4OH$ for $R = NH_2$ or HO- then N-bromosuccinimide, heat for R = Br*

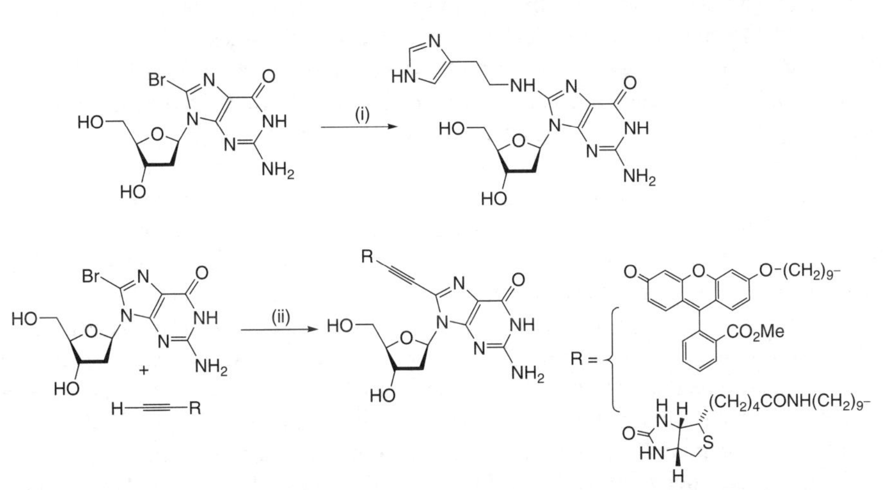


**Figure 3.29** *Syntheses of some C8-substituted purine nucleosides. Reagents: (i) histamine, $Et_3N$, $H_2O$, heat; and (ii) $(PPh_3)_4Pd$, CuI, $Et_3N$, DMF*

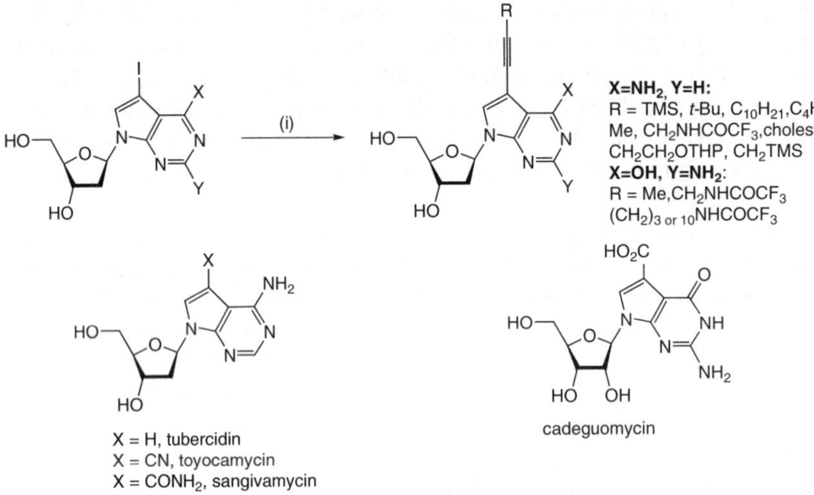


**Figure 3.30** *Syntheses of 7-substituted 7-deaza-2′-deoxyadenosine and 7-deaza-2′-deoxyguanosine nucleosides (upper). Reagents: (i) terminal alkyne, Pd($PPh_3$)$_4$, CuI, $Et_3N$, DMF. Some naturally occuring antibiotic nucleosides (lower)*

The Sonogashira substitution has also been widely employed by Seela in the syntheses of C7-modified 7-deazapurine (pyrrolo[2,3-*d*]pyrimidine) nucleosides[43,44] (Figure 3.30). These analogues have attracted widespread interest since they represent 7-substituted purine nucleosides that can maintain Watson–Crick base pairing while retaining the normal *anti*-conformation about the glycosylic bond (Section 2.1.4). The 5′-triphosphates of 7-deaza-2′-deoxyadenosine, 7-deaza-2′-deoxyguanosine and their 7-substituted analogues are generally excellent substrates for many DNA polymerases and have been widely used in developing DNA sequencing methodology.[67] In addition, 7-deazapurine forms the basis of a number of naturally occurring antibiotics such as **tubercidin**, **toyocamycin**, **sangivamycin** and **cadeguomycin** (Figure 3.30), while several 7-substituted 7-deazaguanosines such as **nucleoside Q** are found in some tRNAs. Furthermore, many 7-substituted 7-deazapurine nucleoside analogues have been shown to stabilize DNA duplexes.[70]

Many nucleosides have been modified at the 4-position of pyrimidine or 6-position of purine, starting from uridine or guanosine respectively. The modification of pyrimidine nucleosides at C-4 can be achieved through nucleophilic substitution of 4-triazolo-pyrimidine nucleoside derivatives, which are stable, isolable compounds that can be transformed into a variety of analogues (Figure 3.31).[71] $O^4$-Methylthymidine is an important analogue formed during DNA damage by alkylating agents, while the highly mutagenic analogues $N^4$-amino- and $N^4$-hydroxy-2′-deoxycytidine are formed by the action of hydrazine and hydroxylamine on DNA (Section 8.4). The reaction of triazolo derivatives with ammonia is particularly useful for converting uracil into cytosine-containing nucleosides, especially for analogues containing modified sugars.[72] Activation at O-4 for subsequent nucleophilic displacement may also be achieved by use of nitrophenoxy-, dinitrophenylthio- and sulfonate esters, although these are often generated and used *in situ* because of their high reactivity.

Nucleosides containing bases with 6-keto functions, such as guanine and hypoxanthine, can also be transformed in a similar way into 6-substituted purine nucleosides. ***O*$^6$-Sulfonate esters** of purines can be displaced by a variety of nucleophiles, although hard nucleophiles such as alkoxide react at sulfur, and so 6-alkoxypurines are made *via* the highly reactive trimethylammonium salt (Figure 3.32).

Thus, syntheses of compounds such as **2-aminopurine** nucleosides (a fluorescent base analogue),[73] **2-amino-6-vinylpurine-2′-deoxyriboside**[74] (allows covalent cross-linking within DNA) and ***O*$^6$-methyl-2′-deoxyguanosine**[75] (an important analogue resulting from DNA alkylation damage) are possible, while a variation on this theme leads to **6-thio-2′-deoxyguanosine**[76] and other analogues (Figure 3.33).

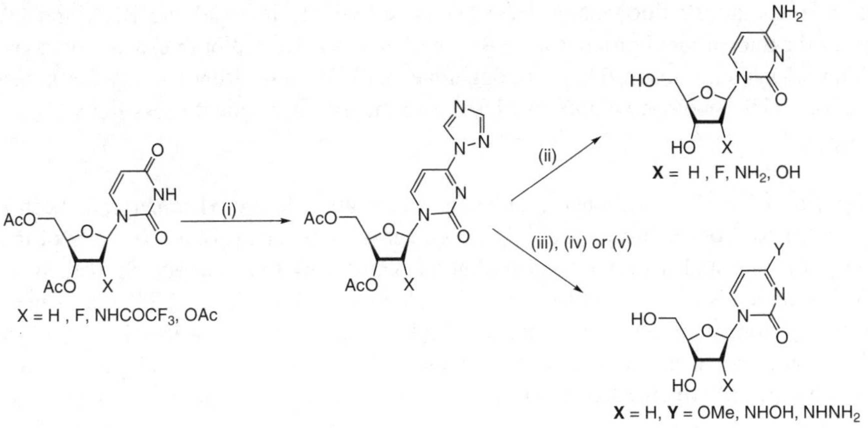


**Figure 3.31** *Conversion of uridine and 2′-modified uridines into 2′-modified cytidines and C-4 modified 2′-deoxyribopyrimidine nucleosides. Reagents: (i) $POCl_3$, triazole, $Et_3N$, $CH_3CN$; (ii) $NH_3$ in dioxan, heat; (iii) DBU in MeOH then $NH_4OH$; (iv) $NH_2OH.HCl$ in pyridine then $NH_4OH$; and (v) $NH_2NH_2$ in EtOH, heat*

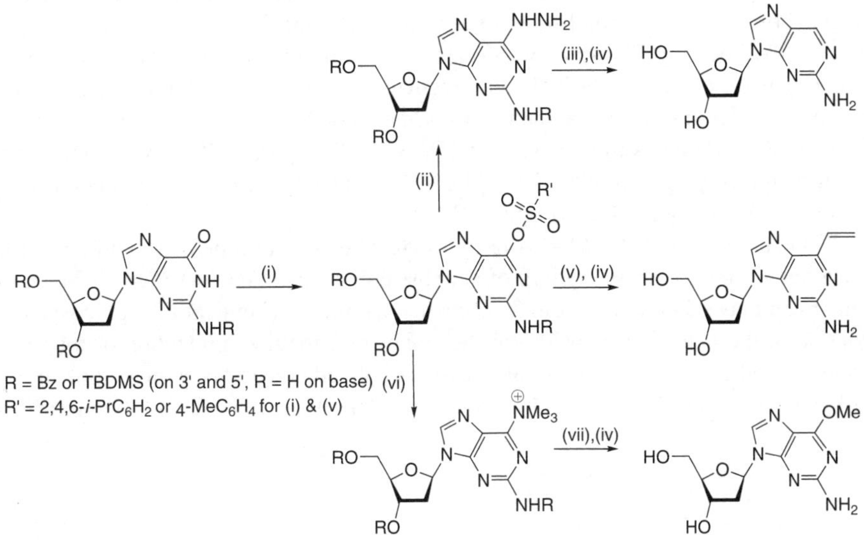


**Figure 3.32** *Syntheses of some 6-substituted purine nucleosides via $O^6$-sulfonate esters. Reagents: (i) arylsulfonyl chloride, 4-dimethylaminopyridine, $Et_3N$, $CH_2Cl_2$; (ii) $NH_2NH_2$ in THF; (iii) $Ag_2O$ in $THF/H_2O$; (iv) Deprotection; NaOMe in MeOH for R = Bz, TBAF in THF for R = TBDMS; (v) $H_2C{=}CHSnBu_3$, $Pd(PPh_3)_4$, LiCl, dioxan; (vi)trimethylamine; and (vii) MeOH, DBU*

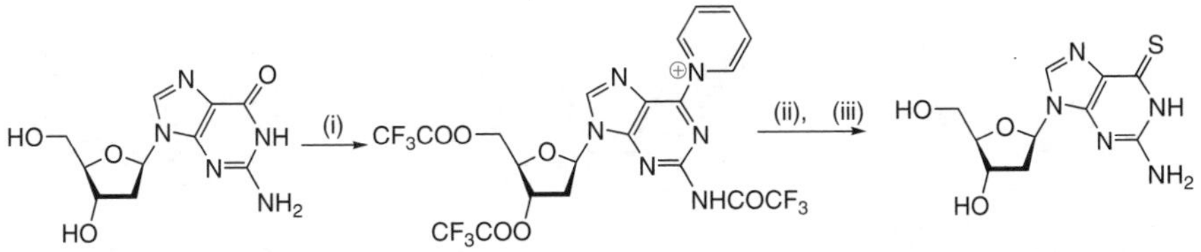


**Figure 3.33** *One-pot conversion of 2′-deoxyguanosine into 6-thio-2′-deoxyguanosine. Reagents: (i) $(CF_3CO)_2CO$ in pyridine; (ii) NaSH in DMF; and (iii) dilute $NH_4OH$*

Oligonucleotides containing thioguanine have been used widely for studying RNA and DNA–protein interactions, as the thiocarbonyl group is a weaker hydrogen bond acceptor than a carbonyl group. In addition, the long wavelength absorption of thioguanine (340–350 nm) allows its use as a spectral probe of conformation, while photoactivation enables the formation of covalent cross-links for studying 3-D structures.[2]

*3.1.4.2 Modified Sugars.* Ribonucleosides in which the 2′-hydroxyl moiety has been replaced by a fluorine,[2,3,72] amino[2,3] or methoxy group[2,3,77] have been used extensively in the study of the properties of RNA and ribozymes and within antisense oligonucleotides by Eckstein and Sproat. Synthetic routes to the former compounds are derived from $O^2$,2′-cyclonucleosides (Figure 3.34). Early work in Todd's group identified cyclonucleosides as intermediates in the conversion of 5′-*O*-acetyl-2′-*O*-tosyluridine into 2′-deoxy-2′-iodouridine using sodium iodide, which gives retention of configuration at C-2. Such **cyclonucleosides** (or **anhydronucleosides**) can be prepared by use of a variety of condensing agents and are stable, isolable compounds.[78] Jack Fox showed that their reactions with a variety of nucleophiles[8] under anhydrous acidic conditions leads to cleavage of the $O^2$-C-2′ ether linkage and the formation of substituted nucleosides with the ribose configuration at C-2′ (Figure 3.34). Hydrolysis under aqueous acidic conditions provides a route to **arabinonucleosides**, (C-2 epimer of ribonucleosides) many of which are biologically active anti-viral compounds. Cyclonucleosides also support the synthesis of 2′-azido and 2′-amino analogues (Figure 3.34),[31] while 2′-modified-2′-deoxyuridines can also be transformed into the corresponding cytidine analogues *via* 4-triazolo derivatives (Figure 3.31).

Purine nucleosides bearing 2′-azido or 3′-azido (and amino) substituents can be prepared from **2′-azido-2′-deoxyuridine** or **3′-azido-2′,3′-dideoxyuridine** using chemical transglycosylation,[31,32] while 2′-fluorinated nucleosides have been obtained by fluorination of 3′,5′- and base-protected *ara*-G and *ara*-A nucleosides using diethylaminosulfur trifluoride (DAST)[79,80] (Figure 3.35). Cyclonucleoside formation has also been used for the preparation of 3′-modified-2′,3′-dideoxynucleosides as exemplified by the synthesis of the anti-viral compound **AZT** (Figure 3.36).[81]

2′-Deoxy-2′-methoxynucleosides (**2′-*O*-methylnucleosides**) of uridine and cytidine can be prepared through alkylation at the 2′-hydroxyl group of suitable precursors (Figure 3.37).[2,3,77] The corresponding adenosine analogue may be obtained from 2′-*O*-methyluridine by chemical transglycosylation[82] (Figure 3.38). Alkylation of the natural nucleosides at the 2′-position requires protection of the reactive lactam functions on uracil and guanine and the use of 3′,5′-*bis*-silylated precursors avoids problems of separating 2-′ and 3′-alkylated products. However, alkylation of the unprotected riboside of 2-amino-6-chloropurine

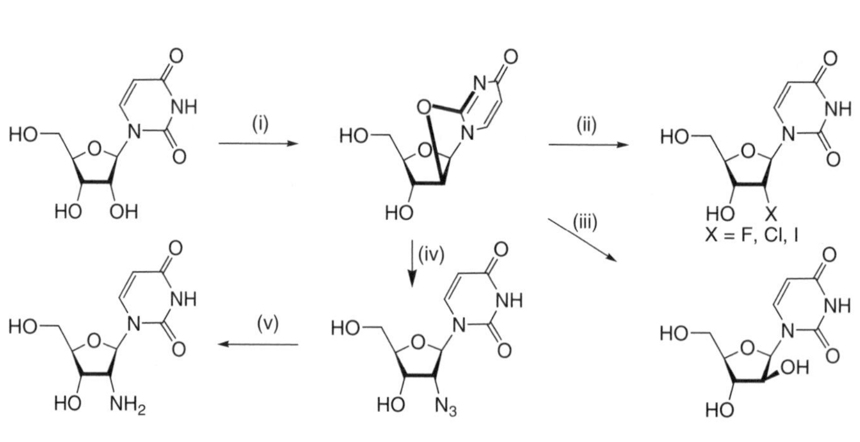


**Figure 3.34** *$O^2$,2′-Cyclonucleosides as precursors to 2′-modified pyrimidine nucleosides. Reagents: (i) $(PhO)_2CO$, $NaHCO_3$, DMF, heat; (ii) HX in dioxan; (iii) $H^+$ aq; (iv) $TMSN_3$, LiF, TMEDA, DMF, heat; and (v) $Ph_3P$, aq $NH_4OH$, dioxan*

produces the 2′-monoalkylated nucleoside together with the 2′,3′-*bis* alkylated by-product. Subsequent hydrolysis provides a simple and efficient route to 2′-*O*-methylguanosine (Figure 3.38).[82]

**2′,3′-Dideoxynucleosides** were first prepared in the groups of Todd and Robins in the 1950s. More recently, many such nucleoside derivatives have been discovered to have important anti-viral properties, particularly effective against HIV (Section 3.4), and they have been used widely in DNA sequencing using the Sanger method (Section 5.1). While glycosylation routes to these analogues are known, they generally give anomeric mixtures of α- and β-nucleosides. A general route to these compounds *via* the corresponding **2′,3′-didehydro-2′,3′-dideoxynucleosides** is shown in Figure 3.39.[83]

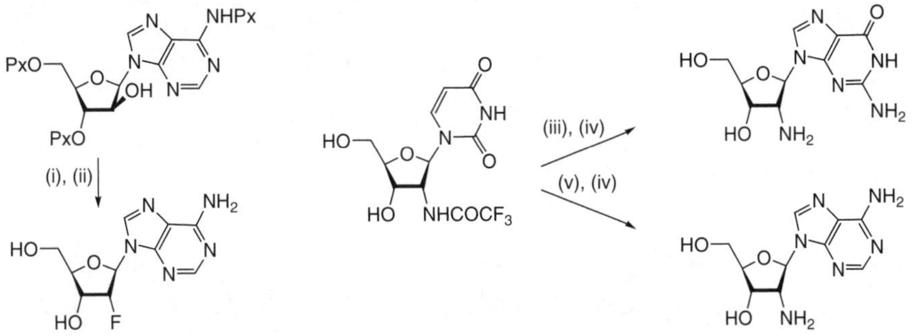


**Figure 3.35** *Preparation of various 2′-modified purine nucleosides. Reagents: (i) DAST, DMF, $CH_3CN$ (px = pixyl/9-phenylxanthyl); (ii) $H^+$; (iii) $N^2$-palmitoyl guanine, bistrimethylsilyl acetamide, reflux; (iv) $NH_3$/MeOH; and (v) $N^6$-octanoyl adenine, bistrimethylsilyl acetamide, reflux*

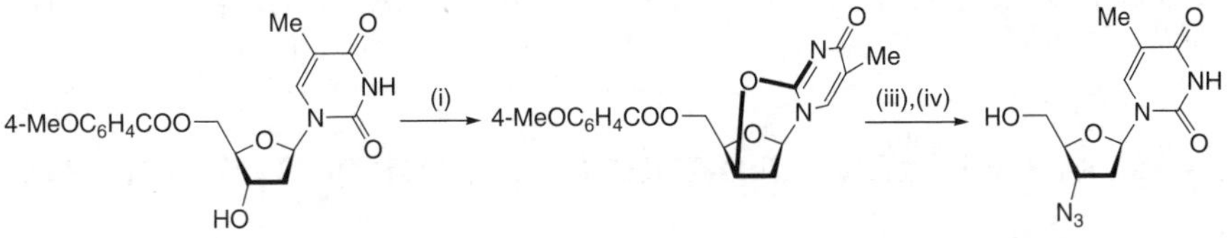


**Figure 3.36** *Synthesis of AZT via $O^2$,3′-cyclonucleoside. Reagents: (i) i-$PrO_2CN{=}NCO_2$i-Pr, $Ph_3P$, DMF; (ii) $LiN_3$, DMF; and (iii) NaOMe in MeOH*

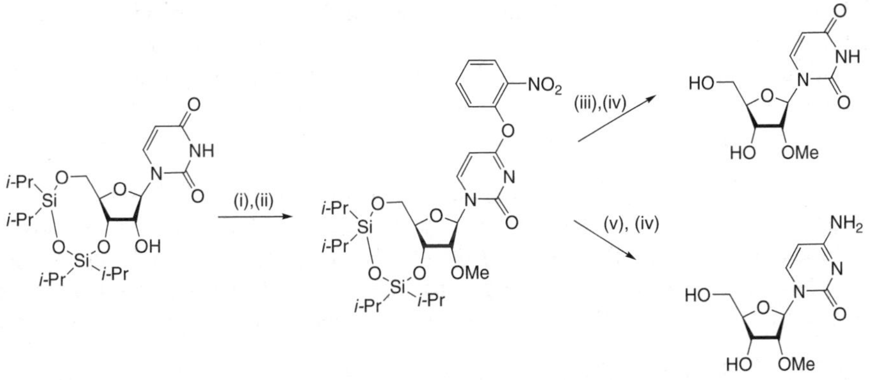


**Figure 3.37** *Syntheses of 2′-OMe modified pyrimidine nucleosides. Reagents: (i) $MeSO_2Cl$,DMAP then 2-nitrophenol; (ii) MeI,$Ag_2O$ in acetone; (iii) 4-$O_2NC_6H_4CH{=}NO^-.(Me_2N)_2C{-}NH_2^+$ then $H_2O$/dioxan; (iv) TBAF in THF; and (v) $NH_3$ in THF*

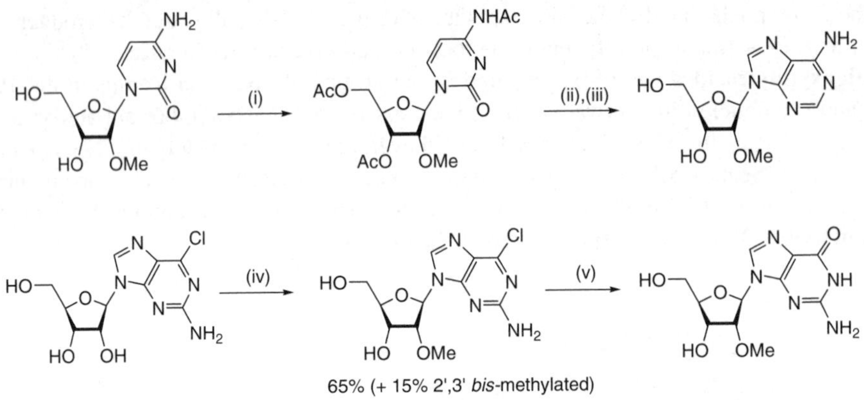


**Figure 3.38** *Syntheses of 2'-OMe modified purine nucleosides. Reagents: (i) $Ac_2O$, DMF, pyridine; (ii) $N^6$-benzoyladenine, bistrimethylsilyl acetamide, TMSOTf, $CH_3CN$, heat; (iii) $NH_3$/MeOH; (iv) MeI, NaH, DMF, −20°C; and (v) 1,4-diazabicyclo[2.2.2]octane (DABCO)/water, heat*

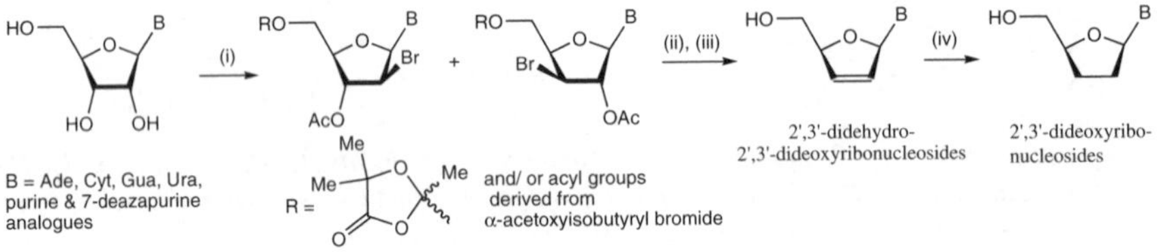


**Figure 3.39** *Syntheses of 2',3'-didehydro-2',3'-dideoxyribonucleosides and 2',3'-dideoxyribonucleosides. Reagents: (i) $Me_2C(OAc)COBr$ in $CH_3CN/H_2O$; (ii) Zn-Cu/DMF or Zn/HOAc/DMF; (iii) $NH_3$/MeOH; and (iv) $H_2$, Pd-C in EtOH*

## 3.2 CHEMISTRY OF ESTERS AND ANHYDRIDES OF PHOSPHORUS OXYACIDS

### 3.2.1 Phosphate Esters

The predominant forms of phosphorus in biology are orthophosphoric acid $(HO)_3P{=}O$, its esters, anhydrides and some amides. Orthophosphates are tetra-substituted at phosphorus, which is in the P(V) oxidation state and has tetrahedral geometry. The bonding can be described by using $sp^3$ hybrid orbitals at phosphorus for the 'single' P—O bonds, which are ~1.6 Å long. In triesters the P—O 'double' bond is shorter, ~1.46 Å, and involves additional π-bonding from $d_\pi$–$p_\pi$ overlap between the phosphorus and oxygen (Figure 3.40). Phosphorus can participate in such bonding simultaneously to more than one oxygen ligand and so any negative charge is delocalized across all unsubstituted oxygen atoms (Figure 3.41). The corresponding π-bonding to neutral nitrogen ligands in phosphoramidates is rather weak, so the nitrogen remains moderately basic.

*3.2.1.1 Phosphate Triesters.* Triesters have all three hydrogen atoms of phosphoric acid replaced by alkyl or aryl groups. They are non-ionic, soluble in many organic solvents, and sufficiently stable to be purified by chromatography. The P=O bond is effectively transparent in the UV region and has an IR absorption at 1280 $cm^{-1}$. When all three ester groups are different, the phosphorus atom is a stereogenic centre, as in $S_P$ methyl ethyl phenyl phosphate (Figure 3.42a), and so optically active triesters can be made.

*3.2.1.2 Phosphate Diesters.* Diesters have two hydrogen atoms replaced by alkyl or aryl groups. The remaining OH group is strongly acidic ($pK_a$ ~1.5). Consequently, phosphate diesters exist as monoanions at pH >2, and are usually water-soluble. The negative charge is shared equally between the two unsubstituted oxygen atoms (Figure 3.42b). When the two ester groups are different, the two unsubstituted

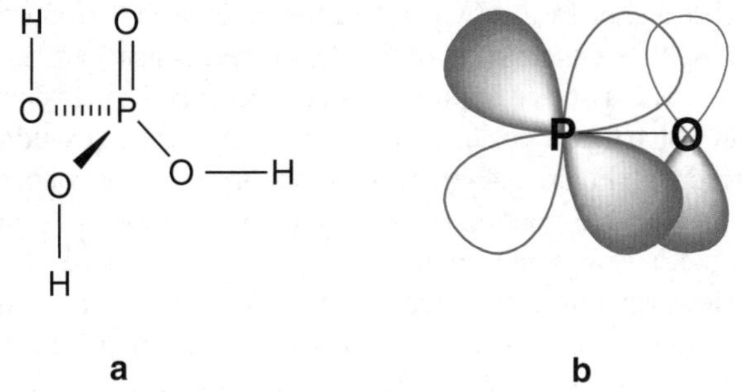

**Figure 3.40** *(a) Orthophosphoric acid; and (b) P—O dπ–pπ bonding*

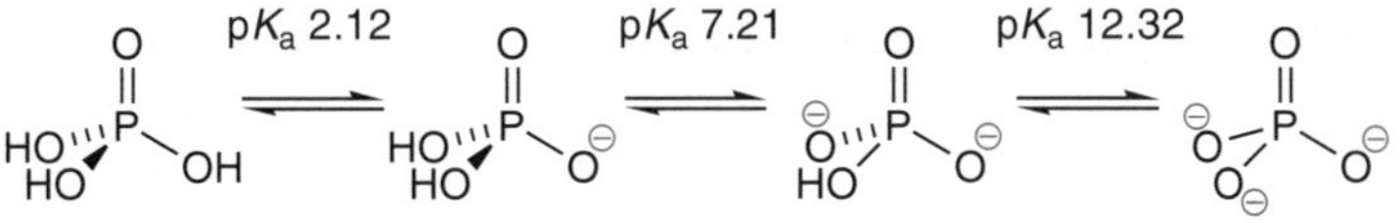

**Figure 3.41** *Orthophosphoric acid and its conjugate bases*

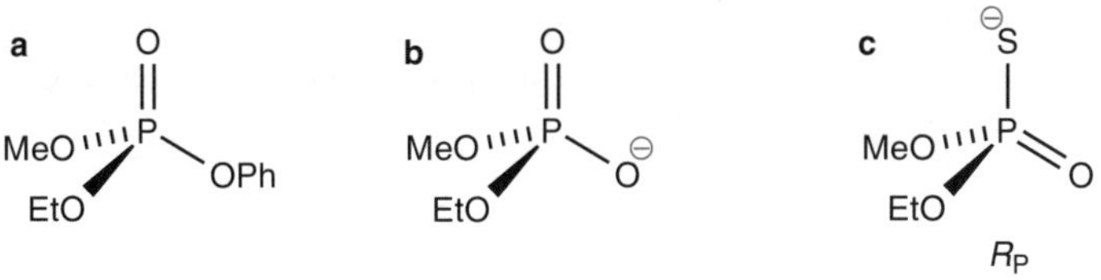

**Figure 3.42** *(a) Chiral phosphate triester; (b) pro-chiral oxygens in a phosphate diester; and (c) chiral phosphothioate diester*

oxygen atoms are non-equivalent (*i.e.* diastereotopic) and the phosphorus atom is a pro-chiral centre. By substituting one of these oxygen atoms by sulfur (Figure 3.42c) or by distinguishing them isotopically, the phosphorus can be made into a stereogenic centre for the stereochemical analyses of substitution reactions.

*3.2.1.3 Phosphate Monoesters.* Monoesters have a single alkyl or aryl group and two ionisable OH groups. These have $pK_{a1} \sim 1.6$ and $pK_{a2} \sim 6.6$, so there is an equilibrium in neutral solution (effectively from pH 5 to pH 8) involving significant concentrations of both the monoanion and dianion. The equivalent oxygen atoms share the negative charge in both monoanions and dianions and there is partial double bonding to each. In the monoanion, the hydrogen atom translocates rapidly between the three oxygen atoms making them all equivalent in solution. These three oxygen atoms are pro-pro-chiral. Thus the use of the three isotopes of oxygen, $^{16}O$, $^{17}O$ (=Ø) and $^{18}O$ (=O) is required for stereodifferentiation and has been widely used in stereochemical synthesis and analysis of substitution reactions of phosphate monoesters (Figure 3.43).

## 3.2.2 Hydrolysis of Phosphate Esters[84]

The great stability of phosphate diesters and monoesters during hydrolysis under physiological conditions is an essential feature of the chemistry of nucleosides and nucleic acids and is intrinsic to life itself. Studies of mechanisms for their hydrolysis have had to be carried out at elevated temperatures (up to 250°C) and often at extremes of pH and the data is then extrapolated to ambient temperature and pH 7. Reactivity can also be enhanced by use of aryl esters, and 4-nitrophenyl esters have been used frequently because of their enhanced reactivity and convenient chromophoric properties.

Both C—O and P—O cleavage pathways (which give the same overall products) are observed. For attack at phosphorus, associative ($S_N2(P)$: either $A_ND_N$ or $A_N + D_N$), mechanisms are more common than dissociative ($S_N1(P)$: $D_N + A_N$) ones. The associative process is best described by invoking a 5-coordinate, trigonal bipyramidal intermediate in which ligand positional interchange, also called **pseudorotation**, is usually slower than breakdown to form products (Figure 3.44a). This leads to inversion of configuration at phosphorus. In a fully dissociative reaction, a planar, 3-coordinate species, often described as a monomeric metaphosphate, would be formed (Figure 3.44c). As it could capture an incoming nucleophile on either face, racemization at phosphorus would result. However, this intermediate is not sufficiently stable to exist in aqueous solution and real dissociative reactions involve an exploded transition state (see Section 3.2.2.4) in which the nucleophile begins to bond with the phosphorus atom before the leaving group has fully departed (see Figure 3.52). In between these two extremes lie concerted displacement reactions (Figure 3.44b), where bond making matches bond breaking and the reaction involves a *penta*-coordinate transition state.

There has been much discussion in terms of the **associative** or **dissociative** character of **transition states for phosphoryl transfer**. The associative/dissociative character is best defined according to the sum of bond

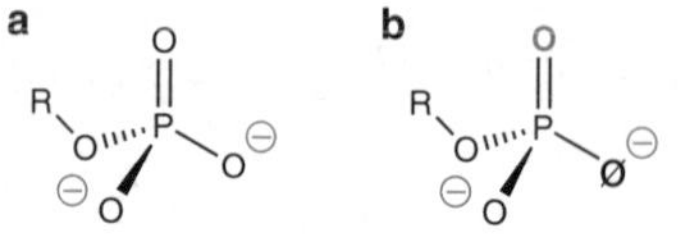

**Figure 3.43** *(a) Pro-pro-chiral oxygens in a phosphate monoester; and (b) phosphate monoester chiral through having three isotopes of oxygen*

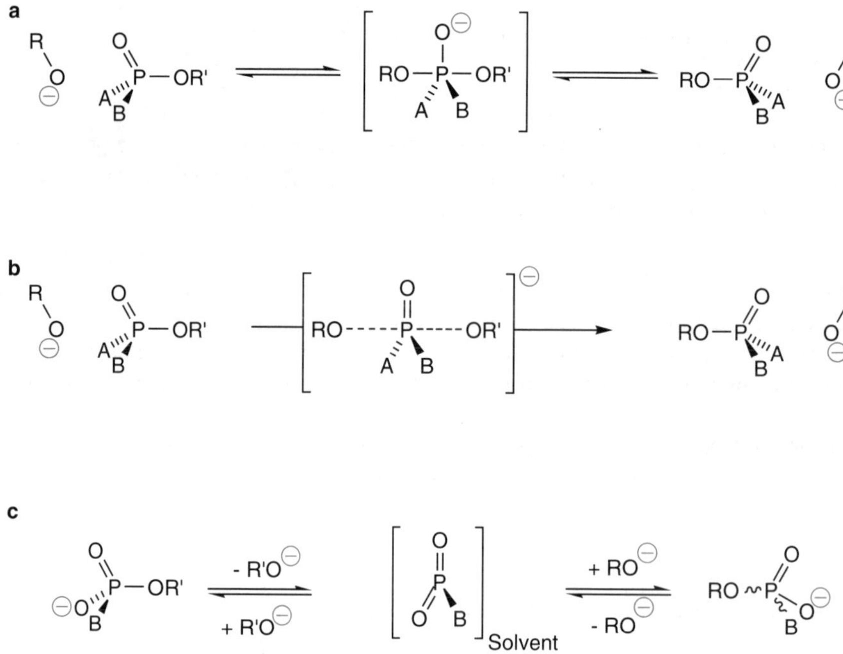

**Figure 3.44** *Mechanisms of displacement reactions for phosphate esters. (a) Addition-elimination ($A_N + D_N$) via a pentacoordinate phosphorane intermediate; (b) synchronous displacement ($S_N2(P)$) via a pentacoordinate transition state; and (c) stepwise displacement ($S_N1(P)$ or $D_N + A_N$) via a solvated metaphosphate intermediate*

formation from phosphorus to the nucleophile and the leaving group. If this is greater in the transition state than in the starting state (*i.e.* greater than one), then the transition state is considered associative (starting to resemble the intermediate in Figure 3.44a); if it is less than one, then the transition state is dissociative in character (starting to resemble the intermediate in Figure 3.44c). This issue is most important for enzymes, which have evolved to stabilize transition states effectively, and considerable progress has been made through the use of 'metaphosphate' mimics, such as $AlF_3$ and $MgF_3$, in X-ray structures of nucleotide complexes with enzymes that utilise ATP or GTP.

*3.2.2.1 Hydrolysis of Alkyl Triesters.* Trimethyl phosphate is hydrolysed in alkaline solution in an $S_N2(P)$ process ($k_{OH^-} = 1.6 \times 10^{-4}\ M^{-1}\ s^{-1}$ at 25°C). With $H_2{}^{18}O$ as solvent, no $Me^{18}OH$ is formed, which shows that the reaction involves exclusively P—O cleavage. Other 'hard' nucleophiles such as $F^-$ react similarly, and indeed, fluoride catalysis of the *trans*-esterification of triesters is a useful process (Figure 3.45). The *intra*-molecular migration of phosphorus in a triester to a vicinal hydroxyl group is especially easy and must be avoided in the synthesis of oligoribonucleotide and inositol phosphate precursors.

Trimethyl phosphate is hydrolysed extremely slowly in neutral and acidic conditions ($k_w = 2 \times 10^{-8}\ s^{-1}$ at 25°C) with C—O cleavage. Soft nucleophiles, such as $RS^-$, $Br^-$ or $I^-$, also dealkylate phosphate triesters with C—O cleavage (Figure 3.46a). Such reactions are typical $S_N2$ processes and show a clear preference for dealkylation in the order of Me > Et > $R_2CH$. This characteristic is particularly well exploited in the thiophenolate deprotection of methyl phosphate triesters used in the phosphodiester chemistry of oligonucleotide synthesis (Section 4.1) (Figure 3.46b). However in this case, some (~20%) oligonucleotide chain cleavage can result through attack of thiophenolate at the 5′-carbon of the 3′-nucleotide unit.

Alkyl phosphate triesters are sensitive to β-elimination processes as has been exploited for the selective deprotection of phosphate triesters in oligonucleotide synthesis (Section 4.1). The **2-cyanoethyl group** possesses an acidic β-hydrogen atom and may be removed by β-elimination mechanism under mildly

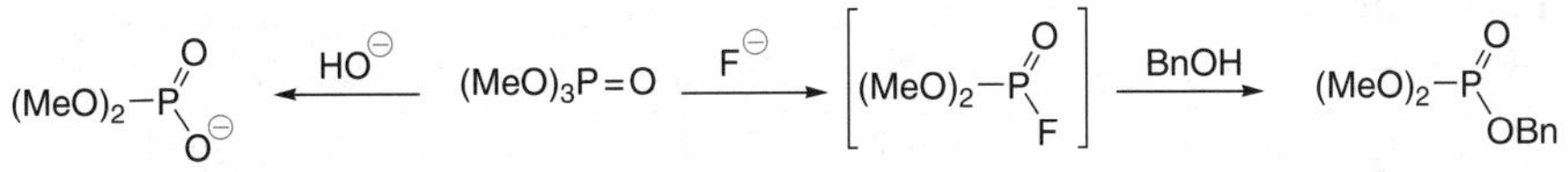


**Figure 3.45** *P—O cleavage reactions with hard nucleophiles for triesters*

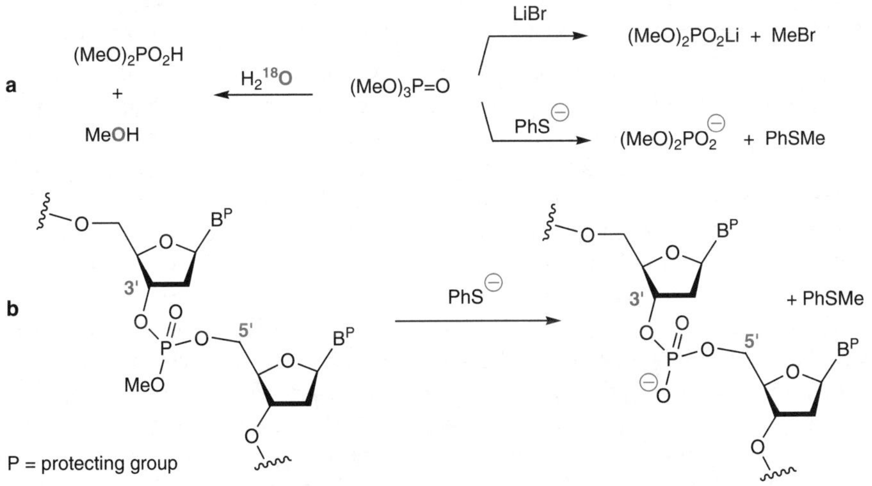


**Figure 3.46** *C—O cleavage reactions of triesters with soft nucleophiles. (a) trimethyl phosphate; (b) a phosphate triester following DNA synthesis*

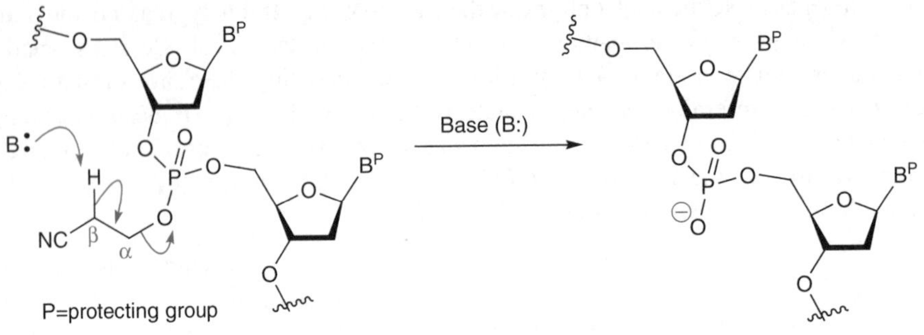


**Figure 3.47** *Selective deprotection of oligonucleotide phosphate triesters by β-elimination*

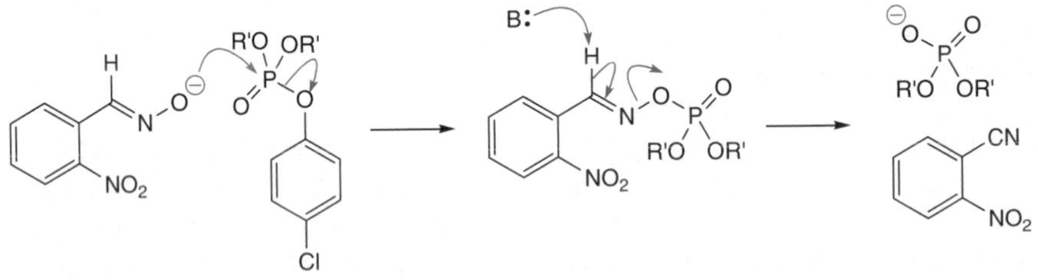


**Figure 3.48** *Selective nucleophilic displacement in an aryl ester*

basic conditions without competing cleavage of the oligonucleotide chain (Figure 3.47). Consequently, it is the protecting group of choice in oligonucleotide synthesis (Sections 4.1 and 4.2). β-Elimination is also important biologically in the base excision repair pathway involving the enzymatic cleavage of phosphate diesters following removal of damaged bases by glycosylase enzymes (Section 8.11.4).

*3.2.2.2 Hydrolysis of Aryl Triesters.* Because aryl phosphates are much more reactive than alkyl phosphate triesters, it is possible to achieve selective, nucleophilic displacement of the phenolic residue in a dialkyl aryl phosphate on account of its better leaving group ability ($pK_a > 5$). One of the best nucleophiles for this purpose is the **oximate anion** (Figure 3.48). Although aryl triesters were used historically during oligonucleotide synthesis, they were replaced by the cyanoethyl group, since this group may be removed in a single step along with the base protecting groups and oligomer cleavage from the solid support (Section 4.1).

*3.2.2.3 Hydrolysis of Phosphate Diesters.*[85,86] At pH > 2, phosphate diesters exist as their monoanions, which are extremely stable kinetically (Table 3.1). Even in strongly alkaline conditions, diesters hydrolyse with predominant C—O cleavage (the extent depending on the nature of the alkyl group) and far more slowly than triesters, since attack at the phosphorus atom is impeded by anion–anion repulsion. The spontaneous (pH-independent) reaction of the monoanion is so slow that it is yet to be quantified for simple phosphate diesters with alkoxy leaving groups. In acidic conditions, their hydrolysis occurs through the neutral species, which are similar to trialkyl phosphates in reactivity. The diaryl esters are rather more reactive under alkaline conditions, as is to be expected for reactions involving a better leaving group, and also allow the pH-independent reaction to be observed (which is very sensitive to the $pK_a$ of the aryloxy leaving group).

This marked stability of the phosphate diester linkage during hydrolysis is a vital feature of the biological role of DNA, where maintenance of the primary structure is required to preserve the genetic code. It is dramatically changed for esters of 1,2-diols, such as the ones found in RNA. Here, the vicinal hydroxyl group enormously enhances the rate of hydrolysis of di- and triesters. Similarly, the cyclic phosphates of 1,2-diols hydrolyse more than $10^7$ times faster than their acyclic or 6- and 7-membered cyclic relatives. This corresponds to a decrease in $\Delta G^{\ddagger}$ of 36 kJ mol$^{-1}$. About 60% of this acceleration is attributed to relief of strain in the five-membered cyclic ester, which has a 98° O—P—O angle and an enhanced enthalpy of hydrolysis of $-20$ kJ mol$^{-1}$.

**Table 3.1** *Some rate constants for the hydrolysis of phosphate esters (25°C) and patterns of bond cleavage*

| | *Phosphate + $H_2O$* $k_w$ *($s^{-1}$)* | *Phosphate monoanion + $H_2O$* $k_w$ *($s^{-1}$)* | *Phosphate + $HO^-$* $k_{OH^-}$ *($M^{-1} s^{-1}$)* |
|---|---|---|---|
| $(MeO)_3PO$ | $2 \times 10^{-8}$ (C—O) | — | $2 \times 10^{-4}$ (P—O) |
| $(MeO)_2PO_2H$ | $5 \times 10^{-10}$ (C—O) | $<2 \times 10^{-14}$ | $7 \times 10^{-12}$ (C—O)* |
| | $1 \times 10^{-10}$ (P—O) | | $1 \times 10^{-15}$ (P—O)*[138] |
| $(MeO)PO_3H_2$ | $1 \times 10^{-10}$ (C—O) | $3 \times 10^{-10}$ (P—O) | $2 \times 10^{-20}$ (P—O) †[139] |
| $(PhO)_3PO$ | — | — | $5 \times 10^{-3}$ (P—O)[140] |
| $(PhO)_2PO_2H$ | — | $5 \times 10^{-15}$ (P—O)[141] | $2 \times 10^{-7}$ (P—O)*[142] |
| $(PhO)PO_3H_2$ | — | $2 \times 10^{-8}$ (P—O)[143] | $1 \times 10^{-13}$ (P—O)†[138] |
| UpU[86] | — | — | $2 \times 10^{-3}$ (P—O)‡ |
| | $7 \times 10^{-5}$ (P—O)§,¶ | $5 \times 10^{-6}$ (P—O)¶ | 5 (P—O)¶ |
| Ethylene phosphate | — | — | $5 \times 10^{-4}$ (P—O)*[144] |

* monoanion.
† pH-independent reaction of dianion, $s^{-1}$.
‡ intramolecular transesterification at phosphate diester monoanion.
§ intramolecular attack by 2′OH.
¶ at 90°C.

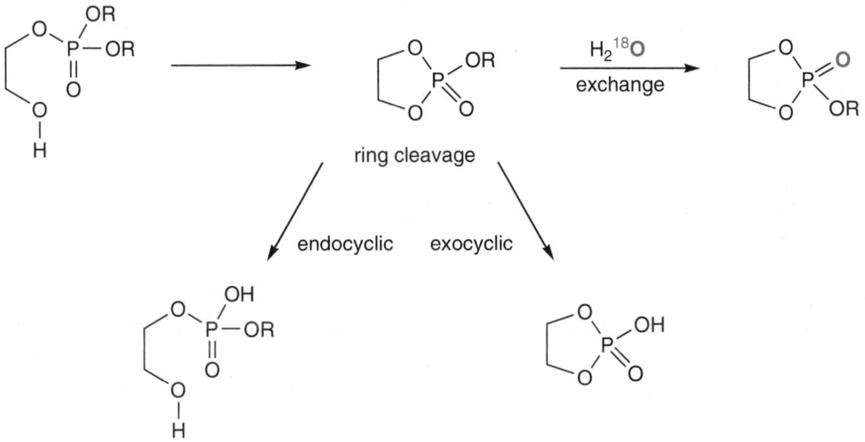


**Figure 3.49** *Accelerated P—O cleavages associated with 5-membered ring phosphate esters in acidic and in alkaline solution*

The essential observation is the acceleration of both ring closure and ring opening. Furthermore, exocyclic P—O bond cleavage is also accelerated in five-membered cyclic esters (Figure 3.49).

How can ring strain accelerate both endocyclic and exocyclic substitution at phosphorus? The hydrolysis of ethylene phosphate shows incorporation of isotopic label from $H_2{}^{18}O$ solvent into P—O bonds in both acidic and alkaline conditions, and these reactions must involve an $A_N + D_N$ (*i.e.* an associative) process (Figure 3.44a). It is generally agreed that a transient *penta*-coordinated phosphorane intermediate is both stabilized and made kinetically more accessible relative to its acyclic counterpart because of the geometry of the five-membered ring. It is then reasonable to invoke topoisomerism (in this case the pseudorotation of the trigonal bipyramidal species) to explain most of the phenomena associated with this remarkably enhanced reactivity (Figure 3.50).

This phenomenon is clearly important in the hydrolysis of RNA by alkali and ribonucleases. In both cases, the 5-membered 2′,3′-cyclic phosphates of nucleosides are formed by the displacement of the

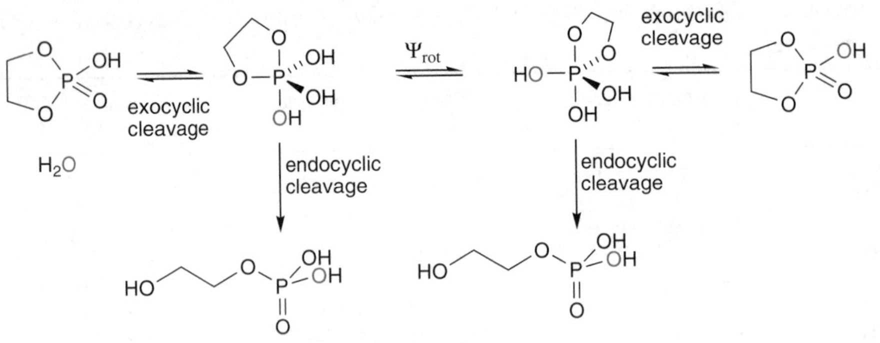


**Figure 3.50** *Role of trigonal bipyramidal pseudorotation ($\psi_{rot}$) in $^{18}O$ isotope (O) incorporation into ethylene phosphate*

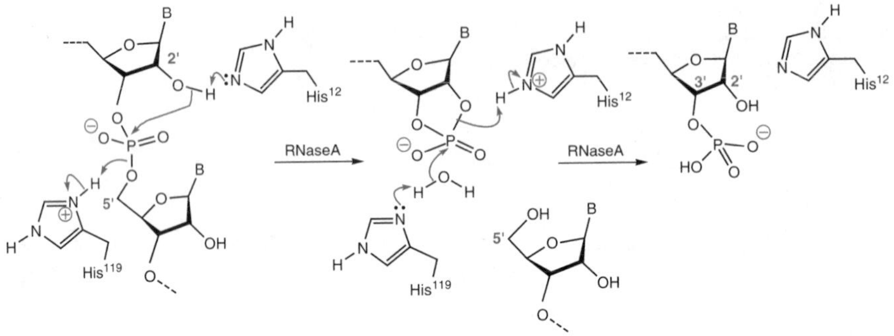


**Figure 3.51** *Ribonuclease A hydrolysis of RNA via 2′,3′-cyclic phosphate. Imidazoles (of His-12 and His-119 residues) act as a general acid and general base*

5′-*O*-nucleoside residue. The enzymatic reaction is completed by the regioselective ring opening of the cyclic phosphate to give only a 3′-nucleoside phosphate (Figure 3.51). By contrast, alkaline hydrolysis leads to a mixture of 2′- and 3′- phosphates. These reactions exhibit overall retention of configuration at the phosphorus centre. This is accounted for by the double inversion of stereochemistry that occurs in the two successive '**in-line**' displacement processes.

It must be emphasised that this remarkable reactivity appears to be exclusive to five-membered cyclic phosphate esters and esters of 1,2-diols. This contrasts with the relative stability of esters of 1,3-diols and 6-ring cyclic phosphates. An important example is 3′,5′-cAMP, whose key role as the second messenger in cell signalling is dependent on its kinetic stability to non-enzymatic hydrolysis.

*3.2.2.4 Phosphate Monoesters.*[87] The hydrolysis of monoalkyl phosphates at very low pH proceeds *via* the conjugate acid, and is similar in mechanism to that of triesters and neutral diesters (Table 3.1). These esters are very resistant to hydrolysis under alkaline conditions where they exist as dianions (and catalysis by hydroxide is never observed). The reaction of the dianion proceeds by P—O cleavage and has many of the characteristics of a dissociative process *via* a hypothetical **metaphosphate** intermediate. A better description invokes the idea of an 'exploded' transition state in which there is very weak bonding to the incoming nucleophile and outgoing leaving group (*i.e.* a concerted reaction which is very dissociative in character). The reaction is extremely sensitive to the p$K_a$ of the leaving group, and alkyl phosphate monoester dianions are even more stable than the corresponding diesters. By contrast, the monoanion shows both an unusually high relative reactivity towards hydrolysis and is very insensitive to the leaving group p$K_a$. This is explained by involving the minor tautomer (where the leaving group O carries the proton) as the reactive ionic form, which then hydrolyses through a similar transition state to the dianion (Figure 3.52). For

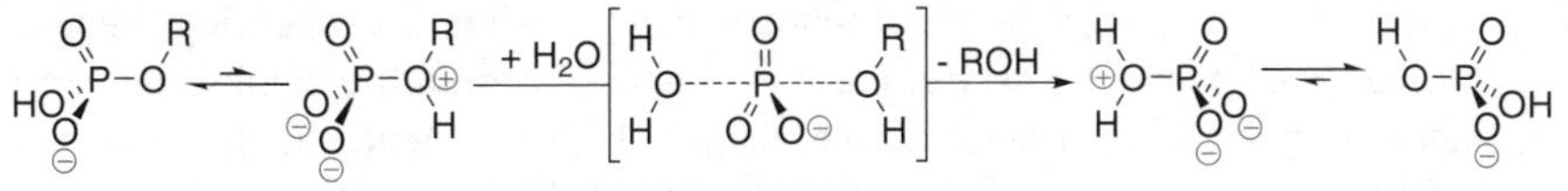

**Figure 3.52** *Hydrolysis of a phosphate monoester monoanion through an 'exploded' transition state*

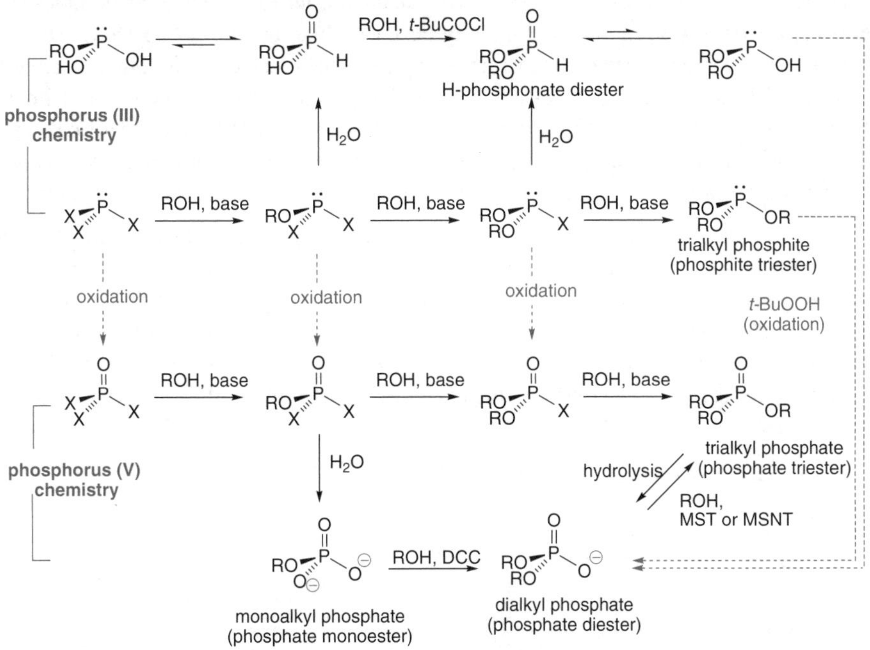


**Figure 3.53** *Formal relationship between exters of P(III) and P(V) oxyacids. DCC = N,N′-dicyclohexylcarbodiimide, MST = mesitylenesulfonyl tetrazolide, MSNT = mesitylenesulfonyl (3-nitrotriazolide)*

very good leaving groups, the difference between mono- and dianion reactivity is small, but for poor leaving groups, the monoanion is by far the more reactive ionic form and accounts totally for the observed reaction even at high pH. Similar phenomena have been analysed for spontaneous hydrolyses of acetyl phosphate, creatine phosphate and ATP (loss of the γ-phosphate), all of which have good leaving groups on a terminal phosphate.

### 3.2.3 Synthesis of Phosphate Diesters and Monoesters

The most common approaches to dinucleoside phosphate ester synthesis use phosphorylation reactions in which a 3′-nucleotide component is converted into a reactive phosphorylating species by a condensing agent. One of the major problems is that the more reactive condensing agents not only activate the phosphate, but may also react with nucleoside bases or the product, leading to unacceptable reduction in yield. The ideal condensing agent should have a high rate of activation of the phosphate species and a negligible rate of reaction with the alcohol component or *N*-protected bases.

*3.2.3.1 Interrelationships of Esters of Phosphorus Oxyacids.* The formal relationship between phosphorus halides (X = Cl) and the mono-, di- and tri-alkyl esters of phosphorus oxyacids is shown schematically (Figure 3.53). Actual reaction conditions have to be controlled carefully to avoid the formation of by-products, especially of alkyl halides. Many of the interconversions shown are best accomplished using nitrogen ligands at phosphorus (X = *i*-$Pr_2N$ or an azole).

Early syntheses of nucleoside phosphate esters worked mainly with mild condensing agents such as **dicyclohexylcarbodiimide (DCC)**. More powerful reagents, such as **arenesulfonyl chlorides**, were introduced next and were improved by building in steric factors. Some valuable references to the early phosphorylation chemistry may be found in a number of reviews.[34,88–90] After 1980, the demand for faster reactions for oligonucleotide synthesis has switched attention to P(III) chemistry.[91] In a more recent variation, the use of H-phosphonates as a 4-coordinate P(III) species[92] has been built upon pioneering studies of Todd in the 1950s.

*3.2.3.2 Syntheses via Phosphate Diesters.*[88,89,93] In the diester route to oligonucleotides (Figure 3.54), the key step is the condensation of a phosphate monoester with an alcohol using DCC. The reactions are slow, but at room temperature there is no formation of triesters. The mechanism is complex: an initial imidoyl phosphate adduct of DCC and the 5′-nucleotide is probably converted into the cyclic trimetaphosphate species before reaction with the 3′-hydroxyl component and subsequent final formation of the phosphate diester (Figure 3.54). Since the reaction of trimetaphosphate with alcohols is relatively slow, DCC was superseded by mesitylenesulphonyl chloride as a faster and more efficient condensing agent.

*3.2.3.3 Syntheses via Phosphate Triesters.*[88,89] The greater reactivity for phosphorylation using arenesulfonyl chloride as activating agent enables the syntheses of triesters from dialkyl phosphates and an alcohol, and so it forms the basis of the first triester syntheses of oligonucleotides. The key step here is the condensation of a suitable nucleotide diester as $(RO)_2POX$, with the 3′-hydroxyl group of a second nucleoside to give a phosphate triester, $(RO)_3PO$. To avoid problems arising from the nucleophilicity of chloride anion, the condensing agents now used are **mesitylenesulfonyl tetrazolide (MST)** or **nitrotriazolide**

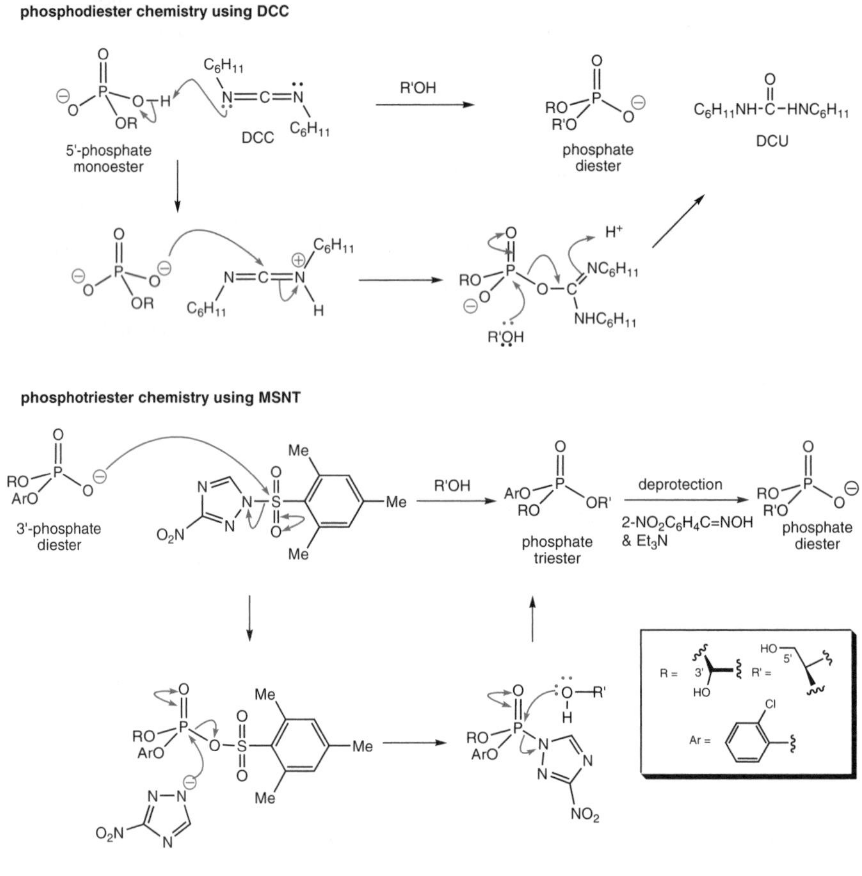


**Figure 3.54** *Mechanisms and reagents of phosphodiester (upper) and phosphotriester (lower) chemistry*

**(MSNT)**. A mixed phosphoryl–sulfonyl anhydride is produced initially, in which the methyl groups of the mesitylene ring provide steric hindrance to reaction at the sulfur atom and ensure condensation at the phosphorus atom. Subsequent complex reactions, which may also involve condensed phosphate intermediates, lead to the triester product (Figure 3.54). The final conversion of the triester into the desired diester uses one of the specific cleavages described earlier (Section 3.2.2).

*3.2.3.4 Syntheses via Phosphite Triesters.*[88,91,94] The P(III) triester route to oligonucleotides initially introduced by Letsinger (Figures 3.53 and 3.55) was designed to exploit the intrinsically greater reactivity shown by $PCl_3$ as compared to $POCl_3$ to achieve faster coupling steps. A major breakthrough was achieved by Caruthers,[95,96] who established the value of alkyl **phosphoramidites** ($X = i\text{-}Pr_2N$) as stable 3′-derivatives of nucleosides, which nonetheless react rapidly and efficiently with nucleoside 5′-hydroxyl groups in the presence of azole catalysts. The resulting product is an unstable phosphite triester, which must be oxidised immediately to give the stable phosphate triester in a process that can be cycled up to 100 times on a solid-phase support. Removal of the phosphate-protecting group affords the phosphate diester (Figure 3.55). Racemization that occurs during reactions carried out with purified *R*p and *S*p diastereoisomers have confirmed the intermediacy of a **phosphorotetrazolide** during formation of the phosphite triester (Figure 3.56).[97]

*3.2.3.5 Syntheses via H-phosphonate Diesters.*[92] The H-phosphonate monoesters of protected 3′-nucleosides are readily prepared using $PCl_3$ and excess imidazole followed by mild hydrolysis. The

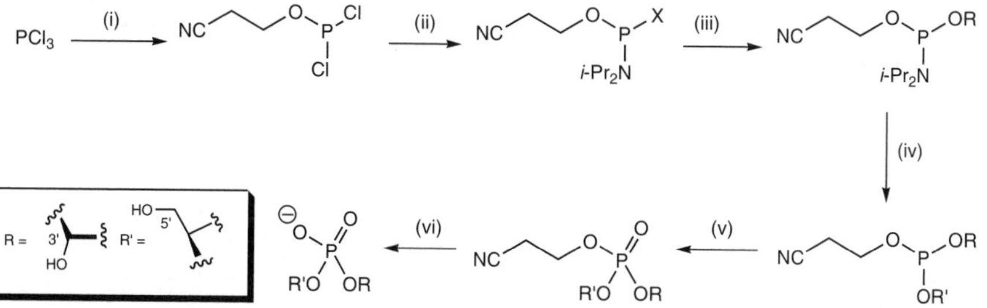


**Figure 3.55** *Synthesis of phosphate triesters and diesters via phosphite triesters. Reagents: (i) 2-cyanoethanol, pyridine, ether −78°C; (ii) i-Pr$_2$NH, ether, −20°C: X = Cl, 2eq. or X = i-Pr$_2$N 4 eq.; (iii) X = Cl: ROH, i-Pr$_2$EtN,THF, or X = i-Pr$_2$N: ROH, i-Pr$_2$EtNH$_2$$^+$ tetrazolide$^-$ in CH$_2$Cl$_2$ or THF; (iv) tetrazole in CH$_3$CN and R′OH; (v) I$_2$ in pyridine/H$_2$O/THF; and (vi) aq NH$_4$OH*

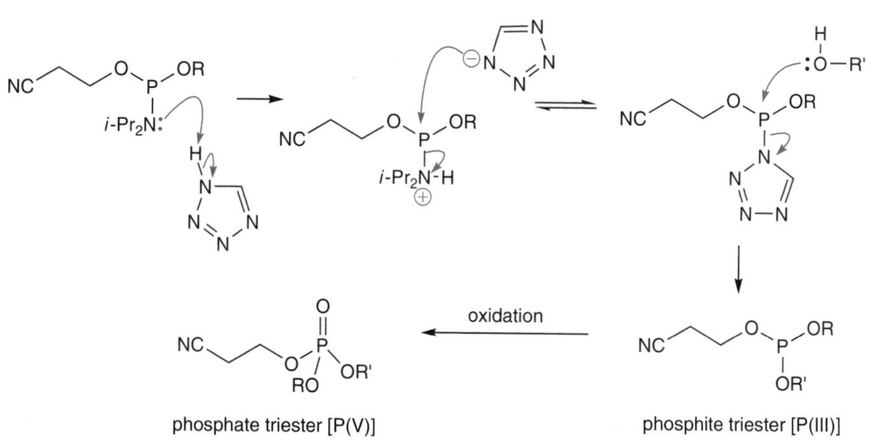


**Figure 3.56** *Synthesis of phosphate triesters using phosphoramidite reagents*

intermediates do not require further protection at phosphorus and are rapidly and efficiently activated by a range of condensing agents, such as pivaloyl chloride (Figure 3.53). This can be carried out before or after the addition of the second nucleoside if excess condensing agent is avoided. A dinucleoside H-phosphonate is rapidly formed in high yield. The procedure can be repeated many times before a single oxidation step finally converts the H-phosphonate diesters into phosphate diesters (Figure 3.57). Studies on the mechanism of condensation indicate that a mixed phosphonate anhydride is the likely intermediate.

*3.2.3.6 Synthesis of Phosphate Monoesters.*[34,90] Monoesters are usually made either from triesters by selective deprotection or by direct condensation of an alcohol with a reactive phosphorylating agent, usually a polyfunctional species such as **phosphoryl chloride** ($POCl_3$). The first formed product is immediately hydrolysed to give the desired monoester.

Triester procedures almost always call for selective protection of the nucleoside hydroxyl groups to leave only the reaction centre free. Reagents such as dibenzyl phosphorochloridate (with deprotection *via* catalytic hydrogenolysis), *bis*-(2,2,2-trichloroethyl) phosphorochloridate (deprotection by zinc reduction) or 1,2-phenylene phosphorochloridate (deprotection by hydrolysis to the diester and then oxidative removal of catechol) give good yields of phosphate monoester (Figure 3.58a). Selective reaction at the 5′-hydroxyl

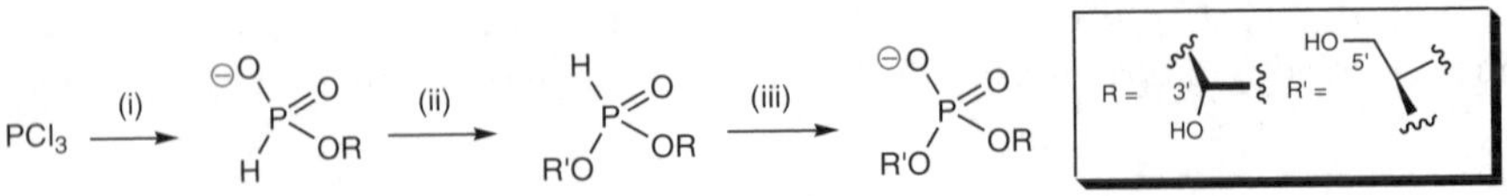


**Figure 3.57** *Synthesis of phosphate diesters via H-phosphonates. Reagents: (i) ROH then aq. $Et_3N$; (ii) $(CH_3)_3COCl$ and R′OH; and (iii) $I_2$ in aq. Pyridine*

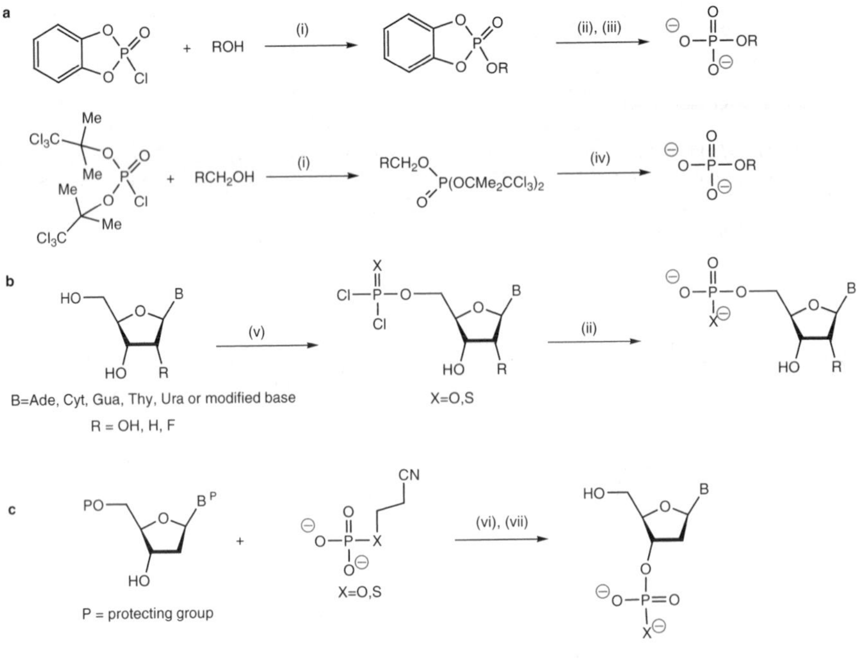


**Figure 3.58** *Syntheses of phosphate monoesters. (a) Triester procedures, (b) using $POCl_3$ or $PSCl_3$, (c) synthesis of 3′-phosphate monoesters. Reagents: (i) base; (ii) aq. $Et_3NH^+$ $HCO_3^-$; (iii) $Br_2/H_2O$ (oxidation); (iv) Zn, $MeCO_2H$; (v) $POCl_3$ [$PSCl_3$ (+collidine for pyrimidine nucleosides)] in $(MeO)_3PO$, 0°C; (vi) DCC, pyridine; and (vii) aq. $NH_4OH$(X=O) or $^tBuNH_2$/pyridine then aq. $NH_4OH$(X=S)*

group may be achieved using bulky phosphorylating agents such as *bis*-(2,2,2-trichloro-1,1-dimethyl) phosphorochloridate (Figure 3.58a).

For a wide range of sugars and bases, direct phosphorylation at the 5′-position of unprotected nucleosides and 2′-deoxyribonucleosides using $POCl_3$ in a trialkyl phosphate solvent system (the **Yoshikawa method**[98]) is probably the simplest and most convenient method. Hydrolysis of the highly reactive phosphorodichloridate intermediate with aqueous buffer gives the monophosphate (Figure 3.58b). No protection of the base or sugar is required, although by-products arising from phosphorylation of the secondary hydroxyl groups of the sugar vary, depending on the nature of the nucleoside. In the same way, isotopic oxygen can be readily introduced by the generation of $P^{18}OCl_3$ *in situ*, while the use of thiophosphoryl chloride ($PSCl_3$) leads to phosphorothioates.[99] In a modification to the procedure, the use of aqueous pyridine in acetonitrile as solvent (the **Sowa–Ouchi method**)[100] can provide yields greater than 80% with over 90% regioselectivity.

Synthesis of nucleoside 3′-phosphate monoesters requires protection of the 5′-hydroxyl group and base and can be achieved, for example, by DCC-mediated condensation with 2-cyanoethyl phosphate. The method also allows the preparation of phosphorothioate monoesters (Figure 3.58c).

## 3.3 NUCLEOSIDE ESTERS OF POLYPHOSPHATES

### 3.3.1 Structures of Nucleoside Polyphosphates and Co-Enzymes

Phosphoric acid can form chains of alternating oxygen–phosphorus linkages, which are relatively stable in neutral aqueous solution (Figure 3.59). The major condensed phosphates of biological importance are pyrophosphoric acid, $(HO)_2P(O)OP(O)(OH)_2$, its esters and esters of tripolyphosphoric acid. The stability of such species can be related to that of the corresponding phosphates after due allowance for (1) the changed stability of the anionic charge and (2) improved leaving group characteristics.

Thus, tetraethyl pyrophosphate is an ethylating agent towards hard nucleophiles and also it is a phosphorylating agent. Tetraphenyl pyrophosphate is exclusively a phosphorylating agent. $P^1,P^2$-Dialkyl pyrophosphates have considerable stability towards hydrolysis at ambient pHs where they exist exclusively in the form of a dianion. This feature is very important for the stability of the pyrophosphate link as a structural feature in many co-enzymes, including NADH, FAD and CoA. Similarly, $P^1,P^3$-diesters of tripolyphosphoric acid are stable components of the 'cap' structure of eukaryotic mRNA (Figure 3.60) and $\boldsymbol{P^1,P^4}$**-diadenosyl tetraphosphate**

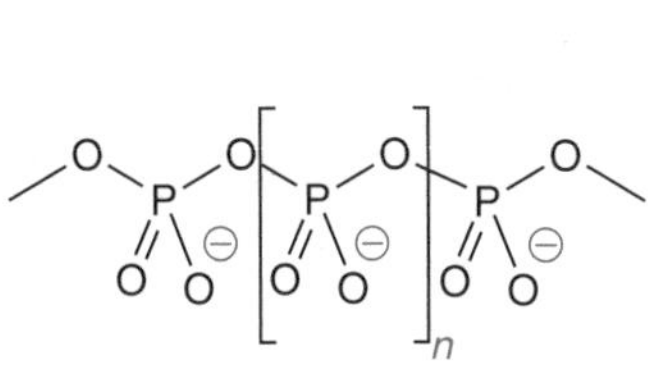

**Figure 3.59** *Structure of polyphosphates*

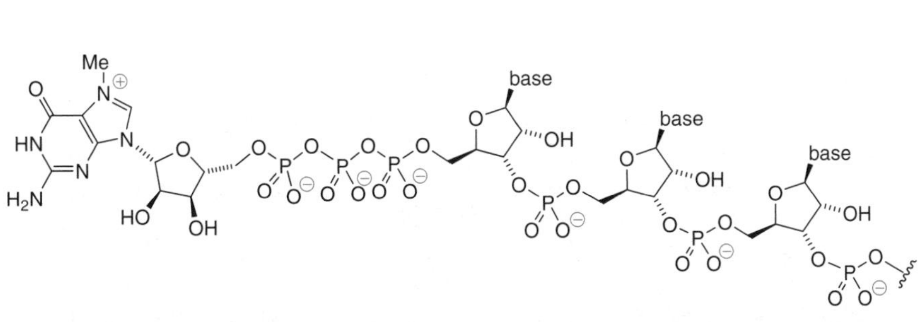

**Figure 3.60** *$P^1$, $P^3$-Dinucleosidyl triphosphate in the 'cap' structure at the 5′-end of eukaryotic mRNA*

($Ap_4A$) and other polyphosphates such as $Ap_3A$ (Figure 3.63) are stable minor nucleotide species found in low concentration in all mammalian tissues.

*3.3.1.1 Monoalkyl Esters.* Monoalkyl esters are the most ubiquitous examples of $P^1$-nucleoside esters of polyphosphates. They include the ribo- and deoxyribo-nucleoside esters of pyrophosphoric acid (NDPs and dNDPs) and of tripolyphosphoric acid (NTPs and dNTPs) (Figure 3.61). These esters are metabolically labile and participate in a huge range of C—O and P—O cleavage processes; thiamine pyrophosphate is a co-enzyme.

Among the minor nucleoside polyphosphates, the 'magic spot' nucleotides MS1 (ppGpp) and MS2 (pppGpp) are species formed by stringent strains of *Escherichia coli* during amino acid starvation.

*3.3.1.2 Dialkyl Esters.* Dialkyl esters are biologically significant for di-, tri-, tetra- and penta-polyphosphoric acids. In every case, the esters are located on the two terminal phosphate residues leaving (as described below) an ionic phosphate at every position, which ensures stability during spontaneous hydrolysis. Several of the co-enzymes such as co-enzyme A, flavine adenine dinucleotide (FAD) and nicotinamide adenine dinucleotide ($NAD^+$) are stable $P^1$,$P^2$-diesters of pyrophosphoric acid (Figure 3.62a). Their

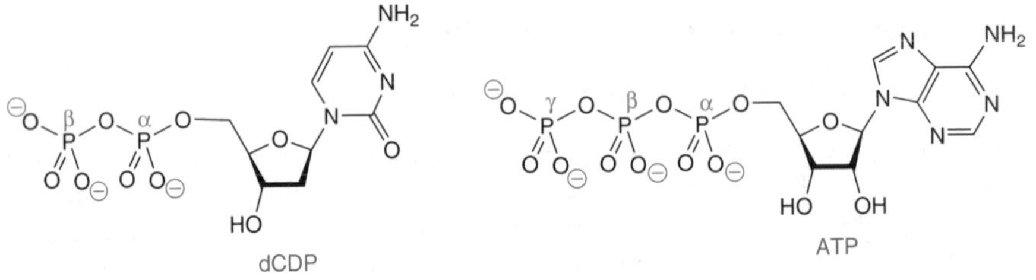


**Figure 3.61** *Structures of nucleoside 5′-di- and tri-phosphates*

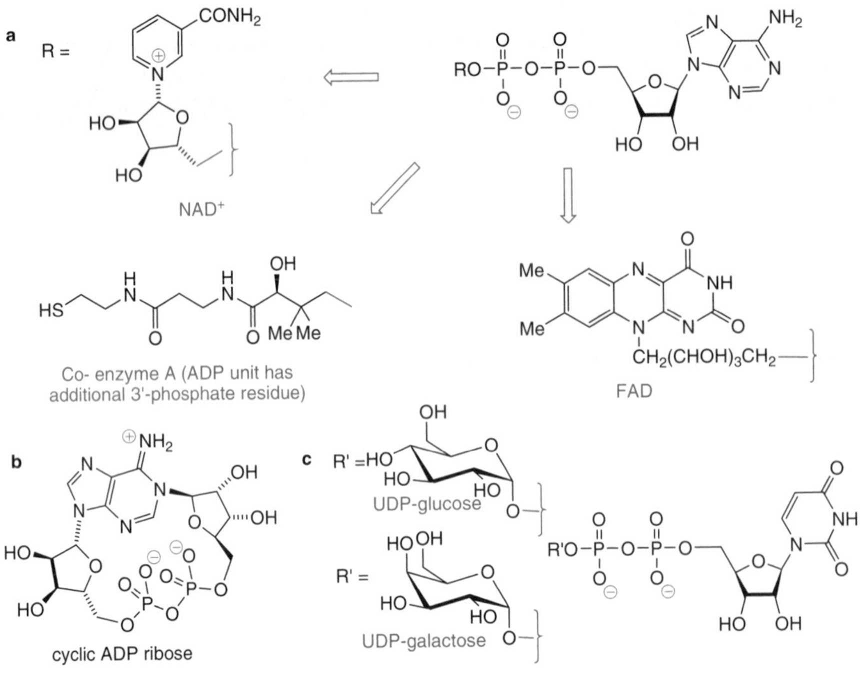


**Figure 3.62** *(a) Structures of three adenosine co-enzymes; (b) Structure of cyclic ADP ribose; (c) Structures of UDP-hexoses*

biosynthesis involves the condensation of ATP with a monoalkyl phosphate, and the pyrophosphate appears to act generally as a structural unit providing coulombic binding to appropriate enzyme residues.

**Cyclic ADP ribose** (Figure 3.62b) is a metabolic product formed by the cyclization of $NAD^+$ to close C-1 of the second ribose onto N-1 of the adenine ring. The instability of this nucleotide appears to originate not from its pyrophosphate diester but from the second glycosylic linkage. It has an important role as a second messenger and is involved in calcium signalling in cells. The active forms of many hexoses are found as pyrophosphate esters of uridine 5′-diphosphate. These include UDP-glucose, UDP-galactose and **UDP-*N*-acetylglucosamine** (Figure 3.62c) that are formed biosynthetically from UTP and hexose-1-α-phosphate. The pyrophosphate ester is a good leaving group and is employed in catabolic processes involving C-1 of the hexose residue.

The $P^1$,$P^4$-dinucleosidyl tetraphosphates, $Ap_4A$ and $Ap_4G$, are found along with $Ap_3A$ in all cells, especially under conditions of metabolic stress (Figure 3.63). They are produced as a result of the phosphorolysis of aminoacyl adenylates, particularly tryptophanyl and lysyl adenylates, with ATP or GTP. Although these minor nucleotides were discovered by Zamecnik in 1966, their purpose remains uncertain. They may have a role in the initiation of DNA biosynthesis, and their analogues inhibit the aggregation of blood platelets. $Ap_3A$ is involved in signalling for apoptosis. Somewhat related structures are found in the 'caps' at the 5′-ends of eukaryotic mRNAs, which have a 7-methylguanosin-5′-yl residue linked to the 5′-triphosphate. Both of these species and their analogues have been targets for synthesis as a means of discovering their biological function.

### 3.3.2 Synthesis of Nucleoside Polyphosphate Esters

All of the naturally occurring nucleoside polyphosphates have at least one negative charge on each phosphate residue. This is because uncharged phosphoryl residues in a string of phosphates are readily hydrolysed spontaneously. As a result, most syntheses have avoided the formation of fully esterified intermediates, though an early synthesis of UTP was achieved (in low yield) by the catalytic hydrogenolysis of its tetrabenzyl ester. Generally, syntheses of monoalkyl esters fall into two classes; they involve C—O bond or P—O—P bond formation.

The exploitation of the alkylating properties of nucleoside 5′-*O*-tosylates towards pyrophosphate or tripolyphosphate anions and their methylene analogues is put to good use in the **Poulter reaction**.[101] This has made direct syntheses of nucleoside 5′-di- and tri-phosphates and their analogues possible (Figure 3.64).

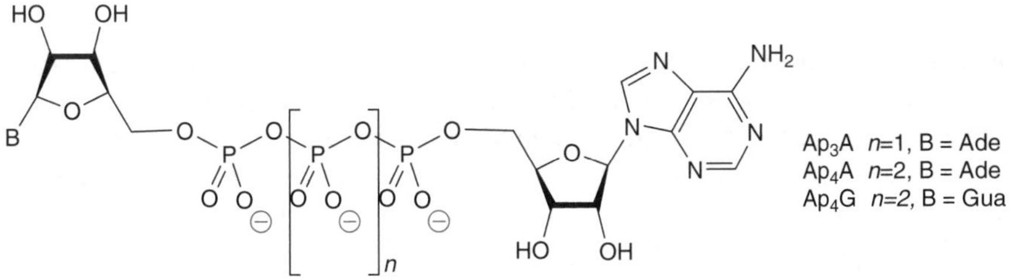


**Figure 3.63** *Structures of $P^1$, $P^3$-Dinucleoside triphosphate and $P^1$, $P^4$-dinucleoside tetraphosphates*

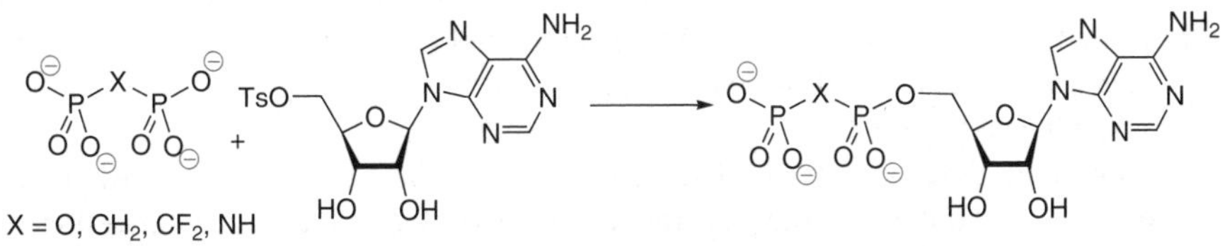


**Figure 3.64** *Synthesis of ADP and its analogues by C—O bond formation*

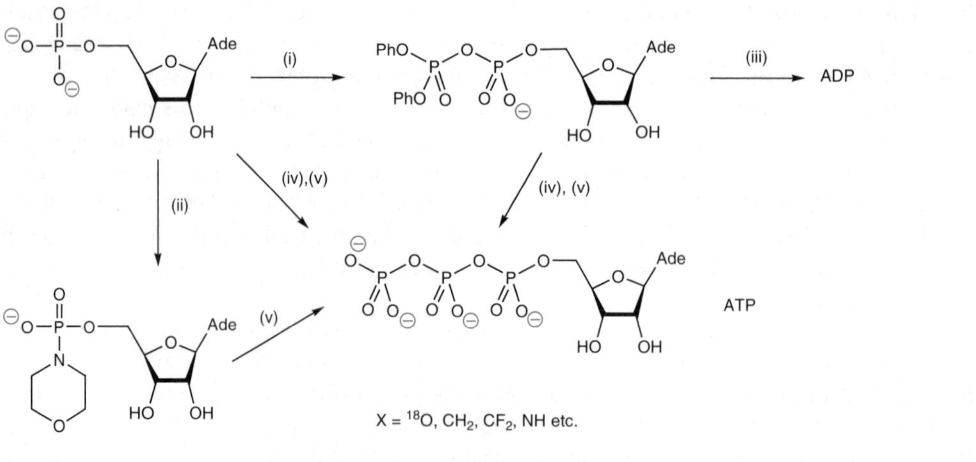


**Figure 3.65** *Synthesis of nucleoside diphosphates and triphosphates and analogues by P—O bond formation. Reagents: (i) $(PhO)_2POCl$ in DMF; (ii) DCC, morpholine, pyridine; (iii) tributylammonium phosphate in DMF; (iv) CDI in pyridine; and (v) tributylammonium pyrophosphate or pyrophosphate analogue in DMF*

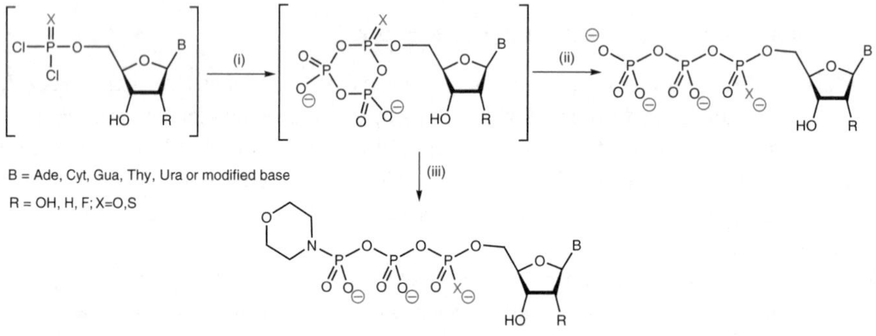


**Figure 3.66** *Syntheses of nucleoside triphosphates and analogues using phosphorus(V) chemistry. Reagents: (i) tetrakis (tributylammonium) pyrophosphate in DMF; (ii) $Et_3NH^+$ $HCO_3^-$; and (iii) morphiline*

A related but more general route to nucleoside 5′-triphosphates or β,γ-substituted analogues thereof involves the reaction of an activated nucleoside monophosphate, which is then able to condense with pyrophosphate or a pyrophosphate analogue[102,103] such as **methylenebisphosphonate**. Among the condensing agents which have been used widely are DCC to prepare the reactive **nucleoside 5′-phosphoromorpholidates** and **carbonyl diimidazole** (Figure 3.65).[34,90,104,105] This procedure is well suited to the introduction of isotopic oxygen into nucleotides in a non-stereochemically controlled fashion, for subsequent use in positional isotope exchange (PIX) studies.

Base and sugar-modified nucleoside 5′-triphosphates and **5′-(1-thio)triphosphates** (**α-thiotriphosphates**) are generally more often made in a simple one-pot reaction exploiting the Yoshikawa phosphorylation. Here a nucleoside-5′-phosphorodichloridate (Figure 3.66) is reacted *in situ* with *tetrakis* (tributylammonium) pyrophosphate.[106] The resulting cyclic triphosphate formed is then hydrolysed to afford linear triphosphates[107] or (1-thio)triphosphates,[99] while hydrolysis by other nucleophiles (*e.g.* morpholine), results in γ-substituted triphosphates (Figure 3.66).

In related methodology developed by Ludwig and Eckstein, the more reactive P(III) reagent, **salicyl chlorophosphite**, is used and this allows the preparation of nucleoside 5′-triphosphates after final oxidation.[108] Modifications also furnish routes to 5′-(1-thio)triphosphates and dithiotriphosphate derivatives,[109,110] and Barbara Shaw has used the procedure for the preparation of nucleoside **5′-(1-borano)triphosphates** (**α-boranotriphosphates**)[111] (Figure 3.67).

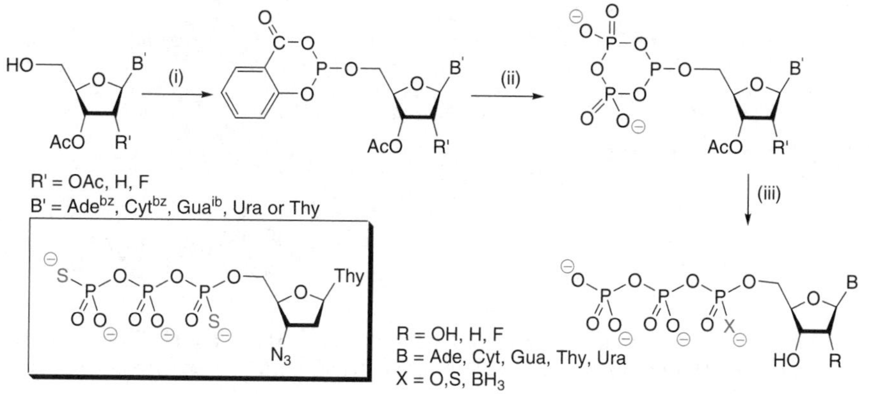


**Figure 3.67** *Syntheses of nucleoside triphosphates and analogues using phosphorus(III) chemistry. Inset AZT-α, γ-phosphorodithioate formed through ring opening of cyclic thiotriphosphate with $Li_2S$ in DMF. Reagents: (i) salicyl chlorophosphite in dioxan/DMF; (ii) tetrakis (tributylammonium) pyrophosphate in DMF; and (iii) X = O: iodine in aq pyridine, then $NH_4OH$; X = S: $S_8$, then $NH_4OH$; X = $BH_3$: borane-i-$Pr_2EtN$ complex, then $NH_4OH$*

All of these methods produce nucleotide phosphorothioates as racemic mixtures, although the diastereoisomers can generally be resolved by reversed phase HPLC (the $S_p$ diastereoisomer always elutes first) and characterized by $^{31}$P NMR (the signal for the α-phosphate of the $S_p$ diastereoisomer is downfield).[108,112]

Nucleotide thioesters have become prime tools for the investigation of the stereochemistry of enzyme-catalysed phosphoryl transfer processes.[112] For example, the ($S_p$) isomer of adenosine 5′-*O*-(1-thio)triphosphate, ATPαS, is readily made from AMPS by the combined action of adenylate kinase and pyruvate kinase (both enzymes can be immobilized on a polymer support for large scale syntheses) and by use of phosphoenol pyruvate with a little ATP to start the cycle. This synthesis illustrates the stereospecificity of adenylate kinase. The $^{31}$P NMR of this product has been used to identify the ($R_p$) and ($S_p$) diastereoisomers of dATPαS, which have been synthesised and separated by HPLC. Such species have been employed *inter alia* as substrates for DNA and RNA polymerases, which only incorporate nucleotide thiotriphosphates of the $S_p$ configuration, to show that polymerases, as well as T4 RNA ligase and adenylate cyclase, operate on adenine nucleotides with inversion of configuration at Pα.[113] For such purposes, ATP has been made with incorporation of either $^{17}$O or $^{18}$O in just about every possible position in the three phosphate residues.[114] The more useful species for nucleic acid chemistry are the α-phosphate substituted nucleotides. These can be made either by *ab initio* synthesis or the stereochemically controlled replacement of sulfur of an α-thiophosphate residue by isotopic oxygen. This transformation is best carried out by controlled bromine oxidation in $^{17}$O- or $^{18}$O-enriched water (Figure 3.68). While this reaction proceeds with inversion of configuration, similar oxidations with *N*-bromosuccinimide or cyanogen bromide have been found to be less stereoselective. An alternative procedure, although less widely applied, has used [$^{18}$O]-styrene oxide, when the substitution of sulfur by oxygen proceeds with exclusive retention of stereochemistry at phosphorus.

The $P^1$,$P^2$-diesters of pyrophosphoric acid are most often made by coupling together two phosphate monoesters using DCC, by a morpholidate procedure or by diphenyl phosphorochloridate.[115–117] A classical example is Khorana's synthesis of co-enzyme A.[118] The same methods have worked well for syntheses of $Ap_4A$ and its analogues, where the use of an excess of activated AMP and limiting pyrophosphate (or one of its analogues) gives acceptable yields of $P^1$,$P^2$-dinucleosidyl tetraphosphate or analogue (Figure 3.69).

Making the $P^1$,$P^2$-dialkyl triphosphates of the mRNA 'cap' structures has called for more sophisticated coupling procedures.[119] This is partly on account of the lability of the glycosylic bond in the 7-MeGuo residues and partly because of the unsymmetrical character of the diester (Figure 3.70). In general, the major problem encountered in the syntheses of all these species has arisen during purification as there appears to be no good alternative to ion-exchange chromatography, although high-performance reverse phase silica chromatography has some uses.

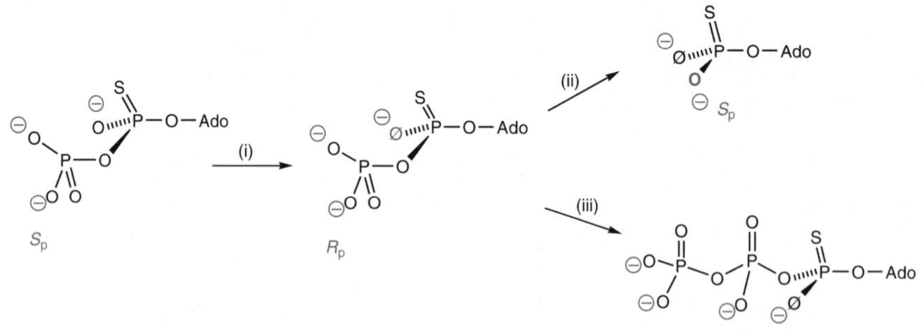


**Figure 3.68** *Syntheses of (Rp)-[α−$^{17}$O] ATP and (Sp) AMP. Reagents: (i) $Br_2$, $H_2{}^{17}O$; (ii) snake venom phosphodiesterase, $H_2{}^{18}O$ (retention); and (iii) pyruvate kinase, $Mg^{2+}$, $K^+$, phosphoenol pyruvate*

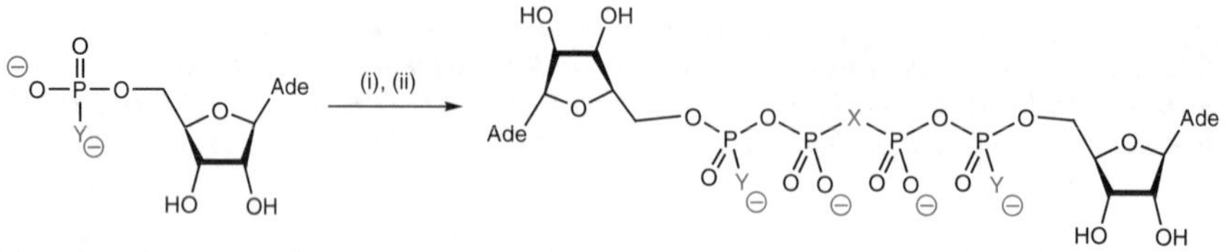


**Figure 3.69** *Syntheses of $Ap_4A$ and some analogues. Reagents: (i) $(PhO)_2POCl$, pyridine; and (ii) $O_3PXPO_3^{4-}$ (X = O, $CH_2$, $CF_2$, NH etc.; Y = O or S)*

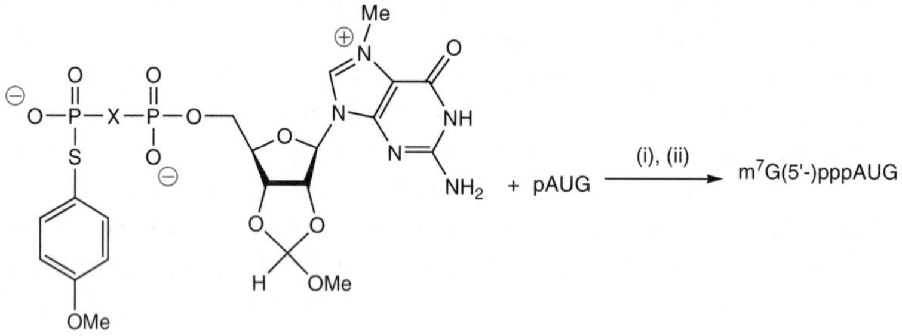


**Figure 3.70** *Synthesis of the 'cap' structure of mRNA. Reagents: (i) $Ag^+$, imidazole, DMF; and (ii) $H_3O^+$*

## 3.4 BIOSYNTHESIS OF NUCLEOTIDES

Nucleotides play a key role as the precursors of DNA and RNA, as activated intermediates in many biosynthetic processes and as metabolic regulators. One particular nucleotide, adenosine 5′-triphosphate (ATP), is an important energy source. For example, an average human turns over 40 kg of ATP per day and can require 0.5 kg $min^{-1}$ during exercise. The biosynthesis of nucleotides involves both constructive (anabolic) and destructive (catabolic) pathways. In the following sections, we will concentrate on only the general principles of nucleotide and nucleic acid metabolism and then show how certain steps are prime targets for biosynthetic interference, especially for the design of anti-cancer and anti-viral agents.

### 3.4.1 Biosynthesis of Purine Nucleotides[120,121]

*3.4.1.1 De novo Pathways.* The key intermediate in the biosynthesis of both pyrimidines and purines is α-D-5-phosphoribosyl 1-pyrophosphate (PRPP), which is formed from α-D-5-ribose 5-phosphate by a

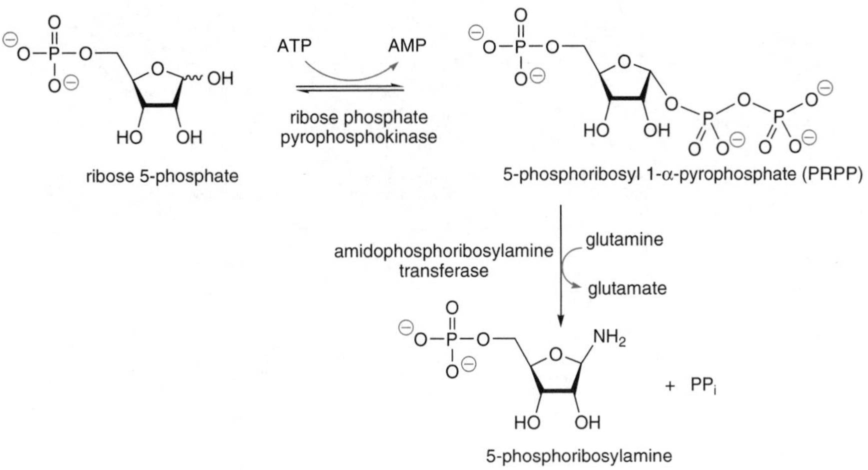


**Figure 3.71** *Biosynthesis of 5-phosphoribosylamine*

reaction catalysed by the enzyme ribose phosphate pyrophosphokinase (Figure 3.71). ATP acts as the donor of pyrophosphate while ribose 5-phosphate comes primarily from the pentose phosphate pathway.

In contrast to pyrimidine nucleotide biosynthesis, where a preformed heterocycle is incorporated intact (Section 3.4.2), in purine nucleotide biosynthesis the purine ring is constructed stepwise. The first irreversible step (the **committed step**) is displacement of pyrophosphate at C-1 or PRPP by ammonia from glutamine to give β-D-5-phosphoribosylamine (Figure 3.71). The reaction proceeds with inversion at C-1 to give the glycosylic bond in the β-configuration. The equilibrium in this reaction is displaced towards the phosphoribosylamine by the hydrolysis of the pyrophosphate co-product.

The five carbon and the remaining three nitrogen atoms of the purine skeleton are derived from six different precursor sources and assembled by nine successive steps (Figure 3.72). These steps are

1. reaction of PRPP with glycine to give glycinamide ribonucleotide;
2. formylation of the α-amino terminus of the glycine moiety by $N^{10}$-formyltetrahydrofolate to give α-*N*-formylglycinamide ribonucleotide;
3. conversion into the corresponding glycinamide with ammonia derived from glutamine;
4. ring closure to give 5-aminoimidazole ribonucleotide;
5. carboxylation of the imidazole C-4 (The carbon atom derived from $CO_2$);
6. condensation with aspartate;
7. elimination of fumarate to give 5-aminoimidazole-4-carboxamide ribonucleotide;
8. formylation of the amino imidazole by $N^{10}$-formyltetrahydrofolate; and
9. ring closure condensation to form inosine 5′-monophosphate (IMP).

Inosine is a nucleoside rarely found in natural nucleic acids except in the 'wobble' position of some tRNAs (Section 7.2.4). In such cases, the inosine comes from adenosine in the preformed tRNA by displacement of adenine by hypoxanthine.

Inosine 5′-monophosphate is used entirely for the production of the natural purine nucleotides, adenosine 5′-monophosphate (AMP) and guanosine 5′-monophosphate (GMP) (Figure 3.73). AMP receives its amino group at C-6 from aspartate in a reaction that utilises GTP as the phosphate donor. GMP is derived in two steps from xanthosine 5′-monophosphate (XMP) with the final amino group being donated by glutamine, and ATP is consumed in the process. In both these pathways, a carbonyl group of an amide is replaced by an amino group to give an amidine. This is a common type of mechanism whereby the amide is phosphorylated by ATP or GTP to its imido-*O*-phosphoryl ester and then the phosphoryl ester displaced by an amine

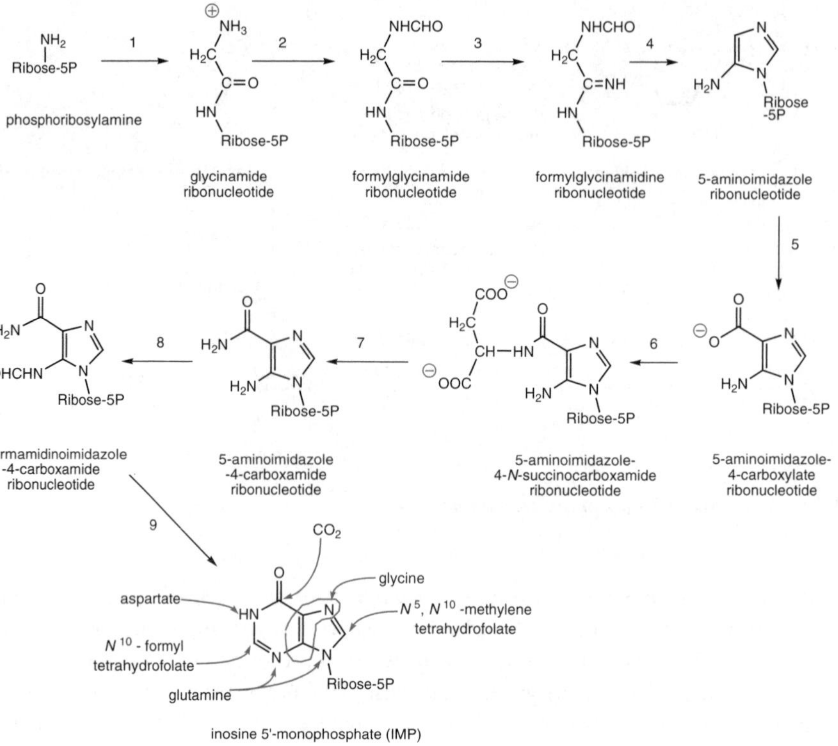


**Figure 3.72** *Formation of the purine ring; biosynthesis of IMP*

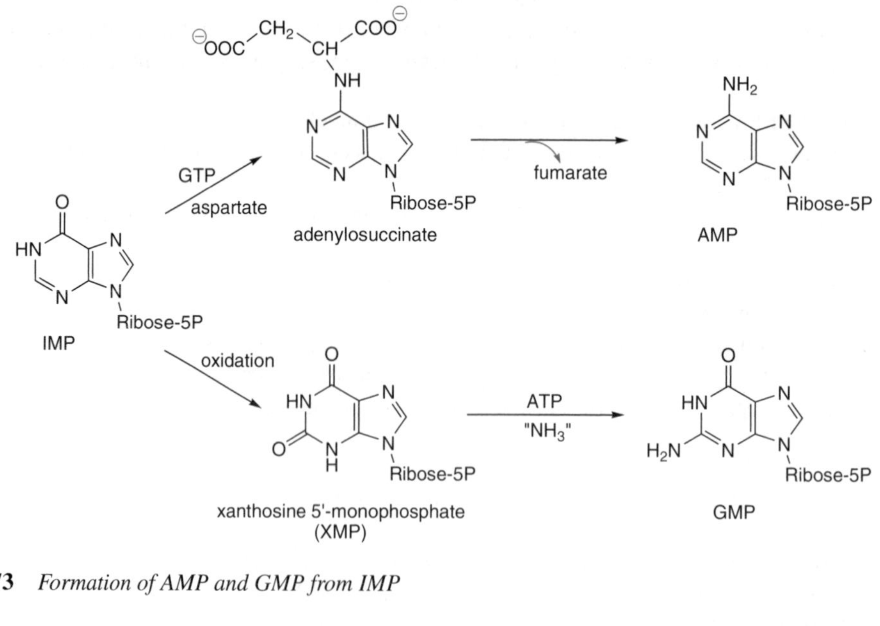


**Figure 3.73** *Formation of AMP and GMP from IMP*

or ammonia. The leaving group can be inorganic phosphate, pyrophosphate or AMP, while the displacing nucleophile can be ammonia, the side chain of glutamine or the $\alpha$-amino group of aspartate (Figure 3.74).

Steps in the biosynthesis of purine nucleotides furnish good examples of a standard control mechanism in metabolic pathways. This is **feedback inhibition**, where an enzyme catalysing an early step in the pathway

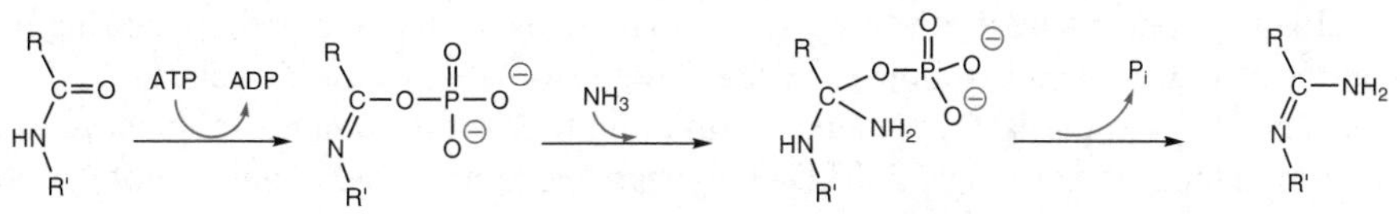


**Figure 3.74** *General mechanism for biosynthetic formation of an amidine from an amide*

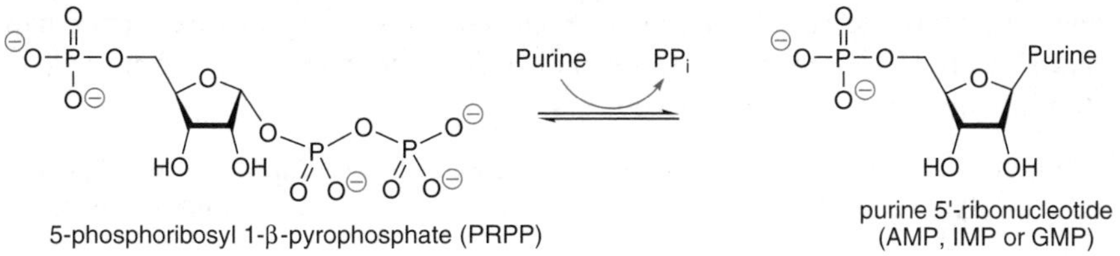


**Figure 3.75** *Salvage biosynthesis of purine ribonucleosides*

is inhibited by the final product of the pathway. For example, the enzyme ribose phosphate pyrophosphokinase (Figure 3.71) is inhibited by AMP, GMP and IMP and this inhibition regulates the production of PRPP. Similarly, the enzyme amidophosphoribosyl transferase, which is responsible for catalysing the committed step, is inhibited by a number of purine ribonucleotides including AMP and GMP, which act synergistically. AMP and GMP also inhibit the conversion of IMP into their own immediate precursors, adenylosuccinate and XMP. A separate control feature is that GTP is required in the synthesis of AMP, while ATP is required in the synthesis of GMP.

*3.4.1.2 Salvage Pathways.* Most organisms use a pathway of nucleotide biosynthesis known as **salvage**. This is advantageous since degradation products of nucleic acids can be recycled rather than destroyed, which is much less wasteful than the energy-demanding reactions of the *de novo* pathways. In some cancer cells or virus-infected cells, extra synthetic capacity is required. Here salvage may become the dominant pathway and hence has become a target for chemotherapeutic inhibitors.

Purine bases, which arise by hydrolytic degradation of nucleotides and nucleic acids, react with PRPP to give the corresponding purine ribonucleotide and pyrophosphate is eliminated (Figure 3.75). The enzyme, adenine phosphoribosyl transferase, is specific for the reaction with adenine; whereas, another enzyme, hypoxanthine-guanine phosphoribosyl transferase (HGPRT), catalyses the formation of IMP and GMP. A deficiency of HGPRT is responsible for the serious Lesch–Nyhan syndrome, which is often characterised by self-mutilation, mental deficiency and spasticity. Here, elevated concentrations of PRPP give rise to an increase in *de novo* purine nucleotide synthesis and degradation to uric acid (Section 3.5).

Another salvage route involves the reaction of a purine (or purine analogue) with ribose 1-phosphate. The reaction is catalysed by a nucleoside phosphorylase (nucleoside phosphotransferase) and the resultant ribonucleoside is then converted into its corresponding 5′-nucleotide by a cellular kinase. Similarly, a deoxynucleoside phosphotransferase produces deoxyribonucleosides from purines and 2-deoxyribose 1-phosphate.

### 3.4.2 Biosynthesis of Pyrimidine Nucleotides

*3.4.2.1 De novo Pathways.* Carbamoyl phosphate is an important intermediate in pyrimidine biosynthesis. It is formed from glutamine and bicarbonate in a reaction catalysed by a carbamoyl phosphate synthetase, and the reaction uses ATP as its energy source (Figure 3.76). The committed step is the subsequent formation of *N*-carbamoyl aspartate from carbamoyl phosphate and aspartate. This step is subject to feedback inhibition by cytidine triphosphate (CTP), which is the final product of the pathway, while the synthesis of carbamoyl phosphate is inhibited by UMP. In the next step, the pyrimidine ring is formed by cyclization and loss of water and is followed by dehydrogenation to give orotic acid.

The enzymes involved in the last three steps form a multi-enzyme complex in eukaryotes (but not in prokaryotes) and are located on a single 200 kDa polypeptide chain. A potent inhibitor of the first enzyme, aspartate transcarbamoylase, is *N*-phosphonoacetyl-L-aspartate (PALA) (Figure 3.77). PALA is an example of a **transition state inhibitor**, which mimics the transition state of a reaction. PALA binds tightly to aspartate transcarbamoylase and has proved to be useful in the production and isolation of the enzyme complex.

Orotate then reacts with PRPP to give orotidylate (Figure 3.78). There is inversion of configuration at C-1′ and the β-nucleotide is formed. The equilibrium of the reaction is once again driven forward by hydrolysis of pyrophosphate. Finally, UMP is produced by decarboxylation. The other pyrimidine nucleotides are derived from UMP after its conversion into UTP (Section 3.4.3).

*3.4.2.2 Salvage Pathways.* The enzyme, orotate phosphoribosyl transferase, which is involved in the production of orotidylate from orotate, can also utilize a number of other pyrimidines that are produced as

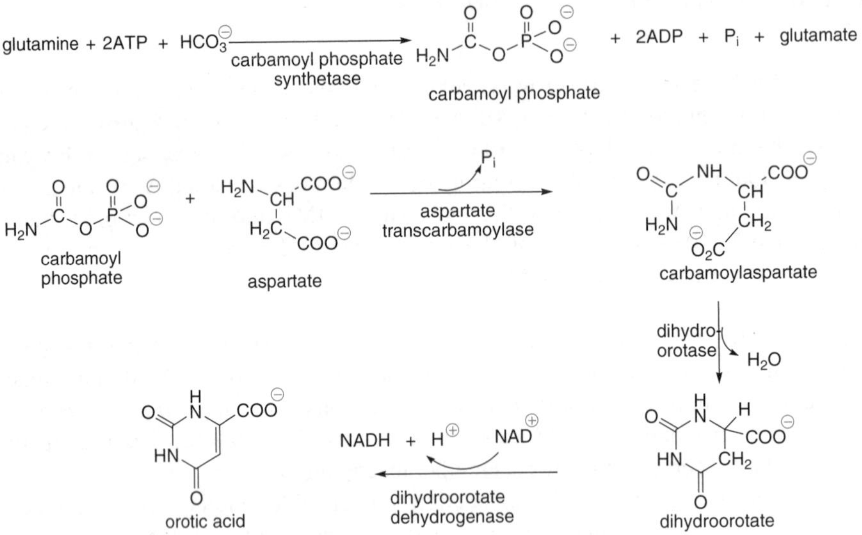


**Figure 3.76** *De novo biosynthesis of pyrimidines; formation of orotate*

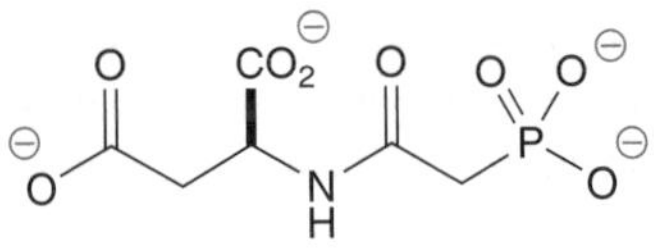

**Figure 3.77** *N-Phosphonoacetyl-L-aspartate (PALA)*

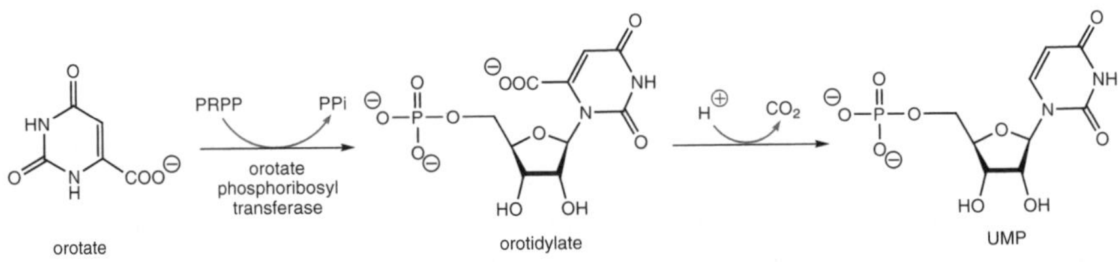


**Figure 3.78** *Formation of UMP from orotate*

a result of hydrolysis of DNA or RNA. In a similar way as in the salvage of purines, phosphorylases catalyse nucleoside formation from a variety of pyrimidines with either ribose 1-phosphate or 2-deoxyribose 1-phosphate. A cellular kinase is also required to convert the nucleoside product into the corresponding 5′-nucleotide. Thus, uridine kinase will accept both uridine and cytidine as substrates while thymidine kinase will accept deoxyuridine as well as deoxythymidine. The fact that many viral thymidine kinases have a reduced specificity for their substrates enables a distinction to be made between normal and virally-infected cells and has led to a strategy for viral interference (Section 3.7.2). Nucleoside transferases will catalyse base exchange between nucleosides exclusively in the 2′-deoxy series.

### 3.4.3 Nucleoside Di- and Triphosphates

The immediate biosynthetic precursors of the nucleic acids are normally the nucleoside triphosphates; whereas, diphosphates can also be used in energy conversions. Diphosphates are obtained from the corresponding monophosphates by means of a specific nucleoside, monophosphate kinase. Adenylate kinase converts AMP into ADP while UMP kinase converts UMP into UDP. Both enzymes use ATP as the phosphoryl donor. Nucleoside triphosphates are interconvertible with diphosphates through nucleoside diphosphate kinase, an enzyme that has a broad specificity. Thus Y and Z (Figure 3.79) can be any of the several purine or pyrimidine ribo- or deoxyribonucleosides.

Cytidine triphosphate is formed from UTP by replacement of the oxygen atom at C-4 by an amino group. In *E. coli* the donor is ammonia, but in mammals the ammonia comes from the amide group of glutamine. In both cases, ATP is required for the reaction (Figure 3.74).

### 3.4.4 Deoxyribonucleotides

Deoxyribonucleotides are formed by the reduction of the corresponding ribonucleotides. The 2′-hydroxyl group of the ribose is replaced by a hydrogen atom in a reaction that takes place at the level of the ribonucleoside 5′-diphosphate. The mechanism is rather complicated. The key enzyme is **ribonucleotide reductase** (ribonucleoside diphosphate reductase) and the electrons required for the reduction of the ribose are transferred from NADPH to sulfhydryl groups at the catalytic site of the enzyme. The enzyme from *E. coli* is a prototype for most eukaryotic reductases. A larger subunit (2 × 86 kDa) binds the NTP substrate and the smaller subunit (2 × 43 kDa) contains a binuclear iron centre and a tyrosyl radical at residue-122. A mechanism based on all the available data is shown (Figure 3.80). The reduction of ribonucleoside diphosphates is controlled by allosteric interactions (an allosteric enzyme is one in which the binding of another substance, usually product, alters its kinetic behaviour) through the use of two allosteric sites that bind a number of nucleoside 5′-triphosphates and lead to a variety of conformations, each with different catalytic properties.

In the event that any dUTP is formed from dUDP, it is rapidly hydrolysed to dUMP by an active dUTPase, which thereby limits the incorporation of dUTP into DNA. Nonetheless, some uracil residues do occur in DNA, which may in part arise through deamination of cytosines. These premutagenic events are repaired by uracil DNA glycosylase (Section 8.11.3). As a result, deoxythymidine 5′-triphosphate (dTTP) is the

$$\text{UMP} + \text{ATP} \xrightleftharpoons{\text{UMP kinase}} \text{UDP} + \text{ADP}$$

$$\text{AMP} + \text{ATP} \xrightleftharpoons{\text{UMP kinase}} 2\ \text{ADP}$$

$$\text{YDP} + \text{ZTP} \xrightleftharpoons[\text{kinase}]{\text{nucleoside diphosphate}} \text{YTP} + \text{ZDP}$$

**Figure 3.79** *Biosynthesis of nucleoside di- and triphosphates*

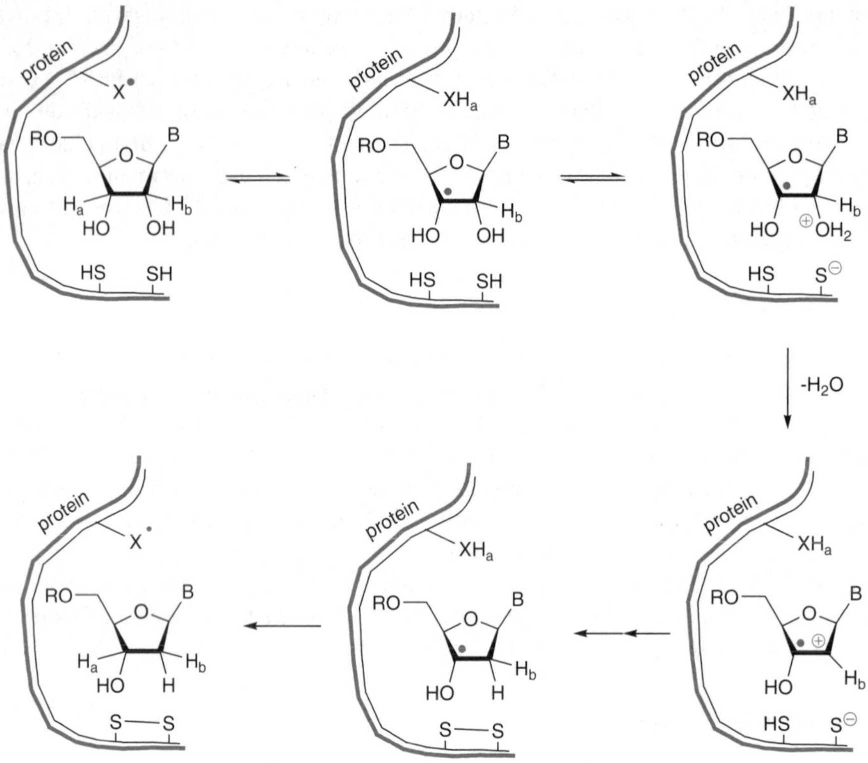


**Figure 3.80** *Postulated mechanism for reduction of nucleotides to deoxynucleotides by E. coli ribonucleotide reductase (X = Tyr$^{122}$–O)*

predominant dioxopyrimidine nucleotide incorporated into DNA. First, deoxythymidine 5′-monophosphate (dTMP) is biosynthesised from dUMP *via* the enzyme **thymidylate synthase**. The methyl group is provided by $N^5$,$N^{10}$-methylenetetrahydrofolate, which also acts as an electron donor (Figure 3.81) and becomes oxidised to dihydrofolate. Tetrahydrofolate is regenerated by dihydrofolate reductase (DHFR) using NADPH as the reductant. These two enzymes are excellent targets for cancer chemotherapy because cancer cells have an increased level of DNA synthesis and, thus, a heavy requirement for dTMP (see 5-fluorouracil and methotrexate in Section 3.7.1).

dTMP is next converted into dTTP in two stages by means of a thymidylate kinase and then a nucleoside diphosphate kinase. In virus-infected cells, the viral thymidine kinase (Section 3.7.2) often also plays the role of a thymidylate kinase.

## 3.5 CATABOLISM OF NUCLEOTIDES

The degradation of nucleotides is of major importance as a target for drug design. RNA is metabolically much more labile than DNA and is constantly being synthesised and degraded. Degradation occurs initially through the action of ribonucleases and deoxyribonucleases, which form oligonucleotides that are further broken down to nucleotides by phosphodiesterases.

Nucleotides are hydrolysed to nucleosides by nucleotidases (and by phosphatases). Of great importance is the final cleavage of nucleosides by inorganic phosphate to base and ribose 1-α-phosphate (or 2-deoxyribose 1-α-phosphate) catalysed by the widely distributed enzyme PNP (Figure 3.82). The ribose phosphate can then be isomerized to ribose 5-phosphate and reused for the synthesis of PRPP. In mammalian tissues, adenosine and deoxyadenosine are resistant to the phosphorylase. AMP is therefore first deaminated by adenylate

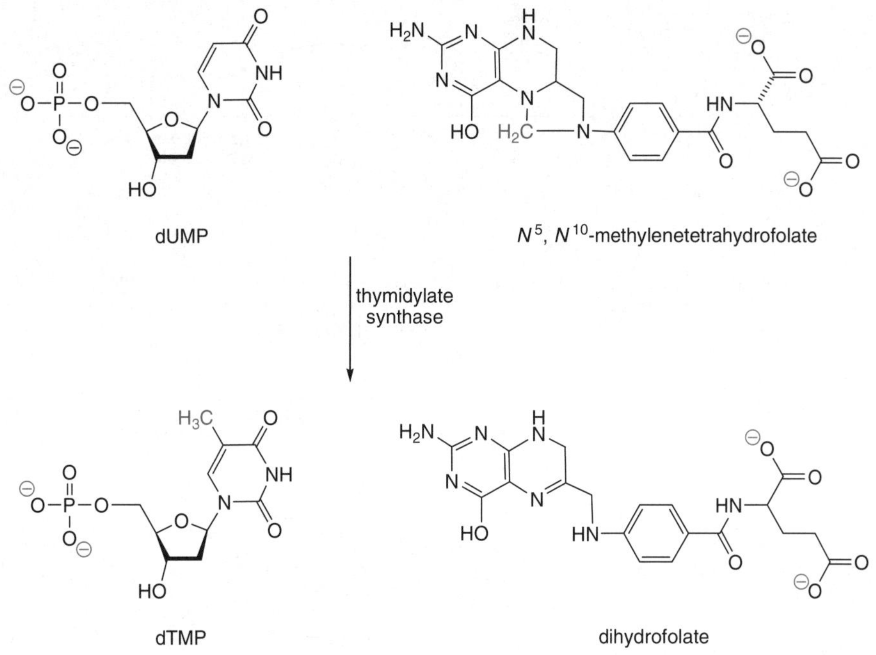


**Figure 3.81** *Formation of dTMP from dUMP*

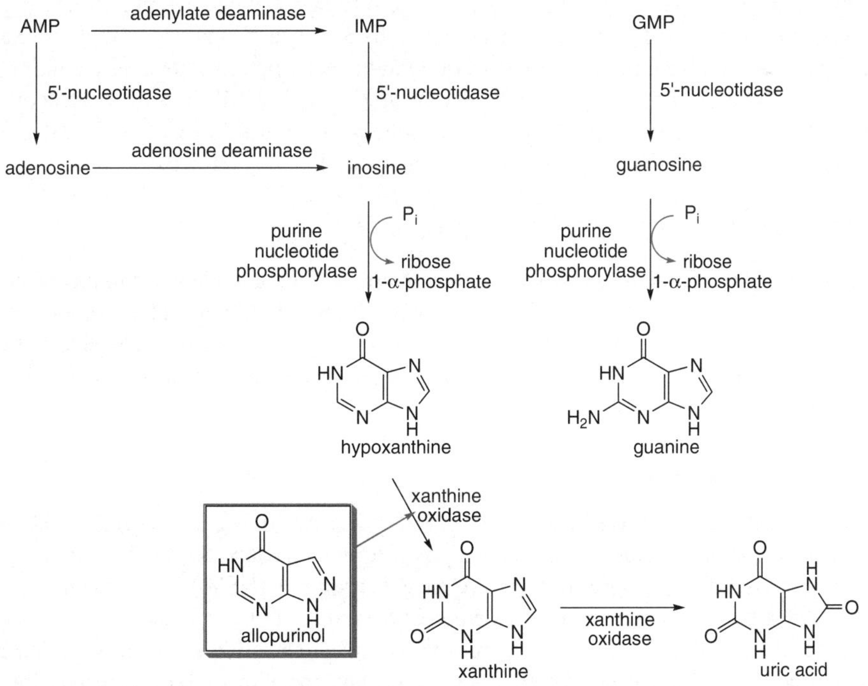


**Figure 3.82** *Catabolism of purine nucleotides*

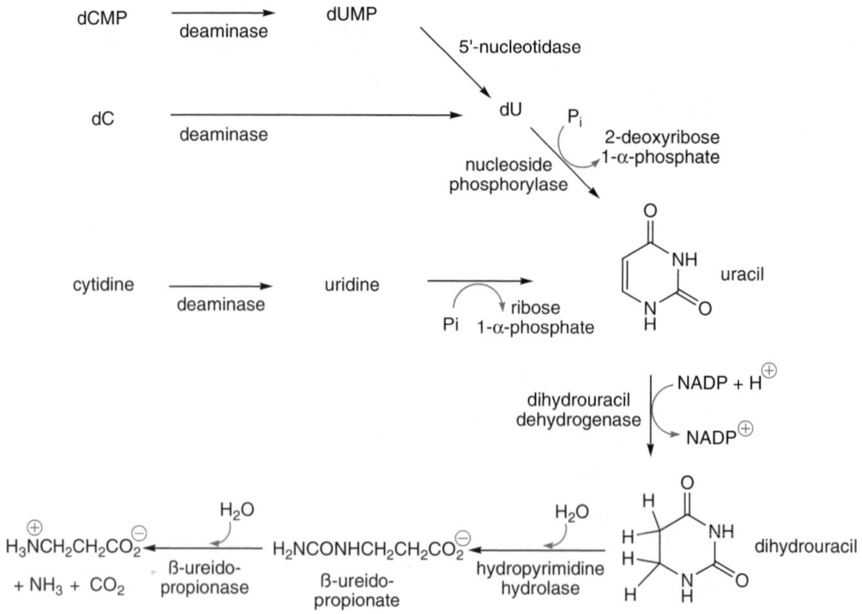


**Figure 3.83** *Catabolism of pyrimidine nucleotides*

deaminase to IMP and adenosine is deaminated to inosine by adenosine deaminase, an enzyme that is thought to be present in elevated levels in leukaemic cells. Oxidation of hypoxanthine is catalysed by xanthine oxidase to give xanthine, which is also the deamination product of guanine. Xanthine is further oxidised to uric acid, which in humans is excreted in the urine. Gout is a painful disease caused by the excessive production of monosodium urate, which is deposited as crystals in the cartilage of joints. **Allopurinol** is an analogue of hypoxanthine that provides an effective treatment of gout by acting as a substrate inhibitor of xanthine oxidase. Since the allopurinol becomes irreversibly bound to the enzyme it is known as a **suicide inhibitor**.

A variety of deaminases convert cytidine, 2′-deoxycytidine and dCMP into the corresponding uracil-containing derivatives. All of these products can be hydrolysed to uracil, which is then degraded reductively (Figure 3.83). Thymine is degraded in a way exactly analogous to uracil.

## 3.6 POLYMERISATION OF NUCLEOTIDES

While the complex series of reactions involved in the polymerisation of nucleotides to form DNA and RNA are described in detail in Chapter 6, we are here primarily concerned with the polymerases as potential targets for chemotherapy, as they are the enzymes responsible for polymerisation of nucleoside 5′-triphosphates into nucleic acids. In each case there is a requirement for a template strand of nucleic acid and an oligoribo- or oligodeoxyribo-nucleotide primer.

### 3.6.1 DNA Polymerases

All cellular polymerases use DNA as a template and polymerise in a 5 → 3 direction (Section 6.6). While polymerases are potential targets for cancer chemotherapy (*e.g.* for intercalators, Section 9.6), much greater scope is available for anti-viral therapy since many viruses (*e.g.* herpesvirus (HSV)) encode their own DNA polymerases, which often have substrate specificities different from those of the cellular enzymes (Section 3.7.2).

One group of RNA-containing viruses, the retroviruses, replicates *via* a double-stranded DNA intermediate. Retroviruses are important since many cause cancer and one of them, **HIV**, is responsible for the disease AIDS. Its RNA genome is first transcribed into DNA by an RNA-dependent DNA polymerase, also known as a **reverse transcriptase** (RT) (Section 6.4.6). In contrast to the cellular polymerases, these RTs are unique to retroviruses and they are also tolerant to a wide range of nucleoside triphosphate analogues, which identifies them as targets for chemotherapy (Section 3.7.2).

### 3.6.2 RNA Polymerases

DNA is transcribed into RNA by RNA polymerases (Sections 6.6.2 and 10.7.2). Several antibiotics are highly potent inhibitors of transcription. **Actinomycin D** is an intercalator that binds tightly and selectively inhibits ribosomal RNA chain elongation. In contrast, rifampicin interacts directly with one of the subunits of RNA polymerase and inhibits initiation of RNA synthesis. *Cis*-Diamminedichloroplatinum (II) (**cisplatin**) has strong anti-tumour activity. It cross links two adjacent guanines present in the same DNA strand at their N-7 positions and interferes with transcription (Section 6.6).

Some viruses encode their own RNA-dependent RNA polymerase. These are also potential targets for chemotherapy, as these enzymes are generally specific for viral RNA. For example, 2′-*C*-methylnucleoside derivatives are terminators of hepatitis C virus (HCV) RNA polymerase and are in clinical trials for HCV treatment.

## 3.7 THERAPEUTIC APPLICATIONS OF NUCLEOSIDE ANALOGUES

At the beginning of the twenty first century, all countries faced the scourges of cancer and of many viral and parasitic diseases. Nucleoside analogues form a substantial core of the clinician's armoury against viral infections and cancer. In the remainder of this chapter we will examine the modes of action of these compounds and also some important non-nucleoside drugs, and assess rational design for anti-cancer and anti-viral therapy.

In the anti-cancer field, our knowledge of metabolic differences between normal and cancer cells is growing, particularly for those proteins that are altered in pattern of regulation during oncogenesis. We also have a better understanding of chromosomal translocations that cause cancers. This is encouraging the development of chemical agents that specifically target cancer cells and, for example, trigger apoptosis (programmed cell death).

Anti-viral chemotherapy has made particularly good progress in the past decade and there are now over twice the number of new anti-viral agents in the clinic since the publication of the second edition of this book. The need for viable anti-cancer and anti-viral chemotherapy is huge and will remain so for some time, not least because of the limitations in the use of anti-viral vaccines. We can take advantage of the understanding of the metabolic pathways of normal, virus-infected and cancer cells gained earlier in this chapter to study the role of many anti-cancer and anti-viral drugs.

### 3.7.1 Anti-Cancer Chemotherapy

Most anti-cancer drugs act by inhibiting DNA synthesis in some way. They exhibit a greater toxicity for faster growing tissues such as bone marrow, gastrointestinal epithelium, hair follicles and gonadal tissue. Many of these drugs cause nausea and vomiting – especially a problem with the alkylating agents and cisplatin. The majority of drugs target DNA directly: many antibiotics form physical complexes that inhibit polymerases and topoisomerases (Section 9.10.2), or generate covalent interactions with DNA and RNA (Section 8.7) or a combination of both of these activities. Some drugs are alkylating agents that react covalently with DNA (Section 8.10). The antimetabolites are an important class of agents designed to impede the supply of monomers for DNA biosynthesis and so arrest cell division. Finally, several alkaloid drugs act by interfering with the cell cycle – as for example with the formation of tubulin or topoisomerases. The use of combinations of these different types of drug in cancer chemotherapy has been a major advance in this field and offers several important advantages.

- Decreased incidence of resistance
- A greater than additive or synergistic effect of the drugs
- Use of drugs with different types of toxic effects reduces overall toxicity or at least the toxicity to any one system.

The following section will focus on antimetabolites while parts of Chapters 8 and 9 will deal with other chemotherapeutic agents. The limited selectivities shown for cancer cells in the examples that follow are sometimes based on slightly different transport properties between cell types, or perhaps on a salvage pathway that is working at a higher level or the change in pH or oxygen tension due to the rapid metabolism of a tumour cell.

*3.7.1.1 Antimetabolites.*[120,122,123] Antimetabolites are structural analogues of naturally occurring compounds that interfere with the production of nucleic acids. They work through a variety of mechanisms including competition for binding sites on enzymes and incorporation into nucleic acids. Antimetabolites inhibit the growth of the most rapidly proliferating cells in the body (*e.g.* bone marrow, G.I. tract, *etc.*). There are three categories of antimetabolites: purine analogues, pyrimidine antimetabolites and antifolates.

**3.7.1.1.1 Thiopurines.** The purine analogues **6-mercaptopurine** and **6-thioguanine** (Figure 3.84) are used in cancer chemotherapy, particularly against childhood acute lymphoblastic leukaemia. These drugs are analogues of hypoxanthine and guanine, respectively. These antipurines can inhibit nucleotide and nucleic acid synthesis, can be incorporated into nucleic acid and can sometimes do both. Most studies indicate that the thiopurines work at multiple sites and that their mechanism of action is a result of combined effects at these different sites. Their biological activity relies on their conversion into the corresponding nucleoside 5′-triphosphates by the salvage enzyme HGPRT. This causes feedback inhibition of amidophosphoribosyl transferase in the synthesis of 5-phosphoribosylamine from PRPP (Figure 3.71) and also prevents IMP being converted into XMP and adenylosuccinate (Figure 3.73). The mononucleotide derivatives are ultimately converted into triphosphates, which can be incorporated into RNA and DNA.

**3.7.1.1.2 Deoxyadenosine Analogues. Cladribine** and the more soluble **fludarabine** (Figure 3.85) are used in the treatment of hairy-cell leukaemias (HCL) and chronic lymphocytic leukaemia (CLL), respectively. Both compounds are typically given intravenously, which in the case of the former compound leads to rapid dephosphorylation by serum phosphatases. Following cellular uptake by the target cells, the free nucleosides are phosphorylated by deoxycytidine kinase and further phosphorylated by cellular kinases to the triphosphates, which are then incorporated in DNA. Once incorporated into DNA, both compounds lead to chain termination of DNA replication and ultimately to cell death. Furthermore, the triphosphates are known to inhibit ribonucleotide reductase, thereby reducing the available pool of natural

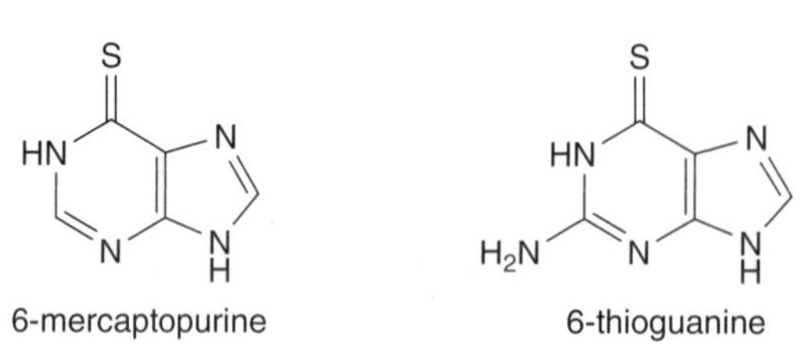


**Figure 3.84** *Anti-cancer drugs 6-mercaptopurine and 6-thioguanine*

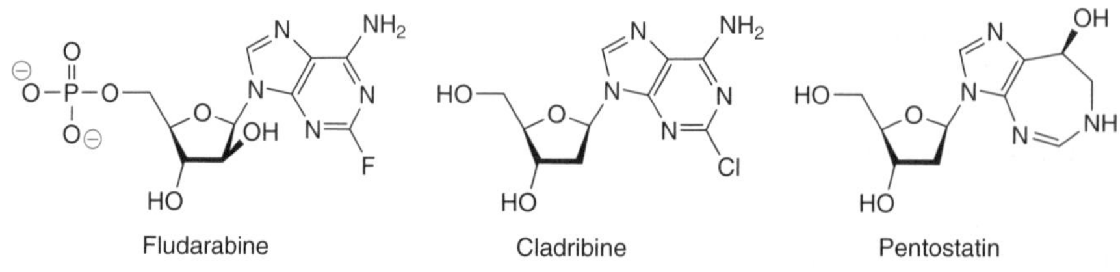


**Figure 3.85** *Anti-cancer drugs based on adenosine*

dNTPs required for DNA synthesis. However, the mechanisms by which these drugs achieve selectivity for their target cells are not clearly understood.

Fludarabine in combination with the alkylating agent cyclophosphamide (see Section 3.7.1.2) has been highly successful in clinical treatments for CLL. Clinical trials of the nucleoside analogue pentostatin (Figure 3.85) together with **cyclophosphamide** are currently in progress. Cytarabine (cytosine arabinoside or *ara*-C) is an analogue of 2′-deoxycytidine in which the 2′-hydroxyl is sterically inverted (Figure 3.85). It is used primarily for the treatment of acute myelocytic leukaemia. *Ara*-C is first converted into its monophosphate (*ara*CMP) by deoxycytidine kinase. The monophosphate then reacts with appropriate kinases to form the di- and triphosphate nucleotides. *Ara*CTP is believed to be the key active component and its accumulation causes potent inhibition of DNA synthesis in many cells. While this nucleotide is a competitive inhibitor of many DNA polymerases, it is also a substrate for some DNA polymerases. It thereby becomes incorporated into DNA, leading to inhibition of chain elongation. Unlike many antimetabolites, the effects of *ara*-C are directed almost exclusively towards DNA and it has little or no effect on RNA synthesis or function. Some evidence indicates that the inhibition of synthesis is secondary to incorporation of *ara*CMP into DNA.

**3.7.1.1.3 Fluorouracil.** 5-Fluorouracil (5FU) must first be converted into the nucleotide to be active as a cytotoxic agent. The 5′ ribonucleotide (5FUMP) is formed *via* several different pathways. 5FUMP is then incorporated into RNA and is also converted into the deoxynucleotide (FdUMP) by ribonucleotide reductase. FdUMP is also be formed by the direct phosphorylation of FdUrd by thymidine kinase.

The formation of FdUMP is crucial for the cytotoxicity of 5FU. This is because FdUMP inhibits the enzyme thymidylate synthetase and so blockades the formation of dTTP, one of the four essential constituents of DNA. Thymidylate synthetase catalyses the methylation of dUMP in a multi-step process that involves formation of a ternary complex between dUMP, methylenetetrahydrofolate and the enzyme. This complex reacts further by loss of a proton from position-5 of the uracil ring. FdUMP (Figure 3.86) also forms such a ternary complex but its breakdown would require loss of a fluorine cation. Thus the complex is sufficiently stable so that the enzyme cannot turnover. DNA synthesis is thus inhibited until the drug is removed and *de novo* enzyme synthesis begins. dFUMP is thus a suicide inhibitor. Dan Santi's extensive work on this intermediate has shown that a cysteine residue in the enzyme is the nucleophile that adds to position-6 of the pyrimidine ring (Figure 3.86).

As FdUrd is converted into FdUMP directly in a single step, it is a potent inhibitor of dTMP synthetase and is often effective in the low nanomolar concentration range. On the other hand 5FU, though less expensive, is only effective at micromolar concentration where further active metabolites are formed including 5FUTP. This can be incorporated into RNA in place of UTP and it affects the function of both rRNA and mRNA. Although 5FU and some of its pro-drugs are widely used in the treatment of common solid tumours, they show little selectivity and are therefore toxic, causing suppression of the immune system.

**3.7.1.1.4 Antifolates[124]–Methotrexate.** The importance of folates in tumor cell growth was demonstrated by Farber in 1948, when aminopterin was shown to produce remissions in leukaemia. Antifolates produced both the first striking remissions in leukaemia and the first cure of a solid tumour, choriocarcinoma.

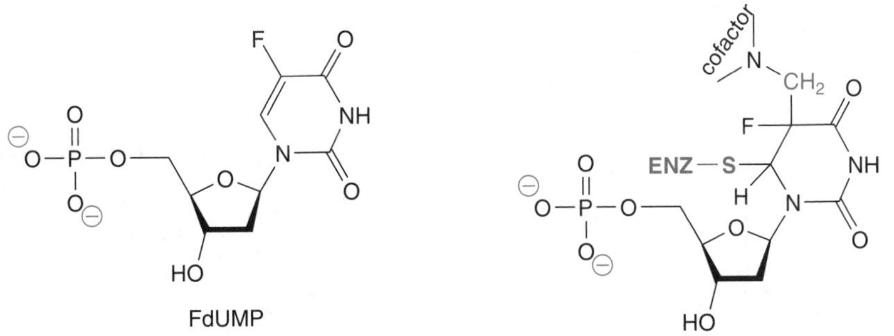


**Figure 3.86** *Structure of FdUMP (left) and its suicide complex with thymidylate synthetase (right)*

Although aminopterin was the first clinically useful folate, methotrexate was soon introduced in therapy and it has become the major folate used in cancer therapy (Figure 3.87).

Folic acid is an essential growth factor that leads to a series of tetrahydrofolate cofactors that provide one-carbon groups for the synthesis of RNA and DNA precursors, such as thymidylate and purines. Folic acid is reduced in two successive steps by DHFR using NADPH to give tetrahydrofolate, which is the active form. Thus, the enzyme DHFR is the primary site of action of most folate analogues such as methotrexate. Methotrexate has a high affinity for the tumor cell DHFR and so blocks the formation of tetrahydrofolate needed for thymidylate and purine biosynthesis. Cell death probably results from inhibition of DNA synthesis. Methotrexate is only partially selective for tumour cells and is toxic to all rapidly dividing normal cells, such as those of the gastrointestinal epithelium and bone marrow.

### *3.7.1.2 Alkylating Agents*

**3.7.1.2.1 Cyclophosphamide.** Many anti-cancer alkylating agents are both mutagenic and carcinogenic, but are used in chemotherapy under controlled circumstances.[125] One example, cisplatin, is thought to cross link DNA (Section 8.5.4) though its selectivity for tumour cells is not understood. Cyclophosphamide is a masked nitrogen mustard and is one of the clinically most useful drugs. It can be administered orally or parenterally. It is thought to cross-link DNA and interferes with replication. Cyclophosphamide was originally designed to be preferentially activated in tumour tissues, as they were believed to contain elevated levels of phosphatases. It is now known that this does not happen, but the drug does undergo metabolic activation in the liver catalysed by cytochrome P-450 microsomal enzymes. The nitrogen mustard metabolite formed is active as a cytotoxic agent while a second metabolic product, acraldehyde, is believed responsible for cystitis, a side effect caused by cyclophosphamide. Ifosphamide, a structural analogue of cyclophosphamide, is also metabolized by the cytochrome P-450 system and undergoes transformations similar to those for cyclophosphamide. However, a smaller proportion of ifosphamide is converted into undesirable products and thus larger doses can be used clinically as compared to cyclophosphamide (Figure 3.88).

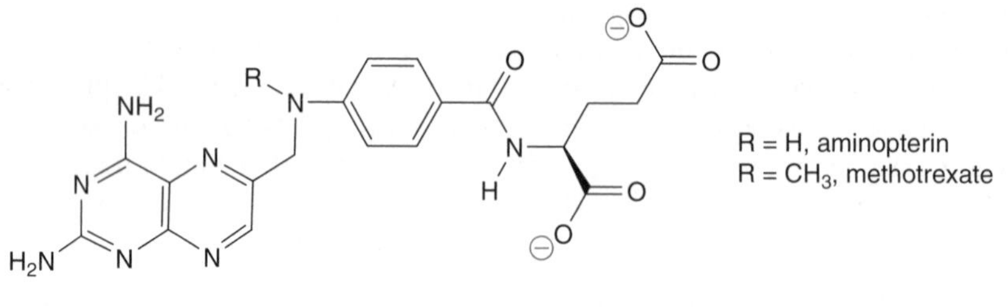


**Figure 3.87** *Folic acid analogues*

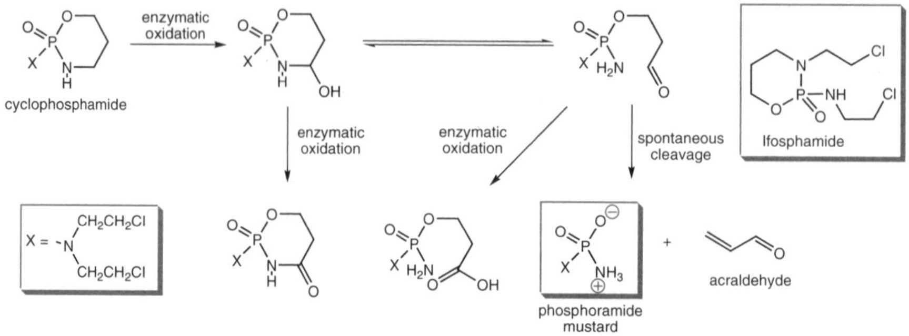


**Figure 3.88** *Metabolic activation and deactivation of cyclophosphamide. Structure of Ifosphamide*

Other important alkylating agents used in cancer treatment include the nitrosoureas carmustine (BCNU) and lomustine (CCNU) which alkylate the O-6 and N-1 positions of guanine bases to give $N^1,O^6$-ethenoguanine (Figure 3.89),[126] which reacts to give an interstrand cross link with cytosine. Other alkylating agents such as temozolomide (Figure 3.89) alkylate the O-6-position of guanines, which causes G→A mutations during DNA replication. $O^6$-Alkylguanine DNA alkyltransferase is a protein that can repair such DNA damage and is expressed at elevated levels in some tumours. As a result, it is a key target for inhibition by a number of compounds.[126] Both $O^6$-benzylguanine and $O^6$-(4-bromothenyl)guanine (Figure 3.89) have been used clinically in combination with chemotherapeutic regimes that employ alkylating agents.

In all the above compounds, the therapeutic target is inhibition of DNA replication, whether by preventing synthesis of precursors or by alkylating the DNA itself. As most of the effective drugs are also toxic, patients receiving cancer chemotherapy are unusually susceptible to viral and bacterial infections.

## 3.7.2 Anti-Viral Chemotherapy[127]

The life cycle of a virus involves a combination of its own enzymes and those of the host cell. Thus the design of anti-viral agents can be more directly targeted than that of anti-cancer agents. Since most virus classes are unrelated to each other and have unique replication cycles, there appears little chance for the discovery of wide-activity anti-viral agents comparable to the broad-spectrum antibiotics, such as the β-lactams. Different approaches have been successful for retroviruses such as HIV and DNA viruses such as HSV and hepatitis B virus (HBV).

*3.7.2.1 Retrovirus Inhibitors.* Much effort has been expended on finding a chemotherapeutic agent to alleviate the symptoms of AIDS. The HIV is a member of the lentivirus family (a sub-class of retrovirus) and its reverse transcriptase has been an obvious target. Virtually all the compounds currently used for the treatment of HIV infections, or in advanced research, belong to one of four main classes:

- nucleoside/nucleotide reverse transcriptase inhibitors (NRTIs);
- non-nucleoside reverse transcriptase inhibitors (NNRTIs);
- protease inhibitors (PIs); and
- host-cell receptor based therapeutic agents.

**3.7.2.1.1 Nucleoside Reverse Transcriptase Inhibitors.** There are around 40 anti-viral compounds in clinical use (Table 3.2), with over half of them being used in the treatment of HIV patients. The majority are nucleoside analogues effective against HIV or HSV infections. The key structural feature of a nucleoside analogue as a chain terminator for RT is the absence of a 3′-OH function. Thus after incorporation

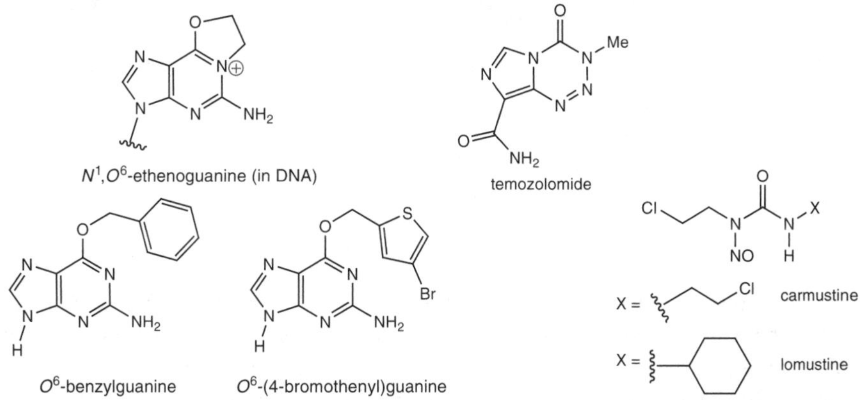


**Figure 3.89** *Structures of $N^1,O^6$-ethenoguanine, $O^6$-benzylguanine, $O^6$-(4-bromothenyl)guanine and various DNA alkylating agents*

**Table 3.2** *Anti-viral compounds in clinical use*

| *Drug* | *Virus* |
|---|---|
| Amantadine, Rimantadine | Influenza A |
| Oseltamivir, Zanamivir | Influenza A and B |
| Zidovudine (AZT), Retrovir® | HIV |
| Didanosine (ddI), Videx® | HIV |
| Zalcitabine (ddC), Hivid® | HIV |
| Stavudine (d4T), Zerit® | HIV |
| Lamivudine (3TC), Epivir®, Zeffix® | HIV and HBV |
| Abacavir, Ziagen® | HIV |
| Emtricitabine, Emtriva® | HIV and HBV |
| Tenofovir disoproxil (oral prodrug of PMPA) | HIV and HBV |
| Nevirapine, Delavirdine, Efavirenz (non-nucleoside RT inhibitors NNRTIs) | HIV |
| Saquinavir, Ritonavir, Indinavir, Nelfinavir, Amprenavir, Lopinavir, Atazanavir | HIV protease inhibitors |
| Enfuvirtide | HIV viral entry inhibitor |
| Adefovir dipivoxil (oral prodrug of PMEA) | HIV and HBV |
| Acyclovir (ACV), Zovirax® | Herpes Simplex Virus (HSV-1 and HSV-2) and Varicella Zoster Virus (VZV) |
| Valaciclovir, Zelitrex®, Valtrex® oral prodrug of acyclovir) | HSV-1, HSV-2 and VZV |
| Penciclovir, Denavir®, Vectavir® | HSV-1, HSV-2 and VZV |
| Famciclovir, Famvir® | HSV-1, HSV-2 and VZV |
| Idoxuridine, Herpid®, Idoxene®, *etc.* | HSV-1, HSV-2 and VZV |
| Trifluridine (TFT), Viroptic® | HSV-1, HSV-2 and VZV |
| Brivudin (BVDU), Zostex®, Zerpex®, *etc.* | HSV-1 and VZV |
| Ganciclovir (DHPG), (GCV), Cymevene®, Cytovene® | HSV-1 and HSV-2 cytomegalovirus (CMV) |
| Valganciclovir (VGCV), Valcyte® | HSV-1, HSV-2 and CMV |
| Foscarnet, Foscavir® | HSV-1, HSV-2, VZV, CMV and HIV |
| Cidofovir (HPMPC), (CDV), Vistide® | Herpes viruses, papilloma-, polyoma-, adeno- and poxviruses |
| Fomiversen, Vitravene® | CMV retinitis |
| Ribavirin | Influenza A and B, measles, respiratory syncytial virus (RSV) and adenovirus |

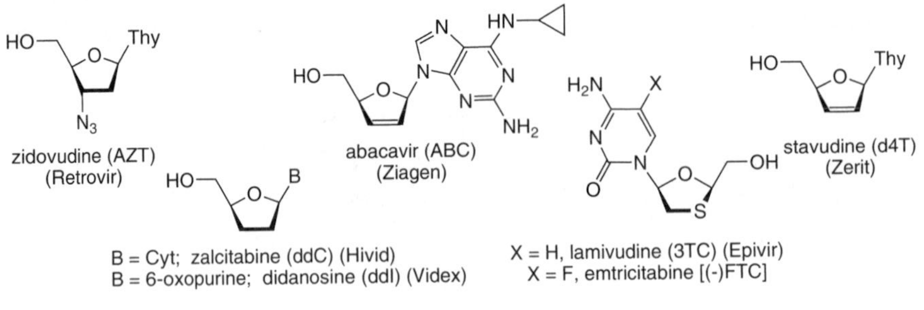


**Figure 3.90** *Structures of some licensed nucleoside-based reverse transcriptase inhibitors*

at the growing 3′-end, this moiety cannot become phosphorylated by the next dNTP monomer. One of the first compounds to be widely used was AZT (Figure 3.90) as it was found to be the least toxic of several analogues in clinical use. A range of effective NRTIs is shown in Figure 3.90.

The active species for all nucleoside analogues of this type is the 2′-deoxynucleoside 5′-triphosphate. Since the retrovirus does not encode its own kinase and because it is difficult to get highly anionic nucleotides into cells, it is necessary for the native human cellular thymidine kinase to perform the initial phosphorylation steps within the infected cell. Sometimes a non-specific enzyme (*e.g.* pyruvate kinase) can perform this task. Mammalian cellular kinases are usually highly selective in their substrate requirements and this greatly restricts the acceptability of potential chain terminator nucleosides. AZT has only a small structural modification and is a good substrate for the host kinase. Thus its triphosphate is available as a substrate for the viral RT and so causes chain termination. Unfortunately, the high dose levels needed for AZT (around 1 g per day) give rise to considerable host toxicity. This is probably because the triphosphate of AZT is to some extent a substrate for the DNA polymerase of the host cell, and thus can contribute to the observed toxic side effects of the drug, including bone marrow suppression.

A number of other nucleosides, including dideoxynucleoside analogues, such as **2′,3′-dideoxyinosine** (**ddI**, Figure 3.90) and **2′,3′-dideoxycytidine** (**ddC**, Figure 3.90), are also approved for the treatment of HIV infection. They share the same problems of toxicity and requirement for phosphorylation by host enzymes with AZT. The use of **nucleotide prodrugs** has greatly improved the efficacy of these and other nucleoside analogues that are not good substrates for phosphorylation *in vivo*.[128] Such prodrugs are non-ionic, which aids their cellular uptake, and they are converted enzymatically or spontaneously into their monophosphates after entering the cell. Phosphorylation to the bioactive triphosphate forms then follows.

The use of **combination therapy** in the treatment of HIV is becoming increasingly common to combat problems of drug resistance. This typically involves the use of AZT together with a second anti-HIV compound such as lamivudine (Figure 3.90). The rationale for this approach is that the use of a combination of drugs that are synergistic and have no overlapping toxicity can reduce toxicity, improve efficacy and prevent drug resistance from arising.

The prolonged use of AZT in AIDS patients leads to the development of drug-resistant HIV strains because the drug is not 100% effective in killing the virus and mutants resistant to AZT survive and proliferate. Single mutations at residue-184 of the RT in HIV cause high-level resistance to 2′,3′-dideoxy-3′-thiacytidine (3TC, lamivudine, Figure 3.90) that is an important component of triple-drug anti-AIDS therapy. Such mutations contribute to the failure of anti-AIDS combination therapy.

Considerable progress is being made in understanding the nature of drug resistance through analysis of X-ray structures of wild-type and mutated HIV RT complexed with a nucleotide drug and DNA. Arnold[129] has determined crystal structures of the 3TC-resistant mutant HIV-1 RT (M184I) in both the presence and absence of a DNA/DNA template-primer. In the absence of a DNA substrate, the wild-type and mutant structures are very similar. However, comparison of structures of M184I mutant and wild-type HIV-1 RT with and without DNA shows that the template-primer is repositioned in the M184I/DNA binary complex and there are other smaller changes in residues in the dNTP-binding site. These structural results support a model that explains the ability of the 3TC-resistant mutant M184I to incorporate dNTPs but not the nucleotide analogue 3TCTP. The same model can also explain the 3TC resistance of analogous hepatitis B polymerase mutants.

**3.7.2.1.2 Non-Nucleoside Reverse Transcriptase Inhibitors.**[130] The structure of HIV-1 RT has been solved by X-ray crystallography (Figure 10.30). The active form of the enzyme is a heterodimer having one polymerase active site and one RNaseH active site. Several potent and specific inhibitors of HIV-1 RT were discovered in the early 1990s that are not nucleosides (NNRTIs) and probably do not require kinase metabolism to generate an active form (Figure 3.91). One of them, nevirapine (Figure 3.91c), has been co-crystallised with the transcriptase and its binding site on the enzyme is seen to be a hydrophobic pocket guarded by two tyrosine residues close to the polymerase active site. Thus NNRTIs result in allosteric inhibition of the enzyme rather than the competitive inhibition that results from the nucleoside-based inhibitors binding to the active site and so binding of an NNRTI can inhibit reverse transcription directly.

Although, the structures of these non-nucleoside inhibitors are very diverse, they are all believed to bind in a similar (though not identical) location. They show little toxicity and have a very high anti-HIV activity in cell culture. However, HIV-1 rapidly becomes resistant to these drugs, in most cases, owing to the selection of strains containing an RT mutated at one or both of the tyrosines, while retaining infectivity. In addition to nevirapine, two other NNRTIs are also in clinical use are – delavirdine and efavirenz (Figure 3.91).

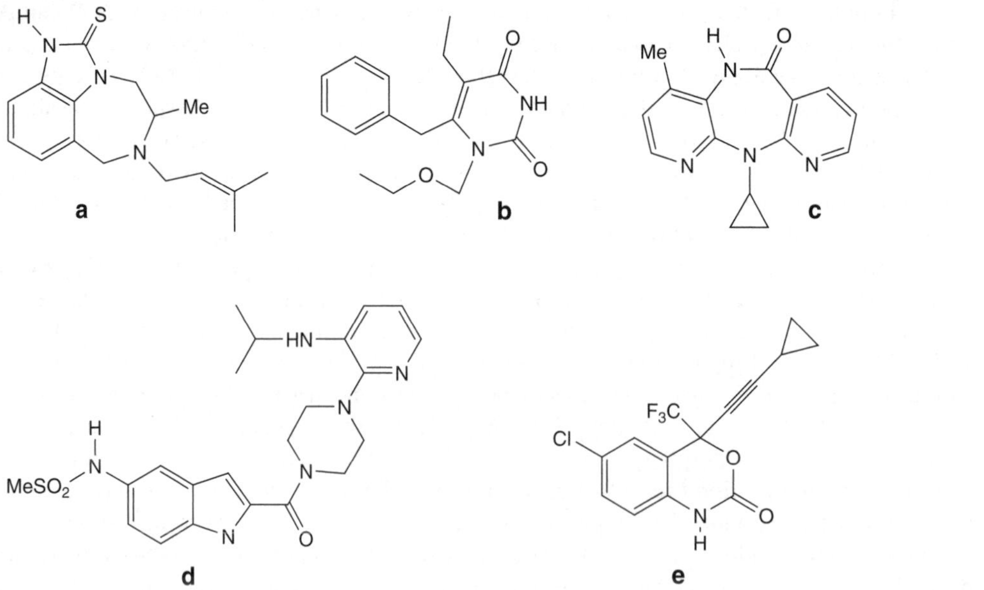


**Figure 3.91** *Non-nucleoside HIV reverse transcriptase inhibitors: (a) TIBO class; (b) HEPT class; (c) Nevirapine; (d) Delavirdine; (e) Efavirenz*

Large randomized and cohort studies on asymptomatic patients have demonstrated that non-nucleoside reverse-transcriptase inhibitors are at least as effective as protease inhibitors as part of initial triple-drug therapy. The effectiveness of specific highly active antiretroviral therapy (HAART) combinations has provided support for the use of triple therapy with zidovudine, lamivudine (Combivir) and efavirenz (ZDV/3TC/EFV).[131]

Various new NRTIs and NNRTIs have been developed that possess improved metabolic characteristics (*i.e.* phosphoramidate and *cyclo*saligenyl pronucleotides by-passing the first phosphorylation step of the NRTIs) or increased activity ('second' or 'third' generation NNRTIs (*i.e.* TMC-125, DPC-083)) against those HIV strains which are resistant to the 'first' generation NNRTIs.

**3.7.2.1.3 Protease Inhibitors.** The HIV protease is a vital enzyme in the HIV life cycle that cleaves the transcribed *gag-pol* protein into three HIV polypeptides: the RT, integrase and *gag* polypeptides. Six HIV protease inhibitors have been approved of which saquinavir and ritonavir are two widely used analogues. The use of combination therapies against HIV that employ two nucleoside-based RT inhibitors together with a PI agent are now proving to be highly effective, as for example, the combination of zidovudine (AZT) and stavudine (d4T) along with a PI such as indinavir. It appears that triple-drug regimens containing two NRTIs with a PI, a NNRTI or a third NRTI may provide comparable activity.[132]

**3.7.2.1.4 Host-Cell Receptor Based Therapeutic Agents.** A virus initiates infection by attaching to a specific receptor on the surface of a susceptible host cell. Agents that inhibit such virus–receptor interactions in the case of HIV and its CD4 receptor on T lymphocytes are in clinical trial for AIDS. However, approaches of this type do not involve the use of nucleoside or nucleotide analogues and will not be discussed further.

Finally, we have to recognise that host-cell toxicity is a major problem. Many nucleoside drugs are directed at enzymes (such as polymerases) that have counterparts in host cells and it is often inevitable that they may interfere with a fundamental biochemical pathway in a normal cell. Because such toxicity is frequently unacceptable for non-life-threatening disease states, extensive testing is inevitable, time consuming, and inordinately expensive. In contrast, 'fast-tracking' of new candidate drugs is available for life-threatening

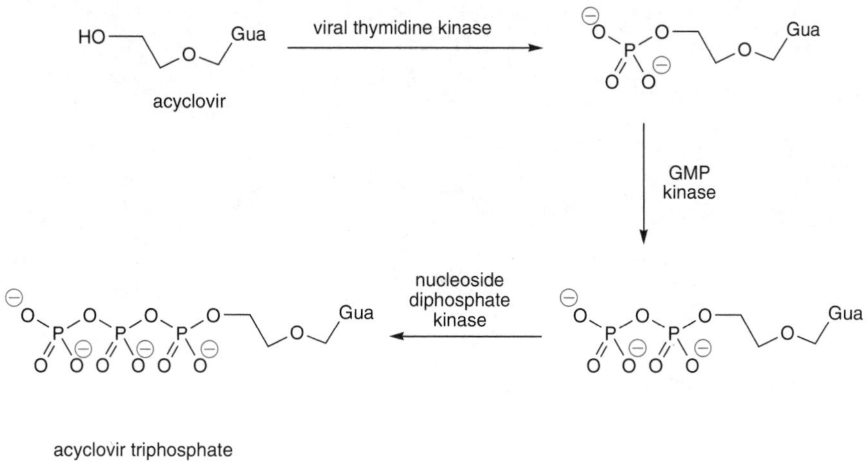


**Figure 3.92** *Enzymatic phosphorylation of acyclovir*

viral infections, such as HIV. For example, the anti-AIDS drug AZT (Figure 3.90) was marketed within a year of the report of its *in vitro* properties.

*3.7.2.2 Anti-DNA Virus Drug Design.* Herpesviruses are double-stranded DNA viruses that cause a variety of diseases in humans: cold sores, eye infections (keratitis), genital sores, chickenpox, shingles and glandular fever (infectious mononucleosis). They all exhibit latency, which means that following infection of a cell, the virus produced can go into a latent state in nerve endings from where it can be reactivated by various stimuli (stress, UV light, other viral infections, *etc.*). Since it has so far been impossible to destroy the virus in the latent state (*i.e.* prevent it from replicating), antiviral chemotherapy must be directed first against primary infection and then against subsequent recurrent episodes.

Herpesviruses code for many enzymes involved in their own replication and metabolism. These are sufficiently different from the corresponding ones in the host cell to give an opportunity for selective interference. For example, HSVs rely largely on the salvage pathway for the production of dTTP for DNA synthesis and so the viruses encode their own thymidine kinase, TK. The specificity of the viral TK is not as great as that of the host cell and it phosphorylates a wide range of nucleoside analogues that, once activated, inhibit viral replication. HSVs also code for their own DNA polymerase, which has a different specificity from the cellular polymerases and hence presents a target for selective attack.

**3.7.2.2.1 Acyclovir and Related Acyclonucleosides.**[133] Acyclovir (Figure 3.92) is effective against HSVs, and its metabolic conversion into the active form is remarkable. Although acyclovir is a purine nucleoside analogue lacking C-2′ and C-3′ of the sugar ring, it is specifically phosphorylated at the position equivalent to the 5′-hydroxyl group by the thymidine kinase of the HSV. Not surprisingly, no such metabolism occurs in an uninfected cell. However, in the virally infected cell, 5′-phosphorylated acyclovir is now recognised by the host-cell guanylate kinase and is taken to the diphosphate, from which a nucleoside diphosphate kinase produces the 5′-triphosphate. This is now a substrate for the HSV-encoded DNA polymerase and it is incorporated into viral DNA. Since the analogue has no 3′-hydroxyl group, it is a chain terminator and thus stops the synthesis of viral DNA (Figure 3.92).

Developments aimed at enhancing the oral bioavailability of acyclovir have resulted in the discovery of valaciclovir (Figure 3.93), the L-valyl ester of acyclovir. This acts as prodrug of the parent nucleoside and, as it has increased solubility, up to threefold higher plasma levels of acyclovir can be achieved.

The acyclonucleoside **ganciclovir** (Figure 3.93) is active against both HSV and cytomegalovirus (CMV) infections. In the latter case, the initial phosphorylation is carried out by a CMV-encoded protein kinase.

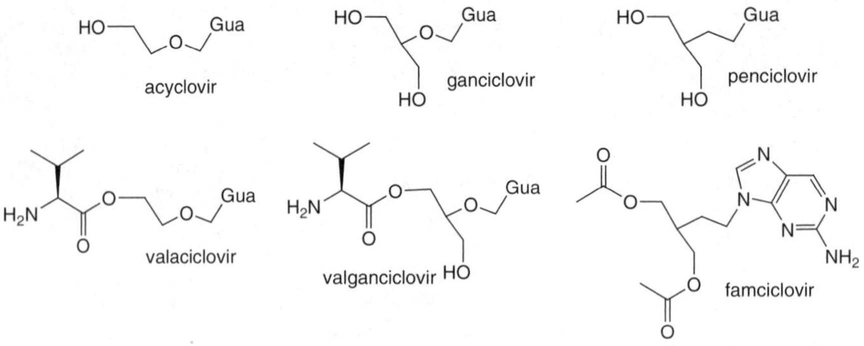


**Figure 3.93** *Structures of acyclonucleosides used as anti-viral agents*

The inconvenience of the need to administer this drug intravenously for CMV infections can also be overcome by using the L-valyl ester prodrug, **valganciclovir** (Figure 3.93) that can be administered orally.

**Penciclovir** (Figure 3.93) is also used as an anti-herpes drug, particularly for recurrent cold sores and has the same mechanism of action as the other acyclonucleosides. However, penciclovir has even poorer oral bioavailability than acyclovir. Fortunately an oral prodrug, **famciclovir** (Figure 3.93), has been developed, which is converted into penciclovir through oxidation by xanthine oxidase in the gut, followed by removal of the acetyl groups by esterases in the liver.

**3.7.2.2.2 Nucleotide Analogues.** The analogue **(*S*)-9-(3-hydroxy-2-phosphonylmethoxypropyl) adenine (HPMPA)** (Figure 3.94), discovered by the groups of Hóly and DeClercq in the 1980s, was the first nucleotide analogue to show antiviral properties. Deletion of the 5′-oxygen creates a phosphonic acid analogue of a nucleoside 5′-phosphate that is stable to nucleotidases. **Tenofir disoproxil** (Figure 3.94) is a non-ionic oral prodrug that is converted into its bioactive form (an analogue of HPMPA) following hydrolysis *in vivo* by carboxyl esterases and spontaneous cleavage of the phosphonate esters. It is currently used in the treatment of HIV. **Adefovir dipivoxil** (Figure 3.94), a prodrug of **PMEA**, is used for the treatment of chronic hepatitis B infections; while the phosphonate derivative **cidofovir** (Figure 3.94), is licensed for used in the treatment of HSV infections, and particularly CMV retinitis in AIDS patients. It also shows activity against a range of other herpes infections and has considerable potential for the treatment of adeno-, papilloma- and poxvirus infections. In each case, it is necessary for the phosphonic acid to be accepted as a substrate for a nucleotide kinase to generate the analogues of the 5′-diphosphate and 5′-triphosphate sequentially (Figure 3.92).

**3.7.2.2.3 5-Substituted-pyrimidine 2′-Deoxyribonucleosides.**[134–136] 5-Iodo-2′-deoxyuridine (Figure 3.95a) was discovered by Prusoff in the 1960s and was the first anti-viral nucleoside to be marketed against HSV and **VZV** infections as the drug **idoxuridine**. The mode of action of this nucleoside is still not known, although it is incorporated into both cellular and viral DNA. It is likely that the toxicity of this drug arises because both viral and cellular kinases can phosphorylate it and, therefore, it is further metabolized in infected and non-infected cells.

**5-Vinyl-2′-deoxyuridine** (Figure 3.95b) is in many orders of magnitude more potent than the 5-iodo derivative against HSV *in vitro*. While this compound is very toxic in cell culture, in animals it is neither toxic nor does it have anti-viral properties. This is because the nucleoside is a very good substrate for nucleoside phosphorylase, an enzyme that is absent from many tissue culture cell lines. The enzyme cleaves it to give the heterocyclic base and 2-deoxyribose 1-pyrophosphate, neither of which has anti-viral properties. From this example we learn that nucleoside analogues in this series must be resistant to nucleoside phosphorylase in order to possess anti-viral activity. **(*E*)-5-(2-Bromovinyl)-2′-deoxyuridine** (**BVDU**; Figure 3.95c) is even more effective ($IC_{50}$ 0.001 mg ml$^{-1}$) against HSV-1 and VZV but less so against HSV-2. This is

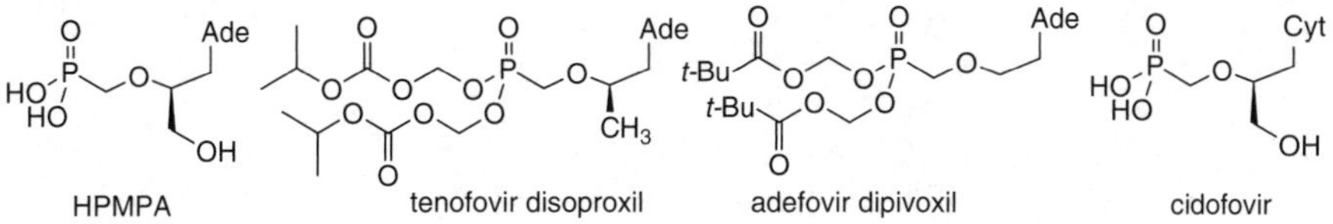


**Figure 3.94** *Structures of acyclonucleoside phosphonates used as anti-viral agents*

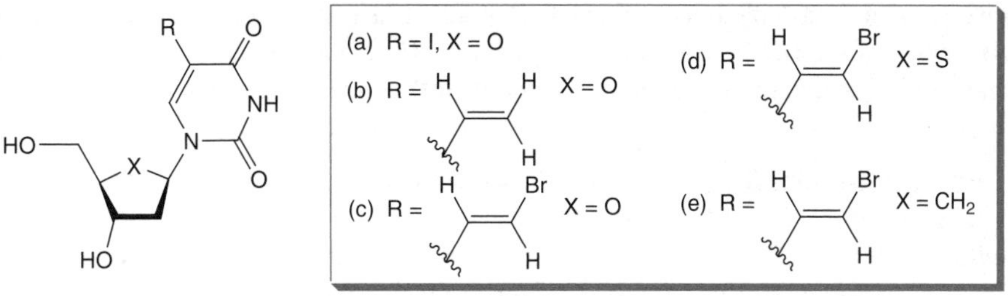


**Figure 3.95** *Anti-viral pyrimidine 5-substituted 2′-deoxyuridines*

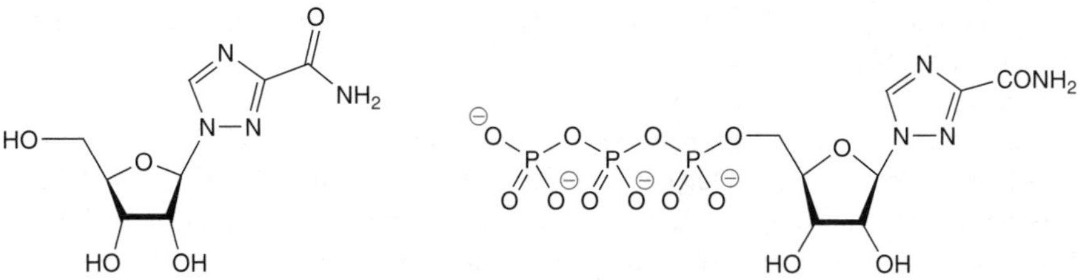


**Figure 3.96** *Ribavirin (left) and its triphosphate (right)*

because the nucleoside is a substrate for the viral **thymidine kinase** but not for the host-cell TK. This viral thymidine kinase is also a thymidylate kinase and produces the BVDU 5′-diphosphate, but only in HSV-1 infected cells because diphosphate formation does not occur efficiently with the HSV-2 encoded enzyme. BVDU is a substrate for nucleoside phosphorylase but it sufficiently avoids degradation to show useful clinical activity. The base (*E*)-5-(2-bromovinyl)uracil is an inhibitor for pyrimidine-5,6-dihydroreductase, the first enzyme in pyrimidine catabolism and thus can actually be salvaged and the 2′-deoxyribonucleoside can be regenerated. BVDU triphosphate is an inhibitor of the virally encoded DNA polymerase. A nucleoside phosphorylase-resistant analogue of BVDU (Figure 3.95d) has been described which has substantially greater activity *in vivo* because it has a much longer half-life in serum. The carbocyclic analogue of BVDU (Figure 3.95e) also has significant activity.

**3.7.2.2.4 Ribavirin.** The broad spectrum anti-viral activity of **ribavirin** (1-β-D-ribofuranosyl-1,2,4-triazole-3-carboxamide; Figure 3.96) was first described in 1972.[137] Since then, it has been studied in more animals and against more viruses than any other anti-viral agent. It is also apparently active in cell culture against about 85% of all virus species studied and shows little or no cellular toxicity. Until recently its main clinical use in the USA was in aerosol form against respiratory syncytial virus in young children. Very recently it has been approved as a combination treatment against chronic hepatitis C infections. It is also known to be effective against Lassa fever and influenza and its potential for the treatment of severe acute respiratory syndrome (SARS) is currently under investigation.

The mode of action of ribavirin is somewhat controversial. The most abundant form of ribavirin in cells is the triphosphate (Figure 3.96) and this was originally thought to inhibit inosine monophosphate

**Figure 3.97** *Anti-viral analogues of pyrophosphate. (a) Phosphonoacetic acid; (b) Phosphonoformic acid*

dehydrogenase, resulting in a depletion of cellular GTP pools. This in turn means that ribavirin 5′-triphosphate is an effective competitive inhibitor of the viral-specific RNA polymerase for some viruses. Ribavirin 5′-triphosphate is also known to inhibit the viral-specific mRNA-capping enzymes, guanyl transferase and $N^7$-methyl transferase, so that viral protein synthesis is interrupted.

**3.7.2.2.5 Phosphonoformic Acid.** **Phosphonoacetic acid** (PAA, Figure 3.97a) was discovered to have antiherpetic activity *in vitro* following random screening in 1973. Two years later it was shown to be a selective inhibitor of the virally encoded DNA polymerase, and the related **phosphonoformic acid** (**PFA**, Figure 3.97b) was subsequently found to be an even stronger inhibitor of this enzyme. PFA (foscarnet, Foscavir®) is used clinically for the treatment of CMV retinitis in AIDS patients. Both PAA and PFA are analogues of pyrophosphate, a product of polymerases, and presumably bind to the corresponding site on the enzyme and thus prevent replication. One problem with compounds of this sort is that they require no prior activation and therefore the difference in affinity between the virus-encoded and the host-cell polymerase determines their effectiveness.

Hepatitis B virus has a partially double stranded DNA genome with an ORF that codes for a DNA polymerase that is also an RT. It thus presents a very focused target for anti-viral nucleoside development. A wide range of nucleoside analogues have been used for HBV treatment (Table 3.2 and Figures 3.90 and 3.94) illustrating the breadth of modification to the 2-deoxy-D-ribose that has been explored by chemical synthesis. They reveal that the DNA polymerase of HBV has a preference for the L- over the D-enantiomers of some dNTP analogues, an advantage enhanced by the fact that L-enantiomers are often less toxic and more stable to metabolism than their D-counterparts.

There is still much to be done in this area of research. The rational design of novel inhibitors will continue to rely on knowledge of the details of viral replication at the molecular level. The number of structures of important virus-target enzymes will steadily rise. But success in designing analogues that are effective inhibitors *in vitro* and then converting such knowledge into a useful drug still calls for intact delivery of the drug at an effective concentration to the desired location and with minimum toxicity. Orally active forms of drugs are increasingly desirable for non-lethal infections.

## REFERENCES


1. J. Kurreck, Antisense technologies – improvement through novel chemical modifications. *Eur. J. Biochem.*, 2003, **270**, 1628–1644.
2. S. Verma and F. Eckstein, Modified oligonucleotides: synthesis and strategy for users. *Ann. Rev. Biochem.*, 1998, **67**, 99–134.
3. B.E. Eaton and W.A. Pieken, Ribonucleosides and RNA. *Ann. Rev. Biochem.*, 1995, **64**, 837–863.
4. E. Fischer and B. Helferich, Synthetische glucosine der purine. *Chem. Ber.*, 1914, **47**, 210–235.
5. J. Davoll, A.R. Lythgoe and A.R. Todd, Experiments on the synthesis of purine nucleosides. Part XIX. A synthesis of adenosine. *J. Chem. Soc.*, 1948, 967–969.
6. J. Davoll, A.R. Lythgoe and A.R. Todd, Experiments on the synthesis of purine nucleosides. Part XX. A synthesis of guanosine. *J. Chem. Soc.*, 1948, 1685–1687.
7. J. Davoll and B.A. Lowy, A new synthesis of purine nucleosides. The synthesis of adenosine, guanosine and 2,6-diamino-9-β-D-ribofuranosylpurine. *J. Am. Chem. Soc.*, 1951, **73**, 1650–1655.

8. J.J. Fox, N. Yung, I. Wempen and J. Doerr, Pyrimidine nucleosides III. On the synthesis of cytidine and related pyrimidine nucleosides. *J. Am. Chem. Soc.*, 1957, **79**, 5060–5064.
9. K.A. Watanabe, D.H. Hollenberg and J.J. Fox, Nucleosides LXXXV. On mechanisms of nucleoside synthesis by condensation reactions. *J. Carbohydr., Nucleosides, Nucleotides*, 1974, **1**, 1–37.
10. E. Diekmann, K. Friedrich and H.-G. Fritz, Dideoxy-ribonucleoside durch schmelzkondensation. *J. Prakt. Chem.*, 1993, **335**, 415–424.
11. T. Sato, in *Synthetic Procedures in Nucleic Acid Chemistry.* W.W. Zorbach and R.S. Tipson (eds), Wiley Interscience, New York, 1968, 264.
12. J.T. Witkowski and M.J. Robins, Chemical synthesis of the 1,2,4-triazole nucleosides related to uridine, 2′-deoxyuridine, thymidine, and cytidine. *J. Org. Chem.*, 1970, **35**, 2635–2641.
13. G.E. Hilbert and T.B. Johnson, Researches on pyrimidines. CXVII. A method for the synthesis of nucleosides. *J. Am. Chem. Soc.*, 1930, **52**, 4489–4494.
14. G.E. Hilbert and T.B. Johnson, Researches on Pyrimidines. CXV. Alkylation on nitrogen of the pyrimidine cycle by application of a new technique involving molecular rearrangements. *J. Am. Chem. Soc.*, 1930, **52**, 2001–2007.
15. J. Pliml and M. Prysta, Hilbert-Johnson reaction of 2,4-dialkoxy-pyrimidines with halogenoses. *Advan. Heterocycl. Chem.*, 1967, **8**, 115–142.
16. T. Nishimura, B. Shimizu and I. Iwai, A new synthetic method of nucleosides. *Chem. Pharm. Bull.*, 1963, **11**, 1470–1472.
17. L. Birkhofer, A. Ritter and H.P. Kuelthau, Disilylated carboxamides. *Angew. Chem. Int. Ed. Engl.*, 1963, **75**, 209.
18. E. Wittenburg, *Z. Chem.*, 1964, **4**, 303.
19. U. Niedballa and H. Vorbrüggen, Synthesis of nucleosides. 9. General synthesis of N-glycosides. I. Synthesis of pyrimidine nucleosides. *J. Org. Chem.*, 1974, **39**, 3654–3660.
20. H. Vorbrüggen, K. Krolikiewicz and B. Bennua, Nucleoside syntheses. 22. nucleoside synthesis with trimethylsilyl triflate and perchlorate as catalysts. *Chem. Ber. Recl.*, 1981, **114**, 1234–1255.
21. H. Vorbrüggen, Some recent trends and progress in nucleoside synthesis. *Acta Biochim. Pol.*, 1996, **43**, 25–36.
22. H. Vorbrüggen and C. Ruh-Pohlenz, in *Handbook of Nucleoside Synthesis*. H. Vörbruggen and C. Ruh-Pohlenz (eds), Wiley, New York, 2001, 15–24.
23. H. Vorbrüggen and G. Hofle, Nucleoside syntheses. 23. On the mechanism of nucleoside synthesis. *Chem. Ber.*, 1981, **114**, 1256–1268.
24. H. Vorbrüggen and C. Ruh-Pohlenz, in *Handbook of Nucleoside Synthesis*. H. Vörbruggen and C. Ruh-Pohlenz (eds), Wiley, New York, 2001, 38–46.
25. M. Hoffer, α-Thymidin. *Chem. Ber.*, 1960, **93**, 2777.
26. A.J. Hubbard, A.S. Jones and R.T. Walker, An investigation by $^1$H-NMR spectroscopy into the factors determining the beta-alpha ratio of the product in 2′-deoxynucleoside synthesis. *Nucleic Acids Res.*, 1984, **12**, 6827–6837.
27. J.N. Freskos, Synthesis of 2′-deoxypyrimidine nucleosides via copper(I) iodide catalysis. *Nucleosides Nucleotides*, 1989, **8**, 549–555.
28. H. Vorbrüggen and C. Ruh-Pohlenz, in *Handbook of Nucleoside Synthesis*. H. Vörbruggen and C. Ruh-Pohlenz (eds), Wiley, New York, 2001, 33–38.
29. H. Vorbrüggen and C. Ruh-Pohlenz, in *Handbook of Nucleoside Synthesis*. H. Vörbruggen and C. Ruh-Pohlenz (eds), Wiley, New York, 2001, 25–33.
30. J. Boryski, Transglycosylation reactions of purine nucleosides. A review. *Nucleosides Nucleotides*, 1996, **15**, 771–791.
31. M. Imazawa and F. Eckstein, Facile synthesis of 2′-amino-2′-deoxyribofuranosylpurines. *J. Org. Chem.*, 1979, **44**, 2039–2041.
32. M. Imazawa and F. Eckstein, Synthesis of 3′-azido-2′,3′-dideoxyribofuranosylpurines. *J. Org. Chem.*, 1978, **43**, 3044–3048.

33. J.R. Hanrahan and D.W. Hutchinson, The enzymatic-synthesis of antiviral agents. *J. Biotechnol.*, 1992, **23**, 193–210.
34. D.W. Hutchinson, in *Comprehensive Organic Chemistry*, vol 5, D.H.R. Barton and W.D. Ollis (eds), Pergamon Press, Oxford, 1979, 105–145.
35. T.A. Krenitsky, J.L. Rideout, E.Y. Chao, G.W. Koszalka, F. Gurney, R.C. Crouch, N.K. Cohn, G. Wolberg and R. Vinegar, Imidazo[4,5-*c*]pyridines (3-deazapurines) and their nucleosides as immunosuppressive and antiinflammatory agents. *J. Med. Chem.*, 1986, **29**, 138–143.
36. T. Utagawa, H. Morisawa, S. Yamanaka, A. Yamazaki and Y. Hirose, Enzymatic synthesis of nucleoside antibiotics. 6. Enzymatic synthesis of virazole by purine nucleoside phosphorylase of enterobacter-aerogenes *Agr. Biol. Chem.*, 1986, **50**, 121.
37. M.G. Stout, D.E. Hoard, M.J. Holman, E.S. Wu and J.M. Siegel, Preparation of 2′-deoxyribonucleosides via nucleoside deoxyribosyl transferase. *Methods Carbohydr. Chem.*, 1976, **7**, 19–29.
38. D. Betbeder, D.W. Hutchinson and A.O. Richards, The stereoselective enzymatic synthesis of 9-β-D-2′-deoxyribofuranosyl 1-deazapurine. *Nucleic Acids Res.*, 1989, **17**, 4217–4222.
39. B.R. Baker, in *The CIBA Foundation Symposium on the Chemistry and Biology of the Purines*. G.E.W. Wolstenholme and C.M. O'Connor (eds), Churchill, London, 1957, 120.
40. R.A. Lessor and N.J. Leonard, Synthesis of 2′-deoxynucleosides by deoxygenation of ribonucleosides. *J. Org. Chem.*, 1981, **46**, 4300–4301.
41. M.J. Robins and J.S. Wilson, Nucleic-acid related compounds.32. Smooth and efficient deoxygenation of secondary alcohols – a general procedure for the conversion of ribonucleosides to 2′-deoxynucleosides. *J. Am. Chem. Soc.*, 1981, **103**, 932–933.
42. Z. Kazimierczuk, H.B. Cottam, G.R. Revankar and R.K. Robins, Synthesis of 2′-deoxytubercidin, 2′-deoxyadenosine, and related 2′-deoxynucleosides via a novel direct stereospecific sodium-salt glycosylation procedure. *J. Am. Chem. Soc.*, 1984, **106**, 6379–6382.
43. F. Seela, Base-modified nucleosides and oligonucleotides: synthesis and application. *Collect. Czech. Chem. Commun. Special Issue*, 2002, **5**, 1–15.
44. F. Seela, N. Ramzaeva and H. Rosemeyer, *Houben–Weyl, Methods of Organic Synthesis*. vol E9b, Thieme, Stuttgart, 1997, 304–550.
45. F. Seela, B. Westermann and U. Bindig, Liquid-liquid and solid-liquid phase-transfer glycosylation of pyrrolo 2,3-D-pyrimidines – stereospecific synthesis of 2-deoxy-β-D-ribofuranosides related to 2′-deoxy-7-carbaguanosine. *J. Chem. Soc. Perkin Trans. 1*, 1988, 697–702.
46. D.M. Brown and R.C. Ogden, A synthesis of pseudouridine. *J. Chem. Soc.*, 1981, 723–725.
47. B.A. Schweitzer and E.T. Kool, Aromatic non-polar nucleosides as hydrophobic isosteres of pyrimidine and purine nucleosides. *J. Org. Chem.*, 1994, **59**, 7238–7242.
48. L.A. Agrofoglio, I. Gillaizeau and Y. Saito, Palladium-assisted routes to nucleosides. *Chem. Rev.*, 2003, **103**, 1875–1916.
49. L.B. Townsend, Imidazole nucleosides and nucleotides. *Chem. Rev.*, 1967, **67**, 553–563.
50. S.-I. Nagatsuka, T. Ohgi and T. Goto, Synthesis of wyosine (nucleoside Yt), a strongly fluorescent nucleoside found in *Torulopsis utilis* tRNAPhe and 3-methylguanosine. *Tetrahedron Lett.*, 1978, **29**, 2579–2582.
51. G. Shaw and R.N. Warrener, Synthesis of 2-thiouridine. *Proc. Chem. Soc.*, 1957, 351.
52. S. De Bernado and M. Weigele, C-Nucleoside antibiotics 2. Synthesis of oxazinomycin (minimycin). *J. Org. Chem.*, 1977, **42**, 109–112.
53. M.-I. Lim, R.S. Klein and J.J. Fox, Synthesis of the pyrrolo[3,2-*d*]pyrimidine C-nucleoside isostere of inosine. *Tetrahedron Lett.*, 1980, **21**, 1013–1016.
54. R. Noyori, T. Sato and Y. Hayakawa, A stereocontrolled general synthesis of C-nucleosides. *J. Am. Chem. Soc.*, 1978, **100**, 2561–2563.
55. F. Burlina, A. Favre, J.L. Fourrey and M. Thomas, An expeditious route to carbocyclic nucleosides: (−)- aristeromycin and (−)-carbodine. *Bioorg. Med. Chem. Lett.*, 1997, **7**, 247–250.

56. S. Ohira, T. Sawamoto and M. Yamato, Synthesis of (−)-neplanocin-a *via* C—H insertion of alkylidenecarbene. *Tetrahedron Lett.*, 1995, **36**, 1537–1538.
57. W.B. Choi, S. Yeola, D.C. Liotta, R.F. Schinazi, G.R. Painter, M. Davis, M. Stclair and P.A. Furman, Synthesis, anti-Human-Immunodeficiency-Virus, and anti-Hepatitis-B Virus activity of pyrimidine oxathiolane nucleosides. *Bioorg. Med. Chem. Lett.*, 1993, **3**, 693–696.
58. H. Gao and A.K. Mitra, Synthesis of acyclovir, ganciclovir and their prodrugs: A review. *Synthesis*, 2000, 329–351.
59. M.J. Robins, in *Nucleoside Analogues: Chemistry, Biology, and Medical Applications*. R.T. Walker, E. DeClercq and F. Eckstein (eds), Plenum Press, New York, 1979, 165–192.
60. L.B. Townsend, in *Nucleoside Analogues: Chemistry, Biology, and Medical Applications*. R.T. Walker, E. DeClercq and F. Eckstein (eds), Plenum Press, New York, 1979, 193–223.
61. B. Luy and J.P. Marino, Measurement and application of $^{1}$H-$^{19}$F dipolar couplings in the structure determination of 2′-fluorolabeled RNA. *J. Biomol. NMR*, 2001, **20**, 39–47.
62. P. Bachert, Pharmacokinetics using fluorine NMR *in vivo*. *Prog. Nucl. Magn. Reson. Spectrosc.*, 1998, **33**, 1–56.
63. J.X. Yu, V.D. Kodibagkar, W.N. Cui and R.P. Mason, $^{19}$F: A versatile reporter for non-invasive physiology and pharmacology using magnetic resonance. *Curr. Med. Chem.*, 2005, **12**, 819–848.
64. S.E. Lee, A. Sidorov, T. Gourlain, N. Mignet, S.J. Thorpe, J.A. Brazier, M.J. Dickman, D.P. Hornby, J.A. Grasby and D.M. Williams, Enhancing the catalytic repertoire of nucleic acids: a systematic study of linker length and rigidity. *Nucleic Acids Res.*, 2001, **29**, 1565–1573.
65. K. Sakthivel and C.F. Barbas, Expanding the potential of DNA for binding and catalysis: Highly functionalized dUTP derivatives that are substrates for thermostable DNA polymerases. *Angew. Chem. Int. Ed-Engl.*, 1998, **37**, 2872–2875.
66. J.L. Ruth, in *Oligonucleotides and Analogues*. F. Eckstein (ed), OUP, New York, 1991, 255–281.
67. J.P. Anderson, B. Angerer and L.A. Loeb, Incorporation of reporter-labeled nucleotides by DNA polymerases. *Biotechniques*, 2005, **38**, 257–264.
68. C. McGuigan, A. Brancale, G. Andrei, R. Snoeck, E. De Clercq and J. Balzarini, Novel bicyclic furanopyrimidines with dual anti-VZV and -HCMV activity. *Bioorg. Med. Chem. Lett.*, 2003, **13**, 4511–4513.
69. D.E. Bergstrom and J.L. Ruth, Preparation of C-5 mercurated pyrimidine nucleosides. *J. Carbohydr., Nucleosides, Nucleotides*, 1977, **4**, 257–269.
70. N. Ramzaeva and F. Seela, Duplex stability of 7-deazapurine DNA: oligonucleotides containing 7-bromo- or 7-iodo-7-deazaguanine. *Helv. Chim. Acta*, 1996, **79**, 1549–1558.
71. K.J. Divakar and C.B. Reese, 4-(1,2,4-triazol-1-yl) and 4-(3-nitro-1,2,4-triazol-1-yl)-1-(β-D-2,3,5-tri-*O*-acetylarabinofuranosyl)pyrimidin-2(1h)-ones – valuable intermediates in the synthesis of derivatives of 1-(β-D-arabinofuranosyl)cytosine (Ara-C). *J. Chem. Soc.-Perkin Trans., 1*, 1982, 1171–1176.
72. K.W. Pankiewicz, Fluorinated nucleosides. *Carbohydr. Res.*, 2000, **327**, 87–105.
73. B.A. Connolly, in *Oligonucleotides and Analogues*. F. Eckstein (ed), OUP, New York, 1991, 155–183.
74. F. Nagatsugi, K. Uemura, S. Nakashima, M. Maeda and S. Sasaki, 2-aminopurine derivatives with C6-substituted olefin as novel cross-linking agents and the synthesis of the corresponding β-phosphoramidite precursors. *Tetrahedron*, 1997, **53**, 3035–3044.
75. B.L. Gaffney and R.A. Jones, Synthesis of $O^{6}$-alkylated deoxyguanosine nucleosides. *Tetrahedron Lett.*, 1982, **23**, 2253–2256.
76. P.P. Kung and R.A. Jones, One-flask syntheses of 6-thioguanosine and 2′-deoxy-6-thioguanosine. *Tetrahedron Lett.*, 1991, **32**, 3919–3922.
77. B.S. Sproat and A.I. Lamond, in *Oligonucleotides and Analogues*. F. Eckstein (ed), OUP, New York, 1991, 49–86.
78. J.J. Fox, Pyrimidine nucleoside transformations *via* anhydronucleosides. *Pure Appl. Chem.*, 1969, **18**, 223–255.

79. D.B. Olsen, F. Benseler, H. Aurup, W.A. Pieken and F. Eckstein, Study of a hammerhead ribozyme containing 2′-modified adenosine residues. *Biochemistry*, 1991, **30**, 9735–9741.
80. F. Benseler, D.M. Williams and F. Eckstein, Synthesis of suitably-protected phosphoramidites of 2′-fluoro-2′-deoxyguanosine and 2′-amino-2′-deoxyguanosine for incorporation into oligoribonucleotides. *Nucleosides Nucleotides*, 1992, **11**, 1333–1351.
81. A.V.R. Rao, M.K. Gurjar and S.V.S. Lalitha, Discovery of a Novel Route to β-thymidine – a precursor for anti-AIDS compounds. *J. Chem. Soc. Chem. Commun.*, 1994, 1255–1256.
82. L. Beigelman, P. Haeberli, D. Sweedler and A. Karpeisky, Improved synthetic approaches toward 2′-*O*-methyl-adenosine and guanosine and their *N*-acyl derivatives. *Tetrahedron*, 2000, **56**, 1047–1056.
83. M.J. Robins, J.S. Wilson, D. Madej, N.H. Low, F. Hansske and S.F. Wnuk, Nucleic-acid related compounds.88. Efficient conversions of ribonucleosides into their 2′,3′-anhydro, 2′-(and 3′)-deoxy, 2′,3′-didehydro-2′,3′-dideoxy, and 2′,3′-dideoxynucleoside analogs. *J. Org. Chem.*, 1995, **60**, 7902–7908.
84. G.R.J. Thatcher and R. Kluger, Mechanism and catalysis of nucleophilic substitution in phosphate esters. *Adv. Phys. Org. Chem.*, 1989, **25**, 99–265.
85. E.V. Anslyn and D.M. Perreault, Unifying the current data on the mechanism of cleavage-transesterification of RNA. *Angew. Chem. Int. Ed. Engl.*, 1997, **36**, 432–450.
86. M. Oivanen, S. Kuusela and H. Lönnberg, Kinetics and mechanisms for the cleavage and isomerisation of the phosphodiester bonds of RNA by Brønsted acids and bases. *Chem. Rev.*, 1988, **98**, 961–990.
87. A.C. Hengge, in *Comprehensive Biological Catalysis: A Mechanistic Reference*, vol 1. M. Sinnott (ed), Academic Press, New York, 1988, 517–542.
88. D.M. Brown, in *Methods in Molecular Biology*, vol 20. S. Agarwal (ed), Humana Press Inc., Totowa, NJ, 1993, 1–17.
89. C.B. Reese, The chemical synthesis of oligo- and poly-nucleotides by the phosphotriester approach. *Tetrahedron*, 1978, **34**, 3143–3179.
90. L.A. Slotin, Current methods of phosphorylation of biological molecules. *Synthesis*, 1975, **11**, 737–752.
91. S.L. Beaucage and R.P. Iyer, Advances in the synthesis of oligonucleotides by the phosphoramidite approach. *Tetrahedron*, 1992, **48**, 2223–2311.
92. J. Stawinski and A. Kraszewski, How to get the most out of two phosphorus chemistries studies on H-phosphonates. *Acc. Chem. Res.*, 2002, **35**, 952–960.
93. S. Narang, DNA synthesis. *Tetrahedron*, 1983, **39**, 3–22.
94. C.B. Reese, The chemical synthesis of oligo- and poly-nucleotides: a personal commentary. *Tetrahedron*, 2002, **58**, 8893–8920.
95. M.D. Matteucci and M.H. Caruthers, Nucleotide chemistry.4. Synthesis of deoxyoligonucleotides on a polymer support. *J. Am. Chem. Soc.*, 1981, **103**, 3185–3191.
96. S.L. Beaucage and M.H. Caruthers, Deoxynucleoside phosphoramidites – a new class of key intermediates for deoxypolynucleotide synthesis. *Tetrahedron Lett.*, 1981, **22**, 1859–1862.
97. W.J. Stec, G. Zon, W. Egan and B. Stec, Automated solid-phase synthesis, separation, and stereochemistry of phosphorothioate analogs of oligodeoxyribonucleotides. *J. Am. Chem. Soc.*, 1984, **106**, 6077–6079.
98. M. Yoshikawa, T. Kato and T. Takenishi, Studies of phosphorylation. III. Selective phosphorylation of unprotected nucleosides. *Bull. Chem. Soc. Jpn.*, 1969, **42**, 3505–3508.
99. A. Arabshahi and P.A. Frey, A simplified procedure for synthesizing nucleoside 1-thiotriphosphates – dATP(α)S, dGTP(α)S, UTP(α)S, and dTTP(α)S. *Biochem. Biophys. Res. Commun.*, 1994, **204**, 150–155.
100. T. Sowa and S. Ouchi, Facile synthesis of 5′-nucleotides by the selective phosphorylation of a primary hydroxyl group of nucleosides with phosphoryl chloride. *Bull. Chem. Soc. Jpn.*, 1975, **48**, 2084–2090.

101. V.J. Davisson, D.R. Davis, V.M. Dixit and C.D. Poulter, Synthesis of nucleotide 5′-diphosphates from 5′-*O*-tosyl nucleosides. *J. Org. Chem.*, 1987, **52**, 1794–1801.
102. G.M. Blackburn, D.E. Kent and F. Kolkmann, The synthesis and metal-binding characteristics of novel, isopolar phosphonate analogs of nucleotides. *J. Chem. Soc.-Perkin Trans., 1*, 1984, 1119–1125.
103. T.C. Myers, K. Nakamura and J.W. Flesher, Phosphonic acid analogs of nucleoside phosphates. I. The synthesis of 5′-adenylyl methylenediphosphonate, a phosphonic acid analog of adenosine. *J. Am. Chem. Soc.*, 1963, **85**, 3292–3295.
104. K.-H. Scheit, in *Nucleotide Analogues. Synthesis and Biological Function.* K.-H. Scheit (ed), Wiley Interscience, New York, 1980, 96–141.
105. K.-H. Scheit, in *Nucleotide Analogues. Synthesis and Biological Function.* K.-H. Scheit (ed), Wiley Interscience, New York, 1980, 195–218.
106. K. Burgess and D. Cook, Syntheses of nucleoside triphosphates. *Chem. Rev.*, 2000, **100**, 2047–2059.
107. J. Ludwig, in *Biophosphates and Their Analogues – Synthesis, Structure, Metabolism and Activity.* K.S. Bruzik and W.J. Stec (eds), Elsevier, Amsterdam, 1987, 131–133.
108. J. Ludwig and F. Eckstein, Rapid and efficient synthesis of nucleoside 5′-O-(1-thiotriphosphates), 5′-triphosphates and 2′,3′-cyclophosphorothioates using 2-chloro-4H-1,3,2-benzodioxaphosphorin-4-one. *J. Org. Chem.*, 1989, **54**, 631–635.
109. J. Ludwig and F. Eckstein, Stereospecific synthesis of guanosine 5′-O-(1,2-dithiotriphosphates). *J. Org. Chem.*, 1991, **56**, 5860–5865.
110. J. Ludwig and F. Eckstein, Synthesis of nucleoside 5′-*O*-(1,3-dithiotriphosphates) and 5′-*O*-(1,1-dithiotriphosphates). *J. Org. Chem.*, 1991, **56**, 1777–1783.
111. B.R. Shaw, M. Dobrikov, X. Wang, J. Wan, K.Z. He, J.L. Lin, P. Li, V. Rait, Z.A. Sergueeva and D. Sergueev, *Therapeutic Oligonucleotides*, S. Cho-chung, A.M. Gewirtz and C.A. Stein (ed) vol 1002. New York Academy of Science, New York, 2003, 12–29.
112. F. Eckstein, Nucleoside phosphorothioates. *Ann. Rev. Biochem.*, 1985, **54**, 367–402.
113. F. Eckstein and J.B. Thomson, *DNA Replication*, vol 262. Academic Press, San Diego, 1995, 189–202.
114. J.R. Knowles, Enzyme catalysed phosphoryl transfer reactions. *Ann. Rev. Biochem.*, 1980, **49**, 877–919.
115. G.M. Blackburn, G.E. Taylor, G.R.J. Thatcher, M. Prescott and A.G. McLennan, Synthesis and resistance to enzymatic hydrolysis of stereochemically-defined phosphonate and thiophosphate analogs of $P^1$,$P^4$-bis(5′-adenosyl) tetraphosphate. *Nucleic Acids Res.*, 1987, **15**, 6991–7004.
116. N.B. Tarussova, T.I. Osipova, P.P. Purygin and I.A. Yakimova, The synthesis of $P^1$, $P^3$-bis(5′-adenosyl)triphosphate, $P^1$,$P^4$-bis(5′-adenosyl)tetraphosphate and its phosphonate analog with the use of carbonyl derivatives of nitrogen-containing heterocycles. *Bioorg. Khim.*, 1986, **12**, 404–407.
117. G.M. Blackburn, F. Eckstein, D.E. Kent and T.D. Perrée, Isopolar vs isosteric phosphonate analogs of nucleotides. *Nucleosides Nucleotides*, 1985, **4**, 165–167.
118. J.G. Moffatt and H.G. Khorana, The total synthesis of coenzyme A. *J. Am. Chem. Soc.*, 1961, **83**, 663–675.
119. M. Sekine, S. Nishiyama, T. Kamimura, Y. Osaki and T. Hata, Chemical synthesis of capped oligoribonucleotides, $m^7$g5′pppAUG and $m^7$g5′pppAUGCC. *Bull. Chem. Soc. Jpn.*, 1985, **58**, 850–860.
120. R.I. Christopherson, S.D. Lyons and P.K. Wilson, Inhibitors of *de novo* nucleotide biosynthesis as drugs. *Acc. Chem. Res.*, 2002, **35**, 961–971.
121. J.M. Berg, J.L. Tymoczko and L. Stryer, in *Biochemistry*, 5th edn. J.M. Berg, J.L. Tymoczko and L. Stryer (eds), Freeman, New York, 2002, 693–714.
122. P.M.J. Burgers, E.V. Koonin, E. Bruford, L. Blanco, K.C. Burtis, M.F. Christman, W.C. Copeland, E.C. Friedberg, F. Hanaoka, D.C. Hinkle *et al.*, Eukaryotic DNA polymerases: Proposal for a revised nomenclature. *J. Biol. Chem.*, 2001, **276**, 43487–43490.
123. C.M. Galamarini, J.R. Mackey and C. Dumontet, Nucleoside analogues and nucleobases in cancer treatment. *Lancet Oncol.*, 2002, **3**, 415–424.

124. S. Miura and S. Izuta, DNA polymerases as targets of anticancer nucleosides. *Curr. Drug Targets*, 2004, **5**, 191–195.
125. I.M. Kompis, K. Islam and R.L. Then, DNA and RNA synthesis: Antifolates. *Chem. Rev.*, 2005, **105**, 593–620.
126. D.R. Newell, How to develop a successful cancer drug – molecules to medicines or targets to treatments? *Eur. J. Cancer*, 2005, **41**, 676–682.
127. S.L. Gerson, Clinical relevance of MGMT in the treatment of cancer. *J. Clin. Oncol.*, 2002, **20**, 2388–2399.
128. E. DeClercq, Antiviral drugs in current clinical use. *J. Clin. Virol.*, 2004, **30**, 115–133.
129. J.S. Copperwood, G. Gumina, F.D. Boudinot and C.K. Chu, in *Recent Advances in Nucleosides: Chemistry and Chemotherapy*. C.K. Chu (ed), Elsevier, Amsterdam, 2002, 91–147.
130. S.G. Sarafianos, K. Das, A.D. Clark, J.P. Ding, P.L. Boyer, S.H. Hughes and E. Arnold, Lamivudine (3TC) resistance in HIV-1 reverse transcriptase involves steric hindrance with beta-branched amino acids. *Proc. Natl. Acad. Sci. USA*, 1999, **96**, 10027–10032.
131. E. De Clercq, Non-nucleoside reverse transcriptase inhibitors (NNRTIs): Past, present, and future. *Chem. Biodivers.*, 2004, **1**, 44–64.
132. C. Orkin, J. Stebbing, M. Nelson, M. Bower, M. Johnson, S. Mandalia, R. Jones, G. Moyle, M. Fisher and B. Gazzard, A randomized study comparing a three- and four-drug HAART regimen in first-line therapy (QUAD study). *J. Antimicrob. Chemother.*, 2005, **55**, 246–251.
133. J.A. Bartlett, R. DeMasi, J. Quinn, C. Moxham and F. Rousseau, Overview of the effectiveness of triple combination therapy in antiretroviral-naive HIV-1 infected adults. *AIDS*, 2001, **15**, 1369–1377.
134. A. Hóly, in *Recent Advances in Nucleosides: Chemistry and Chemotherapy*. C.K. Chu (ed), Elsevier, Amsterdam, 2002, 167–238.
135. E. DeClercq and R.T. Walker, Synthesis and antiviral properties of 5-vinylpyrimidine nucleoside analogs. *Pharmacol. Ther.*, 1984, **26**, 1–44.
136. E. DeClercq, *Recent Advances in Nucleosides: Chemistry and Chemotherapy*. Elsevier, Amsterdam, 2002, 433–454.
137. R.W. Sidwell, Jt. Witkowsk, L.B. Allen, R.K. Robins, G.P. Khare and J.H. Huffman, Broad-spectrum antiviral activity of virazole – 1-β-D-ribofuranosyl-1,2,4-triazole-3-carboxamide. *Science*, 1972, **177**, 705–706.
138. N.H. Williams and P. Wyman, Base catalysed phosphate diester hydrolysis. *Chem. Commun.*, 2001, 1268–1269.
139. C. Lad, N.H. Williams and R. Wolfenden, The rate of hydrolysis of phosphomonoester dianions and the exceptional catalytic proficiencies of protein and inositol phosphatases. *Proc. Natl. Acad. Sci. USA*, 2003, **100**, 5607–5610.
140. P.W. Barnard, C.A. Bunton, D.R. Llewellyn, C.A. Vernon and V.A. Welch, The reactions of organic phosphates. Part IV. Oxygen exchange between water and orthophosphoric acid. *J. Chem. Soc.*, 1961, 2670–2676.
141. A.J. Kirby and M. Younas, Reactivity of phosphate esters – diester hydrolysis. *J. Chem. Soc. B*, 1970, 510–513.
142. E.T. Kaiser and K. Kudo, Alkaline hydrolysis of aromatic esters of phosphoric acid. *J. Amer. Chem. Soc.*, 1967, **89**, 6725–6728.
143. A.J. Kirby and A.G. Varvoglis, The reactivity of phosphate esters. Monoester hydrolysis. *J. Am. Chem. Soc.*, 1967, **89**, 415–423.
144. J. Kumamoto, J.R. Cox and F.H. Westheimer, Barium ethylene phosphate. *J. Am. Chem. Soc.*, 1956, **78**, 4858–4860.

CHAPTER 4

# Synthesis of Oligonucleotides

## CONTENTS



## 4.1 SYNTHESIS OF OLIGODEOXYRIBONUCLEOTIDES

An oligonucleotide is a single-stranded chain consisting of a number of nucleoside units linked together by phosphodiester bridges. Generally in oligonucleotide synthesis, phosphodiesters are formed between a 3′-hydroxyl group bearing a phosphate derivative and a 5′-hydroxyl group of another nucleoside (Section 4.1.4). In the context of nucleic acids, the prefix ‘oligo’ is usually taken to denote a few nucleoside residues, while the prefix ‘poly’ means many. However, it has become a common practice to refer to all chemically synthesised nucleic acid chains as oligonucleotides, even if they are in excess of 100 residues in length. The term polynucleotide is more often taken to mean single-stranded nucleic acids of less-defined length and sequence, often obtained by a polymerisation reaction, for example, polycytidylic acid, polyC.

### 4.1.1 Overall Strategy for Chemical Synthesis

Nucleic acids are sensitive to a wide range of chemical reactions (see Chapter 8), and relatively mild reaction conditions are required for their chemical synthesis. The heterocyclic bases are prone to alkylation, oxidation and phosphorylation and the phosphodiester backbone is susceptible to hydrolysis. In the case of DNA, acidic hydrolysis occurs more readily than alkaline hydrolysis because of the lability of the glycosylic bond, particularly in the case of purines (depurination, see Section 8.1). Such considerations limit the range of chemical reactions in oligodeoxyribonucleotide synthesis to (1) mild alkaline hydrolysis; (2) very mild acidic hydrolysis; (3) mild nucleophilic displacement reactions; (4) base-catalysed elimination reactions; and (5) certain mild redox reactions (*e.g.* iodine or Ag(I) oxidations and reductive eliminations using zinc).

The key step in the synthesis of oligodeoxyribonucleotides is the specific and sequential formation of internucleoside 3′→5′ phosphodiester linkages. The main nucleophilic centres on a 2′-deoxyribonucleoside are the 5′- and 3′-hydroxyl groups and, in the case of dC, dG and dA, the exocyclic amino groups. To form a specific 3′–5′ linkage between two nucleosides, the nucleophilic centres not involved in the reaction must be protected. The first 5′-unit requires a **protecting group** on the 5′-hydroxyl as well as on the **nucleobase**, whereas the second 3′-unit requires protection of the 3′-hydroxyl as well as the nucleobase. In the example of joining a 5′-dA unit to a 3′-dG unit (Figure 4.1), $R^1$ and $R^2$ protect the 5′-dA and $R^3$ and $R^4$ protect the 3′-dG. One of the two units requires phosphorylation or phosphitylation on the unprotected hydroxyl group and is then joined to the other nucleoside in a **coupling reaction**. The resulting dinucleoside monophosphate is now fully protected. Usually the phosphate group carries a protecting group $R^5$, introduced during the phosphorylation (phosphitylation) step, such that the internucleotide phosphate is a **triester.** To extend the chain, one of the two terminal hydroxyl-protecting groups $R^1$ or $R^3$ must be selectively removed to which a new protected nucleoside unit may be attached.

Where $R^1$ and $R^3$ are conventional protecting groups, oligonucleotide synthesis is referred to as **solution-phase**. Solution-phase synthesis has largely been superseded by a **solid-phase** method, where either $R^1$ or $R^3$ is an insoluble polymeric or inorganic support (Section 4.1.4). Whereas extension of the chain in solution-phase synthesis is possible in either the 3′→5′ or 5′→3′ directions, in solid-phase synthesis the oligonucleotide can be extended only in one direction. The conventional protecting group removed prior to each coupling step ($R^1$ or $R^3$, whichever is not the solid-support) is a **temporary** protecting group. $R^2$, $R^4$, $R^5$ and the solid support are all **permanent** protecting groups, and must remain stable throughout the oligonucleotide synthesis. They are only removed at the end of the synthesis to generate the final deprotected oligonucleotide.

### 4.1.2 Protected 2′-Deoxyribonucleoside Units

The most convenient way to assemble an oligonucleotide is to utilise preformed deoxynucleoside phosphate [P(V)] or phosphite [P(III)] derivatives as building blocks, and to couple these sequentially to a terminal

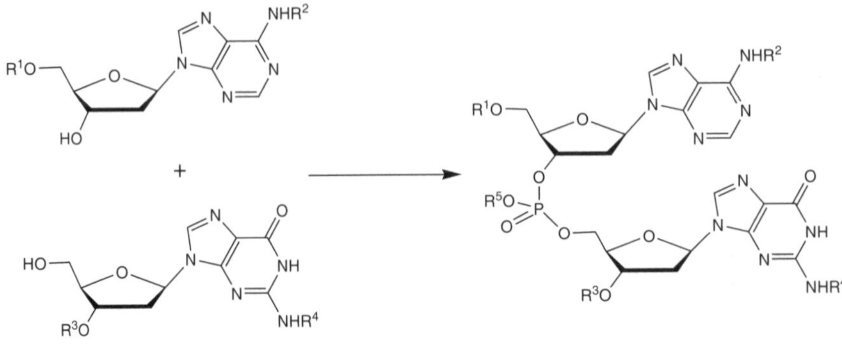


**Figure 4.1** *Joining of a 5′-dA unit to a 3′-dG unit*

nucleoside attached to a **solid support**. Since the 5′-hydroxyl group is a more effective nucleophile than the secondary 3′-hydroxyl group, the phosphate/phosphite group is best placed on the 3′-position. To achieve this selectively it is necessary to protect the nucleobase exocyclic amino groups and the 5′-hydroxyl group.

*4.1.2.1 Nucleobases.* Permanent protecting groups for the exocyclic amino groups of adenine, cytosine and guanine have been used for many years in oligonucleotide synthesis.[1] Acyl protecting groups were chosen, since they are stable for long periods during mildly basic and acidic conditions used during oligonucleotide synthesis, and are removed with concentrated ammonia at the end of the synthesis (Section 4.1.4). The benzoyl group is used to protect both adenine and cytosine, while isobutyryl is used to protect guanine (Figure 4.2). Thymidine does not require protection since it does not have an exocyclic amino group. While these acyl protecting groups are still suitable for oligonucleotide synthesis today, new chemistries and new nucleoside building blocks have been introduced, which require milder deprotection conditions at the end of the synthesis. For example, a matched set of phenoxyacetyl (PAC) for dA, isopropylphenoxyacetyl for dG and acetyl for dC can be removed by treatment with 0.05 M potassium carbonate in methanol at room temperature within a few hours.

When nucleosides are prepared for incorporation into oligonucleotides, it is usual to protect nucleobase exocyclic amino groups first. There are two common methods for the synthesis of acylated nucleosides, **per-acylation** and **transient protection** (Figure 4.3). The per-acylation method involves use of an excess of acylating agent such that the hydroxyl groups and the exocyclic amino groups are each acylated (bis-acylated in the case of the amino groups), and then the hydroxylic and one of the amino acyl groups are removed selectively under mild basic conditions. The selectivity arises because of the greater stability of amides compared to esters (and bis-amides) at high pH. In the transient protection route, the nucleoside is treated with trimethylsilyl chloride (TMSCl), which reacts selectively with the hydroxyl groups. Treatment with benzoyl chloride is then selective for the exocyclic amino group (again the bis-acylated product may be formed). The silyl protecting groups are removed under basic conditions to give the desired $N^6$-benzoyl-2′-deoxyadenosine. This method may also be used for protection of 2′-deoxycytidine.

In the case of 2′-deoxyguanosine protection, the reaction may be carried out with isobutyric anhydride by either per-acylation or the transient protection route. However, in the case of dG, the $O^6$-position is susceptible to reaction under certain conditions, particularly with coupling agents and phosphorylating agents, or in the synthesis of G-rich oligonucleotides. Under these conditions it is necessary to protect the $O^6$-position using alkyl or aryl protecting groups. However, such protection is not necessary in the case of the phosphoramidite method (Section 4.1.3). Another common protecting group for dG is the dimethylformamidine group that is readily introduced using dimethylformamide dimethylacetal.

*4.1.2.2 5′-Hydroxyl Group.* By far the most common protecting group for the 5′-hydroxyl group is the 4,4′-dimethoxytrityl group (DMT) (Figure 4.4). The DMT group is readily removed under acidic conditions. It is introduced onto the 5′-hydroxyl group of *N*-acylated nucleosides with DMT–Cl in the presence of a base such as pyridine or 4-dimethylaminopyridine. Reaction occurs principally at the 5′-hydroxyl rather than at the secondary 3′-hydroxyl group because of steric effects. The DMT group is removed during oligonucleotide

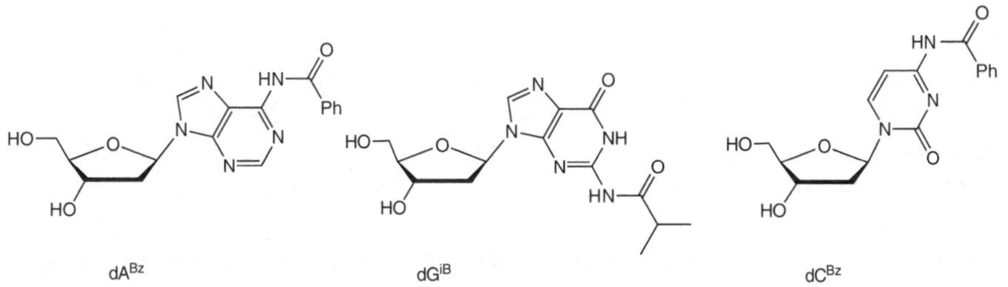


**Figure 4.2** *Common protecting groups for the heterocyclic bases of dA, dG and dC*

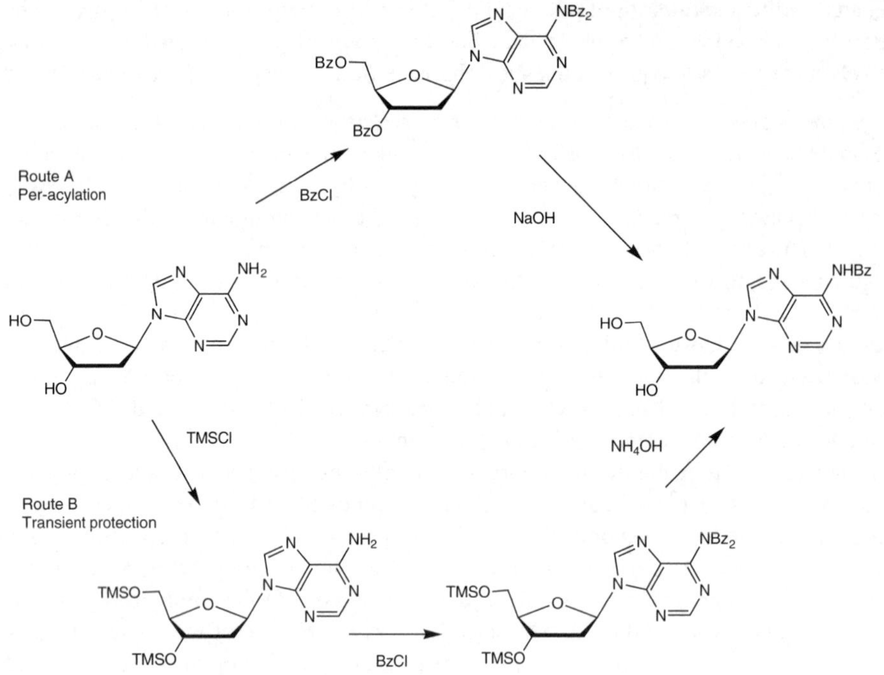


**Figure 4.3** *Routes to $N^6$-benzoyl-2′-deoxyadenosine*

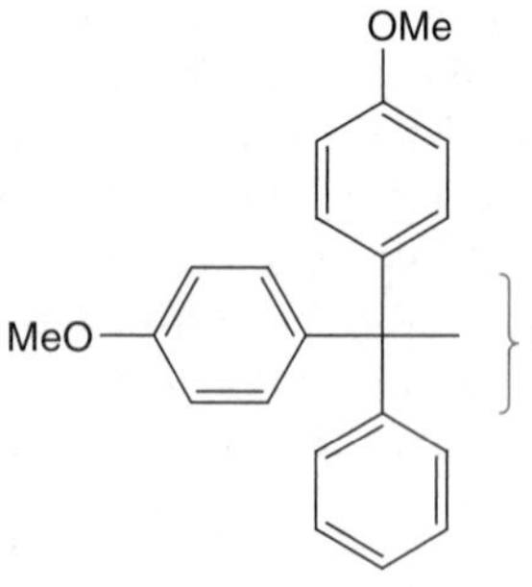


**Figure 4.4** *The 4,4′-dimethoxyphenylmethyl (dimethoxytrityl, DMT) group*

synthesis with either dichloroacetic acid or trichloroacetic acid in non-aqueous solvent, conditions that prevent other side reactions, such as depurination. During deprotection, the bright orange-red DMT cation is liberated and is used as a measure of the yield of coupling of that nucleoside unit (Section 4.1.4).

*4.1.2.3 Introduction of Phosphate.* In the original chemistry developed by Khorana and co-workers (phosphodiester, Section 4.1.3), deoxynucleoside 5′-phosphates were used as building blocks. In other chemistries developed more recently 5′-*O*-dimethoxytrityl-(*N*-acylated)-2′-deoxynucleosides are phosphorylated or phosphitylated at the 3′-hydroxyl group (Figure 4.5). In these cases the products of synthesis after assembly of the oligonucleotide are phosphate triesters, where the internucleoside phosphate carries a protecting group. In **phosphotriester** chemistry [P(V)] the best protecting groups are aryl (usually mono- or di-chlorophenyl derivatives). This is because an aryl phosphodiester is a much more reactive deoxynucleoside building block than an alkyl phosphodiester in a coupling reaction. For example, 5′-*O*-dimethoxytrityl-$N^6$-benzoyl-2′-deoxyadenosine gives the corresponding 3′-*O*-(2-chlorophenyl) phosphodiester by reaction

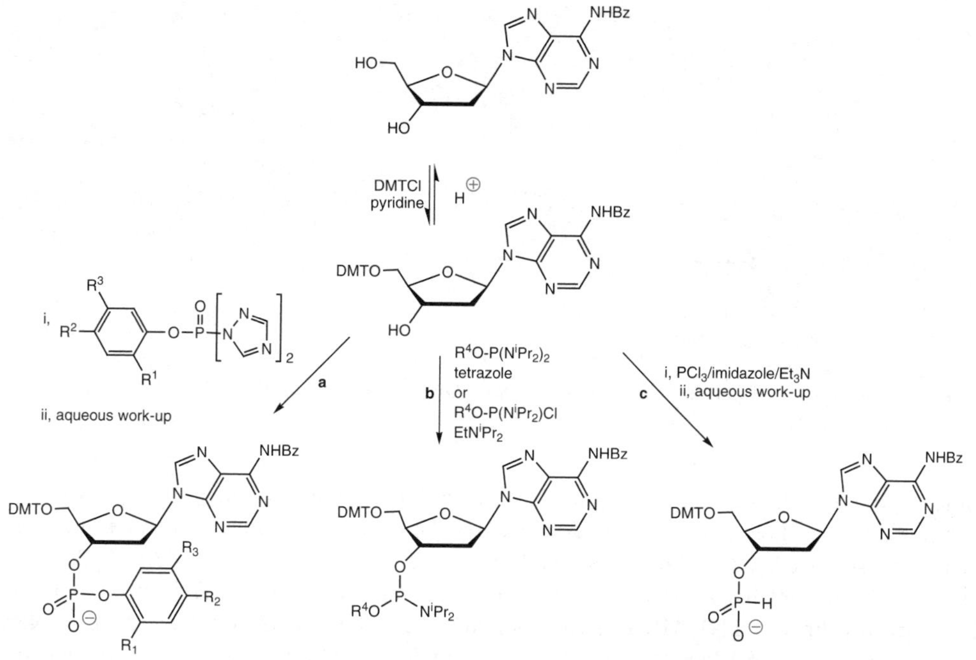


**Figure 4.5** *Introduction of a 3′-phosphate by (a) phosphorylation, (b) phosphitylation, and (c) H-phosphonylation. $R^1$, $R^3$ = H, $R^2$ = Cl, 4-chlorophenyl; $R^2$, $R^3$ = H, $R^1$ = Cl, 2-chlorophenyl; $R^2$ = H, $R^1$, $R^3$ = Cl, 2,5-dichlorophenyl; $R^4$ = methyl or 2-cyanoethyl*

with 2-chlorophenyl phosphoro-*bis*(triazolide) (Figure 4.5, Route **a**). Despite this being a bifunctional phosphorylating agent it acts as a monofunctional one in the absence of any stronger catalyst.

In **phosphate-triester** chemistry [P(III)] both aryl and alkyl phosphates are highly reactive species. Here, a methyl group or 2-cyanoethyl group is the preferred protecting group because they can be removed conveniently and selectively at the end of the synthesis (Section 4.1.4). Again a bifunctional reagent is used in a monofunctional manner, but to obtain a sufficiently stable product a phosphoramidite is prepared (Route **b**). The monofunctional chlorophosphoramidite can also be used.

**H-Phosphonate** chemistry does not require protection of the phosphate group, since the internucleoside H-phosphonate linkage in an oligonucleotide is stable to the conditions used in the assembly of the oligonucleotide. In a sense, a proton is the protecting group. A 2′-deoxyribonucleoside 3′-H-phosphonate is prepared by the reaction of a deoxynucleoside with phosphorus trichloride and imidazole or triazole in the presence of a basic catalyst, such as *N*-methylmorpholine, followed by an aqueous work-up (Route **c**).

### 4.1.3 Ways of Making an Internucleotide Bond

The development of an efficient method for forming an internucleotide bond was for many years the most central issue in oligonucleotide synthesis.[2] The problem was solved by the development of phosphite triester chemistry (phosphoramidite) and, to some extent, H-phosphonate chemistry. However, an understanding of earlier phosphodiester and phosphotriester chemistry is important (see Section 3.2.3).

*4.1.3.1 Phosphodiester.* In the pioneering gene syntheses by Khorana and colleagues in the 1960s and 1970s (see Section 5.4.1),[3] oligonucleotide synthesis involved coupling a 5′-protected deoxynucleoside derivative with a 3′-protected deoxyribonucleoside-5′-phosphomonoester (Figure 4.6). The coupling agent (triisopropylbenzenesulfonyl chloride, TPS) activates the phosphomonoester by a complex reaction mechanism that gives a powerful phosphorylating agent, which reacts with the 3′-hydroxyl group of the

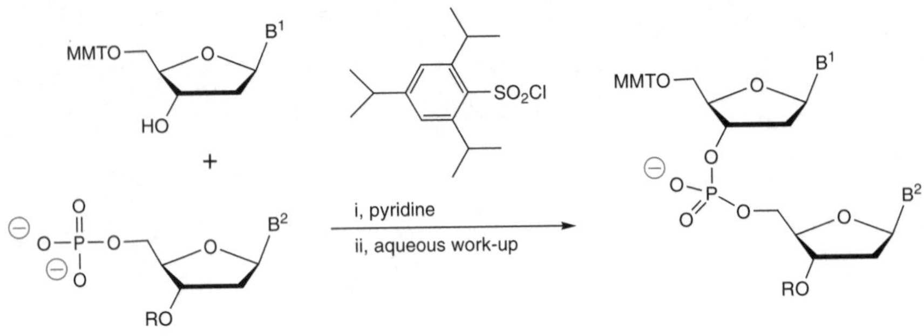


**Figure 4.6** *Formation of an internucleotide bond by the phosphodiester method. B = T, $C^{Bz}$, $A^{Bz}$ or $G^{iB}$. MMT is monomethoxytriphenylmethyl*

5′-unit to yield a dinucleoside phosphodiester. The main drawback is that the product phosphodiester is also vulnerable to phosphorylation by the activated deoxyribonucleoside phosphomonoester to give a trisubstituted pyrophosphate derivative. An aqueous work-up is necessary to regenerate the desired phosphodiester. Extension of the chain involves removal of the 3′-protecting group with alkali (for R = acetyl) or fluoride ion (for R = *tert*-butyldiphenylsilyl, TBDPS) and coupling with another deoxyribonucleoside 5′-phosphate derivative. To prepare oligonucleotides beyond five units, preformed blocks containing two or more deoxyribonucleotide residues must be coupled. Such blocks require significant effort to synthesise and contain unprotected phosphodiesters that undergo considerable side reactions. The synthetic products of coupling reactions require lengthy purification. Thus, synthesis of an oligonucleotide of 10–15 residues (the effective limit of the method) took upwards of 3 months. Although in the late 1970s phosphodiester chemistry was successfully applied to solid-phase synthesis (Section 4.1.4), the low yields intrinsic to phosphodiester chemistry remained.

*4.1.3.2 Phosphotriester.* Although this chemistry was first applied to solution-phase synthesis, it proved particularly successful when applied to solid-phase synthesis in the early 1980s.[2] A 5′-*O*-(chlorophenyl phosphate) is coupled to a deoxynucleoside attached at its 3′-position to a solid support (Figure 4.7). The coupling agent (mesitylenesulfonyl 3-nitro-1,2,4-triazolide, MSNT) is similar to that used in phosphodiester synthesis, except that 3-nitrotriazolide replaces chloride. The coupling agent activates the deoxyribonucleoside 3′-phosphodiester and allows reaction with the hydroxyl group of the support-bound deoxyribonucleoside. The rate of reaction can be enhanced by addition of a nucleophilic catalyst such as *N*-methylimidazole. This participates in the reaction by forming a more activated phosphorylating intermediate (an *N*-methylimidazolium phosphodiester), since the *N*-methylimidazole is a better leaving group. The product is a phosphodiester and accordingly is protected from further reaction with phosphorylating agents. The yield is therefore much better than in the case of a phosphodiester coupling, but phosphotriester chemistry could only be used satisfactorily after the development of selective reagents for cleavage of the aryl protecting group. To extend the chain, the DMT group is removed by the treatment with acid to liberate the hydroxyl group for further coupling. Note the direction of extension is 3′→5′, in contrast to solid-phase phosphodiester chemistry.

Two side reactions give rise to limitations. During coupling there is a competitive reaction (about 1%) of sulfonylation of the 5′-hydroxyl group by the coupling agent. This limits the efficiency of phosphotriester coupling to 97–98%, and thus also the length of oligonucleotide attainable to about 40 residues. More seriously, deoxyguanosine residues are subject to both phosphorylation and nitrotriazole substitution at the O-6-position unless the O-6-position is protected. $O^6$-Phosphorylation is particularly serious since this is not easily reversible (in contrast to phosphitylation) and leads to chain branching and eventually chain degradation.

The phosphotriester method is particularly useful for large-scale (multi-gram) synthesis of short oligonucleotides. Here the solid support is usually replaced by an acetyl or benzoyl group for solution phase stepwise synthesis, or by a soluble polymeric carrier.

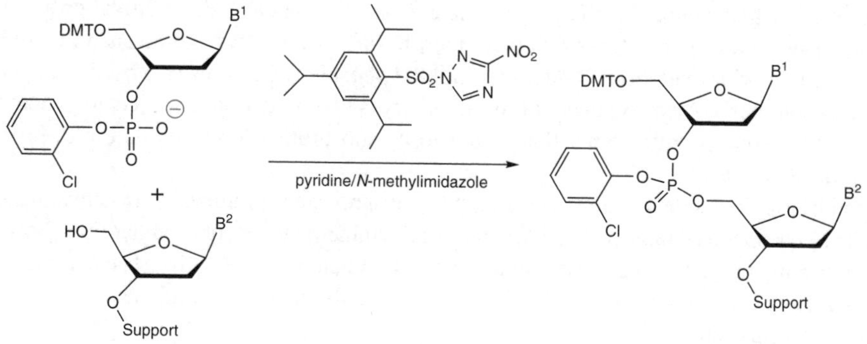


**Figure 4.7** *Formation of an internucleotide bond by the solid-phase phosphotriester method*

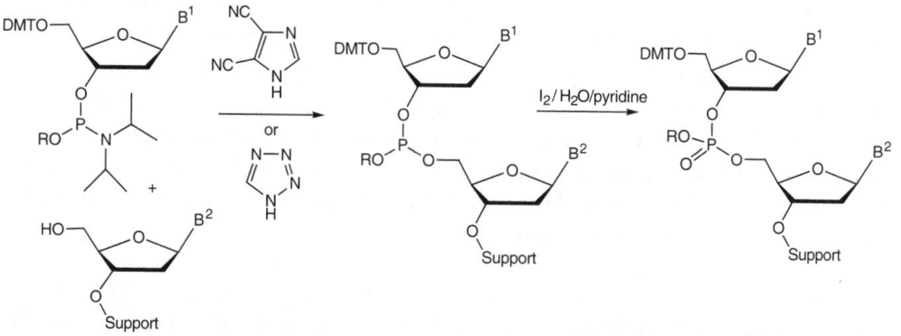


**Figure 4.8** *Formation of an internucleotide bond by the solid-phase phosphoramidite method. R = methyl or 2-cyanoethyl*

*4.1.3.3 Phosphite Triester.* The development of phosphite triester (or phosphoramidite) chemistry by Caruthers and co-workers in the early 1980s transformed oligonucleotide synthesis into an efficient and automated process.[4,5] The crux of this chemistry is a highly efficient coupling reaction between a 5′-hydroxyl group of a support-bound deoxyribonucleoside and a 5′-DMT-(*N*-acetylated)-deoxyribonucleoside 3′-*O*-(*N,N*-diisopropyl *O*-alkyl phosphoramidite (the alkyl group being methyl or 2-cyanoethyl) (Figure 4.8). In early development of this chemistry, a chlorophosphite was used in place of the *N,N*-diisopropylphosphoramidite, but was found to be unstable on storage. By contrast, a phosphoramidite is considerably less reactive and requires protonation on nitrogen to make the phosphoramidite into a highly reactive phosphitylating agent. A weak acid (such as tetrazole or 4,5-dicyanoimidazole) can do this without causing loss of the DMT group. The product of coupling is a dinucleoside phosphite, which must be oxidised with iodine to the phosphotriester before proceeding with chain extension.

The efficiency of coupling is extremely high (>98%) and the only major side reaction is phosphitylation of the $O^6$-position of guanosine. Fortunately, after coupling, treatment with acetic anhydride and *N*-methylimidazole (introduced to cap off any unreacted hydroxyl groups) completely reverses this side reaction. Solid-phase phosphoramidite chemistry may be used for synthesis of oligodeoxynucleotides up to 150 residues in length and to prepare products on a scale from micrograms to many grams.

*4.1.3.4 H-Phosphonate.* Although the origins of this chemistry lie with Todd and co-workers in the 1950s, the potential in oligonucleotide synthesis emerged more recently.[2] A deoxyribonucleoside 3′-*O*-(H-phosphonate) is essentially a tetra-coordinated P(III) species, preferring this structure to the tautomeric tri-coordinated phosphate monoester. Activation is achieved with a hindered acyl chloride (*e.g.*

pivaloyl chloride), which couples the H-phosphonate diester to a nucleoside hydroxyl group (Figure 4.9). The resultant H-phosphonate diester is relatively inert to further phosphitylation, such that the chain may be extended without prior oxidation. Oxidation of all the phosphorus centres is carried out simultaneously at the end of the synthesis. An advantage of this chemistry is that oxidation is subject to general base catalysis and this allows nucleophiles other than water to be substituted during oxidation to give a range of oligonucleotide analogues.

Unfortunately, a serious side reaction occurs if an H-phosphonate is premixed with activating agent before coupling. The H-phosphonate rapidly dimerises to form a symmetrical phosphite anhydride. Subsequent reaction of this with a hydroxyl group gives rise to a branched trinucleotide derivative. The complete elimination of this side reaction, even under optimal conditions, is probably impossible and may account for the lower yields obtained by this route.

### 4.1.4 Solid-Phase Synthesis

The essence of solid-phase synthesis is the use of a heterogeneous coupling reaction between a deoxynucleoside derivative in solution and another residue bound to an insoluble support. This has the advantage that a large amount of the soluble deoxynucleoside derivative can be used to force the reaction to high yield. The support-bound product dinucleotide is removed from the excess of reactant mononucleoside derivative simply by filtration and washing. Other reactions are also carried out heterogeneously and reagents removed similarly. This process is far faster than a conventional separation technique in solution and easily lends itself to mechanisation. Protocols and full details of the chemistry are available.[6,7] There are four essential features of solid-phase synthesis.

*4.1.4.1 Attachment of the First Deoxynucleoside to the Support.* Of the many types of support that have been used for solid-phase synthesis of oligonucleotides, only controlled pore glass (CPG) and polystyrene have proved to be generally useful. CPG beads are ideal in being rigid and non-swellable. They are manufactured with different particle sizes and porosities and they are chemically inert to reactions involved in oligonucleotide synthesis. Currently, 500–1000 Å porosities are favoured, the latter for synthesis of chains longer than 80 residues. The silylation reactions involved in functionalisation of glass (introduction of reactive

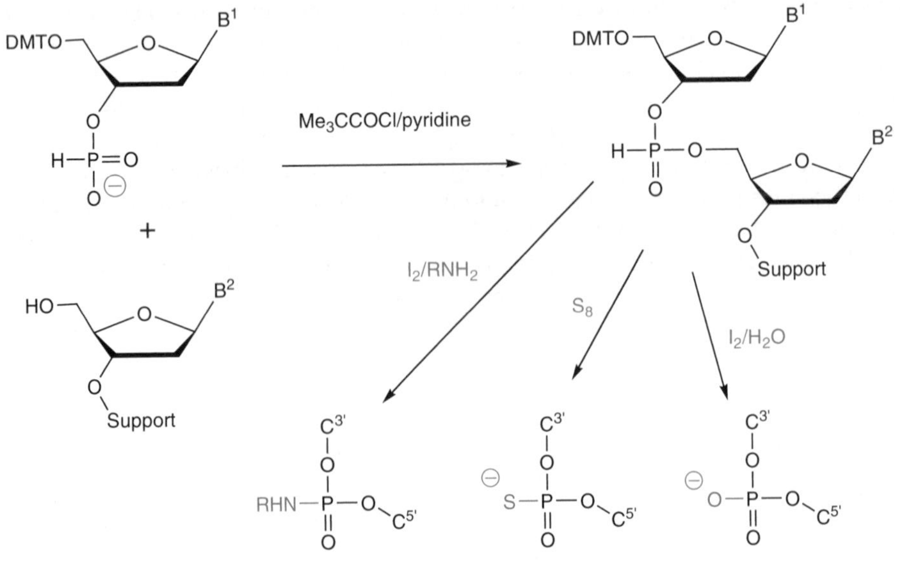


**Figure 4.9** *Formation of an internucleotide bond by the solid-phase H-phosphonate method*

sites) are beyond the scope of this Chapter. It is sufficient to note that a long spacer is used to extend the sites away from the surface and ensure accessibility to all reagents. One type of spacer is illustrated (Figure 4.10). The loading of amino groups on the glass is best kept within a narrow band of 30–80 $\mu$mol g$^{-1}$, below which the reactions become irreproducible and above which they are subject to steric crowding between chains. Highly cross-linked polystyrene beads have the advantage of good moisture exclusion properties, and allow efficient oligonucleotide synthesis on an extremely small scale (10 nmole).

The 3′-terminal deoxyribonucleoside of the oligonucleotide to be synthesised is attached to the solid support *via* an ester linkage by conversion of the protected 5′-*O*-DMT derivative into its corresponding active succinate ester, which is subsequently reacted with amino groups on the support (Figure 4.10). An assembled oligonucleotide is released from the support by treatment with ammonia. Several other types of derivatised solid supports are now available, which are obtainable through reagent suppliers.

*4.1.4.2 Assembly of Oligonucleotide Chains.* Assembly of the protected oligonucleotide chain is carried out by packing a small column of deoxynucleoside-loaded support and flowing solvents and reagents through in predetermined sequence. Columns containing only a few milligrams (10 nmole) up to tens of grams (1 mmole or more) can be used. Small-scale assembly is usually accomplished by use of a commercial DNA Synthesiser. Machine specifications vary, but the basic steps for oligonucleotide synthesis are as shown in Figure 4.11.

Step 1. Detritylation (removal of the 5′-DMT group) is carried out with dichloroacetic or trichloroacetic acid in dichloromethane. The orange colour from the dimethoxytrityl cation liberated from this step is compared by intensity in a UV–Visible spectrometer to obtain the **coupling efficiency** of the previous step.

Step 2. Activation of the phosphoramidite occurs when it is mixed with coupling agent (4,5-dicyanoimidazole, tetrazole or a derivative such as *S*-ethyl thiotetrazole) in acetonitrile solution (see Figure 3.56).

Step 3. Addition of the activated phosphoramidite to the growing chain.

Step 4. Capping is a safety step introduced to block chains that have not reacted during the coupling reaction and also limits the number of failure sequences. A fortuitous benefit of this step is that phosphitylation of the O-6-position of guanosine is reversed. This is carried out using a mixture of two solutions: acetic anhydride/2,6-lutidine and *N*-methylimidazole each in tetrahydrofuran (THF).

Step 5. Oxidation of the intermediate phosphite to the phosphate triester is achieved with iodine and water in THF. Pyridine or 2,6-lutidine is added to neutralise the hydrogen iodide liberated.

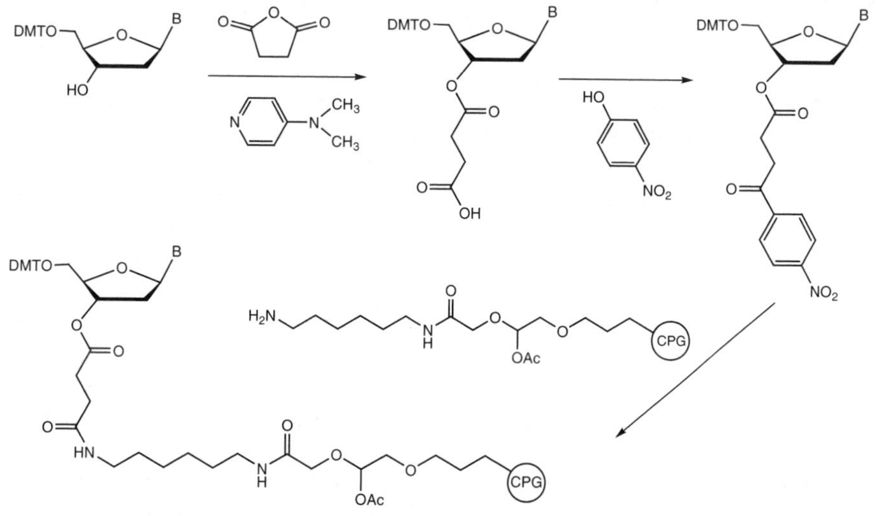


**Figure 4.10** *Attachment of a 5′-protected nucleoside to a solid support of controlled pore glass (CPG) functionalised by a long chain alkylamine*

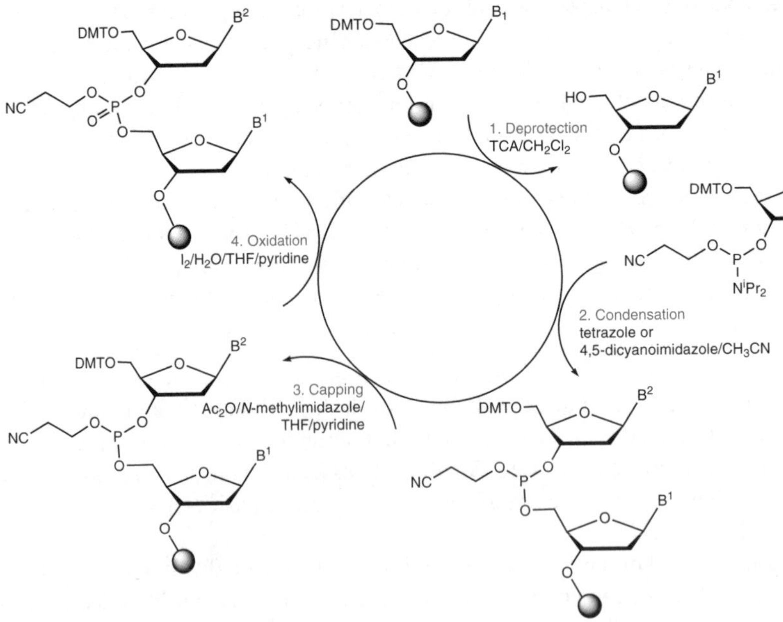


**Figure 4.11** *Basic steps in a cycle of nucleotide addition by the phosphoramidite method*

This cycle is repeated the requisite number of times for the length of the oligonucleotide required, with each deoxynucleoside phosphoramidite added in the desired sequence. Synthesis by this method is carried out in the 3′→5′ direction.

The traditional method for the synthesis of oligonucleotides outlined above is in the 3′→5′ direction. However, there are applications where it is desirable to reverse this direction of synthesis, for example, when oligonucleotides are required with their 5′-end attached to a support such as on a microarray chip or to a bead. In such cases, synthesis in the 5′→3′ direction has been made possible by the use of 5′-phosphoramidite building blocks. The overall chemical strategy for synthesis remains unchanged, but the functional groups on the 3′- and 5′-hydroxyl groups are exchanged.

*4.1.4.3 Deprotection and Removal of Oligonucleotides from the Support.* Unless there is a need to purify the oligonucleotides by reversed phase chromatography (see Section 4.1.4.4), the 5′-DMT group must first be removed using the same conditions as those used during oligonucleotide synthesis. If phosphoramidite chemistry has been used, then deprotection and removal of the oligonucleotide from the solid support is carried out in a single step. The solid support-bound oligonucleotide is treated with concentrated ammonia for 30 min and the column is then washed with a further portion of ammonia solution. This serves to cleave the oligonucleotide from the solid support. Nucleobase and phosphate protecting groups (2-cyanoethyl is removed from the phosphate by a β-elimination reaction, Figure 3.47) are then removed by heating an ammoniacal solution at 50°C overnight. Shorter deprotection times and lower temperatures are used when the mild-deprotection groups (PAC, *etc.*) are used. If methyl phosphoramidites are used, then the methyl group may be removed by treatment with thiophenolate ion (generated with thiophenol and triethylamine) prior to treatment with ammonia (Figure 3.46b). Lyophilisation of the ammonia solution gives the crude oligonucleotide.

In phosphotriester chemistry, the phosphate aryl group is selectively displaced by use of *syn*-2-nitrobenzaldoximate ion or by 2-pyridine-carbaldoximate ion (Figure 3.48). The product of this reaction undergoes elimination in the presence of water. Removal of the base protecting groups and cleavage of the oligonucleotide from the solid support is then carried out with ammonia as described above.

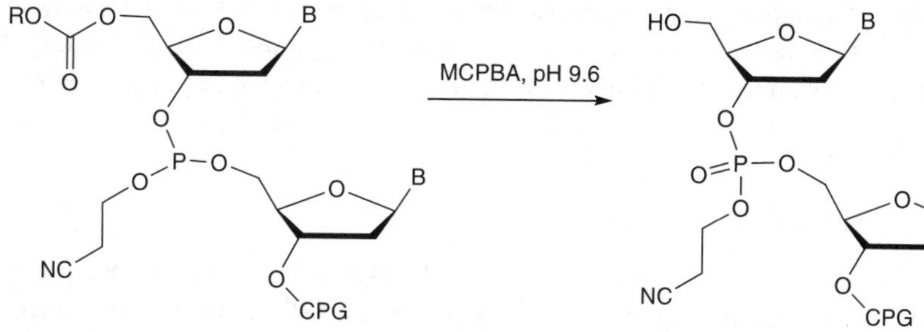


R = 4-chlorophenyl or 3-(trifluoromethyl)phenyl

**Figure 4.12** *Simultaneous oxidation of the internucleotide linkage and removal of the 5′-aryloxycarbonyl protecting group carried out with m-chloroperbenzoic acid (MCPBA) in the presence of lithium hydroxide and 2-amino-2-methyl-1-propanol at pH 9.6 by the phosphoramidite method*

*4.1.4.4 Purification of the Oligonucleotides.* The average yield for each step during an oligonucleotide synthesis is usually in excess of 98%, but for a long oligonucleotide this will correspond to a significant quantity of impurities and truncated oligonucleotides. The efficient removal of these impurities is an important process in the synthesis of oligonucleotides, and powerful separation methods have been developed for purification of microgram to milligram quantities of oligonucleotides.

**4.1.4.4.1 Polyacrylamide Gel Electrophoresis (PAGE).** PAGE separates oligonucleotides according to their unit charge difference (see Section 11.4.3). Oligonucleotides are applied to thick gels (1–2 mm) and after electrophoresis, the presence of the oligonucleotides may be detected with short wavelength (254 nm) UV light and the appropriate band cut out. The oligonucleotide may then be removed from the gel either by a soaking buffer or by electro-elution, followed by a desalting step using either a desalting column or dialysis. This method of purification is suitable for oligonucleotides of any length: short oligonucleotides being separated in a high percentage polyacrylamide gel (*e.g.* 20%) and longer oligonucleotides separated using lower polyacrylamide gel concentrations.

**4.1.4.4.2 High Performance Liquid Chromatography (HPLC).** HPLC is particularly suitable for purification of oligonucleotides. Ion exchange chromatography resolves predominantly by charge difference, and can be used both analytically and preparatively for oligonucleotides up to about 100 residues long. Reversed phase HPLC separates according to hydrophobicity, but the elution profile is less predictable than ion exchange chromatography. A common and more reliable method is to purify oligonucleotides before removal of the 5′-terminal DMT group, where the oligonucleotide will be resolved from the shorter non-DMT containing impurities. The 5′-DMT group is then cleaved after purification, and may be removed by a reversed-phase desalting cartridge or on a small gel filtration column.

There have been a number of recent protecting group strategies and improved reagents for the synthesis cycle devised to improve the yields of oligonucleotides even further. For example, in a recent method developed by Caruthers, aryloxycarbonyl protection is used for the 5′-hydroxyl group and DMT protection for the nucleobase exocyclic amino groups (Figure 4.12). After coupling in the usual manner, treatment of the extended chain with peroxy anions at pH 9.6 simultaneously cleaves the 5′-carbonate protection and oxidises the internucleoside phosphite linkage. In this way the number of steps in the synthesis cycle is reduced to two, resulting in a shorter nucleotide addition cycle.

## 4.2 SYNTHESIS OF OLIGORIBONUCLEOTIDES

The development of effective chemical methods for the synthesis of oligoribonucleotides has been slower than for oligodeoxyribonucleotides, largely because of the need to find a suitable protecting group for the additional 2′-hydroxyl group in ribonucleosides. Three effective methods for the synthesis of RNA are now available. Two of the methods involve silyl-type protecting groups at the 2′-position while DMT is used at the 5′-position. The third utilises silyl protection at the 5′-position and an acid-labile group to protect the 2′-position.

### 4.2.1 Protected Ribonucleoside Units

*4.2.1.1 Hetereocyclic Base Protection.* As with DNA synthesis, the exocyclic amino groups of adenine, guanine and cytosine need protection. Acyl groups are still the method of choice. Benzoyl and acetyl protecting groups are commonly used, since they are readily removed during the ammonia deprotection at the end of the synthesis. However, with the newer 2′-*O*-protecting group strategies it is often desirable to have amino-protecting groups that are removed under milder conditions. Therefore, PAC or dimethylaminomethylene are often used for adenine and guanine, while acetyl, though more stable than PAC, is also employed frequently. For coupling using phosphotriester chemistry, additional protection for the $O^6$- and $O^4$-positions of guanine and uracil respectively is necessary, since these positions are susceptible to reaction during phosphorylation.

*4.2.1.2 Hydroxyl Group Protection.* The 2′-hydroxyl group needs to be protected with a group that is stable throughout the synthesis and which can be removed selectively at the end without side-reactions. To introduce a protecting group at the 2′-hydroxyl group, orthogonal protection of both 5′- and 2′-positions is needed. In addition during such synthesis, there is danger of migration of protecting groups between the 2′- and 3′-hydroxyl positions, which can occur under both acidic and basic conditions and which makes the separation and purification of the desired 2′-protected nucleoside difficult.

Currently there are three main types of RNA phosphoramidite building blocks based on different O-2′-protecting groups (Figure 4.13). After initial 5′-protection, one of these protecting groups is introduced into a ribonucleoside selectively at the 2′-position and then phosphitylation of the 3′-hydroxyl group is carried out as described for 2′-deoxyribonucleosides.

**4.2.1.2.1 TBDMS.** The *tert*-butyldimethylsilyl (TBDMS) group is moderately stable to the acidic conditions used during sequential deprotection of the 5′-DMT group that is used in chain assembly, but

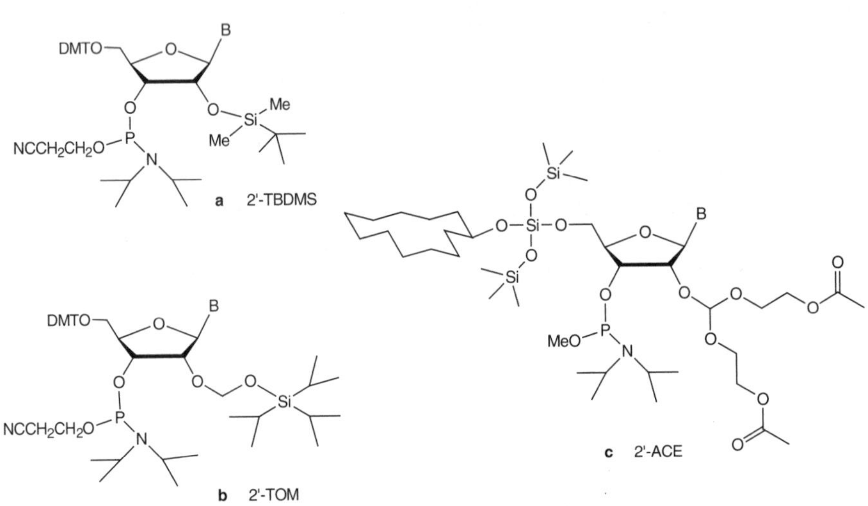


**Figure 4.13** *Standard building blocks used for RNA synthesis, (a) tert-butyldimethylsilyl (TBDMS) phosphoramidite, (b) triisopropylsilyloxymethyl (TOM) phosphoramidite, and (c) tris(acetoxyethyl) orthoformate (ACE) phosphoramidite*

can be removed by treatment with fluoride ion at the end of assembly (Section 4.2.2). TBDMS chemistry is useful both on small and larger production scale (Figure 4.13a).[8]

**4.2.1.2.2 TOM.** Triisopropylsilyloxymethyl (TOM) is a silyl-protected acetal, which has the advantage that no 2′↔3′ migration occurs under the usual basic conditions of introduction.[9] The nucleobase protecting groups are *N*-acetyl and the 5′-hydroxyl group is protected with DMT. TOM chemistry may have an advantage over TBDMS of slightly improved overall yields in RNA synthesis, but this view is not universally held (Figure 4.13b).

**4.2.1.2.3 ACE.** The 2′-*O*-[*bis*[2-(acetyloxy)ethoxy]methyl] (ACE) protecting group is entirely different from the other two RNA chemistries in being a **protected protecting group**. It is a protected orthoester.[10] The nucleobases are protected by acyl groups ($N^4$-acetyl-C, $N^6$-benzoyl-A and $N^2$-isopropyl-G), but at the 5′-position there is a cyclododecyloxy-*bis*(trimethylsiloxy)silyl (SIL) group rather than the usual acid-labile DMT group (Figure 4.13c). The ACE protecting group becomes acid-labile once the two acetyl groups of ACE have been removed (Section 4.2.2). The 3′-phosphoramidite has a methyl protecting group instead of the usual 2-cyanoethyl group, because the latter is unstable under the fluoride ion conditions needed to remove the 5′-SIL protecting group. ACE chemistry is currently only used for relatively small-scale syntheses on a polystyrene support (glass or silica is not compatible with the 5′-deprotection conditions) but has become particularly popular for siRNA synthesis (see Section 5.7.2).

## 4.2.2 Oligoribonucleotide Synthesis

*4.2.2.1 Assembly.* The assembly cycle for each of the three different RNA chemistries follows a similar overall route to that for DNA assembly (Section 4.1.4) and involves

(i) Deprotection of the 5′-protecting group. For TBDMS and TOM chemistry, this involves removal of DMT groups with di- or trichloroacetic acid, but for ACE chemistry, deprotection of the 5′-silyl group (SIL) is accomplished with triethylamine.3HF.
(ii) Coupling of the 3′-phosphoramidite to the free 5′-hydroxyl group using *S*-ethylthio tetrazole as activator.
(iii) Capping of any unreacted 5′-hydroxyl groups (acetic anhydride/2,6-lutidine and *N*-methylimidazole in MeCN, same reagents as for DNA assembly).
(iv) Oxidation with iodine/pyridine/water (same reagent as for DNA assembly).

*4.2.2.2 Deprotection and Purification.* Deprotection and removal of oligonucleotides from the solid support also uses procedures similar to those described for DNA synthesis (Section 4.1.4), but the reagents and conditions depend on the choice of protection strategy.

**4.2.2.2.1 TBDMS Chemistry.** 5′-Deprotection of DMT groups is usually carried out first by acidic treatment, while oligoribonucleotides are still attached to the support. Subsequent ammonia treatment then results in cleavage of the linkage of the oligonucleotide to the solid support simultaneously with the removal of nucleobase and phosphate (2-cyanoethyl) protecting groups. Mild methanolic ammonia (PAC protection) or aqueous ammonia (dimethylaminomethylene protection) at room temperature is suitable. Lastly, removal of the 2′-TBDMS group is effected by treatment with 1 M tetrabutylammonium fluoride (TBAF) in THF for 16–24 h or with triethylamine·(3HF). Final purification of deprotected oligoribonucleotides is carried out by polyacrylamide gel electrophoresis or by HPLC on ion exchange columns (Section 4.1.4). For 'DMT-on' purification, steps two and three are performed and the DMT-protected oligoribonucleotide is purified by HPLC and stored. The DMT group is removed by mild acid treatment before use.

**4.2.2.2.2 TOM Chemistry.** 5′-Deprotection of DMT groups is the same as for the TBDMS route. Cleavage from the solid support and removal of nucleobase protecting groups is effected with a 1:1 mixture of 40% aqueous methylamine and 33% ethanolic methylamine at room temperature overnight or 6 h at 35°C. 2′-TOM deprotection uses 1 M TBAF/THF. Removal of the 2′-hemiacetal occurs with the addition of 1 M Tris buffer. Purification is similar to that for TBDMS chemistry.

**4.2.2.2.3 ACE Chemistry.** Removal of the phosphate methyl ester is effected first by use of 1 M disodium 2-carbamoyl-2-cyanoethylene-1,1-dithiolate. Cleavage of the oligonucleotide from the solid support and removal of the nucleobase protecting groups is carried out with 40% aqueous methylamine at 55°C for 10 min, which also cleaves the acetyl groups from the ACE protecting group, rendering it acid-labile. Alternatively, the oligoribonucleotide can be desalted after release from the support and stored with the 2′-ACE protecting group intact. The ACE group may then be removed under mild acidic conditions just before use since the by-products from that deprotection are all volatile. Oligoribonucleotides may be purified by HPLC or by gel electrophoresis at either 2′-protected or deprotected stages (Section 4.1.4).

## 4.3 ENZYMATIC SYNTHESIS OF OLIGONUCLEOTIDES

Oligonucleotides of less than 50 residues are not usually prepared enzymatically because their chemical synthesis is very efficient and capable of producing sufficient quantities for most purposes. However, there are some occasions when it is desirable to synthesise oligonucleotides enzymatically. In particular enzymatic synthesis is used frequently to incorporate the triphosphate of a nucleoside analogue onto the 3′-end of a chemically-synthesised DNA primer in a primer-extension reaction. Further, RNA transcription is usually less expensive than chemical synthesis, especially on larger scale, and is more efficient than chemical synthesis for lengths of RNA of 50 residues or more.

### 4.3.1 Enzymatic Synthesis of Oligodeoxyribonucleotides

Numerous nucleoside and related analogues have been synthesised and their properties studied in enzymatic reactions (see Sections 3.1 and 3.7). Most commonly, such analogues are converted into a phosphoramidite or H-phosphonate and incorporated into an oligonucleotide (Section 4.4) so that their properties within a **template** may be studied. Alternatively, the analogue may be converted into a 5′-triphosphate derivative and then incorporated at the 3′-end of an oligodeoxyribonucleotide **primer** in a **primer-extension reaction** (Figure 4.14). In each case, a short (typically 18–24 nucleotide) primer is annealed to a template and extension carried out in the presence of deoxyribonucleoside triphosphates and a DNA polymerase (see Section 3.6.1) (*e.g.* exonuclease-deficient Klenow fragment, or *Taq* DNA polymerase, see Section 5.2.2). To visualise the reaction it is necessary first to label the primer, typically by addition of 5′-$^{32}$P-phosphate with T4 polynucleotide kinase (see Section 5.3.3) or by incorporation of a fluorescent label onto the 5′-end of the primer during chemical synthesis.

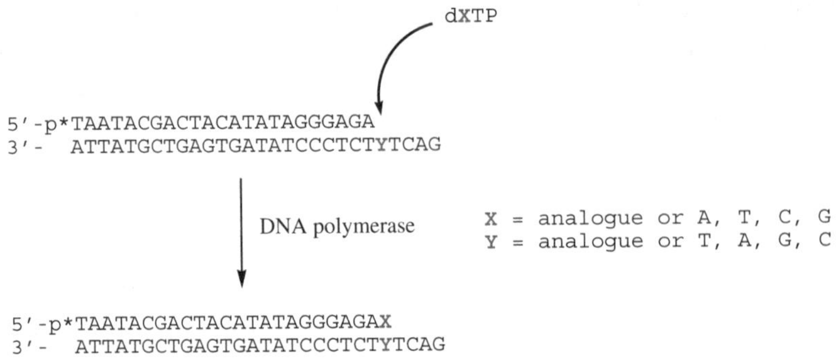


**Figure 4.14** *Primer extension reactions with DNA (or RNA) polymerases may be used to study the incorporation of nucleoside analogues as their 5′-triphosphates (X = analogue) or their properties when placed in a DNA template (Y = analogue)*

## 4.3.2 Enzymatic Synthesis of Oligoribonucleotides

*4.3.2.1 Transcription by T7 RNA Polymerase.* A powerful method of enzymatic RNA synthesis makes use of the RNA polymerase (see Sections 3.6.2 and 10.7.2) from bacteriophage T7 to copy a synthetic DNA template.[11] The template is prepared from two chemically synthesised oligodeoxyribonucleotides. Upon annealing, a duplex is formed corresponding to the base-pairs −17 to +1 of the T7 promoter sequence. Position +1 is the site of initiation of transcription, which in natural DNA would be in a fully base-paired duplex. For short RNA transcripts of 10–60 residues, it is possible to use a bottom strand that carries a single-stranded 5′-extension corresponding to the complement of the desired oligoribonucleotide. Transcription of this template *in vitro* with T7 RNA polymerase and nucleoside triphosphates gives up to 40 μmol of transcript per micromole of template (Figure 4.15).

Unfortunately there are limitations to this method. There are significant variations in the yield of RNA run-off transcripts, especially depending on the sequence from +1 to +5 in the template. In some cases there can be a high proportion of abortively-initiated transcripts. Transcription of higher efficiency and reliability is often obtained by the use of a fully double-stranded DNA template, either by chemical synthesis of both strands or by transcription of a plasmid DNA where the desired sequence is cloned 3′- to a T7-promoter and linearised by cutting with a restriction enzyme (see Section 5.3.1). Run-off transcription takes place up to the end of the DNA duplex at the restriction site.

A second problem is that in some cases a non-template-encoded nucleotide may be added to the oligoribonucleotide or the main product may be one nucleotide shorter than expected. An ingenious solution to this problem is to engineer the desired sequence within the plasmid 3′- to the T7-promoter and flanked by other sequences which, when transcribed, fold into self-cleavage domains,[12] as for example for the hammerhead (5′-flank) and hepatitis delta virus (3′-flank) ribozymes (see Section 7.6.2). During transcription the transcribed RNA folds and cleaves itself to give unique 5′- and 3′-ends.

To obtain oligoribonucleotides lacking the 5′-triphosphate, whichever transcription method is used, it is possible to initiate transcription by including in the reaction a high proportion of rGpG or the nucleoside rG, which is incorporated at the 5′-end of the transcript.

*4.3.2.2 Joining of Oligoribonucleotides.* An RNA ligase from the bacteriophage T4 (RNA ligase 1) catalyses the joining of a 5′-phosphate group of a **donor** molecule (minimum structure pNp) to a 3′-hydroxyl group of an **acceptor** oligonucleotide (minimum structure NpNpN) (Figure 4.16).[13]

The enzyme exhibits a high degree of preference for particular nucleotide sequences, favouring purines in the acceptor and a pyrimidine at the 5′-terminus of the donor, although there are substantial variations depending on the exact sequences of each. To prevent other possible ligation reactions, the acceptor

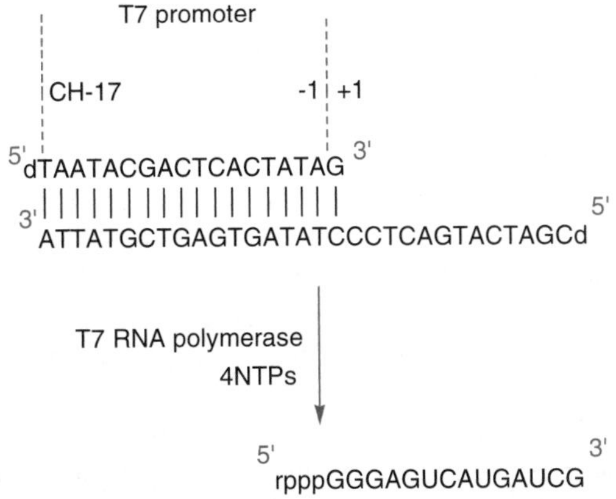


**Figure 4.15** *Use of T7 RNA polymerase to transcribe synthetic DNA templates*

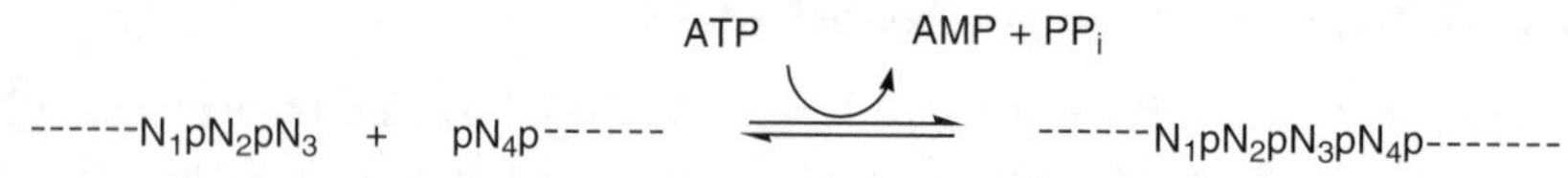


**Figure 4.16** *Joining oligoribonucleotides by use of RNA ligase*

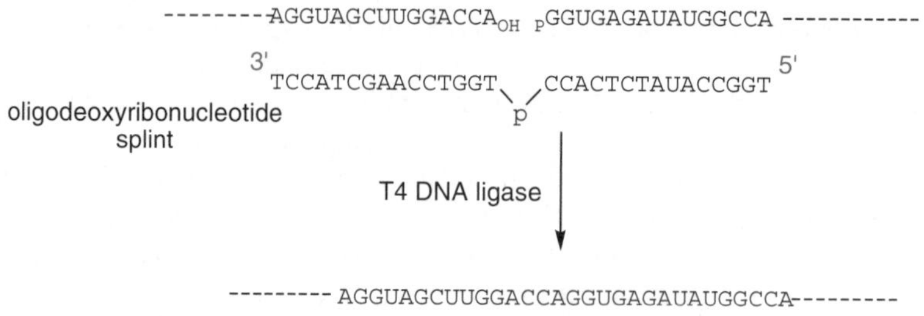


**Figure 4.17** *Joining of oligoribonucleotides by use of T4 DNA ligase*

carries no terminal phosphate whereas the donor is phosphorylated at both ends. The 3′-phosphate of the donor acts essentially as a protecting group. After joining it can be removed by treatment with alkaline phosphatase to generate a free 3′-hydroxyl group and thus a new potential acceptor. A particularly useful application is in the $^{32}$P-labelling of RNA, where T4 RNA ligase is used to catalyse the addition of [$^{32}$P]pCp to the 3′-end of the RNA.

Another method for joining RNA involves the use of a DNA ligase from bacteriophage T4 (normally used to join DNA, see Section 5.3.5) to unite two oligoribonucleotides or segments of RNA in the presence of a complementary oligodeoxyribonucleotide splint.[14] Both donor and acceptor oligoribonucleotides can be obtained by T7 RNA polymerase transcription or by chemical synthesis. In the example shown (Figure 4.17), the donor may be prepared by transcription with an rGpG or rG primer (this section) and then 5′-phosphorylated by the use of ATP and T4 polynucleotide kinase. Advantages of this method of ligation include a high sensitivity for acceptor oligoribonucleotides of the correct sequence (*i.e.* incorrect n + 1 long acceptor transcripts are not joined) and the lack of a need for 3′-protection of the donor oligoribonucleotide. The method has proved useful in incorporation of rG analogues at the joined site.

## 4.4 SYNTHESIS OF MODIFIED OLIGONUCLEOTIDES

### 4.4.1 Modified Nucleobases

Among the many research enterprises that involve modified oligonucleotides, the synthesis of nucleobase-modified oligonucleotides is probably the largest group.[15–19] Phosphoramidites of deoxyribo- or ribonucleosides, containing a number of modified nucleobases, are commercially available for incorporation into synthetic DNA or RNA by standard solid-phase synthesis (Section 4.1) (Figure 4.18). Among numerous applications, certain modified bases are used to increase the stability of a DNA duplex. For example, **5-propynyl-dU** extends the π-structure of the nucleobase and allows improved stacking with neighbouring bases within a DNA duplex. **7-Deaza-dG** (Figure 4.18b) is an analogue in which the N7 nitrogen atom is replaced by a methine (CH) group. Thus, it is very useful for understanding the role of the Hoogsteen edge of a G residue in the recognition of DNA by drugs and enzymes within the major groove of a synthetic DNA duplex (see Chapters 9 and 10). It is also used as a triphosphate analogue in place of dGTP for improving DNA sequencing (see Section 5.1) where a long run of dG residues would be formed in a sequencing reaction that would result in unusual structures, such as G-quartets (see Section 2.3.7). **5-BromodU** and **5-iododU** derivatives undergo photolytic cross-linking reactions and are useful for DNA–protein cross-linking. Similarly **4-thioU** is useful for RNA–RNA and RNA–protein photocross-linking. **2-Aminopurine** is an example of a fluorescent base with a high quantum yield that is useful for probing the conformation of RNA structures (Figure 4.18e).[20]

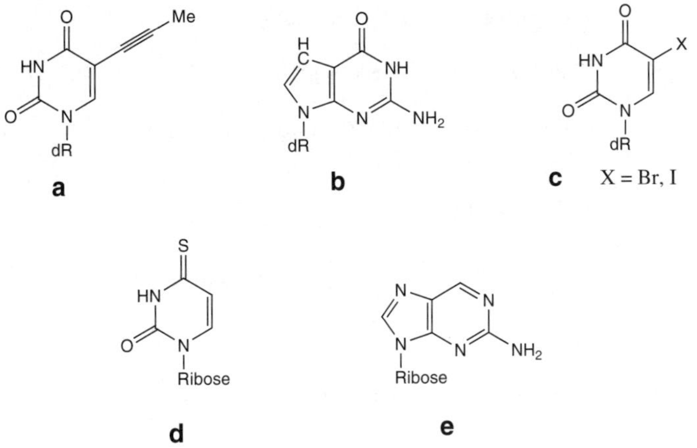


**Figure 4.18** *Nucleoside analogues used in structural studies involving oligonucleotides. (a) 5-propynyl-2′-deoxyuridine, (b) 7-deaza-2′-deoxyguanosine, (c) 5-halo-2′-deoxyuridine, (d) 4-thiouridine, and (e) 2-aminopurine riboside*

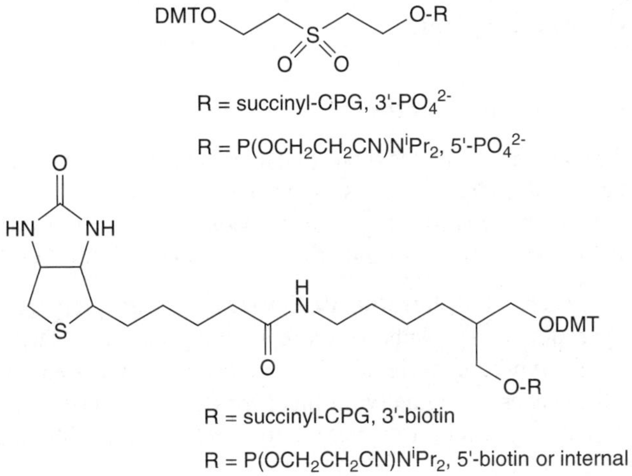


**Figure 4.19** *Oligonucleotide terminal modifiers*

### 4.4.2 Modifications of the 5′- and 3′-Termini

There are a number of reagents useful for attachment to the termini of synthetic oligonucleotides during chemical synthesis.[21,22] For example, phosphoramidite-building blocks are available for the synthesis of oligonucleotides bearing either 5′- or 3′-phosphate groups (Figure 4.19). An important class of modifiers are **reporter groups**. For example, fluorophores are useful in fluorescence studies for quantification or localisation of an oligonucleotide, in fluorescence resonance energy transfer (**FRET**) studies (Section 11.1.2) and in automated DNA sequencing (see Section 5.1.2). Another important reporter group is **biotin**, which may be incorporated, for example, during solid-phase synthesis as a phosphoramidite reagent or on the 3′-end of an oligonucleotide through attachment to the solid support (Figure 4.19). Biotinylated oligonucleotides have the advantage that they can be separated from other biomolecules by the extremely tight interaction of biotin with **streptavidin**, for example, with streptavidin-coated beads or micro-titre plates.

**Linkers** have become increasingly important units for conjugation of oligonucleotides to other biomolecules, particularly those linkers that have the capability of generating a terminal amino, thiol or carboxylate group following oligonucleotide synthesis and a subsequent linker deprotection reaction (Figure 4.20).

Amino linker $R^1NH(CH_2)_6O\text{-}R^2$

Thiol linker $TrS(CH_2)_6O\text{-}R^2$

Disulfide linker $DMTO(CH_2)_6\text{-}S\text{-}S\text{-}(CH_2)_6O\text{-}R^2$

Carboxylate linker

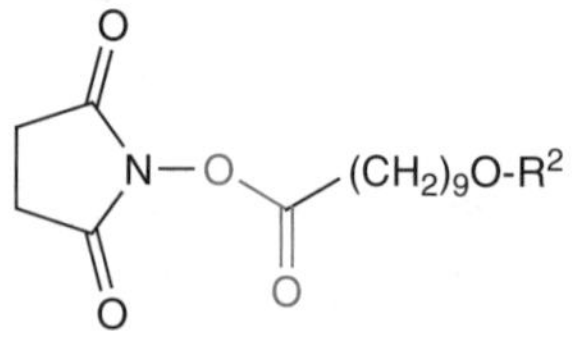


$R^1$ = monomethoxytrityl or trifluoroacetyl
$R^2$ = oligonucleotide

**Figure 4.20** *Linkers for oligonucleotide conjugation*

Oligonucleotide **conjugates** are formed by reaction with a complementary reactive group on the biomolecule, such as a peptide, for example to give amide, disulfide or thioether linkages depending on the types of functionalities involved in the conjugation.

## 4.4.3 Backbone and Sugar Modifications

Many modifications to the oligonucleotide backbone (the internucleotide linkage and/or sugar moiety) have found applications in the use of oligonucleotides as antisense agents (see Section 5.7.1) or in synthetic siRNA (see Section 5.7.2) for the control of gene expression.[23–25] The most common backbone modifications are described below, noting their advantages and disadvantages for their use.

*4.4.3.1 Phosphorothioates.* **Phosphorothioate** linkages were first prepared by Fritz Eckstein. They have a non-bridging oxygen atom of a phosphodiester replaced by sulfur.[18,19] They can be prepared during solid phase phosphoramidite synthesis by replacement of the oxidation step with a sulfurisation step. While elemental sulfur ($S_8$) was used originally for this purpose, the sulfurisation step is now carried out more rapidly and conveniently by use of a reagent such as 3*H*-1,2-benzodithole-3-one-1,1-dioxide (the **Beaucage reagent**)[26] (Figure 4.21).

The replacement of an oxygen atom by sulfur results in a mixture of two diastereoisomers at phosphorus and these are designated ($R_P$) and ($S_P$).* For an oligonucleotide containing a single phosphorothioate linkage, separation of the two diastereoisomers is usually possible by HPLC. For multiple sulfur substitutions, separation by HPLC is not possible, and the required pure diastereoisomer must be synthesised by stereospecific phosphorothioate chemistry developed by Wojciech Stec.[27]

Nucleotide phosphorothioates are isopolar and isosteric with phosphates, and generally only one of the two diastereoisomers is a substrate for native polymerases. ($S_P$)-α-Thiotriphosphates (Section 3.3.2) with a complete DNA polymerase lead to pure ($R_P$) phosphorothioate linkages, since the polymerase extension reaction proceeds with inversion of configuration (Figure 4.21). However, ($R_P$)-α-thiotriphosphate nucleotides can be accepted as poorer substrates by DNA polymerases using manganese or by the Klenow fragment of Pol-1 (see Section 5.1.1).

Phosphorothioate modifications in oligonucleotides have been particularly valuable in antisense applications for clinical use (see Section 5.7.1). They are more resistant to both exo- and endonucleases and are therefore used to enhance the stability of oligonucleotides in cells and in sera. One disadvantage is that

* Stereochemistry at a thiophosphate is defined according to the CIP convention with priority $S > O3' > O5' > O(=P)$.

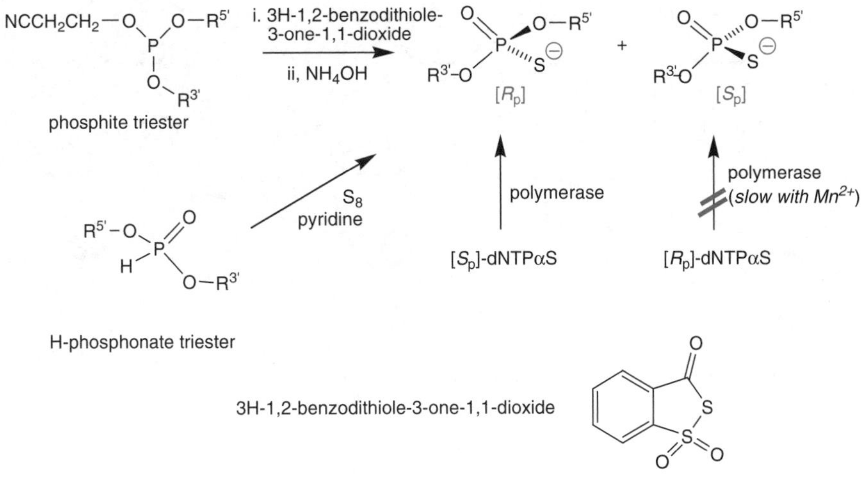


**Figure 4.21** *Synthesis of oligonucleotide phosphorothioates*

phosphorothioate-modified oligodeoxynucleotides with mixed phosphorothioate stereochemistry bind more weakly to complementary RNA targets than do regular phosphate oligomers. Also, in the case of uniform phosphorothiate oligonucleotides, there can be a loss of specificity of binding to nucleic acids while some non-specific binding to proteins may be observed. The challenge of the chemical synthesis of homochiral all-($R_P$) and all-($S_P$) oligomers has been accomplished by Wojciech Stec[28,29] leading to the conclusive result that oligomers having all-($R_P$) phosphorothioate linkages bind more tightly to RNA than do their phosphate counterparts. Single-site stereospecific modifications have been used in mechanistic studies involving oligonucleotides, for example in studies of the cleavage reaction of ribozymes (see Section 7.6.2).

Replacement of a **bridging** oxygen atom by sulfur is more difficult to achieve synthetically. Internucleotide coupling reactions involving sulfur are more difficult since the sulfur atom is less nucleophilic for phosphorus. Nevertheless, oligonucleotides in which the 3′- or 5′-bridging oxygen has been replaced by sulfur have been prepared and used in mechanistic studies.

*4.4.3.2 Phosphorodithioates.* **Phosphorodithioate** linkages have both the non-bridging oxygen atoms replaced by sulfur. Such linkages are non-chiral and are completely resistant to cleavage by all known nucleases. Caruthers has developed a synthesis of phosphorodithioate oligonucleotides that couples a 2′-deoxyribonucleoside 3′-phosphorothioamidite to a support-bound nucleoside 5′-hydroxyl group and is followed by a sulfurisation step (Figure 4.22). The 2-benzoylthioethyl group is removed by ammonia deprotection at the end of the synthesis.[30] Although this method can incorporate a phosphorodithioate linkage at any position in an oligonucleotide, phosphorodithioates are used infrequently because they bind to complementary oligonucleotides with reduced discrimination and also bind to various proteins.

*4.4.3.3 Methylphosphonates.* **Methylphosphonates** are uncharged analogues of phosphodiester anions in which a non-bridging oxygen atom of the phosphate group has been replaced by a methyl group (Figure 4.23a) (several other alkyl or aryl groups attached to phosphorus have also been used). Oligonucleotides containing methylphosphonate modifications are prepared from 3′-*O*-methylphosphonamidite nucleoside building blocks using conditions similar to standard phosphoramidite synthesis.[31] The methylphosphonate is chiral at phosphorus, so a mixture of isomers occurs and the synthesis of defined stereoisomers has been accomplished. Methylphosphonate diester linkages have enhanced stability to exo- and endonucleases and duplexes containing them have elevated $T_m$s. However, as this modification results in a loss of the phosphate anionic charge, poor aqueous solubility and aggregation of oligonucleotides can result from multiple methylphosphonate substitutions.

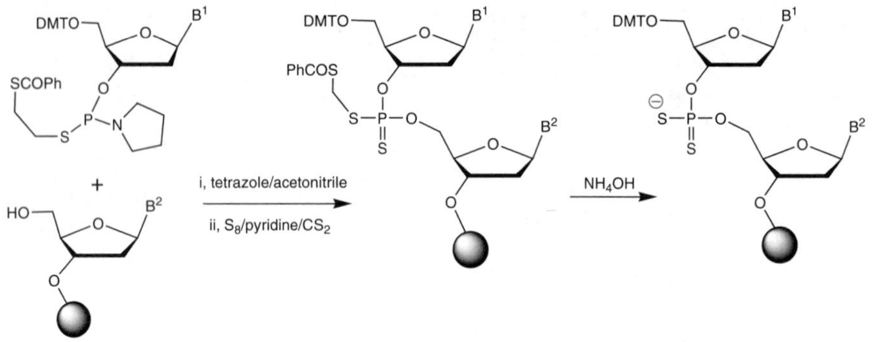


**Figure 4.22** *Solid-phase synthesis of oligonucleotide phosphorodithioates*

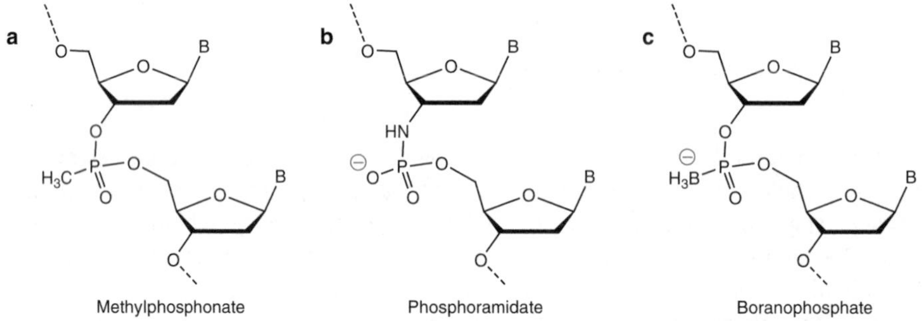


**Figure 4.23** *Structures of (a) the methylphosphonate internucleotide linkage, (b) the N3′-P phosphoramidate internucleotide linkage, and (c) the boranophosphate linkage*

*4.4.3.4 Phosphoramidates.* **Phosphoramidates** are internucleotide linkages in which either the 3′- or 5′-oxygen of the phosphodiester is replaced by an amino group. Much of the work with 3′-phosphoramidates was pioneered by Sergei Gryaznov.[32] In general these are now prepared by coupling a nucleoside 5′-phosphoramidite to a solid support-bound 3′-deoxy-3′-amino nucleoside to form an N3′-P-phosphoramidate linkage (Figure 4.23b).[33] Oligomers with phosphoramidate linkages show enhanced resistance to snake venom phosphodiesterases and give significantly higher $T_m$s for duplexes with complementary DNA and RNA strands. The internucleotide phosphonamidite linkage can be sulfurised to form a phosphorothioamidate and oligodeoxynucleotides with N3′-P5′-amidate linkages are useful steric block antisense reagents because their duplexes with RNA are not recognised by RNase H (see Section 5.7.1).

*4.4.3.5 Other Internucleotide Modifications.* One interesting analogue involves the use of **boranophosphate** internucleotide linkages first described by Barbara Shaw.[34] In the boranophosphate linkage, a non-bridging oxygen atom is replaced by a borano group ($BH_3^-$) (Figure 4.23c). This also creates a P-chiral centre. A boranophosphate is isoelectronic and isosteric with a natural phosphate, but it has increased lipophilicity. Boranophosphate-modified oligonucleotides can induce RNase H-mediated cleavage of complementary RNA and they have enhanced resistance to nucleases.

**2′–5′ linked oligoadenylates** (Figure 2.39) are an important class of naturally occurring oligoribonucleotide in which consecutive nucleotide units have 2′–5′ linkages. 2′–5′ Oligoadenylates are prepared by 2–5A synthetase from ATP in interferon-treated cells, and play a key role in mediating the antiviral effect of interferon.

*4.4.3.6 2′-Modifications.* Of the many 2′-modifications,[35] the **2′-*O*-methyl ribonucleoside** is the most well known (Figure 4.24). 2′-*O*-Methyloligoribonucleotides are more stable in binding complementary DNA or RNA than are oligodeoxyribonucleotides because the 2′-*O*-methyl sugar adopts a C3′-*endo*

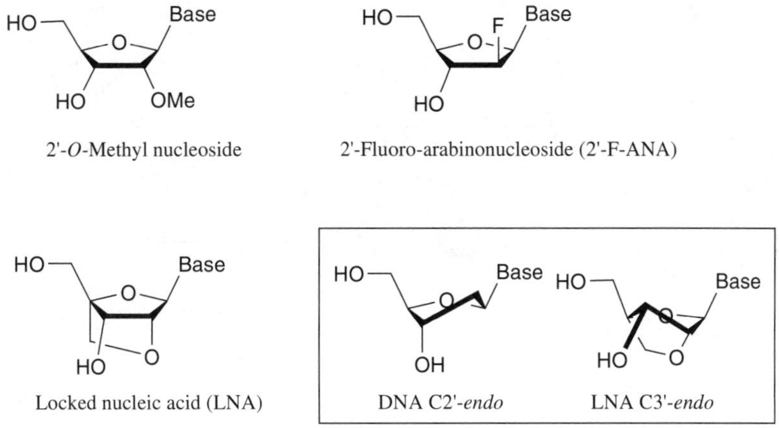


**Figure 4.24** *Structures of sugar modifications: 2′-O-methyl, 2′-O-F-ANA and locked nucleic acids (LNA). The conformation of the LNA sugar is compared to that of DNA*

ribose conformation (see Section 2.1.1). In contrast to RNA, there is a considerable increase in the stability of the oligonucleotide towards exonuclease degradation. Thus, 2′-*O*-methyl modifications are particularly useful in the 3′- and 5′-flanking regions of oligonucleotide gapmers and in steric block applications (see Section 5.7.1), with or without additional phosphorothioate modifications. The 2′-*O*-methoxyethyl (MOE) modification has similar uses.

2′-Deoxy-2′-fluoro-β-D-ribofuranosides can be considered close analogues of the natural β-D-ribose found in RNA since the sugar favours a C3′-*endo* pucker and A-type conformation when hybridised with RNA. However, such 2′-fluoro-containing oligonucleotides are not substrates for RNase H in duplexes with RNA. However, 2′-fluoronucleotides are one of a number of analogues being explored for use in synthetic siRNA (see Section 5.7.2) and in aptamer applications (see Section 5.7.3). By contrast, the 2′-deoxy-2′-fluoro-β-D-arabinonucleoside oligomers (2′-F-ANA, Figure 4.24) are substrates to direct cleavage by RNase H.[36]

Since modifications at the 2′-position are generally very well tolerated in oligonucleotide duplexes, the 2′-position has been widely used to attach a large variety of substituents, such as fluorophores, into oligonucleotides, either using the 2′-hydroxyl group or *via* a 2′-amino-2′-deoxy modification.

*4.4.3.7 Locked Nucleic Acids (LNA).* **LNA**s, also known as BNA (Figure 4.24), were first described by Takeshi Imanishi[37] and Jesper Wengel,[38] LNA has a methylene bridge between the 2′-oxygen and the C4′-carbon, which results in a locked 3′-*endo* sugar conformation, reduced conformational flexibility of the ribose ring and an increase in the local organisation of the phosphate backbone. The entropic constraint in LNA results in significantly stronger binding of LNA to complementary DNA and RNA. LNA-modified oligonucleotides have considerably enhanced resistance to nuclease degradation and they have proven to be effective in antisense strategies when used in flanking regions of gapmers or in steric block applications (see Section 5.7.1).

*4.4.3.8 Peptide Nucleic Acids (PNA).* **PNA**s were first introduced by Peter Nielsen and have normal nucleobases attached to a peptide-like backbone that is built from 2-aminoethylglycine units (Figure 4.25). As a result, PNA is electrically neutral but has excellent natural DNA and RNA recognition properties.[39,40]

PNA is synthesised by sequential **solid phase synthesis** similar to the methods employed in peptide synthesis and using protected PNA building blocks. In one system, the PNA unit has an acid-labile *t*-butyloxycarbonyl (tBoc) group for *N*-protection and an active ester activation of the carboxylic group. Additional benzyloxycarbonyl protecting groups are removed at the end of PNA assembly by treatment with HF. In a second method, involving milder chemistry, a 9-fluorenylmethoxycarbonyl (Fmoc) amino protecting group is removed by treatment with 20% piperidine/DMF while nucleobase protection uses a

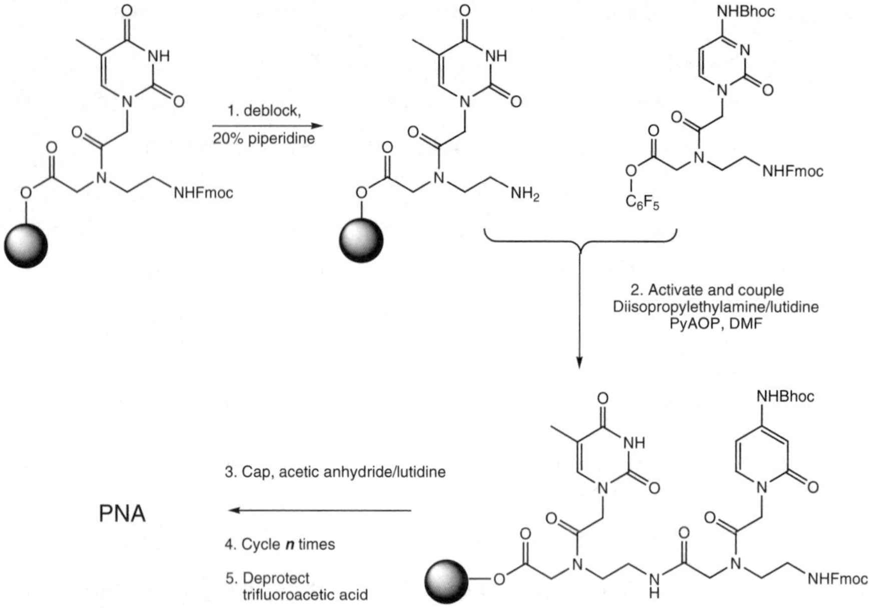


**Figure 4.25** *Synthesis of PNA by the Fmoc method. Bhoc: benzhydryloxycarbonyl, PyBOP: 7-azabenzotriazol-1-yloxytris (pyrrolidino)phosphonium hexafluorophosphate*

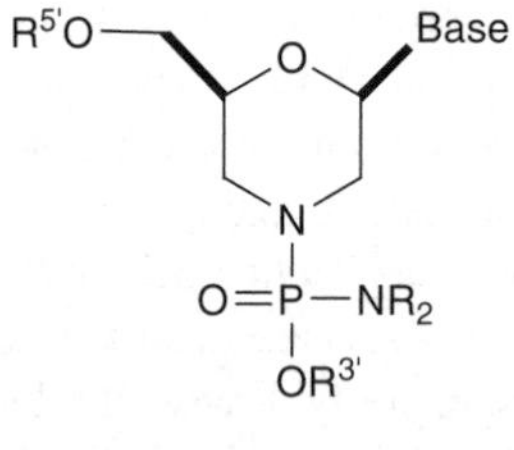


Morpholino phosphorodiamidate nucleoside

**Figure 4.26** *Structure of a morpholino phosphorodiamidate nucleotide residue*

benzhydryloxycarbonyl (Bhoc) group that is cleaved with aqueous trifluoracetic acid (Figure 4.25).[41] In both these systems, the coupling reactions are similar to those used in solid phase peptide synthesis to form an amide bond. Since unmodified PNA is rather insoluble in water, it is usual to incorporate a few cationic amino acids (especially lysines) to aid solubility. One advantage of PNA is that amino acids or peptides can be synthesised as direct conjugates with the DNA analogue. Such PNA–peptide conjugates are being explored in antisense applications for direct delivery of PNA into cells.

PNA forms particularly strong hybrids with DNA and RNA oligonucleotides and the inter-base distance in PNA when bound to such oligonucleotides is approximately the same as in the natural nucleotide strand. When bound to RNA, RNase H is not induced and therefore PNA is only used in steric block antisense approaches (see Section 5.7.1) or in microarray diagnostic applications (see Section 5.5.4). In addition, when targeted at DNA duplexes, PNA is able to displace one strand of the duplex to form a PNA:DNA:PNA triplex (see Section 2.3.6).

*4.4.3.9 Phosphorodiamidate Morpholino Modifications.* One final modification that has been used is the double replacement of the pentose by a morpholino-group and the phosphate non-bridging oxygen by an amino group to give a phosphorodiamidate morpholino (PMO) linkage (Figure 4.26). This type of modification has a number of advantages that have warranted its use in steric block antisense applications.[42] Such morpholino modifications give oligonucleotide analogues that are electrically neutral, show enhanced binding to DNA and RNA, are completely nuclease resistant, and have lower toxicity and greater specificity than phosphorothioate modifications. They have been used successfully in gene knockdown experiments by microinjection into cells and embryos.

## REFERENCES


1. S.L. Beaucage, in *Current Protocols in Nucleic Acids Chemistry*, Vol. 1, E.W. Harkins (ed). Wiley, 2005, 2.1.
2. C.B. Reese, The chemical synthesis of oligo- and polynucleotides: a personal commentary. *Tetrahedron*, 2002, **58**, 8893–8920.
3. H.G. Khorana, Total synthesis of a gene. *Science*, 1979, **203**, 614–625.
4. M.H. Caruthers, Gene synthesis machines – DNA chemistry and its uses. *Science*, 1985, **4723**, 281–285.
5. M.H. Caruthers, Chemical synthesis of DNA and DNA analogues. *Acc. Chem. Res.*, 1991, **24**, 278–284.
6. S. Agrawal, *Protocols for oligonucleotides and analogs*. Humana Press, Totowa, New Jersey, 1993.
7. E.W. Harkins, *Current Protocols in Nucleic Acids Chemistry*. Wiley, 2004.
8. M.J. Gait, C.E. Pritchard and G. Slim, in *Oligonucleotides and Analogues: A Practical Approach*, F. Eckstein (ed). Oxford University Press, Oxford, UK, 1991, 25–48.
9. S. Pitsch, P.A. Weiss, L. Jenny, A. Stutz and X. Wu, Reliable synthesis of oligoribonucleotides (RNA) with 2′-*O*-[(triisopropylsilyl)oxy]methyl (2′-*O*-tom)-protected phosphoramidites. *Helv. Chim. Acta*, 2001, **84**, 3773–3795.
10. S.A. Scaringe, F.E. Wincott and M.H. Caruthers, Novel RNA synthesis method using 5′-silyl-2′-orthoester protecting groups. *J. Am. Chem. Soc.*, 1998, **120**, 11820–11821.
11. J.F. Milligan, D.R. Groebe, G.W. Witherell and O.C. Uhlenbeck, Oligoribonucleotide synthesis using T7 RNA polymerase and synthetic DNA templates. *Nucleic Acids Res.*, 1987, **15**, 8783–8798.
12. S.R. Price, C. Oubridge, G. Varani and K. Nagai, in *RNA: Protein Interactions. A Practical Approach*, C.W.J. Smith (ed). OUP, Oxford, 1998, 37–74.
13. T. Middleton, W.C. Herlihy, P. Schimmel and H.N. Munro, Synthesis and purification of oligoribonucleotides using T4 RNA ligase and reverse phase chromatography. *Anal. Biochem.*, 1985, **144**, 110–117.
14. M.J. Moore and P.A. Sharp, Site-specific modification of pre-mRNA: the 2′-hydroxyl groups at the splice sites. *Science*, 1992, **256**, 992–997.
15. P. Herdewijn, Heterocyclic modifications of oligonucleotides and antisense technology. *Antisense Nucl. Acid Drug Dev.*, 2000, **10**, 297–310.
16. P. Hensley, Defining the structure and stability of macromolecular assemblies in solution: the re-emergence of analytical ultracentrifugation as a practical tool. *Structure*, 1996, **4**, 367–373.
17. I. Luyten and P. Herdewijn, Hybridisation properties of base-modified oligonucleotides within the double and triple helix motif. *Eur. J. Med. Chem.*, 1998, **33**, 515–576.
18. F. Eckstein, Nucleoside phosphorothioates. *Ann. Rev. Biochem.*, 1985, **54**, 367–402.
19. F. Eckstein and G. Gish, Phosphorothioates in molecular biology. *Trends Biol. Sci.*, 1989, **14**, 97–100.
20. *Specialist Periodical Reports: Organophosphorus Chemistry*. Royal Society of Chemistry, Cambridge, 2004, Vol. 33.
21. S.L. Beaucage and R.P. Iyer, The functionalization of oligonucleotides via phosphoramidite derivatives. *Tetrahedron*, 1993, **49**, 1925–1963.
22. M. Manoharan, Oligonucleotide conjugates as potential antisense drugs with improved uptake, biodistribution, targeted delivery, and mechanism of action. *Antisense Nucl. Acid Drug Dev.*, 2002, **12**, 103–128.
23. B.S. Sproat, Chemistry and applications of oligonucleotide analogues. *J. Biotechnol.*, 1995, **41**, 221–238.
24. R.P. Iyer, A. Roland, W. Zhou and K. Ghosh, Modified oligonucleotides-synthesis, properties and applications. *Curr. Opin. Mol. Ther.*, 1999, **1**, 344–358.

25. C. Leumann, DNA analogues: from supramolecular principles to biological properties. *Bioorg. Med. Chem.*, 2002, **10**, 841–854.
26. R.P. Iyer, W. Egan, J.B. Regan and S.L. Beaucage, 3H-1,2-Benzodithiole-3-one 1,1-dioxide as an improved sulfurizing reagent in the solid-phase synthesis of oligodeoxyribonucleoside phosphorothioates. *J. Am. Chem. Soc.*, 1990, **112**, 1253–1254.
27. W.J. Stec and A. Wilk, Stereocontrolled synthesis of oligo(nucleoside phosphorothioate)s. *Angew. Chem. Int. Ed.*, 1994, **33**, 709–722.
28. P. Guga and W.J. Stec, *Synthesis of phosphorothioate oligonucleotides with stereodefined phosphorothioate linkages*. Wiley, Hoboken, NJ, 2003.
29. M. Boczkowskaa, P. Guga and W.J. Stec, Stereodefined phosphorothioate analogues of DNA: relative thermodynamic stability of the model PS-DNA/DNA and PS-DNA/RNA complexes. *Biochemistry*, 2002, **41**, 12483–12487.
30. W.T. Wiesler and M.H. Caruthers, Synthesis of phosphorodithioate DNA via sulfur-linked base-labile protecting groups. *J. Am. Chem. Soc.*, 1996, **61**, 4272–4281.
31. P.S. Miller, M.P. Reddy, A. Murakami, K.R. Blake, S.-B. Lin and C.H. Agris, Solid-phase synthesis of oligodeoxyribonucleoside methylphosphonates. *Biochemistry*, 1986, **25**, 5092–5097.
32. J.-K. Chen, R.G. Schultz, D.H. Lloyd and S.M. Gryaznov, Synthesis of oligodeoxyribonucleotide N3′-P5′ phosphoramidates. *Nucleic Acids Res.*, 1995, **23**, 2661–2668.
33. J.S. Nelson, K.L. Fearon, M.Q. Nguyen, S.N. McCurdy, J.E. Frediant, M.F. Foy and B.L. Hirschbein, N3′-P5′ oligodeoxyribonucleotides phosphoramidates: a new method of synthesis based on a phosphoramidite amin-exchange reaction. *J. Org. Chem.*, 1997, **62**, 7278–2287.
34. J.S. Summers and B.R. Shaw, Boranophosphates as mimics of natural phosphodiesters in DNA. *Curr. Med. Chem.*, 2001, **8**, 1147–1155.
35. S.M. Frier and K.-H. Altmann, The ups and downs of nucleic acid duplex stability: structure-stability studies on chemically modified DNA:RNA duplexes. *Nucleic Acids Res.*, 1997, **25**, 4429–4443.
36. M.J. Damha, C.J. Wilds, A. Noronha, I. Brukner, G. Borkow, D. Arion and M.A. Parniak, Hybrids of RNA and arabinonucleic acids (ANA abd 2′-F-ANA) are substrated of ribonuclease H. *J. Am. Chem. Soc.*, 1998, **120**, 12976–12977.
37. S. Obika, D. Nanbu, Y. Hari, K. Morio, in Synthesis of 2′-*O*,4′-*C*-methylenuridine and -cytidine. Novel bicyclic nucleosides having a fixed C-3′,-endo sugar puckering. *Tetrahedron Lett.*, Y. Ishida and T. Imanishi (eds), 1997, **38**, 8735–8738.
38. S.K. Singh, P. Nielsen, A.A. Koshkin and J. Wengel, LNA (locked nucleic acids): synthesis and high-affinity nucleic acid recognition. *J. Chem. Soc. Chem. Commun.*, 1998, 455–456.
39. P.E. Nielsen, M. Egholm, Berg and O. Buchardt, Sequence-selective recognition of DNA by strand displacement with a thymine-substituted polyamide. *Science*, 1991, **5037**, 1497–1500.
40. M. Egholm, O. Buchardt, L. Christensen, C. Behrens, S.M. Freier, D.A. Driver, R.H. Berg, S.K. Kim, B. Norden and P. Nielsen, PNA hybridizes to complementary oligonucleotides obeying the Watson–Crick hydrogen bonding rules. *Nature*, 1993, **365**, 566–568.
41. S.A. Thomson, J.A. Josey, R. Cadilla, M.D. Gaul, C.F. Hassman, M.J. Luzzio, A.J. Pipe, K.L. Reed, D.J. Ricca, R.W. Wiethe *et al.*, Fmoc mediated synthesis of peptide nucleic acids. *Tetrahedron*, 1995, **51**, 6179–6194.
42. J. Summerton, D. Stein, S.B. Huang, P. Matthews, D.D. Weller and M. Partridge, Morpholino and phosphorothioate antisense oligomers compared in cell-free and in cell systems. *Antisense Nucl. Acid Drug Dev.*, 1997, **7**, 63–70.

CHAPTER 5

# Nucleic Acids in Biotechnology

## CONTENTS

## 5.1 DNA SEQUENCE DETERMINATION

### 5.1.1 Principles of DNA Sequencing

There are two major ways of determining the sequence of a DNA molecule. These methods were developed in the laboratories of Sanger and of Gilbert for which each received a Nobel Prize in 1980. Both methods rely upon sequencing only one strand at a time.

*5.1.1.1 Sanger DNA Sequencing.* In the traditional method of **Sanger DNA Sequencing**,[1,2] the DNA to be sequenced acts as a template and a new strand of DNA is synthesised enzymatically by use of either the **Klenow fragment of DNA polymerase I**, which lacks the 3′-5′-exonuclease, or the DNA polymerase from bacteriophage T7[3] (Figure 5.1). The method depends on obtaining specific termination of the reaction at just one nucleotide base type to generate a mixture of shorter sequences.

To terminate the polymerisation at a specific point, a small amount of one of four 2′,3′-**dideoxynucleoside** 5′-triphosphates is added. These can be incorporated into a growing DNA strand, but since they possess no 3′-hydroxyl group, they are unable to accept the addition of any extra nucleotides. They are thus **chain terminators**. The addition of a small amount of one of these, together with all four of the normal 2′-deoxyribonucleoside 5′-triphosphates to a polymerisation reaction gives rise to a series of oligonucleotides, each terminated by a dideoxynucleotide. Four reactions are carried out in parallel, each with a different dideoxynucleoside triphosphate (A, G, C and T). Separation of the oligonucleotide extension products from each individual reaction is achieved by **polyacrylamide gel electrophoresis** under denaturing conditions (Section 11.4.3) to generate a **sequencing ladder**. For visualisation by autoradiography, one of the unmodified deoxynucleoside triphosphates is radiolabelled with $^{32}$P (or with $^{35}$S *via* use of an α-thio triphosphate, Section 3.3.2). For example in Figure 5.1, there are two fragments generated in the ddATP reaction, two in the ddGTP reaction, one in the ddCTP reaction and three in the ddTTP reaction. The order of fragments up the gel represents the sequence of the extension product from 5′ to 3′. The complement of this 'read' sequence is that of the template.

In practice, it is necessary to elongate a short primer that has already been annealed to the template, since DNA polymerases can only elongate existing hybrids. For this purpose, the DNA fragment to be sequenced is usually sub-cloned into a vector (Section 5.2) that has known sequences flanking the insertion site. Chemically synthesised oligonucleotides (typically 17–25 nucleotides in length) that correspond to one or the other side of the insert are annealed to the sub-clone of DNA and the dideoxy-sequencing reactions are carried out on these templates. The polymerisation reaction can proceed on double-stranded templates, with one strand being displaced by the elongated primer. More usually, single-stranded templates are used, such as the viral DNA from bacteriophage M13-derived recombinants. Two hundred to three hundred nucleotides can be sequenced routinely by this approach for each set of reactions.

Modern sequencing polymerases are derivatives of the **thermostable polymerase** from *Thermus aquaticus* (Taq). This allows sequence data to be obtained from a very few copies of DNA template by carrying out amplification cycle sequencing in a similar manner to PCR (Section 5.2.2).

*5.1.1.2 Maxam and Gilbert Sequencing.* This now rarely used method relies upon radioactive labelling of only one end of the DNA.[4,5] The labelled DNA is then subjected to four separate, partial, base-selective, chemical modification (or for G + A, depurination) reactions (Table 5.1). These reactions allow

**Table 5.1** *Base-selective cleavage reactions for sequencing DNA*

| *3′-Cleavage adjacent to* | *Modification* | *Reagent* | *Strand breakage* |
|---|---|---|---|
| G | Methylation | Dimethyl sulfate | 1 M piperidine (at 90°C for 30 min) |
| G + A | Depurination | 88% Formic acid | 1 M piperidine (at 90°C for 30 min) |
| T + C | Base ring-opening | Hydrazine | 1 M piperidine (at 90°C for 30 min) |
| C | Base ring-opening | Hydrazine, high salt | 1 M piperidine (at 90°C for 30 min) |

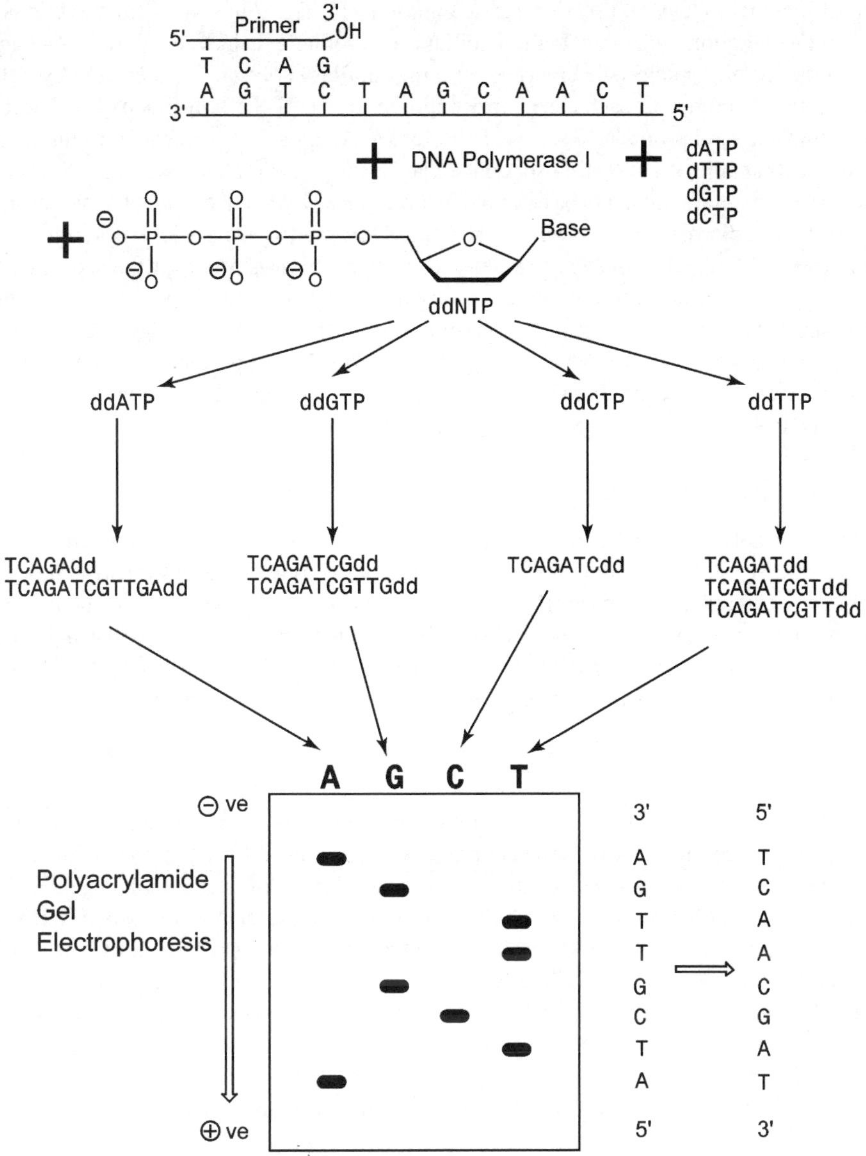


**Figure 5.1** *The principles of classical DNA sequencing. A primer oligodeoxynucleotide is annealed to the DNA template to be sequenced (top) and four separate extension reactions are carried out in the presence of DNA polymerase I, the four deoxynucleoside triphosphates (one usually $^{32}$P-labelled and a single 2′-3′-dideoxynucleoside triphosphate) to produce a series of truncated extension products (middle). The products of the reactions are separated by denaturing polyacrylamide gel electrophoresis and the gel autoradiographed to obtain a DNA sequence ladder (bottom)*

the DNA to become sensitive at these sites to cleavage by alkaline hydrolysis. The fragments created are then separated by polyacrylamide gel electrophoresis in very much the same way as for Sanger sequencing.

### 5.1.2 Automated Fluorescent DNA Sequencing

Machines have now been developed to separate and identify the products of dideoxy-sequencing reactions. Here a fluorescent label is built into a set of four alternative dideoxynucleotide chain terminators each with a different dye attached (Section 8.5.4, Figure 8.11). The four sequencing reactions are carried out together in one reaction and subjected to capillary electrophoresis in free solution. As each length of fluorescently labelled oligonucleotide emerges from the capillary column, it is detected by a fluorescence detector, the particular colour in each case corresponding to one of the four dideoxynucleotides. A computer analyses the data and produces a series of fluorescent signals that correspond to the read sequence. Such machines are capable of generating sequence reads of 500–1000 residues.

Sequencing machines have proved to be essential in large-scale DNA sequencing of genomes. For example, the DNA sequences of the yeast *Saccharomyces cerevisiae*, the fruitfly *Drosophila melanogaster* and *Homo sapiens* have been determined in this way. In genome sequencing, it is usual for the sequence to be determined three to ten times from different clones or fragments, including from both strands of the DNA, to obtain higher accuracy. Powerful computer programmes are then able to determine overlaps between fragments, and align the sequences against genome maps (Section 6.5.2). The positions of genes, introns and alternative splicing patterns can be predicted and genomes compared between different organisms to obtain knowledge of the RNA transcripts as well encoded proteins.

### 5.1.3 RNA Sequencing by Reverse Transcription

It is possible to carry out base-specific chemical treatments or digestions by nuclease enzymes to determine an RNA sequence directly. However, a more common procedure is to use a **reverse transcriptase** enzyme to make a cDNA copy of the single-stranded RNA. The reaction is initiated by use of an oligodeoxyribonucleotide primer from a known part of the sequence. This cDNA can then be amplified by the **polymerase chain reaction** (PCR) (Section 5.2.2) and sequence determined by standard DNA sequencing. The method only gives information regarding the base sequence and not regarding RNA modifications.

## 5.2 GENE CLONING

**Cloning** is the technique of growing large quantities of genetically identical cells or organisms that are derived from a single ancestor (clones). **Gene cloning** is an extension of this whereby a particular gene, group of genes or a fragment of DNA is selected from a mixed population (often a complete genome) and amplified to a huge extent. This can be carried out by insertion of the chosen DNA into a **vector DNA** and introduction of the hybrid (**recombinant DNA**) into cells by **transformation** (transfection). The cells containing the recombinant DNA are propagated and each cell in a colony contains an exact copy (or copies) of the gene 'cloned' in the vector. Cloning and recombinant DNA technology are well documented in established textbooks and manuals.[6,7]

Nowadays, a separate and complementary approach that uses PCR can achieve the same objective in a fraction of the time, of amplification of DNA segments without involving living cells.

### 5.2.1 Classical Cloning

*5.2.1.1 Vectors.* Several classes of vector exist into which a foreign DNA can be inserted and amplified. The major classes are **plasmids, bacteriophage** and **cosmids and bacterial or yeast artificial chromosomes** (YACs). Prokaryotic plasmids are almost always circular double-stranded DNAs that contain antibiotic resistance genes as markers and a variety of restriction sites that can be used for insertion of the foreign DNA. A large number of plasmids are available for use in *E. coli*. The most useful bacteriophage is λ, which has been engineered in many ways to accept inserts of many different sizes and types, up to approximately 20,000 base pairs. Because the transformation frequency is high, screening is easy. Cosmids are large plasmids that contain the packaging site for bacteriophage λ DNA. Therefore they can either be packaged into phage particles or they can be replicated as plasmids. Since the amount of DNA that can be packaged in a λ phage particle is 50,000 base pairs, the potential size of cosmid inserts is very large. Cosmids have been used in **chromosome walking** (see later this section), but they are somewhat more difficult to manipulate than is λ. Artificial chromosomes are vectors containing the constituents of natural yeast chromosomes, namely a prokaryotic origin of DNA replication in the case of **bacterial artificial chromosomes** (BACs) or a **centromere** (the region required for correct segregation of daughter chromosomes during mitosis and meiosis) and two **telomeres** (the ends of the chromosome) in the case of YACs. BACs accept inserts of up to about 150,000 bp and YACs can accept inserts of many hundreds or thousands of base pairs. Both types replicate inside their host cells in exactly the same way as a natural chromosomes. This carries the disadvantage that only one YAC molecule can exist per yeast cell. BACs are preferred nowadays over YACs because YACs have problems with instability of the inserts and an unacceptably high occurrence of chimeric inserts derived from more than one genomic region.

The choice of vector is dependent on the ease of screening, the transformation efficiency, the insert size and the ease of isolating the DNA after cloning. If the average insert size is large, then less recombinants are needed to obtain a representative recombinant **library**. Vector DNA use allows not only an original cloning to be carried out, but also sub-cloning into more amenable fragments later in a project.

*5.2.1.2 DNA Inserts.* Often the source of the nucleic acid is a cDNA copy of messenger RNA (mRNA), which has been generated using the enzyme reverse transcriptase. Sometimes the DNA of interest can be synthesised chemically (Section 4.1). More often now it is the product of PCR (Section 5.2.2). Often, the DNA to be cloned is inserted as a duplex into a restriction site of the vector DNA, after treatment of the vector with that restriction enzyme, by use of the enzyme DNA ligase (Section 5.3.5, Figure 5.2). Usually the insert DNA is a restriction fragment with termini compatible with the vector ends. Sometimes oligonucleotide **linkers** need to be joined to the insert. These are self-complementary synthetic duplex oligonucleotides that specify a recognition site for a restriction enzyme. Linkers are ligated to the fragment to be cloned and then treated with the restriction enzyme, thus generating new termini, which are now identical or compatible for joining to the cleaved vector. In all cases, the joined DNA is then transfected into a host cell line.

*5.2.1.3 Identification of Clones.* The rate-limiting step in classical cloning is often the identification of the correct clone from a huge excess of other molecules. Typically, a vast number of visually identical bacterial colonies or bacteriophage plaques are generated on the surface of agar in Petri dishes. To identify the very few clones of interest, several approaches can be used. Often, a copy of the entire set of recombinants is made by touching a nitrocellulose or nylon filter membrane to the agar surface. The 'master' copy agar plate is stored away until the location of the required clone has been established on the filter copy.

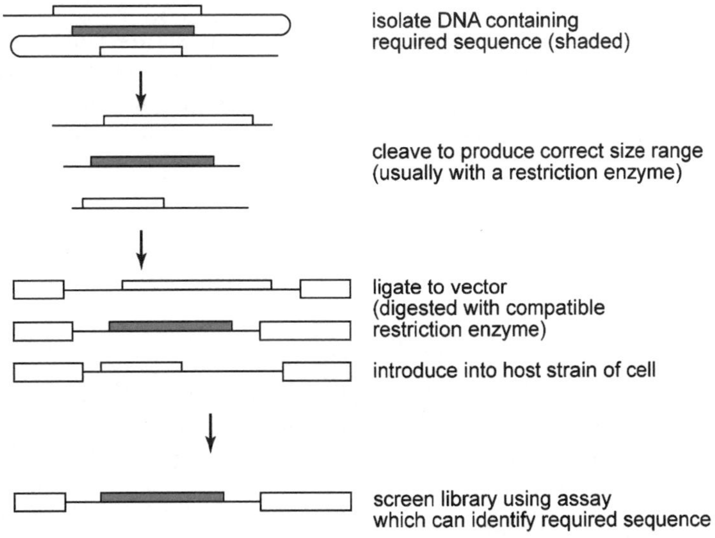

isolate DNA containing required sequence (shaded)
cleave to produce correct size range (usually with a restriction enzyme)
ligate to vector (digested with compatible restriction enzyme)
introduce into host strain of cell
screen library using assay which can identify required sequence

**Figure 5.2** *Basic cloning procedure*

In some cases, nucleic acid **hybridisation** (annealing of complementary strands, Section 5.5) is useful to find the desired clone. This is particularly true if a related sequence, such as that from another species, has already been cloned. Alternatively, one can deduce the DNA sequence from the corresponding protein sequence (if available) and chemically synthesise **oligonucleotide probes** complementary to part of the target DNA for screening of clones.

A complication here is that most amino acids are encoded by more than one nucleotide triplet. The result is that many different oligonucleotides need to be made to be sure of using the correct one. The number can be reduced by choosing a region of protein sequence containing the less ambiguous amino acids such as methionine and tryptophan which are specified by a single codon (TGG and ATG respectively). It is also possible to synthesize a mixture of oligonucleotides with two, three or four bases at the points where ambiguity is present, since the first two bases are often invariant for a particular amino acid (Figure 5.3; see also Figure 7.25). Lastly, several different regions of a protein can be used to derive a battery of probes all of which can be used to screen the library. In this way artefacts can be discounted. The probe is labelled (either by radioactivity or by use of a non-radioactive reporter molecule) and a solution of the probe is incubated with the DNA clones on the filter. After careful washing of the filter, only that probe which is exactly complementary to the desired sequence is left attached to the filter and positive clones can be identified by autoradiography or by visualisation of the reporter molecule. The hybridisation conditions used in such experiments are often crucial to a successful outcome.

In other cases, antibody screening of the filter copies can be used to detect the required clone. Of course this can only succeed if an antibody to the polypeptide product of the required gene is available and the vector into which the gene has been cloned contains appropriate transcriptional and translational regulatory sequences for the expression of the cloned gene as protein.

Another way of screening libraries is by the use of PCR. This is particularly powerful when screening artificial chromosome libraries of complete genomes. First, the individual clones in the library are arrayed into individual tubes (typically wells in a 24 × 16 multi-well plate). Next, a set of pooled subsets of the library is prepared. For example, a complete library of 10 multi-well plates might be split into ten pools, each comprising the complete contents of a single plate. PCR reactions on these ten samples would narrow down the target clone to a single multi-well plate. Further pools can be prepared, such as a particular row number for every plate (24 pools) or a particular column number (16 pools). In such a circumstance, these three sets of PCRs, totalling 10 + 24 + 16 = 50 reactions, would identify a single well in the library containing the clone producing the PCR product.

There are a number of shortcuts to molecular cloning which are sometimes useful.

*5.2.1.4 Transposon Tagging.* A previously cloned transposon (Section 6.6.5) is used to create mutations in the required gene (Figure 5.4). The transposon can then itself be used as a molecular 'tag' to isolate the gene by hybridisation (the transposon and its surrounding DNA must both be isolated by this method). Note that the detection method is based entirely upon the mutant phenotype and therefore no knowledge of the structure or biochemical function of the gene or gene product is needed.

*5.2.1.5 Microdissection.* It is possible physically to dissect and clone the required part of the chromosome (provided the chromosomal location of the gene of interest is known). Chromosomes may be separated from one another by pulsed-field gel electrophoresis or by fluorescence-activated sorting.

| | | | | | | |
|---|---|---|---|---|---|---|
| Glu | Asp | Ile | Trp | Lys | Lys | Phe |
| GAG | GAT | ATT | TGG | AAA | AAA | TTC |
| A | C | C | | G | G | T |
| | | A | | | | |

**Figure 5.3** *Example of a mixed sequence oligonucleotide incorporating each alternative base which is used in gene cloning to probe for the gene encoding this peptide*

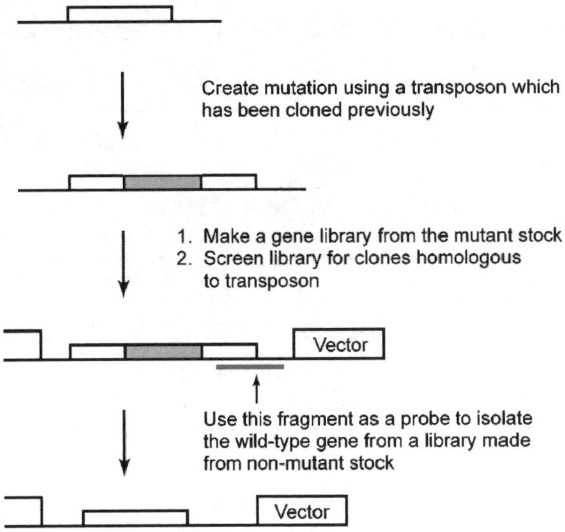


**Figure 5.4** *Transposon tagging*

*5.2.1.6 Chromosome Walking.* If an overlapping series of clones can be isolated, it is possible to use one clone to isolate the next in line and thus 'walk' along the DNA to the required sequence. This is a very time-consuming process, but nevertheless has been used frequently.

*5.2.1.7 Chromosome Jumping.* This is an extension of chromosome walking that proceeds by larger steps and ignores the large DNA stretches in the middle of each step (hence the word 'jumping').

### 5.2.2 The Polymerase Chain Reaction

A DNA or gene of interest is best isolated now directly from the total DNA of the organism in question by use of PCR.[8] Since the complete DNA sequences of many organisms (human, mouse, *etc.*) are now known, it is easy to design and synthesise chemically a pair of oligonucleotide primers of 20–30 residues flanking the region to be amplified, with each complementary to a different DNA strand (Figure 5.5). If it is required to clone the DNA into a vector, the oligonucleotide primers each can contain a restriction site at the 5′-ends, whereas the 3′-ends are complementary to the ends of the sequence to be cloned.

The target duplex DNA is denatured by heat and annealed to the primers, which are in vast excess to prevent the target DNA strands renaturing with each other. The two primers are next elongated on the separated target DNA template strands by use of a thermostable DNA polymerase, for example, Taq DNA polymerase from the thermophilic bacterium *Thermus aquaticus*, to give a twofold amplification. Since the primers are derived from different DNA strands, each newly synthesised strand now contains a binding site for the primer used for copying of the *other* strand. A second round of denaturation by heat, annealing and primer extension results in a fourfold amplification. After 20 rounds of amplification, $2^{20}$ copies of the original target DNA are formed. Such a powerful technique can produce as much DNA as can be made by classical cloning methods. For cloning, the DNA is either treated with both restriction enzymes to create sticky ends, and the product joined to a similarly treated vector with T4 DNA ligase, just as in classical cloning (Section 5.2.1), or cloned without restriction digestion. In the latter case, the PCR products are often cloned into a special cleaved vector with single 3′-T overhanging bases at the cleavage sites, because PCR products usually contain a single A base overhang at both 3′-ends, which is added in a template-independent manner. But of course, it is not essential to clone the DNA, since the PCR reaction can be repeated again if further amounts of DNA are required.

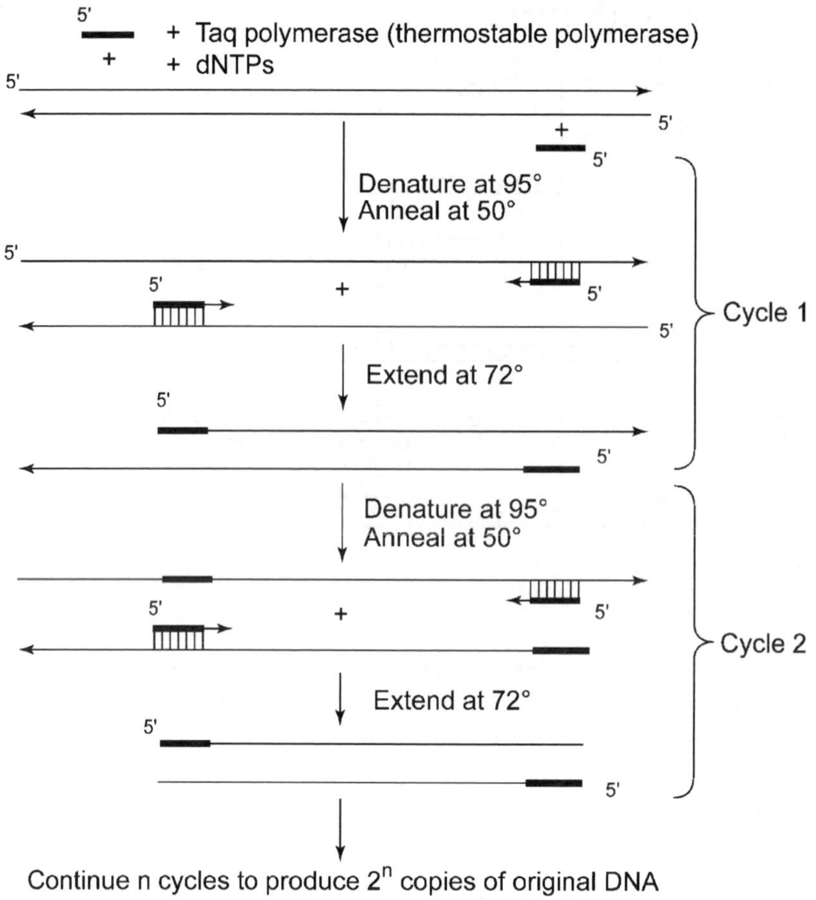


**Figure 5.5** *Steps involved in the polymerase chain reaction (PCR)*

The PCR method has also found widespread application in forensics for the detection and amplification of very tiny amounts of DNA, for example at crime scenes, and in DNA mutation detection.

## 5.3 ENZYMES USEFUL IN GENE MANIPULATION

Gene cloning would not have become possible without the discovery and isolation of a range of enzymes that act on DNA to enable manipulation of particular sections. One important class of enzymes is the DNA polymerases (Section 3.6.1). Other enzymes are involved in cutting the DNA, adding or removing a terminal phosphate or joining DNA fragments.

### 5.3.1 Restriction Endonucleases

Bacteria require a system to prevent foreign DNA from being replicated. This is provided by restriction endonucleases, which recognise and bind to DNA sequences at specific sites and make a double-stranded cleavage (Section 10.5.1). There are three types of restriction and modification systems, termed I, II and III in order of their discovery. The names of the enzymes are usually based on the names of the bacteria from which they are isolated. More than 3600 enzymes have been identified to date (http://rebase.neb.com/rebase/rebase.html).

**Table 5.2** *Some restriction endonucleases and their recognition sequences. N signifies any nucleotide. Cleavage sites indicated*▼▲

| *Type* | *Enzyme* | *Recognition site* |
|---|---|---|
| Type I | *EcoK* | A A C $(N)_6$ G T A C<br>T T G $(N)_6$ C A C G |
| | *EcoB* | T G A $(N)_8$ T G C T<br>A C T $(N)_8$ A C G A |
| Type II | *EcoRI* | G▼A A T T C<br>C T T A A▲G |
| | *SmaI* | C C C▼G G G<br>G G G▲C C C |
| | *PstI* | C T G C A▼G<br>G▲A C G T C |
| | *Sau3A1* | ▼G A T C<br>C T A G▲ |
| | *NotI* | G C▼G G C C G C<br>C G C C G G▲C G |
| | *BglI* | G C C N N N N▼N G G C<br>C G G N▲N N N N C C G |
| | *MnlI* | C C T C $(N)_7$▼<br>G G A G $(N)_7$▲ |
| Type III | *EcoPI* | A G A C C<br>T C T G G |

Type I consists of a large enzyme complex containing subunits encoding endonuclease, methylase and several other activities. The recognition sequence comprises a trinucleotide and a tetranucleotide separated by about six non-specific base pairs (Table 5.2), but the endonucleolytic cleavage site can be up to 7000 base pairs distant.

Type II systems have independent endonucleases and methylases that act on the same DNA sequence. These sequences are generally **palindromic** (*i.e.* they have a twofold axis of symmetry) and the cleavage sites are usually within or very close to the recognition sites (Section 10.5.1). In some cases, the symmetrical recognition sequence is interrupted (*e.g. BglI*), while a few enzymes recognise an asymmetric sequence and cleave at a defined distance (*e.g. MnlI*). Restriction enzymes cleave both strands of the DNA either symmetrically to give **blunt ends** or asymmetrically to give **sticky ends**. A vast range of enzymes with different specificities has been isolated from a wide variety of organisms and type II restriction enzymes are highly useful tools in recombinant DNA research, and the products of cleavage (restriction fragments) can often be rejoined using DNA ligase (Section 5.3.5).

The type III system shares features in common with both type I and type II. There are two independent polypeptides, one of which acts independently as a methylase, but both are required for specific endonucleolytic activity. In the case of *EcoPI*, for example, the recognition sequence is an asymmetric pentanucleotide and the cleavage site is 25 bp downstream (Table 5.2).

### 5.3.2 Other Nucleases

Almost every organism contains a wide variety of nucleases, of which some are involved in the salvage of nucleotides and some feature as intrinsic activities of proteins used in replication and repair processes. Apart from non-specific nucleases, such as DNase I, and ribonucleases, there are several other nucleases that are used in the manipulation of DNA and RNA (Table 5.3).

**Table 5.3** *Some endonucleases and their activities*

| *Nuclease* | *Origin* | *Activities* |
|---|---|---|
| Exonuclease III | *E. coli* | (1) ss *exo*-cleavage from 3′-ends of dsDNA<br>(2) *endo*-cleavage for apurinic DNA<br>(3) RNase H<br>(4) 3′-phosphatase |
| Exonuclease VII | *E. coli* | ss *exo*-cleavage from 5′- or 3′-end of ssDNA |
| Bal31 | *Alteromonas espejiana* | (1) ss *exo*- and *endo*-cleavage from 5′- or 3′-end of dsDNA<br>(2) ssDNA *endo*-cleavage |
| S1 | *Aspergillus oryzae* | ssDNA or RNA *exo*- and *endo*-cleavage |
| Lambda exonuclease | Infected *E. coli* | ss *exo*-cleavage from 5′-end of dsDNA |
| Phosphodiesterase I | Bovine spleen | ss *exo*-cleavage from 5′-end of ssDNA or RNA |
| Phosphodiesterase II | *Crotalus adamanteus* (or other snakes) | ss exo-cleavage from 3′-end of ssDNA or RNA |

### 5.3.3 Polynucleotide Kinase

A polynucleotide kinase isolated from bacteriophage T4 catalyses the transfer of the γ-phosphate of ATP to the 5′-hydroxyl terminus of DNA, RNA or an oligonucleotide in a reaction that requires magnesium ions. The enzyme is particularly useful for introducing a radioactive label on to the end of a polynucleotide, where the phosphate donor is γ-$^{32}$P-ATP. Both single- and double-stranded polynucleotides can be phosphorylated, although recessed 5′-hydroxyl groups in double-stranded DNA, such as those obtained by cleavage with certain restriction enzymes, are poorly phosphorylated. This sort of polynucleotide kinase activity, though not found in bacteria, has been found in some mammalian cells. The T4 enzyme is the only well-characterised kinase that has polynucleotides as substrates. This T4 protein also has a 5′-phosphatase activity which is unusually specific for a 3′-phosphate of a nucleoside or polynucleotide.

### 5.3.4 Alkaline Phosphatase

Phosphatases catalyse the hydrolysis of phosphate monoesters to produce inorganic phosphate and the corresponding alcohol. Most phosphatases are non-specific. Alkaline phosphatases are found in bacteria, fungi and higher animals (but not plants) and will remove terminal phosphates from polynucleotides, carbohydrates and phospholipids. The *E. coli* enzyme is a dimer of molecular weight about 89 kDa, requires a zinc (II) ion, and is allosterically activated by magnesium ions. During dephosphorylation of the substrate, its phosphate is transferred to a serine residue on the enzyme located in the sequence Asp-Ser-Ala. This same sequence is found in mammalian alkaline phosphatases (the calf intestinal enzyme is particularly well characterised) and it is similar to the active centre of serine proteases. Acidic phosphatases are also common, but these do not usually operate on polynucleotides as substrates.

### 5.3.5 DNA Ligase

A ligase is an enzyme that catalyses the formation of a phosphodiester linkage between two polynucleotide chains.[9] In the case of DNA ligases, a 5′-phosphate group is esterified by an adjacent 3′-hydroxyl group and there is concomitant hydrolysis of pyrophosphate in $NAD^+$ (bacterial enzymes) or ATP (phage and eukaryotic enzymes). Particularly efficient joining takes place when the phosphate and hydroxyl groups are held close together within a double helix, typically where the joining process seals a 'nick' and creates a perfect duplex (Figure 5.6). This situation occurs both in gene synthesis (Section 5.4) and in recombinant DNA technology (Section 5.2) in ligation of identical 'sticky ends' formed by cleavage with a restriction endonuclease. *E. coli* and phage T4 DNA ligases are well-characterised enzymes which have an important role in DNA replication (Section 6.6.4). T4 DNA ligase will join blunt DNA duplex ends when used at high concentrations and

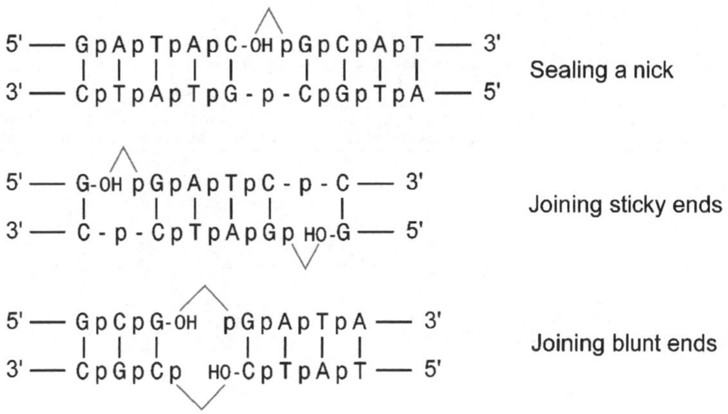


**Figure 5.6** *Joining reactions carried out by DNA ligase*

it will also catalyse the joining of two oligoribonucleotides in the presence of a complementary splint oligodeoxyribonucleotide.

## 5.4 GENE SYNTHESIS

### 5.4.1 Classical Gene Synthesis

The principles of gene assembly were developed 35 years ago by Khorana and his colleagues.[10]

*5.4.1.1 5′-Phosphorylation.* To join the 3′-end of one oligonucleotide to the 5′-end of another, a phosphate group must be attached to one of the ends. This is most easily accomplished at a 5′-end either chemically or enzymatically. The chemical procedure involves reaction of the 5′-hydroxyl group of a protected oligonucleotide, while still attached to a solid support with a special phosphoramidite derivative (*e.g.* $DMTO(CH_2)_2SO_2(CH_2)_2OP(N^{iPr})_2OCH_2CH_2CN$) (Section 4.4.2, Figure 4.19). The DMT group is removed by acidic treatment and during subsequent ammonia deprotection both the 2-cyanoethyl and hydroxyethylsulfonylethyl groups are removed to liberate the 5′-phosphate. Alternatively, and so as to introduce a $^{32}P$-radiolabel, phosphorylation is carried out enzymatically using T4 polynucleotide kinase (Section 5.3.3) to transfer the γ-phosphate of ATP to the 5′-end of an oligonucleotide.

*5.4.1.2 Gene Assembly.* Figure 5.7 shows schematically the construction of a gene coding for a small bovine protein caltrin (a protein believed to inhibit calcium transport into spermatozoa).[11] Each synthetic oligonucleotide is denoted by the position of the arrows. These are arranged such that annealing (heating to 90°C and slow cooling to ambient temperature) of all ten oligonucleotides simultaneously gives rise to a contiguous section of double-stranded DNA, the sequence of which corresponds to the desired protein sequence. In this example, the oligonucleotides are 24–38 residues long, but chains of 80 residues or more have been used in gene synthesis.

Oligonucleotides C2–C9 are previously phosphorylated such that, for example, the 5′-phosphate group of C3 lies adjacent to the 3′-hydroxyl group of C1. The duplex is only held together by virtue of the complementary base pairing between strands. The enzyme T4 DNA ligase (Section 5.3.5) is now used to join the juxtaposed 5′-phosphate and 3′-hydroxyl groups (the caret marks denote the joins). The overlaps are such that each oligonucleotide acts as a splint for joining of two others.

Note that oligonucleotides C1 and C10 are not phosphorylated. Each end corresponds to a sequence that would be generated by cleavage by a restriction enzyme (Section 5.3.1). Lack of a phosphate group prevents these self-complementary ends from joining to themselves during ligation. The ends are later joined to a vector DNA, previously cleaved by the same two restriction enzymes, to give a closed circular duplex ready for transformation and cloning in *E. coli* (Section 5.2).

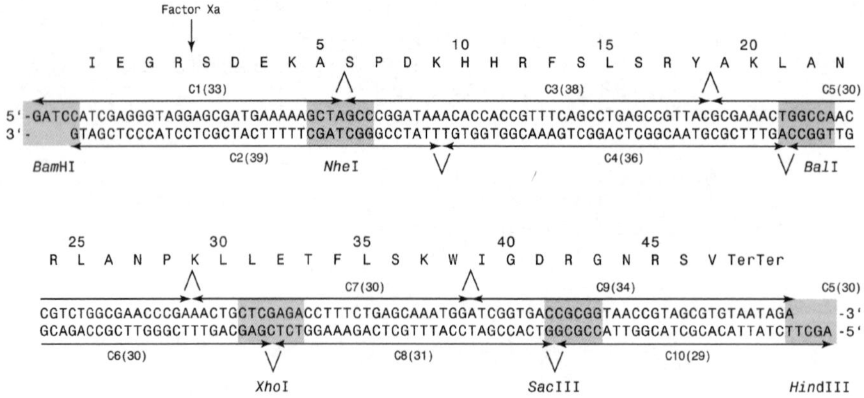


**Figure 5.7** *A synthetic gene for bovine caltrin. Oligonucleotides used for the gene assembly are indicated by arrows and caret marks denote points of ligation. The amino acid sequence is shown above. Restriction enzyme recognition sites are shaded*
(Adapted from Ref. 11; © (1987), with permission from Oxford University Press)

Other features of the synthetic gene are internal restriction enzyme sites (shaded), which can be introduced artificially merely by judicious choice of codons specifying the required amino acid sequence. This particular gene is designed without a methionine initiation codon, since the protein is intended for expression as a fusion with another vector-encoded protein. This fusion can be cleaved to generate caltrin by treatment with the proteolytic enzyme factor *Xa* (an enzyme important in the blood-clotting cascade and whose natural substrate is prothrombin), since the synthetic gene has been designed to include a section encoding the tetrapeptide recognition sequence for this enzyme.

### 5.4.2 Gene Synthesis by the Polymerase Chain Reaction

There are numerous procedures for gene synthesis that involve use of PCR (Section 5.2). A particularly simple version known as **recursive PCR** has been used for the preparation of large genes such as that for human lysozyme. Oligonucleotides are synthesised 50–90 residues long but, unlike the classical approach, only their ends have complementarity (Figure 5.8). Overlaps of 17–20 bp are designed to have annealing temperatures calculated to be in the range 52–56°C. A computer search ensures that no two ends are similar in sequence. Recursive PCR is carried out in the presence of all oligonucleotides simultaneously with cycles of heating to 95°C, cooling to 56°C and primed DNA synthesis at 72°C using the four deoxynucleoside triphosphates and the thermostable Vent DNA polymerase derived from *Thermococcus litoralis*. In initial cycles (step 1), each 3′-end is extended using the opposite strand as a template to yield sections of duplex DNA. In further cycles (steps 2–5), one strand of a duplex is displaced by a primer oligonucleotide derived from one strand of a neighbouring duplex. Finally (step 6), a high concentration of the two terminal oligonucleotides drives efficient amplification of the complete duplex. Success is due to the useful characteristics of Vent DNA polymerase, which has both a strand displacement activity and an active 3′–5′ proofreading activity that reduces the chances of incorrect nucleotide incorporation.

## 5.5 THE DETECTION OF NUCLEIC ACID SEQUENCES BY HYBRIDISATION

Molecular cloning is only the beginning of the study of a gene. Often it is important to study the same gene from a variety of different individuals. For example, much can be learned from structural analysis of a series of mutants in the gene. While it is possible to molecularly clone the gene from each mutant individual, it is often much easier simply to analyse the uncloned nucleic acid, for example by PCR amplification (Section 5.2.2) and DNA sequencing (Section 5.1). It is also important to be able to detect the RNA

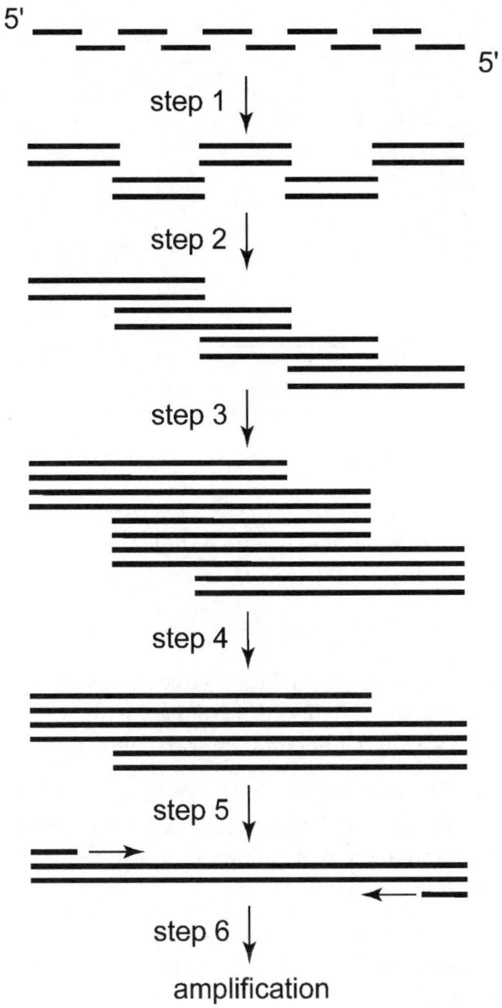


**Figure 5.8** *Gene synthesis by recursive PCR. Bars represent oligonucleotides and their extension products after PCR*

encoded by a gene and to determine its levels of transcription and tissue specificity. Methods exist for both of these purposes and both depend upon the ability of a single-stranded nucleic acid to pair specifically with its complementary strand.

### 5.5.1 Parameters that Affect Nucleic Acid Hybridisation

Hybridisation is the annealing of two single strands of a nucleic acid to form a duplex. Duplex strength is measured by observation of the **melting temperature** ($T_m$) and is affected by several parameters.

*5.5.1.1 Base Composition (%GC).* Since G–C base pairs have three hydrogen bonds, they are stronger than A–T base pairs, which have only two. Thus duplexes with higher G–C content have higher melting temperatures.

*5.5.1.2 Temperature (T).* The rate of association of single-stranded DNA into a duplex varies markedly with temperature (Figure 5.9: see also Section 2.5.1). The shape of this curve is governed by two

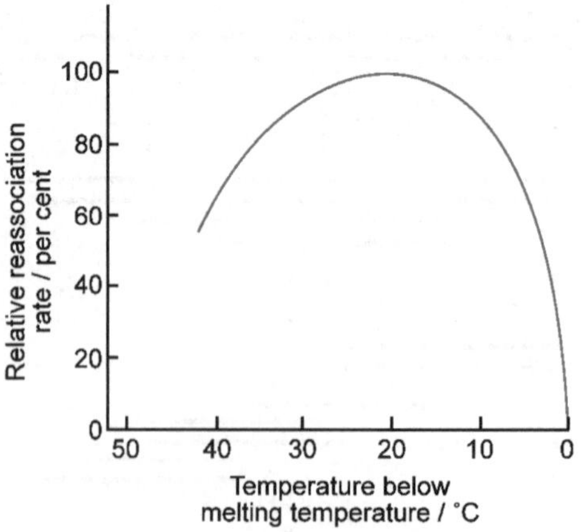


**Figure 5.9** *Dependence of the reassociation rate of DNA upon temperature*

factors. At low temperatures, the re-association rate is determined by the difference in free energy between the unassociated and the transition state.

$$k = Ze^{-E_a/RT} \tag{5.1}$$

where $k$ is the re-association rate constant, $E_a$ the activation free energy, $R$ the gas constant and $T$ the absolute temperature. At higher temperatures, the stability of the duplex is markedly reduced until eventually it is unstable and the hybrid melts. Thus there is a fall off in re-association rate as this point is approached.

*5.5.1.3 Monovalent Cation Concentration (M).* The melting temperature of a hybrid (Section 2.5.1) is reduced at lower salt concentration because cations help to stabilise a duplex. Divalent cations such as magnesium are much more effective in stabilisation of hybrids, but are less frequently used in hybridisation studies (Section 9.3).

*5.5.1.4 Duplex Length (L).* The melting temperature of a duplex shorter than a few hundred base pairs is length dependent. In practice, these four factors can be combined into an empirical equation giving the melting temperature $T_m$ of a hybrid DNA.

$$T_m = 69.3 + 0.41(\%GC) + 18.5\log 10\ \mathrm{M} - 500\ \mathrm{L}^{-1}/°\mathrm{C} \tag{5.2}$$

Web-based algorithms are available now for calculating $T_m$ from knowledge of the various parameters (*e.g.* see http://www.basic.northwestern.edu/biotools/OligoCalc.html). Use of hybridisation temperatures from 10 to 20°C below the calculated $T_m$ of the hybrid is optimal in practice to ensure annealing of strands.

For synthetic oligonucleotide probes of 15–20 residues, the calculation of $T_m$ is simplified to 2°C per dA·dT and 4°C per dG·dC base pair in 1 M sodium chloride solution. This is known as the **Wallace rule**. In the case of the quaternary ammonium salt tetramethylammonium chloride (3 M TMAC), the $T_m$ of a duplex is independent of base composition and is thus directly proportional to its length. This is of practical value for example in cloning applications that involve hybridisation of mixed sequence oligonucleotides (Section 5.2.2).

## 5.5.2 Southern and Northern Blot Analyses

It is possible to use nucleic acid hybridisation to detect uncloned genomic DNA. Genomic DNA is immobilised on a nitrocellulose or nylon filter, in basically the same way as described for gene cloning (Section 5.2.1). The gene of interest is detected on the filter by hybridising a complementary nucleic acid strand labelled either with radioactivity or an affinity label such as biotin, which can be detected with great sensitivity. Of course, if the DNA is just spotted onto the filter, all that is seen is a spot whose intensity reflects the

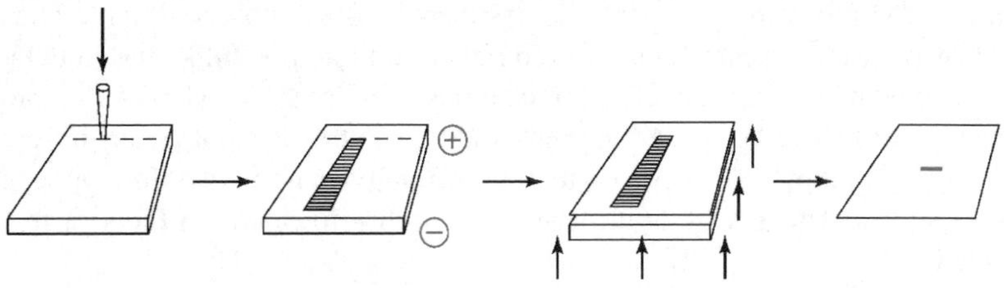
load restriction-digested DNA onto agarose gel
electrophorese DNA
transfer DNA out of gel onto filter
probe filter for specific sequence of interest

**Figure 5.10** *Southern blot analysis*

concentration of the corresponding gene in the sample. This latter technique is called **dot blotting** and is very useful in this limited respect. However, if the DNA is fractionated before transfer, much more information is acquired. **Southern blot analysis** (named after its inventor Ed Southern) involves fractionation of DNA by gel electrophoresis, followed by transfer of the DNA out of the gel onto a filter (Figure 5.10). The filter is then probed for the gene of interest as before. In the commonest case, the DNA has been digested with restriction enzymes and the result is the detection of those restriction fragments that are homologous to the gene probe. In this way, **restriction maps** of genes can be derived from genomic DNA without resort to cloning.

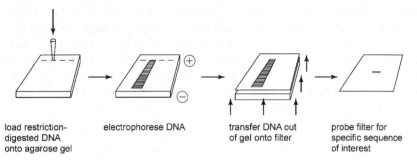

This technology can be extended to the study of RNA (**Northern blot analysis**). RNA can be electrophoresed in gels and immobilised on filters, provided it is denatured by treatment with formaldehyde. It can then be detected in the same way as DNA. Unfortunately, RNA cannot be cut into large defined fragments with the same ease as DNA, so such an approach is more limited. It is particularly useful for the determination of the sizes of RNAs and their tissue specificities, the latter approach relying upon the isolation of RNAs from different tissues. Northern blot analysis is often used to determine the transcribed regions in a stretch of DNA. By this approach, a battery of different restriction fragments, which together span the DNA of interest, are separately used to probe a Northern blot. Those DNA fragments that are transcribed detect bands in the Northern blot. The complementary approach (by use of radioactive RNA to probe restriction-digested DNA) is only possible if the transcripts arising from the DNA are particularly abundant.

## 5.5.3 DNA Fingerprinting

The notion that human characteristics can be inherited is long established. However, the ABO blood-group system can still only be used to classify people into just four types (groups A, B, AB and O). Moreover, such serological and protein markers are all too readily degraded in aged forensic samples. Clearly, the solution to such limitations lies in the direct examination of the genetic material itself. Even before the DNA revolution, it was evident that the 3 billion base pairs that make up the human genome must contain a huge number of sites of heritable variation and ought to support truly positive biological identification. Moreover, DNA is surprisingly tough and bits can survive in typeable form for remarkably long periods.

Genetic fingerprinting was developed in 1984 by accident. It was at first an academic curiosity, but then moved speedily into real-life casework where it established that molecular genetics could really provide an entirely new dimension to biological identification.[12,13] This technology has changed the lives of thousands of people involved in criminal investigations, paternity disputes, immigration challenges, identification of victims of mass disasters and the like. The analysis of human DNA has been of prime importance though there are tremendous applications in non-human DNA analysis, in particular the use of animal and plant DNA-typing and the field of 'microbial forensics', which has expanded as a response to the threat of bio-terrorism.

*5.5.3.1 Super Markers.* Alec Jeffreys started a search for hypervariable regions in human DNA in the 1980s.[13] He found the answer in **minisatellites**. These are regions of DNA consisting approximately 30 base pairs repeated over and over again tens or hundreds of times, and with different alleles varying in the number of stutters. The problem was how to access them. Jeffreys observed that a chance-studied minisatellite tucked away inside a human gene looked rather familiar, not unlike the stutters in the few other minisatellites described in the literature. The implications were clear – a hybridisation probe consisting of this DNA sequence motif shared by different minisatellites should latch onto many different minisatellites simultaneously, giving unlimited access to these potentially extremely informative genetic markers. Minisatellites are simply detected by hybridisation of probes to Southern blots of restriction-enzyme-digested genomic DNA.

*5.5.3.2 Stumbling upon DNA Fingerprinting.* In September 1984, Jeffreys tested a range of samples that included DNA from a human father/mother/child trio. The results provided multiple, highly variable DNA fragments. While mother and father were obviously different, the child seemed to be a union of the DNA patterns of the parents.[14] Improved technology was able to resolve large numbers of extremely variable DNA fragments containing these minisatellites (Figure 5.11), not just in humans but in other organisms as well. In humans, the banding patterns are individual-specific, with essentially zero chance of matching even between close relatives or members of an isolated inbred community. For any individual, the patterns are constant, irrespective of the source of DNA. The multiple markers that make up a DNA fingerprint are inherited in a simple Mendelian fashion, with each child receiving a random selection of about half of the father's bands and half of the mother's. Happily, the term 'DNA fingerprint' was chosen rather than the more accurate description 'idiosyncratic Southern blot minisatellite hybridisation profile' (Section 5.5.2).[13]

*5.5.3.3 The Evolution of Forensic Genetics.* The amount of variation currently accessible in DNA is extremely informative. Sequence variations between different minisatellite loci allows probes to detect many independent minisatellites simultaneously, yielding the hypervariable multi-band patterns known as DNA fingerprints.[14,15] By use of only a single probe, the match probability is estimated to be $<3 \times 10^{-11}$, while two probes together give a value of $<5 \times 10^{-19}$. This is so low that the only individuals sharing DNA fingerprints are monozygotic twins.[14] At the same time, a method known as differential lysis was developed[15] that selectively enriches sperm concentration in vaginal fluid/semen mixtures, thereby avoiding the problem of the victim's DNA (which is in great excess) masking that of a rapist.

*5.5.3.4 Single-Locus Probes.* Although use of DNA fingerprinting persisted for some years in paternity testing, criminal casework soon concentrated on the use of specific cloned minisatellites. Each of these 'single-locus probes' (SLPs) revealed only a single, highly polymorphic, restriction fragment length polymorphism, thereby simplifying interpretation. Typically, four SLPs were used successively to probe a Southern blot, yielding eight hyper-variable fragments per individual. SLPs were used in the first DNA-based criminal investigation in the UK in 1986.[16]

*5.5.3.5 Profiling DNA.* DNA fingerprints are excellent for some applications, but not for forensic investigations that have to identify the origin of a biological sample with as much certainty as possible.[12] This is because fingerprint patterns are complex and their interpretation is readily open to challenge in court, they are not easy to computerise, and they require significant amounts of good quality DNA, equivalent to that obtained from a drop of fresh blood. The solution to these problems was simple – the isolation and cloning of minisatellites. Each cloned minisatellite, used as a hybridisation probe, produces a much simpler pattern of just two bands per person, corresponding to the two alleles in an individual (Figure 5.11c). Such simple profiles can be obtained using considerably less DNA (one hair root is enough), and the estimated lengths of the DNA fragments easily support database construction. These **DNA profiles** have exposed the true variability of human minisatellites, some showing 100 or more different length alleles in human populations.

DNA profiles are not individual-specific no matter how variable the minisatellite is between unrelated people. This is particularly true for siblings, who have a one in four chance of sharing exactly the same profile. Nevertheless, by typing DNA sequentially, typically with a battery of five different minisatellites, excellent levels of individual specificity are obtained, leading to routine match frequencies of one in a billion with DNA profiling.

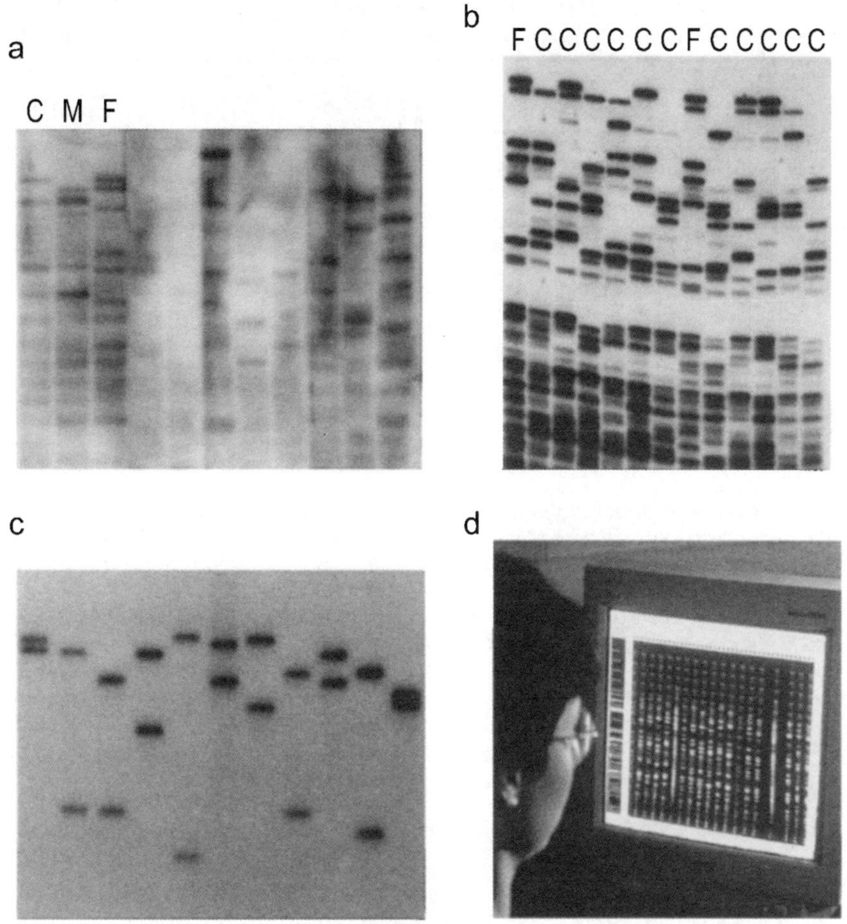


**Figure 5.11** *The evolution of DNA typing systems. (a) The very first DNA fingerprints with a family group at left (M, mother; F, father; C, child) plus DNA from various non-human species. (b) Improved DNA fingerprints from a single family with the father (analysed twice) and his 11 children. Note how DNA fingerprints readily distinguish even close relatives and how bands in the missing mother can be easily identified as bands in the children that are not present in the father. (c) Simpler DNA profiles of unrelated people. (d) DNA profiling using PCR-amplified microsatellites. Several microsatellites are amplified at the same time and the resulting profiles are displayed on a computer and automatically interpreted for databasing*
(Courtesy of Orchid-Cellmark[13]; A.J. Jeffreys, *Genetic Fingerprinting, The Darwin Lectures 2003*, Darwin College, University of Cambridge, 2003, 49–67. © (2003), with permission from Cambridge University Press)

*5.5.3.6 PCR-Based Methods.* The discovery of **short tandem repeats** (STRs) together with the introduction of automated sequencing technology has led to the current powerful systems for the identification of individuals. Human forensic casework is now carried out using commercially developed **autosomal STR multiplexes** (single-tube PCR (Section 5.2.2) that amplify multiple loci) and is established worldwide because of its advantages of high discriminating power, sensitivity, ability to resolve simple mixtures, speed and automation. The resulting reduced cost has paved the way for the creation of national STR DNA databases (http://www.cstl.nist.gov/div831/strbase/). For example, the UK National DNA Database contained some 2.5 million reference profiles and about 200,000 crime-scene profiles as at July 2004 (http://www.forensic.gov.uk/forensic/news/press_releases/2003/NDNAD_Annual_Report_02-03.pdf). Automated sequencing equipment for multiplex analysis typically uses multi-channel capillary electrophoresis systems that detect fluorescently labelled PCR products. These are combined with robotics

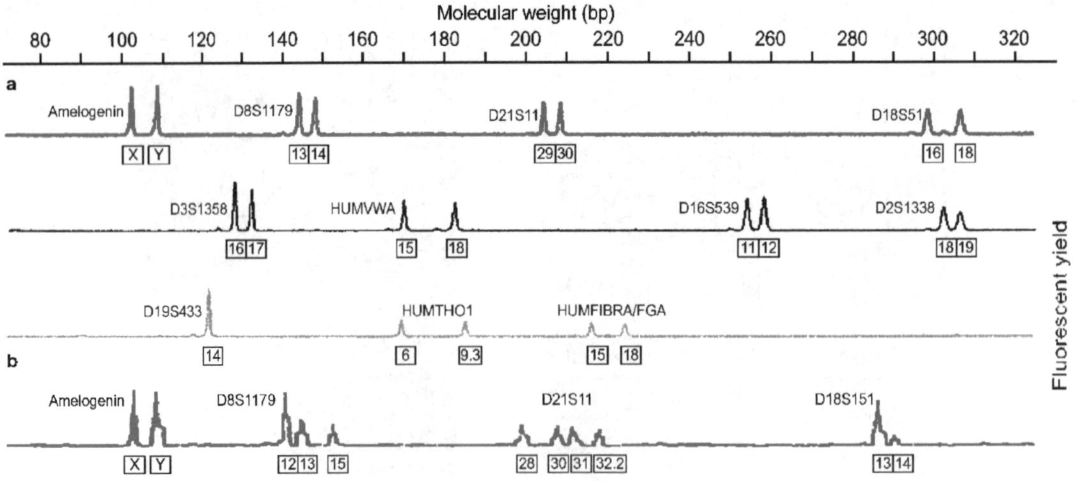


**Figure 5.12** *Electropherograms illustrating autosomal short tandem repeat (STR) profiles. (a) An electropherogram of the second-generation multiplex 'SGM Plus' profile from a male, including X- and Y-specific amelogenin products of 106 and 112 bp, respectively. Most short tandem repeats (STRs) are heterozygous and the alleles are evenly balanced. Numbers beneath STR peaks indicate allele sizes in repeat units. The STR profile (displayed here as red, black and grey) uses a four-colour fluorescent system, with the fourth channel being used for a size marker (not shown). (b) A typical mixture from two individuals (red channel only shown). Mixtures can only be identified if the alleles of the minor component are above the background 'noise' in an electropherogram (in practice a ratio of ~1:10) and can usually be resolved by inspection. In this example, the contributions are in even proportions – for example, D21S11 shows four alleles where the peaks are approximately equal in height, whereas D18S51 shows two peaks in a 3:1 ratio. The X- and Y-specific amelogenin peaks are of approximately equal height, indicating that this is a mixture from two males*
(Adapted from Ref. 12; © (2004), with permission of Macmillan Publishers Ltd)

and laboratory information management systems, including bar coding of samples, to reduce operator errors. A typical electropherogram output is illustrated (Figure 5.12).

A 'second-generation multiplex' (SGM) has further included a PCR assay targeted at the XY-homologous amelogenin genes[17] to reveal the sex of a sample donor. An additional four loci were added to the multiplex, now renamed 'SGM Plus10' (Figure 5.12), giving it a match probability of less than $10^{-13}$.

Although some differences in practice between individual national jurisdictions remain, there has been rapid development and near-universal acceptance of this new DNA-based technology in forensic genetics.[18]

### 5.5.4 DNA Microarrays

DNA microarrays are now one of the most widely used tools in functional genomics.[19] They are providing biology with the equivalent to the chemist's periodic table – a classified inventory of all the genes for a living organism. Oligonucleotide **microarrays**, also known as **DNA chips**, are miniature parallel analytical devices containing libraries of oligonucleotides robotically spotted (printed) or synthesised *in situ* on solid supports (glass, coated glass, silicon or plastic). The major DNA-chip technologies are distinguished by the sizes of the DNA fragments arrayed, by methods of arraying, by their chemistry and linkers for attaching DNA to the chip, and by hybridisation and detection methods.

Microarrays work by exploiting the ability of a given cDNA or mRNA test sample to hybridise to the DNA template from which it originated. By use of a two-dimensional (2-D) array containing very many DNA samples, the expression levels of hundreds or thousands of genes within a cell can be determined quickly by measuring the amount of cDNA or mRNA bound to individual sites on the array. The precise amount of mRNA bound to each locus gives a profile of gene expression in the cell. Alternatively, comparative binding of a test and a standard probe provides an immediate signal of the presence or absence of a particular sequence. Ultimately, such studies promise to expand the size of existing gene families, reveal new patterns of coordinated gene expression across gene families, and uncover entirely new categories of genes.

*5.5.4.1 Technical Foundations.* Two technologies are central to the production and use of DNA microarrays. The first is the fabrication of tens to hundreds of thousands of polynucleotides at high spatial resolution in precise locations on a 2-D surface. The second involves the measurement of molecular hybridisation events on the array using laser fluorescence scanning. By use of one of three different methodologies, DNA is synthesised, spotted or printed onto the support, which is usually a glass microscope slide, but can also be a silicon chip or a nylon membrane. The DNA sequences in a microarray are attached to the support in a fixed way, so that the location of each spot in the 2-D grid identifies a particular sequence. The spots themselves are either oligodeoxynucleotides, DNA or cDNA.

*5.5.4.2 Use of DNA Microarrays.* The five steps for carrying out a microarray experiment are

- **DNA chip** preparation using the chosen target DNAs,
- making a hybridisation solution containing a mixture of fluorescently labelled cDNAs,
- incubating the hybridisation mixture of fluorescently labelled cDNAs with the DNA chip,
- detecting bound cDNA using laser technology and data storage in a computer, and
- data analysis using computational methods.

*5.5.4.3 Microarray Preparation.* The first chip technology came in 1984 from the work of Stephen Fodor in the California-based company, Affymetrix, and is based on **photolithography**. A synthetic linker with a photochemically removable protecting group is bonded to a flat glass substrate. Light is then directed through a photolithographic mask to specific areas on the surface to produce localised photodeprotection (Figure 5.13). The first of a series of DNA phosphoramidite monomers (Section 4.1.2), also having a 5′-(α-methyl-2-nitropiperonyloxycarbonyl), photochemically labile protecting group[20] (Figure 5.14a) is

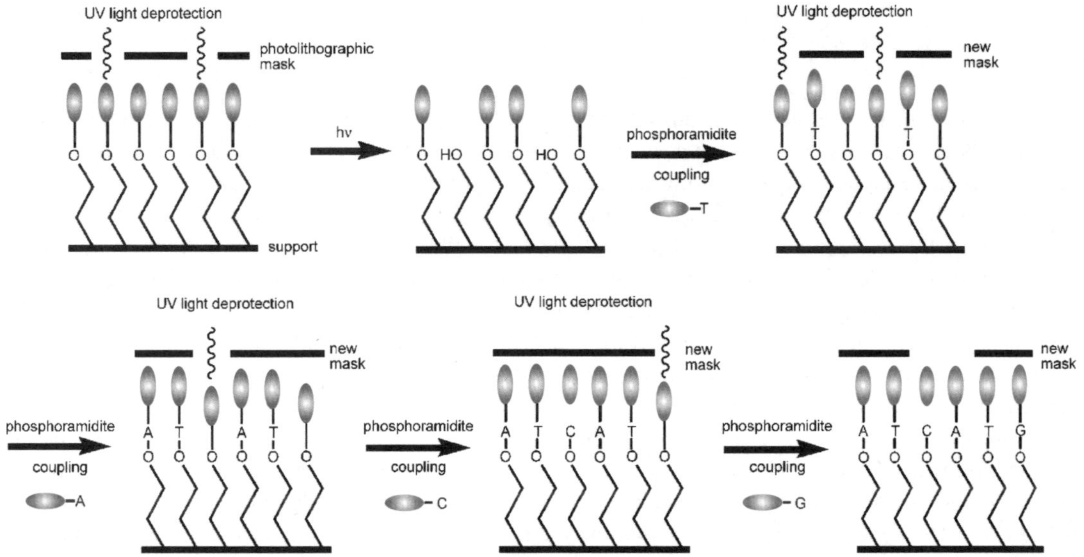


**Figure 5.13** *Light directed oligonucleotide synthesis. Derivatised solid support has hydroxyl groups protected with a photolabile group. Light is directed through a mask to effect selective deprotection. The first dT-phosphoramidite with 5′-photolabile protection is introduced. A new mask enables deprotection of a second set of spots on the array which are then linked to the second nucleotide, dA. Repetition of this procedure for next dC and finally dG completes the cycle for the first nucleotides in the oligomer array. The cycle is then repeated with new masks to install the second nucleotides in the array, and so on*

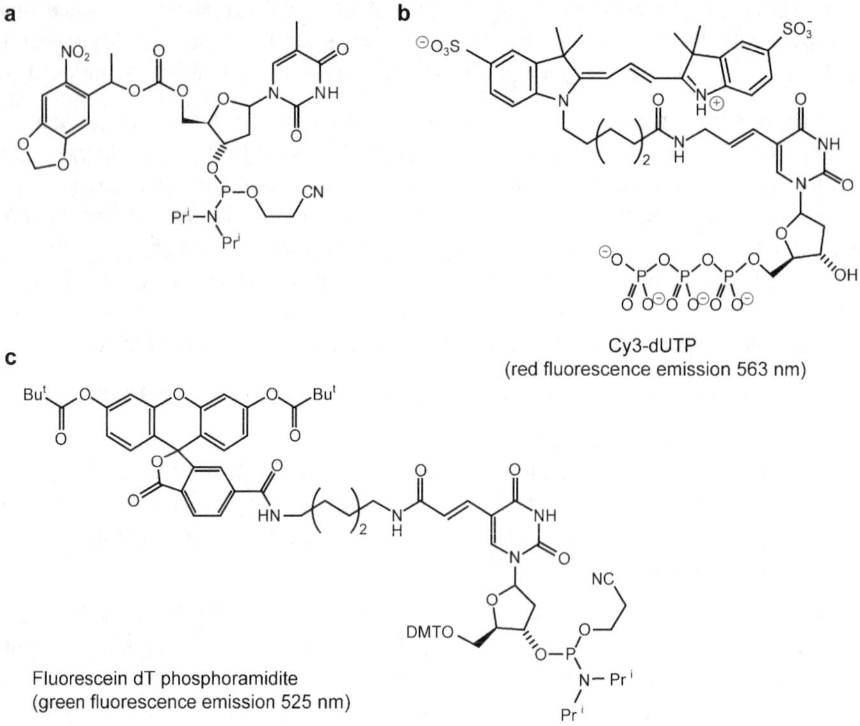


**Figure 5.14** *(a) Deoxythymidine 3′-phosphoramidite with a 5′-photolabile α-methyl-2-nitropiperonyloxycarbonyl protecting group. (b) dUTP analogue linked through the 5-methyl group to a cyanine dye having red fluorescence emission. (c) Deoxyuridine 3′-phosphoramidite linked via C-5 to a protected fluorescein dye having green light emission*

incubated with the surface and chemical coupling occurs only at those sites that have been irradiated in the preceding step. Light is next directed at further regions of the substrate by a new mask, and the reaction sequence is successively repeated for the second, third and fourth of the four monomers, A, C, G and T to complete the first cycle. A second complete cycle lays down the second nucleotide in the oligomer. Further repetitions of this cycle provide the full set of $4^N$ polydeoxyribonucleotides of length $N$, or any subset, in just $N$ complete cycles. Thus, for a given reference sequence, a DNA array can be designed that consists of a highly dense collection of DNA single-stranded oligomers, usually around 25 residues long. This photolithographic process enables construction of arrays with extremely high information content. Large-scale commercial methods permit approximately 300,000 oligodeoxynucleotides to be synthesised on small $1.28 \times 1.28$ cm arrays, while versions with $10^6$ probes per array are being developed.

In a separate development, Patrick Brown at Stanford University developed a cDNA spotting method that is suited to the display of single-DNA fragments, often greater than several hundred base pairs in length.[21] cDNA samples (about 15 ng) are micro-spotted robotically onto a glass (or nylon membrane) surface that has been treated chemically to provide primary amino groups. Droplets (=1 nL) are located ~200 μm apart and the DNA in the spots is covalently bonded to the surface by UV irradiation to link the surface amino groups to thymidine residues.

A third robotic methodology has been developed by Rosetta Inpharmatics that uses ink-jet printer technology to perform classical oligodeoxyribonucleotide synthesis based on the four-step dimethoxytrityl protecting group chemistry (Section 4.1.2).[22] In a fourth approach, Agilent Labs in collaboration with Marvin Caruthers have developed a two-step microarray synthesis cycle to halve the number of steps required to build oligodeoxyribonucleotides on glass surfaces. An entirely new carbonate-based protecting group chemistry enables deprotection and oxidation in a single step, reducing time and cost for microarray synthesis.[23]

*5.5.4.4 Microarray Analysis.* How does one analyse the information encoded in thousands of individual gene sequences on a small glass or silicon chip? The process is based on hybridisation probing, a technique that uses fluorescently labelled nucleic acid molecules as 'mobile probes' to identify complementary DNA sequences using base pair recognition. The DNA probes to be hybridised to the array are labelled by incorporating fluorescently tagged nucleotides (such as Cy3-dUTP, Figure 5.14b) during oligo-primed reverse transcription of mRNA. Alternatively, they can be chemically tagged by 5′-end labelling (Figure 5.14c). Different green and red fluorophores are used to label cDNAs from control (reference) and experimental (test) RNAs. The labelled cDNAs are then mixed together prior to hybridisation to the array so that relative amounts of a particular gene transcript in the two samples are determined by measuring the signal intensities detected for both green and red fluorophores. Because the arrays are constructed on a rigid surface (glass), they can be inverted and mounted in a temperature-controlled hybridisation chamber. When the fluorescent mobile probe, DNA, cDNA or mRNA, locates a complementary sequence on the chip, it will lock onto that immobilised target, and the probe is identified by fluorescence microscopy. The fluorescent tag on the probe is excited by a laser and the digital image of the array is captured. These data are then stored in a computer for analysis. Thus, for example, cDNA from a normal cell and a diseased cell can be separately labelled with green and red fluorescent markers to enable comparative analysis. The location and intensity of both colours shows whether the gene, or a mutant, is present in either the control and/or sample DNA (Figure 5.15). It can also provide an estimate of the expression level of the gene(s) in the sample and control DNA.

*5.5.4.5 Types of Microarray.* There are three basic types of samples used to construct DNA microarrays, two are genomic and the other is 'transcriptomic', for measuring mRNA levels. They differ in the kind of immobilised DNA used to generate the array and, ultimately, the kind of information that is derived from the chip. The target DNA used will also determine the type of control and sample DNA that is used in the hybridisation solution.

**5.5.4.5.1 Changes in Gene Expression Levels.** Determining the level, or volume, at which a particular gene is expressed is called **microarray expression analysis**, and the arrays used in this kind of analysis are called '**expression chips**'. The immobilised DNA is cDNA derived from the mRNA of known

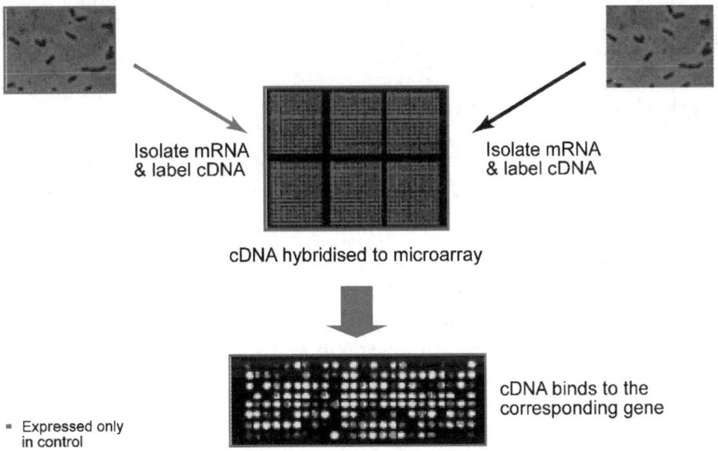


**Figure 5.15** *Microarray analysis to compare the hybridisation of expressed genes in a control cDNA sample (left) and in a mutant (or diseased) cDNA sample (right) to an immobilised reference gene set. Red dots show where the gene is expressed only in the control. Grey dots (green channel) show where the gene is expressed only in the mutant. White dots (normally green plus red = yellow) show here the gene is expressed in both control and mutant sample. Absence of a dot indicates that the gene is not expressed in either DNA sample*

genes, and the control and sample DNA hybridised to the chip is cDNA derived from the mRNA of for example, normal and diseased tissue. If a gene is overexpressed in a certain disease state, then more sample cDNA, as compared to control cDNA, will hybridise to the spot representing that expressed gene. Expression analysis is valuable in drug development, drug response and therapy development.

**5.5.4.5.2 Genomic Gains and Losses.** A technique called **microarray comparative genomic hybridisation** (CGH) is used to look for genomic gains and losses or for a change in the number of copies of a particular gene involved in a disease state. In microarray CGH, large pieces of genomic DNA provide the target DNA, and each spot of target DNA in the array has a known chromosomal location. The hybridisation mixture contains fluorescently labelled genomic DNA harvested from both normal (control: green) and diseased (sample: red) tissue. If the number of copies of a particular target gene has increased, a large amount of sample DNA will hybridise to the corresponding loci on the microarray, whereas comparatively small amounts of control DNA will hybridise to the same spots. As a result, those spots containing the disease gene will fluoresce red with greater intensity than they will fluoresce green. CGH is used clinically for tumour classification, risk assessment and prognosis prediction.

**5.5.4.5.3 Mutations in DNA.** Detection of mutations or polymorphisms in a gene sequence employs the DNA of a single gene as the immobilised target. In such arrays, the target sequence at a given locus will differ from that of other spots in the same microarray sometimes by only one or a few specific bases. A type of sequence commonly used in such analyses are **single nucleotide polymorphisms (SNPs)**. SNPs have a single genetic change within a person's DNA sequence. The analysis of such a target microarray requires genomic DNA derived from a normal sample for use in the hybridisation mixture. An SNP pattern associated with a particular disease can be used to test an individual to determine whether he or she is susceptible to that disease. Such '**mutation/polymorphism analysis**' is commonly used in drug development, therapy development and tracking disease progression.

*5.5.4.6 Microarray Data Management.* Data management technology is critical for the efficient use of microarray results, but is beyond the scope of this book. The Gene Expression Omnibus (GEO: www.ncbi.nlm.nih.gov/geo/) is an online resource for the storage and retrieval of gene expression data from any organism or artificial source.

Personalised drugs, molecular diagnostics, integration of diagnosis and therapeutics are the long-term medical promises of microarray technology. For the future, DNA microarrays offer hope for obtaining global views of biological processes – simultaneous readouts of all the body's components – by providing a systematic way to survey DNA and RNA variation.

### 5.5.5 *In Situ* Analysis of RNA in Whole Organisms

Hybridisation can be used to detect transcripts in a cell or organism. Cells and organisms smaller than about 1 mm are fixed (the macromolecules are immobilised) by treatment with a fixative, such as formaldehyde, glutaraldehyde or methanol/acetic acid. Larger organisms are normally sliced into thin sections before fixation. The fixed specimens are then probed with radioactively or fluorescently labelled nucleic acid in the same way as for a Southern blot. Synthetic 2′-*O*-methyloligoribonucleotides are particularly good probes of mRNA in cells, because they are resistant to cellular nucleases (Sections 3.1.4.2 and 4.4.3.6). By use of microscopy (Section 11.5), the locations of RNAs can be determined at the cellular or even the sub-cellular level.[24,25]

## 5.6 GENE MUTAGENESIS

### 5.6.1 Site-Specific *In Vitro* Mutagenesis

The process of engineering specific changes in a DNA sequence is termed as ***in vitro* mutagenesis**. It is an invaluable tool for modification of a DNA sequence in a pre-determined manner to study its biological function. In classical mutagenesis, alterations are created randomly and the effects of each mutation need to be screened separately, which is time consuming. Now more directed methods are standard, where the DNA is first cloned for ease of manipulation and then deletions, insertions or replacements made.

*5.6.1.1 Deletions.* Deletions can be created at restriction sites (Section 5.3.1) by cleavage with the corresponding enzyme and then by treatment for a short period with an exonuclease enzyme. For example, the exonuclease *Bal* 31 is used to remove both double- and single-stranded DNA from both ends. Alternatively, the enzyme *Exo* III is used to generate single-stranded ends followed by treatment with SI nuclease to trim the created single strands (Table 5.3). Re-ligation of the two new double-stranded ends generates deletion mutations of the parent DNA. This method has the serious limitation that deletions can only be made around restriction sites.

A more general deletion method involves use of synthetic oligonucleotides (Figure 5.16). In this procedure, an oligonucleotide complementary to the desired site of deletion on the DNA, but not containing the nucleotides required for deletion, is used as a primer for synthesis of a second DNA strand. In the process of cloning, mutant DNA segregates from wild type DNA and clones containing mutant the deletion can be selected.

One problem associated with this technique is that bacteria will often attempt to repair the mutagenised strand because the *in vivo*-generated DNA strand is methylated. This can result in low yields of the mutated sequence. Eckstein has developed a reliable method that involves incorporation of **phosphorothioate**-modified nucleotides (Section 4.4.3) into the *in vitro*-generated strand. Such nucleotides are more resistant to nuclease degradation, with the result that the unmutagenised DNA strand can be removed by exonuclease digestion and the gap filled to generate the mutation in both strands (Figure 5.17). Deletion mutants can also be generated by this method by use of PCR (Section 5.2.2).

Deletion mutants can also be generated by PCR (Figure 5.18). This method relies upon the fact that PCR primers are tolerant of primer-template mismatch to create a mutation at the priming site in an analogous way to that shown in Figure 5.17. Unfortunately, this raises the problem that PCR-based mutagenesis can only make a mutated site at an end of the PCR fragment. However, this problem can be solved by generating two PCR products sharing a common central mutated region (Figure 5.18). Denaturation and annealing of these two products, followed by extension of the duplex with *Taq* DNA polymerase yields a larger product with the mutated site in the centre.

*5.6.1.2 Insertions.* Insertions may be generated by ligation of a synthetic oligonucleotide duplex into a restriction site after cutting with the appropriate restriction enzyme. Sequence additions at other sites can

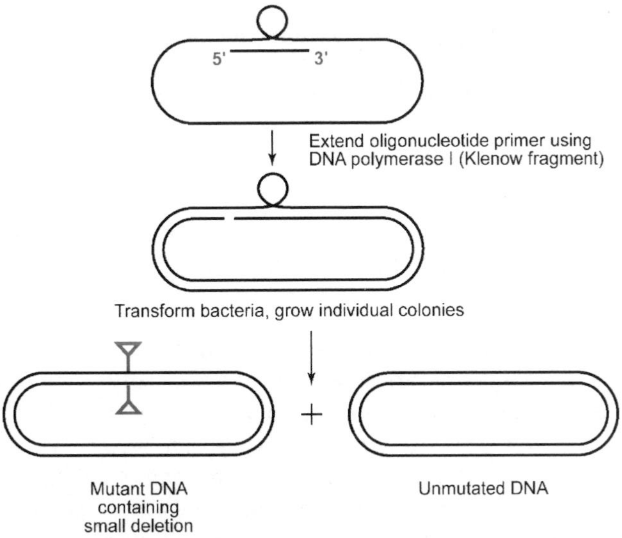


**Figure 5.16** *Oligonucleotide site-directed deletion mutagenesis*

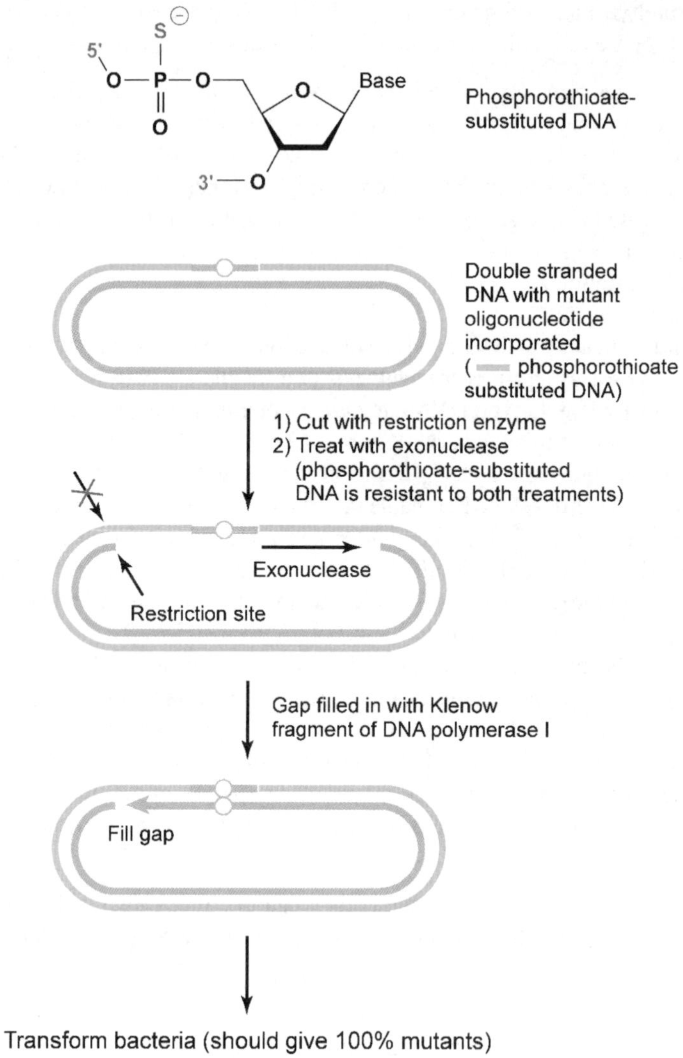


**Figure 5.17** *The use of phosphorothioate-modified nucleotides for* in vitro *mutagenesis*

be achieved by means of site-directed mutagenesis using oligonucleotides in an analogous way to that described for deletions (Figures 5.17 and 5.18), but in this case the synthetic oligonucleotide primer contains the desired additional nucleotide(s).

*5.6.1.3 Replacements.* A common type of mutation is that which maintains the same number of nucleotides but where part of a sequence is replaced. This is particularly useful for single-base alterations that lead to a change of amino acid codon. Expression of the mutated gene leads to the production of a protein with a single amino acid alteration (**protein engineering**). One replacement method is to introduce a small deletion at a restriction site followed by ligation into the gap of an oligonucleotide duplex of the same size but of different sequence. A more general approach involves use of a synthetic oligonucleotide primer in an analogous way to the introduction of deletions and insertions, but with the same number of nucleotides in the mutant strand as wild type (Figures 5.17 and 5.18).

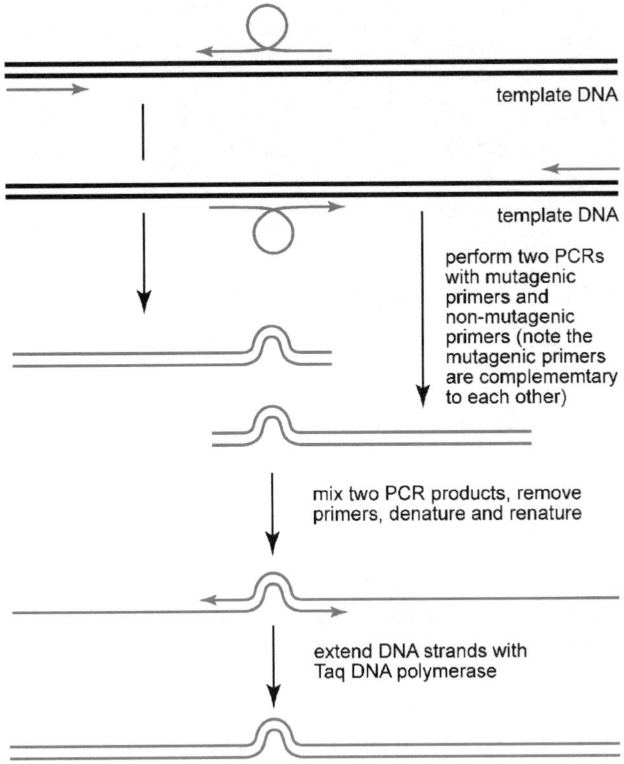
template DNA
template DNA
perform two PCRs with mutagenic primers and non-mutagenic primers (note the mutagenic primers are complemementary to each other)
mix two PCR products, remove primers, denature and renature
extend DNA strands with Taq DNA polymerase

**Figure 5.18** *The creation of DNA deletions by PCR*

### 5.6.2 Random Mutagenesis

**Random mutagenesis** is a method of introduction of multiple mutations into a DNA sequence but in arbitrary sequence positions. Random mutagenesis was achieved classically by reaction of DNA with chemicals (Chapter 8) to establish its biological role. More recently, random mutagenesis has been used to evolve DNA sequences that code for proteins with new or enhanced properties and in the evolution of new DNA catalysts (Section 5.7.3). The introduction of a number of random mutations into a DNA sequence generates a library of new DNA entities, which can then be used to identify those sequences with the required functionality. There are two main methods of random mutagenesis.

In **error-prone *PCR***, the aim is to modify the usual PCR protocol (Section 5.2.2) in order to deliberately introduce mutations. There are several ways in which this may be done. For example, the use of a polymerase which lacks proof-reading ability (such as *Taq* DNA polymerase) allows errors inherent in DNA synthesis to go uncorrected. Other changes to help reduce the **fidelity** of the polymerase include a lower annealing temperature in the PCR cycle, low or unequal dNTP concentrations and/or use of a large number of PCR cycles (60–80), which allows amplification of erroneous copies. Another common method of increasing polymerase infidelity is to increase the $Mg^{2+}$ concentration (up to 10 mM) in the PCR reaction or to replace $Mg^{2+}$ by $Mn^{2+}$ (typically 0.05–0.5 mM). Mutated products can be amplified by further rounds of PCR.

A second method for introduction of mutations is the use of **nucleotide analogues** into the nascent DNA during PCR. When the nucleotide analogue is copied in subsequent rounds of PCR, it is not recognised by the polymerase as a normal nucleotide and an incorrect deoxyribonucleotide is inserted opposite the analogue. Examples of such analogues includes 2′-deoxyinosine, 5-fluoro-2′-deoxyuridine, 8-oxo-2′-deoxyguanosine and the degenerate pyrimidine analogue dP (Figure 5.19).

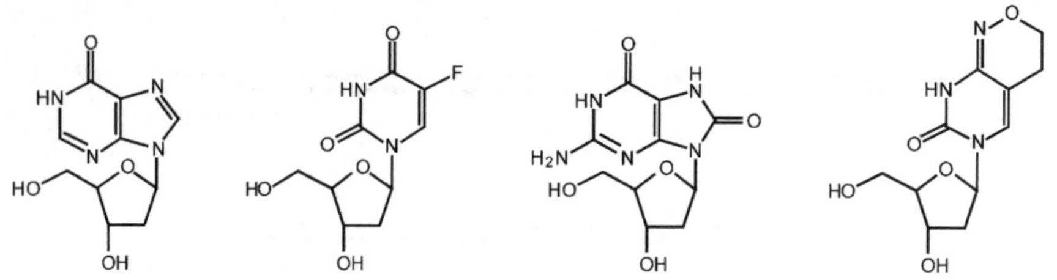

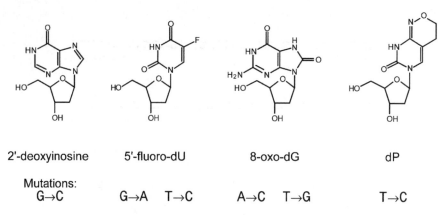

2'-deoxyinosine | 5'-fluoro-dU | 8-oxo-dG | dP

Mutations: G→C | G→A T→C | A→C T→G | T→C

**Figure 5.19** *Nucleoside analogues used in error-prone PCR mutagenesis*

### 5.6.3 Gene Therapy

The ability to introduce new or altered genes into a mammalian genome has tremendous implications. For example, it may prove possible to cure some genetic diseases by introducing a healthy copy of a gene into an afflicted individual. It is already possible to introduce into mammals genes that encode economically or medically important polypeptides such as insulin growth hormones and interferon. The intention is that the animal either grows faster or produces large amounts of protein, which can be harvested. We will address the ways in which this can be carried out, leaving the ethical questions raised by this issue to others.

There are three major ways of introducing DNA into mammalian germ tissue such that the progeny of the recipient will carry the gene. The first involves microinjection of DNA solutions into the nucleus of an egg by means of an extremely fine capillary. Such a technique works very well with a mouse egg but is more difficult with other mammals, such as sheep, where it is extremely hard to see the nucleus. In this way, **transgenic** animals have been created which carry functioning genes from another organism.

The second method involves the use of retrovirus-based vectors (Figure 5.20). As described in Section 6.4.6, retroviruses can infect a cell and then insert their DNA into its chromosomes. The gene to be introduced into the host is ligated into the genome of the retrovirus. The retroviral DNA is then introduced into a cultured cell line, which is capable of producing all components of a retrovirus except for the viral RNA (such a cell culture is called a **helper cell** line). This cell line will then package the recombinant virus stock into virus particles that can be harvested from the culture medium. Helper cells are necessary because the presence of the insert in the retroviral genome disrupts some of the normal retroviral genes needed for viral production. The harvested recombinant virus stock is then used to infect an early embryo, which is then replaced into a donor mother. During growth, some cells of the embryo become infected by the virus and the retroviral gene, including the gene insert, becomes stably inserted into the DNA of these cells. Because not all cells become infected, the animal is a **chimera**. However, if the germ cells of this animal contain proviral DNA then its offspring will retain the recombinant in every cell of its body.

In addition to introduction of a new gene coding for a protein, it is also possible to introduce *via* a retroviral vector a gene that codes for a specialised RNA (*e.g.* antisense RNA (Section 5.7.1), short interference RNA (Section 5.7.2) or an RNA that folds into ribozyme or an aptameric structure (Section 5.7.3) that can act in *trans* to interact with and block the function of another RNA.

The third method for introducing DNA into the mammalian germ line relies upon the existence of cultured cell lines, which can become germ cells if injected into early embryos. This approach is particularly useful in the mouse, where such cells, **embryonal carcinoma cells**, can be grown in dishes. The gene of interest can be introduced into these cells, which are then injected into embryos.

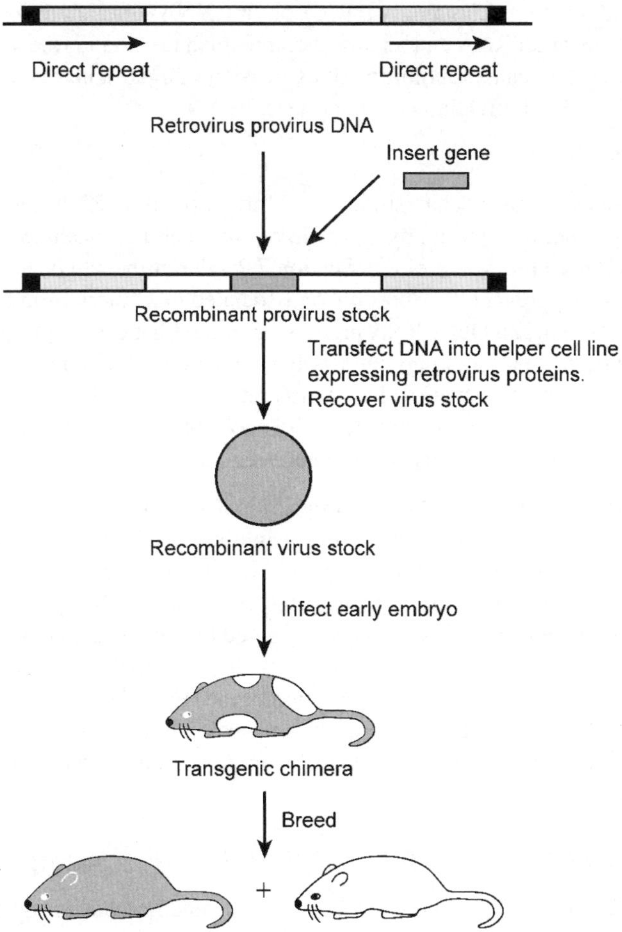


**Figure 5.20** *Creation of transgenic animals by use of retroviral vectors*

## 5.7 OLIGONUCLEOTIDES AS REAGENTS AND THERAPEUTICS

The ability to synthesise DNA and RNA oligonucleotides of defined sequence rapidly (Sections 4.1 and 4.2), including a range of nucleotide analogues (Section 4.4), has led to a large number of applications as therapeutic and diagnostic agents.[26] Many applications involve the principle of recognition of a linear sequence of RNA or DNA by the oligonucleotide. For example, antisense, steric block and short interfering RNA (siRNA) all involve targeting of RNA within cells to form **duplexes** as a means of control of gene expression[27] (Sections 5.7.1 and 5.7.2). Synthetic oligonucleotides have been used also to form **triplexes** (Sections 2.3.6 and 9.10.1) with double-stranded DNA to block gene expression,[28] but this principle has so far not led to therapeutic products. Other types of application include ***in vitro* selection** and design of oligonucleotides that recognise and bind to nucleic acids structures, to proteins or to small molecule ligands (Section 5.7.3).

### 5.7.1 Antisense and Steric Block Oligonucleotides

In 1979, Zamecnik and Stephenson[29] were the first to show that a synthetic oligonucleotide could be used to block specific gene expression in Rous Sarcoma Virus. This pioneering work led to the study of oligonucleotides and their analogues as therapeutic agents. This field is commonly known as 'antisense', since the principle of biological activity usually involves either degradation or steric blocking of the sense strand (the coding strand) of RNA (commonly mRNA or viral RNA) through formation of an exactly base paired duplex between the target RNA and an added complementary strand (the antisense strand) (Figure 5.21). Formation of the duplex causes inhibition of expression of a particular gene within cells or *in vivo* and the aim is to do this without affecting any other gene.[30,31]

### 5.7.1.1 Basic Mechanisms

**5.7.1.1.1 Steric Block.** This mechanism involves formation of an RNA–DNA duplex to physically block the RNA and to prevent recognition by a protein or other cellular machinery. For example, binding of an oligonucleotide close to the 5′-cap site[32] (Section 7.2.1) or at the site of initiation of translation in mRNA[33] (Section 7.3.3) may prevent the ribosome or associated machinery from binding to the RNA and initiating translation (Figure 5.22). Other RNA processing events that can be interfered with sterically by duplex formation include nuclear splicing[34] and polyadenylation[35] (Section 7.2.1), both of which are required for the processing of most mammalian gene transcripts and which involve numerous steps of RNA–protein recognition. In the case of viral RNA, it is possible to block recognition of essential RNA binding proteins that are required for virus-specific gene regulation.

**5.7.1.1.2 Induction of RNase H.** Although steric block activity requires stoichiometric amounts of complementary added oligonucleotide, a more potent inhibitory effect can often be obtained through recognition of an RNA–oligonucleotide duplex by the ubiquitous cellular enzyme Ribonuclease H (RNase H). The normal function of this enzyme is to help the removal of RNA primers in DNA replication (Section 6.6.3). However, when an RNA sequence is targeted in cells by a complementary oligodeoxynucleotide, RNase H-induced cleavage can occur rapidly, usually close to the centre of the targeted RNA section.[36] The loss of intact RNA leads to rapid degradation of the RNA. Thus in the case of mRNA, there is a concomitant reduction in the level of the encoded protein expressed. Most regions of an mRNA can usually be targeted by such oligonucleotides, including 3′- and 5′-untranslated regions.

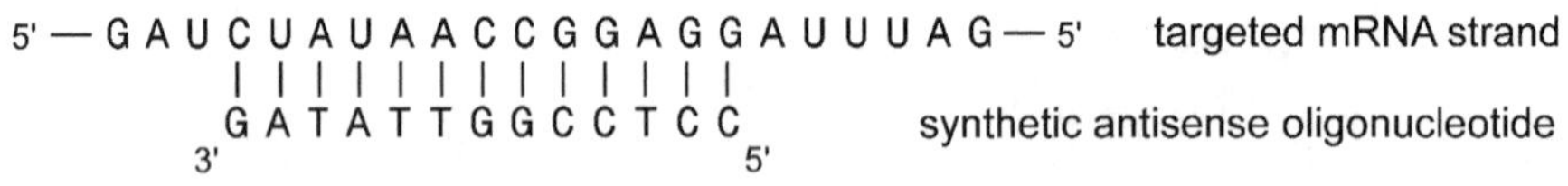


**Figure 5.21** *Duplex formed by an antisense oligodeoxyribonucleotide and a target mRNA*

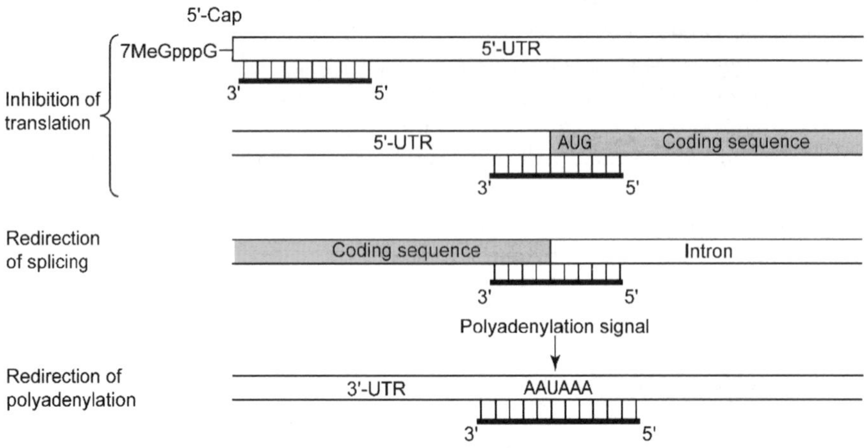


**Figure 5.22** *Three alternative mechanisms of steric block action of antisense oligonucleotides acting upon RNA*

*5.7.1.2 Optimal Oligonucleotide Characteristics.* There are many factors that influence cellular or *in vivo* antisense activity. In practice, oligonucleotide optimisation is carried out by experimentation through use of *in vitro*, cell-based and ultimately *in vivo* assays, although some general principles can be used in oligonucleotide design.

**5.7.1.2.1 Duplex Stability.** For intracellular antisense activity, an oligonucleotide must be of sufficient length to form a strong duplex with its RNA target at 37°C under cellular conditions. In general, binding strength increases as a function of length as well as the number of G:C pairs (Section 5.5.1). In addition, the type of nucleotide analogues used and their placement within the oligonucleotide are also crucial and those nucleoside analogues that adopt an RNA-like, 3′-*endo* sugar conformation (such as 2′-*O*-methylribonucleosides) tend to result in increased binding strength, since there is a tendency to form a more compact A-helix (Section 2.2.3). It is important also that the oligonucleotide does not form unusual secondary structures (such as G quadruplexes, Section 2.3.7) that may hinder duplex formation. Another important consideration is whether the target RNA site is easily accessible, *i.e.* does not exhibit tight RNA secondary or tertiary structure or is not strongly bound by cellular proteins. In this regard, experimental approaches to target choice are often more reliable than RNA structure prediction.

**5.7.1.2.2 Specificity.** For unique sequence recognition within the human genome (*i.e.* no other likely exact match for an oligonucleotide of typically mixed composition), a minimum length of around 12 nucleotides is usually required. However, the longer the chosen sequence, the greater the chance for the oligonucleotide to form a mismatched duplex with an incorrect RNA sequence. This is particularly of concern in the case of RNase H induction where an incorrect RNA may be cleaved in addition to that targeted, leading to side effects. In practice, a compromise between duplex stability and target specificity limits oligonucleotide length usually to 12–25 residues.

**5.7.1.2.3 Nuclease Stability.** Unmodified single-stranded DNA and RNA oligonucleotides are degraded very fast by cellular nucleases in cells and serum. **3′-Exonucleases** are the most prevalent, such that minimally the 3′-end of an antisense oligonucleotide must be protected, usually by chemical modification. But 5′-exonucleases as well as endonucleases are also present in cells, and thus for therapeutic applications, nuclease protection of each internucleotide linkage by inclusion of analogues is often thought necessary.

**5.7.1.2.4 Cellular Uptake.** A significant difficulty is that oligonucleotides and their analogues rarely penetrate cells in culture without co-addition of a carrier or cell delivery agent. For example, popular delivery agents for many cultured mammalian cell lines are **cationic lipids**, which can form complexes with negatively charged oligonucleotides to help cell association, uptake through the endosomal pathway and subsequent release into the cytosol by endosome destabilisation. Oligonucleotides are able to enter cell nuclei readily once they have been released into the cytosol. Oligonucleotides are usually administered *in vivo* without carrier, and here there may be special mechanisms available for cell uptake, but this remains a difficult and controversial subject of study.

**5.7.1.2.5 Pharmaceutical Considerations.** One positive feature of many clinically investigated oligonucleotides to date is their relative lack of toxicity during systemic delivery into animals and man. By contrast, a major concern has been the frequently observed, rapid clearance through the kidney, which is typical of many macromolecules. It is thus not surprising that the greatest success to date for therapeutic oligonucleotides has been in local or topical administration. Pharmaceutical development remains a significant challenge in terms of reaching the required tissue or organ, efficacy of action at the site and the maintenance of a therapeutic dose at manageable and affordable concentration levels. Many studies continue that focus on investigations of new nucleotide analogue types and combinations, conjugates and formulations[31].

*5.7.1.3 Nucleotide Analogues Used in Antisense Applications.* The most potent antisense oligonucleotides to date have been those shown to induce RNase H cleavage within cellular models and several have been taken to clinical trials.[26] Strong recognition by RNase H requires there to be a contiguous stretch of minimally 6–10 residues of 2′-deoxyribonucleosides where internucleotide linkages are phosphodiesters or the close analogue **phosphorothioate** (Section 4.4.3). Phosphorothioates are considerably more resistant than phosphodiesters to nuclease degradation and are well tolerated in humans. Thus, first generation therapeutic oligonucleotides contained only 2′-deoxyribonucleotide phosphorothioates, such as the clinically approved drug Vitravene, which is a 26-mer used for treatment of CMV-induced retinitis in AIDS patients.

Second-generation oligonucleotides employ the principle of a **gapmer**. Such oligonucleotides contain a section of 6–10 residues (usually centrally placed) of 2′-deoxyribonucleoside phosphorothioates, but the flanking regions on each side comprised of other analogues that enhance binding to the RNA target and further increase the oligonucleotide stability to nuclease, but which do not direct RNase H cleavage. Such analogues are generally ribonucleoside analogues, such as 2′-*O*-methyl, 2′-*O*-methoxyethyl or locked nucleic acids (LNA), where the sugar conformation is 3′-*endo*. Overall, gapmers are recognised and direct RNA cleavage by RNase H. Whereas several first generation antisense oligonucleotides, such as ISIS 3521 (Affinatak) targeted to the mRNA for protein kinase C α,[37] failed clinical trials, there is more hope of clinical benefit for some higher potency gapmers against viral and cancer targets (Figure 5.23).[26,31]

For steric block applications, there is no restriction in principle to the type or placement of an analogue within a sequence as long as other antisense considerations are met. The variety of analogues that have been investigated is very large. In some cases analogues can be combined in one oligonucleotide to give **mixmers**. In addition to 2′-*O*-methyl and 2′-*O*-methoxyethyl ribonucleotides described above, other important analogues used in steric block applications fall into two classes: (a) those that contain a phosphate group, such as LNA, tricyclo DNA, 3′-amino phosphoroamidate and phosphorothioamidate, and (b) non-phosphate containing analogues such as peptide nucleic acids (PNA) and morpholinodiamidates. In class (b) it was hoped that the absence of the negative charges on the oligonucleotide would enhance cell uptake, but this is not the case. Attachment of a cationic or other cell penetrating peptide appears to improve cell uptake, but the universality of this approach is still under study. A steric blocking, phosphorothioamidate oligonucleotide targeted to the essential RNA involved in the enzyme telomerase (Section 6.6.5) is moving close to clinical trials as an anti-cancer agent.[38]

*5.7.1.4 Non-Duplex Therapeutic Activities of Single-Stranded Oligonucleotides.* Recently, other biological activities of oligonucleotides have been found that are sequence-dependent but which are independent of duplex formation with an RNA target.

**5.7.1.4.1 Immune Modulation.** Single-stranded oligodeoxynucleotide phosphodiesters and phosphorothioates that contain the dinucleotide sequence CpG can trigger an immune response when administered to humans and animals.[39,40] The response is mediated through binding to a '**toll-like receptor**' TLR9 that is present in cytosolic vesicles and the binding stimulates signalling pathways that activate transcription factors. By contrast, double-stranded RNA and siRNA (Section 5.7.2) binds to another receptor TLR7 and may stimulate a different immune response. The context of the CpG determines the immune modulation specificity, such that mouse TLR9 prefers CpG when flanked at 5′ by two purines and at 3′ by two pyrimidines, while human TLR9 is recognised optimally by GTCGTT and TTCGTT sequences. Such activities are now recognised to have contributed being harnessed for therapeutic applications and as vaccine

**First Generation**
ISIS 3521 (Affinatak) GsTsTsCsTsCsGsCsTsGsGsTsGsAsGsTsTsTsCsA

**Second Generation**
ISIS 9606 GsTsTsCsTsCsGsCsTsGsGsTsGsAsGsTsTsTsCsA

s = phosphorothioate linkage
underline = 2'-O-methoxyethylnucleoside

**Figure 5.23** *First and second generation clinically used phosphorothioate-containing oligodeoxyribonucleotides* adjuvants. Immune stimulation of this or similar type has also been used to explain *in vivo* activities of some supposed antisense oligonucleotides that contain CpG sequences.

**5.7.1.4.2 DNA Aptamers.** Oligodeoxynucleotides have been selected (Section 5.7.3) that bind specific cellular proteins. Such **aptamers** can fold into unusual structures (such as a G quadruplex, Section 2.3.7). A chemically synthesised DNA aptamer that binds strongly to vascular endothelial growth factor has been formulated as a conjugate with polyethylene glycol and has recently been given regulatory approval (Macugen) for treatment of patients with the wet form of age-related macular degeneration, a common cause of blindness due to abnormal blood vessel growth.[41]

## 5.7.2 RNA Interference

In the late 1990s, gene silencing by double-stranded RNA was observed in plants in the laboratory of David Baulcombe[42] and in the worm *C. elegans* in that of Craig Mello.[43] Within a very few years gene silencing activities were found in many diverse organisms and have gone on to be harnessed as powerful diagnostic reagents in genome research and as potential therapeutics.[44–46] **RNA interference** (**RNAi**) probably evolved from the need for eukaryotic cellular defence against foreign (*e.g.* viral) duplex RNA or DNA that is transcribed into RNA. Distinct but overlapping pathways have been elucidated that represent different forms of genetic regulation (Figure 5.24).

Primary RNA transcripts in the nucleus has been found to contain hundreds of endogenous sequences known as **microRNAs** (**miRNAs**) that can fold into hairpins that contain imperfect matches (Section 7.5.3). The transcripts are processed by the nuclear complex **Drosha**, which contains an RNase III activity that cleaves such RNAs to produce hairpins of about 70 residues (pre-microRNAs). After export to the cytosol, the

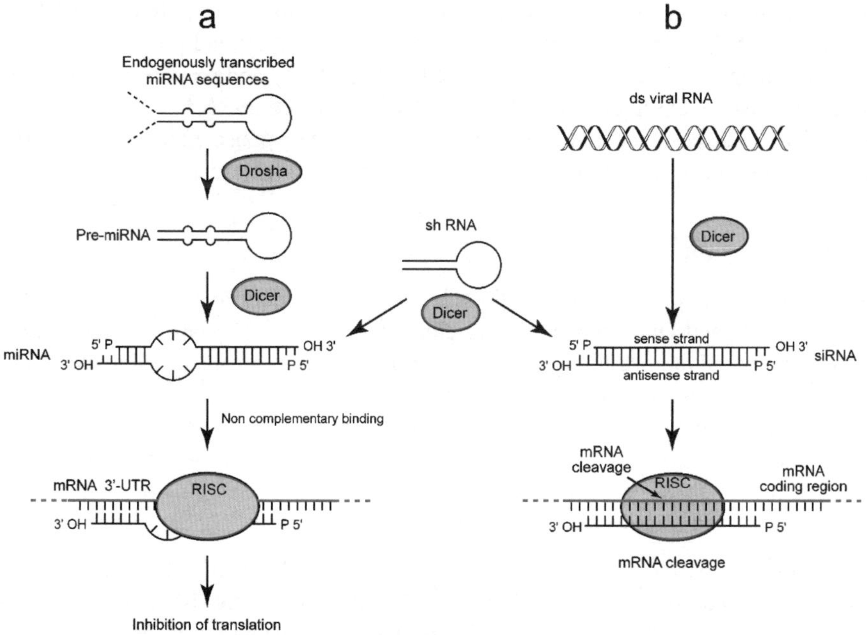


**Figure 5.24** *Mechanisms of action involved in RNA interference (RNAi). Pathway (a) shows steps in the processing of microRNA (miRNA) eventually leading to inhibition of translation. Pathway (b) shows the processing of double-stranded viral RNA to form short interfering RNA (siRNA) and eventual cleavage of mRNA by the RNA-induced silencing complex (RISC) complex. ShortRNA (shRNA) (centre) can in principle enter either pathway*

hairpins are further processed by the enzyme complex **Dicer** (Dcr-1),[47] a multi-subunit protein complex that also contains an RNase III activity, to give imperfectly paired RNA duplexes of about 21–23 residues (miRNAs). These are recognised by the RNA-induced silencing complex (**RISC**), which directs one of the two RNA strands to bind to a selected sequence in the 3′-untranslated region of a gene (a microRNA recognition element) to form an imperfect complement, which results in a block to translation (Figure 5.24, pathway a).

A second pathway is triggered by the introduction into a cell of double-stranded RNA, such as viral RNA. A second DICER variant, Dcr-2, is responsible for processing this duplex RNA into 21–23 residue perfect duplexes (**siRNA**). SiRNAs are then utilised by the RISC complex to direct one strand (antisense or **guide**) to form a duplex with an exact complement on a mRNA and cleave the phosphodiester bond precisely between residue 10 and 11 counting from the 5′-end of the complement, as directed by a member of the Argonaute family (Ago 2), an RNA endonuclease within RISC (Figure 5.24, pathway b). The other strand (sense or **passenger**) is discarded and then degraded.

Thomas Tuschl and colleagues[48] found that when synthetic siRNAs of 21 residues were transfected into mammalian cells, the RISC-dependent cleavage pathway could be triggered, thus allowing site-specific cleavage of mRNA and subsequent inhibition of gene expression. Synthetic siRNAs have now become used widely as reagents for specific gene inhibition in many cell types and seem to be applicable to almost all genes. Generally two-nucleotide 3′-overhangs are added on each strand for optimal activity, similar to those found following natural DICER cleavage of duplex RNA. Although siRNA duplexes appears to be stable to nuclease degradation for hours to days within cells, unlike single-stranded RNA, much effort has been expended on investigation of the tolerance to incorporation of analogues or conjugates that might have advantages *in vivo* to aid stability or pharmacology.[49] The sense strand appears to be highly tolerant of chemical modification, but the antisense strand, which is the one introduced by the RISC complex to pair with mRNA, is less so. A recent demonstration of efficacy in a transgenic mouse model promises that modified siRNAs may have therapeutic value.[50]

A third RNAi pathway is triggered by introduction into cells of short hairpin RNAs (**shRNAs**) of around 29-base pairs (Figure 5.24).[51] Such shRNAs are recognised by DICER and processed to give siRNAs of high potency, presumably because they are generated endogenously and may be more readily utilised by RISC. ShRNAs are also potential precursors of imperfectly matched miRNA.

Although siRNAs have been shown to generate some immune response effects, it is not clear at present whether these will present significant problems or not for their *in vivo* use.

### 5.7.3 *In Vitro* Selection

*5.7.3.1 Principles of In Vitro Selection (SELEX).* The advent of *in vitro selection* or **SELEX** (systematic evolution of ligands by exponential enrichment) in the early 1990s by the groups of Gold,[52] Szostak and Joyce marked the beginning of a new age in the design of functional nucleic acid molecules as both ligands for given targets and as catalysts. SELEX is a combinatorial technique in which nucleic acids with specific properties, such as binding with high affinity to a given target molecule (an **aptamer**) or catalysis of a chemical reaction, are selected from a pool of typically $10^{12}$–$10^{15}$ RNA or DNA molecules of randomised sequence.[53,54] The technique exploits the wide range of structures that single-stranded nucleic acids can adopt and mimics the natural processes of evolution.

*5.7.3.2 Selection of Aptamers.* The basic principle of SELEX (Figure 5.25) involves the creation by automated chemical synthesis of an initial oligonucleotide library consisting of an internal random nucleotide sequence flanked by 5′- and 3′-tails of a constant sequence which act as primer binding sites for subsequent amplification of the library by PCR (Section 5.2.2). The random sequence, typically 10–100 nucleotides long, is generated by delivering a mixture of all four nucleoside phosphoramidites simultaneously during automated synthesis (Section 4.1). Since the library contains just a few copies of each sequence, it is first amplified by PCR in which one of the oligonucleotide primers carries a biotin modification. To act as ligands to a specific target molecule, the nucleic acids within the library must be free to fold into a wide

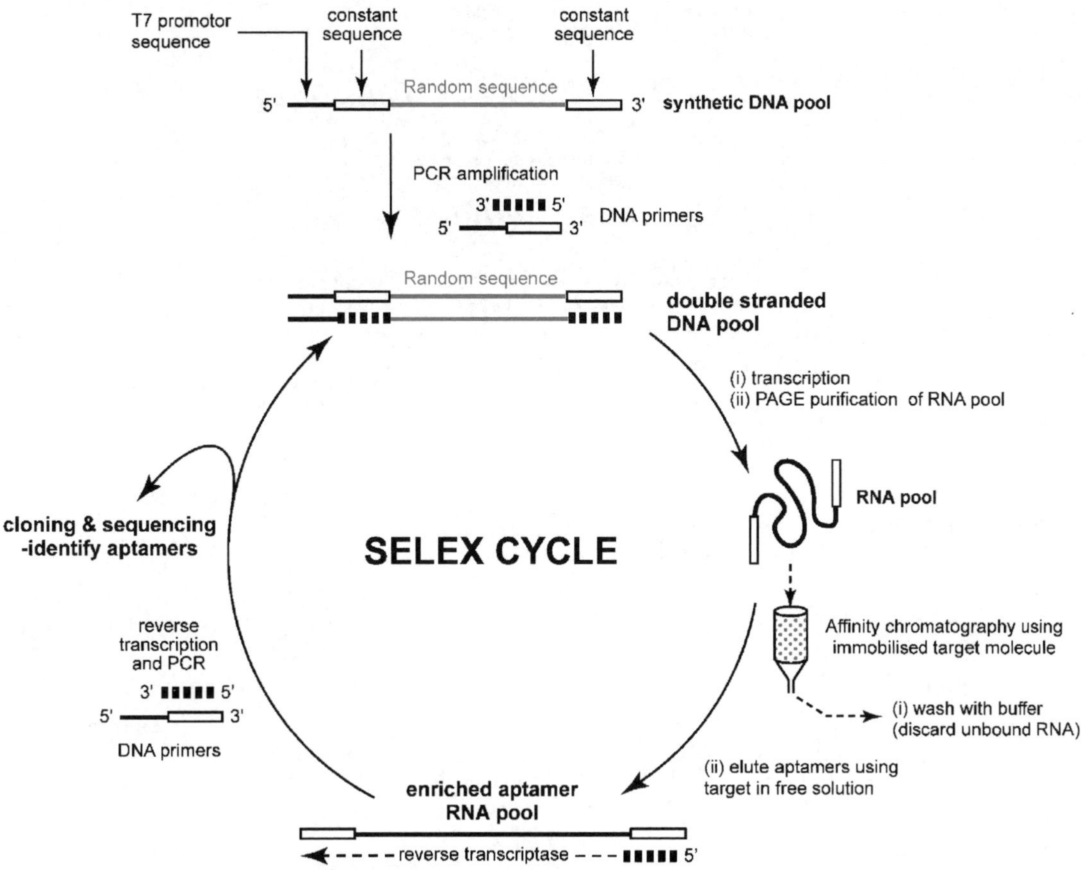


**Figure 5.25** *The generation of RNA aptamers using SELEX (systematic evolution of ligands by exponential enrichment)*

range of 3-D structures and hence must be single stranded. Thus the selection of RNA aptamers, the duplex DNA library contains a T7 RNA promoter sequence upstream of the 5′-constant sequence. This enables a single-stranded RNA pool to be generated through transcription by use of T7 RNA polymerase (Figure 5.25). A single-stranded DNA pool required for the selection of DNA aptamers is produced by capture of only the 5′-biotinylated strand from the DNA duplex on streptavidin-derivatised beads (Figure 5.26).

The selection of DNA or RNA aptamers, which can bind to a given target molecule is achieved by use of affinity chromatography. Thus the target molecule is immobilised on a solid support usually within a small column and a solution of the nucleic acid pool is passed through. Unbound nucleic acids are eluted by simple washing of the column with a suitable buffer. Sequences that have some affinity for the target are bound by the column and subsequently eluted by washing with a buffer that contains the free target molecule. For RNA aptamers, a cDNA library is then generated from the bound fraction using enzyme reverse transcriptase, which is then amplified by PCR (Figure 5.25). For DNA aptamers, the aptamer-containing fraction is subjected directly to PCR (Figure 5.26). In subsequent rounds of SELEX, the stringency of the washing protocols is increased or the concentration of immobilised target is reduced such that the affinity chromatography step leads to an enrichment of high-affinity binding sequences within the library. Typically about ten cycles of SELEX are carried out after which perhaps about 100 or so different sequences remain. These aptamers are then cloned into a vector and characterised by DNA sequencing.

SELEX has been used to identify a wide range of aptamers to a variety of diverse molecular targets, for example ions, small molecules such as organic dyes, nucleotides and their bases, amino acids, co-factors, antibiotics, transition state analogues, as well as peptides and proteins.[55] The remarkable selectivity of

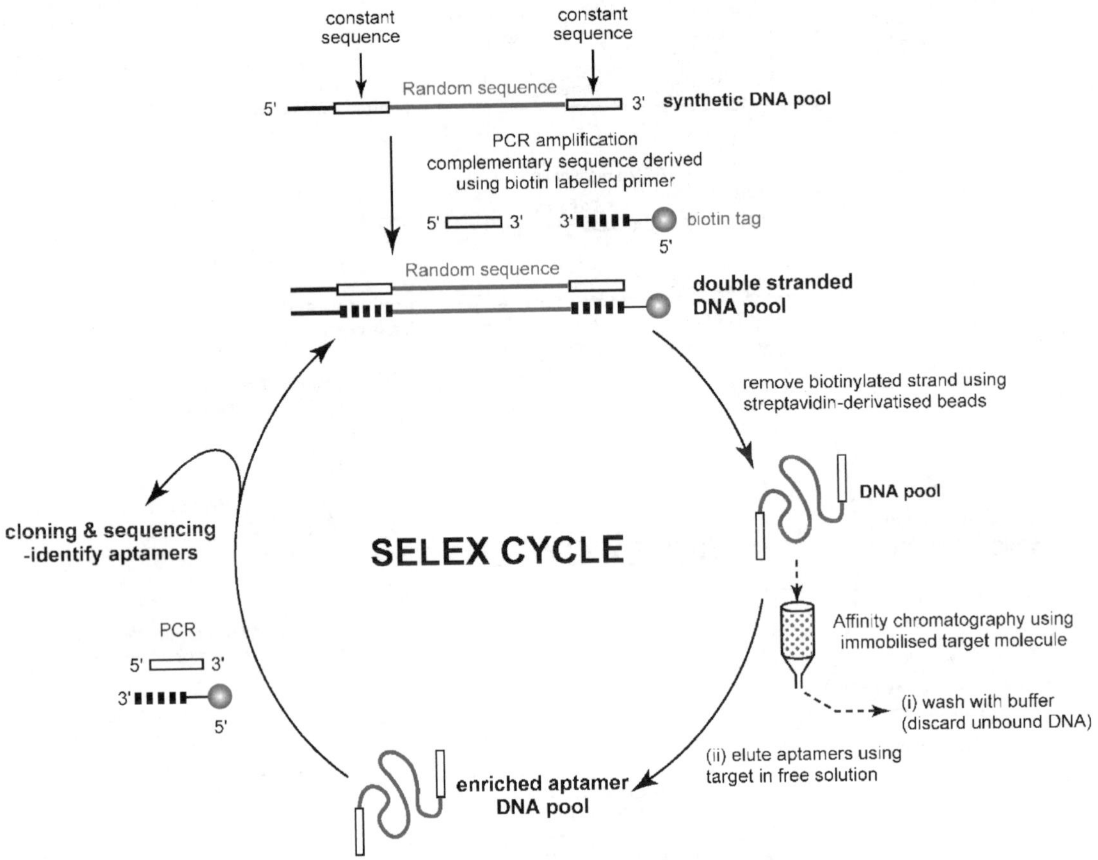


**Figure 5.26** *The generation of DNA aptamers using SELEX*

aptamers for their target molecule is nicely illustrated by one such example that binds with high affinity ($K_D$ 0.6 μM) to theophylline (1,3-dimethylxanthine) but binds the highly similar structure caffeine (1,3,7-trimethylxanthine) about $10^4$ fold less efficiently.

The selection of an aptamer with high affinity for the blood-clotting protein thrombin was the first example of a nucleic acid ligand designed to bind to a protein target that does not normally interact with DNA.[56] Such aptamers show affinities between 25 and 200 nM and contain a highly conserved 14–17 base consensus sequence. NMR of the 15-mer aptamer d(GGTTGGTGTGGTTGG) and X-ray crystallography as a complex with thrombin have revealed that the oligonucleotide forms a DNA quadruplex structure (Sections 2.3.7 and 9.10.2). Such aptamers have potential clinical application as anti-coagulants.[54]

*5.7.3.3 Selection of Nucleic Acid Catalysts.* SELEX has also been exploited to generate nucleic acid-based catalysts for a wide range of chemical reactions. Examples including RNA cleavage, DNA cleavage, DNA ligation, DNA phosphorylation, porphyrin metalation, DNA capping, DNA depurination, amide bond formation and the Diels–Alder reaction and the ability for stereochemical control during catalysis highlight the potential of SELEX in the area of synthetic organic chemistry.[53,57–59] Initial selections require protocols in which the chemical reaction is intramolecular (in *cis*), for example, self-cleavage, self-alkylation or self-phosphorylation. However, the analogous intermolecular reaction (in *trans*) with a separate substrate molecule is generally of more practical value.

In an example derived from the work of Santoro and Joyce,[60] a DNA catalyst (**DNAzyme**) capable of the specific cleavage of an HIV RNA target sequence was identified (Figure 5.27). Here a synthetic DNA library containing a central randomised region of 50 nucleotides flanked by 5′ and 3′ constant sequences

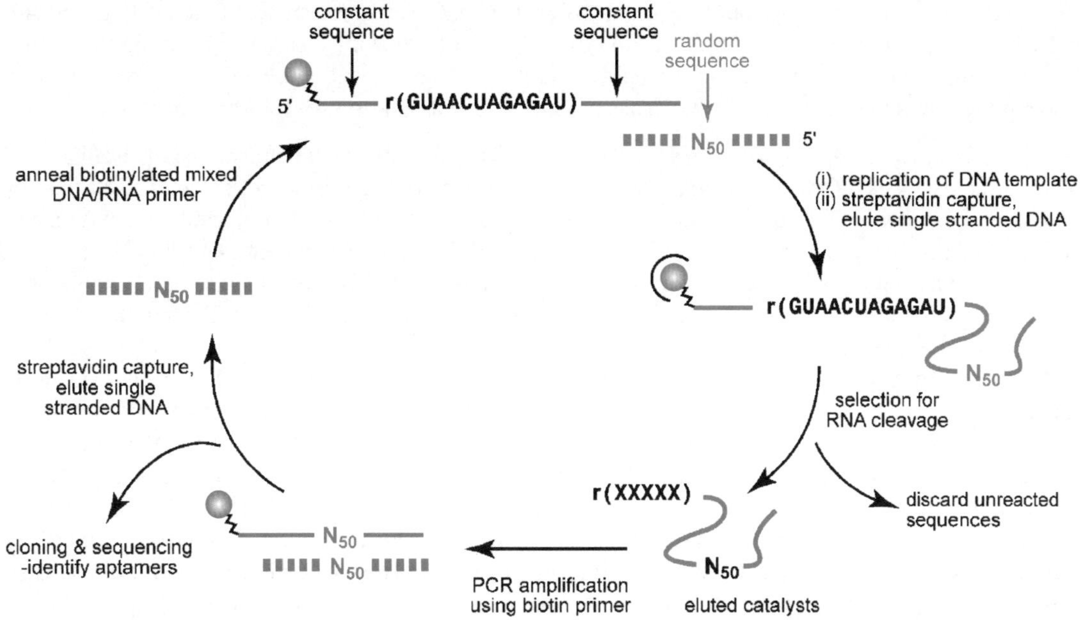


**Figure 5.27** *The generation of a DNAzyme to catalyse the sequence-specific cleavage of RNA*

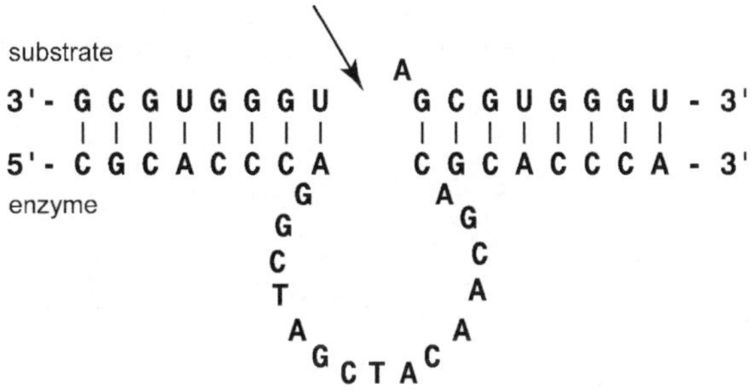


**Figure 5.28** *Composition of a 10–23 catalytic motif. The catalytic domain of 15 nucleotides is flanked by two substrate recognition domains that can be varied providing Watson–Crick base-pairing is maintained. The arrow denotes the site of cleavage*

was first copied using a synthetic 5′-biotinylated mixed DNA/RNA primer. The primer contained an embedded 12 nucleotide RNA sequence and a 3′-DNA tail complementary in sequence to the 3′-terminus of the DNA library. Single-stranded DNA containing both enzyme (within the 50 nucleotide random sequence) and target RNA sequence was then obtained following capture by streptavidin-coated beads, denaturation and removal of the non-biotinylated template. This allows folding and interaction of the enzyme and substrate portions of the immobilised sequence. In the presence of magnesium ion co-factor, the active sequences undergo self-cleavage within the RNA target section. The released nucleic acid sequences are then amplified by PCR where one primer is biotinylated. Single-stranded templates are produced by streptavidin capture of the biotinylated strand ready for the next round of SELEX.

The DNAzyme sequences identified in this work can be simplified and shortened such that cleavage of a separate substrate strand (*trans* cleavage) can occur. The '10–23' enzyme derived from this work comprises of a catalytic core of 15 nucleotides flanked by two substrate recognition domains in which Watson–Crick base pairing occurs (Figure 5.28). The structure of the '10 –23' enzyme can be modified to recognise and cleave different target sequences providing that Watson–Crick recognition between the enzyme and substrate is maintained. Recently a '10–23' DNAzyme capable of achieving cleavage rates of up to 10 $min^{-1}$ under certain metal ion, concentration and pH conditions has been identified, while a related DNAzyme also reported by Joyce[53], the '8–17' motif, has been exploited as a biosensor for $Pb^{2+}$ ions.

*5.7.3.4 Modified Nucleotides Used in SELEX.* The diversity of functional groups available for binding and catalysis offered by the four natural nucleotides is limited. Furthermore, the $pK_a$ values of the nucleobases are not ideal to permit electrostatic transition-state stabilisation or general acid–base mechanisms at appropriate neutral pH. Consequently, both sugar and base modified nucleotides have been exploited in SELEX, but such modified nucleotides must be substrates for T7 RNA polymerase or a suitable DNA polymerase to replace their natural analogues during replication or transcription of the template DNA or RNA pool respectively.

Analogues such as 5-(1-pentynyl)-2′-deoxyuridine (Figure 5.29, structure c) have been used in the selection of aptamers designed to bind thrombin. These aptamers display similar binding to thrombin as the unmodified 15-mer aptamers, but have different structures and do not function if the analogue is replaced by 2′-deoxyuridine or thymidine. The potential use of aptamers as therapeutics has led to attempts to increase their stability towards degradation by nuclease enzymes. Thus 2′-fluoro- and 2′-amino modified ribonucleotides have been employed in SELEX.[54] These analogues impart a considerably increased stability to RNA in respect of both chemical and enzymatic hydrolysis. Complete resistance to nucleases can be achieved if the aptamers are comprised of mirror image or L-RNA (**spiegelmers**).[61]

A recent variation suitable for the isolation of very high affinity nucleic acid ligands is **photo-SELEX**.[62] Here a 5-bromo-UTP, -2′-fluoro-UTP or -dUTP is employed as one of the nucleotide units and the affinity chromatography step is followed by a brief UV laser irradiation step at 308 nm, during which the single-stranded nucleic acid ligand is cross-linked to a proximal electron-rich amino acid in a protein target. The complexes are purified by SDS–PAGE and following proteolysis with proteinase K, the aptamer functions as

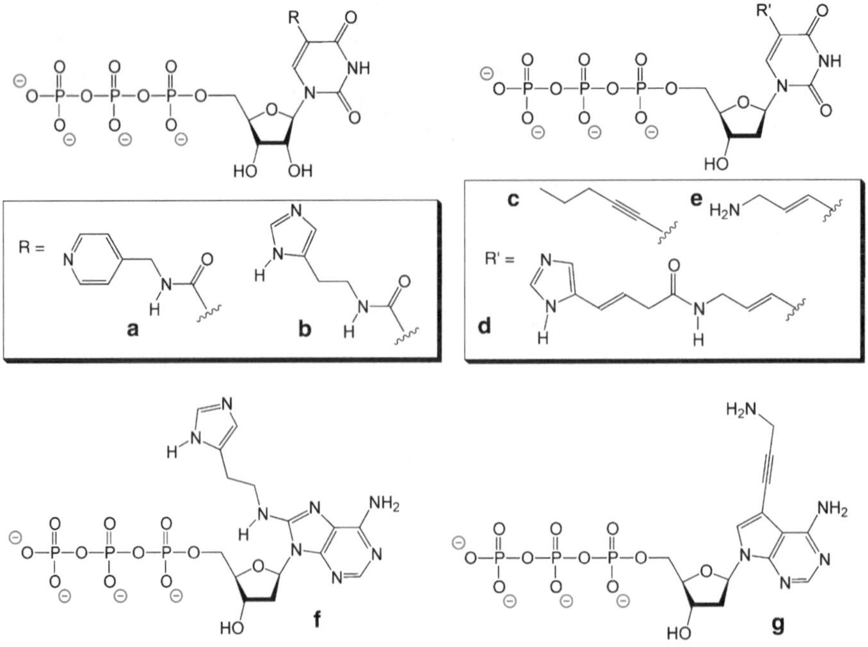


**Figure 5.29** *A selection of modified nucleoside triphosphates that have been used in the selection of nucleic acid aptamers and catalysts*

a template for *Taq* DNA polymerase (DNA aptamer) or reverse transcriptase (RNA aptamer) in further rounds of SELEX. The technique has been used to isolate aptamers with nM and pM affinities to the HIV-1 Rev protein and basic fibroblast growth factor (bFGF)[63] respectively. Only high affinity aptamers have the precise orientation of functional groups to permit cross-linking, and a harsh washing step removes all non-cross-linked proteins interacting with the immobilised aptamer. This reduces background signals due to binding of non-cognate proteins and allows a highly reliable diagnostic assay for proteins attached to a micro-chip.[63]

The incorporation of base modified nucleosides with potential catalytic groups was first demonstrated by Eaton in which a C-5-modified UTP bearing an appended pyridyl or imidazole function (Figure 5.29, structures a and b) was employed in the selection of an RNA capable of acting as a Diels Alderase or an amide synthase.[53] More recently imidazole- and amine-modified nucleotides have been employed to select DNAzymes capable of the sequence-specific cleavage of RNA (Figure 5.29, structures d, f and g).[55,64,65] For example, Perrin[64] and Williams[65] selected DNAzymes functionalised with imidazolyl and amino functional groups that catalyse the sequence specific cleavage of RNA in the absence of metal ions with rate enhancements of about $10^5$ compared to the uncatalysed reaction. Such DNAzymes display the functional side chains that are utilised by protein ribonuclease RNaseA in metal-independent RNA cleavage.

*5.7.3.5 Riboswitches.* While the catalytic potential of natural RNA has been known since the mid-1980s, it is only recently that a natural biological role for RNA aptamers has been revealed. Such naturally occurring aptamers or '**riboswitches**' have been found within the leader sequences of several metabolic genes where they have important roles in regulating both transcription and translation of the respective gene.[66] The function of these riboswitches is to assess cellular levels of certain metabolites, which in turn control expression of that gene. Thus the riboswitch functions in much the same way as synthetic aptamer that can recognise and bind to a small molecule target. The flavin mononucleotide (FMN)-sensing riboswitch found in bacteria was one of the first such examples.[66] FMN and flavin adenine dinucleotide (FAD) are synthesised from riboflavin (vitamin B2). The enzymes that are responsible for riboflavin biosynthesis from GTP are derived from five genes that comprise the riboflavin operon. The first of these genes contains a 300 nucleotide untranslated region which, upon binding of FMN or FAD, changes conformation so as to cause the termination of transcription. In this work, a number of other riboswitches have been described, including examples, which recognise thiamine pyrophosphate, adenosylcobalamin, *S*-adenosyl methionine, lysine and guanine.[66]

## 5.8 DNA FOOTPRINTING

**Footprinting** is a method for determining the precise DNA sequence of bases that is the site for attachment of a particular DNA enzyme or binding-protein or of a DNA-binding drug. DNA footprinting utilises a DNA cleaving agent, which can be either a nuclease or a chemical reagent. The agent must be able to cut DNA non-selectively at every exposed base pair while such DNA cleavage is inhibited at the site where the protein or drug binds to DNA. Thus a 'footprint' of the target sequence is identified as the region where no cutting is observed.

The steps in a DNA 'footprinting' experiment are

- a fragment of dsDNA containing the target sequence (usually 200–300 base pairs) is labelled at the 5′-ends with $^{32}P$ and then the label is removed preferentially from one end (*e.g.* the 3′-end of a gene) by a suitable restriction endonuclease (Section 5.3.1);
- this dsDNA fragment is incubated with the DNA binding-protein so that the protein protects the target region of DNA from DNase I digestion (Section 5.3.2);
- limited DNase I digestion is carried out, so that there is about one cut per strand and the sites of cleavage are randomly distributed among the accessible sites;
- the resulting DNA fragments are analysed by gel electrophoresis (Section 11.4.3), and give a ladder that has a 'footprint' region where there are no cuts, corresponding to the binding site (Figure 5.30). If a control track is generated from a Maxam–Gilbert G + A chemical sequencing reaction (Section 5.1.1) using the same probe as template, then the exact footprint sequence can be read out by comparing the location of the blank with the sequencing reaction.

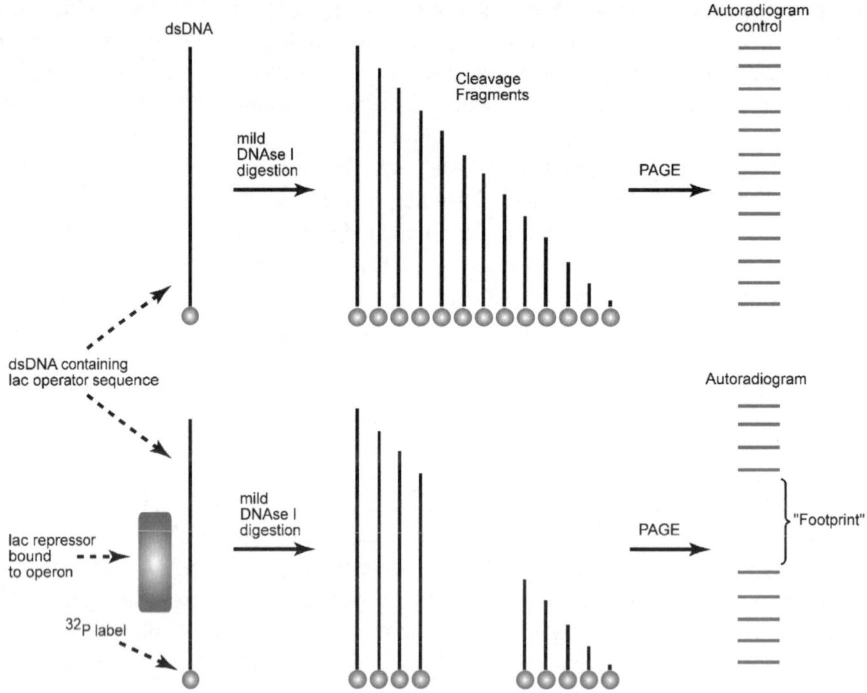


**Figure 5.30** *Scheme illustrating DNA footprinting for lac repressor protein binding to dsDNA containing the lac operator sequence. DNase I cuts DNA molecules randomly. Only one strand is 5′-end labelled with $^{32}P$. For polyacrylamide gel electrophoresis see Section 11.4.3*

Footprinting was first used by Galas[67] to determine the binding sequence for the *lac* repressor protein that established the operator sequence: d(CACCTTAACACTAACCTCTTGTTAAAG)-5′. It is now possible to identify stronger and weaker protein binding and to differentiate between affinities for each of the two DNA strands.

Since protein binding *in vitro* may not accurately reflect binding-site occupancy in the cell nucleus, methods have been developed for DNA footprinting *in vivo*. The GA-LMPCR *in vivo* footprinting system employs **dimethyl sulfate** (0.3–1.5%) to methylate nuclear DNA in whole cells suspended in phosphate buffer. Methylation occurs mainly at guanine N-7 in the major groove (Section 8.5.3) with further methylation at adenine N-3 in the minor groove. Incubation of the protein-free genomic DNA at 90°C and pH 7.0 for 15 min followed by treatment with 1 M NaOH for 30 min at 90°C leads to specific cleavage at methylated G and A sites (Maxam–Gilbert G > A procedure) (Section 5.1.1) Guanine-specific cleavage can also be accomplished by piperidine treatment of methylated DNA. The cleaved strands are then amplified using ligation-mediated PCR and analysed by PAGE, as above. Such methods have been used for the detection of upstream regulatory sequences, known as locus control regions.[68,69]

For some purposes, cleavage of the DNA is better achieved by chemical means and one of the most successful reagents has been the **hydroxyl radical**: Fenton's reagent. The cleavage system used is ferrous ammonium sulfate (1 mM) in conjunction with ascorbic acid (10 mM) and hydrogen peroxide (0.3%) at room temperature for 2 min. This works by generating hydroxyl radicals that abstract a hydrogen atom from the deoxyribose leading to phosphate diester cleavage at that residue (Section 8.9.1).[70]

In addition to its use in studies of protein binding to DNA, footprinting has been widely employed, for example, for investigating the selectivity of drug binding to DNA (Chapter 9) and for conformational analysis of triple helix formation.[71]

## REFERENCES

1. F. Sanger, A. Nicklen and A.R. Coulson, DNA sequencing with chain-terminating inhibitors. *Proc. Natl. Acad. Sci. USA*, 1977, **74**, 5463–5467.
2. F. Sanger, Sequences, sequences and sequences. *Ann. Rev. Biochem.*, 1988, **57**, 1–28.
3. C.W. Fuller, Modified T7 DNA polymerase for DNA sequencing. *Methods Enzymol.*, 1992, **216**, 329–354.
4. A.M. Maxam and W. Gilbert, A new method for sequencing DNA. *Proc. Natl. Acad. Sci. USA*, 1977, **74**, 560–564.
5. A.M. Maxam and W. Gilbert, Sequencing end-labeled DNA with base-specific chemical cleavages. *Methods Enzymol*, 1980, **65**, 499–560.
6. J. Sambrook and D.W. Russell, *Molecular Cloning: a Laboratory Manual*. Cold Spring Harbor Press, New York, 2000.
7. T.A. Brown, *Genomes 2*. Wiley, New York, 2000.
8. M.J. McPherson, P. Quirke and G.R. Taylor, *PCR. A Practical Approach*. Oxford University Press, Oxford, 1991.
9. S. Shuman and B. Schwer, RNA capping enzyme and DNA ligase: a superfamily of covalent nucleotidyl transferases. *Mol. Microbiol.*, 1995, **17**, 405–410.
10. H.G. Khorana, Total synthesis of a gene. *Science*, 1979, **203**, 614–625.
11. S. Heaphy, M. Singh and M.J. Gait, Cloning and expression in *E. coli* of a synthetic gene for the bacteriocidal protein caltrin/seminalplasmin. *Protein Eng.*, 1987, **1**, 425–431.
12. M.A. Jobling and P. Gill, Encoded evidence: DNA in forensic analysis. *Nat. Rev. Genet.*, 2004, **6**, 739–751.
13. A.J. Jeffreys, Genetic fingerprinting. *Nat. Med.* 2005, **11**, 1035–1039.
14. A.J. Jeffreys, V. Wilson and S.-L. Thein, Individual-specific 'fingerprints' of human DNA. *Nature*, 1985, **316**, 76–79.
15. P. Gill, A.J. Jeffreys and D.J. Werrett, Forensic application of DNA 'fingerprints'. *Nature*, 1985, **318**, 577–579.
16. N. Rudin and K. Inman, *An Introduction to Forensic DNA Analysis*. 2nd edn. CRC Press, Boca Raton, FL, 2002.
17. K.M. Sullivan, A. Mannucci, C.P. Kimpton and P. Gill, A rapid and quantitative DNA sex test: fluorescence-based PCR analysis of X–Y homologous gene amelogenin. *Biotechniques*, 1993, **15**, 636–641.
18. J.M. Butler, *Forensic DNA Typing: Biology and Technology Behind STR Markers*. Academic Press, New York, 2001.
19. E.M. Southern, K. Mir and M. Shchepinov, Molecular interactions on microarrays. *Nat. Genet.*, 1999, **21**(Suppl. 1), 5–9.
20. G.H. McGall, A.D. Barone, M. Diggelmann, S.P.A. Fodor, E. Gentalen and N. Ngo, The efficiency of light-directed synthesis of DNA arrays on glass substrates. *J. Am. Chem. Soc.*, 1997, **119**, 5081–5090.
21. P.O. Brown, Genome scanning methods. *Curr. Opin. Genet. Dev.*, 1994, **4**, 366–373.
22. T.R. Hughes, M. Mao, A.R. Jones, J. Burchard, M.J. Marton, K.W. Shannon, S.M. Lefkowitz, M. Ziman, J.M. Schelter, M.R. Meyer *et al.*, Expression profiling using microarrays fabricated by an ink-jet oligonucleotiode synthesizer. *Nat. Biotech.*, 2001, **19**, 342–347.
23. A.B. Sierzchala, D.J. Dellinger, J.R. Betley, T.K. Wyrzykiewicz, C.M. Yamada and M.J. Caruthers, Solid-phase oligodeoxynucleotide synthesis: a two-step cycle using peroxy anion deprotection. *J. Am. Chem. Soc.*, 2003, **125**, 13427–13441.
24. D.P. Bratu, C. B.-K., M.M. Mhlanga, F.R. Kramer and S. Tyagi, Visualizing the distribution and transport of mRNAs in living cells. *Proc. Natl. Acad. Sci. USA*, 2003, **100**, 13308–13313.
25. R.W. Dirks, C. Molenaar and H.J. Tanke, Visualizing RNA molecules inside the nucleus of living cells. *Methods*, 2003, **29**, 51–57.

26. J.B. Opalinska and A.M. Gewirtz, Nucleic-acid therapeutics: basic principles and recent applications. *Nat. Rev. Drug Discovery*, 2002, **1**, 503–514.
27. L.J. Scherer and J.J. Rossi, Approaches for the sequence-specific knockdown of mRNA. *Nat. Biotechnol.*, 2003, **21**, 1457–1465.
28. M. Faria, C.D. Wood, L. Perrouault, J.S. Nelson, A. Winter, M.R.H. White, C. Hélène and C. Giovannangeli, Targeted inhibition of transcription elongation in cells mediated by triplex-forming oligonucleotides. *Proc. Natl. Acad. Sci. USA*, 2000, **97**, 3862–3867.
29. P.C. Zamecnik and M.L. Stephenson, Inhibition of Rous sarcoma virus replication and transformation by a specific oligodeoxynucleotide. *Proc. Natl. Acad. Sci. USA*, 1978, **75**, 280–284.
30. P. Sazani, M.M. Vacek and R. Kole, Short-term and long-term modulation of gene expression by antisense therapeutics. *Curr. Opin. Biotech.*, 2002, **13**, 468–472.
31. J. Kurreck, Antisense technologies. Improvement through novel chemical modifications. *Eur. J. Biochem.*, 2003, **270**, 1628–1644.
32. B.F. Baker, S.S. Lot, T.P. Condon, S. Cheng-Flourney, E.A. Lesnik, H.M. Sasmor and C.F. Bennett, 2′-*O*-(2-methoxy)ethyl-modified anti-intercellular adhesion molecule 1 (ICAM-1) oligonucleotides selectively increase the ICAM-1 mRNA level and inhibit formation of the ICAM-1 translation initiation complex in human umbilical vein endothelial cells. *J. Biol. Chem.*, 1997, **272**, 11994–12000.
33. M. Faria, D.G. Spiller, C. Dubertret, J.S. Nelson, M.R.H. White, D. Scherman, C. Hélène and C. Giovannangeli, Phosphoramidate oligonucleotides as potent antisense molecules in cells *in vivo*. *Nature Biotech.*, 2001, **19**, 40–44.
34. D.R. Mercatante and R. Kole, Control of alternative splicing by antisense oligonucleotides as a potential chemotherapy: effects on gene expression. *Biochim. Biophys. Acta*, 2002, **1587**, 126–132.
35. T.A. Vickers, J.R. Wyatt, T. Burckin, C.F. Bennett and S.M. Freier, Fully modified 2′-MOE oligonucleotides redirect polyadenylation. *Nucl. Acids Res.*, 2001, **29**, 1293–1299.
36. J.J. Toulmé, C. Boiziau, B. Larrouy, P. Frank, S. Albert and R. Ahmadi, in *DNA and RNA Cleavers and Chemotherapy of Cancer and Viral Diseases*, B. Meunier (ed). Kluwer Academic Publishers, The Netherlands, 1996, 271–288.
37. R.A. McKay, L.J. Miraglia, L.L. Cummins, S.R. Owens, H. Sasmor and N.M. Dean, Characterization of a potent and specific class of antisense oligonucleotide inhibitor of human protein kinase C-α expression. *J. Biol. Chem.*, 1999, **274**, 1715–1722.
38. A. Asai, Y. Oshima, Y. Yamamoto, T. Uochi, H. Kusaka, S. Akinaga, Y. Yamashita, K. Pongracz, R. Pruzan, E. Wunder *et al.*, A novel telomerase template antagonist (GRN163) as a potential anticancer agent. *Cancer Res.*, 2003, **63**, 3931–3939.
39. S. Agrawal and E.R. Kandimella, Medicinal chemistry and therapeutic potential of CpG DNA. *Trends Mol. Med.*, 2002, **8**, 114–121.
40. S. Agrawal and E.R. Kandimella, Antisense and siRNA as agonists of Toll-like receptors. *Nat. Biotech.*, 2004, **22**, 1533–1537.
41. E.S. Gragoudas, A.P. Adamis, E.T. Cunningham, M. Feinsod and D.R. Guyer, Pegaptanib for neovascular age-related macular degeneration. *New Engl. J. Med.*, 2004, **351**, 2805–2816.
42. A.J. Hamilton and D.C. Baulcombe, A species of small antisense RNA in post-transcriptional gene silencing in plants. *Science*, 1999, **286**, 950–952.
43. A. Fire, S. Xu, M.K. Montgomery, S.A. Kostas, S.E. Driver and C.C. Mello, Potent and specific genetic interference by double-stranded RNA in *Caenorhabditis elegans*. *Nature*, 1998, **391**, 806–811.
44. G. Hannon and J.J. Rossi, Unlocking the potential of the human genome with RNA interference. *Nature*, 2004, **431**, 371–378.
45. O.A. Kent and A.M. MacMillan, RNAi: running interference for the cell. *Org. Biomol. Chem.*, 2004, **2**, 1957–1961.
46. S.W. Jones, D.S. P.M. and M.A. Lindsay, siRNA for gene silencing: a route to drug target discovery. *Curr. Opin. Pharmacol.*, 2004, **4**, 522–527.
47. M. Tijsterman and R.H.A. Plasterk, Dicers at RISC: the mechanism, of RNAi. *Cell*, 2004, **117**, 1–4.

48. S.M. Elbashir, J. Harborth, W. Lendeckel, A. Yalcin, K. Weber and T. Tuschl, Duplexes of 21-nucleotide RNAs mediate RNA interference in cultured mammalian cells. *Nature*, 2001, **411**, 494–498.
49. M. Manoharan, RNA interference and chemically modified small interfering RNAs. *Curr. Opin. Chem. Biol.*, 2004, **8**, 1–10.
50. J. Soutschek, A. Akinc, B. Bramlage, K. Charisse, R. Constien, M. Donoghue, S.M. Elbashir, A. Geick, P. Hadwiger, J. Harborth *et al.*, Therapeutic silencing of an endogenous gene by systemic administration of modified siRNAs. *Nature*, 2004, **432**, 173–177.
51. P.J. Paddison, A.A. Caudy, E. Bernstein, G.J. Hannon and D.S. Conklin, Short hairpin RNAs (shRNAs) induce sequence-specific silencing in mammalian cells. *Genes Dev.*, 2002, **16**, 948–958.
52. L. Gold, B. Polisky, O. Uhlenbeck and M. Yarus, Diversity of Oligonucleotide Functions. *Ann. Rev. Biochem.*, 1995, **64**, 763–797.
53. G.F. Joyce, Directed evolution of nucleic acid enzymes. *Ann. Rev. Biochem.*, 2004, **73**, 791–836.
54. D.S. Wilson and J.W. Szostak, *In vitro* selection of functional nucleic acids. *Ann. Rev. Biochem.*, 1999, **68**, 611–647.
55. S.W. Santoro, G.F. Joyce, K. Sakthivel, S. Gramatikova and C.F. Barbas, RNA cleavage by a DNA enzyme with extended chemical functionality. *J. Am. Chem. Soc.*, 2000, **122**, 2433–2439.
56. L.C. Bock, L.C. Griffin, J.A. Latham, E.H. Vermaas and J.J. Toole, Selection of single-stranded-DNA molecules that bind and inhibit human thrombin. *Nature*, 1992, **355**, 564–566.
57. C. Frauendorf and A. Jaschke, Catalysis of organic reactions by RNA. *Angew. Chem. Int. Ed.*, 1998, **37**, 1378–1381.
58. A. Jaschke, C. Frauendorf and F. Hausch, *In vitro* selected oligonucleotides as tools in organic chemistry. *Synlett*, 1999, **6**, 825–833.
59. A. Jaschke and B. Seelig, Evolution of DNA and RNA as catalysts for chemical reactions. *Curr. Opin. Chem. Biol.*, 2000, **4**, 257–262.
60. S.W. Santoro and G.F. Joyce, A general purpose RNA-cleaving DNA enzyme. *Proc. Natl. Acad. Sci. USA*, 1997, **94**, 4262–4266.
61. D. Eulberg and S. Klussmann, Spiegelmers: biostable aptamers. *ChemBiochem*, 2003, **4**, 979–983.
62. K.B. Jensen, B.L. Atkinson, M.C. Willis, T.H. Koch and L. Gold, Using *in vitro* selection to direct the covalent attachment of human immunodeficiency virus type 1 Rev protein to high-affinity RNA ligands. *Proc. Natl. Acad. Sci. USA*, 1995, **92**, 12220–12224.
63. E.N. Brody, M.C. Willis, J.D. Smith, S. Jayasena, D. Zichi and L. Gold, The use of aptamers in large arrays for molecular diagnostics. *Mol. Diag.*, 1999, **4**, 381–388.
64. L. Lermer, Y. Roupioz, R. Ting and D.M. Perrin, Toward an RNaseA mimic: a DNAzyme with imidazoles and cationic amines. *J. Am. Chem. Soc.*, 2002, **124,** 9960–9961.
65. A.V. Sidorov, J.A. Grasby and D.M. Williams, Sequence-specific cleavage of RNA in the absence of divalent metal ions by a DNAzyme incorporating imidazolyl and amino functionalities. *Nucl. Acids Res.*, 2004, **32**, 1591–1601.
66. E. Nudler and A.S. Mironov, The riboswitch control of bacterial metabolism. *Trends Biochem. Sci.*, 2004, **29**, 11–17.
67. D.J. Galas, The invention of footprinting. *Trends Biochem. Sci.*, 2001, **26**, 690–693.
68. E.C. Strauss and S.H. Orkin, *In vivo* interactions at hypersensitive site 3 of the human β-globin locus control region. *Proc. Natl. Acad. Sci. USA*, 1992, **89**, 5809–5813.
69. I.L. Cartwright and S.E. Kelly, Probing the nature of chromosomal DNA-protein contacts by *in vivo* footprinting. *Biotechniques*, 1991, **11**, 188–196.
70. W.J. Dixon, J.J. Hayes, J.R. Levin, M.F. Weidner, B.A. Dombroski and T.D. Tullius, Hydroxyl radical footprinting. *Meth. Enzymol.*, 1991, **208**, 380–413.
71. K.R. Fox and M.J. Waring, High-resolution footprinting studies of drug-DNA complexes using chemical and enzymatic probes. *Meth. Enzymol.*, 2001, **340**, 412–430.

CHAPTER 6

# Genes and Genomes

## CONTENTS

## 6.1 GENE STRUCTURE

The primary function of polymeric nucleic acids in all living organisms is the storage and transmission of genetic information. Every living thing on Earth is constructed from a genetic blueprint encoded by its nucleic acid genome. For all independently living organisms, this blueprint is comprised of DNA. Less complex entities, such as viruses, which rely on hosts to live and reproduce themselves, may use RNA instead. This chapter will describe how this genetic information is stored, replicated, repaired and copied into the functional products on which life depends.

The basic unit of genetic information is the **gene**. Genes were described originally in 1865 by Mendel as apparently indestructible factors, which specify traits of an organism such as colour or shape. The pioneering work of Avery and co-workers[1] showed that genes are in fact comprised of nucleic acid and a 'Golden Age' of molecular biology in the 1950s and 1960s laid the foundation for our present day understanding of gene structure and expression.[2] The modern definition of a gene is a discrete nucleic acid that encodes an RNA or protein that has biological function. It is important to note here that not all genes encode proteins. Many genes encode functional RNAs, such as transfer RNAs, ribosomal RNAs (rRNAs) or spliceosomal RNAs (Sections 2.4 and 7.3).

Gene structure is remarkably diverse. The only property shared by all genes is the presence of a nucleic acid region that encodes a functional component. There are three dominant types of gene structure seen in living cells (Figure 6.1). The first and simplest (Figure 6.1a) consists of a single uninterrupted coding region flanked by signals necessary for starting and stopping the transcription of the gene into RNA. The former signal is known as a transcriptional promoter and the latter as a transcriptional terminator. The second type of gene structure (Figure 6.1b) commonly found in prokaryotes, such as in the bacterium *Escherichia coli*, dispenses with individual promoters and terminators and pools genes together into a cluster called an **operon**[3] under the control of a single **promoter**. The third major type of gene structure found (Figure 6.1c) is the interrupted gene,[4–6] where the internal region is split into segments, which either are present in the mature functional RNA gene product (**exons**[6]) or removed during RNA splicing (Section 7.2.2) and destroyed (**introns**[6]). This seemingly bizarre organisation points back to the origin of genes in that small segments of DNA, representing discrete units of function, are thought to have gradually become assembled into the exons of more complex genes that now code for multi-domain proteins.[7]

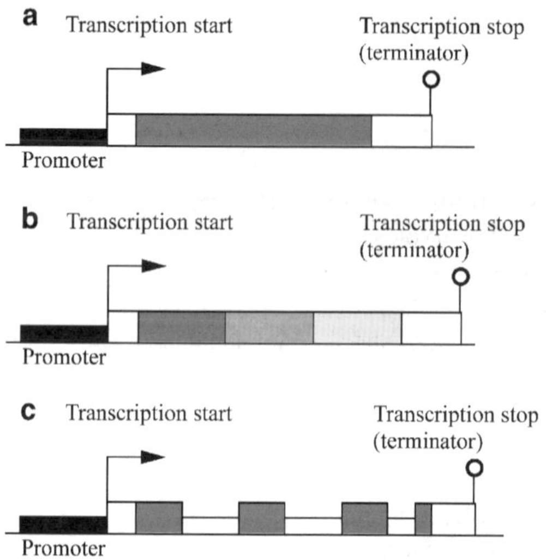


**Figure 6.1** *Basic gene structures. (a) A gene with its promoter and terminator. (b) An operon containing several genes under the control of a single promoter. (c) An interrupted gene containing exons (red shaded boxes) and introns (uncoloured smaller boxes). Red shaded regions are protein coding*

### 6.1.1 Conventional Eukaryotic Gene Structure – The β Globin Gene as an Example

Most of the genes in eukaryotes belong to the third class described above and the great majority of these encode proteins. The pathway from the gene to its protein product and the structural relationships between the gene and its gene products are exemplified by the β globin gene (Figure 6.2), which is found in all vertebrate animals. The gene is first transcribed in the cell nucleus to produce a precursor RNA that contains all the gene's introns. The 5′-end of the precursor RNA corresponds to the transcription start but the 3′-end extends past the eventual terminus of the mature messenger RNA (**mRNA**) product. Such a precursor RNA is typically unstable and is quickly processed into a mature mRNA by removal of its introns and by cleavage at its 3′-end, followed by the addition of a few hundred adenosine bases to produce a '**poly A tail**'.[8] The mRNA includes start and stop sites for translation. Therefore, an mRNAs always contains extra nucleic acid sequences at both its 5′- and 3′-end that are not converted into protein (shown in white in the RNA in Figure 6.2). The mature mRNA is exported from the nucleus[9] to the cytoplasm of the eukaryotic cell where it is translated into protein.

### 6.1.2 Complex Gene Structures

The large majority of protein-encoding genes have the general structures shown in Figure 6.1, with variations in overall size and number of exons. However, there are many examples of more complicated gene structure (Figure 6.3).

*6.1.2.1 Alternative Promoters.* It is relatively common to find a single gene, which contains more than one promoter.[10] An example is the alcohol dehydrogenase gene of the fruit fly *Drosophila* (Figure 6.3a). Typically, the different promoters function either in different tissues of the organism, at different developmental stages or in response to different stimuli. **Alternative promoters** therefore provide a way of varying

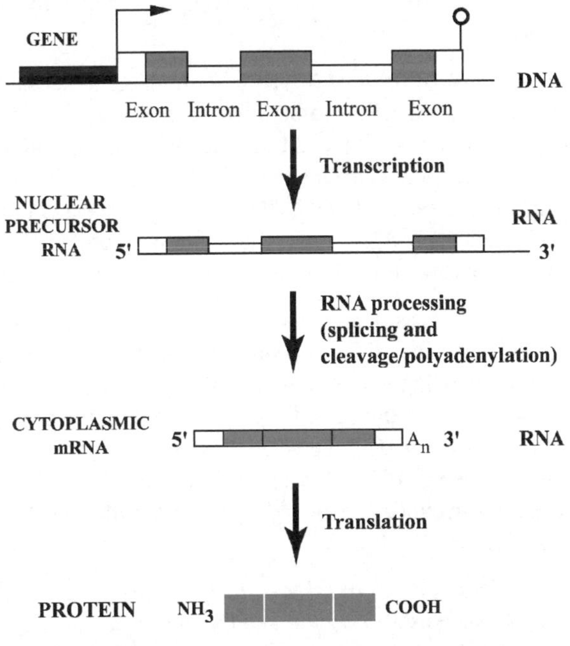


**Figure 6.2** *Relationships between a typical eukaryotic protein-coding gene and its gene products. Red shaded regions are translated into protein*

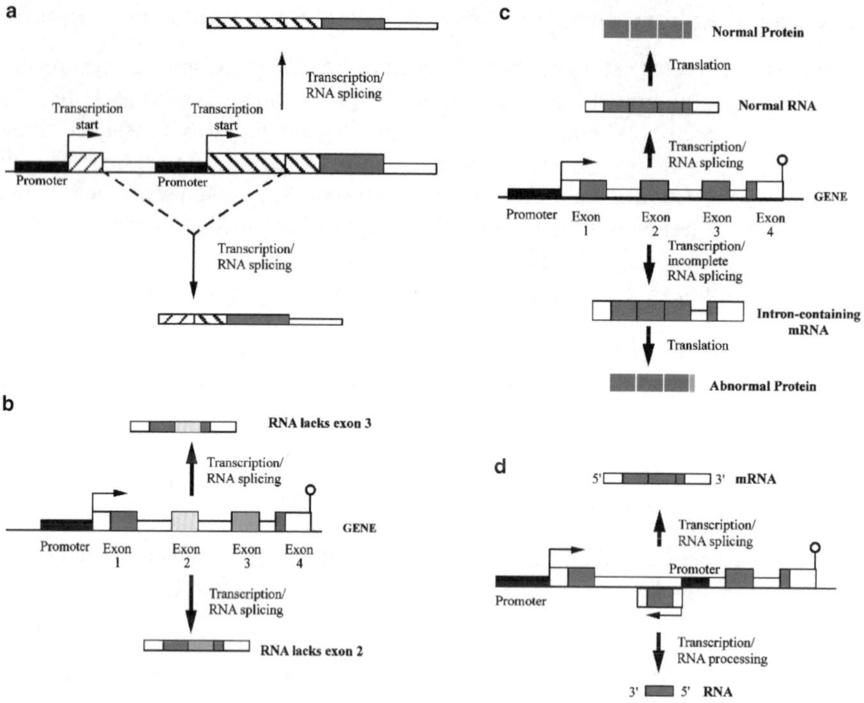


**Figure 6.3** *Complex gene structures. (a) Alternative promoters for a single gene. (b) Optional exon usage. (c) Intron omission. (d) A gene within the intron of another gene*

the amount of gene product produced (in this case two corresponding enzymes). Often, different promoters use a different splicing site.

*6.1.2.2 Alternative Exons and Optional Splicing.* A single gene does not necessarily use all of its exons to produce a gene product (Figure 6.3b). Sometimes an exon may be omitted during **RNA splicing**. If this exon is protein coding, the proteins produced from the two different mRNAs differ from each other. In this way, a single gene can produce more than one protein.[11] Another way that an encoded protein sequence can be altered is by an intron being missed out in the RNA splicing process (Figure 6.3c). This results in an mRNA containing an intron. This is likely to terminate the synthesised protein, either by introducing a translational frame shift or a stop codon (Section 7.3.1), leading to a protein truncated at its carboxyl terminus.

*6.1.2.3 Genes Within Genes.* Occasionally, two genes can be found within the same section of DNA sequence. Most commonly, a small gene, such as a small RNA-encoding gene, can be found within the intron of a conventional protein-encoding gene (Figure 6.3d). In such circumstances, the two genes do not seem to be expressed in the same cells at the same time, thus avoiding the problem of head-to-head collisions between RNA polymerases transcribing the two DNA strands simultaneously. Certain small RNA encoding genes, such as small nucleolar RNA (snoRNA) genes, may be found within introns in the same transcriptional orientation as the surrounding gene.[12] In these cases, the snoRNA is cut out of the precursor RNA by RNA processing (Section 7.5.2).

*6.1.2.4 The Complexity of Some Genes in Higher Eukaryotes.* A significant number of the genes in a variety of higher organisms, such as *Drosophila* and humans, are highly complex and very large. For example, in some of the major RNAs encoded by the *Ubx* gene of *Drosophila* (Figure 6.4) there are multiple promoters, optional or alternative introns and multiple polyadenylation sites, which are combined in

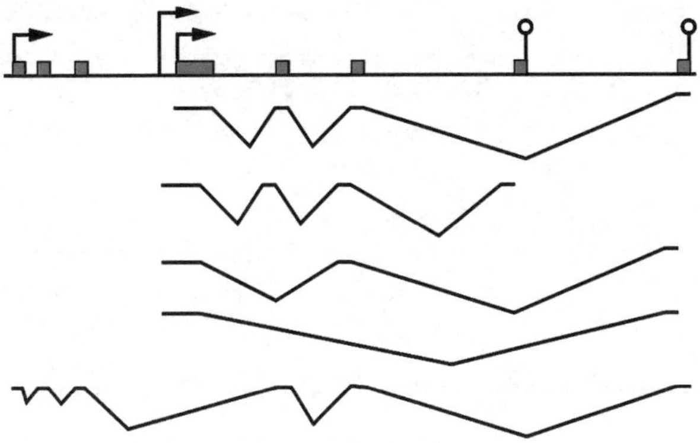

**Figure 6.4** *The Ubx gene of Drosophila – an example of complex gene structure*

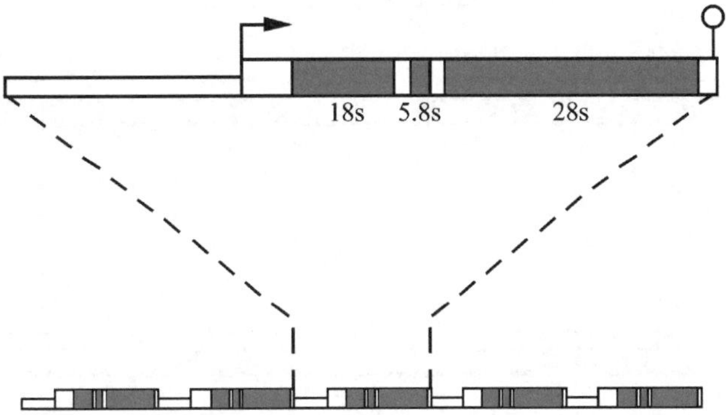


**Figure 6.5** *Structure of a tandemly repeated gene family, the human rRNA genes. Red shaded regions are expressed as mature RNA, white boxed regions are transcribed and removed from the RNA precursor by RNA processing*

a huge gene to produce a highly complex set of different proteins that function in different parts of the anatomy of an organism.

## 6.2 GENE FAMILIES

Most organisms, particularly the more complex ones, contain more than one copy of a given gene. For example, even simple prokaryotes such as *E. coli* contain several genes encoding rRNA. This is probably necessary to ensure that sufficient amounts of the gene products are produced. In eukaryotes it is more common to find multi-copy genes than true **single copy genes**. Human DNA has nearly 300 rRNA genes, located in five clusters on different chromosomes.[13,14] Each of these repeated sequences is virtually identical to the others. **rRNA gene clusters** (Figure 6.5) are comprised of tandemly repeated units, each unit containing a 28S, 5.8S and 18S gene, all driven from a single promoter. This structure resembles the operons of prokaryotes (Figure 6.1).

There are examples of high copy number for other types of gene. The silk moth *Bombyx mori* has hundreds of genes encoding the chorion (the egg shell). In this case however, there is far more complexity in the sequences of the genes (Figure 6.6). It seems that an ancestral gene pair has proliferated and diversified to produce the multiplicity of different but related genes seen today.[15] This is an extreme example of a very common process in gene and genome evolution. A comparison of haemoglobin genes across the vertebrates shows a gradual process of duplication and diversification (Figure 6.7). Haemoglobin contains

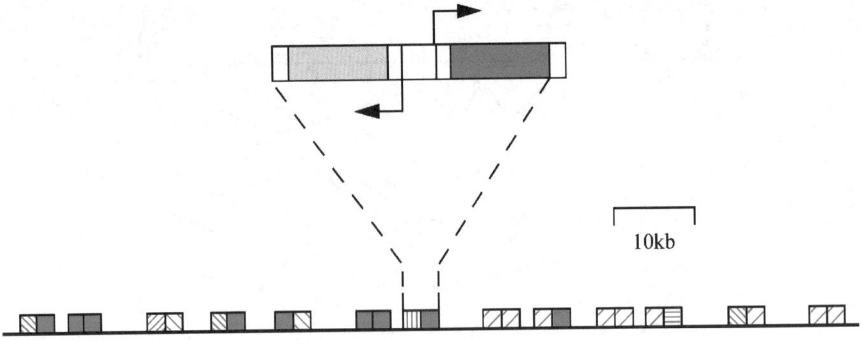


**Figure 6.6** *The silkmoth chorion multi-gene family: a 2-gene unit, where each gene is transcribed in the opposite orientation and which has been amplified multiple times. Shadings indicate sequence divergence that has occurred following amplification*

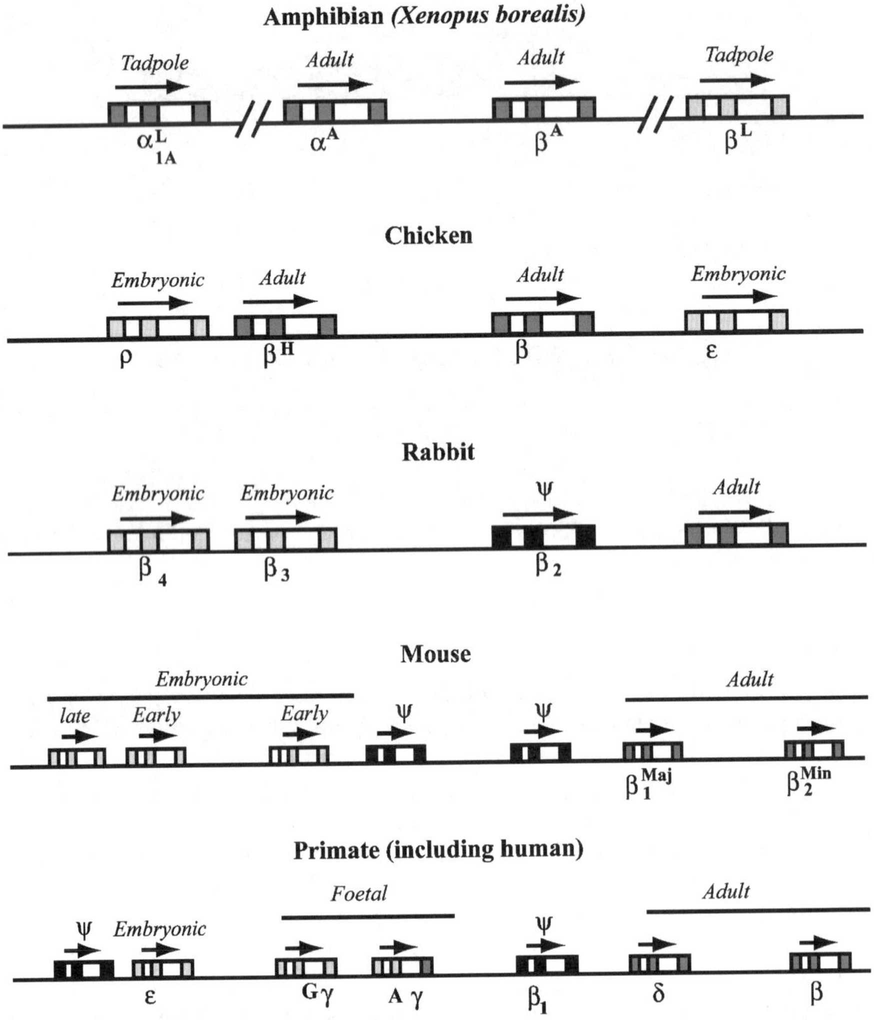


**Figure 6.7** *The evolution of the β globin gene cluster. Transcription orientation is indicated by arrows. Exons are shaded. α: an a globin-like gene. β: a β globin-like gene. Diagonal lines in the amphibian cluster indicate a longer distance between the genes. ψ: pseudogene*

two proteins, α globin and β globin. In the amphibian *Xenopus laevis* the genes for both globin types are found in the same region of the genome. In birds the two gene families have split from each other and this is also seen for mammals, where there has been an increase in gene number and complexity. Mammals need to supply their unborn young with oxygen and this cannot happen efficiently unless the foetal globin can sequester oxygen from the adult globin. For this to happen, the protein sequence, and hence the DNA sequence, of foetally expressed globin must diverge from that of the adult. Thus natural selection has promoted the proliferation and diversification of haemoglobin genes.

The β globin locus of humans shows several other interesting features (Figure 6.7). First, there are two foetal genes (Gγ and Aγ) that encode identical proteins. This is probably the result of an evolutionarily recent duplication of the foetal gene in this lineage. Second, we see several examples of defective genes carrying frame shifts and stop codons. These pseudogenes[16] may be the result of gene evolution having taken a wrong path. **Pseudogenes** are very common in mammals but surprisingly quite rare in plants. Third, each β globin gene has the same transcriptional orientation. This is also true for all vertebrates and is a consequence of the mechanism of **gene duplication**, which involves unequal **homologous recombination** between repeated sequences flanking the genes (Figure 6.8).

## 6.3 INTERGENIC DNA

In prokaryotes there is very little extra DNA besides that encoding genes. However, the genomes of eukaryotes are very different. For example, a typical stretch of the maize genome contains relatively few genes and a large amount of repetitious DNA (Figure 6.9).[17] But this is not a fixed rule; for example,

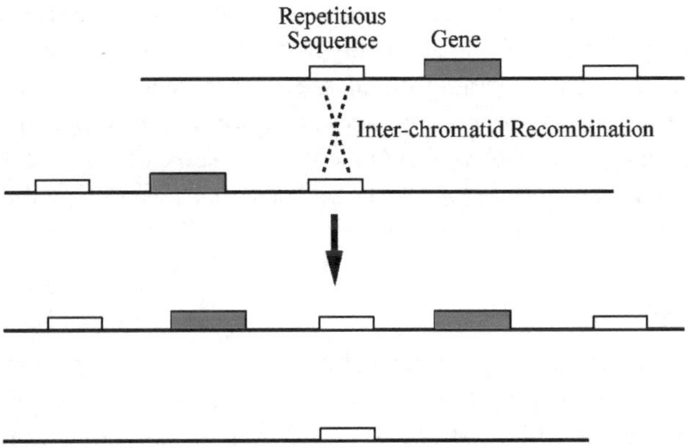


**Figure 6.8** *Gene duplication and deletion promoted by unequal exchange between flanking repetitious DNAs*

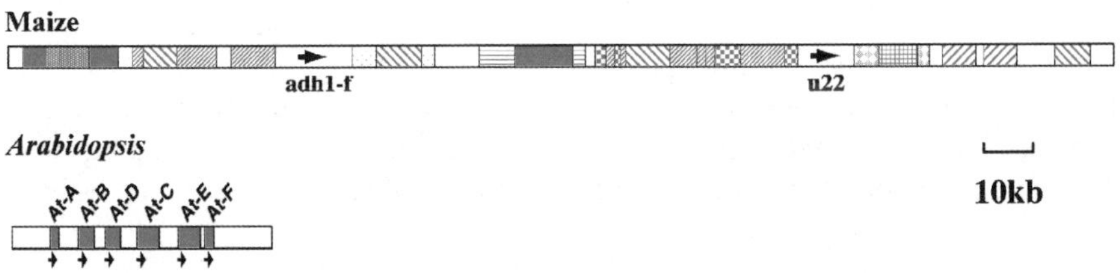


**Figure 6.9** *Intergenic DNA in plants. Maize, a plant with a large genome, has complex sets of retrotransposon insertions (Section 6.10.3) between two genes (indicated by arrows). Arabidopsis thaliana has six genes with relatively little intergenic DNA*

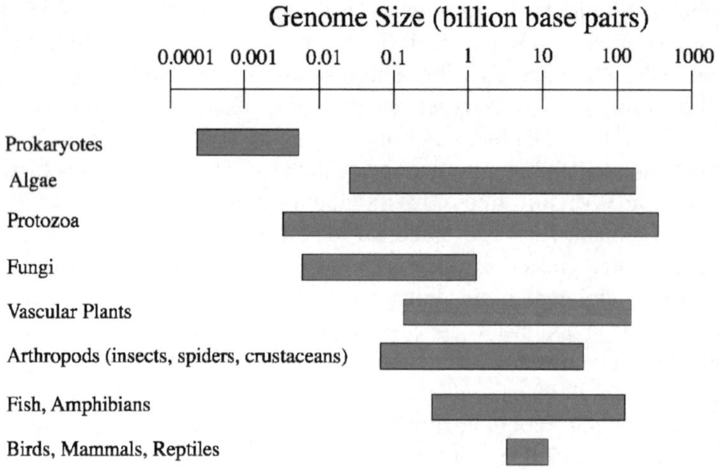


**Figure 6.10** *Variation in genome size across the five kingdoms of life*

another plant, the weed *Arabidopsis thaliana*, has far less intergenic **repetitious DNA** (Figure 6.9).[18] In fact, the **intergenic DNA** of eukaryotes is much more susceptible to change than their genes.

For some genomes such as maize the repetitious DNA can outweigh the DNA devoted to genes, leading to 'genomic obesity'.[19] This observation explains an old puzzle for molecular geneticists, namely that the sizes of eukaryotic genomes are extremely variable and there is no obvious correlation between genome size and evolutionary level (Figure 6.10). For example, one amphibian may possess 50-fold more genetic information than another. The puzzle has been called the **C-value paradox**.

Two major classes of repetitious DNA make up the majority of this so-called 'junk DNA'.[20] The first is **satellite DNA**, which is comprised of seemingly endless tandem repeats of a simple sequence. For example, the fruit fly *Drosophila virilis* possesses a huge number of ACAAACT repeats that together add up to almost a quarter of the entire genome. Such DNA is also found in **heterochromatin**[21] (Section 6.4.2), particularly at the **centromeres** of chromosomes (Section 6.4.5). The second major class of repetitious DNA is **transposable elements**, particularly **retrotransposons** (Section 6.8.3).

## 6.4 CHROMOSOMES

The DNA of an organism is arranged on one or more **chromosomes**. The number of chromosomes per species is invariant but can vary a lot between related species. Each chromosome is comprised of a double-stranded DNA molecule, packed together with a set of associated proteins and other components into a complex called **chromatin**.

### 6.4.1 Eukaryotic Chromosomes

All eukaryotes contain at least two chromosomes. There is no clear correlation between the chromosome number and the type of organism. For example, the yeast *Saccharomyces cerevisiae* has 16 chromosomes per haploid cell, the fruit fly *Drosophila melanogaster* has 4 and the human has 23. Most multi-cellular eukaryotes contain mostly diploid cells and for these cells the chromosome number is doubled. All eukaryotic nuclear chromosomes studied to date contain simple linear double-stranded DNA.

### 6.4.2 Packaging of DNA in Eukaryotic Chromosomes

The DNA of a eukaryotic chromosome must fit into a space far smaller than its total length. For example, a human chromosome has around 3–10 cm of DNA that must fit into a cell nucleus a thousand times smaller.

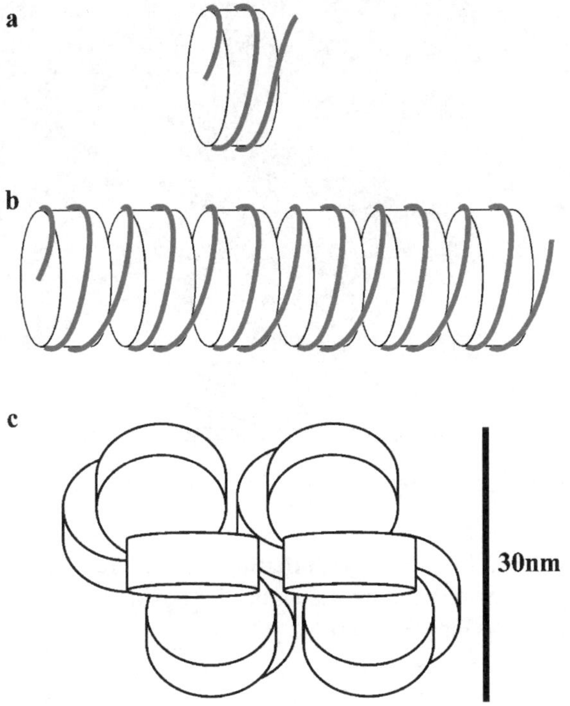


**Figure 6.11** *Nucleosomes and chromatin packing. Nucleosome proteins are shown as a disc with DNA wrapped around. (a) The nucleosome. (b) The 30 nm fibre. (c) The 300 nm chromatin fibre*

This is achieved by successive levels of packaging of the DNA with proteins.[22] The first level is the winding of DNA around a complex of basic proteins called **histones** to form the **nucleosome**[23] (Figure 6.11, see also Section 10.6.1). There are four different histone proteins in the nucleosome, Histones 2A, 2B, 3 and 4, and two molecules of each are used in each nucleosome. The nucleosome has a flattened cylindrical structure, with two turns of the DNA molecule around each monomer.[24] The nucleosomes themselves are wound again to form a 30 nm fibre, which has a helical periodicity and which contains six nucleosomes per turn. Other proteins, including different histones, participate in this second level of packaging. There are further levels of packaging, which are poorly understood at present. The 30 nm fibre is drawn into looped domains, which are condensed further into a 300 nm chromatin fibre. The familiar visible condensed chromosomes seen in spreads of cells in metaphase (*i.e.*, undergoing division) are further condensed from this (Section 2.6.1).

The packaging of DNA in chromatin has profound effects on gene expression. DNA that is tightly packaged is inaccessible to the machinery for gene expression and 'domains' of similar gene expression, which span multiple genes, are defined by particular boundary DNA elements and the protein complexes, which bind to them. These effects of DNA packaging are covered below (Section 6.6.2).

Chromatin in cells fixed to microscope slides can be stained by a variety of compounds to reveal structural features related to the level of chromatin condensation. This kind of analysis is particularly revealing when fully condensed chromosomes (in metaphase) that are about to undergo **segregation** into daughter cells are visualised (Figure 6.12a). Such analysis shows that certain regions of chromosomes are very tightly wound into a dense structure, which is called **heterochromatin**. The regions surrounding **centromeres** (see below) are often heterochromatic and other defined heterochromatic regions are characteristic to the particular chromosomes containing them. Additionally, the Y-chromosomes of mammals are made up almost entirely of heterochromatin. **Heterochromatin** was long thought to be free of genes and to consist wholly of non-coding highly repetitious DNA. We now know that genes do indeed reside in heterochromatin. For example, the fine structure of the giant polytene chromosomes of *Drosophila* (Figure 6.12b), shows closely

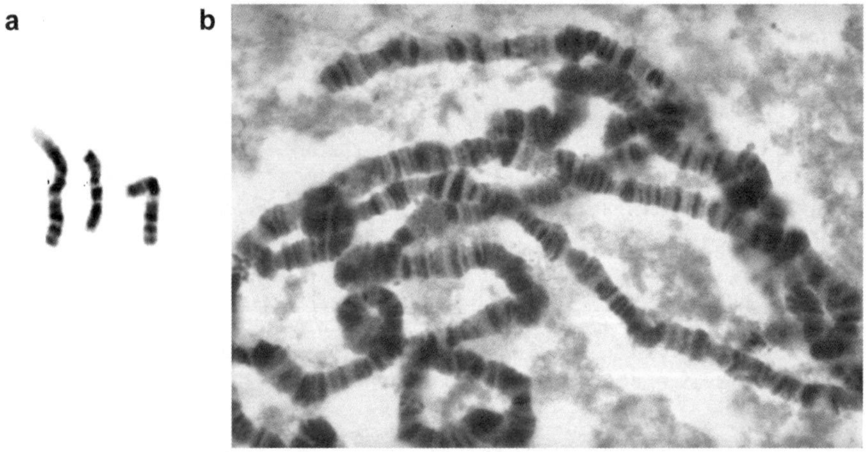


**Figure 6.12** *Chromosome banding. (a) Three human metaphase chromosomes. (b) Drosophila polytene chromosomes*

interspersed stretches of high and low density **chromatin**, which are responsible for the beautiful banding pattern seen in these chromosomes. The much cruder bands seen in mammalian chromosome spreads are caused by a similar phenomenon (Figure 6.12a).

### 6.4.3 Prokaryotic Chromosomes

Most prokaryote chromosomes contain circular double-stranded DNA but some of them are linear, like those of eukaryotes. Prokaryotic DNA is associated with **DNA gyrase, DNA topoisomerase**[25] (Section 2.3.5) and packaging proteins into a nucleoprotein complex, which is analogous to eukaryotic chromatin but the details of which are dissimilar.

### 6.4.4 Plasmid and Plastid Chromosomes

Both prokaryotes and eukaryotes contain extra DNA besides that belonging to their regular chromosomes. Bacteria contain a wide variety of plasmids, which are smaller double-stranded DNA. Most, but not all are circular. In *E. coli* these plasmids are not absolutely essential for the life of the host but they carry genes that can be useful, particularly those conferring resistance to antibiotics. In other prokaryotes, plasmids can be more important. For example, the spirochaete *Borrelia burgdorferi*, the causative agent of Lyme disease, carries many linear and circular plasmids. Long term culture of this prokaryote results in loss of some of these plasmids and concomitant loss of infectivity.

Eukaryotes contain **plastids**, the most prominent of which constitute the genomes of the **mitochondria** of virtually all eukaryotes and the **chloroplasts** of plants and algae. These organelles and their associated genomes are the descendants of ancient prokaryotes that either invaded or were engulfed by the ancestors of their present day hosts. Mitochondrial genomes are typically circular and carry genes required for the generation of ATP from respiration, together with some genes needed for their translation. Strangely, the human mitochondrial genome is smaller than that of *S. cerevisiae* (17 kb compared to 75 kb). Chloroplast genomes are generally 100–200 kb long and contain genes for the light-harvesting complex, which drives photosynthesis.

### 6.4.5 Eukaryotic Chromosome Structural Features

Eukaryotic chromosomes need to be replicated faithfully, with no loss of DNA from their ends. After replication they need to separate (**segregate**) into the daughter cells. The preservation of the ends of chromosomes depends on structures called **telomeres** and segregation requires **centromeres**.

*6.4.5.1 Centromeres.* **Centromeres** can be found at different regions depending on the particular chromosome. A **metacentric chromosome** has its centromere near the middle of the chromosome and a **telocentric chromosome** has one at the telomere. The centromere has been intensively studied by genetic, molecular and microscopic analysis. Genetic and molecular analysis in the yeast *S. cerevisiae* has identified a minimal structure necessary for centromere function (Figure 6.13).[26,27] Two regions, CDE1 and CDE3 have important conserved sequences necessary for centromere function. The first of these regions binds to a protein called CBF1 and the second binds to a complex of three proteins, namely CBF3b, NDC10 and CTF13. The DNA sequence CDE2 that separates these regions is about 80 bp in length, has approximately 90% A/T content, and binds to the MIF2 protein. The centromeres of more complex eukaryotes are much larger than this. Additionally, centromeric regions tend to accumulate even more repetitious DNA than the rest of the genome. Repetitious DNA presents serious problems in sequence determination and computer-based structural analysis and is often difficult to clone.

*6.4.5.2 Telomeres.* **Telomeres** are also essential chromosomal components.[28] In their absence, the chromosome shortens until essential genes are lost and the cell dies. The extreme ends of chromosomes do not have complex structures; they are simply double-stranded DNA of repeating sequence, exemplified by the common sequence shown in Figure 6.14. Telomeres counter their natural tendency to become shorter at their ends by generating new copies of these repeats. The mechanisms whereby this occurs are described below (Section 6.6.5).

### 6.4.6 Viral Genomes

Viruses are parasites that can only replicate inside a host cell.[29] Probably all organisms can act as hosts to viruses. Viruses can have genomes made up of DNA or RNA (Table 6.1). The simplest **virus** has only a short nucleic acid encoding a handful of genes, which is packaged into a protein particle. More complicated

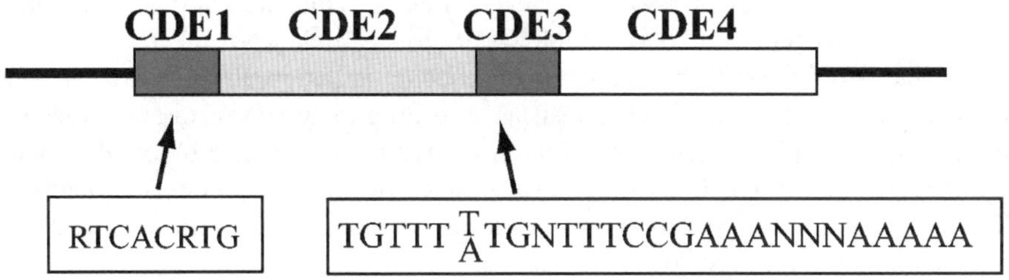


**Figure 6.13** *Structure of the yeast centromere*

```
TTGGGGTTGGGGTTGGGGTTGGGG   3'
AACCCCAACCCCAACCCCAACCCC   5'
```

**Figure 6.14** *Sequence of the human telomere repeat*

**Table 6.1** *Eukaryotic viruses*

| *Genome* | *Type* | *Example* | *Size (base pairs)* | *Structure* |
|---|---|---|---|---|
| ds DNA | Poxvirus | Smallpox | 250,000 | Linear |
| ssDNA | Parvovirus | AAV | 2000 | Circular |
| ss/ds DNA | Hepadnavirus | Hepatitis B | 3000 | Circular |
| dsRNA | Reovirus | Reovirus | 25,000 | Linear |
| ssRNA + strand | Picornavirus | Poliovirus | 7000 | Linear |
| ssRNA − strand | Myxovirus | Influenza | 12,000 | Linear (several pieces) |

viruses may have many enzymes within the particle and hundreds of genes encoded by the nucleic acid. Viral genomes are always very compact, with almost every nucleotide devoted to genes.

*6.4.6.1 The Viral Life Cycle.* First, a virus must enter its host cell. The simpler viruses then uncoat completely and the DNA enters the nucleus where parts of its genome are transcribed. More complex viruses, such as poxviruses preserve an internal core structure inside the cell, which stays in the cytoplasm. In poxviruses, the host cell's **RNA polymerase** components migrate to the cytoplasm.

For most viruses, there is more than one stage of viral infection. For the simpler viruses there is an early phase (pre-DNA replication) and a late phase (post-DNA replication). For DNA viruses the early phase involves the transcription of 'early' genes, the jobs of which are to make sure the cell is not in a resting phase and to coordinate the switching on of viral DNA replication and transcription of 'late' genes that typically encode the components of the virus particle. For more complex viruses there are several phases, for example herpesviruses have three distinct phases.

*6.4.6.2 RNA Viruses.* RNA viruses all encode their own polymerases for replication, because host cells do not contain enzymes capable of copying RNA. (+) Strand viruses contain RNA that can act as an mRNA for the production of the viral polymerase that is responsible for synthesis of a minus strand. The +/− duplex (or replication intermediate) is then transcribed to generate further (+) strand, to thus produce more mRNA as well as viral RNA. (−) Strand viruses cannot act as mRNA when they enter the cell, so they must carry polymerase inside their virus particles. This polymerase then replicates the (−) strand, usually to form a +/− RNA duplex as before, which is then copied to produce **mRNA**.

One of the most interesting and important classes of RNA virus is the **retroviruses**.[30] These are (+) strand viruses that depart from a normal life cycle by going through a DNA intermediate. The AIDS retroviruses HIV 1 and 2 are major pathogens, which are responsible for the deaths of millions of people. Retroviruses carry an RNA-dependent DNA polymerase (**reverse transcriptase**) that is responsible for the synthesis of a double-stranded DNA copy of the viral RNA. This DNA copy is inserted into the chromosomal DNA by another virus-encoded enzyme (**integrase**). The integrated DNA is then transcribed to produce (+) strand RNA, which then can be processed to become mRNA or virus particle RNA.

An important feature of viruses is their high rate of sequence mutation. In some cases the consequences are serious, since such mutation can lead to resistance to drug treatment or to antibodies raised by the human immune system. Rapid sequence mutation in a virus is often due to the viral polymerase being more error-prone than the cellular **DNA polymerase**. Since the viral life cycle is typically measured in hours or a few days, a virus will go through many successive replications of its genome during an infection, each one of which can give rise to new mutations.

## 6.5 DNA SEQUENCE AND BIOINFORMATICS

The remarkable advances in cloning and sequencing technologies since the 1970s have made it reasonably easy to sequence very long stretches of DNA. Whole genome sequences are being deposited in sequence databases at an accelerating rate.[13,14] The enormous amount of data involved in a genome sequence must be stored in such databases and analysed. To get an idea of the volume of information, the complete sequence of a single, smaller than average-sized human gene, the β globin gene is shown (Figure 6.15). One thousand such genes are represented below and 30 such sets are roughly equivalent to the complete human gene complement. This constitutes about 1/60th of the entire human genome, with repetitious DNA making up the rest.

### 6.5.1 Finding Genes

The acquisition of complete genome sequences allows potential access to every gene in the organism. To realise this goal, genes must be found within the 'haystack' of non-coding DNA. **Bioinformatics** provides reasonably reliable search algorithms to predict gene structure, particularly for location of **exons** within

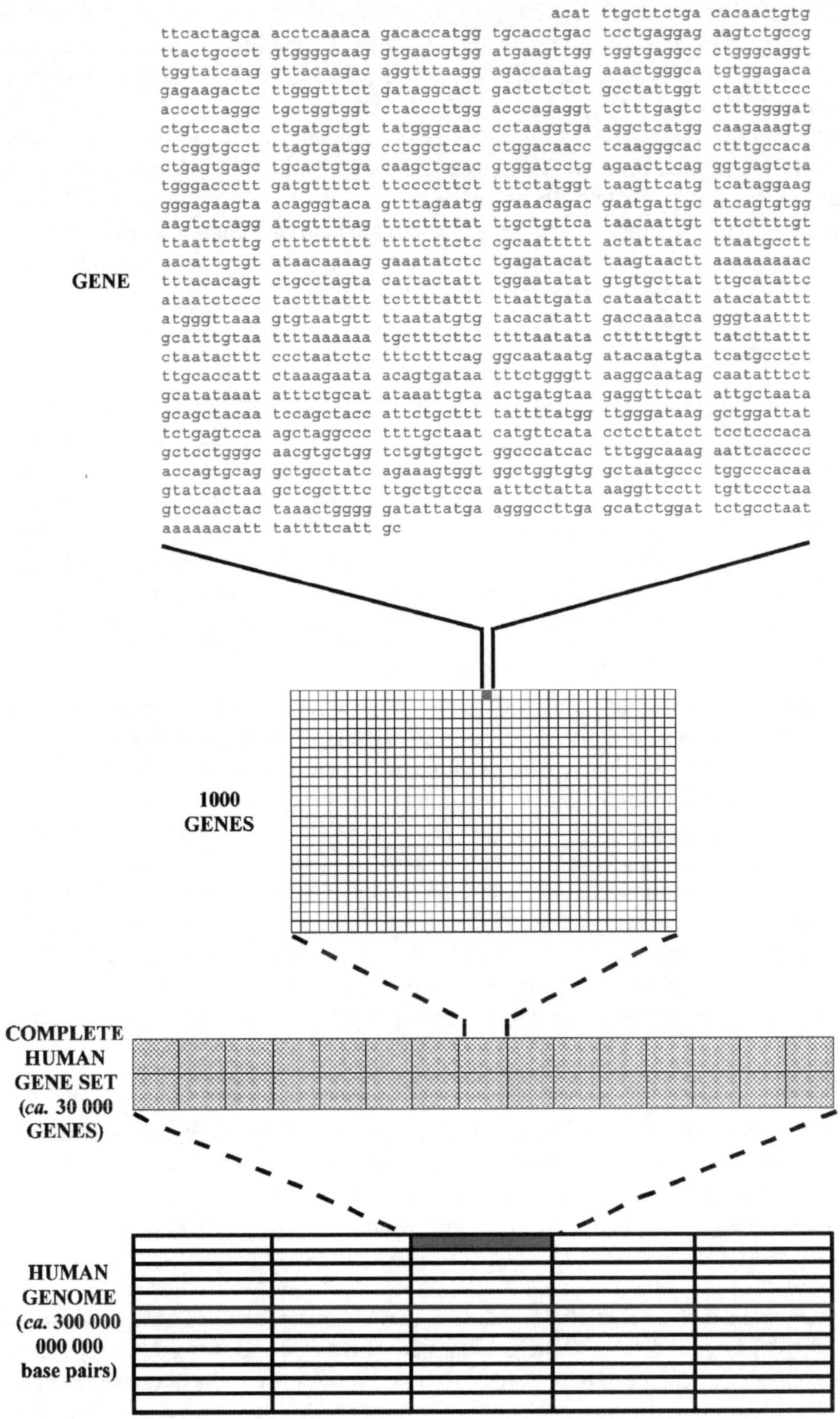


**Figure 6.15** *The information content of the human genome*

genomic sequences. But when these are carefully tested on well-characterised genomic regions containing known genes, they often fall short, either by missing exons, by predicting exons to be complete genes or by predicting exons where none exist. Such failings are particularly pronounced for complex genes such as *Ubx* (Figure 6.4).

One way to help find the locations of genes in DNA is by sequence analysis of large numbers of transcribed sequences, because in general only genes are transcribed, and by comparison with the complete genome sequence. Typically, a **cDNA library** is made by reverse transcription from the RNA of the organism. Then thousands of individual sequences are determined. To find the rare RNAs, often single sequencing experiments are carried out (typically *ca.* 500 bp), on a large number of subclones, instead of determination of the complete sequence for relatively few RNAs.

### 6.5.2 Genome Maps

Once the genomic sequence is reasonably well ordered into accurate, large contiguous pieces (**contigs**), which eventually extend to whole chromosomes, the cDNA sequences can be mapped onto the respective genome. Such maps are useful in identifying genes, which may be associated with important traits, such as the predisposition to inherited diseases. The gene map obtained from such a study can be aligned against other important maps, showing the extents of large insert clones (**BAC**, **YAC** clones, *etc.*, see Section 5.2.1). BAC and YAC contigs are obtained by sequence analysis of the ends of randomly selected large insert clones and then by a search of previously acquired data for identical sequences.

### 6.5.3 Molecular Marker Maps

Another important map, which can be aligned against the genome and cDNA maps, is a **molecular marker** map. A molecular marker is any difference in DNA sequence observed at a precise genomic location between two individuals of an organism, for example two human beings. Such differences represent just a tiny fraction of the huge amount of genetic variation in a species and mostly lie in non-coding DNA that is not subject to natural selection to preserve its sequence. Molecular markers are useful research tools in that they can be mapped genetically in the same way as visible traits, such as Mendel's pea seed traits. They can also be physically mapped on genomic DNA. Indeed, there are now hundreds of times more molecular markers mapped, both genetically and physically, on the human genome than there are genes identified. Genetic markers that are tightly linked to particular gene variants (**alleles**) can be of medical importance. For example, whether a baby carries a defective cystic fibrosis gene can now be assessed by a simple marker assay on DNA isolated from a pinprick of blood, rather than having to clone and sequence the gene itself.

### 6.5.4 Molecular Marker Types

The first types of molecular markers are **restriction fragment length polymorphisms (RFLPs)**. These DNA sequence variants (usually **point mutations** or small insertions or deletions) result in the creation or destruction of a **restriction enzyme** cleavage site (Section 5.3.1). Such mutations sometimes alter the restriction map of the genomic region in which they reside. Such DNA alterations can be detected either by **Southern blot analysis** (Section 5.5.2) or more often nowadays by PCR (Section 5.2.2) followed by restriction digestion of the amplified DNA to reveal the polymorphic restriction site.

Two more important molecular marker types in use now are **microsatellites**[31] and **single nucleotide polymorphisms (SNPs)** (Section 5.5.3).[32] Microsatellites are also called simple sequence repeats (SSRs). SSRs contain a varying number of repeats of typically 2–3 base pairs. At a given locus (genomic region), one individual might have six repeats of the dinucleotide GT whilst another may have nine such repeats. These differences are revealed by DNA amplification of the region containing the repeat and by determination of its length. SNPs are merely single nucleotide changes in a given genomic region, e.g. a G substitution by an A at position 543. Much effort is currently being invested in finding cheap and efficient methods for identifying such simple SNPs.

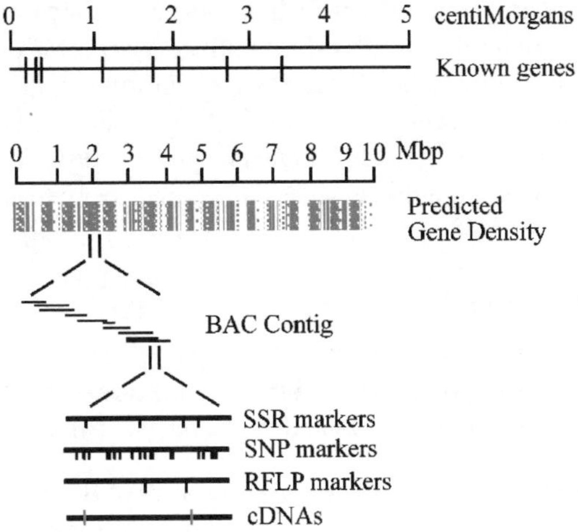


**Figure 6.16** *Composite maps for genomics. A genetic map (top) is aligned against a sequence-derived gene density map, a BAC contig physical map and four physical maps (bottom). Positions of molecular markers and cDNAs on the maps are shown as vertical lines*

### 6.5.5 Composite Maps for Genomes

It is common to show schematically a composite map for particular regions of the human genome (*e.g.*, Figure 6.16), which combines the various maps described above. The seven maps in the figure include a genetic map (at the top), a physical map representing the gene density predicted from the complete genome sequence, a BAC contig map, three molecular marker maps and a cDNA map. Each of these can be linked to each other. For example, the exact location of an SSR marker relative to a cDNA clone on the genomic sequence can be determined by a database search. Additionally, every molecular marker can be mapped genetically and placed on the genetic map. Eventually, every gene can be assigned to a recognised cDNA. This is a crucial requirement, since often-genetic traits give little or no clue to the gene responsible for them. For example, a predisposition to lung cancer could derive from a multiplicity of genes.

## 6.6 COPYING DNA

Since DNA is the source of genetic information from which every living organism is constructed, it must be copied faithfully and transmitted into the daughter cells to preserve the genetic integrity of the lineage. DNA must also be transcribed faithfully into its corresponding RNA products to construct and maintain the function of the organism.

### 6.6.1 A Comparison of Transcription with DNA Replication

Both **DNA replication** and **transcription** are copying processes and both occur in the nucleus of cells from the same DNA template, often at the same time. But they are fundamentally different in their mechanisms. In transcription only a small subset of the genetic information needs to be transcribed into RNA namely just those genes whose gene products, proteins or structural RNAs, are needed at a particular stage of cell life. Different genes need to be expressed at different levels. One gene's RNA may perhaps be present as a single copy in the cytoplasm of a particular cell, a second gene may be expressed as thousands of RNAs and a third gene may be totally switched off. Thus, transcription needs to be extremely versatile, to cater for huge differences in expression profiles of thousands of genes, in hundreds of cell types, as well as the need to alter transcription profiles multiple times during cell development. However, transcription does not need to be extremely accurate, because a cell can still operate fully even if 0.1% of its RNAs are not functional.

In contrast, DNA replication requires that all of the genetic information in the cell be copied, and copied only once, into a single daughter molecule that is as identical to the parent molecule as possible. DNA replication is therefore extremely accurate, with rounds of proofreading and error correction, as well as checks to make sure that a DNA strand only becomes copied once per cell cycle.

## 6.6.2 Transcription in Prokaryotes

**Transcription** involves the copying of a gene into an RNA molecule. Several phases are involved in this process, namely, initiation of transcription, **elongation**, **termination** and **RNA processing** (Section 7.2). There are many similarities between prokaryotes and the eukaryotes in these processes.

*6.6.2.1 Prokaryotic RNA Polymerases.* There are two types of **RNA polymerase**. Viral polymerases are simple, single subunit enzymes in the range of $1-2 \times 10^4$ Da in mass. These polymerases can only initiate transcription from one or a very small number of very similar promoters. In contrast, all RNA polymerases that transcribe cellular genes are large, multi-subunit enzymes, which are more versatile in their ability to recognise different promoters.

In prokaryotes, a single RNA polymerase is responsible for the synthesis of all RNA. The complete enzyme (**holoenzyme**) has a molecular mass approximately $4.8 \times 10^5$ Da. It is pentameric in structure and is comprised of two a subunits and two related B subunits, b and b′, together with an associated unit called σ (sigma). There are several **sigma factors** available[33] and these modulate the specificity of the RNA polymerase for different promoters (see below).

*6.6.2.2 Prokaryotic Transcriptional Initiation.* **Transcriptional initiation** is the first event in copying the DNA template into RNA.[34] This occurs at a specific region of the gene, called the **promoter**. *E. coli* RNA polymerase that lacks a sigma factor (the core enzyme) has a relatively weak affinity for all DNA, with no great preference for promoter regions. The function of σ is to make sure that RNA polymerase binds stably to DNA only at promoters. RNA polymerase containing a sigma factor (holoenzyme) has a far lower affinity for DNA in general but a higher affinity for promoter regions in particular. Different promoters show large differences in their affinities for the holoenzyme, with 'strong' promoters having far higher affinities.

**6.6.2.2.1 Steps in Prokaryotic Transcriptional Initiation.** The process of transcriptional initiation has been elucidated *in vitro* using purified RNA polymerase holoenzyme and a DNA template. Four distinct stages are observed (Figure 6.17). First, the core enzyme binds to a region from about 40 bases upstream (the **−35 box**) to about +20 bases downstream of the transcription start site to form a **closed promoter complex**. At this point the DNA template is still an intact double helix. Second, the RNA polymerase moves downstream and a limited region of the helix at another conserved sequence (the **Pribnow** or **−10 box**) is unwound to form an open promoter complex. Third, the polymerase begins to synthesise a short RNA molecule on the template DNA strand at the start site. Usually, several abortive short RNAs of between two and nine nucleotides are synthesised before the polymerase succeeds in clearing the promoter. At this point the σ factor detaches from the holoenzyme.

**6.6.2.2.2 Promoter Identification.** Promoters, both prokaryotic and eukaryotic, have been identified in one or a combination of the following ways:

(i) *Consensus searches.* Many promoters are aligned with each other and conserved regions are thus identified. The Pribnow and −35 boxes were originally identified in this way.
(ii) *Mutation analysis.* Naturally occurring or mutagen-induced mutations that affect transcriptional initiation are examined by sequence analysis of the promoters to determine the molecular basis for the mutations.

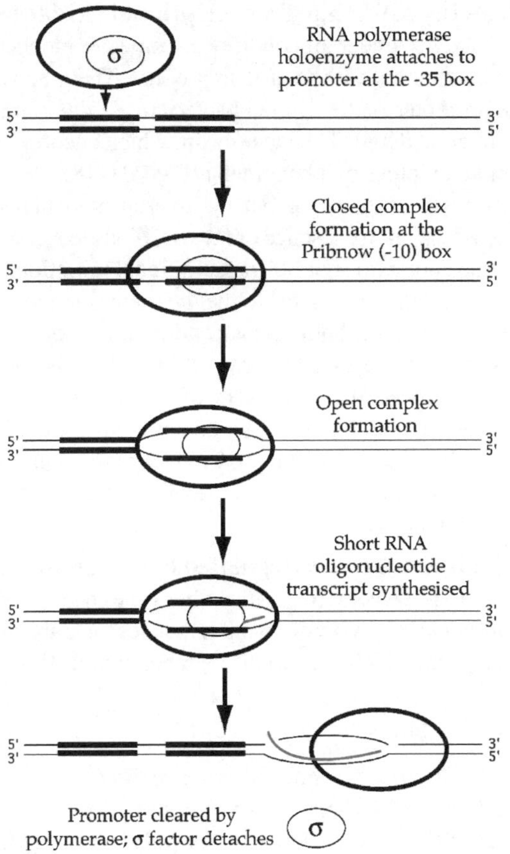


**Figure 6.17** *Initiation of transcription in E. coli. Sigma (σ) factor. The RNA transcript is shown in red*

(iii) *Deletion analysis*. Sub-regions of the DNA template are excised and the effects of this deletion on transcriptional initiation are observed either in an *in vitro* reaction or in living cells.

(iv) *RNA start site mapping*. The location of the 5′-end of the RNA transcript is determined on its DNA template.

(v) *Footprinting*. The binding site of RNA polymerase on the DNA template is determined.

**6.6.2.2.3 Promoter Structure in *E. coli*.** Hundreds of *E. coli* promoters have been studied. The majority of these show some sequence similarities, especially short conserved stretches. The strongest consensus is the **Pribnow box**, a 6 bp sequence similar to the consensus TATAAT, but very few promoters contain this exact sequence. The percentage chances of finding these bases in any given Pribnow box are: T80 A95 T45 A60 A50 T96. The other conserved sequence found in many *E. coli* promoters, the −35 box, has the following consensus: T82 T84 G78 A65 C54 A45.

Both the −35 and Pribnow boxes are very sensitive to sequence change. In general, mutations that make the sequence less like the consensus tend to weaken the promoter and *vice versa*. The strongest promoter is a combination of the two consensus boxes. It is important to remember that not all promoters need to be strong. Different genes need to be transcribed at different rates in an organism. In keeping with the observations from *in vitro* transcription studies (Figure 6.17) mutations in the −35 box alter the rate of closed complex formation, not the conversion into open complex, whereas mutations in the Pribnow box do not affect closed complex formation and have the opposite effect.

**6.6.2.2.4 Regulatory Proteins Affecting Transcriptional Initiation in *E. coli*.** Promoters do not necessarily have the same activity under all conditions. Some are induced and/or repressed under different conditions, such as the need or not for the protein product. These activities are mediated by regulatory proteins that bind to the promoter region. In the bacterium *E. coli*, paradigms for such inducers and repressors have been studied in great detail. The **lac operon**, which encodes the enzymes for metabolising lactose, shows both induction and repression phenomena (Figure 6.18).[3,35]

*E. coli* cell uses lactose as a source of sugar but the operon is switched off in its absence. This is achieved by the ***lac* repressor**, which, in the absence of lactose, binds to a control region (the ***lac* operator**), downstream of the *lac* operon promoter, to shut off transcriptional initiation. However if lactose is present in the cell, one of its metabolites, allolactose, binds to the *lac* repressor and blocks its ability to bind to the operator. Furthermore, in cells containing both lactose and glucose, the *lac* operon is virtually inactive (glucose is a more attractive source of energy than lactose). This effect is mediated by the **catabolite activator protein** (CAP),[35] which activate the promoters of several genes, which encode enzymes that metabolise sugars other than glucose. A cell containing glucose has low cyclic AMP (cAMP) levels, leading to a loss of the ability of CAP to bind to its target. This leads to an almost complete switch off of all these operons. If glucose falls below a threshold level, cAMP levels rise, the CAP protein attaches to its binding site, and the operons are induced.

**6.6.2.2.5 Promoter Specificity in *E. coli* is Regulated by Different σ Factors.** Most *E. coli* genes are transcribed with the aid of a single σ factor (σ 70) but other genes need to be turned on under specific circumstances. For example, heat shock or nitrogen starvation induces the transcription of a series of genes that have different promoters. Such promoters have variant −35 boxes and 'Pribnow-like' boxes (Table 6.2).[33]

### 6.6.3 Transcription in Eukaryotes

*6.6.3.1 Eukaryotic RNA Polymerases.* Eukaryotes have three different nuclear **RNA polymerases**, RNA polymerases I, II and III, as well as separate enzymes for their chloroplasts and/or mitochondria. **RNA polymerase I** transcribes rRNA exclusively.[36] **RNA polymerase II** transcribes all protein-coding

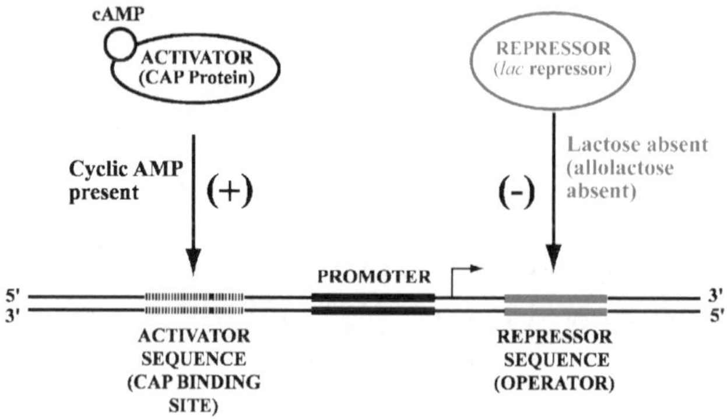


**Figure 6.18** *Activators and repressors of transcriptional initiation in E. coli – the lac operon. Activation and repression of transcription are indicated by (+) and (−) respectively*

**Table 6.2** *Different σ factors in E. coli*

| *σ factor* | *Use* | *−35 sequence* | *Gap (base pairs)* | *−10 sequence* |
|---|---|---|---|---|
| σ70 | General | TTGACA | 16–18 | TATAAT |
| σ32 | Heat shock | CNCTTGAA | 13–15 | CCCCATNT |
| σ54 | Nitrogen starvation | CTGGNA | 6 | TTGCA |

genes and many small nucleoprotein RNA (snRNA) genes. Finally, **RNA polymerase III** transcribes the rest of the small RNAs, particularly tRNAs and 5S RNAs.

The complete nuclear RNA polymerase enzymes are multi-subunit structures of around $5 \times 10^5$ Da in mass. Each polymerase is composed of two major subunits, usually about $2 \times 10^5$ and $1.4 \times 10^5$ Da, respectively. These correspond to the β and β′ subunits of *E. coli* RNA polymerase. Additionally, eukaryotic RNA polymerases contain up to ten smaller subunits of between $1 \times 10^4$ Da and $9 \times 10^4$ Da. Several of these subunits are shared between different types of RNA polymerase.

*6.6.3.2 Transcriptional Initiation for RNA Polymerase II.* **RNA polymerase II** is the most interesting and important of the three nuclear RNA polymerases, since it transcribes all protein-encoding genes. Initiation of transcription by RNA polymerase II is a highly complex process that is a major factor controlling the levels of mRNAs produced and is thus key to regulating the levels of the tens of thousands of different cellular proteins. In addition to the basic transcriptional initiation machinery,[37] there are a multitude of positive and negative regulators available.[38] Many features of eukaryotic promoters are shared with prokaryotes, in particular the basic concepts of induction and repression mediated by proteins.

**6.6.3.2.1 RNA Polymerase II Promoter Structure and the Basal Transcription Machinery.** As for prokaryotes, there is a conserved box at about 25 base pairs upstream from the transcriptional initiation site. This box, the **TATA box**, has the consensus TATAAATA. Deletion of this region in some cases damages promoter strength, for example in the case of the β globin promoter or many yeast promoters. In other cases, TATA box removal does not abolish transcriptional initiation but destroys its specificity, leading to multiple staggered transcriptional initiation sites. Thus, the TATA box has different functions in different genes.

The TATA box binds a protein complex called **transcription factor IID** (TFIID).[37] This is a multimeric protein, one constituent of which is **TATA binding protein** (TBP).[39] TBP is also a constituent of transcription factors for RNA polymerases I and III, despite the fact that these act on promoters that lack TATA boxes.

The initial steps of transcription complex assembly do not involve the polymerase at all. Instead, TFIID binds first to the TATA box followed by two other factors, before the RNA polymerase enters the complex (Figure 6.19). Then several other factors bind to assemble the basal transcription machinery and, finally to complete transcriptional initiation, the carboxy-terminal domain of the $2 \times 10^5$ Da RNA polymerase subunit is phosphorylated.

**6.6.3.2.2 Regulatory Proteins Affecting RNA Polymerase II Transcriptional Initiation in Eukaryotes.** A large number of transcription factors affect transcriptional initiation.[38] For example, **transcription factor** genes constitute at least 10% of the total gene number in the genomes of *Arabidopsis thaliana* and *Homo sapiens*. Consequently, it is not surprising that there are no consensus boxes common to all protein-coding genes. Instead boxes are often specific to a particular class of genes that are transcribed under similar conditions, in an analogous way to that described for *E. coli* (Table 6.2). For example, the seven heat shock genes of *Drosophila* are induced by elevated temperature. All of these genes share a region of homology approximately 70 bp upstream of the transcription start site (the lower case letters are less well conserved):

**C T g G A A t N T T C t A G a**

If several copies of this box are inserted next to a gene lacking a promoter **heat inducible transcription** is observed. But instead if a mutated version of the box is inserted, transcription is abolished. Therefore, the 'heat shock box' is necessary and sufficient to confer heat inducibility to a gene. The heat shock response in eukaryotes is mediated by a transcription factor called heat shock activator protein (HAP). This protein is always present in cell nuclei but does not induce heat shock gene transcription at ambient temperature. Under these conditions, RNA polymerase II can bind to a heat shock promoter but stutters and only makes short RNA transcripts, in a rather similar way to that described above for *E. coli* (Figure 6.17). On heat shock, HAP forms a trimer and binds to the heat shock boxes, leading to successful transcriptional initiation.

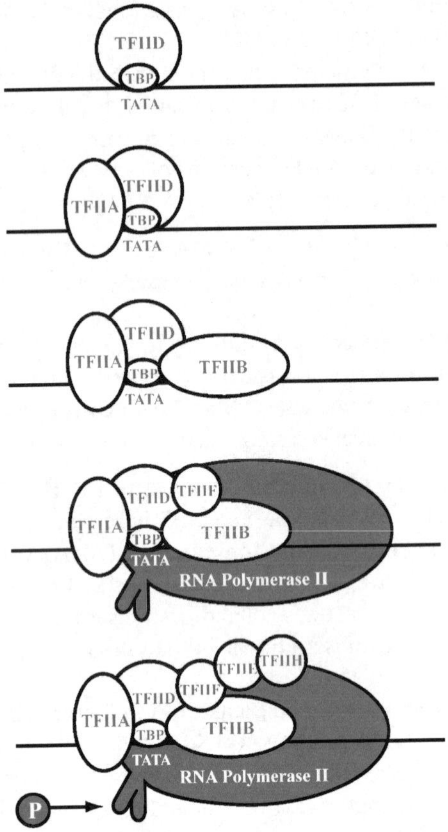


**Figure 6.19** *Assembly of the basal transcriptional initiation complex for RNA polymerase II in eukaryotes. TF: Transcription factor. TATA, TATA box. TBP, TATA binding protein. Circled P indicates protein phosphorylation of RNA polymerase II*

**6.6.3.2.3 Complex RNA Polymerase II Promoters.** A typical example of a eukaryotic promoter that has complex tissue specificity for gene expression occurs in the *Drosophila* gene, *even-skipped* (Figure 6.20). Altogether, 11 regions of the gene control transcriptional initiation in a variety of ways. Each is associated with a different requirement of the protein product, such as position in the body (stripes), cell type (neurons, muscle, anal plate ring). Additionally, most of the specificities (shown by the shading in Figure 6.20) are sub-divided. For example, each of the four neuronal control elements confers transcriptional induction in a different sub-set of the neurons of the fly.

**6.6.3.2.4 Transcriptional Enhancers.** The complete region controlling transcriptional initiation in the *even-skipped* gene is $9 \times 10^3$ bp long, which demonstrates that transcriptional control elements can act at great distances. It is thought that the intervening DNA between a distant control element and its basal transcription machinery is looped out (Figure 6.21). Such distant positive control elements have been termed **enhancers** and their negatively regulating analogues are called **silencers**. Like the other promoter elements, enhancers and silencers work by binding specific transcription factors, which then interact with the transcription machinery (Figure 6.21).

**6.6.3.2.5 Transcriptional Insulators.** An enhancer or silencer can act on more than one transcriptional start site (Figure 6.22). So the cell must be able to prevent enhancers and silencers from working on

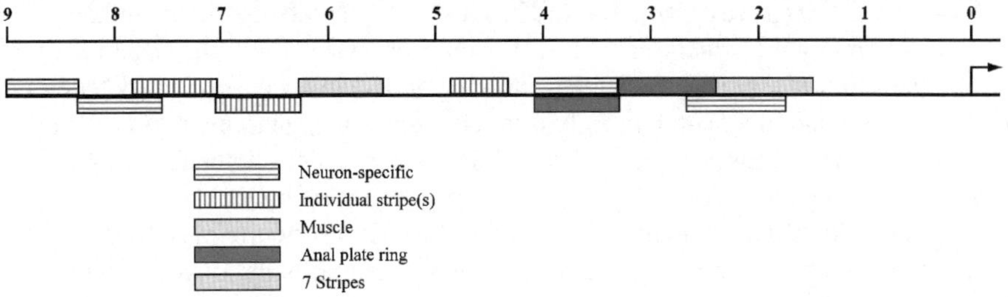


**Figure 6.20** *Structure of a complex, upstream transcriptional control gene – the even-skipped gene of Drosophila. Coloured, shaded and hatched boxes indicate transcriptional control elements necessary for expression in the cell types indicated. The scale is in kilobases upstream form the transcriptional start site (shown by an arrow)*

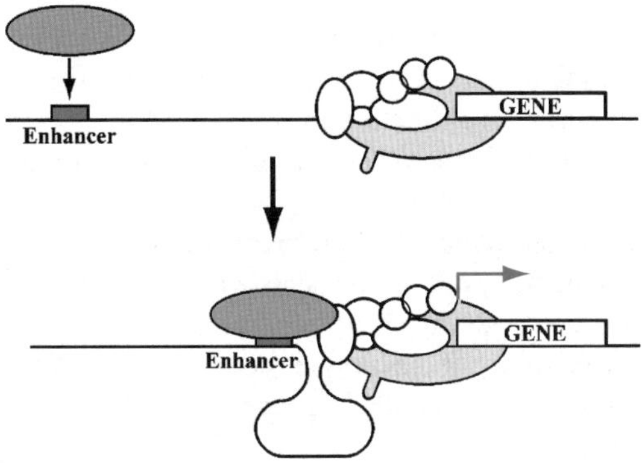


**Figure 6.21** *Transcriptional enhancers act at a distance by looping out intervening DNA*

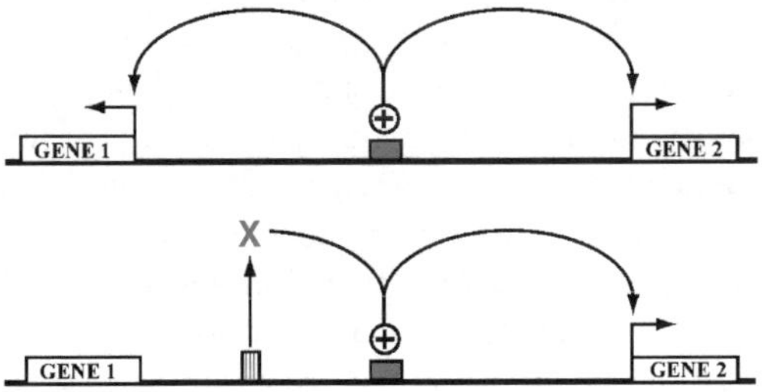


**Figure 6.22** *Transcriptional enhancers and insulators. Circled (+) indicates positive enhancer activity. X indicates blocking of an enhancer activity by an intervening insulator*

every transcription unit in the chromosome and thus causing transcriptional chaos. This is achieved by use of yet another control element called an **insulator**.[40] Insulators confine the effects of enhancers and silencers to domains of effect (Figure 6.22). Insulators are believed to exert their effect by binding specific proteins and by acting as a barrier to the migration of chromatin structure. When DNA is wrapped tightly in nucleosomes (Section 6.4.2), then the genes contained within it are inaccessible to the transcription machinery and cannot be expressed. Since a single chromosome contains many regions with differing levels of chromatin condensation, there must be a mechanism for keeping these regions in the right places. Insulators are believed to be an important component of this mechanism.

**6.6.3.2.6 Chromatin Structure and Gene Expression.** **Nucleosomes** modulate the accessibility of genes to the transcription machinery in at least two different ways. First, the exact position of individual nucleosomes in promoter regions can change, allowing or denying access of promoter elements to their transcription factors. Second, the overall density of nucleosomes in a genomic region can alter, leading to **chromatin packing** or unpacking, which can have global effects on promoter accessibility.

Transcribed genes in a cell nucleus are more susceptible to digestion by the nuclease DNAse I than those not undergoing transcription. This demonstrates that the chromatin structure must loosen on transcription. Sometimes, this **'active' chromatin** structure persists in a gene that is no longer being transcribed. This shows that active chromatin may be a prerequisite for transcriptional activation but it is not sufficient. The regions of DNase I sensitivity can extend a thousand base pairs or more away from the transcribed region, suggesting that there are active domains within chromosomes. These domains may be determined by where they are attached to a **nuclear scaffold**, also called **nuclear matrix**,[41] which are comprised mainly of histone H1 and topoisomerase proteins (Section 2.6.2).

Certain sites within the transcribed regions are even more susceptible to cleavage by DNase I and are therefore termed **nuclease hypersensitive sites**.[42] This hypersensitivity is presumably a consequence of the nucleosome–DNA interactions. These sites often correspond to promoter regions. For example, in a developing chick embryo, the adult β globin gene becomes nuclease hypersensitive before transcription begins, which implies that a change in chromatin structure must have already occurred. Such hypersensitivity is not seen in tissues that never express the gene. For example, no β globin genes ever become hypersensitive in developing brain tissue.

A key factor that affects chromatin packing is **histone acetylation**.[43] The histone components of the nucleosome are basic proteins that have many lysine amino acids which bind the phosphate backbone of the DNA double helix. Some of these lysine residues can become acetylated by nuclear histone acetyltransferase enzymes, which leads to a reduced affinity of the histones both for the DNA and also each other. Intriguingly, some proteins known to affect transcriptional initiation have turned out also to be histone acetyltransferases.

**6.6.3.2.7 DNA Methylation and Gene Expression.** Many of the CG dinucleotides in animals and CNG trinucleotides (where N can be any nucleotide) in plants carry methyl groups at position 5 of the cytosine residues.[44] This position lies in the major groove of the double helix and does not disturb either the helix structure or the base pairing within it (Section 2.2.1). Normally, both C residues of each strand of the duplex are methylated. It is the only common covalent modification to DNA in eukaryotes and is found less in lower than in higher eukaryotes. For example, *Drosophila* has nearly no DNA methylation and yeast has none. **DNA methylation** is detected by the inability of some restriction endonucleases (Section 5.3.1) to cleave methylated DNA when a cleavage recognition site is present. For example, *Hpa* II cleaves CCGG but cannot digest CmCGG. Other enzymes that recognise the same site (isoschizomers) may be unaffected by methylation, or example *Msp* I cuts CmCGG. Unfortunately, not all methylated sites can be detected in this way, because many are not within restriction sites.

Many constitutively active genes (*i.e.*, those whose expression are never switched off) possess many more CG dinucleotides than do inducible genes. These '**CpG-rich islands**' are generally undermethylated throughout the life of the organism. When methylated DNA is replicated in the cell, the newly synthesised DNA strands are unmethylated. A DNA methylation complex scans DNA and, if it finds such hemimethylated sites in the DNA duplex, it methylates the other strand at the appropriate site.

DNA methylation can have a dramatic effect on gene expression. In general, DNA methylation is associated with non-expressed regions of the genome. For example, the majority of detectable methylation sites for the embryonic β-like globin genes become unmethylated in expressing tissue. In adult tissues, after the switch from embryonic gene expression to adult, the embryonic genes become partially methylated, and in tissues not expressing globin they are fully methylated.

Thus if an unmethylated segment of the mouse β globin locus, containing both the foetal γ and adult β genes (Figure 6.23), is introduced into cultured mouse cells, both genes are expressed. If instead a methylated γ-globin gene is introduced next to an unmethylated β globin gene, the γ gene is no longer active

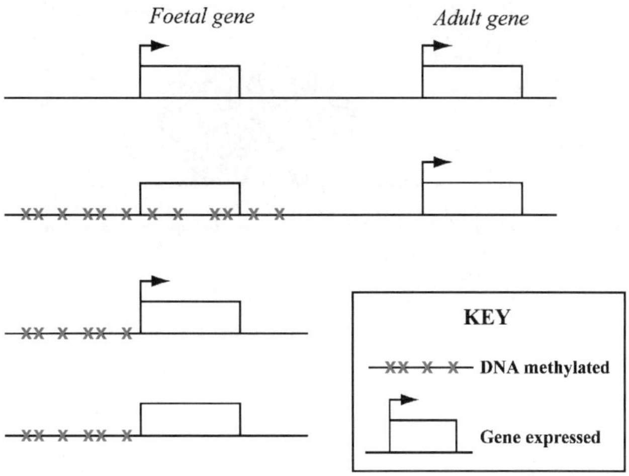


**Figure 6.23** *DNA methylation and gene expression*

but the β gene remains expressed. Further more detailed methylation experiments show that methylation of the transcriptional control region of the γ gene leads to gene inactivation but methylation of the body of the gene does not inhibit gene expression. Thus, at least for some RNA polymerase II-transcribed genes, demethylation of promoter regions is necessary for transcriptional activation but methylation of internal regions appears to be unimportant.

Intriguingly, there seems to be a link between DNA methylation and histone deacetylation (see above). One of the proteins found in the histone deacetylation polyprotein complex is Me-CpG binding protein. This implies that DNA methylation controls gene expression by inducing histone deacetylation and consequently chromatin compaction.

*6.6.3.3 Transcriptional Initiation for RNA Polymerases I and III.* RNA polymerase I transcribes a single type of gene, namely the rRNA gene cluster. Upstream regions of such gene clusters are important in their gene expression but the sequences have diverged so much between organisms that we cannot easily identify 'homology boxes'. This may be due to the highly repetitious nature of the rRNA gene cluster.

Transcriptional initiation for RNA polymerase I, involves protein components of similar complexity to that for RNA polymerase II. However, the RNA polymerase I promoter is simpler and contains a single type of upstream control element as well as the core promoter region surrounding the transcription start site. A dimeric transcription factor called **upstream binding factor** (UBF), together with at least three other factors, binds to both regions.

For most genes transcribed by RNA polymerase III no conserved upstream regions are discernible. Instead the transcriptional control elements for most RNA polymerase III genes reside unusually within the genes themselves.[45] Transcription factors TFIIIA ($4 \times 10^4$ Da) and TFIIIC bind to these internal promoters (Figure 6.24). The binding of these two factors is required for binding of a third factor, TFIIIB, upstream of the start site. It is TFIIIB that aids RNA polymerase III binding.

## 6.6.4 DNA Replication

*6.6.4.1 Introduction.* Before a cell divides it must have already created an exactly duplicated set of chromosomes so that both daughter cells can carry a set of genes identical to those in the parental cell. The basis for this DNA replication is carried within the DNA itself. First, the DNA double helix carries two complementary copies of the genetic information encoded within it (one on each strand). Secondly, Watson–Crick

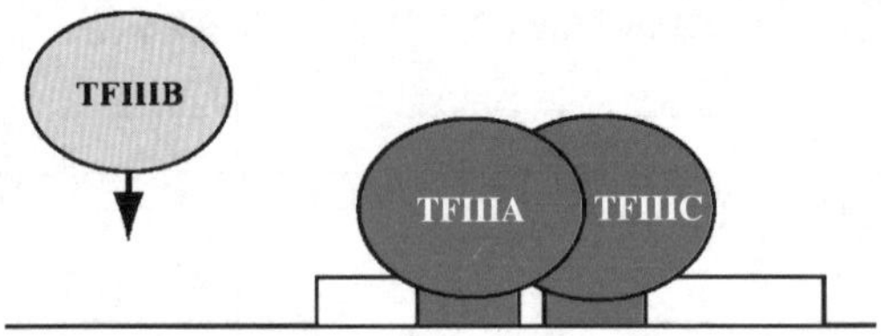


**Figure 6.24** *DNA control elements and transcription factors involved in transcriptional initiation by RNA polymerase III. TF: Transcription factor*

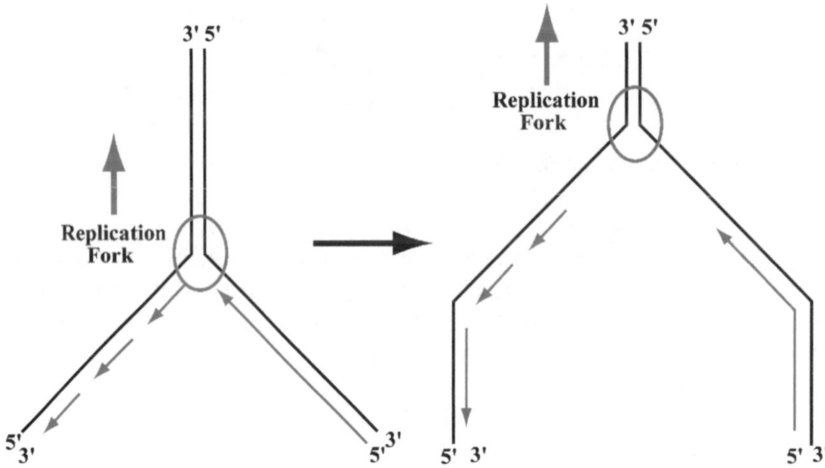


**Figure 6.25** *The replication fork and the polarity of DNA replication. Black lines indicate old DNA and red lines indicate newly synthesised DNA. The direction of movement of the replication fork and the direction of synthesis of DNA are both shown by red arrows*

base pairing determines the identity of the nucleotide to be added at each step during the replication process. In addition to DNA, many other components are required to ensure faithful copying of the DNA.

**6.6.4.1.1 DNA Topology.** DNA cannot be copied unless the complementary DNA strands are first unwound. One problem is that because one strand is wound round the other, unwinding one region by pulling the two strands apart leads to an increase in the number of superhelical turns (supercoils) in another adjacent region (Section 2.3.5). The DNA replication machinery therefore needs to relieve these extra turns as replication proceeds. This problem is solved by the use of a **DNA topoisomerase**[25] to relax these extra supercoils as they are generated during replication.

**6.6.4.1.2 Strand Polarity.** The two DNA strands in a double helix have opposite polarities (Section 2.2). Every enzyme in nature that copies DNA or RNA into DNA or RNA does so by adding single nucleotides to the 3′-end of the elongating strand (*i.e.*, replication proceeds in a 5′ to 3′ direction). If DNA replication is to proceed in a given direction along a duplex, a '**replication fork**' must migrate in that direction.[46] Only one DNA strand can be copied in that direction, the '**leading strand**' and the new DNA elongated continuously. The other strand (the '**lagging strand**') must be copied in the reverse direction and is elongated discontinuously[47] (Figure 6.25).

**6.6.4.1.3 Semi-Conservative DNA Replication.** In principle there are at least three different ways that DNA could be copied (Figure 6.26). The correct mechanism was shown by Meselson and Stahl

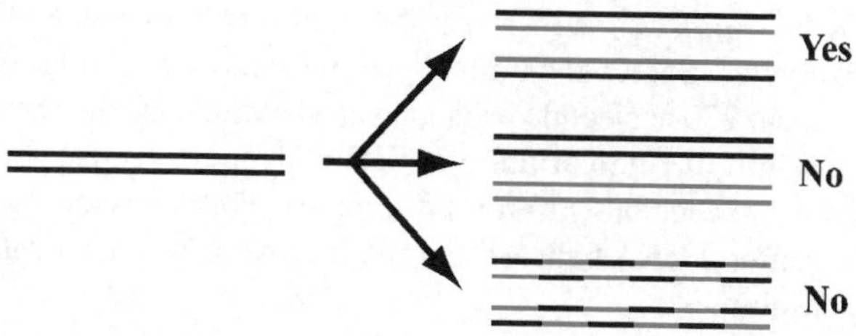


**Figure 6.26** *DNA replication is semi-conservative. Only the top mode of DNA replication (semi-conservative replication) is observed for replication of prokaryote and eukaryote chromosomes*

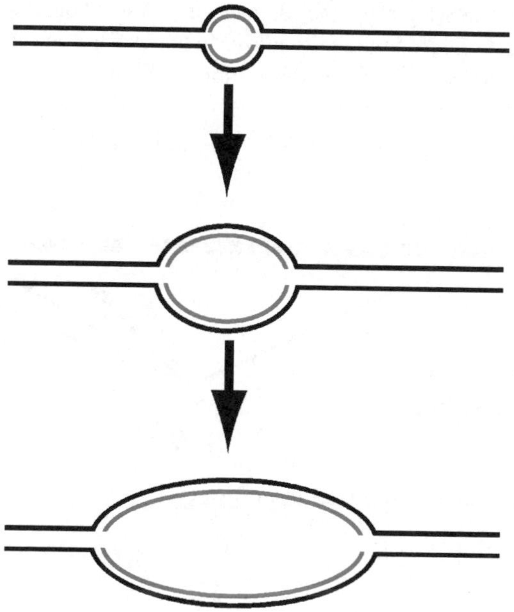

**Figure 6.27** *DNA replication for eukaryotic and prokaryotic chromosomes is bi-directional. Newly synthesised DNA is shown in red*

to involve the preservation of one of the two parent strands in each of the two newly synthesised duplexes. This is called **semi-conservative replication**, because half of the strands generated are old and half are new.

**6.6.4.1.4 Origins and Direction of DNA Replication.** In bacteria, fungi and viruses, DNA replication starts at distinct **origins**, but in higher eukaryotes such sites are far less easy to identify.[48] Replication of all chromosomes proceeds in both directions, creating 'bubbles' that can be seen in electron micrographs (Figure 6.27).

*6.6.4.2 Priming of DNA Replication.* The enzymes responsible for making a complementary copy of a DNA are called **DNA polymerases**. But they can only elongate an existing duplex. Therefore, another enzyme is needed to initiate the synthesis of a new strand on the DNA template. This enzyme is a specialised RNA polymerase called a **primase**,[49] which occurs in both prokaryotes and eukaryotes. A short RNA oligonucleotide is synthesised first on the DNA template strand and then a DNA strand is synthesised from the 3′-end of this short RNA molecule. The unwanted RNA primer is removed by a specialised ribonuclease called **RNase H**, which digests only the RNA strand within an RNA–DNA duplex.

*6.6.4.3 Initiation of DNA Replication.* Most of our current knowledge of how DNA replication is initiated comes from prokaryotes. Bacterial chromosomes have individual origins of replication. In *E. coli* this is called ***oriC***. *OriC* is about 250 nucleotide pairs long and has the structure shown in Figure 6.28. Various proteins interact with the **origin of replication** to initiate copying of the DNA. DnaA protein molecules bind first to a 9 bp motif, then to each other to form a complex. This leads to unwinding of the helix in the A/T-rich 13 bp motifs. Thereafter, a **DNA helicase**, the DnaB protein, in concert with the DnaC protein, binds to form a **pre-priming complex**.

*6.6.4.4 DNA Elongation.* During the elongation process (Figure 6.29), the DnaB helicase protein migrates along the duplex, attached to the 'lagging strand', breaking base pairs as it goes.[50] Every turn of the double helix that is removed in this way generates an extra turn 'upstream' of the fork, which is relieved by the enzyme DNA topoisomerase (Section 2.3.5). The newly generated DNA single strands are protected by **single strand binding proteins** (SSBs) from DNA damage or unwanted binding to other nucleic acids or proteins. These are later displaced as DNA polymerase moves in to make the new complementary lagging strand.

The leading strand can be copied without any discontinuity (Figure 6.30) but the lagging strand requires a new primer every thousand or so nucleotides. This is synthesised by the primase and then the DNA polymerase (DNA polymerase III in *E.* coli) takes over, extending the primer for about 1000 nucleotides before

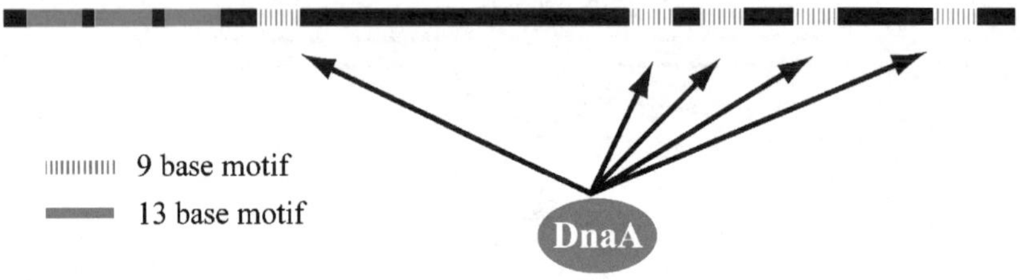


**Figure 6.28** *Structure of the E. coli origin of DNA replication OriC*

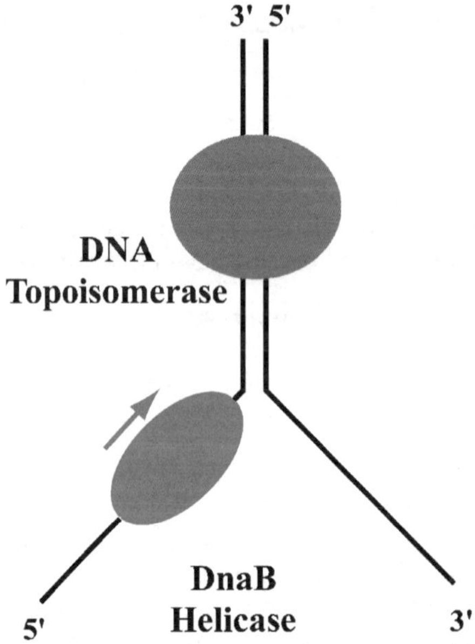


**Figure 6.29** *Elongation of DNA synthesis involves DnaB helicase and DNA topoisomerase*

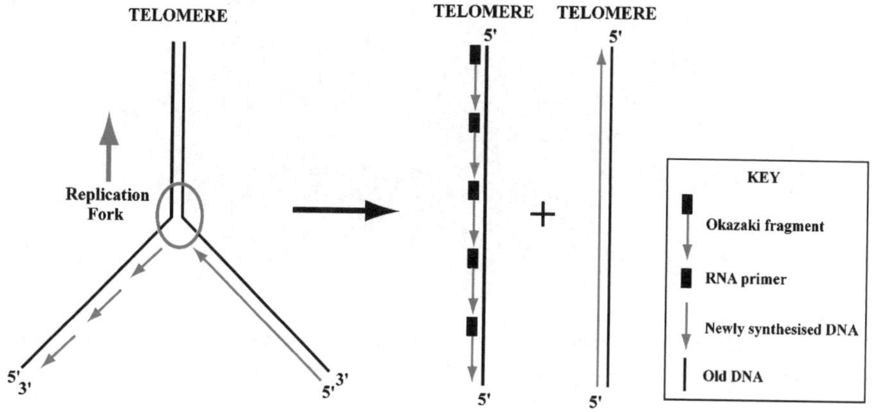


**Figure 6.30** *DNA replication and Okazaki fragments. Synthesis of the leading strand of DNA is continuous and discontinuous for the other strand, resulting in generation of Okazaki fragments*

the cycle repeats. The RNA–DNA fragments generated in this way are called **Okazaki fragments** after their discoverer.[47]

The Okazaki fragments become joined up by DNA polymerase I, which possesses a 5′–3′ exonuclease activity that degrades the RNA primers and replaces them by DNA copied from the template strand. This leaves a gap in the phosphodiester backbone (absence of a single phosphodiester bond), which is closed by the enzyme DNA ligase.

*6.6.4.5 Termination of DNA Replication.* DNA replication terminates when another replication fork or the telomere is reached (Section 6.4.4). In *E. coli*, a circular genome, replication always terminates within a defined region roughly opposite *oriC*. This is mediated by DNA binding proteins called Replication Terminator Proteins (TUS) that allow replication forks to proceed through them in one direction only, thus trapping them at the termination region.

### 6.6.5 Telomerases, Transposons and the Maintenance of Chromosome Ends

Eukaryotes have linear chromosomes that pose a problem for the DNA replication machinery. When the DNA polymerase elongation complex reaches the end of the chromosome, the leading strand can *perhaps* be completed in its entirety (assuming that the DNA polymerase is happy to replicate these final nucleotides although it is falling off the end of the chromosome). However, the lagging strand must at least lack the region corresponding to the template for the short RNA primer, and thus DNA replication cannot proceed in the reverse direction by priming from beyond the telomere (Figure 6.30). Furthermore, the telomeres are susceptible to degradation by nuclease action and have been shown to progressively shorten in somatic (non-germ line) cells.

Almost all telomeres contain short tandemly repeated sequences (Section 6.4.4 and Figure 6.14).[28] In most organisms, including all mammals, cells preserve the integrity of their telomeres by synthesis of further copies of the repeated sequence by a special DNA polymerase called **telomerase**. The telomerase has an RNA component that contains a short sequence that acts as a template for telomerase to extend the 3′-end of one DNA strand (Figure 6.31). This enzyme is thus a highly specialised reverse transcriptase, since it synthesises DNA from an RNA template. Interestingly, in some other organisms, notably *Drosophila*, a different mechanism for telomere maintenance exists. In these species, a specialised long interspersed element **(LINE) transposable element** (Section 6.8.3) transposes into the region and thus extends the telomere in order to counteract degradation.

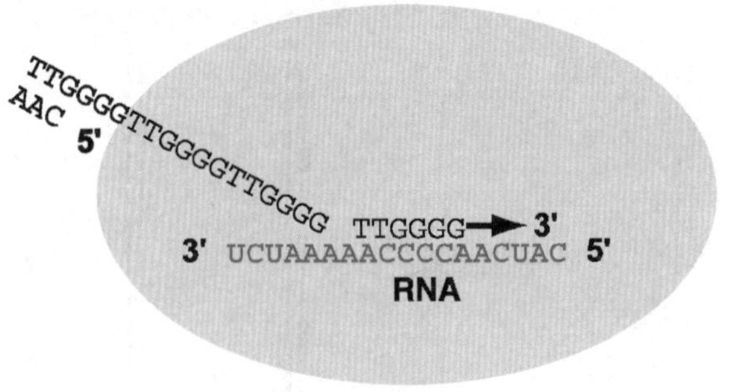


**Figure 6.31** *Reverse transcriptase activity of telomerase preserves telomeres. Telomeric DNA is shown in black and the template RNA, which is copied by the telomerase enzyme, in red*

## 6.7 DNA MUTATION AND GENOME REPAIR

Every organism is constantly subjected to a barrage of mutagenic agents, including ionising radiation and chemical mutagens (Chapter 8). Therefore, highly efficient mechanisms of DNA repair are needed to maintain the genome integrity.[51] A number of complex machineries have evolved for recognising and correcting the many different types of damage caused to DNA.

### 6.7.1 Types of DNA Mutation

Figure 6.32 shows some major types of DNA damage. DNA damage may result in an altered nucleotide, which is read by the DNA replication machinery as a different base, resulting in a change of DNA sequence (**point mutation**). If the mutation is within a coding sequence, it may give rise also to a change in the protein sequence. If the mutation is within a non-coding sequence, it may affect control functions. Other mutation types include (i) the addition of extra bases in a 'microsatellite' sequence by **polymerase slippage** during DNA replication, (ii) removal of the base from the ribose backbone (almost always a purine base, depurination, Section 8.1) leading to loss of the ribose and a break in the DNA strand, (iii) modification of the base or the ribose by alkylating agents (Section 8.10) leading either to base mispairing, and consequent point mutation, or inhibition of DNA replication, (iv) cross-linking of bases (particularly adjacent thymines; Figure 6.32 (see also Figure 8.30) leading to major errors during DNA replication, and (v) single or double stranded breaks.

### 6.7.2 Mechanisms of DNA Repair

*6.7.2.1 Direct Repair.* Most DNA damage can only be repaired by the removal of the complete nucleotide and often some surrounding sequence, followed by insertion of the correct nucleotides (Figure 6.33). However, some lesions can be corrected *in situ*. Simple nicks in one of the phosphodiester backbones are corrected by the enzyme DNA ligase. Also, certain alkyl groups can be removed at specific positions from particular bases, for example *O*-6 methyl groups can be removed from guanosine bases by *O*-6-methylguanine-DNA methyltransferase (Section 8.11.1).

*6.7.2.2 Excision Repair.* **Excision repair** is a very important mechanism in most organisms (Section 8.11.2).[52,53] One form of excision repair, used in situations where only a single base is slightly damaged, involves a two-step process, whereby the damaged base is excised first by a **DNA glycosylase**. An **endonuclease** then cleaves the sugar phosphate backbone, leaving a single base gap, which is repaired by DNA polymerase. In more extensively damaged DNA, the entire nucleotide(s) and a short region around them are removed by an endonuclease complex. Once again, the gap created is filled in by DNA polymerase and the remaining nick sealed by **DNA ligase**.

a

| Mutation | Example | Common Cause(s) |
|---|---|---|
| Point Mutation | G/C → G/T | DNA replication error, deamination |
| Microsatellite variation | GTGT/CACA → GTGTGT/CACACA | Polymerase slippage |
| Depurination | GTGT/CACA → GT_T/CACA | Heat |
| Alkylation | G/C → G(Me)/T | Alkylating agent |
| Thymine Dimer Formation | TT/AA → T^T/AA | UV irradiation |
| Strand breaks | GTGT/CACA → GT GT/CA CA | X irradiation |

b

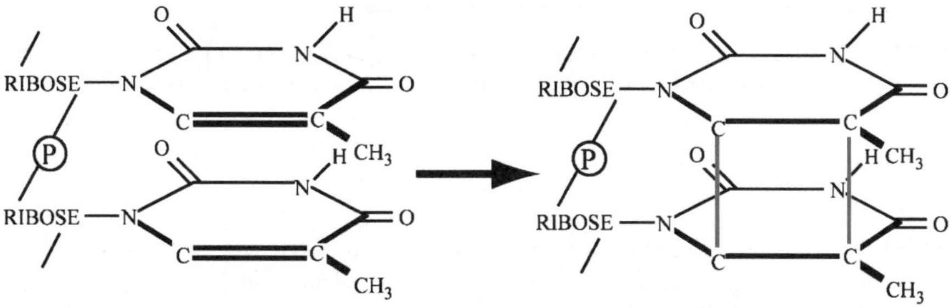


**Figure 6.32** *Types of DNA damage encountered by the DNA repair machinery. (a) Common types of DNA damage, together with examples and their cause(s). (b) Formation of a cyclobutyl adduct (commonly called a thymine dimer)*

*6.7.2.3 Mismatch Repair.* DNA polymerases[54] are extremely accurate at copying DNA templates ($10^{-7}$ error rate for *E. coli*) but they are not perfect. In part, this great accuracy involves a mechanism called **proofreading**. The major DNA polymerases of prokaryotes and eukaryotes possess a 3′–5′ exonuclease activity, which removes any nucleotide that has not been correctly base paired with the template during the extension reaction. Nevertheless, very occasionally, an incorrect nucleotide is inserted into a new DNA strand by DNA polymerase. This generates a sequence mismatch, which is corrected by **mismatch repair**.[55] The sequence mismatch causes a small irregularity in the double helix, which leads to a loss of base pairing around the mismatch. A short region on one of the two strands must be removed and the lesion filled in by DNA polymerase. How does the repair machinery know which strand to remove? Many organisms, such as *E. coli* and humans, have DNA methylation (Section 6.6.3), which distinguishes old DNA from newly synthesised DNA. Other organisms lacking DNA methylation, such as yeast and *Drosophila*, must use another way to recognise recently synthesised DNA, but this is not yet understood.

*6.7.2.4 Repairing Double-Stranded DNA Breaks.* A break in both DNA strands is extremely dangerous for the cell. If such damage is not repaired the exposed ends might become degraded, leading to a deletion of DNA at the break point. Alternatively, an entire chromosome segment may be lost along with its genes or one chromosome segment translocated on to the telomere of another chromosome. In humans

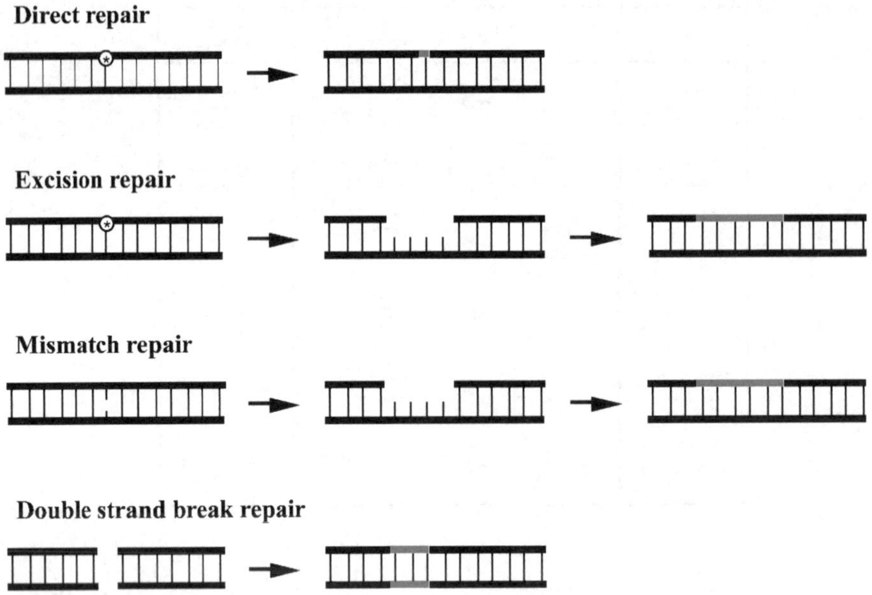


**Figure 6.33** *Major pathways for DNA repair. Circled * indicates damage to a base or ribose. Longer vertical lines indicate base pairs and shorter lines show unpaired bases. Thick horizontal lines indicate the ribose backbone*

double-stranded DNA breaks are repaired by DNA ligase in concert with a multi-subunit complex containing **DNA protein kinase** and two so-called Ku proteins (Ku70 and Ku80).[56] The protein complex brings the broken ends together, a few bases are removed at the ends and the break is repaired (Figure 6.34).

## 6.8 DNA RECOMBINATION

### 6.8.1 Homologous DNA Recombination

One of the seminal discoveries of genetics was the observation that the different genes on the parental chromosome pairs are shuffled before donation to the offspring. How do these 'beads on a string' become assorted? The answer is by **homologous recombination**. A diploid cell in the germinal lineage (the lineage will give rise to haploid germ cells or gametes) contains two copies of a given chromosome. Segments from one chromosome can recombine with corresponding segments of the other (Figure 6.35). As we shall see below this recombination is ultimately dependent on the DNA sequence homology between the two chromosomes.

What advantages does homologous recombination provide? It enables the host organism to sort alleles (differing copies of the same gene) into novel groups. If a copy of a particular gene in a fruit fly has randomly acquired a mutation that yields a more efficient enzyme, then Darwinian natural selection can operate on that gene, provided it is not shackled to all the other genes on the chromosome. Thus, favourable and unfavourable alleles can be shuffled randomly and then the many combinations in the population can be tested by natural selection. Another advantage which recombination provides is the ability to repair a damaged gene in an otherwise favourable chromosome. If no ability to assort different alleles existed then a single unfavourable mutation in a chromosome would consign the whole of it to oblivion.

*6.8.1.1 The Mechanism of Homologous Recombination.* Homologous recombination is linked to DNA replication but does not occur during replication. Rather, it takes place between intact double helices. Genetic and DNA sequence analysis has shown that recombination is accurate to a single base

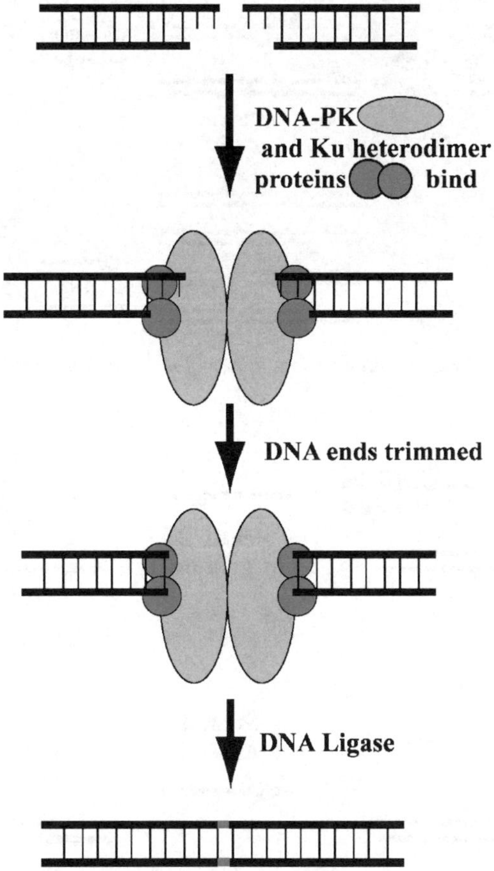


**Figure 6.34** *Repair of double-stranded DNA breaks in eukaryotes*

**Figure 6.35** *Homologous recombination involves the reciprocal exchange of DNA between chromosomes*

pair. The inference from this is that base pairing is involved during the process. Damage to DNA stimulates recombination, suggesting strongly that homologous recombination is initiated from broken DNA strands. Presumably, under normal circumstances the cell creates such breaks enzymatically to promote homologous recombination.

**6.8.1.1.1 The Holliday Junction.** Numerous models have been proposed to explain the mechanism of homologous recombination. A key intermediate in almost all of these is the **Holliday junction**, named after its proposer (Figure 6.36).[57] One of the key properties of this structure is its ability to migrate along the DNA helices by a process called **branch migration**. A reversal of the Holliday junction formation gives a strand swapping event. Finally, DNA replication fixes the new arrangement in the daughter chromosomes.

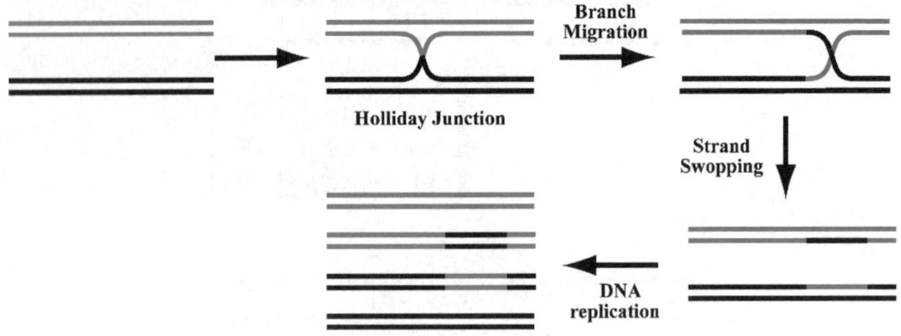


**Figure 6.36** *The Holliday junction is an intermediate in homologous recombination*

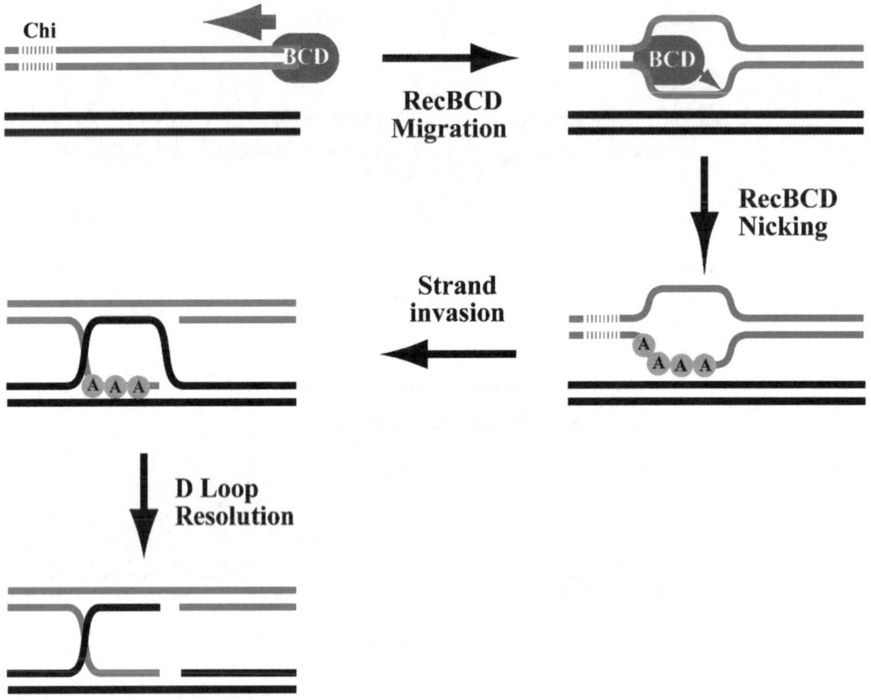


**Figure 6.37** *Involvement of the RecBCD and RecA proteins in the generation of the Holliday junction*

The first step in the likely mechanism for the generation of the Holliday junction (Figure 6.37) involves formation of a single stranded nick in one of the DNA duplexes. This leads to strand invasion and formation of a D loop with the displaced strand. The enzymes in *E. coli* that can catalyse this single strand nicking and strand invasion process are known.[58,59] Nicking is achieved by the **RecBCD** enzyme complex, a large protein complex of about $3 \times 10^5$ Da mass. RecBCD can only bind to a free DNA duplex end, but once bound, it moves along the duplex, unwinding the helix as it goes and rewinding the DNA behind it. If RecBC encounters a specific sequence termed a **Chi site** as it moves along the DNA, it cuts 56 bases 3′ to it. RecBC continues to unwind the DNA but the rewinding is prevented by the nick. This leaves a single-stranded region that participates in strand invasion. This process is catalysed by the **RecA protein**.[60] RecA binds to the single-stranded region and inserts it into a DNA duplex with which it is homologous. In this way, a combination of RecA and RecBCD proteins can catalyse the formation of a Holliday junction.

The Holliday junction is a topologically symmetrical structure. A few simple manipulations of it, which involve no breaking of covalent or hydrogen bonds, results in two possible fates, namely recombination or strand swapping (Figure 6.38). Enzymes which recognise and cleave Holliday junctions are known, such as bacteriophage T4 endonuclease 7. So if a Holliday junction is formed, it is a plausible substrate for recombination. The choice between which bonds are cleaved in Figure 6.38 determines whether a strand swap or a recombination event occurs. This choice may be influenced by the DNA sequence at the junction, both because the enzymes may display sequence specificity for cleavage and also because the three dimensional structure of the junction is not a simple tetrahedron but is distorted by the sequence.

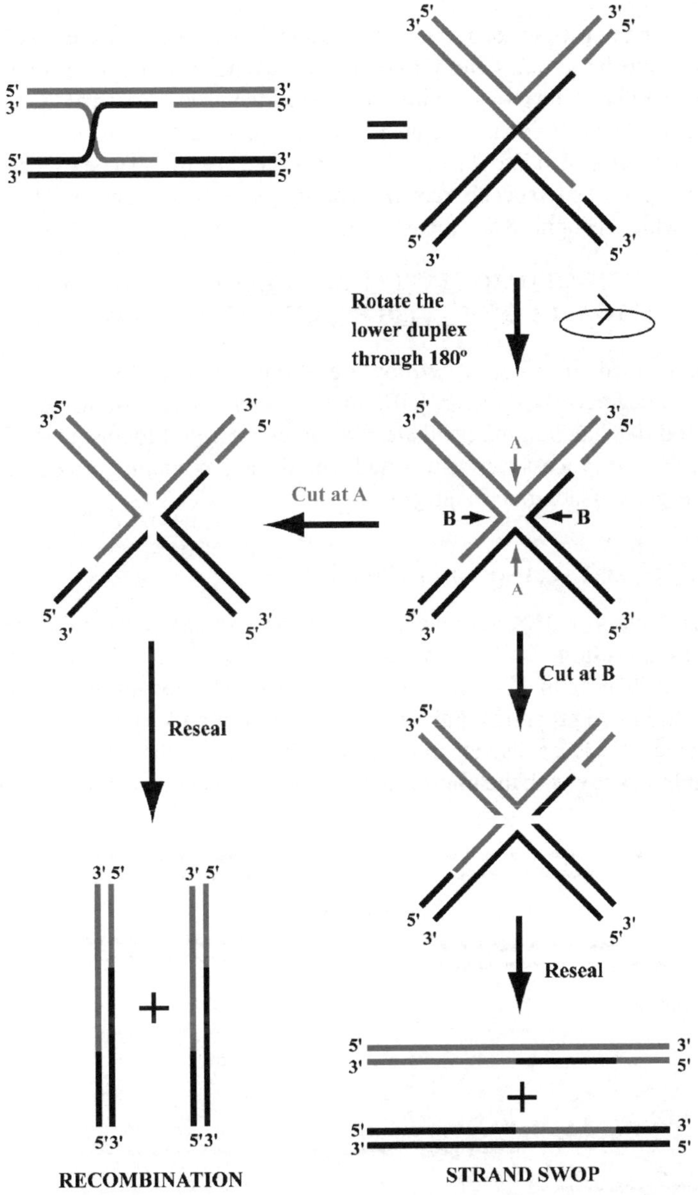


**Figure 6.38** *The Holliday junction can yield recombination or strand swapping. Endonucleolytic cleavage at A or B yields a recombination event or a strand swap, respectively*

*6.8.1.2 Some Implications of Recombination.* We have already seen that recombination offers advantages to the organism by assorting alleles and repairing genes. What other advantages does it give? One major effect is the potential for expanding or contracting the number of copies of genes by **unequal exchange** (Figure 6.8), leading to the evolution of multi-gene families. Another consequence of recombination is **gene conversion**.[57] The normal product of homologous recombination is the swapping of DNA between two duplexes (Figure 6.39). Sometimes however, two copies of one duplex are produced whereas the other is destroyed. Gene conversion provides an evolutionary mechanism for genes to edit and correct one another, while remaining unchanged themselves.

### 6.8.2 Site-Specific Recombination

Recombination is sometimes used to control gene expression.[61] One well studied example is the integration of bacteriophage λ DNA into the chromosome of its host, *E. coli*. When bacteriophage λ virus infects *E. coli*, two outcomes are possible. Either, lytic growth of the virus results in the destruction of the host cell and release of many virus particles or the bacteriophage remains dormant in the cell. In the latter case, called **lysogeny**, the viral DNA becomes integrated into the *E. coli* chromosome and thus becomes a part of its host genome (Figure 6.40). The process is a **site-specific recombination** event, which is catalysed by a λ-encoded enzyme, called an integrase, which recognises the recombination sites (which are called *att* sites) on λ and *E. coli*:

5′ CTGGTTCAGCTTTTTTATACTAAGTTGGCAT 3′ λ
5′ TGAAGCCTGCTTTTTTATACTAACTTGAGCG 3′ *E. coli*

This process can be made to occur *in vitro* by use of only the two DNAs, $Mg^{2+}$, λ integrase and an *E. coli* protein called integration host factor (IHF). In this way it has been shown that the integration event, which involves a Holliday junction intermediate, resembles the type I topoisomerase-catalysed reactions (Section 2.3.5). Lastly, this process can be reversed with the aid of another protein called an excisionase, encoded by the λ *xis* gene, in addition to integrase.

### 6.8.3 Transposition and Transposable Elements

There is another major class of DNA rearrangement, which is not dependent on sequence homology. This is the process of **transposition**, which involves the movement of DNA into new locations in the host genome.[62,63] DNAs which move in this way are collectively termed **transposable elements** or **transposons**, although the latter term more properly applies only to a subset of prokaryotic transposable elements.

In transposition (Figure 6.41), a discrete transposable element becomes inserted into a target site, which bears no significant homology with the transposable element sequence. A short stretch of DNA sequence,

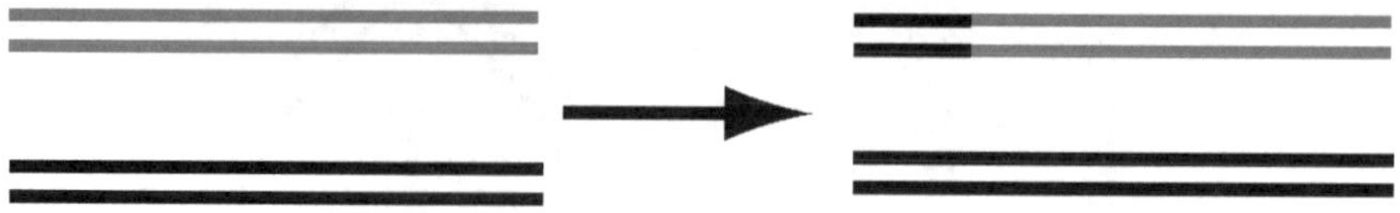

**Figure 6.39** *Gene conversion*

**Figure 6.40** *Bacteriophage λ inserts into the E. coli chromosome by homologous recombination at att sites*

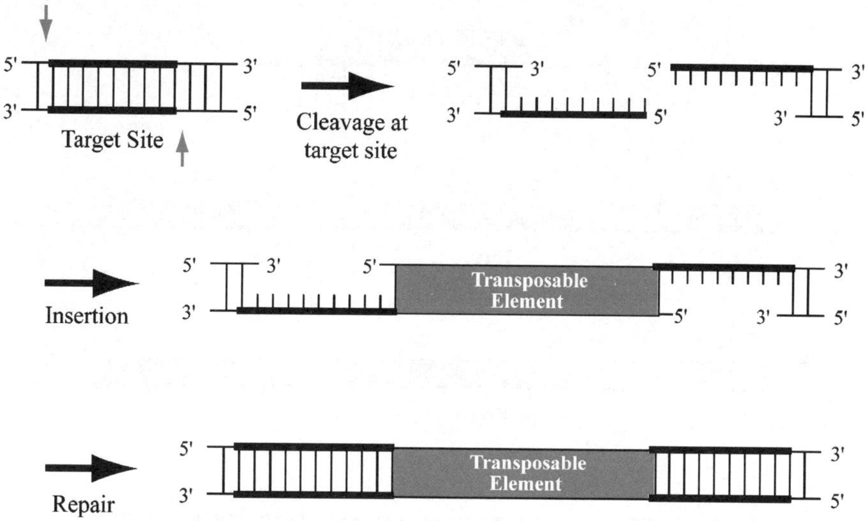


**Figure 6.41** *Transposable elements create small duplications of target sequence when they insert into the chromosome*

typically 4–12 nucleotide pairs, but constant for a given transposable element, is duplicated as a result of the insertion. The transposable element itself moves as a unit. Thus, further copies of the same sequence are found in other genomic locations. Also important to most transposition processes are small inverted repeats at each end of the transposable element in addition to at least one gene inside the transposable element which encodes an enzyme (usually called a **transposase**) that catalyses the transposition process. The inverted repeats constitute the recognition site for the transposase.

Transposition depends on endonucleolytic cleavage by the transposase enzyme at both the ends of the transposable element and at the target site (Figure 6.41).[61] The target site duplication arises as a result of staggered nicking of the target DNA by the transposase. For a given transposable element the target sequence may be effectively random but some elements have preferred DNA motifs for insertion within chromosomal regions.

*6.8.3.1 Prokaryotic Transposable Elements.* There are several major classes of transposable element known in prokaryotes.[62] The simplest are termed insertion sequence elements (**IS elements**; Figure 6.42). These contain just a single transposase gene and terminal repeats. More complicated prokaryotic transposable elements comprise a central region flanked by long repeats, which are either direct or inverted with respect to each other. The long repeats are IS elements or derivatives thereof. These more complex elements are called transposons. An example of a transposon with direct repeats is Tn9 and one with inverted repeats is Tn10 (Figure 6.42). Finally, large transposons without internal repeats, for example Tn3, are also known.

The central regions of transposons contain antibiotic resistance genes and it is the transposition of these mobile DNAs that causes the rapid dissemination of antibiotic resistance between different strains and species of bacteria. They are therefore of considerable medical importance. It is presumed that complex transposons evolved from the fortuitous juxtaposition of two IS elements. In the laboratory the IS elements which comprise a complex transposon can sometimes be mobilised independently. Such events, however, are much less common than transposition of the entire genetic element.

**6.8.3.1.1 The Transposition Mechanism.** Transposition in bacteria can result in two fundamentally different outcomes.[61] In the first case, the transposable element simply leaves one site on the one chromosome and enters another ('**cut and paste**'). In the second case, the transposon is preserved at its original location and a new element appears at a distant site ('**copy and paste**'). Thus, transposition in the latter

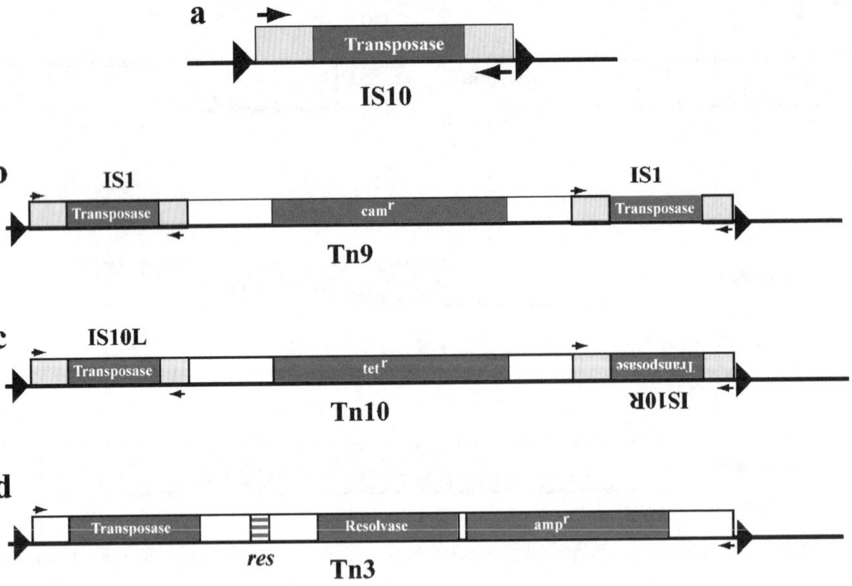


**Figure 6.42** *Structures of prokaryotic transposable elements*

instance is replicative. Some transposons, including IS1, can use both mechanisms, indicating that these two processes are mechanistically linked. The way in which this is achieved is shown in Figure 6.43.

The **Shapiro transposition intermediate** (Figure 6.43) is as important to models of transposition as the Holliday junction is to homologous recombination mechanisms. It is important to note that replicative transposition results in the donor and recipient strands being joined to form a structure called a **cointegrate**. This can be seen during transposition from one plasmid to another and necessitates a homologous recombination step to separate the two duplexes (Figure 6.44). This latter step is catalysed by a transposon-encoded **resolvase** enzyme. The resolvase stimulates recombination at a specific site (called *res*) inside the transposon by endonucleolytic cleavage in a similar manner to that used by λ integrase (Section 6.8.2). Furthermore, the sequence of *res* is similar to that of the core sequence used by λ

*res* GATAATTTATAATAT
*att* GCTTTTTTATACTAA

A final point to note from Figure 6.43 is that during non-replicative transposition the donor chromosome is left broken. In prokaryotes this often results in the loss of the chromosome. In addition to simple transposition, transposons can catalyse rearrangements of the DNA surrounding them. Deletion or inversion of flanking DNA is often seen. Such rearrangements are abortive by-products of the normal transposition process (Figure 6.45).

*6.8.3.2 Eukaryotic Transposons.* There exists a wide variety of different types of transposable element in eukaryotes.[63] In common with their prokaryote relatives, most **eukaryotic transposable elements** possess small inverted terminal repeats and generate small direct repeats of the target site during integration. Perhaps all eukaryotes contain at least one type and the variety of size and structure is bewildering. However, two major classes are apparent. **Class I transposable elements** all use RNA intermediates for transposition and **Class II elements** all use only DNA transposition intermediates. Class II elements are evolutionarily related to IS elements and will be considered first.

**6.8.3.2.1 Eukaryotic Class II Transposable Elements: The P Element.** The fruit fly *D. melanogaster* is host to a wide variety of transposable elements, which together comprise at least 10%

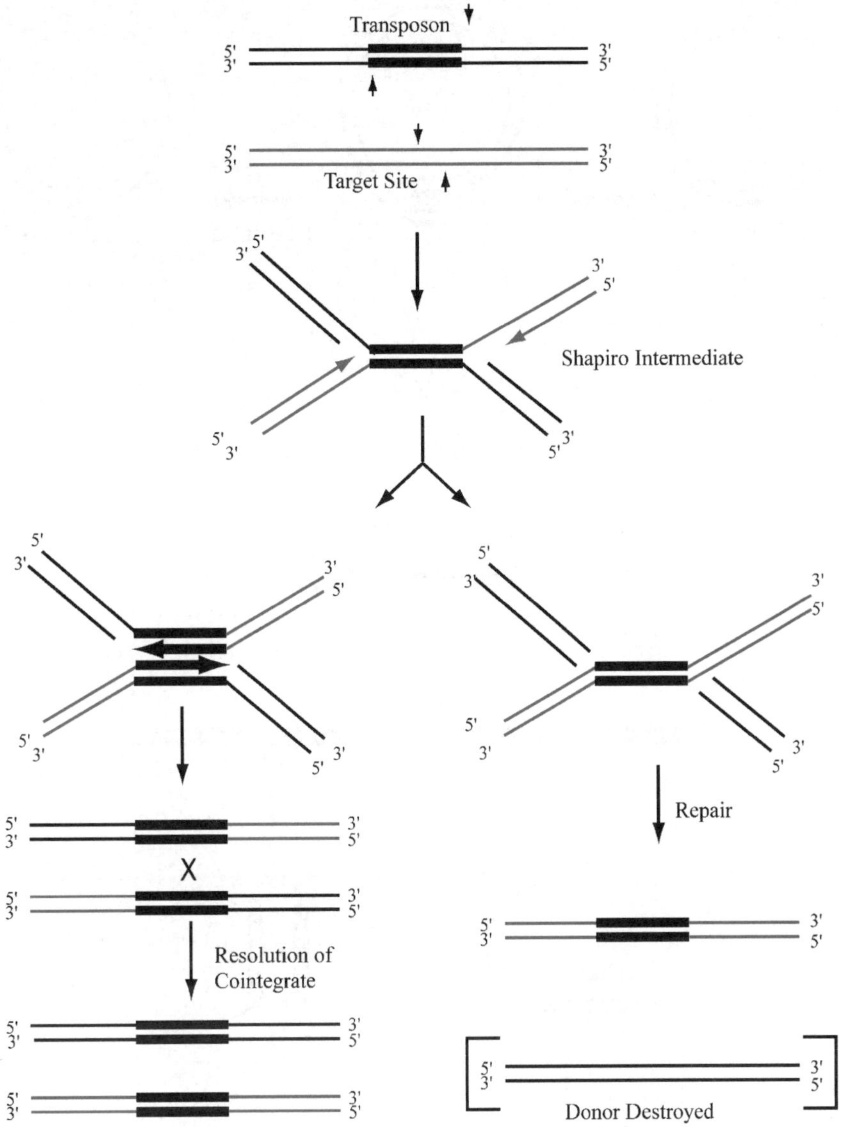


**Figure 6.43** *The Shapiro intermediate is central to the mechanism of illegitimate recombination. Either replicative or non-replicative transposition may result from this structure*

of its total genome. One of the most interesting and best-studied elements in *Drosophila* is the P element. This transposable element is structurally simple and very similar to IS elements. It has a single transposase gene, this time interrupted by three introns (Figure 6.46). It transposes by a 'cut and paste' mechanism effectively identical to that displayed by IS elements.

The P element is only transposed in the germ cells of the female fly and only if the female lacks intact P elements but her mate has them. No other combination of mating produces movement of the transposable element. Interestingly, P element RNA is found in all flies containing these elements and is made in all their tissues. If this RNA encodes the transposase, what prevents transposition in all tissues of flies containing P elements? It turns out that the protein synthesised in somatic tissue (the entire body minus the germ cells) of the fly only contains sequence information from the first three exons of the P element (Figure 6.46). This

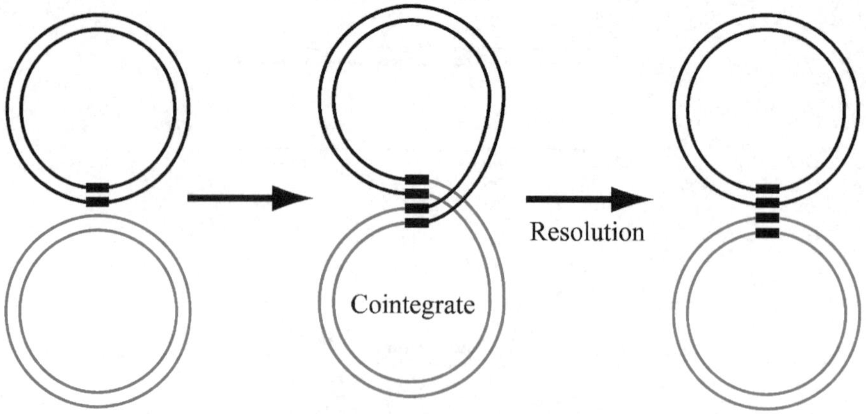


**Figure 6.44** *Cointegrate structures are intermediates in replicative transposition*

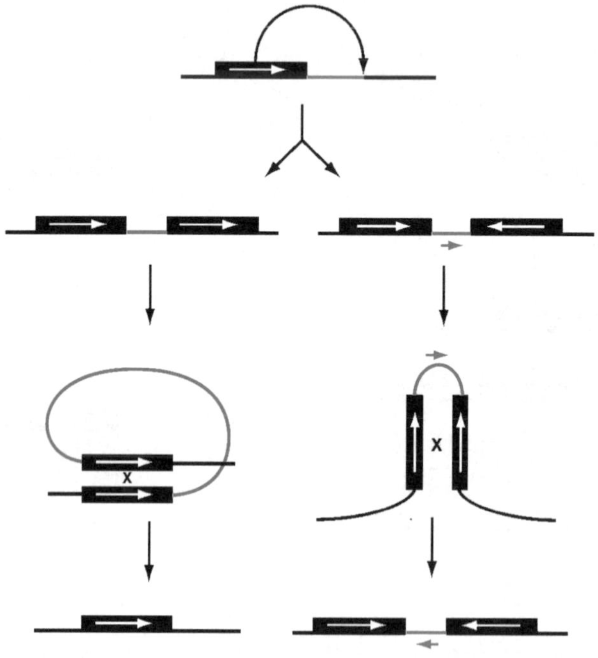


**Figure 6.45** *Deletion or inversion of DNA surrounding adjacent transposable element insertions*

truncated protein is unable to catalyse transposition and is actually a repressor of the transposase. The mRNA which gives rise to this defective protein retains intron 3 of the transposase gene, leading to premature translational termination of the transposase open reading frame within the intron. In germ cells, intron 3 is efficiently removed from the mRNA, active transposase is translated and the result is transposition of the P elements.

**6.8.3.2.2 Eukaryotic Class I Transposable Elements.** Eukaryotic Class I transposable elements fall into three major categories, namely **retroviruses, retrotransposons** and **dispersed pseudogenes**[64] (Figure 6.47). Retroviruses and retrotransposons are closely linked evolutionarily and structurally. The key distinction between the two is that retroviruses are capable in principle of producing an extracellular infectious virus. In practice, there are very many defective retroviruses, some of which have been called retrotransposons.

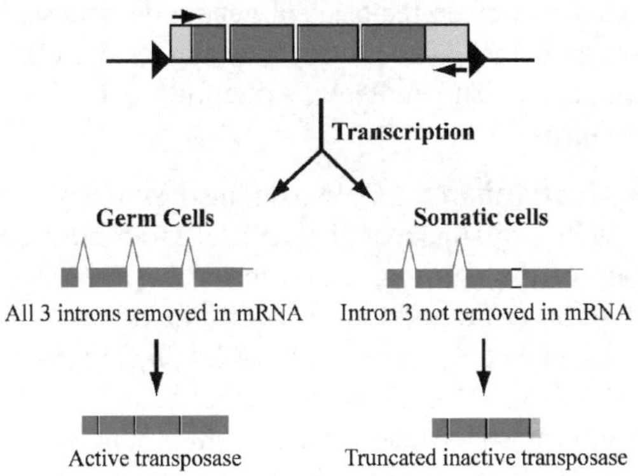


**Figure 6.46** *The control of transposition of the Drosophila P element. mRNA splicing limits the synthesis of active transposase to the germ cells*

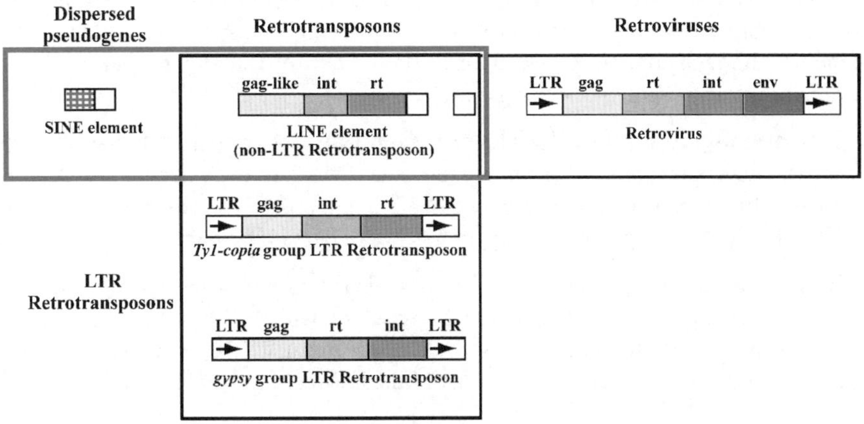


**Figure 6.47** *Classification of Class I transposable elements. Large outlined boxes indicate overlapping classifications for retrotransposons, retroposons and dispersed pseudogenes. Smaller shaded and hatched boxes indicate gene conservation among different types of Class I elements*

Retrotransposons are the dominant class of transposable element found in eukaryotes. Two types of true retrotransposon are known, **LTR retrotransposons** and **non-LTR retrotransposons**, more popularly known as long interspersed elements (**LINEs**).[65] Dispersed pseudogenes are defective derivatives of cellular genes that have never encoded the proteins needed for successful transposition, but which are probably moved to new chromosomal locations by the transposition machinery of retroviruses or retrotransposons. They are thus genetic 'hitch-hikers'.

The best-known dispersed pseudogenes are the short interspersed elements (**SINEs**) and the prototype example of these in humans is the **Alu family**.[66] Both SINEs and LINEs have also been classified as '**retroposons**' (Class I elements which use a common transposition mechanism called target-primed reverse transcription). Unfortunately, the retroposon classification, although well established, conflicts with the retrotransposon classification, because LINE elements are both retrotransposons and retroposons.

LINEs are probably the most primitive retrotransposon group and may be ancestral to the other groups. They contain a gene encoding the protein component of a capsid-like intracellular structure (*gag*) required for the transposition process. Some non-LTR retrotransposons also encode an endonuclease, which is involved in transposition. Finally, all LINEs contain a reverse transcriptase gene, encoding the enzyme that converts the element's RNA into DNA.

LTR retrotransposons are sub-divided on the basis of gene order and sequence similarity into the Ty1-*copia* group and the *gypsy* group, named after prototype examples discovered in *Drosophila* and yeast.[64] Both groups are also found in plants, fish, amphibia and reptiles but neither has been described in mammals (including humans) or in birds.

**6.8.3.2.3 Transposition Mechanisms.** Only the transposition cycle for LTR retrotransposons will be discussed here, since it is much better understood than LINE transposition mechanisms (Figure 6.48).[64] First, an integrated DNA copy of the retrotransposon in the host genome is transcribed into an RNA. Transcription is initiated within one LTR and terminated in the other. It thus contains the entire genetic information of the transposable element. Some of this RNA is spliced before export into the cytoplasm and some is exported unspliced. Both spliced and unspliced RNAs can be translated into the various protein products encoded by the retrotransposon. Most of the protein produced comprises the components of an intracellular virus-like particle. This process is similar to that seen for many retroviruses. A subset of the full-length unspliced retrotransposon RNAs is encapsidated into this particle, along with two enzymes necessary for transposition, namely reverse transcriptase and integrase.

Reverse transcription is carried out by the encapsidated reverse transcriptase. The process is completed when the RNA has been converted into a linear unintegrated DNA. Insertion of this DNA into a new chromosomal location is catalysed by the integrase protein. As a rule, retrotransposons insert at random sites into their host genome, although some LTR retrotransposons of *Drosophila* and *Saccharomyces* have preferences for particular insertion sequences or genomic regions.

**6.8.3.2.4 Retrotransposons and the Human Genome.** Approximately half of human genome consists of repeated elements interspersed among the genes. Much of this interspersed DNA is comprised of **Class I transposable elements**, predominantly **SINEs** (approximately 1 million copies or 10–15% of the genome), **LINEs** (15–20% of the genome) and pseudogenes of human retroviruses (*ca.* 1% of the genome).[67]

**6.8.3.2.5 Retroviruses.** Retroviruses were first identified as agents involved in the onset of cancer about 80 years ago.[30] More recently the AIDS epidemic has been shown to be due to the HIV retrovirus. In the early 1970s it was discovered that retroviruses could replicate their RNA genomes *via* conversion into DNA, which becomes stably integrated in the DNA of the host cell. It is only comparatively recently that

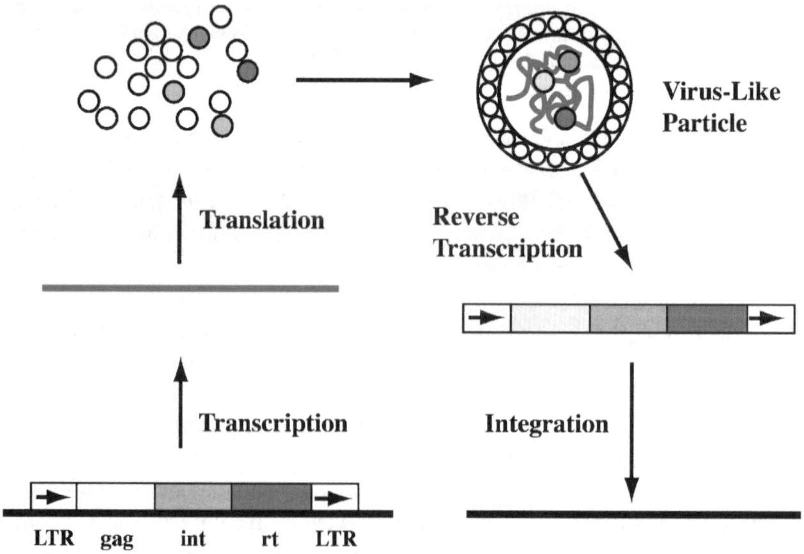


**Figure 6.48** *The life cycle of LTR retrotransposons. All intermediates are intracellular. Shaded boxes and circles indicate genes and their protein products respectively*

retroviruses have been recognised as particularly specialised forms of eukaryotic transposon. In effect they are retrotransposons, which usually can leave the host cells and infect other cells. This ability is donated by an extra gene (*env*) encoding a glycoprotein, which coats the virus, allowing entry and exit from the host cell. The integrated DNA form (or provirus) of the retrovirus is almost identical to a retrotransposon.

## REFERENCES

1. O.T. Avery, C.M. MacLeod and M. McCarty, Studies on the chemical nature of the substance inducing transformation of pneumococcal types. *J. Exp. Med.*, 1944, **79**, 137–158.
2. J. Cairns, G.S. Stent and J.D. Watson, *Phage and the Origins of Molecular Biology.* Cold Spring Harbor Laboratory Press, Cold Spring Harbor, NY, 1966.
3. F. Jacob and J. Monod, Genetic and regulatory mechanisms in the synthesis of proteins. *J. Mol. Biol.*, 1961, **3**, 318–356.
4. G.M. Cooper and G.E. Hausman, *The Cell.* Sinauer Associates, Sunderland, MA, 2004, 142–143.
5. R. Breathnach, J.L. Mandel and P. Chambon, Ovalbumin gene is split in chicken DNA. *Nature*, 1977, **270**, 314–319.
6. W. Gilbert, Why genes in pieces? *Nature*, 1978, **271**, 501.
7. A. Stoltzfus, D.F. Spencer, M. Zuker, J.M. Logsdon Jr. and W.F. Doolittle, Testing the exon theory of genes: The evidence from protein structure. *Science*, 1994, **265**, 202–207.
8. D.F. Colgan and J.L. Manley, Mechanism and regulation of mRNA polyadenylation. *Genes Dev.*, 1997, **11**, 2755–2766.
9. E.P. Lei and P.A. Silver, Protein and RNA export from the nucleus. *Dev. Cell.*, 2002, **2**, 261–272.
10. T.A. Ayoubi and W.J. Van De Ven, Regulation of gene expression by alternative promoters. *FASEB J.*, 1996, **10**, 453–460.
11. A.J. Lopez, Alternative splicing of pre-mRNA: developmental consequences and mechanisms of regulation. *Ann. Rev. Genet.*, 1998, **32**, 279–305.
12. E. Enerly, Ø. L. Mikkelesen, M. Lyamouri and A. Lambertsson, Evolutionary profiling of the U49 *snoRNA* gene. *Hereditas*, 2003, **138**, 73–79.
13. International Human Genome Sequencing Consortium, Initial sequencing and analysis of the human genome. *Nature*, 2001, **409**, 860–921.
14. J.C. Ventor and 273 others, The sequence of the human genome. *Science*, 2001, **291**, 1304–1351.
15. T.H. Eickbush and F.C. Kafatos, A walk in the chorion locus of *Bombyx mori*. *Cell*, 1982, **29**, 633–643.
16. P.F.R. Little, Globin pseudogenes. *Cell*, 1982, **28**, 683–684.
17. P. SanMiguel, A. Tikhonov, Y.K. Jin, N. Motchoulskaia, D. Zakharov, A. Melake-Berhan, P.S. Springer, K.J. Edwards, M. Lee, Z. Avramova and J.L. Bennetzen, Nested retrotransposons in the intergenic regions of the maize genome. *Science*, 1996, **274**, 765–768.
18. M. Rossberg, K. Theres, A. Acarkana, R. Herrero, T. Schmitt, K. Schumacher, G. Schmitz and R. Schmidt, Comparative sequence analysis reveals extensive microcolinearity in the lateral suppressor regions of the tomato, *Arabidopsis*, and *Capsella genomes. Plant Cell*, 2001, **13**, 979–988.
19. J.L. Bennetzen and E.A. Kellogg, Do plants have a one-way ticket to genomic obesity? *Plant Cell*, 1997, **9**, 1509–1514.
20. G. Martin, D. Wiernasz and P. Schedl, Evolution of *Drosophila* repetitive-dispersed DNA. *J. Mol. Evol.*, 1983, **19**, 203–213.
21. M.L. Pardue and J.G. Gall, Chromosomal localization of mouse satellite DNA. *Science*, 1970, **168**, 1356–1358.
22. K.E. Van Holde and J. Zlatanovai, Chromatin higher order structure: Chasing a mirage. *J. Biol. Chem.*, 1995, **270**, 8373–8376.
23. R.D. Kornberg, Chromatin structure: a repeating unit of histones and DNA. *Science*, 1974, **184**, 868–871.
24. K. Luger, A.W. Mader, R.K. Richmond, D.F. Sargeant and T.J. Richmond, Crystal structure of the nucleosome core particle at 2.8Å resolution. *Nature*, 1997, **389**, 251–260.

25. J.J. Champoux, DNA topoisomerases: Structure, function and mechanism. *Ann. Rev. Biochem.*, 2001, **70**, 369–413.
26. J. Carbon, Yeast centromeres: Structure and function. *Cell*, 1984, **37**, 351–353.
27. D. Kippling and P.E. Warburton, Centromeres, CEN-P and Tigger too. *Trends Genet.*, 1997, **13**, 141–144.
28. C.W. Grieder and E.H. Blackburn, Telomeres, telomerases and cancer. *Sci. Am.*, 1996, **274**, 80–85.
29. I. Liljas, Viruses. *Curr. Opin. Struct. Biol.*, 1996, **6**, 151–156.
30. J.M. Coffin, S.H. Hughes and H.E. Varmus, *Retroviruses.* Cold Spring Harbor Laboratory Press, Cold Spring Harbor, NY, 1997.
31. A.J. Jeffreys, V. Wilson and S.L. Thein, Hypervariable 'minisatellite' regions in human DNA. *Nature*, 1985, **314**, 67–73.
32. A. Chakravarti, Single nucleotide polymorphisms: to a future of genetic medicine. *Nature*, 2001, **409**, 822–823.
33. C.A. Gross, C. Chan, A. Dombroski, T. Gruber, M. Sharp, J. Tupy and B. Young, The functional and regulatory roles of sigma factors in transcription. *Cold Spring Harbor Symp. Quant. Biol.*, 1998, **63**, 141–155.
34. S. Busby and R.H. Ebright, Promoter structure, promoter recognition and transcription activation in prokaryotes. *Cell*, 1994, **79**, 743–746.
35. S. Busby and R.H. Ebright, Transcription activation by catabolite activator protein (CAP). *J. Mol. Biol.*, 1999, **293**, 199–213.
36. A. Sentenac, Eukaryotic RNA polymerases. *CRC Crit. Rev. Biochem.*, 1985, **18**, 31–90.
37. D.B. Nikolov and S.K. Burley, RNA polymerase II transcription initiation: a structural view. *Proc. Natl. Acad. Sci. USA*, 1997, **94**, 15–22.
38. D.S. Latchman, *Eukaryotic transcription factors*. Academic Press, London, 1995.
39. B.F. Pugh, Control of gene expression through regulation of the TATA-binding protein. *Gene*, 2000, **255**, 1–14.
40. T.I. Gerasimova and V.G. Corces, Chromatin insulators and boundaries: effects on transcription and nuclear organization. *Ann. Rev. Genet.*, 2001, **35**, 193–208.
41. R. Berezney, M. Mortillaro, H. Ma, X. Wei and J. Samarabandu, The nuclear matrix: a structural milieu for genomic function. in *International Reviews of Cytology*, vol 162A, R. Berezney and K.W. Jeon (eds). Academic Press, New York, 1995, 1–65.
42. D.S. Gross and W.T. Garrard, Nuclease hypersensitive sites in chromatin. *Ann. Rev. Biochem.*, 1988, **57**, 159–197.
43. J.T. Kadonaga, Eukaryotic transcription: an interlaced network of transcription factors and chromatin modifying machines. *Cell*, 1998, **92**, 307–313.
44. K.D. Robertson and P.A. Jones, DNA methylation: past, present and future directions. *Carcinogenesis*, 2000, **21**, 461–467.
45. Y. Huang and R.J. Maraia, Comparison of the RNA polymerase III transcription machinery in *Schizosaccharomyces pombe*, *Saccharomyces cerevisiae* and human. *Nucleic Acids Res.*, 2001, **29**, 2675–2690.
46. S. Waga and B. Stillman, The DNA replication fork in eukaryotic cells. *Ann. Rev. Biochem.*, 1998, **67**, 721–751.
47. T. Ogawa and T. Okazaki, Discontinuous DNA replication. *Ann. Rev. Biochem.*, 1980, **49**, 421–457.
48. S.P. Bell and A. Dutta, DNA replication in eukaryotic cells. *Ann. Rev. Biochem.*, 2002, **71**, 333–374.
49. D.N. Frick and C.C. Richardson, DNA primases. *Ann. Rev. Biochem.*, 2001, **70**, 39–80.
50. T.M. Lohman and K.P. Bjornson, Mechanism of helicase-catalysed DNA unwinding. *Ann. Rev. Biochem.*, 1996, **65**, 169–214.
51. E.C. Friedberg *et al.*, *Trends Biochem. Sci.*, 1995, **20**(10), 381–439 (11 articles).
52. W.L. de Laat, N.G.J. Jaspers and J.H.J. Hoeijmakers, Molecular mechanism of nucleotide excision repair. *Genes Dev.*, 1999, **13**, 768–785.

53. A. Sancar, DNA excision repair. *Ann. Rev. Biochem.*, 1996, **65**, 43–81.
54. U. Hubscher, G. Maga and S. Spadari, Eukaryotic DNA polymerases. *Ann. Rev. Biochem.*, **7**, 133–163.
55. B.D. Harfe and S. Jinks-Robertson, DNA mismatch repair and genetic instability. *Ann. Rev. Genet.*, 2000, **34**, 359–399.
56. W.S. Dynan and S. Yoo, Interaction of Ku protein and DNA-dependent protein kinase catalytic subunit with nucleic acids. *Nucleic Acids Res.*, 1998, **26**, 1551–1559.
57. R. Holliday, A mechanism for gene conversion in fungi. *Genet. Res.*, 1964, **5**, 282–304.
58. S.C. West, Enzymes and molecular mechanisms of genetic recombination. *Ann. Rev. Biochem.*, 1992, **61**, 603–640.
59. S.C. Kowalczykowski and A.K. Eggleston, Homologous pairing and DNA strand exchange proteins. *Ann. Rev. Biochem.*, 1994, **63**, 991–1043.
60. C.M. Radding, Helical interactions in homologous pairing and strand exchange driven by RecA protein. *J. Biol. Chem.*, 1991, **266**, 5355–5358.
61. B. Hallet and D.J. Sherratt, Transposition and site-specific recombination: adapting DNA cut-and-paste mechanisms to a variety of genetic rearrangements. *FEMS Microbiol. Rev.*, 1997, **21**, 157–178.
62. J.A. Shapiro (ed), *Mobile Genetic Elements*, Academic Press, London, 1983.
63. M.G. Kidwell and D. Lisch, Transposable elements as sources of variation in animals and plants. *Proc. Natl. Acad. Sci. USA*, 1997, **94**, 7704–7711.
64. A.J. Flavell, Retroelements, reverse transcriptase and evolution. *Comp. Biochem. Physiol. B*, 1995, **110B**, 3–15.
65. E.M. Ostertag and H.H. Kazazian Jr., Biology of mammalian L1 retrotransposons. *Ann. Rev. Genet.*, 2001, **35**, 501–538.
66. C.W. Schmid and W.R. Jelinek, The Alu family of dispersed repetitive sequences. *Science*, 1982, **216**, 1065–1070.
67. R. Lower, J. Lower and R. Kurth, The viruses in all of us: characteristics and biological significance of human endogenous retrovirus sequences. *Proc. Natl. Acad. Sci. USA*, 1996, **93**, 5177–5184.

CHAPTER 7

# RNA Structure and Function

## CONTENTS



## 7.1 RNA STRUCTURAL MOTIFS

RNA is the most versatile macromolecule in nature. The linear sequence of an RNA can encode large amounts of complex information that is subsequently transformed into functional proteins. However, many

RNA sequences also contain sufficient information to fold themselves into specific shapes with distinct chemical properties. Thus, RNA is unique amongst biopolymers in that it encodes genetic information, provides structural scaffolding, recognizes and transports other molecules, and carries out many forms of chemical catalysis in the cell.

### 7.1.1 Basic Structural Features of RNA

It is remarkable that the diverse capabilities of RNA, and many of its distinctions from DNA, stem from a few simple chemical differences (Section 2.4). The most important distinction in RNA is the ribose sugar, which bears a **2′-hydroxyl group** (Figure 2.2). This simple modification confers unique conformational features, specific hydration and electrostatic properties in the RNA polymer, and provides a set of hydrogen-bond donors and acceptors along the RNA backbone. A second difference is the **uracil** nucleobase in RNA, which lacks the major groove 5-methyl group of thymidine in DNA.

Primarily as a consequence of the ribose sugar conformation, RNA duplexes are more compact and have geometrical features that differ from B-DNA. This is because ribose nucleotides tend to adopt the **C3′-endo sugar pucker** (Figure 2.11), which draws the flanking phosphates close together (5.9 Å), resulting in compact **A-form duplexes** that contain 11 base pairs per turn. The two grooves in A-form duplexes differ from those in B-form DNA. The "major" groove of A-form helices is very narrow and deep, while the "minor" groove is wide and flat (Figure 2.40). As a result, ligands employ a different set of strategies for recognizing RNA and DNA.

The major groove of RNA has a markedly negative **electrostatic** potential (Figure 7.1), which tends to draw small, positively charged ions and side-chains into the major groove.[1] In DNA, the minor groove

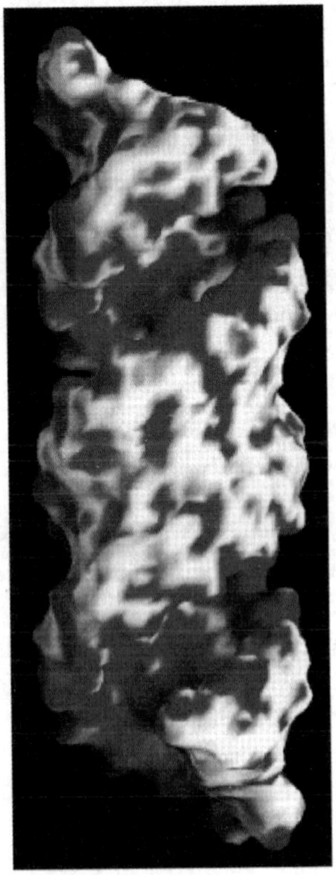

**Figure 7.1** *The electrostatic surface potential of an RNA helix. Red color indicates the region of greatest negative potential, seen particularly in the RNA major groove. Blue color indicates region of most positive electrostatic potential, and white approximates neutrality*
(Reprinted from Ref. 1. © (1999), with permission from Macmillan Publications Ltd)

behaves in this fashion. The **hydration** properties of RNA duplexes are also distinctive.[2] Complex networks of water molecules associate with both nucleobase and backbone atoms, resulting in a rich array of functionalities for molecular recognition (Figure 7.2).

### 7.1.2 Base Pairings in RNA

Unlike DNA, RNA structures accommodate a variety of **alternative base pairings**.[3] In addition to the canonical Watson–Crick G:C and A:U base pairs (Figure 2.8), a large variety of other base-pair combinations are observed. The most common alternative pairings are the **G:U wobble pair** and the **$A^+$:C pair** (Figure 7.3, see also Figure 2.9). The G:U pair is notable in that it provides a large hydrogen-bond donor group (exocyclic amino) in the minor groove of RNA duplexes, which is an important recognition element for proteins and other ligands. The G:U pair occurs in many biological contexts, including the codon–anticodon interactions that form the genetic code between tRNA and mRNA (Section 7.3). In the $A^+$:C pair, the adenine N-1 is protonated as a result of a shift in $pK_a$ that can sometimes occur within folded RNA structures. Shifts in $pK_a$ diversify the function of the four nucleobases and result in altered pairing as well as chemical capabilities. The $A^+$:C pair is observed within some RNA loops and in the core of many catalytic RNAs (ribozymes, Sections 7.2 and 7.6).

Purine nucleotides can also readily pair with one another. **G:A pairs** are common at the termini of RNA helices, in loops and in folds that comprise RNA tertiary structure. While there are many types of G:A, G:G, and A:A pairs, sheared G:A pairs are the most common (*e.g.*, Figure 7.3), which include a **Hoogsteen** interaction that involves the N-7 and 6-amino groups in the major groove edge of the A base (Figure 7.3).

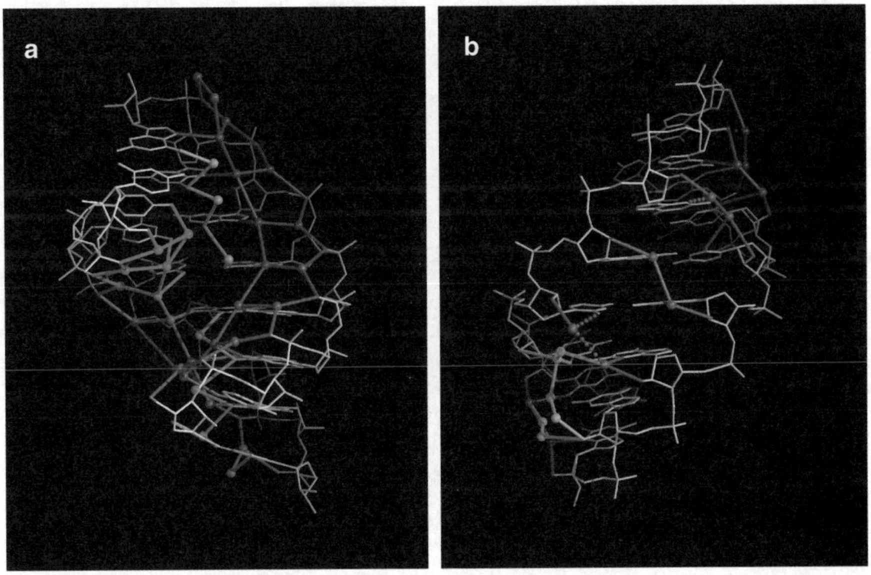


**Figure 7.2** *Hydration of RNA in the major (a) and minor (b) grooves in the crystal structure of the RNA duplex* $[r(CCCCGGGG)]_2$. *In the major groove, water pentagons along strand 1 are red, along strand 2 they are green, and water bridges between O-6 atoms of guanines are yellow and cyan in the top and bottom halves of the duplex, respectively. In the minor groove, water bridges between 2′-hydroxyl groups from opposite strands within base-pair steps are red. Bridges between 2′-hydroxyl groups within base-pair planes and sharing a water molecule with the former bridges are cyan. Two water molecules and a phosphate oxygen from an adjacent duplex link the hydroxyl groups of residues G7 and C11 and are colored yellow*

(Reprinted from Ref. 2. © (1996), with permission from the American Chemical Society)

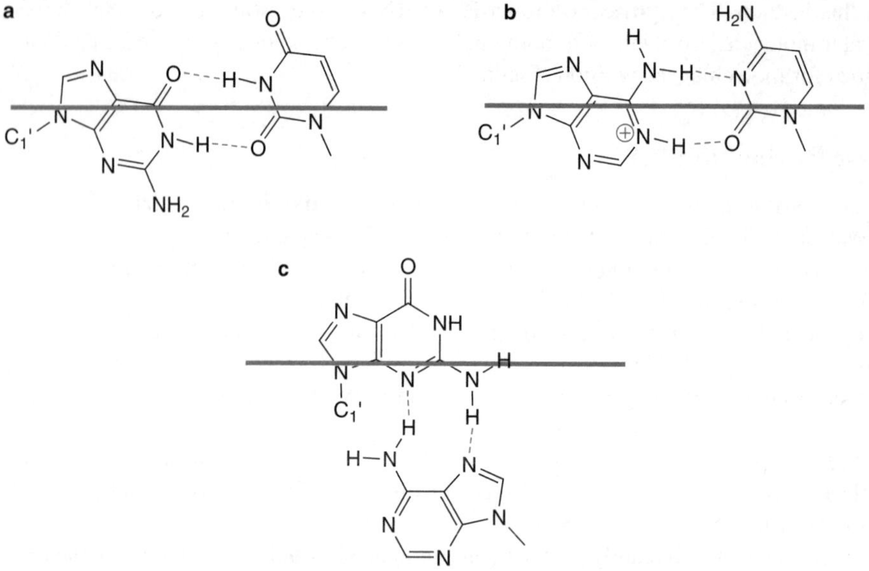

**Figure 7.3** *Common forms of alternative base pairing in RNA. The pink line indicates the base plane axis for a wobble pairing. (a) G–U wobble pair. (b) A$^+$–C pair. (c) Sheared form of the G–A pair*

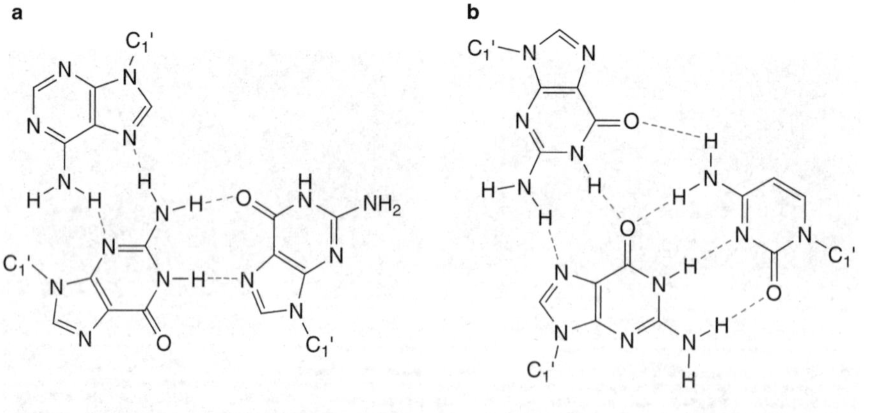

**Figure 7.4** *Examples of the unusual triple interactions observed in RNA. (a) One type of GGA triplex (GGA N-7-imino, carbonyl-amino; N-3-amino, amino-N-7) observed in tRNA bound to a synthetase.[4] (b) A type of GGC triplex (GGC amino(N-2)-N-7, imino-carbonyl, carbonyl-amino(N-4); Watson–Crick) observed in the 50S ribosome[5]*

## 7.1.3 RNA Multiple Interactions

In addition to base pairing, RNA nucleotides commonly participate in multiple interactions that involve both the bases and the sugars. These interactions are sometimes observed individually within a tertiary structure,[3] or they can be linked together to form extended triplexes and quadruplexes (see also Sections 2.4.5 and 2.3.7, respectively). Multiplexes such as the square guanosine tetraplex **(G-quartet, G-tetrad)** are similar in both RNA and DNA (Section 2.3.7). However, a greater diversity of triple and quadruple interactions is observed in RNA molecules. For example, an all-purine **triple interaction** that is dominated by Hoogsteen contacts is observed in the complex of a seryl tRNA synthetase with its cognate tRNA (Figure 7.4a).[4] The 50S

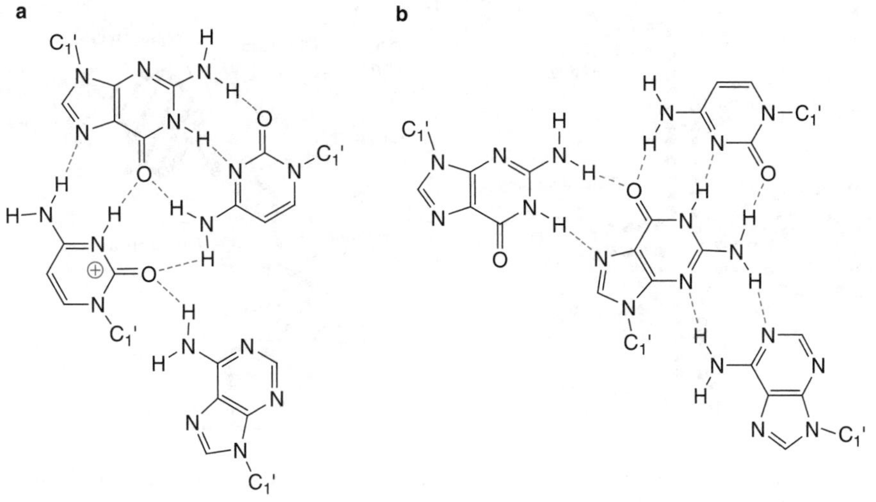


**Figure 7.5** *Examples of quadruple interactions observed in RNA. (a) An ACGC quadruplex (ACGC amino-carbonyl, amino(C)-N-7(G):N-3(+)C-carbonyl(G): carbonyl(C)-amino(C); Watson–Crick) from a frame-shifting pseudoknot*[6]. *(b) The GGCA quadruplex (GGCA imino-N-7, carbonyl-amino; Watson–Crick; N-3-amino(G), amino-N-1(G)) from a dye-binding aptamer*[7]

ribosomal subunit contains a striking diversity of RNA triple interactions, one of which involves a network of hydrogen bonds that are shared amongst all three nucleotides (Figure 7.4b).[5]

Highly folded RNA molecules can contain **quadruple interactions** or even larger arrays of hydrogen-bonded bases. For example, an interesting quadruple variation is observed in the structure of a frameshifting viral pseudoknot, that includes a protonated cytosine N-3 (Figure 7.5a).[6] Another remarkable quadruple interaction occurs within an *in vitro*-selected **aptamer**, which consists of a central G base encircled by hydrogen bonds to three other bases (Figure 7.5b).[7] An aptamer is an RNA or DNA molecule that has been selected from a pool of random sequence, based on its ability to bind with high affinity to a particular ligand (Section 5.7.3).[8]

### 7.1.4 RNA Tertiary Structure

*7.1.4.1 Tertiary Structural Motifs.* Although RNA molecules are commonly represented as two-dimensional secondary structures, most RNAs are folded into compact and defined tertiary structures that are necessary for function (Figure 7.6). Specific types of tertiary structures are seen for tRNA, rRNA, snRNA, certain introns, and ribozymes.[9,10] But even mRNA can adopt complex tertiary structures, particularly in the untranslated terminal regions (UTRs), which can be essential for proper gene expression.

One of the most important contributions to the three-dimensional form of a folded RNA is **coaxial stacking** between adjacent sets of short RNA duplex,[11] for example as seen in a **kissing hairpin complex** Figure 7.7. Among the most common substructures for stabilizing long-range interactions in RNA is the **tetraloop-receptor** motif.[9] These involve a specific arrangement of base stacking and hydrogen bonds between the **GNRA tetraloop** (a highly conserved loop with a defined structure) and a conserved stem-loop sequence (Figure 7.8).

The **A-minor motif** is a ubiquitous interaction that involves contact between an adenosine and the minor groove of a Watson–Crick base pair, thereby forming a type of triple interaction.[12] A-minor motifs are observed in self-splicing introns and throughout the ribosome.[10] They can occur in isolation but are often arranged in stacked arrays that may confer additional stabilization (Figure 7.9).

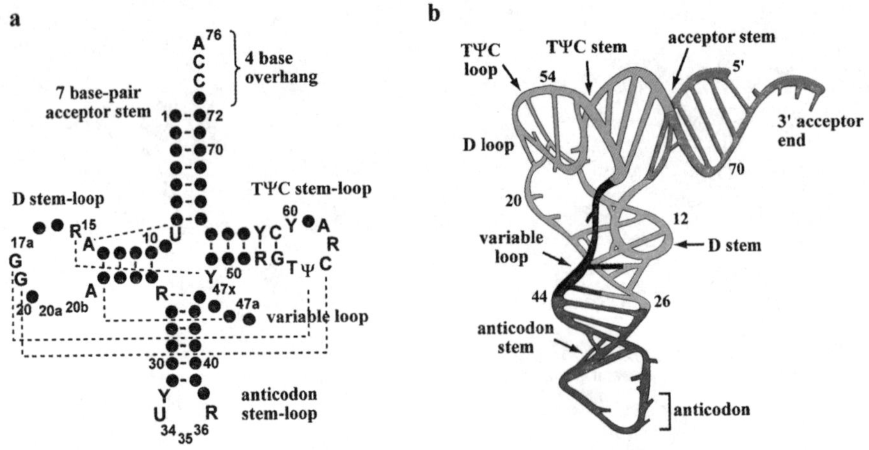


**Figure 7.6** *The structure of tRNA. (a) Secondary structure. Bold dots indicate positions of paired bases. The identity of conserved bases and functions of tRNA regions are indicated. Dashed lines represent tertiary interactions in the three-dimensional structure. (b) Tertiary structure. Colours and labels differentiate the various functional domains of the tRNA*

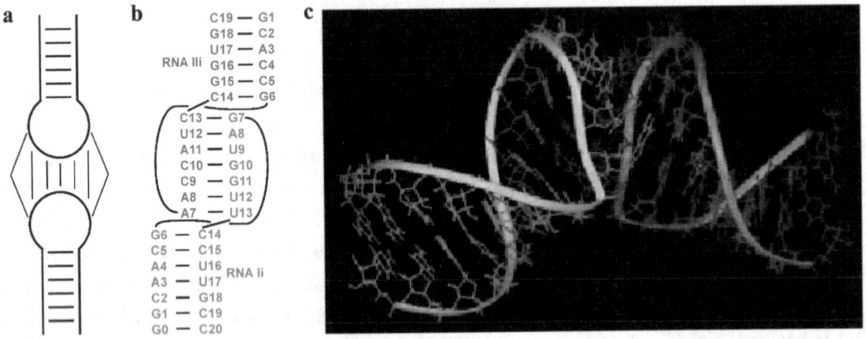


**Figure 7.7** *The secondary and tertiary structure of a loop–loop interaction. Coaxial stacking is preserved in a "kissing loop" complex by the formation of a bend in the overall structure. (a) Schematic of a loop–loop interaction. (b) Secondary structure of the ColE1 loop–loop interactions. (c) NMR structure of the complex*
(Reprinted for Ref. 74. © (1998), with permission from Elsevier)

Because of their ability to serve as both hydrogen-bond donors and acceptors, 2′-OH groups can interdigitate at the interface between two RNA duplexes, resulting in **ribose zippers**. This essential mode of RNA packing has been observed in the crystal structures of many large RNA molecules, and it is a core motif within the hepatitis delta ribozyme active site (Figure 7.10).[13]

**7.1.4.1.1 Metal Ions in RNA Tertiary Structure.** Most RNA tertiary structures are stabilized by **metal ions**, particularly $Mg^{2+}$ and $K^{+}$.[14–16] Metal ions can interact with RNA in at least four ways (Figure 7.11).

1. They can non-specifically screen the charge of the polyanionic backbone, thereby reducing repulsion between RNA strands.
2. They can bind to specific RNA sites and, without forming specific interactions, provide local stability to regions of strongly negative electrostatic potential.

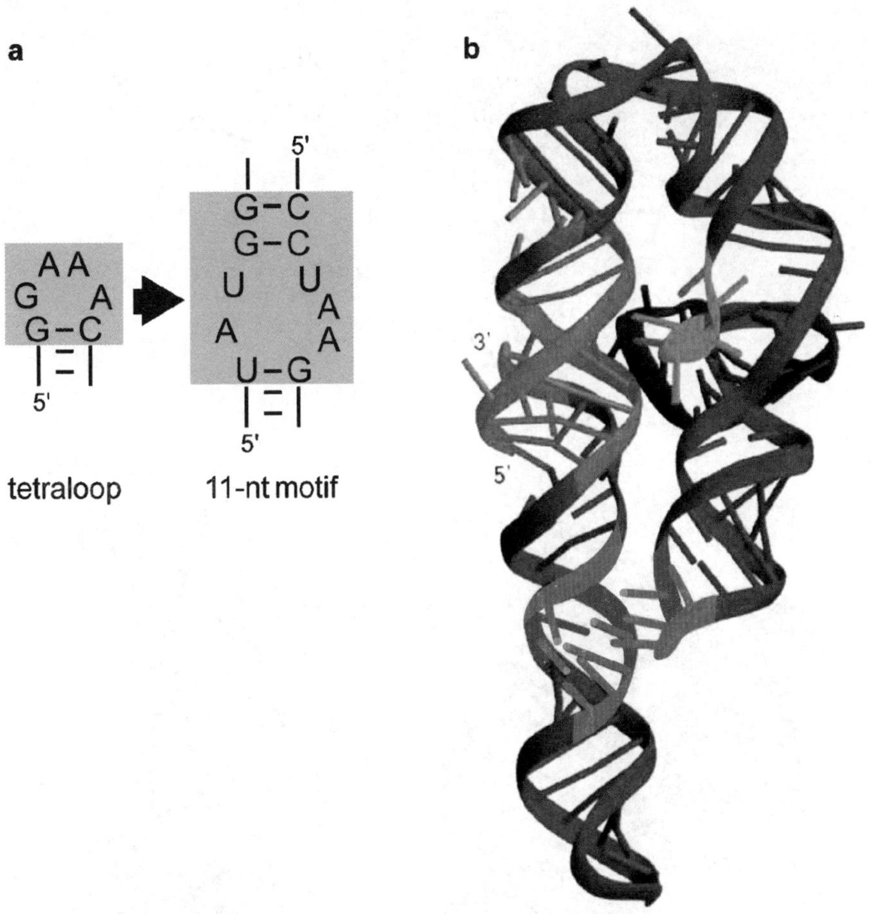


**Figure 7.8** *The tetraloop–receptor interaction in RNA tertiary structure. (a) A secondary structural diagram of the tetraloop–receptor interaction. (b) Ribbon diagram from the crystal structure of the P456 domain of a group I intron,[7] with the tetraloop–receptor interaction highlighted in red*

3. They can interact with RNA through "outer-sphere" contacts that are mediated by coordinated water molecules (Figure 7.11a).
4. They can coordinate directly with RNA functional groups, particularly phosphate oxygens and base heteroatoms (Figure 7.11b).

Because all four of these mechanisms are operative in most large RNA structures, it is often hard to define the role of a particular metal ion. However, the locations of metal ions can be identified through crystallography, NMR, or by biochemical studies involving metal-ion replacement, such as $Mg^{2+}$ by ions of the lanthanide Terbium(III).[17]

A variety of RNA substructures serve as metal-ion binding sites. **A-platforms** are created by adjacent A bases that lie side-by-side rather than stacking on one another (Figure 7.12a). The resultant flat plane readily stacks on other nucleotides and results in a motif that is stabilized by direct interactions with a potassium ion (Figure 7.12b).[18,19]

*7.1.4.2 RNA Folding Pathways.* How RNA molecules go from an unfolded to a folded state is called the **RNA folding** problem.[20] It is remarkable that an RNA sequence contains sufficient information to

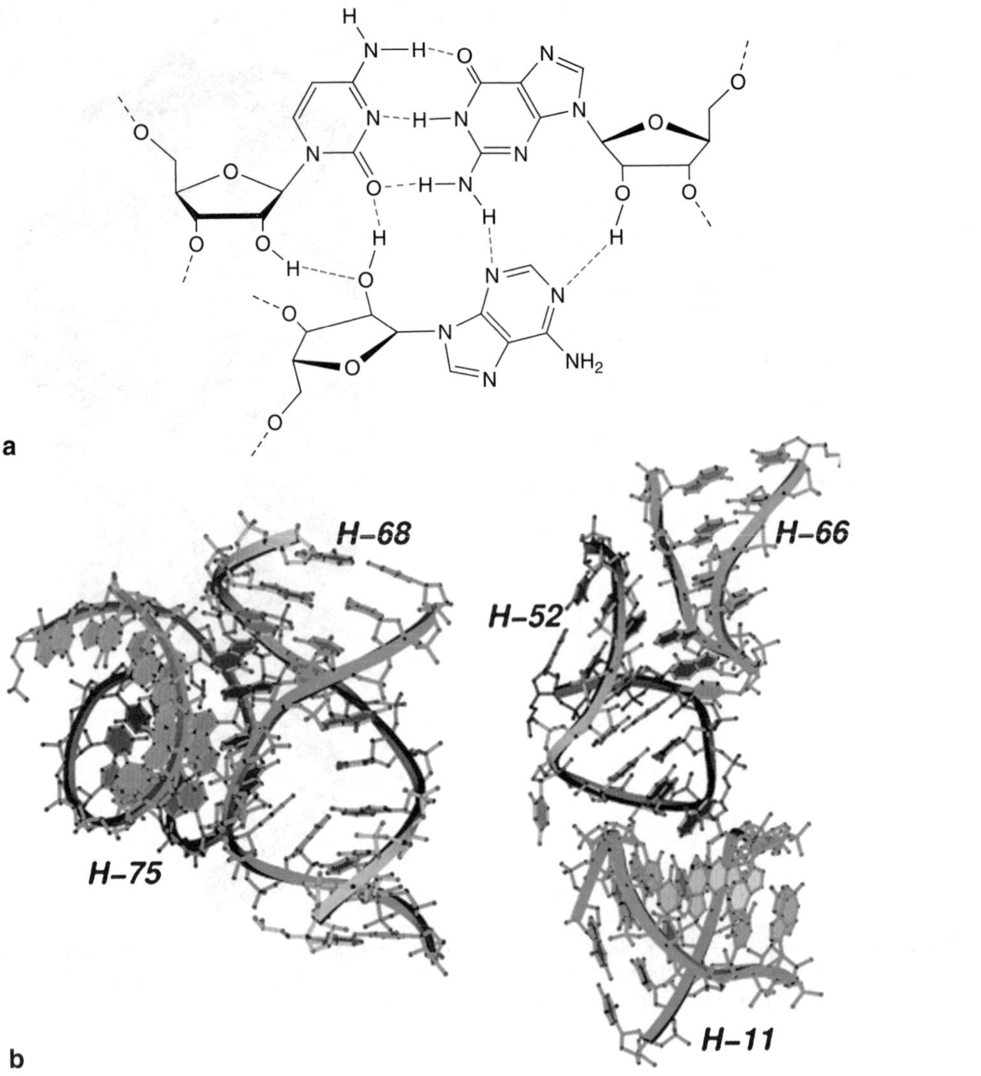


**Figure 7.9** *A-minor motifs and ribosomal RNA packing. (a) A class of A-minor motif, showing adenosine docked into the minor groove of a G–C pair. (b) A-minor interactions (adenosine in red) in the 50S ribosomal subunit*
(Reprinted from Ref. 10. © (2001), with permission from the National Academy of Sciences, USA)

assemble usually into a unique folded structure. However, this process is very sensitive to reaction conditions, such as temperature changes, and requires cations to overcome electrostatic repulsion.

An early event in RNA folding is the formation of secondary structure. This is promoted by cations, including monovalent ions, since duplex formation is stimulated by simple charge screening. Tertiary structure formation involves the tight packing of RNA strands and the formation of cavities. Most RNA molecules begin to form tertiary structure upon interaction with divalent cations, particularly $Mg^{2+}$. RNA molecules tend to fold in a hierarchical manner in which one domain of the molecule precedes the formation of other domains. This is particularly easy to observe in cases where the folding intermediate is stable (as in group I intron folding), but it can still occur in cases where the intermediate is transient. In some

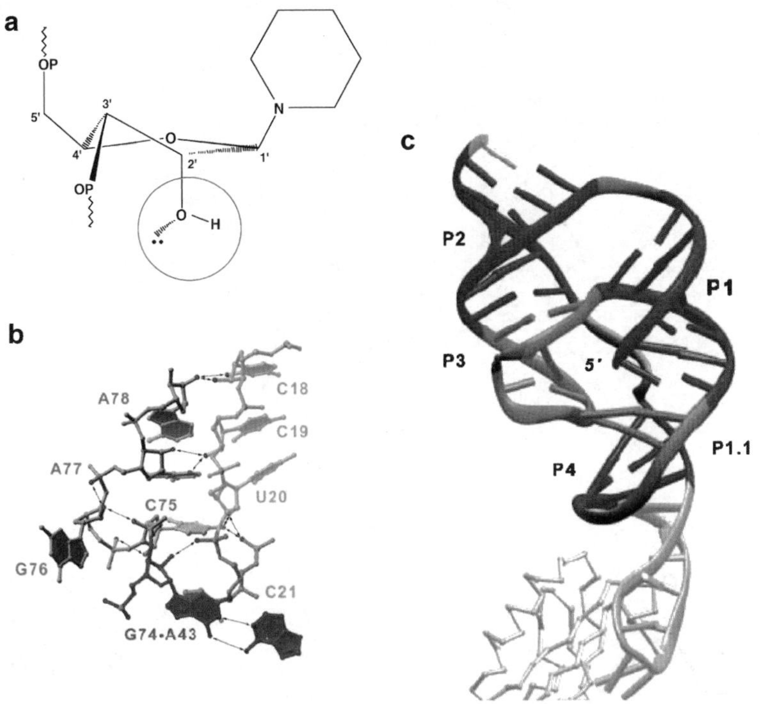


**Figure 7.10** *The ribose-zipper motif in RNA packing. (a) The 2′-OH group is bifunctional, serving as both H-bond donor and acceptor. (b) and (c) Interdigitation of ribose residues in the core of the hepatitis delta ribozyme*
(Reprinted from Ref. 13. © (1998), with permission from Macmillan Publications Ltd)

cases, folding intermediates contribute to the formation of the native folded structure. In other cases, folding intermediates are inhibitory "kinetic traps," or misfolded intermediates, that delay formation of the native state. **Misfolding** by RNA is considered to be a more serious problem than in proteins because RNA secondary structural elements are often highly stable. As a result, many RNA molecules traverse a "rough folding landscape," in which misfolded states are prevalent, and in some cases as stable as the native state. Thus when RNA is handled *in vitro*, it often needs to be refolded by careful denaturation and renaturation procedures. *In vivo*, **RNA chaperone proteins** are likely to assist in proper RNA folding.[21]

*7.1.4.3 The Architecture of RNA Tertiary Structures.* Once formed, RNA tertiary structures are often stable, globular assemblies, which can be visualized by crystallography or through the creation of molecular models that are based on biochemically obtained distance constraints. NMR has also been useful in the elucidation of smaller RNA tertiary structures.

The first high-resolution crystal structure of a nucleic acid molecule was obtained for tRNA in 1972.[22,23] Its secondary structure resembles a cloverleaf (Figure 7.6a), but in its folded form it is L-shaped (Figure 7.6b). The tRNA structure revealed several examples of non-Watson–Crick base pairs, base-triples, and 2′-OH tertiary interactions.

Twenty years later, substantial advances in RNA crystallography resulted in structure determination of the hammerhead ribozyme[24] and a large stable subdomain (P456) within a group I intron RNA.[9] The **P456 structure** demonstrated that a large RNA (160 nucleotides in this case) could be folded *in vitro*, crystallized, and its structure solved by use of conventional methods (Figure 7.8). Since then, the crystal structures of many other large RNAs and ribozymes have been solved (Figures 7.9 and 7.20),[13,25–27] resulting in a wealth of RNA tertiary-structure information. This success has been capped by high-resolution crystal

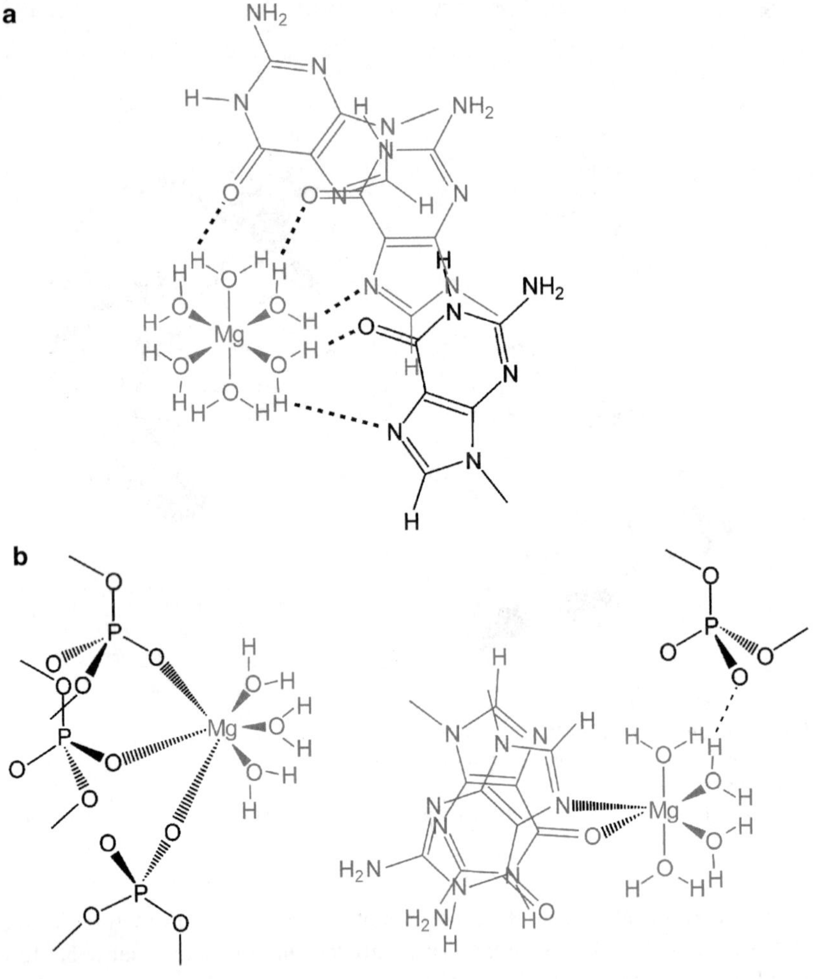

**Figure 7.11** *Metal ions in RNA tertiary structure. Direct interactions between RNA and divalent cations can take two forms: (a) Outer-sphere and (b) Inner-sphere*

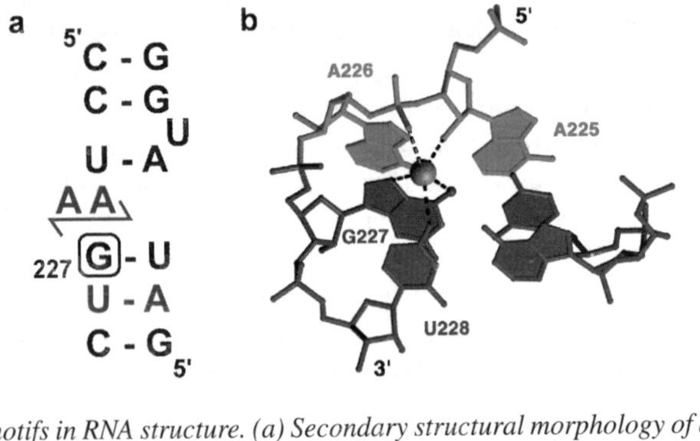


**Figure 7.12** *A-platform motifs in RNA structure. (a) Secondary structural morphology of an A-platform. (b) Tertiary structure of the platform. The specifically bound $K^+$ ion is shown as a gold sphere*
(Reprinted from Ref. 18. © (1998), with permission from Macmillan Publications Ltd)

structures of the complete 70S ribosome,[28] a complex of many proteins and 3 RNAs, as well as the 30S and 50S ribosomal subunits (Section 7.3.3).[5,29]

## 7.2 RNA PROCESSING AND MODIFICATION

RNA molecules are transcribed from DNA by **RNA polymerase** enzymes, which initiate transcription after binding to specific DNA **promoter** sequences. By use of the adjacent DNA antisense strand as a template, polymerases synthesize RNA **transcripts** from the 5′- to the 3′-terminus (Section 6.6). Each organism has at least one, and often several, distinct RNA polymerases. After it has been transcribed, an RNA molecule is typically not functional until it has undergone **RNA processing**.[30] The attachment of modifications, the removal of long sequences, and in some cases changes to the base sequence itself, are often required before an RNA can be transported to the proper cellular compartment and carry out its function.

There are many different types of RNA in the cell. For example, messenger RNA (**mRNA**) encodes protein sequences, transfer RNA (**tRNA**) acts at the ribosome to decode mRNA information to specify particular amino acids, ribosomal RNAs (**rRNA**) assemble into the ribosome where protein is manufactured, small nuclear RNAs (**snRNAs**) tailor other RNAs to the proper size, and microRNAs (**miRNAs**) are tiny sequences that bind and regulate the function of other RNAs. The cellular localization and biological function of RNA molecules dictate the type of processing that they undergo. For the sake of simplicity, the processing of eukaryotic mRNA will be our major focus.

### 7.2.1 Protecting and Targeting the Transcript: Capping and Polyadenylation

As the 5′-terminus of a new mRNA emerges from a eukaryotic RNA polymerase, it is immediately protected by attachment of a **trimethyl G cap** (Figure 7.13). During this process, the terminal nucleotide of the transcript is joined by a 5′–5′ linkage to a guanosine triphosphate. The guanosine itself is methylated, and often hypermethylated, to distinguish it from other guanosines in the cell. As a result of **capping**, the nascent transcript is protected from 5′-exonucleases that would otherwise digest the RNA as it emerged.

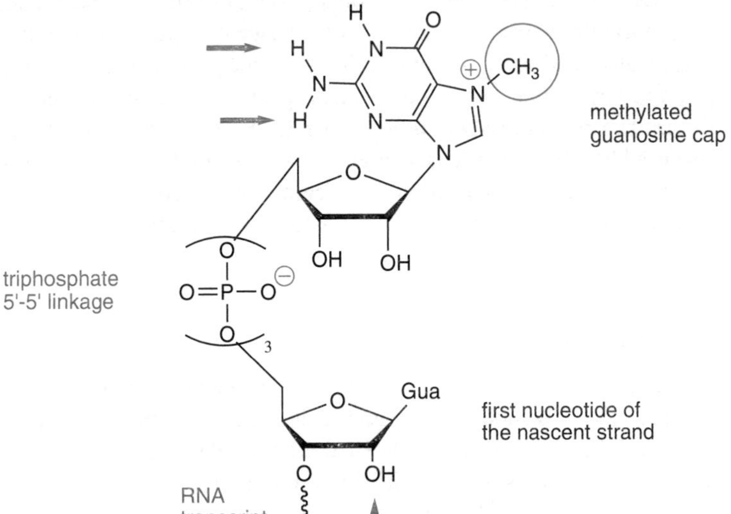


**Figure 7.13** *The cap structure of eukaryotic mRNA molecules. A red circle indicates the site of methylation on N-7 of guanosine, while arrows indicate common sites of additional methylation. Note the unusual 5′-5′ triphosphate linkage that joins the two guanosine nucleotides of the cap*

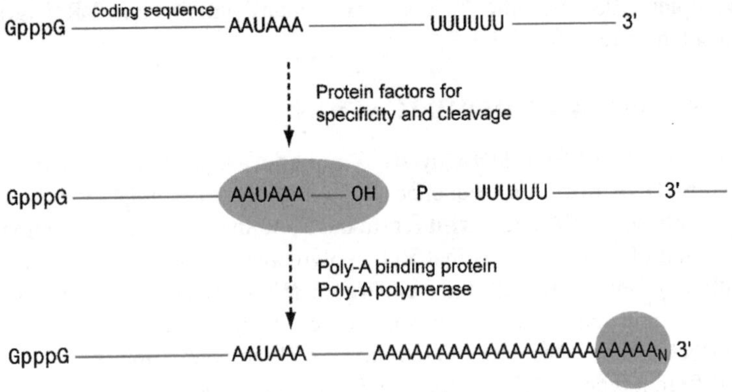


**Figure 7.14** *Polyadenylation. Red lines indicate continuous mRNA sequence. Pink shapes represent poly-A polymerase and its associated proteins*

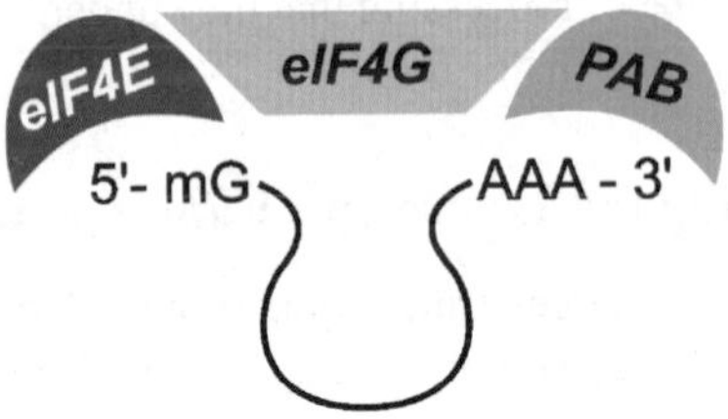


**Figure 7.15** *The cyclization of eukaryotic mRNAs. Because factor eIF4G acts as a bridging molecule (grey), the 5′-end (bound to the eukaryotic initiation factor eIF4E, red) and the 3′-end (bound to poly-A binding protein, PAB, pink) are effectively connected in many eukaryotic messages (dark black line)*

Similarly, once an mRNA transcript has been completely synthesized, its 3′ terminus must be protected through the process of **polyadenylation**[31] (Figure 7.14). Nascent transcripts contain a polyadenylation signal sequence that is located near the 3′-terminus. This sequence is bound by a set of specificity factors and cleaved through a mechanism that remains unclear. A specialized enzyme called **poly-A polymerase** then extends the 3′-terminus by successive addition of adenosine residues, resulting in a **poly-A tail**. Once this tail has been added, the mRNA is recognized by export factors and transported across the nuclear membrane to the cytoplasm. Many mRNA molecules are subsequently "circularized" by proteins that bridge the 5′-cap and poly-A tail, and this plays a role in subsequent translation by the ribosome (Figure 7.15).[32]

### 7.2.2 Splicing and Trimming the RNA

In addition to protection of their termini, many RNA molecules undergo additional processing events. To understand these, it is helpful to consider the schematic diagram of a **eukaryotic pre-mRNA**, which is defined as a new RNA transcript that has not yet been altered in sequence. A typical pre-mRNA contains an abundance of extra sequence that is not translated into protein (Figure 7.16, see also Figure 6.2). For example, most eukaryotic mRNA molecules contain long sequences at each terminus. These **untranslated regions (UTRs)** do not encode protein and they fold into specialized structures that help regulate translation and other functions of the message. The mRNA between the UTRs is divided into short segments of coding RNA (**exons**), which contain sequences that will ultimately encode protein and which are separated by long stretches of "junk" RNA (**introns**) which do not encode protein, but which can have other

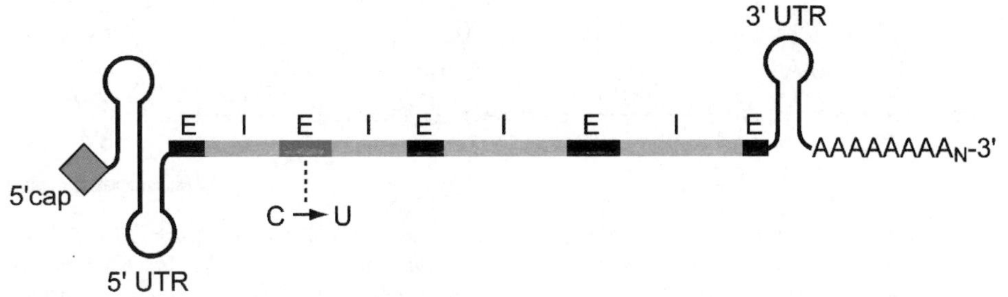


**Figure 7.16** *The structure of a eukaryotic precursor mRNA. The untranslated regions (UTRs) are shown as stem-loop structures and the 5′-cap is a diamond. Exons are black, introns are pink, and an edited exon (undergoing a C→U transversion) is shown in red. The poly-A tail is indicated at the 3′-end*

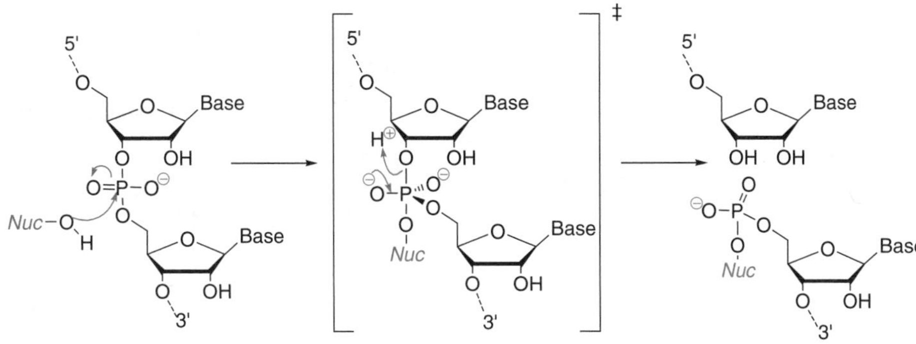


**Figure 7.17** *Reaction mechanism for the spliceosome, group II introns, group I introns, and RNase P. During spliceosomal and group II intron splicing, the nucleophile (Nuc-OH) is the 2′-OH of a specific bulged adenosine within the spliced intron. For group I intron splicing, the nucleophile is the 3′-OH of a guanosine moiety. For RNase P, and for alternative reactions by group II introns, the nucleophile is water*

functions. Remarkably, in higher eukaryotes the exons are very short (~200 nt) while the introns can be very long (>1000 nucleotides) (see Section 6.1).

*7.2.2.1 Nuclear pre-mRNA Splicing.* Before an mRNA can be functional and provide a proper template, obviously the introns must be removed and the exons stitched together so that the processed mRNA contains a coherent coding sequence that is specific for a particular protein. Thus, mature mRNA is much shorter than the precursor transcript. This **pre-mRNA splicing** is carried out by the **spliceosome**, which is a dynamic ribonucleoprotein (RNP) machine that specifically recognizes the sequences at exon/intron boundaries (**splice sites**), and carries out the chemical reactions for cutting and pasting the exons together.[33] The spliceosome is composed of five highly conserved RNA molecules (the small nuclear RNAs or snRNAs U1, U2, U4, U5, and U6) and a host of specialized proteins that bind RNA or remodel it through the consumption of ATP. Splicing proceeds through two sequential *trans*-esterification reactions, each of which involves an $S_N2$ reaction at phosphorus (Figure 7.17). The nucleophile during the first step of splicing is the 2′-hydroxyl group of an adenosine within the intron (the branch-point A), which attacks a specific sequence at the 5′-splice site and releases a 3′-hydroxyl leaving group. During the second step of splicing, this 3′-OH group attacks the 3′-splice site, thereby ligating the exons and releasing an intron lariat molecule (Figure 7.18).

Most pre-mRNA transcripts from mammals contain numerous exons, with ten being an approximate average (Section 6.1). Depending on the tissue and developmental stage of the organism, these exons can

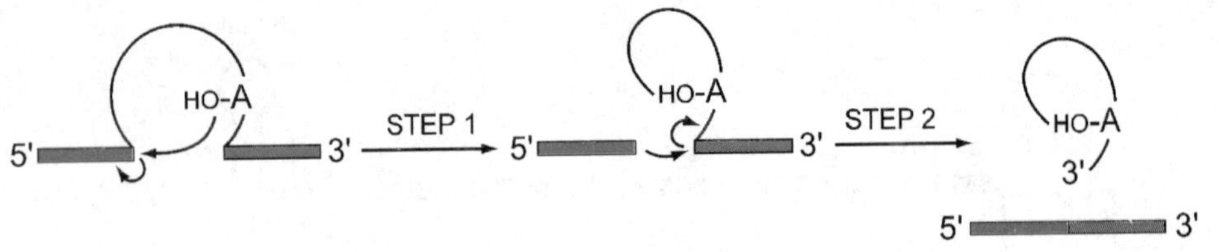


**Figure 7.18** *The two steps of RNA splicing catalyzed by the spliceosome and group II introns. The 5′-exon is shown in red, the 3′-exon is shown in grey, and the nucleophilic adenosine is indicated. The intron is shown as a black line. Note that the lariat structure is connected by 2′–3′–5′ linkages to the adenosine*

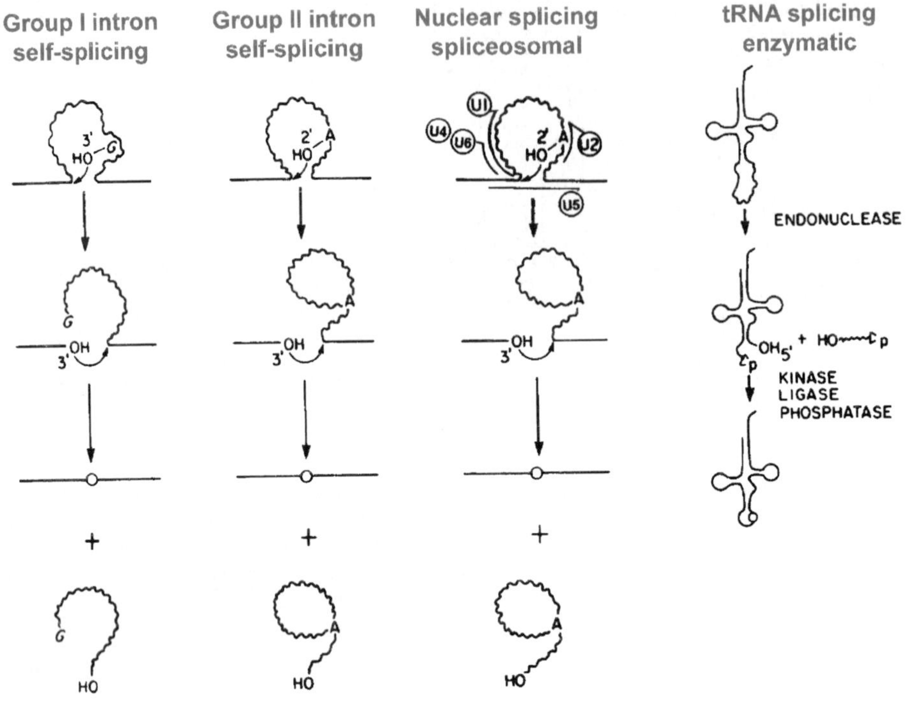


**Figure 7.19** *The four classes of RNA splicing*

either be stitched together sequentially, or certain exons can be skipped and left out of the mature message (see Figure 6.4). As a result of this process, called **alternative splicing**, a single pre-mRNA gene can generate many different types of proteins, thereby providing a form of combinatorial diversity that is not genetically encoded.[34]

*7.2.2.2 Self-Splicing and Other Splicing Pathways.* While most eukaryotic splicing is carried out by the spliceosome, there are specialized genes and introns that are spliced through different mechanisms (Figure 7.19). For example, certain tRNA genes contain introns in the anticodon, and these are removed by the sequential action of protein endonucleases and ligases. But perhaps the most remarkable pathways for splicing involve introns that are inherently reactive, and which can splice themselves out of flanking exons without the aid of spliceosomal machinery. These **self-splicing introns** fall into two categories, the **group I and group II introns**.[35,36] The discovery of these autocatalytic RNA molecules, or **ribozymes**, along with other families of catalytic RNA molecules, was one of the most exciting developments in 20th century biochemistry (Section 7.6.2).

Group I and group II introns are common in lower eukaryotes, such as fungi and yeast, although the latter is abundant in plants as well. Remarkably, group I and group II introns have also been found in bacteria, and constitute the only type of intron found so far among prokaryotic organisms. Group I introns (together with RNase P) were the very first autocatalytic RNA molecules to be discovered.[37] During experiments on the splicing of a ribosomal gene in the protozoan *Tetrahymena thermophila*, Thomas R. Cech and colleagues found that the rRNA gene repeatedly spliced in control reactions to which enzymatic extracts had not been added. In fact, the only cofactors required for group I intron splicing were found to be the nucleotide guanosine and $Mg^{2+}$ ions. Subsequent mechanistic study has revealed that group I introns fold into an elaborate three-dimensional structure that positions both splice sites and binds a guanosine molecule for use as a nucleophile during the first step of splicing (Figure 7.20).[25] As in spliceosomal processing, group I intron splicing is the result of two sequential $S_N2$ *trans*-esterifications that ultimately release intron and ligate the exons (Figures 7.17 and 7.18). Both the folding of the molecule and subsequent catalysis require $Mg^{2+}$ ions, which has been shown to play an important and general role in the **tertiary folding** of RNA molecules.

Group II introns are highly abundant in the organellar genes of plants, fungi, and yeast (Figure 7.21). They have been subjects of particular interest because their mechanism of splicing is so closely related to that of the spliceosome. Like the latter, group II introns utilize a 2′-hydroxyl group of a bulged adenosine

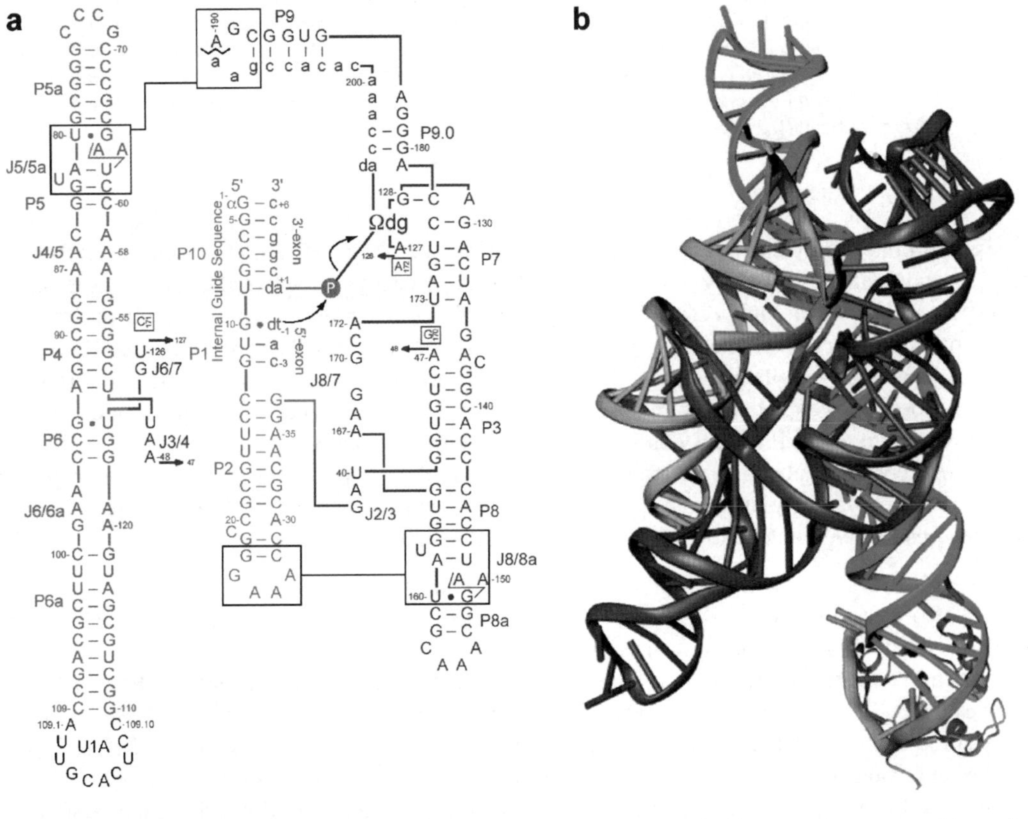


**Figure 7.20** *The structure of a group I intron. (a) The secondary structure and (b) a crystal structure of the tertiary structure for the Azoarchus group I intron. Corresponding colours indicate specific domains of the intron. In this structure, the intron and both exons are intact, revealing all active-site components and their relative locations*
(Reprinted from Ref. 25. © (2004), with permission from Macmillan Publications Ltd)

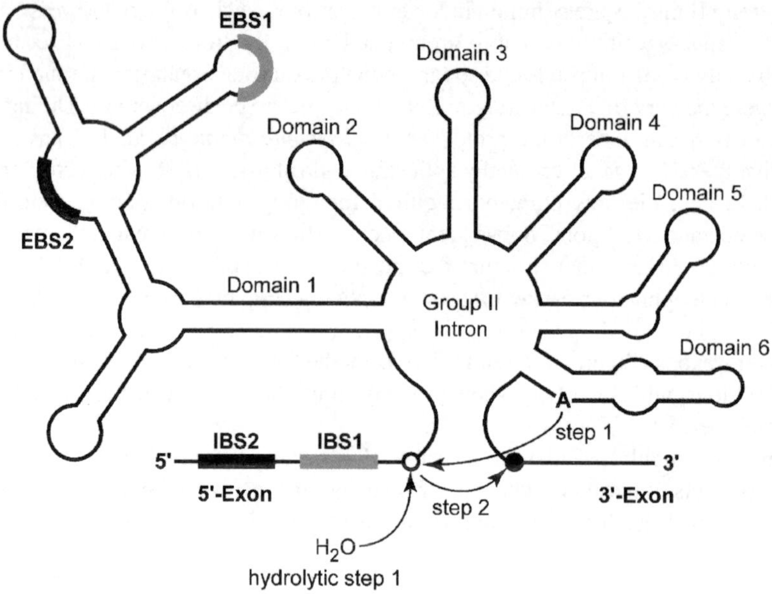


**Figure 7.21** *Schematic of the secondary structure for a group II intron. The EBS1-IBS1 and 2 pairings (thick grey and black lines, respectively) and domain numberings are shown. The 5′-exon is recognized by pairings between EBS1 and IBS1 (grey pairs with grey), and EBS2 and IBS2 (black pairs with black). Step 1 can proceed via attack of the bulged A residue in Domain 6 or through attack of water (hydrolytic step 1). In the second step of splicing, the liberated 5′-exon is ligated to the 3′-exon, thus releasing a lariat intron (see Figure 7.18)*

as the nucleophile during the first step of splicing.[36,38] However, they do not require spliceosomal components or protein enzymes to carry out the chemical steps of catalysis. Unlike the spliceosome, group II introns can also splice through a second pathway in which water serves as the nucleophile during the first step of splicing **(hydrolysis)**, thereby releasing a linear intron (Figure 7.21). Excised group II introns can behave as **infectious mobile elements**, which **reverse-splice** into DNA and thereby spread throughout a genome (or between genomes) (Figure 7.22). Indeed, it has been proposed that all eukaryotic introns may have derived from group II introns that proliferated, degenerated, and were then taken over by the evolution of a spliceosomal apparatus. Group II introns therefore represent a distinctive class of **transposon**, in which ribozyme catalysis plays a role in the mechanism of genetic mobility.[39] Notably, group I introns are also **transposable elements**, although their mechanism for mobility differs (*cf.* Section 6.8.3).

*7.2.2.3 Excision of Terminal Sequences: RNase P and RNase III.* In addition to removal of a sequence from the middle of a transcript, there are also mechanisms for removal of terminal RNA sequences. Many RNA molecules, such as pre-tRNA, have terminal leader sequences that must be excised. The 5′-terminal leader of pre-tRNA is removed by an enzyme called ribonuclease P (**RNase P**), which catalyzes the $S_N2$ attack of a water molecule on the scissile phosphate.[40] In bacteria, RNase P consists of a RNP complex in which the RNA component is sufficient for catalysis. In eukaryotes, RNase P is more complex and requires additional protein components for reactivity. In fact, RNase P is the only "ribozyme" in nature that functions as a true enzyme with multiple turnover in the cell. Most other catalytic RNA molecules are designed to undergo one round of self-cleavage or transposition. There are other enzymes for removal of terminal leader sequences, such as the ribonuclease III family (**RNase III**). Enzymes in this family catalyze a broad spectrum of endonucleolytic reactions on RNA, including the "dicing" of RNA into interfering RNAs (siRNAs) and miRNAs (Section 5.7.2).[41]

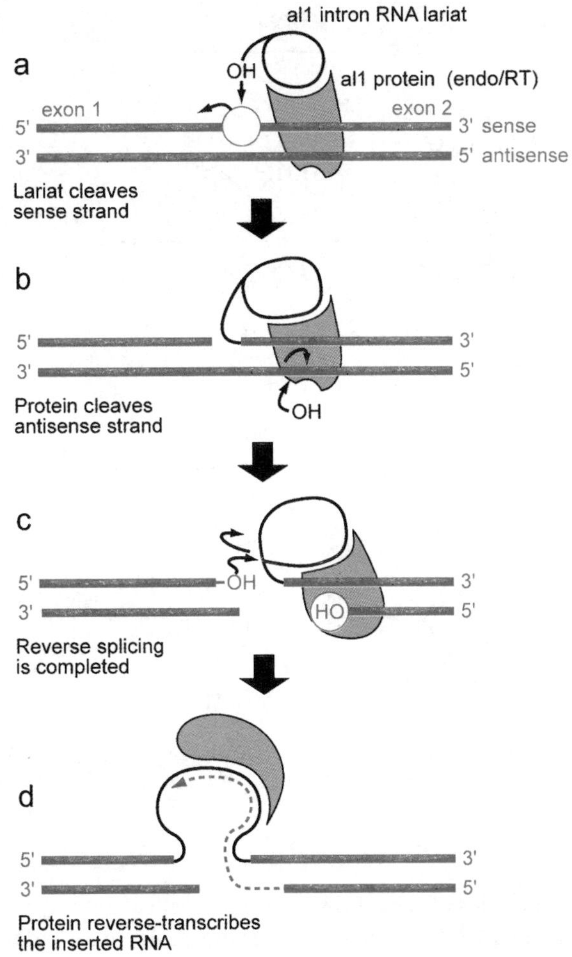


**Figure 7.22** *The mechanism of group II intron insertion and mobility into duplex DNA. Mobility is catalyzed by a ribonucleoprotein particle that contains a lariat group II intron RNA (black line), which is bound to a protein cofactor (grey) that is encoded by the intron itself. Both protein and RNA contain active sites for catalysis of the various steps of intron insertion. (a) After recognition of its target site, the 3′-OH group of the lariat RNA attacks the sense strand of DNA in a reverse-splicing reaction that is catalyzed by the intron. (b) An endonuclease motif within the protein (grey) then attacks the antisense strand. (c) The second step of reverse splicing. (d) Concomitant with or after the second splicing step, a reverse-transcriptase motif makes a DNA copy of the inserted RNA, by use of the cut antisense strand as a primer*

### 7.2.3 Editing the Sequence of RNA

In addition to RNA splicing, there is a second pathway by which RNA is transformed into a different sequence than originally encoded by the parent DNA. Many organisms employ diverse mechanisms for **RNA editing**, during which the identity of individual bases is altered. This can change amino acid identity at a specific position or introduce a new stop codon, thereby resulting in major changes in gene expression.

*7.2.3.1 Transversional Editing.* The mRNA from humans and other higher eukaryotes commonly undergoes **transversional editing**, which changes the identity of individual bases. The most common base changes are **C→U** and **A→I**.[42,43] In the latter case, the inosine residue is read as a guanosine by the translational apparatus and by polymerases that are used to amplify RNA gene products. Transversional editing

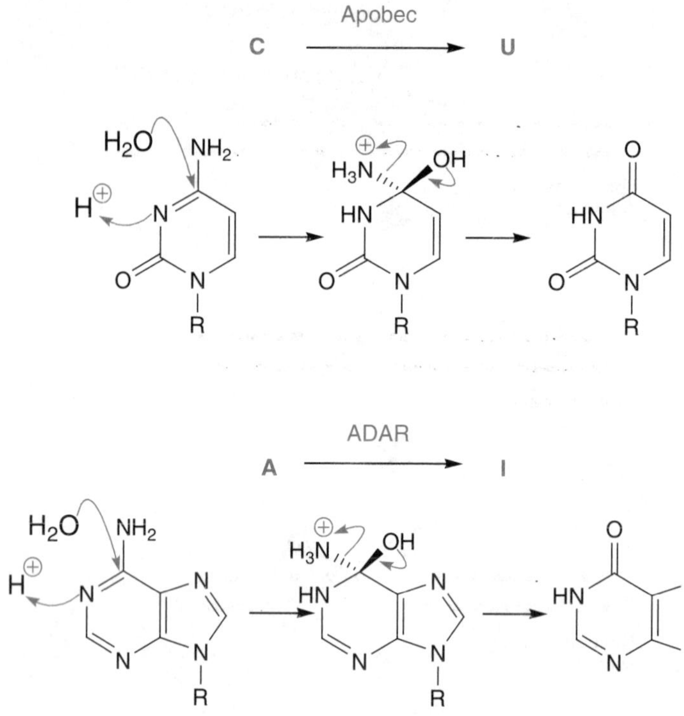


**Figure 7.23** *The deaminase reactions that are catalyzed by RNA editing enzymes Apobec and ADAR. The Apobec enzymes catalyze C→U transversions at specific sites and the ADAR enzymes catalyze A→I (inosine) transversions*

is performed by specialized families of **deaminase enzymes** that catalyze hydrolysis of amino groups on cytidine and adenosine (Figure 7.23). On a much slower timescale, these same reactions occur spontaneously and nonspecifically at A and C, which is one reason why it is challenging to determine the original sequence of DNA or RNA samples that have been extracted from old biological material (>100 years old). However, the deaminases involved in RNA editing have evolved high specificity for their target sequences and, together with additional proteins, they form efficient editing complexes that modify only discrete regions of certain RNA messages.

One example of transversional editing gives rise to variants of the protein **apolipoprotein B**. One form of apolipoprotein B (ApoB-48) is half as long as another common variant (ApoB-100), and the balance between these proteins in different tissues plays a major role in human cardiovascular health. The ApoB-48 variant results from a stop codon in the ApoB-mRNA. However, this stop codon is not DNA-encoded, but results from an RNA editing event, whereby a single cytidine is converted to a uracil by a specialized deaminase protein that has been named **Apobec-1**.[43] The conversion of a glutamine CAA codon into the UAA stop codon results in ApoB-48, while unedited transcripts produce ApoB-100. As in subsequent discoveries of RNA editing, alteration in amino acid identity was only discerned when protein sequences (or cDNA sequences, since they are derived from edited RNAs) were compared to the genomic sequence of the organism. Since the discovery of ApoB editing, numerous other examples of C→U editing have been reported. There is also evidence that Apobec-like enzymes edit the DNA of genes involved in the immune system.[43]

A second class of transversional editing is catalyzed by the **ADAR family** of **adenosine deaminases**.[42] This activity was first noted during biochemical studies on an unusual enzyme that was found to bind to duplex RNA *in vitro* and to convert certain adenosines into inosines. A biological function for this type of enzyme was identified during studies of the human glutamate receptor gene. In neurons, it was noted that certain glutamate receptor subunit proteins (GluR-B) contain the amino acid arginine at a position where the genomic DNA specifies a glutamine codon. The positively charged arginine residue in this protein is essential for proper calcium transport in neuronal tissues and its incorporation was found to result from a transversional RNA editing event. In the case of GluR-B, a specific glutamine codon (CAG) is converted into an arginine codon (CIG) through an **A→I** editing event that is catalyzed by the enzymes **ADAR1** and **ADAR2** (Figure 7.23). It is now known that A→I editing in human tissues is extensive and its catalyzed by a large family of specific ADAR enzymes.[42]

*7.2.3.2 Insertional and Deletional Editing.* The most radical mRNA editing events occur in the mitochondria of trypanosomatids, which are a group of parasitic unicellular eukaryotes. In these organisms, long stretches of uridine are inserted into mRNA, and small numbers of encoded uridines are also removed.[44] The mitochondrial transcript doubles in length! This insertion–deletion editing is catalyzed by a set of endonucleases and ligases that are targeted by specialized **guide RNA molecules (gRNAs)**, which encode the edited sequence. Whilst not as extensive as trypanosomatid editing, certain mRNAs from plants and slime moulds have also been observed to undergo limited insertion, deletion, and even transversional editing. The discovery of RNA editing in almost every type of organism serves as a cautionary tale in the current age of whole-genome sequencing. Thus knowledge of genomic DNA sequence does not necessarily result in accurate information about the sequence of the RNA and protein products.

### 7.2.4 Modified Nucleotides Increase the Diversity of RNA Functional Groups

*7.2.4.1 Major Base Modifications in Mesophiles and Thermophiles.* The information content of nucleic acids is often further diversified by the post-transcriptional attachment of **modifications**, which range in complexity from a simple methyl group to an entire amino acid or isoprenyl moiety. These modifications, which are particularly common in "working RNAs" such as tRNAs and rRNAs, are attached to RNA by a large family of modifying enzymes that are guided to specific target spots by various mechanisms.

DNA is frequently modified at the C-5-position of cytosine and the N-6 position of adenine. Base modifications provide a signal for gene silencing and, in higher eukaryotes, the cytosines in DNA are often more likely to be methylated than not (Figure 7.24). However, the greatest diversity of modifications is found in RNA molecules that are components of large cellular machines such as the ribosome and the spliceosome. Modifications on both base and sugar moieties are common, particularly in thermophiles and hyperthermophiles, where they are believed to enhance the stability of RNA secondary and tertiary structure.[45] Depending on the functional group, nucleotide modifications can radically alter the chemical properties of an RNA molecule, changing electrostatics, hydration, metal-ion binding, molecular recognition, and even the redox properties (Figure 7.24).

*7.2.4.2 Base Modifications in tRNA and rRNA.* tRNA molecules contain numerous modifications, which are involved in diversifying the genetic code, synthetase recognition, and stabilizing tRNA structure.[46] This is exemplified by the remarkable story of **lysidine**, which is a cytidine that has been post-transcriptionally modified at the C-2 position with a lysine amino acid (Figure 7.24). In *E. coli*, isoleucyl tRNA is only recognized and charged by its cognate synthetase enzyme when a lysidine base is present in the tRNA anticodon loop. If lysidine is replaced by cytidine, the tRNA is mis-charged with methionine.[47] Thus, RNA modifications often blur the distinction between nucleic acid and protein, and they contribute in fundamental ways to basic metabolism.

rRNA is heavily modified, particularly in regions that are conserved and critical for function (such as the peptidyl transferase site). In mesophilic eukaryotic ribosomes, certain types of modification are particularly important. For example, a vertebrate ribosome is likely to contain ~100 **pseudouridine** residues (Figure 7.24), which appear to be the dominant form of base modification in mesophilic organisms. Backbone modifications are also observed, particularly in the form of abundant **2′-*O*-methyl groups** (also ~100 in vertebrate ribosomes). Modifications are placed at specific positions through a remarkable process that

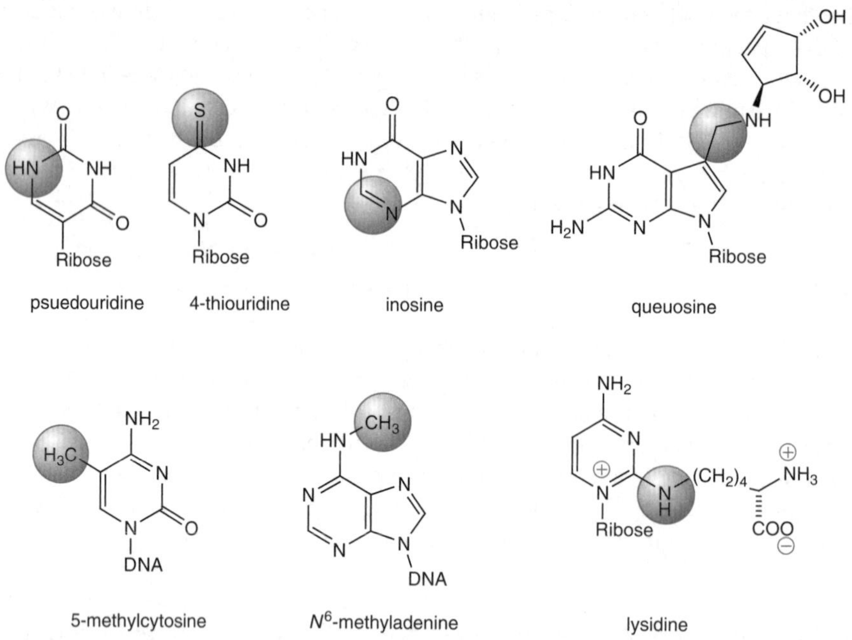


**Figure 7.24** *Common modified bases in RNA and DNA. Hundreds of different base modifications have been observed. A selection of the most common is shown here. The position of modification is highlighted with a pink circle*

involves annealing of short gRNAs (which are actually encoded by specialized introns) that direct the positioning of modifying enzymes.[48]

*7.2.4.3 A Critical Base Modification for Spliceosomal Function.* The U2 snRNA is an integral part of the U2 snRNP that is required for pre-mRNA splicing. A short sequence in the U2 snRNA forms base pairs around the branch-site adenosine on pre-mRNA, thereby specifying the 2′-hydroxyl group of this adenosine as the unique nucleophile for splicing. It is known that a highly conserved pseudouridine pairs immediately next to the intron branch-site and that it is important for pre-mRNA splicing. NMR studies of the U2-mRNA pairing have shown that the pseudouridine causes the branch-site adenosine to flip out of the duplex and to present its 2′-OH group in the proper orientation for nucleophilic attack.[49] When a normal uridine is present at the same position, the branch-site adenosine stacks into the U2-mRNA helix. Thus, a single, subtle modification on U2 snRNA changes the conformational preferences of surrounding nucleotides and may facilitate the process of pre-mRNA splicing.

### 7.2.5 RNA Removal and Decay

Just as it is important to transcribe, process, and modify RNA molecules, it is important for an organism to remove RNA molecules when they are no longer useful. Thus, there are highly regulated pathways for the removal of damaged, improperly processed, and even over-abundant RNAs that might interfere with desired levels of gene expression.

*7.2.5.1 Ribonucleases and the Exosome.* Although nucleases are found in all types of cells and compartments, eukaryotes in particular teem with nucleases that police the structural integrity of RNA, remove invading nucleic acids (such as viral RNAs), and help to regulate the proper quantity of

RNA molecules for appropriate gene expression (Section 5.3).[50] Typical nucleases include the **exonucleases**, which bind to unprotected RNA termini and degrade RNA in either the 3′- or 5′-direction (5′ and 3′-exonucleases). There are also **endonucleases**, which cut in the center of an RNA molecule, often leaving an exposed tail for further degradation by the exonucleases. After splicing, eukaryotic introns are released as lariat molecules, which would be resistant to degradation and recycling were it not for a **debranching enzyme**, which is a specialized nuclease that specifically cuts the 2′–5′ linkage that joins the branch-site to the first nucleotide of an intron. The resultant "linearized" intron can then be degraded by standard nucleases.

While some nucleases diffuse freely throughout a cellular compartment, many of them act in a highly regulated manner, as part of macromolecular machines such as **the exosome**. Processed mRNA molecules are specifically degraded through a carefully orchestrated pathway involving decapping of the 5′-end and loss of the 3′-poly A tail that normally protect mRNA molecules from degradation. Within the exosome, a complex of 3′→5′ exonucleases degrade these unprotected mRNAs. This and other exonucleolytic machinery also degrade mRNAs that have been improperly capped, adenylated or exported, thus providing a form of mRNA quality control.[50]

*7.2.5.2 Nonsense-Mediated Decay and RNA Quality Control.* Transcription, splicing, and other processes involved in RNA synthesis are highly imperfect. Mistakes in these pathways, together with aberrant transcripts from mutated genes and invading viruses lead to deleterious RNA molecules, which, if unchecked, would result in aberrant protein expression in the cell. A major mechanism for destroying these unwanted messages is **nonsense-mediated decay (NMD)**.[51]This pathway is based upon the fact that aberrant transcripts commonly contain stop codons (or "nonsense codons") at inappropriate positions within the RNA sequence. During the process of NMD, RNAs containing these **premature stop codons** are identified and targeted for rapid degradation by exonucleases. Defects in the complex process of NMD have now been linked to numerous important human disorders.

A second important pathway for **RNA quality control** is enforced by the ADAR family of enzymes, which recognize and target duplex RNA molecules that are either produced endogenously or result from viral infection. ADAR enzymes recognize long RNA duplexes and, as in transversional editing, the paired adenosines are converted into inosines. When employed as a form of quality control, this base transversion can result in transcript destabilization and susceptibility to degradation. In a similar (or perhaps related) process, the **RNA interference (RNAi)** machinery also targets RNA duplexes and marks them for destruction (Section 7.5.3).[42]

## 7.3 RNAs IN THE PROTEIN FACTORY: TRANSLATION

In modern organisms, proteins are the dominant macromolecular building materials for cellular function. However, all proteins are the product of a factory that must receive the encoded instructions, gather the amino acid starting materials, and stitch them together in the proper order. But the ribosome is not a mere assembly line for protein synthesis. It is sensitively regulated to produce the correct quantity of protein, to detect problems that arise during synthesis, and to rapidly dispose of defective products. Despite the diversity of life, all organisms utilize similar ribosomal factories to build proteins and modulate the required levels of gene expression.

### 7.3.1 Messenger RNA and the Genetic Code

For protein synthesis, the encoded instructions are contained in **messenger RNA (mRNA)**, which is read like a tape by ribosomes. An mRNA sequence is translated into protein through the **genetic code**, which has the same basic format for all forms of life on earth (Figure 7.25). The code consists of nucleotide triplets (**codons**) that specify the identity and sequential position of amino acids in a protein.[46,52] For example, the sequence 5′-AAA-3′ codes for the amino acid lysine. Each amino acid (and the tRNA to which it is appended) is specified by several different codons (**synonyms**), which differ primarily in the identity of the third position. For example, CCU, CCC, CCA, and CCG all encode the amino acid proline. The ribosome

| first nucleotide (5'-end) | middle nucleotide: U | C | A | G | third nucleotide (3'-end) |
|---|---|---|---|---|---|
| U | Phe | Ser | Tyr | Cys | U |
| | Phe | Ser | Tyr | Cys | C |
| | Leu | Ser | STOP | STOP | A |
| | Leu | Ser | STOP | Trp | G |
| C | Leu | Pro | His | Arg | U |
| | Leu | Pro | His | Arg | C |
| | Leu | Pro | Gln | Arg | A |
| | Leu | Pro | Gln | Arg | G |
| A | Ile | Thr | Asn | Ser | U |
| | Ile | Thr | Asn | Ser | C |
| | Ile | Thr | Lys | Arg | A |
| | Met[a] | Thr | Lys | Arg | G |
| G | Val | Ala | Asp | Gly | U |
| | Val | Ala | Asp | Gly | C |
| | Val | Ala | Glu | Gly | A |
| | Val | Ala | Glu | Gly | G |

[a]AUG is also used as a start signal

**Figure 7.25** *The genetic code. The 64 codons are divided into 16 four-codon boxes. The four codons of a codon box differ in their 3′-terminal nucleotide. Red shows where an amino acid is specified by all four codons of a codon box, pink shows where an amino acid is specified by two (or in one case three) of the four codons, and grey shows where an amino acid is specified by a single codon*

carries out protein synthesis by reading the mRNA sequence that lies immediately downstream of the "start codon" (typically AUG). The codons are then read sequentially, from 5′- to 3′ on the mRNA, until the ribosome reaches a "stop codon" (UAA, UAG, or UGA). Each codon is "read," or **decoded**, by the formation of Watson–Crick pairings with the **anticodon** loop of specific tRNAs that carry **cognate amino acids** into the heart of the ribosome. The sequential arrangement of triplet nucleotides is called the **reading frame**, which can be disrupted by certain types of DNA mutations that alter the mRNA coding sequence and produce nonsense proteins. However, certain mRNAs actually encode multiple proteins through the ribosomal decoding of alternate reading frames (**recoding**).[52]

While the genetic code is remarkably universal from bacteria to higher organisms, there are important exceptions, particularly in mitochondria and in certain lower eukaryotes.[46] For example, the leucine codon (CUG) in *Candida* yeasts has been **reassigned** to encode serine, and the UAA and UAG stop codons have been reassigned to glutamine in diverse ciliates and green algae. In bacteria and eukaryotes, UGA (normally a stop) is often placed in a context that permits pairing with the unusual $tRNA^{sec}$ molecule, by which it is used to encode the 21st amino acid, **selenocysteine**.[53] Modified nucleotides in the anticodon of tRNA can lead to exceptions in the genetic code (such as lysidine, see Section 7.2.4), and even to the incorporation of rare amino acids.

While triplet pairings between tRNA and mRNA have remained a remarkably robust format for information transfer in all organisms, the genetic code has now been expanded artificially in an effort to incorporate unnatural amino acids into proteins and, potentially, to generate entire organisms with new properties. Major success in this area has been achieved by using alternative triplet codons in order to specify unnatural tRNAs that carry modified amino acids.[54] In some cases, new forms of base pairing have been used to generate entirely novel codons, and there have been efforts to expand the genetic code from triplet to quadruplet format.

### 7.3.2 Transfer RNA and Aminoacylation

**tRNAs** are commonly referred to as **adapter molecules**, because the ribosome uses them to translate mRNA triplet codons into protein sequence. tRNA architecture is remarkably conserved throughout all kingdoms of life. It is typically organized into a common secondary structure (the **cloverleaf**) that contains various stems and loops that are essential for tRNA function (Figure 7.6). For example, the seven base-pair **acceptor stem** always terminates with the sequence 5′–CCA-3′, which becomes directly attached to an amino acid upon aminoacylation by **synthetase enzymes**. At the other end of the molecule, the **anticodon loop** contains the three nucleotides that pair with corresponding mRNA codons. All regions of the tRNA molecule play important roles in recognition by proteins, decoding of mRNA, ribosome binding, and in formation of the tertiary structure. For example, the tRNA molecule is not a cloverleaf in solution. Under physiological conditions, it adopts an **L-shaped tertiary fold** that has been visualized by X-ray crystallography and biochemical methods (Figure 7.6). The **D-loop** and the **TψC-loop** (also known simply as the **T-loop**) serve as hinges that permit the L-shaped structure to form. The **variable loop** can be expanded with extra nucleotides, thus explaining why tRNA molecules can vary significantly in size (~74–95 nucleotides), without substantial deviation in their secondary or tertiary structures.

In order for tRNA molecules to function as adapters, they not only form codon–anticodon interactions, they must also carry amino acids into the ribosome so that they can be added to the growing peptide chain. Indeed, tRNA molecules are not even admitted into the ribosome or allowed to pair with mRNA unless they bear "cargo" in the form of an amino acid. This is because only **charged**, or **aminoacylated**, tRNAs are bound to the ribosomal helper protein EF-Tu, which is required for tRNA placement within the ribosome. The 3′-terminus of tRNA is attached to an amino acid through an **aminoacylation reaction** that is catalyzed by a **tRNA synthetase** enzyme (Figure 7.26)[55]. Although synthetases help to stimulate the rate of aminoacylation (which is already a facile chemical reaction), their primary role is in specificity. A given

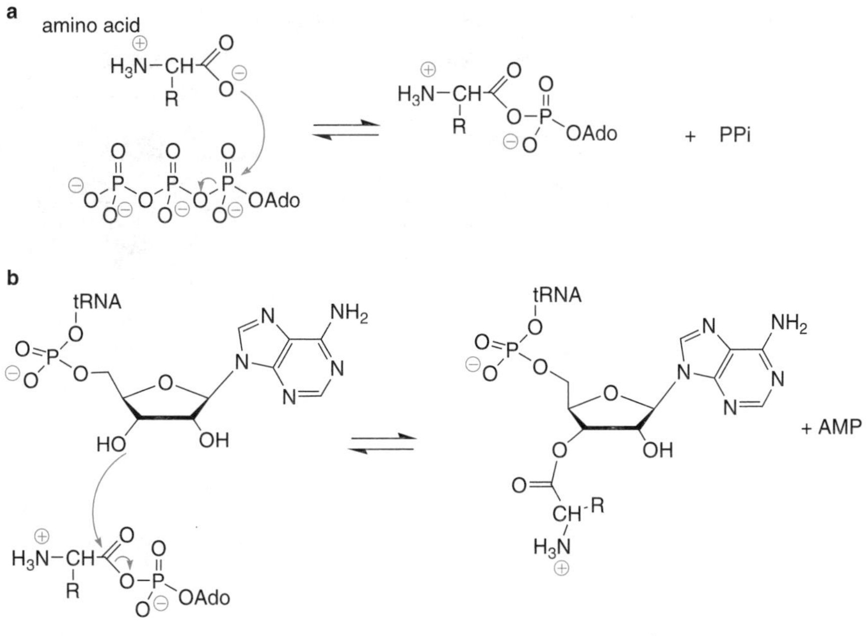


**Figure 7.26** *Steps in aminoacylation of tRNAs catalyzed by aminoacyl-tRNA synthetases. (a) Activation of the amino acid. (b) Transfer of the activated amino acid to the correct tRNA. Note that the reaction of aminoacyl adenylate with tRNA can occur either on the 2′ or 3′-hydroxyl group of the terminal adenosine depending on the particular aminoacyl-tRNA synthetase*

type of tRNA (*e.g.*, tRNA$^{phe}$) is recognized only by a **cognate** synthetase enzyme (*i.e.*, phenylalanine tRNA synthetase), which allows only the proper amino acid to be covalently attached to the 3′-terminus (*i.e.*, phenylalanine). There are two major classes of synthetases, with differing architectures and strategies for tRNA recognition.[55] They are often distinguished by the fact that Class 1 synthetases aminoacylate the 2′-OH of the tRNA acceptor stem, while Class 2 synthetases aminoacylate the 3′-OH. The specificity determinants that govern tRNA–synthethase interactions is a major subject of research, as proper tRNA-synthetase recognition is the foundation of a functional genetic code.

### 7.3.3 Ribosomal RNAs and the Ribosome

The longest, most highly conserved, and most abundant RNA molecules in a cell are the rRNAs. These gigantic transcripts, together with a defined set of ribosomal proteins, assemble to form the two ribosomal subunits (30S and 50S in bacteria; 40S and 60S in eukaryotes) that represent the functional machinery for prokaryotic and eukaryotic ribosomes (Figure 7.27).

The overwhelming majority of ribosomal mass is represented by rRNA, which provides the scaffold for mRNA and tRNA binding, helps translocate them through the ribosomal core, and also catalyzes peptide bond formation. Intrinsic ribosomal proteins (designated S1, S2, *etc.* for proteins of the small subunit and L1, L2, *etc.* for proteins of the large subunit) are particularly important in early stages of subunit assembly, and they contribute subsequently to translation. Additional **translation initiation factors** (*i.e.*, IF1-IF3), **elongation factors** (*i.e.*, EF-Tu and EF-G), and **release factors** (RF) help to facilitate the dynamic process of protein synthesis by the ribosome. Intriguingly, many of these factors (such as EF-Tu, EF-G, and RF) mimic the size, shape, and chemical properties of tRNA molecules, thereby providing important examples of RNA–protein mimicry that is seen in many aspects of RNA biology (Figure 7.28).[56]

The two ribosomal subunits are extremely stable and this has contributed to the recent success in obtaining high-resolution crystal structures of the prokaryotic 30S and 50S RNA–protein particles (**RNPs**) (Figure 7.29, see also the Front Cover). Building on earlier cryo-electron microscopy and biochemical studies

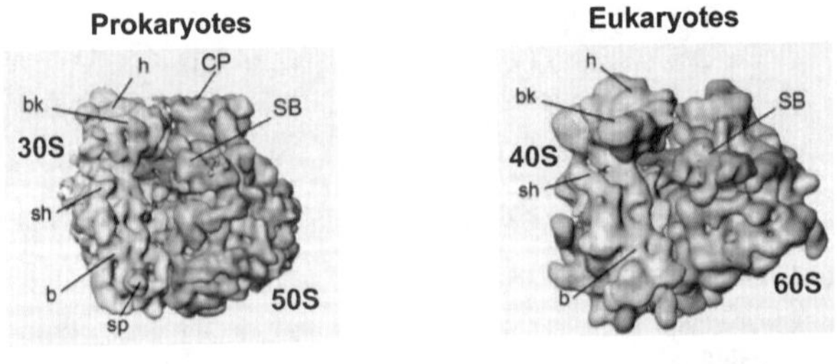


| Parts list | 70S | 80S |
|---|---|---|
| • *intrinsic ribosomal components and their functions:* | 30S subunit (16S rRNA, 21 proteins) initiation<br>50S subunit (23S RNA, 5S RNA, 34 proteins) peptidyl transfer, translocation | 40S subunit (18S rRNA) initiation, translocation<br>60S subunit (5S, 28S, 5.8S rRNAs) peptidyl transfer, translocation |
| • *initiation components* | IF1, IF2, IF3, fMet-tRNAf | eIF4, eIF3, cap binding proteins eIF2-GTP-Met-tRNA |
| • *tRNA shuttling factors* | EF-Tu (brings in the tRNAs) | EF1?, EF1?? (analog of EF-Tu) |
| • *translocation factors* | EF-G (helps translocate tRNAs) | EF2 (analog of EF-G) |
| • *release factors* | RF's (mimic tRNA anticodon) | eRF's |

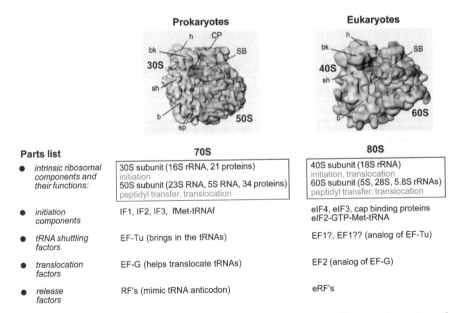

**Figure 7.27** *Two similar designs for the ribosomal factory. Comparison of the prokaryotic and eukaryotic ribosomes and their respective translation cofactors. The structures (above) were obtained by cryo-electron microscopy*
(Reprinted from Ref. 75. © (2001), with permission from Elsevier)

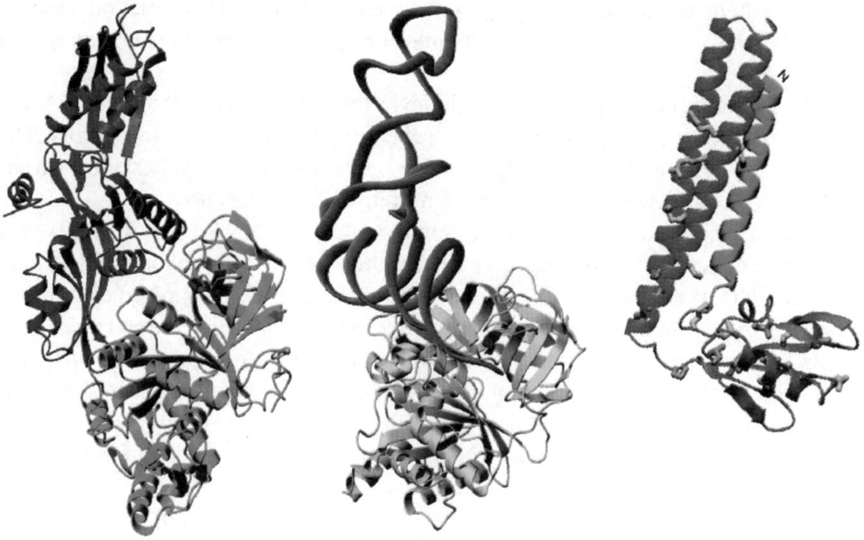


**Figure 7.28** *Proteins mimic the structure and interactions of tRNA. A domain of EF-G (purple) mimics tRNA and binds to the ribosomal A site. The complex between tRNA (red) and EF-Tu (yellow) is shown for comparison. Release factor (RF) closely mimics the L-shape and electrostatic properties of tRNA, even binding the ribosome through an anticodon mimic at the tip of its alpha-helical bundle (blue loops)*
(Reprinted from Ref. 76. © (1999), with permission from AAAS)

**Figure 7.29** *The structures of the ribosomal subunits. Interface views of the 50S (left) and 30S (right) subunits, along with bound tRNAs (yellow, orange, and red, in the A, P, and E site halves of each subunit). The subunits clamp together like the two parts of a shell, enclosing the tRNAs inside (one can visualize the closed form by superimposing the corresponding tRNA molecules)*
(Reprinted from Ref. 28. © (2001), with permission from AAAS)

of the ribosome, the high-resolution structures have provided a wealth of information about RNA tertiary architecture and they have provided new insights into the chemical mechanism of the peptidyl transfer reaction (see below).[5,28,29]

During **translation initiation**, a specialized methionyl tRNA (*N*-formyl methionyl tRNA, or fMet) binds to the AUG initiation codon, together with the 30S subunit and various initiation factors. This **initiation complex** then binds the 50S subunit, thereby forming an active ribosome. Finally, the EF-Tu shuttle protein brings in charged tRNA molecules, and the process of translation commences.

The assembled 70S ribosome contains three binding sites for tRNA: The **A site**, where incoming aminoacylated tRNA molecules are delivered by EF-Tu; the **P-site**, which contains tRNA that is bound to the nascent peptide chain; and the **E-site**, from which uncharged tRNA molecules exit. Similarly, the mRNA traverses and exits through a **tunnel** in the ribosome that helps position the codons during translation. In order to contribute to the peptide chain, each tRNA must **translocate** through the ribosome, visiting the A-site, P-site, and E-sites in turn. Translocation is a dynamic, directional process that is facilitated by motor protein EF-G, which is a tRNA mimic that helps push tRNA from the A-site to the P-site. The process of translocation is best described by the **hybrid states model** (Figure 7.30), which has been confirmed and elaborated by crystallographic analysis of intact, tRNA-bound ribosomes.[57]

The multi-step peptidyl transfer reaction is catalyzed within the 50S subunit. During early stages of the reaction, the incoming amine nucleophile is deprotonated and attacks the activated ester that connects the nascent peptide chain with the P-site tRNA (Figure 7.31). The resultant tetrahedral intermediate then collapses, resulting in deacylated tRNA in the P-site and an expanded peptide chain on the A-site tRNA.[58] High-resolution structural analysis of the 50S subunit has revealed that the active site for peptidyl transfer is strikingly devoid of ribosomal proteins, thereby confirming the long-held view that the ribosome is a ribozyme.[5] While the mechanism of catalysis for peptidyl transfer is still being explored, it is clear that the major functional groups in the active site are nucleotide bases and backbone moieties.

While there are major differences between prokaryotic and eukaryotic translation, much of the ribosomal apparatus is quite similar (Figure 7.27). Bacterial ribosomes jump onto nascent RNA transcripts and begin making protein even as the mRNA is being transcribed. By contrast, eukaryotic ribosomes utilize

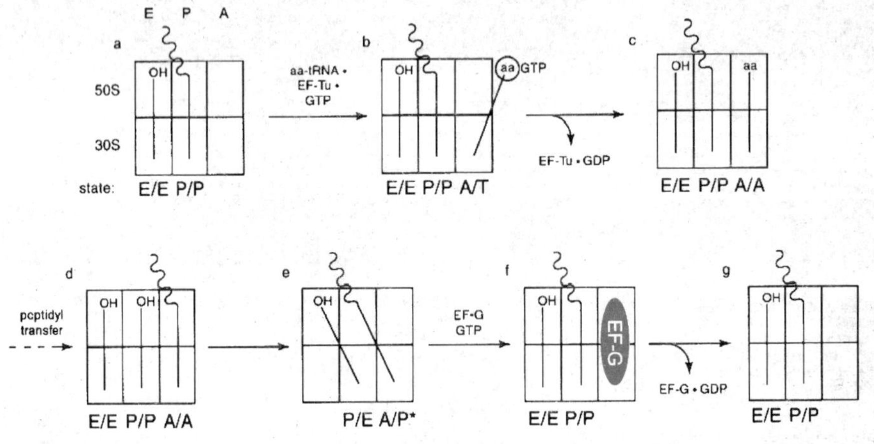


**Figure 7.30** *The hybrid states model for translocation through the ribosome. A charged tRNA (stick with a circle on top) moves sequentially through the A, P, and E sites, respectively, during translocation through the ribosome. According to this model (which has now been confirmed by crystallographic studies), there are "hybrid states" during each stage of translocation, in which a given tRNA is half-way in one site (i.e., the anticodon in the A site) and halfway in another (i.e., the acceptor end in the P site). During translocation, the nascent peptide (a wavy line) is transferred to the amino acid of the incoming tRNA (aa). The tRNAs shift after translocation and release uncharged tRNA (-OH) from the E site*

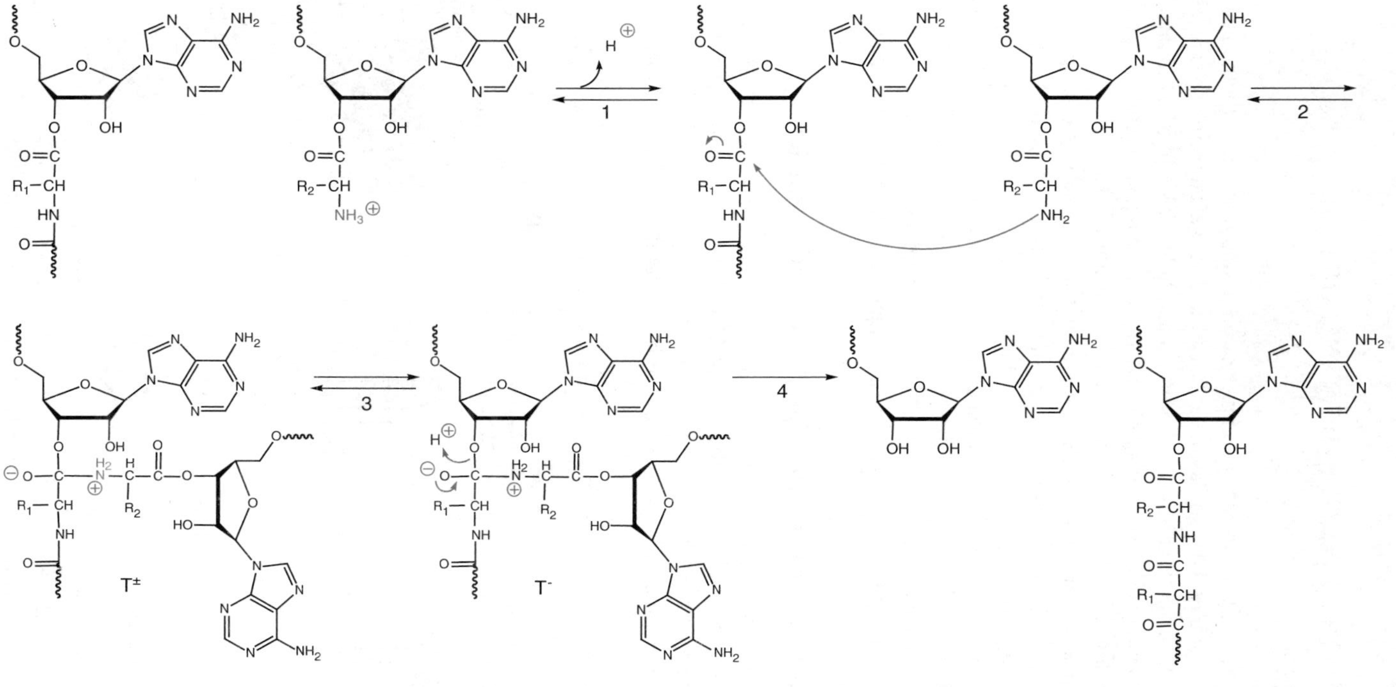

**Figure 7.31** *The peptidyl transfer reaction. The terminal amino group (red) of the incoming tRNA in the A-site is deprotonated and attacks the activated ester at the growing end of the nascent peptide chain (step 2). After resolution of the resulting tetrahedral intermediate (steps 3 and 4), the nascent chain is transferred to the tRNA in the A-site (see Figure 7.30). This peptidyl-tRNA then moves to the P-site so that the process can begin again with a new charged tRNA*

only mature mRNA molecules that have completed the processes of capping, polyadenylation, splicing, editing, and export to the cytoplasm.

## 7.4 RNAs INVOLVED IN EXPORT AND TRANSPORT

### 7.4.1 Transport of RNA

In eukaryotes, RNAs are typically synthesized and processed in one place, usually the nucleus, and then transported to another site, usually the cytoplasm, for function. There are specific sequences or structures in most target RNA molecules that bind transport proteins and help direct an RNA molecule to its functional destination.[59] Like so many other regulatory signals, these targeting structures are located commonly in the **3′-untranslated region (3′-UTR)** of transported RNAs (Figure 7.16).

The 3′-UTR structures that are important for RNA trafficking are highly diverse. Some 3′-UTR targeting sequences are short, such as the 21-nucleotide sequence that binds protein hnRNP A2, which targets transcripts such as the mRNA that codes for a myelin protein. Other 3′-UTR sequences adopt more complex structures, such as those involved in RNA localization during development (*e.g.*, grk mRNA in *Drosophila*), or those that bind the **ZIP-code** family of targeting proteins (*e.g.*, β-actin or Vg1 mRNAs). In almost all cases, the RNA binding proteins that bind these 3′-UTR structures are involved in other processes, such that pre-mRNA splicing, capping, and other events are interdependent and linked to transport of mRNA.

### 7.4.2 RNA that Transports Protein: the Signal Recognition Particle

RNA structures are not only involved in the shuttling of RNA molecules, they are also essential for the transport of proteins. This is exemplified by the role of **7SL RNA** in the **signal recognition particle (SRP)**.[60] The SRP is a RNP complex that is present in all three kingdoms of life (Figure 7.32). Although its complexity has increased with evolution, many of the major components in SRP (such as a conserved

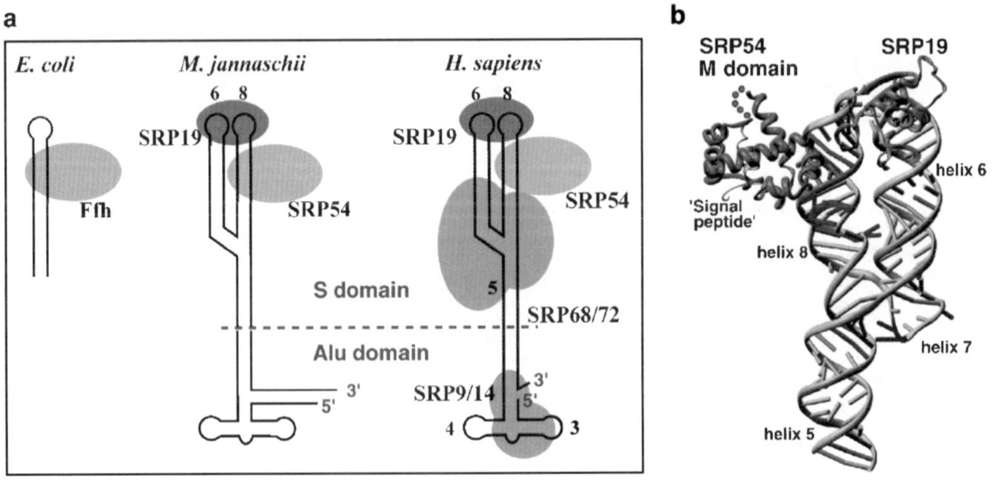


**Figure 7.32** *SRP complexes from the three kingdoms of life. (a) The SRP becomes increasingly complex from eubacteria to archaea, to mammals (left to right). All of them have an "SRP-54"-like component (yellow ellipse) and an "S" domain. Additional components (such as the Alu domain and SRP 9) have been added in more complex organisms*
(Reprinted from Ref. 60. © (2003), with permission from Elsevier)
*(b) Crystal structure of SRP "S" domain RNA in complex with signal peptide and associated proteins*
(Reprinted from Ref. 77. © (2002), with permission from Macmillan Publications Ltd)

RNA core) have remained the same. The role of SRP in the cell is to bind signal peptides that are located at the terminus of nascent membrane and secretory proteins.[61] After forming a complex with the signal sequence, SRP guides the entire ribosome/peptide complex to a receptor site on the endoplasmic reticulum (ER), where the remainder of the protein is synthesized, while it is transported simultaneously through (or into) the ER membrane.

SRP RNA contains an "S" domain that is highly conserved in all kingdoms (Figure 7.32). In Archaea and Eukaryotes, this has been appended to an "Alu domain" RNA. Intriguingly, this Alu domain is the same sequence that is encoded by the eukaryotic mobile genetic element of the same name (Section 7.5.1), and this connection between SRP and genomic plasticity remains a subject of great interest. The SRP RNA is bound by conserved proteins (such as the GTPases SRP54 and SR$\alpha$) that are involved in assembly and function of the particle.

## 7.5 RNAs AND EPIGENETIC PHENOMENA

Genomes and gene expression are constantly being altered by processes that are "outside" the normal processes of DNA replication, RNA metabolism, and protein expression. These "**epigenetic phenomena**" often involve specialized RNA molecules that play a major role in the evolution and metabolism of diverse organisms.

### 7.5.1 RNA Mobile Elements

Genomes are not static environments. Nor do genomes necessarily evolve through small changes that accumulate slowly over time. Rather, genomes often undergo massive changes, the most potent effectors of which are mobile genetic elements, or transposons (Section 6.8.3). While the zoology of mobile elements is diverse,[62] the **retrotransposons** represent a subset of RNA molecules that are of particular interest because of their remarkable mobility mechanisms and their profound influence on eukaryotic genomes. These RNAs assemble with cofactor proteins to form RNPs that encode, or depend upon, **reverse transcriptases (RT)** to produce stable DNA copies of their progeny.

*7.5.1.1 Mammalian L1 and Alu Elements.* A stunning 25% of human DNA, by weight, encodes two RNA molecules (L1, 15%; and Alu, 10%) in millions of copies that have radically altered the organization of mammalian genomes. Although only a fraction of these copies are functional, the mobilization of **L1 and Alu elements** causes significant genomic rearrangement, which can lead to cancer and other diseases.[63]

The L1 gene encodes a large RNA that is ~6000 nucleotides in length. This polyadenylated transcript contains a 5′-UTR and two reading frames (ORF1 and ORF2) that encode proteins essential for L1 transposition. The proteins ORF1p and ORF2p form an RNP by binding to substructures in the L1 RNA. Functional L1 RNPs are then imported into the nucleus, where they attack the host genome through a process that involves reverse transcription of L1 RNA by the ORF2p. RTs such as ORF2p are polymerases that can synthesize DNA from an RNA template. They are important for the replication and genomic integration of many mobile elements and for retroviruses, such as HIV (Section 6.4.6).

The Alu element is a remarkably small RNA (~300 nucleotides) that originated from the 7SL RNA of SRP (Figure 7.32, Section 7.4). Perhaps due to a similarity in the RNA binding properties of SRP and L1 proteins, 7SL RNA is believed to have recruited the L1 transposition apparatus, which allowed it to replicate and proliferate as the Alu mobile element. Alu elements lack ORFs for proteins that stimulate mobility, and therefore they continue to depend on L1 for mobilization and proliferation in the human genome.

*7.5.1.2 Group II Intron Retrotransposons.* A second family of retrotransposon mobilizes a catalytic intron that is commonly found in the organellar genes of plants, fungi, yeast, and also in bacteria. **Group II introns** are self-splicing RNAs (Section 7.2), which contain an ORF that encodes a multifunctional "**maturase**" protein. A group II intron maturase protein usually contains defined segments that are involved in RNA binding, DNA endonuclease activity, and reverse transcription (Figure 7.22). After it is translated, the maturase protein binds its parent intron and stimulates the self-splicing reaction that releases lariat intron. The intron lariat and the maturase then form a stable RNP that is catalytically active for retrotransposition of the intron sequence into duplex DNA.[39]

The transposition reaction initially involves recognition of the DNA target site through interactions with both the maturase and the intron RNA. This is followed by two distinct cleavage events (Figure 7.22). The DNA sense strand is invaded through a **reverse-splicing reaction** that is catalyzed by intron RNA. The antisense strand is cleaved by an endonuclease activity in the maturase protein. Following partial or complete reverse splicing, the maturase RT makes a DNA copy of the intronic RNA, which becomes stably incorporated through DNA repair pathways. Group II integration into duplex DNA represents the first known example of catalytic collaboration between a ribozyme and a protein enzyme, that is active-site functionalities on both components are essential for the reaction. Furthermore, this reaction demonstrated that DNA (and not just RNA) can be the natural substrate for a ribozyme.

### 7.5.2 SnoRNAs: Guides for Modification of Ribosomal RNA

rRNAs contain numerous post-transcriptional modifications (Section 7.2.4). Although nucleotide-modifying enzymes have been known for some time, it was difficult to understand how they are targeted to specific sites on rRNA. The answer has come from studies of the **nucleolus**, which is a cellular organelle that has long been a curiosity because it is literally packed with RNA.

The nucleolus is the manufacturing site for ribosomal subunits, which are exported to the cytoplasm as highly stable RNPs. The raw materials for ribosomes are long pre-rRNAs, which require trimming, modification, and assembly with ribosomal proteins. Prior to assembly, the pre-rRNA is sequence-specifically modified. For example ribose 2′-OH groups are converted into 2′-$OCH_3$ groups, uridines into pseudouridines, and various other modifications are also introduced (Section 7.2.4). This is accomplished by annealing between the rRNA and "**guide RNAs**," which are abundant **small nucleolar RNAs (snoRNAs)**.[48] Modifying enzymes recognize the target nucleotide by measuring the distance between specific base pairings and conserved sequences on the snoRNA (Figure 7.33). Thus, rRNA modification represents a natural

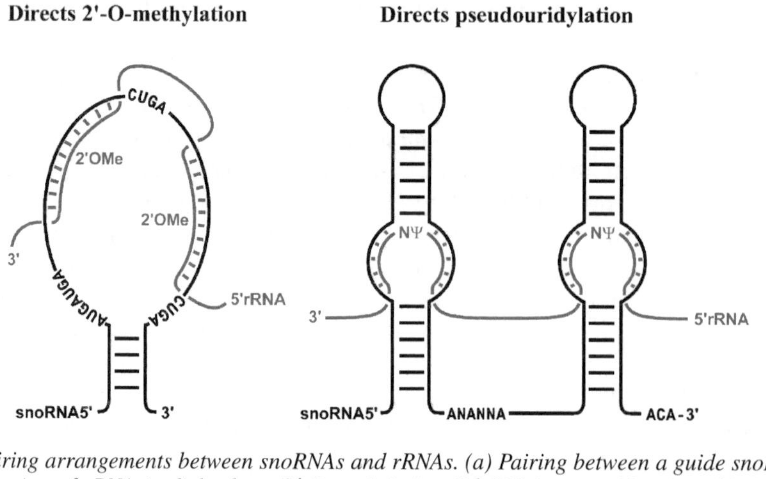


**Figure 7.33** *Pairing arrangements between snoRNAs and rRNAs. (a) Pairing between a guide snoRNA (black) and a region of rRNA (red) leads to 2′-O methylation (2′-OMe) at specific sites. (b) Pairing between a different type of double hairpin guide snoRNA (black) and another region of rRNA (red) leads to specific pseudouridylation events (ψ)*
(Reprinted from Ref. 48. © (1997), with permission from Elsevier)

example of "**antisense**" **targeting**, by which a cellular RNA is manipulated through simple base pairing with a small oligonucleotide.

### 7.5.3 Small RNAs Involved in Gene Silencing and Regulation

Eukaryotic cells contain abundant **small RNAs** of ~22 nucleotides in length. These RNAs are used to control the timing of protein expression, to inhibit the attack of viruses and endogenous transposons, and even to influence the function of DNA, through chromatin silencing.[64]

**miRNAs** are small pieces of single-stranded RNA that base pair with target genes, thereby modulating levels of protein synthesis.[64] The miRNAs are encoded by larger, highly conserved transcripts that contain many long stem-loop structures (Figure 7.34). The functional miRNA molecules are excised by a specialized endonuclease called **Dicer**, which cuts duplex RNA into short segments of ~22 nucleotides.

Invading viruses and transposons may also form RNA duplexes that are cleaved by the Dicer enzyme, which results in short RNA molecules that are called **small interfering RNAs (siRNAs)**.[65] These siRNAs become incorporated into an RNP called the **RNAi-induced silencing complex (RISC)**, which uses the siRNA as a guide for identifying and degrading an invading RNA of complementary sequence. A RISC complex also forms around short miRNAs, and is believed to play a role in translational repression of endogenous gene expression. **RNA interference (RNAi)** is therefore a powerful strategy for host defense and cellular regulation (Section 5.7.2).

RNAi is now widely used for genetic manipulation. To **silence** (eliminate) expression of a particular gene, it is often sufficient to transfect cells with an siRNA that is complementary in sequence to an mRNA target. The siRNA can be chemically synthesized (Section 4.2) or produced from a plasmid (Section 4.3). Once the siRNA is introduced, the cellular RISC machinery responds to the siRNA and causes cleavage of the target RNA (Section 5.7.2).

## 7.6 RNA STRUCTURE AND FUNCTION IN VIRAL SYSTEMS

Many of the most important discoveries about the diversity of RNA function have come from studies on viruses. Due to their small, compact genomes and the rapidity of their evolution, viruses have taken full advantage of the many chemical and structural capabilities of RNA.

### 7.6.1 RNA as an Engine Part: The Bacteriophage Packaging Motor

In addition to the many other attributes of RNA, it can also act as a **molecular motor**. All organisms contain motor enzymes that carry out mechanical work and which undergo conformational changes that are coupled with ATP binding and hydrolysis. These nanomachines are typically made of protein, but in several important cases, they contain essential RNA components. The clearest example of an RNA engine part is the **bacteriophage packaging motor**.[66]

Bacteriophages are a family of viruses that attack bacteria. After replication, many phages (such as phi29) have a remarkable mechanism for **packaging** progeny DNA into the **capsid** shell that will encase a new viral particle. The capsid shell is made up of viral proteins and a collar structure is added, through which DNA is then sucked rapidly and with great force into the capsid shell (Figure 7.35). Indeed, the **phi29 packaging motor** is by far the most powerful molecular motor known, capable of pulling against a load of 50 pN.[67] The collar structure contains a number of important components, one of which is a ring of RNA molecules. Cryoelectron microscopy and biochemical studies have suggested that this RNA ring is an oligomeric structure that is composed of repeating units of the pRNA, which is encoded by the phage (Figure 7.36).[66] Currently, the precise role of pRNA and its rotational movement within the collar structure are undefined. The RNA may play a passive role by creating an anionic corridor that prevents DNA adhesion to the collar, thereby speeding its passage. Alternatively, pRNA may actively translocate DNA

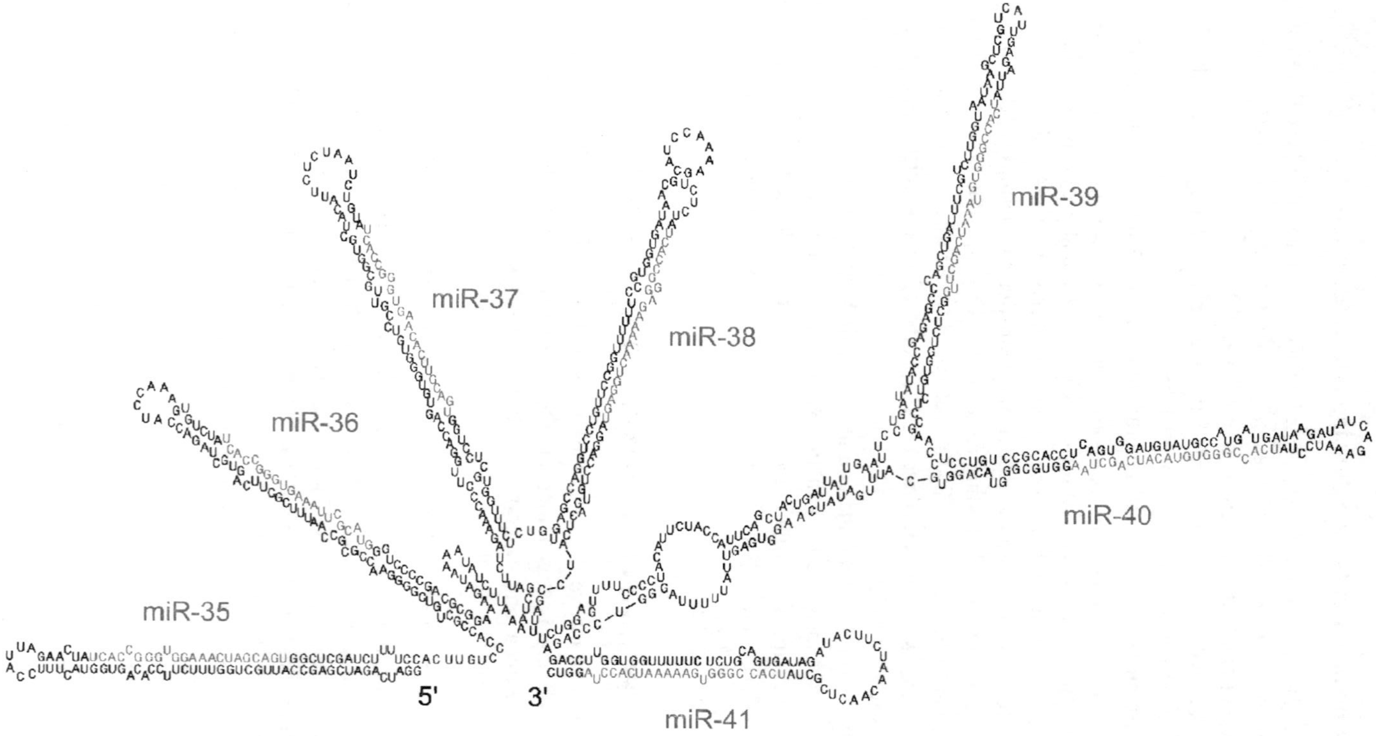


**Figure 7.34** *A microRNA precursor from C. elegans. MicroRNAs can be encoded in large, dendritic operons. Duplex regions encode specific miRNAs (in red). MiRNAs are not always encoded within clusters, particularly in mammals*
(Reprinted from Ref. 78. © (2001), with permission from AAAS)

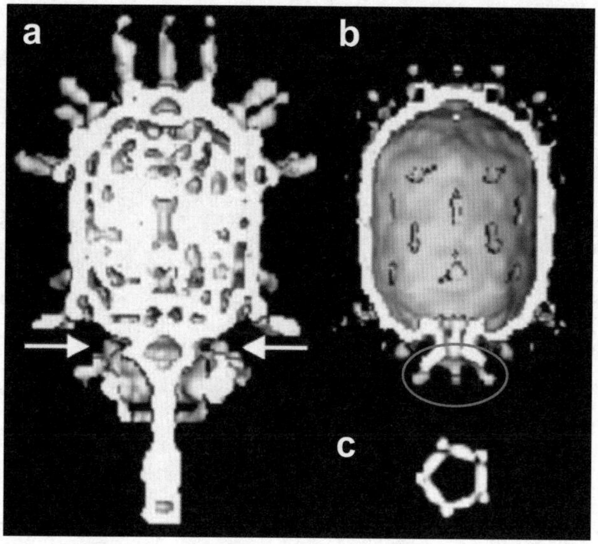


**Figure 7.35** *Cryo-EM representation of the phage packaging motor and its RNA components. (a) The intact phage particle, with arrows indicating the collar region. (b) Cutaway view of the capsid shell with its ring of RNA at the base (circled in red). (c) A top view of the RNA oligomeric ring*
(Reprinted from Ref. 66. © (2000), with permission from Macmillan Publications Ltd)

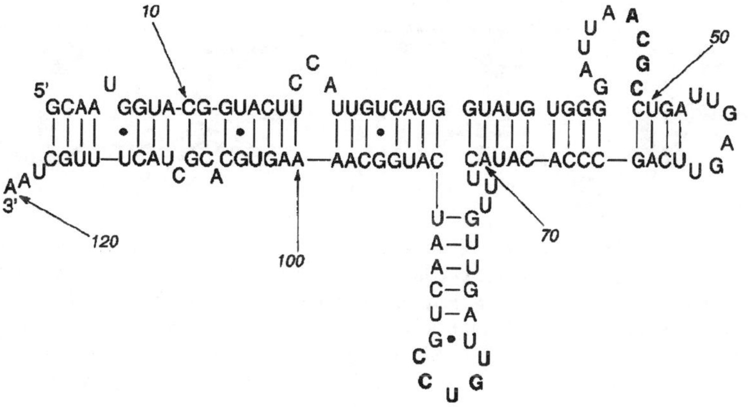


**Figure 7.36** *Secondary structure of monomeric pRNA from phi29 bacteriophage*
(Adapted from Ref. 79. © (2001), with permission from ASBMB)

through the collar by using substructures that engage the DNA bases, much like a gear that engages indentations in a conveyor belt.

## 7.6.2 RNA as a Catalyst: Self-Cleaving Motifs from Viral RNA

Although the phenomenon of RNA catalysis was discovered initially during studies of eukaryotic RNAs, such as the *Tetrahymena* group I intron and RNAse P RNA, other important **ribozymes**, such as hammerhead, hairpin, and hepatitis delta were initially derived from **self-cleaving RNA motifs** that are common in plant and animal viruses (Figures 7.37 and 7.9).[68]

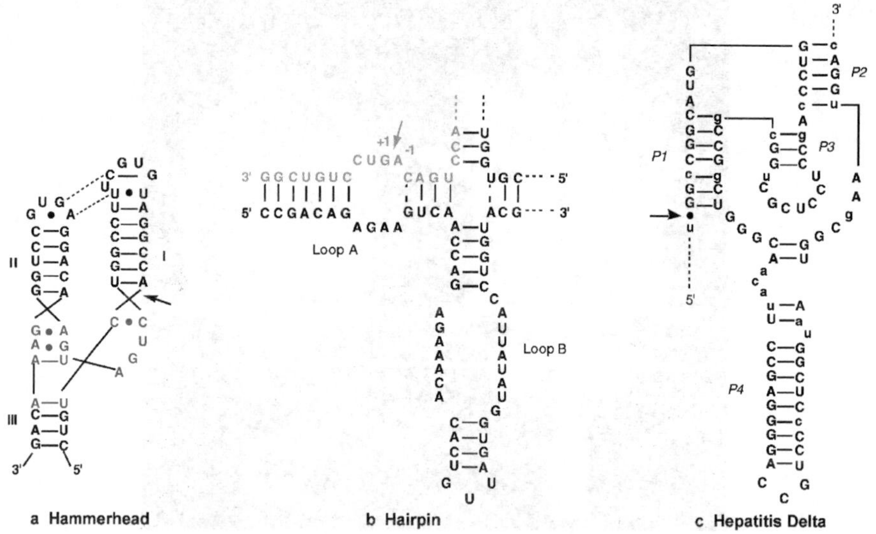


**Figure 7.37** *Three different types of ribozymes derived viroids and viruses. The secondary structures of (a) hammerhead (b) hairpin and (c) hepatitis delta ribozymes are shown, with arrows indicating the sites of cleavage. The catalytic core of the hammerhead ribozyme is shown in red*

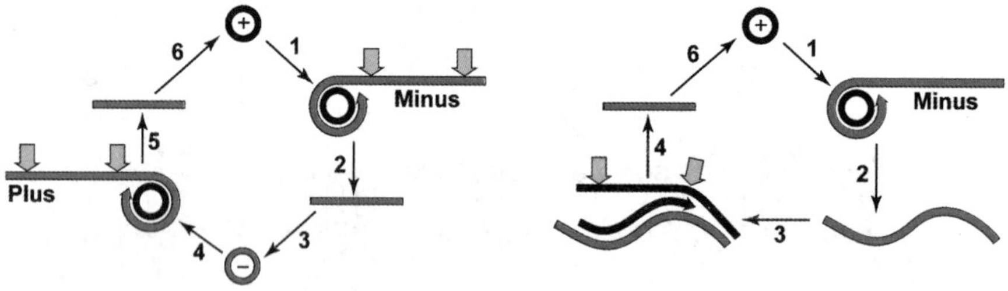


**Figure 7.38** *Mechanisms of rolling circle replication. (a) The positive stranded genome of certain viroids (bold circle around "+") is copied continuously into a long minus-strand RNA (red line) that is cleaved into antigenomic units with a ribozyme (grey arrows). The minus strand RNA circularizes (red circle), is copied into plus strand RNA (black line), which is cleaved into genomic units with another ribozyme (grey arrows, left).*[68] *In pathogens such as the avocado sunblotch virus, the hammerhead ribozyme motif cleaves both (+) and (−) strands, while other viroids/viruses utilize hammerhead or other ribozyme motifs (such as the hairpin). (b) Sometimes the circular (+) strand is copied into a minus strand that does not circularize, but which is copied into a new plus strand that is self-cleaved by ribozyme motifs*

Many plant viruses and animal pathogens undergo **rolling circle replication** from a small circular genome. When DNA or RNA is copied from these templates, the polymerase generates a continuous, repetitive strand that must be cut into pieces that encode single copies of an entire genome (*e.g.*, herpesvirus DNA) or into individual genes (*e.g.*, certain plant viroids and hepatitis delta virus). In the case of **viroids** such as avocado sunblotch virus or tobacco ringspot virus, RNA is transcribed as a continuous strand and the regions between genes then fold into catalytic tertiary structures. These are known as the **hammerhead** and **hairpin** ribozyme motifs, respectively. These motifs undergo **self-cleavage reactions** where the RNA is cleaved into small pieces that encode individual proteins (Figure 7.38).[68] Such ribozymes have been developed as tools for biotechnology.

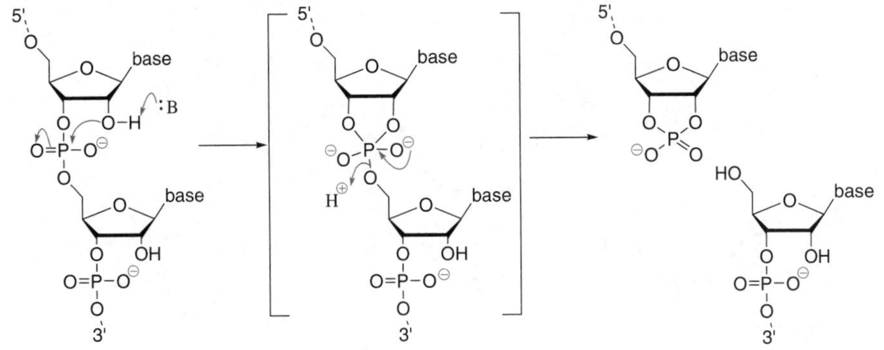


**Figure 7.39** *The mechanism of strand scission by the hammerhead, hairpin, hepatitis delta and Varkud satellite ribozymes*

The viral self-cleaving motifs catalyze sequence-specific RNA cleavage through a simple *trans*-esterification reaction that involves nucleophilic attack of the scissile 2′-OH group on an activated phosphodiester linkage, resulting in products with **2′-3′ cyclic phosphate** and 5′-hydroxyl termini (Figure 7.39). This reaction differs markedly from the self-splicing introns and RNase P in eukaryotes, which stimulate attack by an exogenous nucleophile and which produce a different set of reaction products (Figure 7.17, Section 7.2.2).

While the chemical reaction pathway of self-cleaving motifs is very similar to base-catalyzed RNA hydrolysis (Sections 3.2.2 and 8.1) and to the first part of the mechanism of the cleavage reaction of Ribonuclease A (Figure 3.51), studies on ribozyme constructs have revealed a rich and complex chemistry.[69] Whereas magnesium ions or other divalent cations are important in folding of the ribozyme motifs, some ribozymes, for example the hairpin ribozyme and the hepatitis delta genomic and complementary antigenomic ribozymes (both the sense and antisense strand of this motif are catalytic), do not strictly require divalent cations for the chemical cleavage reaction. Instead, these ribozymes appear to promote strand scission (or ligation, being the reverse reaction) by precise alignment of functional groups within the ribozyme active site.[27] These ribozymes may also stimulate reaction through general acid–base catalysis and electrostatic stabilization (Figure 7.40).[69] Remarkably, certain nucleobases in the active site undergo dramatic p$K_a$ shifts toward neutrality, as exemplified by residue C75 of hepatitis delta virus ribozyme (Figure 7.9), that may allow them to behave much like imidazole moieties of histidine residues within Ribonuclease A.[69]

The catalytic mechanism of the hammerhead ribozyme remains controversial, however. Earlier studies on a smaller RNA section that was thought to be sufficient for cleavage suggested the involvement of magnesium ions in both folding and the catalytic mechanism. We now know that a complete hammerhead is composed of a larger section of RNA (Figure 7.37a) and that two of its loops dock as part of the folding pathway.[70,71] This construct has a much lower magnesium-ion requirement for cleavage than for minimized hammerhead constructs. Further high-resolution structure analysis and biochemical studies may soon resolve whether or not divalent ions are involved in the hammerhead catalytic mechanism.

### 7.6.3 RNA Tertiary Structure and Viral Function

Structured RNA plays a particularly important role in the replication and pathogenicity of viruses. Some of the most important viral threats to human health, such as the flaviviruses (*e.g.*, Yellow Fever), Hepatitis C virus (HCV), Influenza, the coronaviruses (*e.g.*, SARS), and the retroviruses (*e.g.*, HIV) have RNA genomes that contain regulatory elements composed of RNA tertiary structures. Furthermore, all viruses (including those with DNA genomes) produce mRNA molecules that are processed and translated by exploitation of unusual RNA conformations.

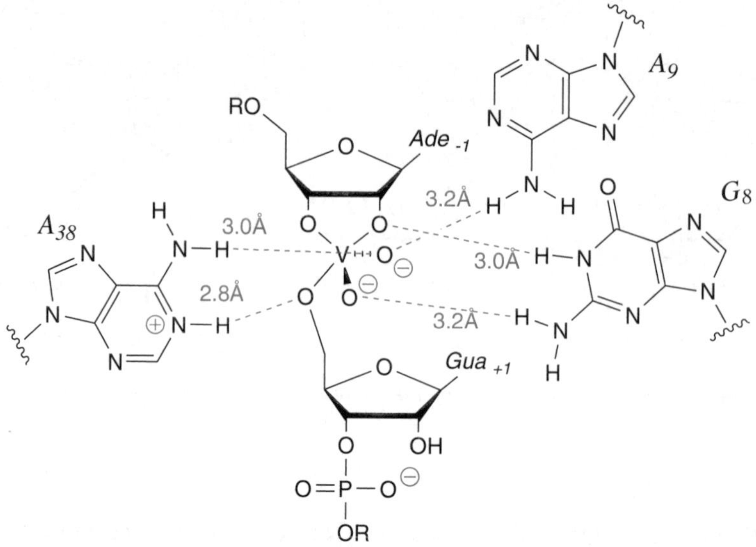


**Figure 7.40** *Transition state of the hairpin ribozyme. A crystal structure of the hairpin ribozyme bound to a transition-state vanadate analogue suggests that reaction is facilitated by precise alignment of nucleobase functional groups*[27]

*7.6.3.1 Pseudoknots that Stimulate Ribosomal Frameshifting and Recoding.* Although the ribosome typically maintains translation **"in-frame"** and reads each sequential triplet codon on mRNA in an orderly fashion (Section 7.3), there are RNA structures that induce **"recoding"** of an mRNA reading frame in order to produce two different proteins from the same RNA sequence. Usually, this is the result of **+1 frameshifting**, whereby the entire frame moves over by 1 nucleotide. This normally occurs at a "**slippery sequence**" (*i.e.*, AAAAAAA), that will bind A- and P-site tRNAs in the same manner, even after the frame shifts by 1. This slippage normally occurs when the ribosome stalls after sensing particular downstream RNA tertiary structures.[72]

Viral genomes are remarkably compact. For example, multiple proteins are commonly encoded by **overlapping frames** of the same gene sequence. Retroviral gene expression depends on a specific type of bent **pseudoknot** (Figure 7.41) that stimulates ribosomal frameshifting and thereby initiates synthesis of viral proteases (pro) and polymerases (pol) from an mRNA sequence that overlaps with the region that encodes structural proteins (gag).

*7.6.3.2 The IRES of Hepatitis C Virus.* Viruses commonly use RNA tertiary structures to "trick" the host translation machinery into making viral proteins. During the first stages of normal eukaryotic translation, an initiation complex (the small ribosomal subunit and various factors) binds the 5′-cap structure on mRNA and recognizes the adjacent start codon. Lacking cap structures, many viruses contain complex tertiary structures at their 5′-termini, immediately upstream of the start codon. These **5′-UTR** structures bind the small ribosomal subunit, allowing it to recognize the proper start codon and initiate the synthesis of viral proteins. Structured elements in viral 5′-UTRs (and even in certain host mRNAs that lack caps) are called **internal ribosome entry sites (IRES)**, and they function by replacing protein factors that are normally required for translation initiation.

One of the best characterized examples of a viral IRES is the 5′-UTR from HCV mRNA.[73] This RNA element is ~330 nucleotides in length, and it contains a four-way junction that adopts a tertiary fold that is essential for ribosomal recognition. When the 40S subunit binds the IRES, it no longer requires eukaryotic initiation factors A, B, G, or E and it initiates translation of the HCV polyprotein at an adjacent start codon (Figure 7.42).

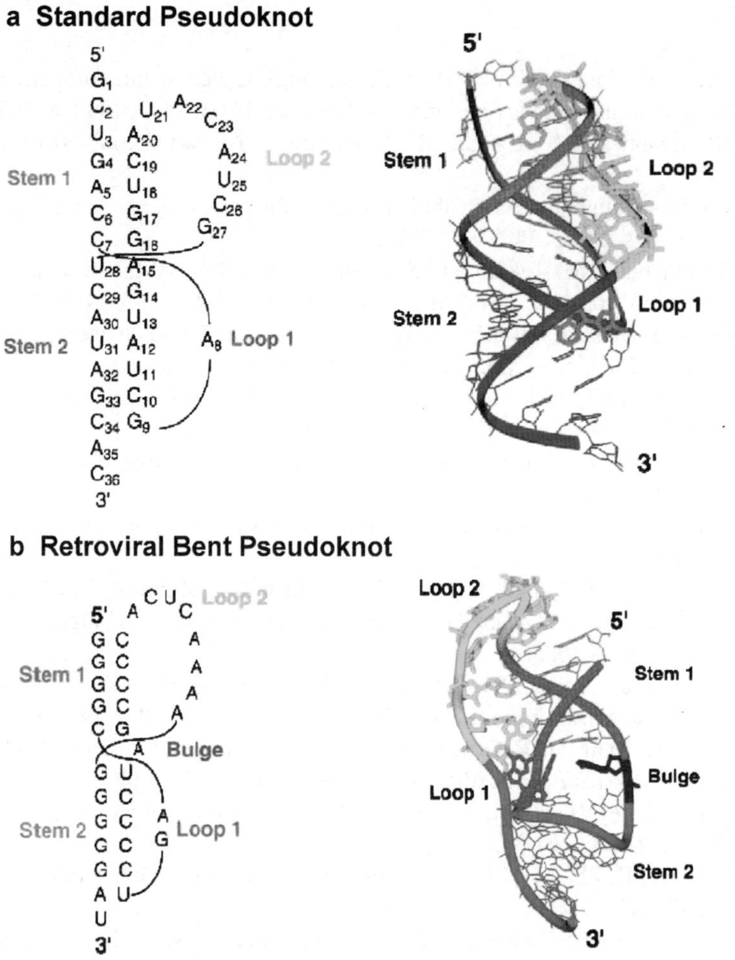


**Figure 7.41** *The secondary and tertiary structures of pseudoknots. (a) Many RNA molecules contain pseudoknots that typically form a straight, coaxial stack of the two helices. (b) The mRNA of retroviruses contains sequences that form an unusual type of pseudoknot. Due to the presence of unpaired nucleotides (bulge) at the junction between stems 1 and 2, these pseudoknots bend, thereby causing a disruption of ribosomal reading frame*
(Reprinted from Ref. 72. © (2000), with permission from Elsevier)

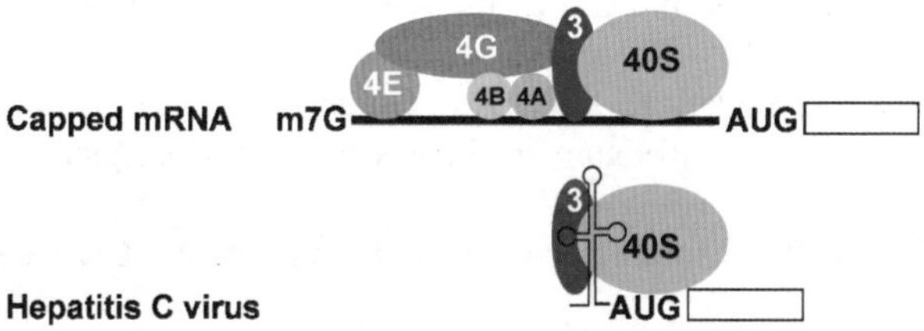


**Figure 7.42** *Translation initiation of eukaryotic mRNAs (top) and Hepatitis C RNAs (bottom). Eukaryotic translation is initiated through binding of the 40S ribosomal subunit (light grey ellipse) to the cap structure (red ellipsoid) at the 5′-end of mRNA molecules. Hepatitis C viral RNA is not capped and therefore it initiates translation by promoting interactions between the 40S subunit and an unusual stem-loop structure, called an internal ribosome entry site (IRES)*

## REFERENCES


1. K. Chin, K. Sharp, B. Honig and A.M. Pyle, Calculating the electrostatic properties of RNA provides new insights into molecular interactions and function, *Nat. Struct. Biol.*, 1999, **6**, 1055–1061.
2. M. Egli, S. Portmann and N. Usman, RNA hydration: a detailed look, *Biochemistry*, 1996, **35**, 8489–8494.
3. N.B. Leontis, J. Stombaugh and E. Westhof, The non-Watson–Crick pairs and their isostericity matrices, *Nucleic Acids Res.*, 2002, **30**, 3497–3531.
4. V. Biou, A. Yaremchuck, M. Tukalo and S. Cusack, The 2.9 Å crystal structure of *T. thermophilus* seryl-tRNA synthetase complexed with tRNA(Ser), *Science*, 1994, **263**, 1404–1410.
5. N. Ban, P. Nissen, J. Hansen, P.B. Moore and T.A. Steitz, The complete atomic structure of the large ribosomal subunit at 2.4 Å resolution, *Science*, 2000, **289**, 905–920.
6. L. Su, L. Chen, M. Egli, J.M. Berger and A. Rich, RNA triplex in the structure of ribosomal frameshifting viral pseudoknot, *Nat. Struct. Biol.*, 1999, **3**, 285–292.
7. C. Baugh, D. Grate and C. Wilson, 2.8 Å crystal structure of the malachite green aptamer, *J. Mol. Biol.*, 2000, **301**, 117–128.
8. L. Gold, B. Polisky, O.C. Uhlenbeck and M. Yarus, Diversity of oligonucleotide functions, *Ann. Rev. Biochem.*, 1995, **64**, 763–797.
9. J.H. Cate, A.R. Gooding, E. Podell, K. Zhou, B.L. Golden, C.E. Kundrot, T.R. Cech and J.A. Doudna, Crystal structure of a group I ribozyme domain reveals principles of higher order RNA folding, *Science*, 1996, **273**, 1678–1685.
10. P. Nissen, J.A. Ippolito, N. Ban, P.B. Moore and T.A. Steitz, RNA tertiary interactions in the large ribosomal subunit: the A-minor motif, *Proc. Natl. Acad. Sci. USA*, 2001, **98**, 4899–4903.
11. F.L. Murphy, Y.-H. Wang, J.D. Griffith and T.R. Cech, Coaxially stacked RNA helices in the catalytic center of the *Tetrahymena* ribozyme, *Science*, 1994, **265**, 1709–1712.
12. E.A. Doherty, R.T. Batey, B. Masquida and J.A. Doudna, A universal mode of helix packing in RNA, *Nat. Struct. Biol.*, 2001, **8**, 339–343.
13. A.R. Ferre-D'amare, K. Zhou and J.A. Doudna, Crystal structure of a hepatitis delta virus ribozyme, *Nature*, 1998, **395**, 567–574.
14. V.K. Misra and D.E. Draper, On the role of magnesium ions in RNA stability, *Biopolymers*, 1998, **48**, 113–135.
15. A.M. Pyle, The role of metal ions in ribozymes. In *Metal Ions in Biological Systems*, H. Sigel and A. Sigel (eds). Marcel Dekker, Inc., New York, 1996, 479–519.
16. R. Shiman and D. Draper, Stabilization of RNA tertiary structure by monovalent cations, *J. Mol. Biol.*, 2000, **302**, 79–91.
17. R.K.O. Sigel, A. Vaidya and A.M. Pyle, Metal ion binding sites in a group II intron core, *Nat. Struct. Biol.*, 2000, **7**, 1111–1116.
18. S. Basu, R.P. Rambo, J. Strauss-Soukup, J.H. Cate, A. Ferre-D'Amare, S.A. Strobel and J.A. Doudna, A specific monovalent metal ion integral to the AA platform of the RNA tetraloop receptor, *Nat. Struct. Biol.*, 1998, **5**, 986–992.
19. J.H. Cate, A.R. Gooding, E. Podell, K. Zhou, B.L. Golden, A.A. Szewczak, C.E. Kundrot, T.R. Cech and J.A. Doudna, RNA tertiary structure mediation by adenosine platforms, *Science*, 1996, **273**, 1696–1699.
20. T.R. Sosnick and T. Pan, RNA folding: models and perspectives, *Curr. Opin. Struct. Biol.*, 2003, **13**, 309–316.
21. R. Schroeder, R. Grossberger, A. Pichler and C. Waldsich, RNA folding *in vivo*, *Curr. Opin. Struct. Biol.*, 2002, **12**, 296–300.
22. S.H. Kim, G. Quigley, F.L. Suddath, A. McPherson, D. Sneden, J.J. Kim, J. Weinzierl, P. Blattman and A. Rich, The three-dimensional structure of yeast transfer RNA: shape of the molecule at 5.5 Å resolution, *Proc. Natl. Acad. Sci. USA*, 1972, **69**, 3746–3750.

23. G.J. Quigley, F.L. Suddath, A. McPherson, J.J. Kim, D. Sneden and A. Rich, The molecular structure of yeast phenylalanine transfer RNA in monoclinic crystals, *Proc. Natl. Acad. Sci. USA*, 1974, **71**, 2146–2150.
24. H.W. Pley, K.M. Flaherty and D.B. McKay, Three-dimensional structure of a hammerhead ribozyme, *Nature*, 1994, **372**, 68–74.
25. P.L. Adams, M.R. Stahley, A.B. Kosek, J. Wang and S.A. Strobel, Crystal structure of an intact self-splicing group I intron with both exons, *Nature*, 2004, **430**, 45–50.
26. A.S. Krasilnikov, X. Yang, T. Pang and A. Mondragon, Crystal structure of the specificity domain of Ribonuclease P, *Nature*, 2003, **421**, 760–764.
27. P.B. Rupert, A.P. Massey, S.T. Siggurdson and A.R. Ferré-D'amare, Transition-state stabilization by a catalytic RNA, *Science*, 2002, **298**, 1421–1424.
28. M.M. Yusupov, G.Z. Yusupova, A. Baucom, K. Lieberman, T.N. Earnest, J.H. Cate and H.F. Noller, Crystal structure of the ribosome at 5.5 Å resolution, *Science*, 2001, **292**, 883–896.
29. B.T. Wimberley, D.W. Broderson, W.M. Clemons, R.J. Morgan-Warren, A.P. Carter, C. Vonrhein, T. Hartsch and V. Ramakrishnan, Structure of the 30S ribosomal subunit, *Nature*, 2000, **407**, 327–339.
30. N.J. Proudfoot, A. Furger and M.J. Dye, Integrating mRNA processing with transcription, *Cell*, 2002, **108**, 501–512.
31. A. Shatkin and J.L. Manley, The ends of the affair: capping and polyadenylation, *Nat. Struct. Biol.*, 2000, **7**, 838–842.
32. A.B. Sachs, P. Sarnow and M.W. Hentze, Starting at the beginning, middle, and end: translation initiation in eukaryotes, *Cell*, 1997, **89**, 831–838.
33. H. Madhani and C. Guthrie, Dynamic RNA–RNA interactions in the spliceosome, *Ann. Rev. Genet.*, 1994, **28**, 1–26.
34. C.W.J. Smith and J. Valcarcel, Alternative pre-mRNA splicing: the logic of combinatorial control, *Trends Biochem. Sci.*, 2000, **25**, 381–388.
35. T.R. Cech, Structure and mechanism of the large catalytic RNAs: group I and group II introns and ribonuclease P. In *The RNA World*, R.F. Gesteland and J.F. Atkins (eds). Cold Spring Harbor Press, Cold Spring Harbor, 1993, 239–270.
36. P.Z. Qin and A.M. Pyle, The architectural organization and mechanistic function of group II intron structural elements, *Curr. Opin. Struct. Biol.*, 1998, **8**, 301–308.
37. K. Kruger, P.J. Grabowski, A.J. Zaug, J. Sands, D.E. Gottschling and T.R. Cech, Self-splicing RNA: autoexcision and autocyclization of the ribosomal RNA intervening sequence of *Tetrahymena*, *Cell*, 1982, **31**, 147–157.
38. C.L. Peebles, P.S. Perlman, K.L. Mecklenburg, M.L. Petrillo, J.H. Tabor, K.A. Jarrell and H.-L. Cheng, A self-splicing RNA excises an intron lariat, *Cell*, 1986, **44**, 213–223.
39. M. Belfort, V. Derbyshire, M.M. Parker, B. Cousineau and A.M. Lambowitz, Mobile introns: pathways and proteins. In *Mobile DNA II*, N.L. Craig, R. Craigie, M. Gellert and A.M. Lambowitz (eds). ASM Press, Washington, DC, 2002, 761–783.
40. J.C. Kurz and C.A. Fierke, Ribonuclease P: a ribonucleoprotein enzyme, *Curr. Opin. Chem. Biol.*, 2000, **4**, 553–558.
41. M.A. Carmell and G.J. Hannon, RNase III enzymes and the initiation of gene silencing, *Nat. Struct. Mol. Biol.*, 2004, **11**, 214–218.
42. B.L. Bass, RNA editing by adenosine deaminases that act on RNA, *Ann. Rev. Biochem.*, 2002, **71**, 817–846.
43. J.E. Wedekind, G.S.C. Dance, M.P. Sowden and H.C. Smith, Messenger RNA editing in mammals: new members of the APOBEC family seeking roles in the family business, *Trends Genet.*, 2003, **19**, 207–216.
44. K. Stuart and A.K. Panigrahi, RNA editing: complexity and complications, *Mol. Microbiol.*, 2002, **45**, 591–596.
45. K.R. Noon, R. Guymon, P.F. Crain, J.A. McCloskey, M. Thomm, J. Lim and R. Cavicchioli, Influence of temperature on tRNA modification in Archaea, *J. Bacteriol.*, 2003, **185**, 5483–5490.

46. P.F. Agris, Decoding the genome: a modified view, *Nucleic Acids Res.*, 2004, **32**, 223–238.
47. T. Muramatsu, K. Nishikawa, F. Nemoto, Y. Kuchino, S. Nishimura, T. Miyazawa and S. Yokoyama, Codon and amino-acid specificities of a transfer RNA are both converted by a single post-transcriptional modification, *Nature*, 1988, **336**, 179–181.
48. C.M. Smith and J.A. Steitz, Sno storm in the nucleolus: a new role for myriad small RNPs, *Cell*, 1997, **89**, 669–672.
49. M.I. Newby and N.L. Greenbaum, Sculpting of the spliceosomal branch site recognition motif by a conserved pseudouridine, *Nature Struct. Biol.*, 2002, **9**, 958–965.
50. R.P. Parker and H. Song, The enzymes and control of eukaryotic mRNA turnover, *Nature Struct. Mol. Biol.,* 2004, **11**, 121–127.
51. L.E. Maquat and G.G. Carmichael, Quality control of mRNA function, *Cell*, 2001, **104**, 173–176.
52. R.F. Gesteland and J.F. Atkins, Recoding: dynamic reprogramming of translation, *Ann. Rev. Biochem.*, 1996, **65**, 741–768.
53. D.L. Hatfield and V.N. Gladyshev, How selenium has altered our understanding of the genetic code, *Mol. Cell. Biol.*, 2002, **22**, 3565–3576.
54. J.W. Chin, T.A. Cropp, J.C. Anderson, M. Mukherji, Z. Zhang and P.G. Schultz, An expanded eukaryotic genetic code, *Science*, 2003, **301**, 964–967.
55. P.J. Beuning and K. Musier-Forsyth, Transfer RNA recognition by aminoacyl tRNA synthetases, *Biopolymers*, 2000, **52**, 1–28.
56. Y. Nakamura, M. Uno, T. Toyoda, T. Fujiwara and K. Ito, Protein tRNA mimicry in translation termination, *Cold Spring Harbor Symp. Quant. Biol.*, 2001, **66**, 469–475.
57. H.F. Noller, M.M. Yusupov, G.Z. Yusupova, A. Baucom, K. Lieberman, L. Lancaster, A. Dallas, K. Fredrick, T.N. Earnest and J.H. Cate, Structure of the ribosome at 5.5 Å resolution and its interactions with functional ligands, *Cold Spring Harbor Symp. Quant. Biol.*, 2001, **66**, 57–66.
58. M.V. Rodnina and W. Wintermeyer, Peptide bond formation on the ribosome: structure and mechanism, *Curr. Opin. Struct. Biol.*, 2003, **13**, 334–340.
59. K.L. Farina and R.H. Singer, The nuclear connection in RNA transport and localization, *Trends Biochem. Sci.*, 2002, **12**, 466–472.
60. K. Nagai, C. Oubridge, A. Kuglstatter, E. Menichelli, C. Isel and L. Jovine, Structure, function and evolution of the signal recognition particle, *EMBO J.*, 2003, **22**, 3479–3485.
61. R.J. Keenan, D.M. Freymann, R.M. Stroud and P. Walter, The signal recognition particle, *Ann. Rev. Biochem.*, 2001, **70**, 755–775.
62. N.L. Craig, R. Craigie, M. Gellert and A.M. Lambowtiz, *Mobile DNA II*, ASM Press, Washington, DC, 2002.
63. J.V. Moran and N. Gilbert, Mammalian LINE-1 retrotransposons and related elements. In *Mobile DNA II*, N.L. Craig, R. Craigie, M. Gellert and A.M. Lambowtiz (eds). ASM Press, Washington, DC, 2002, 836–869.
64. P. Nelson, M. Kiriakidou, A. Sharma, E. Maniataki and Z. Mourelatos, The microRNA world: small is mighty, *Trends Biochem. Sci.*, 2003, **28**, 534–540.
65. G.J. Hannon, RNA interference, *Nature*, 2002, **418**, 244–251.
66. A.A. Simpson, Y. Tao, P.G. Leiman, M.O. Badasso, Y. He, P.J. Jardine, N.H. Olson, M.C. Morais, S. Grimes, D.L. Anderson, T.S. Baker and M.G. Rossman, Structure of the bacteriophage phi29 DNA packaging motor, *Nature*, 2000, **408**, 745–750.
67. D.E. Smith, S.J. Tans, S.B. Smith, S. Grimes, D.L. Anderson and C. Bustamante, The bacteriophage phi29 portal motor can package DNA against a large internal force, *Nature*, 2001, **413**, 748–752.
68. R.H. Symons, Plant pathogenic RNAs and RNA catalysis, *Nucleic Acids Res.*, 1997, **25**, 2683–2689.
69. P.C. Bevilacqua, T.S. Brown, S. Nakano and R. Yajima, Catalytic roles for proton transfer and protonation in ribozymes, *Biopolymers*, 2004, **73**, 90–109.
70. A. Khvorova, A. Lescoute, E. Westhof and S.D. Jayasena, Sequence elements outside of the hammerhead ribozyme catalytic core enable intracellular activity, *Nat. Struct. Biol.*, 2003, **10**, 708–712.

71. J.C. Penedo, T.J. Wilson, S.D. Hayasena, A. Khvorova and D.M.J. Lilley, Folding of the natural hammerhead ribozyme is enhanced by interaction of auxiliary elements, *RNA*, 2004, **10**, 880–888.
72. D.P. Giedroc, C.A. Theimer and P.L. Nixon, Structure, stability and function of RNA pseudoknots involved in stimulating ribosomal frameshifting, *J. Mol. Biol.*, 2000, **298**, 167–185.
73. T.V. Pestova, I.N. Shatsky, S.P. Fletcher, R.J. Jackson and C.U.T. Hellen, A prokaryotic-like mode of cytoplasmic eukaryotic ribosome binding to the initiation codon during internal translation initiation of HCV and classical swine fever virus RNAs, *Genes Dev.*, 1998, **12**, 67–83.
74. A.J. Lee and D.M. Crothers, The solution structure of an RNA loop–loop complex: the ColE1 inverted loop sequence, *Struct. Fold Des.*, 1998, **15**, 993–1005.
75. C.M.T. Spahn, R. Beckmann, N. Eswar, P.A. Penczek, A. Sali, G. Blobel and J. Frank, Structure of the 80S ribosome from *Saccharomyces cerevisiae*. *Cell*, 2001, **107**, 373–386.
76. M. Selmer, S. Al-Karadaghi, G. Hirokawa, A. Kaji and A. Liljas, Crystal structure of *Thermatoga maritima* ribosome recycling factor: a tRNA mimic. *Science*, 1999, **286**, 2349–2352.
77. A. Kuglstatter, C. Oubridge and K. Nagai, Induced structural changes of 7SL RNA during the assembly of human signal recognition particle, *Nat. Struct. Biol.*, 2002, **9**, 740–744.
78. N.C. Lau, L.P. Lim, E.G. Weinstein and D.P. Bartel, An abundant class of tiny RNAs with probable regulatory roles in *C. elegans*, *Science*, 2001, **294**, 858–862.
79. Y. Mat-Arip, K. Garver, C. Chen, S. Sheng, Z. Shao and P. Guo, Three-dimensional interaction of phi29 pRNA dimer probed by chemical modification interference, cryo-AFM, and cross-linking, *J. Biol. Chem.*, 2001, **276**, 32575–32584.

CHAPTER 8

# Covalent Interactions of Nucleic Acids with Small Molecules and Their Repair

## CONTENTS

The simple purpose of this chapter is to provide an outline of the more important examples of covalent interactions of small molecules with nucleic acids. Topics have been chosen as they bear on the modifications of intact nucleic acids, and especially as they relate to mutagenic and carcinogenic effects. While much of the early information has come from studies on nucleosides, more recent work has shown that the net effect of a reagent on an intact nucleic acid in many cases may be quite different from either the sum or the average of its interactions with separate components. Above all, we have to recognise that studies on the more subtle effects of DNA and RNA secondary and tertiary structures on covalent interactions are still in their infancy.

## 8.1 HYDROLYSIS OF NUCLEOSIDES, NUCLEOTIDES AND NUCLEIC ACIDS

Nucleic acids are easily denatured in aqueous solution at extremes of pH or on heating. While the phosphate ester bonds are only slowly hydrolysed (Section 3.2.2), the *N*-glycosylic bonds are relatively labile. Purine nucleosides are cleaved faster than pyrimidines while deoxyribonucleosides are less stable than ribonucleosides.[1] Thus, dA and dG are hydrolysed in boiling 0.1 M hydrochloric acid in 30 min; rA and rG require 1 h with 1 M hydrochloric acid at 100°C, while rC and rU have to be heated at 100°C with 12 M perchloric acid (Figure 8.1). It follows that the glycosylic bonds of carbocyclic nucleoside analogues (Section 3.1.2), which cannot donate electrons from the furanose 4′-oxygen, are much more stable to acidic (and also enzymatic) **hydrolysis** and this property has been used to advantage in many applications.

Formic acid has been used to prepare **apurinic acid**, which has regions of polypentose phosphate diesters linking pyrimidine oligonucleotides. Such phosphate diesters are relatively labile since the pentose undergoes a β-elimination in the presence of secondary amines such as diphenylamine. This gives tracts of **pyrimidine oligomers** with phosphate monoesters at both 3′- and 5′-ends. Total acidic hydrolysis with minimum degradation of the four bases is best achieved with formic acid at 170°C.

DNA is resistant to alkaline hydrolysis but RNA is easily cleaved because of the involvement of its 2′-hydroxyl groups (Section 3.2.2).

## 8.2 REDUCTION OF NUCLEOSIDES

Purine and pyrimidine bases are sufficiently aromatic to resist **reduction** under the mild conditions used, as for example in the hydrogenolysis of benzyl or phenyl phosphate esters. However, hydrogenation with a rhodium catalyst converts uridine or thymidine into **5,6-dihydropyrimidines**.[2] Alternatively, sodium borohydride in conjunction with ultraviolet irradiation gives the same products, which can lead on by further reduction in the dark to cleavage of the heterocyclic ring. Dihydrouridine and 4-thiouridine are easily and selectively reduced in tRNA with sodium borohydride in the dark.[3]

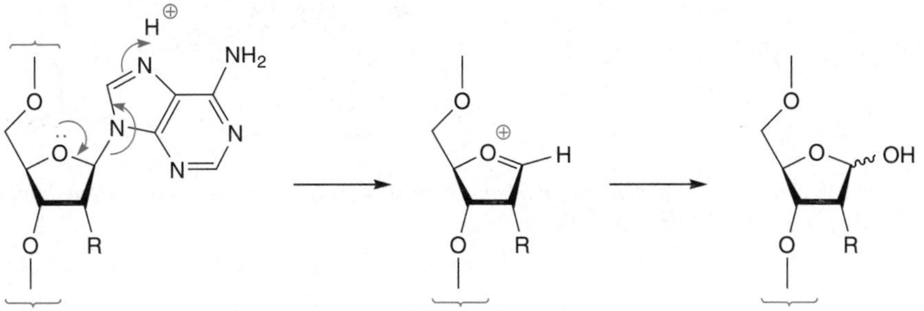


**Figure 8.1** *Mild acidic hydrolysis of purine glycosides in DNA (R = H) and RNA (R = OH)*

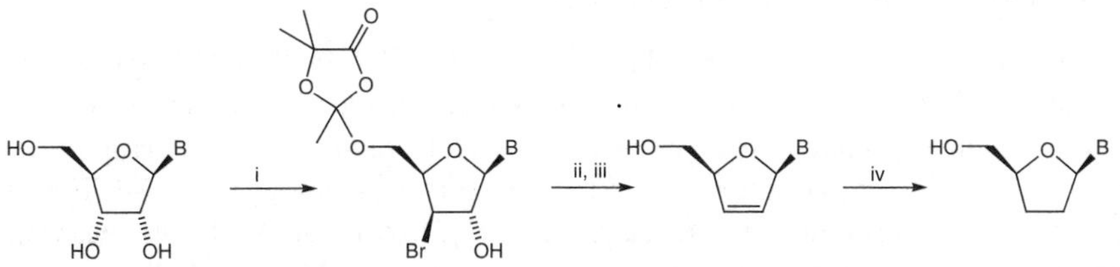


**Figure 8.2** *Synthesis of 2′,3′-dideoxynucleosides by reduction. Reagents: (i) $Me_2C(OAc)COBr$; (ii) $Cr^{2+}$, $(CH_2NH_2)_2$, 75°C; (iii) KOH; and (iv) $H_2$/PdC*

Reduction of ribonucleosides directly to 2′-deoxyribonucleosides can be accomplished by one of the several Barton procedures involving tributyltin hydride (*cf.* Figure 3.11). A nice example is the synthesis of a mixture of 2′- and 3′-deoxyadenosines, which are easily separable.[4] This type of reduction has been widely employed to transform various extensively modified ribonucleosides and their nucleotide analogues into the corresponding deoxyribonucleosides.

**2′,3′-Dideoxyribonucleosides** are valuable for use in DNA sequence analysis and also showed promise for AIDS therapy, both features being related to their chain-terminator activity (Sections 3.7.2 and 5.1). One synthesis involves hydrogenation of 2′,3′-unsaturated nucleosides or an appropriate precursor (Figure 8.2).

## 8.3 OXIDATION OF NUCLEOSIDES, NUCLEOTIDES AND NUCLEIC ACIDS

In general, strong oxidizing agents such as potassium permanganate destroy nucleoside bases. Hydrogen peroxide and organic peracids can be used to convert **adenosine into its N-1-oxide** [5] and **cytidine into its N-3-oxide** while the 5,6-double bond of thymidine is a target for oxidation by **osmium tetroxide**, forming a cyclic osmate ester of the *cis*-5,6-dihydro-5,6-glycol.[6] This reaction is sensitive to steric hindrance and so has been employed to study some details of cruciform structure in DNA (Section 2.3.3). This **thymine glycol** is also formed as a result of ionising radiation (Section 8.9.2).

Recent studies on the oxidation of DNA with hypochlorite and similar oxidants have identified the formation of 8-hydroxyguanine residues as an important mutagenic event (Sections 8.8.3 and 8.9.3).[7]

The pentoses are sensitive to free radicals produced by the interaction of hydrogen peroxide with Fe(III) or by photochemical means, and this causes strand scission (Section 8.9.1). Peter Dervan has made this process sequence-specific *in vitro* by linking radical-generating catalysts to a groove-binding agent (Section 8.8.2) and has also employed it as a '**footprinting**' device by linking an Fe–EDTA complex to an intercalating agent such as methidium (Section 5.8).[8] Other useful oxidative reactions of the pentose moieties are typical of the chemical reactions of primary alcohols and *cis*-glycols. In particular, **periodate cleavage** of the ribose 2′,3′-diol gives dialdehydes.[9] These can be stabilized by reduction to give a ring-opened diol or condensed either with an amine or with nitromethane to give ring-expanded products

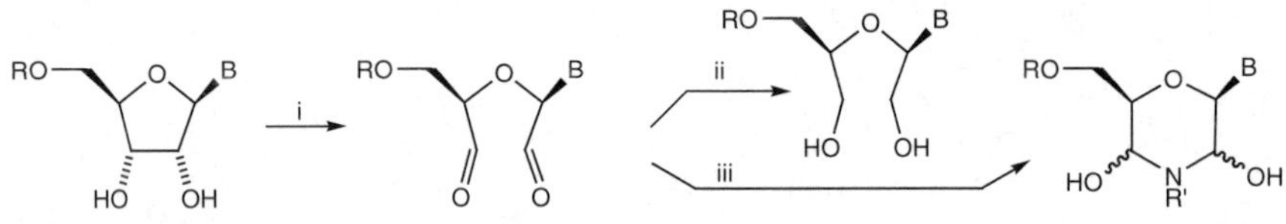


**Figure 8.3** Periodate cleavage *of a 3′-terminal nucleotide (R=RNA) and its subsequent modification. Reagents: (i)* $NaIO_4$*, pH 4.5; (ii)* $NaBH_4$*; and (iii) (R′*$NH_2$*)*

(Figure 8.3). Such procedures have frequently been adopted to make the 3′-terminus of an oligonucleotide inert to 3′-exonuclease degradation.

## 8.4 REACTIONS WITH NUCLEOPHILES

In general, nucleophiles can attack the pyrimidine residues of nucleic acids at C-6 or C-4, while reactions at C-6 of adenine or C-2 of guanine are more difficult. α-Effect nucleophiles, such as hydrazine, hydroxylamine and bisulfite, are especially effective reagents for nucleophilic attack on pyrimidines.

**Hydrazine** adds to uracil and cytosine bases first at C-6 and then reacts again at C-4. The bases are converted into pyrazol-2-one and 3-aminopyrazole, respectively, leaving an *N*-ribosylurea, which can react further to form a sugar hydrazone. These reactions were used in the **Maxam–Gilbert** chemical method of **DNA sequence determination** (now obsolete, Section 5.1), where subsequent treatment of the modified ribose residues with piperidine causes β-elimination of both 3′- and 5′-phosphates at the site of depyrimidination (Figure 8.4).[10]

Cytosine and its nucleosides react with **hydroxylamine**, semicarbazide and methoxylamine under mild, neutral conditions to give $N^4$-substituted products. The mechanism of this process involves reaction with the cytosine cation, as illustrated for hydroxylamine (Figure 8.5). The formation of **$N^4$-hydroxydeoxycytidine** is an important mutagenic event in DNA because this modified base exists to a significant extent in the unusual *imino*-tautomeric form (Section 2.1.2) and thus is capable of base-mispairing with adenine.[11]

A third addition reaction at C-6 of cytosine and uracil residues involves the **bisulfite** anion. While this adds reversibly, the intermediate non-aromatic heterocycles undergo a variety of chemical substitution reactions of which the most important are: (i) transamination of cytosine at C-4 by various primary or secondary amines, (ii) hydrogen isotope exchange at C-5, and (iii) deamination of cytosine to uracil.[12] The third process provides the basis for the mutagenicity and cytotoxicity of bisulfite (which is equivalent to aqueous sulfur dioxide). Such mutations are best carried out at pH 5–6 to bring about the deamination and then at pH 8–9 to eliminate bisulfite (Figure 8.6a, they are the likely basis of bottle-sterilisation by Camden tablets in home-brewing). The *in vitro* incorporation of deuterium at C-5 into cytosine is another substitution reaction that requires the addition of a cysteine-thiol to C-6.

This easy nucleophilic addition of sulfur to C-6 of the pyrimidine ring is a key feature of the biological methylation of pyrimidines. Dan Santi has established that the mechanism of action of **thymidylate synthase** involves addition of a catalytic cysteine to C-6 of the deoxyuridylate in conjunction with electrophilic addition of the methylene group of tetrahydrofolate to C-5.[13] It is this process that underpins the activity of the anti-cancer drug, **5-fluorouracil** in which 5-FU acts as a suicide substrate (Section 3.7.1). In a similar fashion, Rich Roberts has shown that cytosine-specific DNA restriction methylases, such as *M. HhaI,* add a catalytic thiol to C-6 of a deoxycytidine residue in conjunction with transfer of a methyl group from *S*-adenosyl-L–methionine to C-5 (Figure 8.6b).[14]

## 8.5 REACTIONS WITH ELECTROPHILES

### 8.5.1 Halogenation of Nucleic Acid Residues

Uracil, adenine and guanine can be **halogenated directly** by chlorine or bromine and so offer easy routes to 5-chloro-(or bromo-)uridines and 8-chloro-(or bromo-)purines (the latter are readily converted into

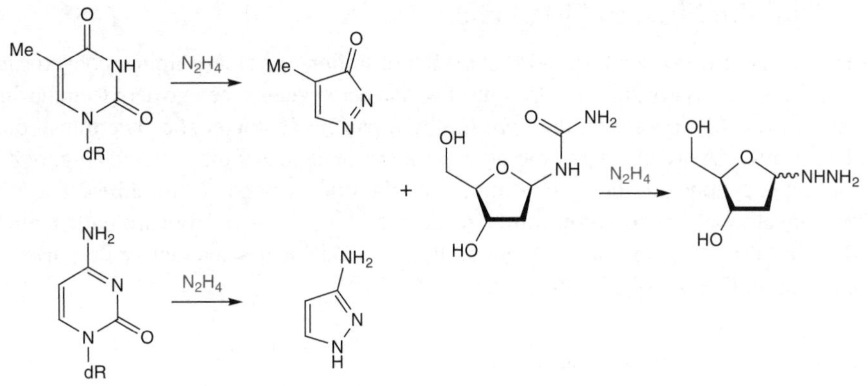

**Figure 8.4** *Hydrazinolysis of pyrimidine nucleosides*

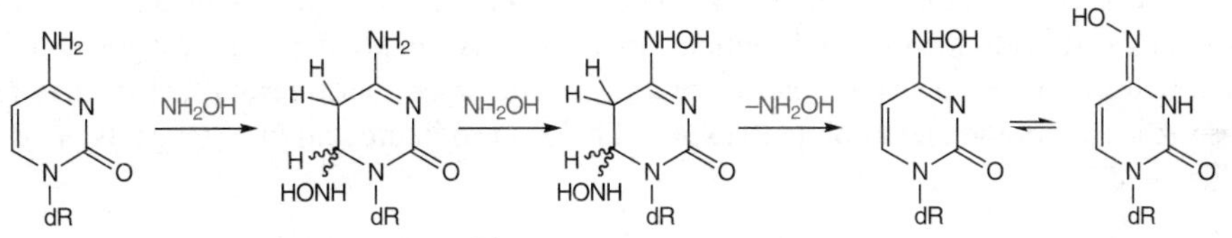

**Figure 8.5** *Reaction of hydroxylamine with deoxycytidine leading to tautomerization of $N^4$-hydroxy-2′-deoxycytidine*

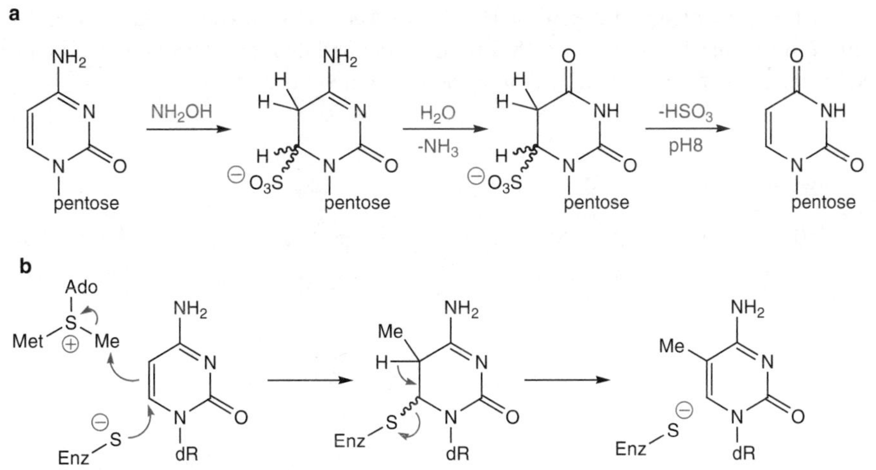

**Figure 8.6** *(a) Mechanism of chemical deamination of cytidine and deoxycytidine catalysed by bisulfite. (b) Schematic mechanism for the restriction methylation of deoxycytidine by S-adenosyl L-methionine catalysed by M Hhal*

8-azidopurine nucleosides for use as photoaffinity labels).[15] It is much more difficult to control the use of elemental fluorine, though fluorine gas has been used in anhydrous acetic acid (care!) to prepare 5-fluorouracil and 5-fluorouridine.[16] 5-Iodouridines are best made by the method described in Section 3.1.4.

### 8.5.2 Reactions with Nitrogen Electrophiles

The standard reaction of **nitrous** acid (as $NO^+$) in the **deamination** of primary amines converts deoxyadenosine into deoxyinosine, deoxycytidine into deoxyuridine and deoxyguanosine into deoxyxanthosine.[17] In each case, the reaction leads to a base with the *opposite* base-pairing characteristic. The transitions dA·dT→dI·dC and dC·dG→dU·dA are characteristic of the mutagenic action of nitrous acid (Figure 8.7).

Aromatic nitrogen cations are the second important class of nitrogen electrophiles. These species are derived either from aromatic **nitro-compounds** by metabolic reduction or from aromatic amines by metabolic oxidation. In both cases, an intermediate hydroxylamine species interacts with purine residues in DNA or RNA either at C-8 or N-7 (Section 8.6.1).

### 8.5.3 Reactions with Carbon Electrophiles

A very large number of reagents form bonds from carbon to nucleic acids. The simplest are species like formaldehyde and dimethyl sulfate. Among the most complex are carcinogens such as benzo[**a**]pyrene, which requires transformation by three consecutive metabolic processes before it can become bound to purine bases in DNA or RNA. Not surprisingly, there is a wide range in selectivity for the sites of attack of these reactive species, some of which have been rationalized in terms of Pearson's **HSAB theory** (HSAB: hard and soft acids and bases). Frontier orbital analysis can provide a more rigorous picture of the problem, but requires a deeper insight into theoretical chemistry. Other relevant factors may relate to the degree of steric access of the electrophile to exposed bases or to intercalation of reagents prior to bonding to nucleotide residues.

*8.5.3.1 Formaldehyde.* Covalent interactions of **formaldehyde** with RNA and its constituent nucleosides take place in a specific reaction of the amino bases. Formaldehyde first adds to the $N^6$-amino group of adenylate residues to give a **6-(hydroxymethylamino)purine** and with guanylate residues to give a **2-(hydroxymethylamino)-6-hydroxypurine**. These labile intermediates can react slowly with a second amino group to give cross-linked products. These have a stable methylene bridge joining the amino groups of two bases. All three possible species, pAdo-$CH_2$-pAdo, pAdo-$CH_2$-pGuo and pGuo-$CH_2$-pGuo, have been isolated from RNA that has been treated with formaldehyde and then hydrolysed with alkali (Figure 8.8). The detailed mechanism of formaldehyde mutagenicity is not yet clear.[18]

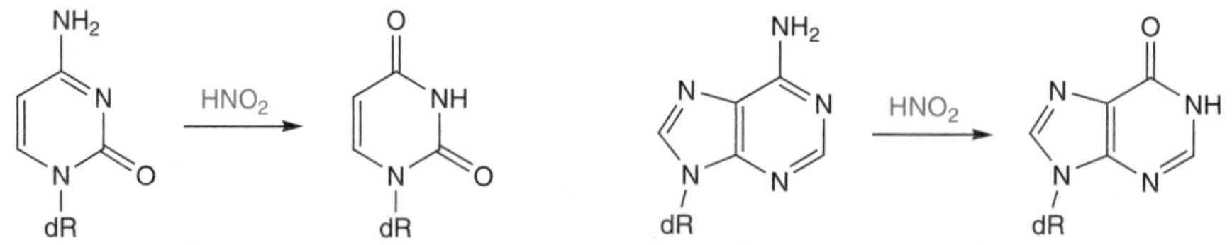


**Figure 8.7** *Pro-mutagenic deamination of dC → dU and of dA → dI by nitrous acid*

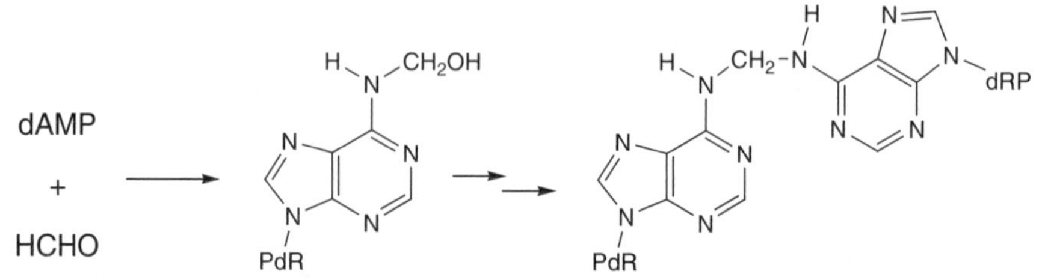


**Figure 8.8** *Formaldehyde mutagenesis of adenine residues*

*8.5.3.2 Alkylating Agents.* Twelve of the nitrogen and oxygen residues of the four nucleic acid bases, in addition to the phosphate oxygen, can be alkylated in aqueous solution at neutral pH. **'Soft' electrophiles**, such as **dimethyl sulfate** (DMS), **methyl methanesulfonate** (MMS) and alkyl halides (such as **methyl iodide**) react in an $S_N2$-like fashion and such alkylation takes place mainly at nitrogen sites with a general selectivity G-N-7 > A-N-1 > A-N-3 > T-N-3. A key measure of 'softness' is a very high ratio of methylation at G-N-7 compared to G-O-6 (typically 250:1).[19] In double-stranded DNA, the major alkylation site for DMS with adenines is at N-3 with lesser substitution at N-7.[20] 'Hard' electrophiles, such as **N-methyl-N-nitrosourea** (MNU) and its ethyl homologue, ENU, are $S_N1$-like alkylating agents. In nucleic acids, MNU **methylation of phosphate esters** can account for up to 50% of total alkylation and also gives higher ratios for G-O-6:G-N-7 products, ranging from 0.08 in liver to 0.15 in brain DNA.[21] Other sites for *O*-alkylation include T-O-2, T-O-4 and C-O-2. The $O^2$-alkylation of ribonucleosides is important for production of modified nucleotides (Section 3.1.4).

In contrast to the C-methylation of nucleic acids by various enzymes, such as thymidylate synthase (Section 3.4), products arising from C-alkylation using electrophilic chemical agents have not been observed.

Many alkylating agents are known to be **primary carcinogens** (agents that act directly on nucleic acids without metabolic activation). An extensive list includes DMS and MMS and their ethyl homologues, **β-propiolactone, 2-methylaziridine**, 1,3-propanesultone and **ethylene oxide**. The list of bifunctional carcinogenic agents includes **bis-chloromethyl ether, bis-chloroethyl sulfide** and **epichlorohydrin** along with such 'first generation' anti-cancer agents as myleran, chlorambucil and cyclophosphamide (Section 3.7.1). In general, 'hard' alkylating agents have been found to be a greater carcinogenic hazard than 'soft' ones (Section 8.10).

*8.5.3.3 Bis-(2-Chloroethyl) Sulfide.* This is the **mustard gas** of World War I as well as of more recent conflicts. It is a typical **bifunctional alkylating agent** in addition to being a proven carcinogen of the respiratory tract. In the early 1960s, Brookes and Lawley showed that it **cross-links** two bases either in the same or in opposite strands of DNA. The typical products isolated (Figure 8.9) have a five-atom bridge between N-7 of one guanine joined either to a second guanine or to adenine-N-1 or to cytosine-N-3. Similar products are formed on alkylation of DNA with 2-methylaziridine. These reagents show little sequence selectivity although nitrogen mustard, $MeN(CH_2CH_2Cl)_2$, shows some preference for alkylation of internal residues in a run of guanines.[22]

*8.5.3.4 Chloroacetaldehyde.* This reagent combines the reactivity of formaldehyde and the alkyl halides. It reacts with adenine and cytosine residues, converting them into **etheno-derivatives**, which have an additional five-membered ring fused on to the pyrimidine ring (Figure 8.10).[23] These modified bases are strongly fluorescent and have been used to probe the biochemical and physiological modes of action of a range of adenine and cytosine species.[24]

*8.5.3.5 Dimethyl Sulfate.* The methylation of deoxyguanosine was also a major feature of the (now obsolete) **Maxam–Gilbert** chemical method for DNA sequence determination (Sections 5.1.1 and 8.4) but is now employed for *in vivo* DNA footprinting (Section 5.8).[25] Following the formation of a

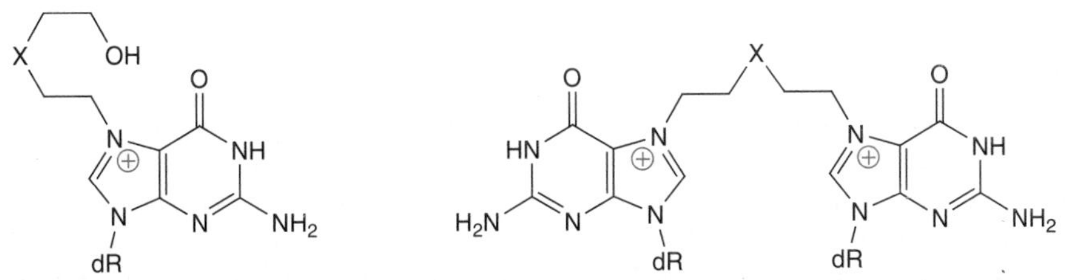


**Figure 8.9** *Mono- and bi-functional products of dG with* sulfur mustard *(X = S) and nitrogen mustard (X = NH or NMe) reagents*

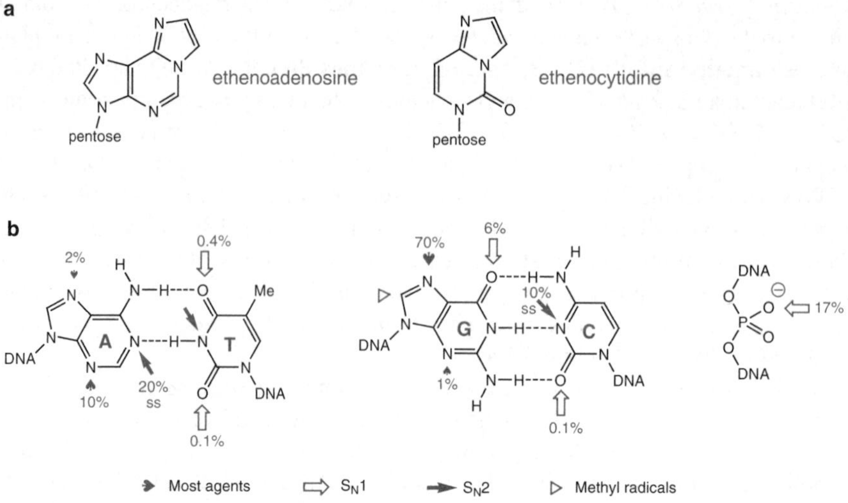


**Figure 8.10** *(a) Alkylation of adenosine and cytidine with chloroacetaldehyde to give fluorescent etheno-derivatives. (b) Sites for the methylation of the DNA bases and sugar-phosphate backbone*[24]

7-methyl-2′-deoxyguanosine residue, treatment of the oligonucleotide with 1 M piperidine at 90°C for 30 min leads to opening of the imidazole ring followed by glycosylic bond cleavage and β-elimination of the phosphate residue. β-Elimination of both phosphates leads to strand cleavage on *both* sides of the dG residue.[10]

## 8.5.4 Metallation Reactions

**Mercury(II) acetate** and chloride readily substitute C-5 of uridine and cytidine nucleosides and nucleotides. The products are easily converted into **organo-palladium** species that are useful intermediates in the synthesis of a range of 5-substituted pyrimidine nucleosides[26] and nucleotides (Section 3.1.4).[27] These include C-5 allyl-, vinyl-, halovinyl- and ethynyl-uridines, all of which have been much studied for possible antiviral activity.

One of the most important recent applications of such chemistry is for the synthesis of **fluorescent chain-terminating dideoxynucleotides**, used in a rapid DNA sequencing (Figure 8.11).[28] 5-Iodo-2′,3′-dideoxyuridine is coupled to *N*-trifluoroacetylpropargylamine using palladium(0) catalysis and the resulting amine is then condensed with a protected succinylfluorescein dye. Deprotection provides the dideoxythymidine terminator species T-526 (which has a fluorescent emission maximum at 526 nm). Related fluorescent derivatives of dideoxycytidine, C-519, dideoxyadenosine, A-512 and dideoxyguanosine, G-505, (the latter are derived by related Heck coupling reactions from 7-deazapurines) provide a complete family of chain-terminators (Figure 8.11). Such modified nucleotides are incorporated with efficiencies comparable to those of unsubstituted ddNTPs and can be used for rapid, single-lane **DNA sequencing** (Section 5.1).

Barnett Rosenberg's discovery of the cytotoxicity of ***cis*-diaminedichloroplatinum(II)** has been carefully developed to make **cisplatin**, the reagent of choice for the successful chemotherapy of **testicular cancer** (and some other cancers), being given FDA approval in 1978. Cisplatin bonds to N-7 in one guanine residue and then links it to N-7 of a second purine. It is selective for d(p**G**p**G**) and d(p**A**p**G**) sequences, but not for d(pGpA) sites and forms predominantly 1,2-intra-strand cross-links. Cisplatin can also bind to two guanines separated by another base, as in d(p**G**pNp**G**pG). Stephen Lippard solved X-ray structures for a model complex of cisplatin with d(p**G**p**G**) in which the platinum is *cis*-linked to N-7 of both guanines and these bases lie in planes almost at right angles.[29]

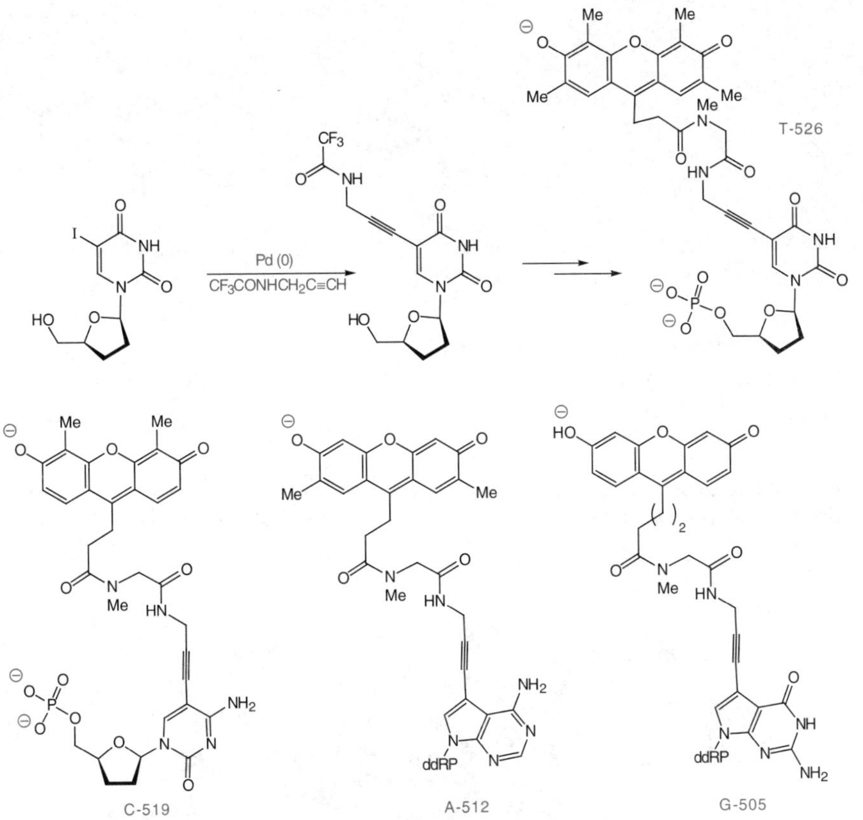


**Figure 8.11** *Synthesis of fluorescent base T-526 using Pd(0) catalysis. Structures of fluorescent dideoxynucleotides related to A, C and G for use in rapid, single-lane DNA sequence analysis*

Intensive development of platinum complexes has identified several analogues of clinical potential including *cis-trans-cis*-ammine(cyclohexylamine)diacetato-dichloroplatinum (Figure 8.12a). More recently, Lippard has solved structures for the complexes of cisplatin, **oxaliplatin** and (Pt(ammine)(cyclohexylamine))$^{2+}$ with the same DNA dodecamer d(CCTCTGGTCTCC).d(GGAGACCACAGG).[31] There is a high degree of homology between these three structures, each of which has an **intra-strand** G-Pt-G link.[30] For oxaliplatin, this lesion induces the duplex to bend toward the major groove (Figure 8.12b). The widened minor groove is shallow and so presents an excellent target for DNA-binding proteins, and these features are also seen in cognate nuclear magnetic resonance (NMR) structures.

## 8.6 REACTIONS WITH METABOLICALLY ACTIVATED CARCINOGENS

Many synthetic chemicals and natural products are known to be carcinogens or mutagens.[32] While these do not react directly with nucleic acids *in vitro*, they are transformed *in vivo* by metabolic processes to give electrophiles that bind covalently to DNA, RNA and also to proteins. Most of these **metabolic transformations** are carried out by the mixed-function **cytochrome P-450 oxidase** (CYP450) enzymes, whose 'proper' function seems paradoxically to be directed towards the detoxification of alien compounds. A well-characterised example is the cytochrome P450 oxidation of vinyl chloride to 2-chlorooxirane which alkylates base residues to give 7-(2-oxoethyl)guanine and other products. The following four classes of metabolically activated compounds are representative of an intensive study of a problem of very grave significance.

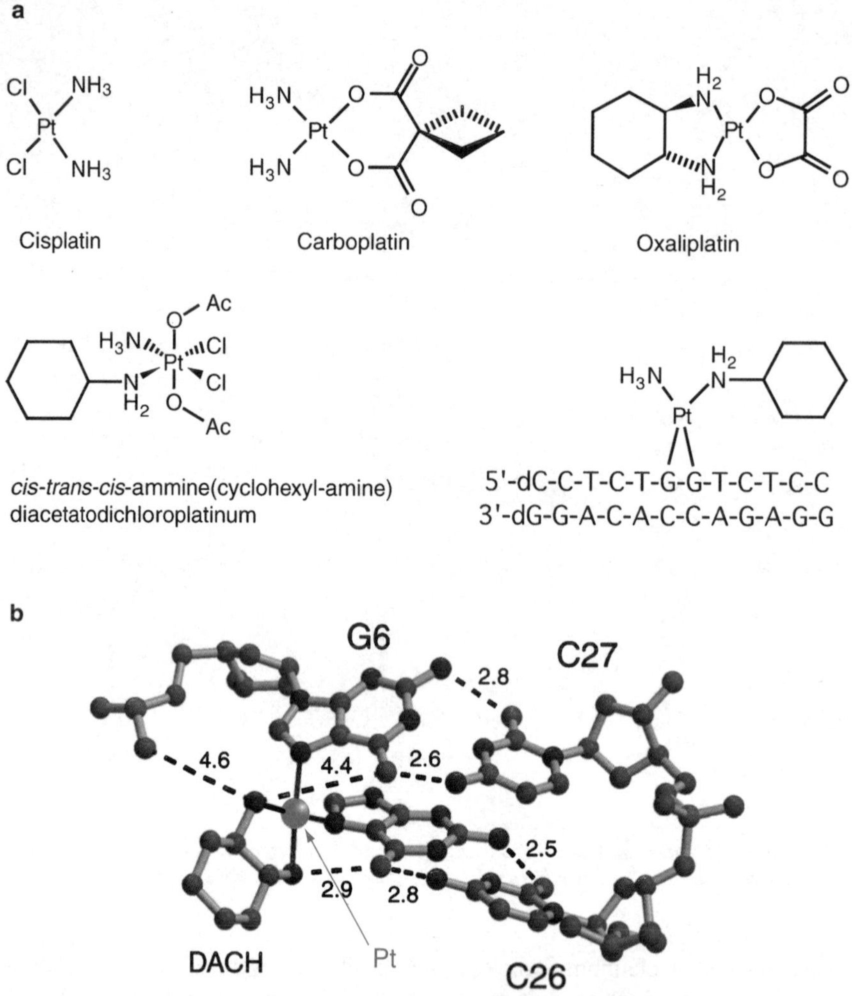


**Figure 8.12** *(a) Structures of cisplatin and some more recent analogues under clinical development. (b) A 2.4 Å molecular structure of oxaliplatin intrastrand cross-link formed with a duplex dodecamer d(CCTCTG-GTCTCC).d(GGAGACCACAGG) showing the G*G* step. Similar intrastrand d(GpG) cross-link structures occur for {cis-Pt($NH_3$)$_2$}$^{2+}$ and {cis-Pt($NH_3$)(Cy$NH_2$)}$^{2+}$ with the same dodecamer. The minor groove is widened and shallow, presenting an excellent target for DNA-binding proteins* (Adapted from Ref. 31. © (2001), with permission from the American Chemical Society)

### 8.6.1 Aromatic Nitrogen Compounds

Investigations of the binding of *N*-aryl carcinogens to nucleic acids began in the 1890s with an investigation of the epidemiology of bladder cancer among workers in a Basel dye factory. The list of chemicals now banned is extensive, but by no means definitive: some examples of proscribed aromatic amines, nitro compounds and azo dyes are illustrated (Figure 8.13).[33]

**Aromatic amines** of this sort are substrates for oxidation by cytochrome P-450 isozymes, which give either phenols, that are inactive and safely excreted, or hydroxylamines. Conjugation of the latter by **sulfotransferase** or **acetate transferase** enzymes converts these proximate carcinogens into **ultimate carcinogens** that can bind covalently to nucleic acid bases, especially to guanine.

The competition between such alternative 'safe' and 'hazardous' metabolic processes is illustrated for **2-acetylaminofluorene** (Figure 8.14a). The corresponding guanine nucleoside adducts have been isolated

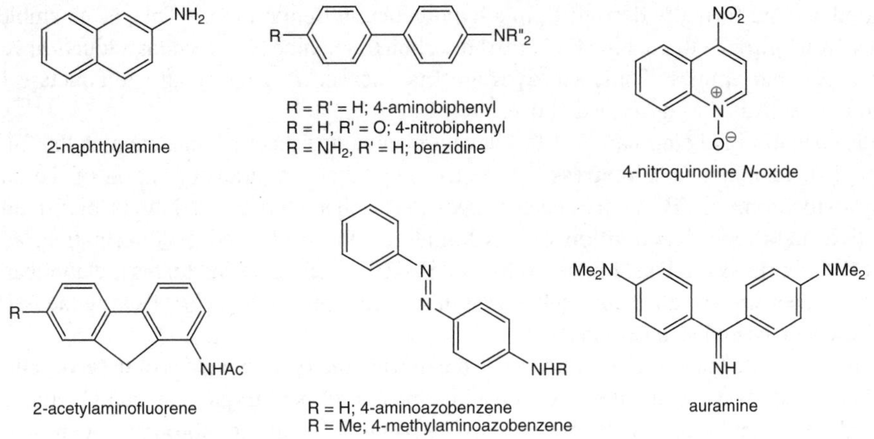


**Figure 8.13** *Examples of N-aryl carcinogens*

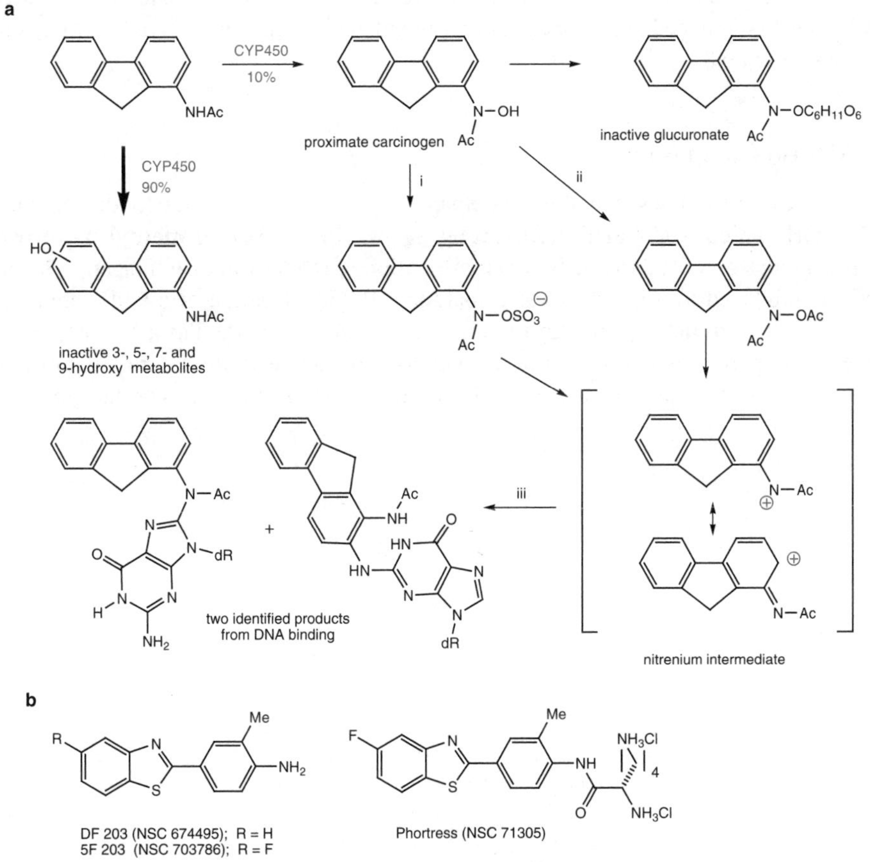


**Figure 8.14** *(a) Metabolic activation of 2-acetylaminofluorene (AAF) and its binding to dG via a hypothetical nitrenium intermediate. (b) Examples of benzothiazolyl anticancer agents. Processes: (i) sulfotransferase; (ii) acetyl transferase; and (iii) binding to DNA in vitro or in vivo*

and identified and are formally derived from a hypothetical nitrenium ion.[34] This, as an ambident cation, bonds either from nitrogen to guanine C-8 or from carbon to guanine N-2. Similar adducts have been identified for many other amines. Thus, azo dyes are first cleaved *in vivo* by an azoreductase to aromatic amines and then activated as described above.

Metabolic oxidation is being harnessed to develop new anti-cancer agents, such as the 2-(4-amino-3-methylphenyl)-benzothiazole, **Phortress.**[35] Selective uptake and biotransformation of the parent amine DF-203 by cytochrome CYP1A1 has been shown to be characteristic of human breast tumour cells. Hydroxylative **metabolic deactivation** of this toluidine can be blocked by fluorination, as in 5F-203 (Figure 8.14b). Phortress is a lysylamide prodrug of 5F-203, in clinical trial, that is metabolically activated to give DNA adducts *in vivo* and cause **cell arrest** in sensitive cells, while cells lacking the ArH signalling pathway show a much-reduced response.

One example of more than usual significance is the **mutagenicity** of two types of **heterocyclic amine** that are found in cooked meats, where they are formed by the pyrolysis of tryptophan and glutamine. Sugimura has identified guanine adducts which are generated by metabolic activation through cytochrome P-450 oxidation and binding to nucleic acids (Figure 8.15).[36] Thus, grilled beef has been estimated to contain nearly 1 ppm of Trp-P-1, while up to 80 ng of this carcinogen has been isolated from the smoke of a single cigarette!

**Aromatic nitro compounds** are found in diesel engine emission, urban air particles and photocopier black toners and some have been identified as mutagens in the Ames' test. They can be reduced to aryl hydroxylamines by anaerobic bacteria in the gut, by xanthine oxidase, or by cytochrome P-450 reductase to give substrates for the processes described above. The best-studied example is **4-nitroquinoline *N*-oxide**, which is first reduced to a hydroxylamine and then binds to DNA *in vivo,* forming characteristic guanine adducts (Figure 8.16).[37]

## 8.6.2 *N*-Nitroso Compounds

Nitrosoureas, nitrosoguanidines and nitrosourethanes hydrolyse to give methyldiazonium hydroxide, Me—N=N—OH, which is a **'hard' methylating agent**. (In the case of **methyl *N*-nitrosoguanidine** (MNNG), thiol groups may catalyse the ***in vivo* methylation of DNA** by this carcinogen). The same methylating species is produced as a result of the cytochrome P-450 oxidation of a wide range of *N*-methylnitrosamines, of which **dimethylnitrosamine** was the first to be identified as a carcinogen in 1956. The common metabolic pattern is hydroxylation of one alkyl residue to form a carbinolamine. This breaks down to give methyldiazohydroxide (Figure 8.17). Many *N*-nitroso compounds of this type have proved to

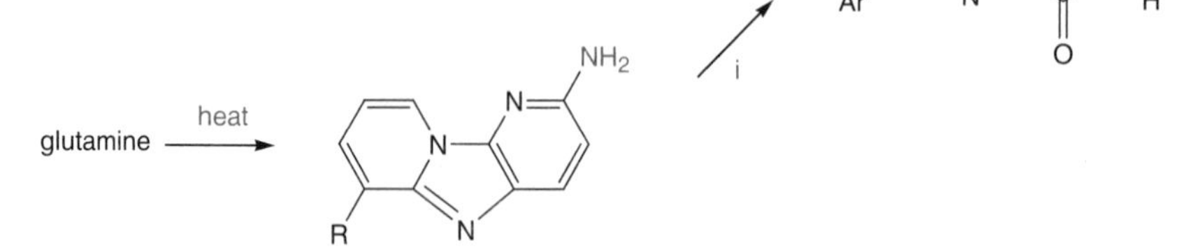


**Figure 8.15** *Metabolic activation of heterocyclic amines from amino acids in cooked food and their binding to DNA; R = H or Me. Processes: (i) cytochrome P-450; (ii) DNA binding; and (iii) hydrolysis*

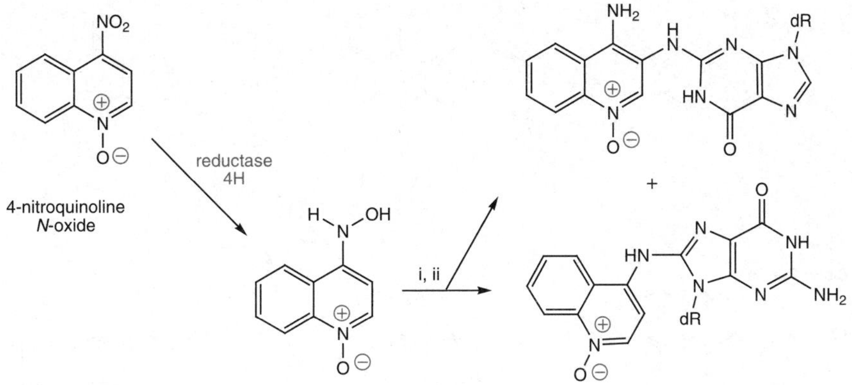


**Figure 8.16** *Reductive activation of 4-nitroquinoline N-oxide and products resulting from its binding to DNA in vivo. Reagents: (i) DNA in vivo; and (ii) hydrolysis*

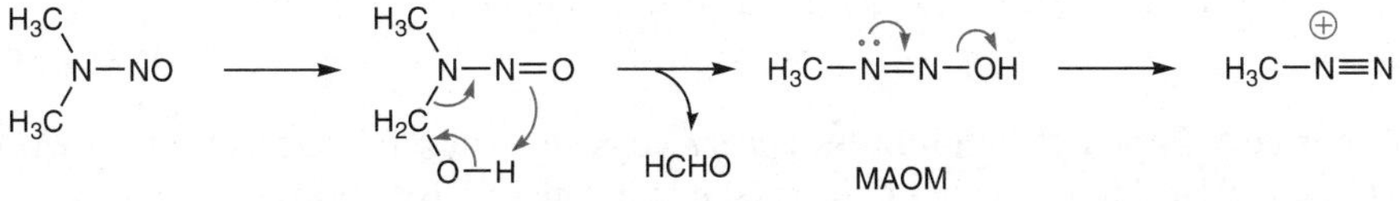


**Figure 8.17** *Cytochrome P-450 oxidation of dimethylnitrosamine and its conversion into methyl azo-oxymethanol (MAOM) en route to DNA methylation*

be carcinogenic in animals and lead to methylation, ethylation, or propylation of DNA, as described above (Section 8.5.3).[21]

### 8.6.3 Polycyclic Aromatic Hydrocarbons

The **polycyclic aromatic hydrocarbons** (PAHs) provided the first example of an industrial carcinogen, **benzo[a]pyrene** (BaP). Its identification marked the initial stage in the molecular analysis of hydrocarbon carcinogenesis, which had begun with Percival Pott's study of scrotal cancer in chimney sweeps in 1775. BaP becomes covalently bound to DNA *in vivo* following a series of three metabolic changes.[38] In the first, **cytochrome CYP1A1** (formerly $P_1$-450) adds oxygen to BaP to give the two enantiomers of **BaP-7,8-epoxide**. Next, these are used as substrates for an **epoxide hydrolase** that converts them into the two enantiomeric *trans*-dihydrodiols. Finally, both diols are again substrates for cytochrome CYP 1A1 and are converted into three of the four possible stereoisomers of the dihydrodiol-9,10-epoxide, BPDE (Figure 8.18).[39]

The carcinogenicity and DNA-binding capability of such **dihydrodiol epoxides** is closely linked to **'Bay Region'** architecture, so called because of the concave nature of this edge of the PAH, which appears to be strongly recognised by the metabolizing enzyme. Among the products that have been characterised are adducts with guanine N-2, guanine N-7, guanine O-6 and adenine N-6. These are formed as a result of a rapid intercalation of the BPDE into $d(A{\cdot}T)_n$-rich parts of the DNA helix, which manifest as a red shift in UV absorption of the hydrocarbon and a negative CD spectrum for the complex. A rate-determining protonation of the C-10 hydroxyl group then leads to the formation of a carbocation that reacts predominantly (90%) with water to give the harmless 7,8,9,10-tetra-ol but less frequently (10%) binds to a proximate nucleic acid base, often a dG residue.

The resulting covalent adducts appear to be of two distinct types. The minor 'site I' adducts have the hydrocarbon still intercalated in an intact DNA helix. The major 'site II' adducts appear to have the hydrocarbon lying at an angle to the helix axis, either in the minor groove of a DNA helix or forming a wedge-shaped intercalation complex. Similar results have been found for **chrysene**, while the Bay Region

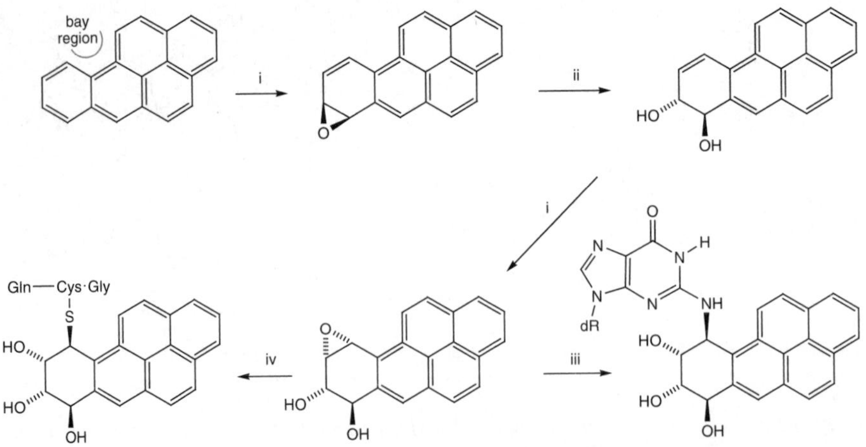


**Figure 8.18** *Metabolic activation of B[**a**]P to BPDE (major stereoisomer illustrated) and its binding to DNA in vivo to give guanine adducts (major product shown). Processes: (i) Cyt $P_1$-450; (ii) epoxide hydrolase; (iii) DNA binding in vivo; and (iv) gluthathione-S-transferase*

dihydrodiol epoxide of **3-methylcholanthrene** appears to be too bulky to intercalate into DNA. Work at the National Institutes of Health has analysed the crystal structure of a BPDE-adenine adduct.[40] A fully synthetic heptadecamer was prepared containing the **BPDE-adenine** base-paired with thymine at a template-primer junction and then complexed with the lesion-bypass DNA polymerase Dpo4 and an incoming nucleotide. Two conformations of the BPDE-adduct are observed: one is intercalated between base-pairs and another is solvent-exposed in the major groove (Figure 8.19). These structures suggest possible mechanisms by which mutations are generated during replication of DNA containing BPDE adducts.

This type of epoxidation is not restricted to synthetic chemicals. One of the most potent groups of carcinogens is the **aflotoxins**, which are fungal products from *Aspergillus flavus*. A dose of less than 1 ppm of aflatoxin B1 can cause lung, kidney and colon tumours in rats and is directly attributable to its oxidation to an epoxide that binds covalently to guanine residues in DNA (Figure 8.20).[41,42]

While both *endo*- and *exo*-epoxides are formed metabolically, only the *exo*-isomer (Figure 8.20) is mutagenic. It seems likely that this metabolite intercalates into the DNA helix with optimal orientation for an $S_N2$ reaction with N-7 of a proximate guanine residue. In contrast, intercalation of the non-mutagenic *endo*-isomer places its epoxide in a non-reactive orientation.

## 8.7 REACTIONS WITH ANTI-CANCER DRUGS

A large number of **'first generation' anti-cancer drugs** were designed to combine a simple alkylating function such as a nitrogen mustard, an aziridine, or an alkanesulfonate ester with another function designed to direct the agent towards the target tissue. Most of these compounds turned out to be less tumour-selective than one might have hoped and, what is worse, many of them have proved to be carcinogens that eventually led to new tumours sometime after the termination of chemotherapy for the original cancer. As a result, their general use is now viewed with some suspicion. However, one success is **Temozolomide**. This compound, invented by Malcolm Stevens and widely used for brain tumours since 1998, is activated by spontaneous hydrolysis and subsequent breakdown to generate the methyl diazonium cation.[43] This is a **'hard' methylating agent** which alkylates guanine residues in the major groove of DNA, prefers runs of guanine tracts, and leads to the formation of ***O*$^6$-methylguanine** in target tissues. The success of Temozolomide likely results from a combination of a slight difference in pH of normal and malignant cells coupled to their reduced ability to repair the *O*$^6$-MeG lesion by ***O*$^6$-alkylguanine-DNA alkyl transferase** (ATase) (Section 8.11.1) in brain tumour and melanoma cells (Figure 8.21).[44]

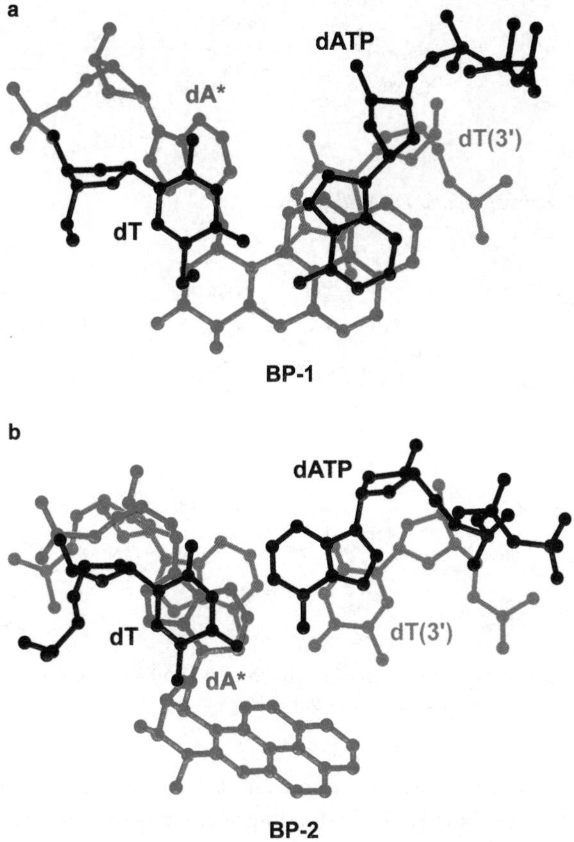


**Figure 8.19** *Distortion of DNA by a BPDE adduct: hydrogen-bond formation in the crystal structure of the dA*(dT and the adjacent replicating base-pair dT·dATP. (a) Looking down the DNA helical axis, the two layers of the base-pair and the PAH adduct are shown, black for the replicating base-pair, red for the dA* adduct and grey for its dT partner. The incoming nucleotide in BP-1 is in the syn conformation. (b) In the BP-2 complex, where the PAH is in the major groove, the adenine base of the dA* is shifted to the major groove, disrupting the normal hydrogen bonds with its dT partner. The location of a normal dA is depicted in grey* (Adapted from Ref. 40. © (2004), with permission from the National Academy of Sciences, USA)

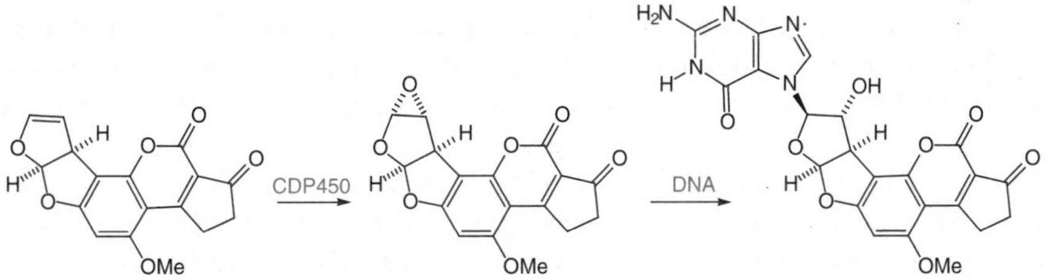


**Figure 8.20** *Metabolic oxidation of aflatoxin $B_1$ and binding of its exo-epoxide to DNA*

A group of **'second generation' anti-cancer agents** has emerged, many of which are natural products, but now augmented by a growing number of rationally designed, synthetic drugs. Their common feature is that they appear to form an initial physical complex with DNA before bonding to it covalently. This heterogeneous group of compounds includes aziridines such as **mitomycin C**[45], several **pyrrolo[1,4]benzodiazepines** and **spirocyclopropanes** such as CC-1065.[46] Their vital purpose is to kill bacteria by disrupting the synthesis

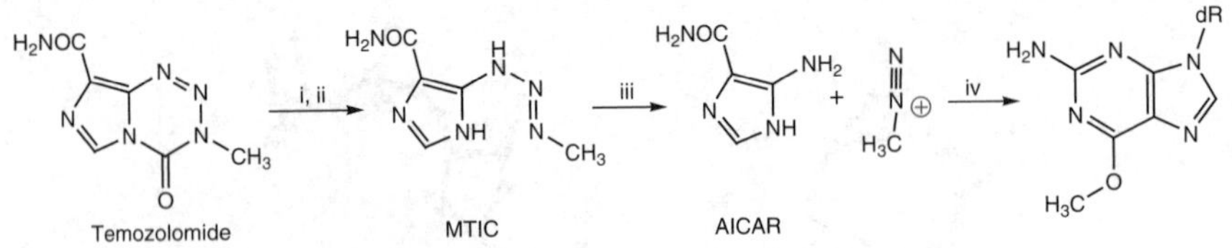


**Figure 8.21** *Temozolomide activation and DNA alkylation*

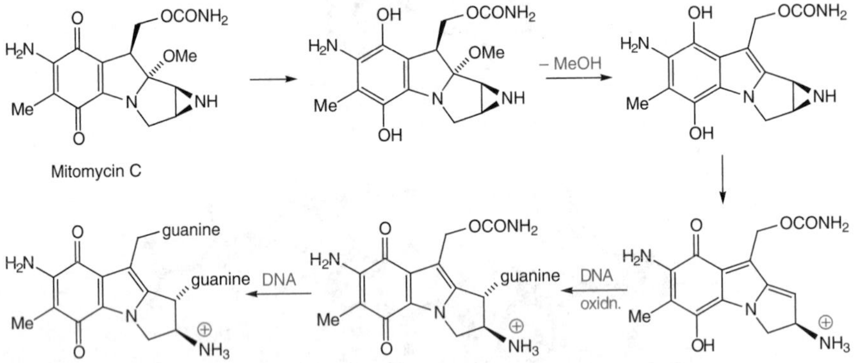


**Figure 8.22** *Activation of mitomycin C by metabolic reduction and bifunctional alkylation of DNA at the 2-amino group of adjacent inter-strand deoxyguanines*

of DNA and RNA, but many of them have also shown useful anti-tumour activity, which must arise from selective toxicity. This can be attributed to DNA-binding specificity or to preferential metabolic activation by tumour cells.

### 8.7.1 Aziridine Antibiotics

An assortment of naturally occurring antibiotics, each having an aziridine ring, has been isolated from *Streptomyces caespitonis*. The most interesting of them in clinical terms is mitomycin C. This compound requires **enzymatic reduction** of its quinone function to initiate the processes that cause it to alkylate DNA. It seems likely that the second step is elimination of methanol that potentiates either monofunctional or **bifunctional alkylation** (Figure 8.22). The antibiotic has been shown to interact with DNA at G-O-6 > A-N-6 > G-N-2 and forms one cross-link for about every ten monocovalent links. The primary process is bonding of the 2-amino group of a guanine residue to C-1 of the reductively activated mitomycin. This reaction shows selectivity for 5′-CG sequences. Cross-linking is completed by alkylation of the 2-amino group of the second guanine to C-10 of the mitomycin (Figure 8.22). This has been accurately analysed by Dinshaw Patel in NMR studies on the adduct of mitomycin C to the hexamer (TACGTA).d(TACGTA) in which the two guanines are crosslinked with the mitomycin molecule positioned in the minor groove of the duplex.[47]

Many drugs that act on DNA exhibit a requirement for **reductive activation**, including adriamycin, daunomycin, actinomycin, streptonigrin, saframycin, bleomycin (Section 9.7) and tallysomycin in addition to mitomycin C. While there is no common factor uniting the chemistry of DNA modification by these agents, the fact that tumour tissues seem to have a higher reducing potential than normal tissue has led to the concept of bioreductive drug activation.

**Carzinophilin A** is also a DNA-alkylating aziridine antibiotic, though it does not appear to need reductive activation.[48] It has been identified as the antibiotic azinomycin B, isolated from *Streptomyces griseofuscus*. **Azinomycin B** operates in the major groove of DNA, causing cross-linking between a guanine residue and a purine residue that is two bases removed in the duplex, as in the sequence d(**G**.Py.Py)·d(C.Pu.**Pu**).

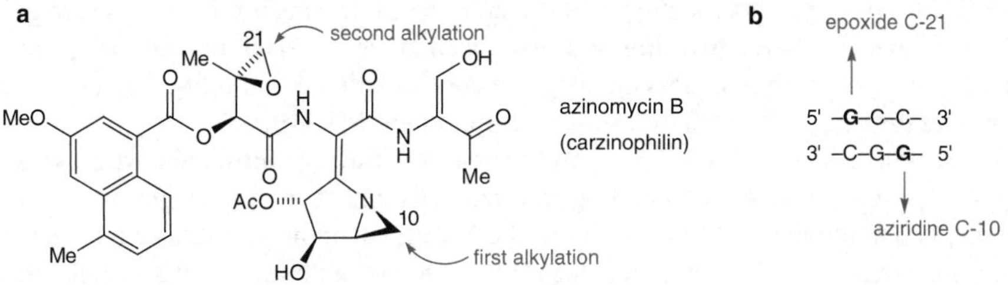


**Figure 8.23** *(a) Structure of azinomycin B (identical with carzinophilin) with sites for nucleophilic attack by N-7 of guanine (purine) residues in opposite strands of DNA. (b) Preferred duplex sequence for its cross-linking to sub-adjacent dG residues*[49]

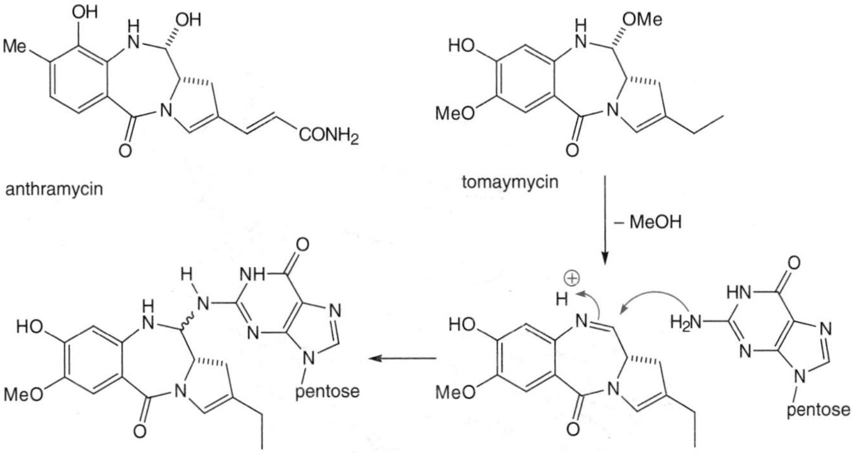


**Figure 8.24** *Binding of P[1,4]B antibiotics to the N$^2$-amino group of guanine*

Robert Coleman has shown that initial alkylation is at N-7 of a dG residue involving the aziridine ring C-10 (Figure 8.23a). It is followed by a slower alkylation of the sub-adjacent purine by the epoxide C-21 as the second alkylating function (Figure 8.23b).[49] The selectivity for the target sequence appears to be determined by the relative nucleophilicity of purines in the major groove while the naphthalene moiety provides hydrophobic binding to DNA without intercalation.

## 8.7.2 Pyrrolo[1,4]Benzodiazepines, P[1,4]Bs

Anthramycin and tomaymycin, along with sibiromycin and neothramycins A and B, are members of the potent P[1,4]B anti-tumour antibiotic group produced by various actinomycetes (Figure 8.24). The first three of these compounds bind physically in the minor groove of DNA where they then form covalent bonds to G-N-2, showing a DNA sequence specificity for 5′-PuGPu sequences.

These P[1,4]Bs appear to interact with DNA in a biphasic process. Initially there is a rapid, non-covalent association that results from a close interaction of the antibiotic with the 'floor' of the minor groove of DNA (Section 10.3.5). Subsequent loss of water or methanol and covalent addition of G-N-2 to C-11 then forms an aminal linkage that is well stabilized by favourable steric and electrostatic interactions.[50]

Lawrence Hurley has established the existence of two distinct tomaymycin-d(ATGCAT)$_2$ species in solution from NMR studies. These have the antibiotic orientated in opposite directions in the minor groove according to its (*R*)- or (*S*)-configuration at C-11. The resulting lesions appear neither to impede Watson–Crick base-pairing nor to distort the B-DNA helix structure so that the two lesions probably pose difficult recognition problems for DNA repair systems (Section 8.11). **Tomaymycin** has been shown to induce greater conformational changes, namely **helix bending** and associated **narrowing of the minor groove**, than does anthramycin. It thus appears that sequence-dependent conformational flexibility may be an important factor in determining the selectivity for DNA sequence binding of P[1,4]Bs.[51]

Richard Dickerson has solved a 2.3 Å X-ray crystal structure of **anthramycin** bonded to a duplex decamer d(C-C-A-A-C-G-T-T-**G**-G). One drug molecule sits within the minor groove at each end of the helix, covalently bound through its C-11 position to the N-2 amine of the penultimate guanine of the chain.[52] The configuration at C-11 is (*S*) for both residues (Figure 8.25). With this configuration, the natural twist of the anthramycin molecule matches the twist of the minor groove whereas a C-11(*R*) drug would fit only into a left-handed helix. The acrylamide tail attached to the five-membered ring extends back along the minor groove toward the center of the helix, binding in a manner reminiscent of netropsin or distamycin. The origin of anthramycin specificity for three successive purines appears to arise not from specific hydrogen bonds but from the low twist angles adopted by purine–purine steps in a B-DNA helix (Section 2.2.4).

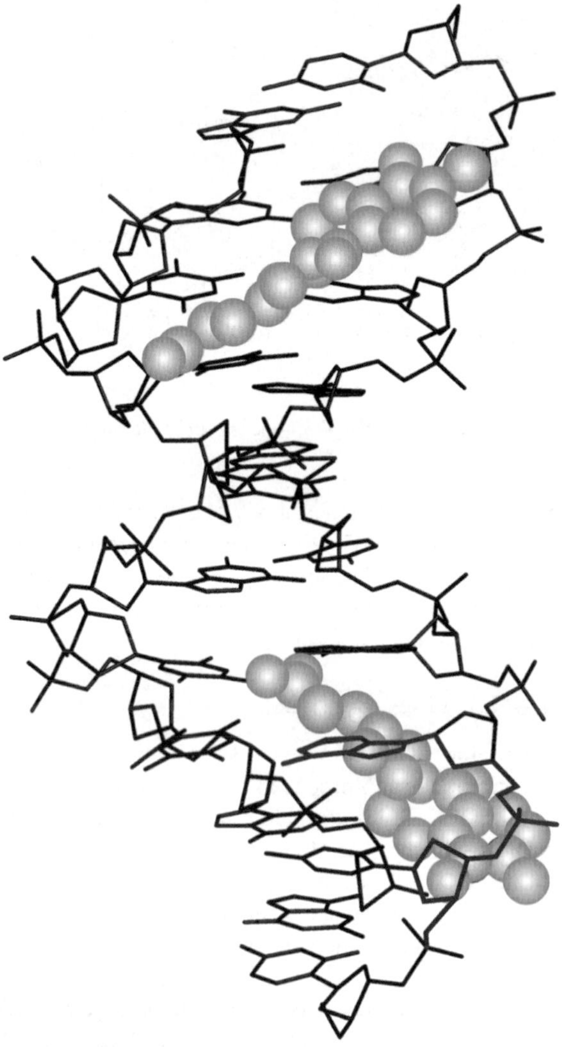

**Figure 8.25** *Crystal structure of a covalent DNA-drug adduct: anthramycin bound to C-C-A-A-C-G-T-T-G-G and a molecular explanation of specificity*[52] (Adapted from a figure kindly provided by M. Kopka.)

### 8.7.3 Enediyne Antibiotics

A range of clinically significant anti-cancer drugs can mediate **oxygen-dependent cleavage** of the ribose phosphate backbone of DNA. They can be broadly assigned into three classes.

- Generators of reactive carbon radicals
- Photo generators of hydroxyl radicals
- Metal-mediated activators of $O_2$.

The first class contains the **enediyne antibiotics**, whose interaction with DNA is more specific than that of many alkylating agents and is irreversible. The second class includes antibiotics such as tetrazomine and quinocarcin, where redox chemistry ultimately results in the reduction of oxygen to superoxide and leads to 'nicking' of DNA. The third class is well represented by the bleomycins, which are discussed in Chapter 9 as compounds that interact reversibly with DNA (Section 9.7).[53]

The structure of the chromophore of the antibiotic **neocarzinostatin**, NCS, was established in 1985 and was soon followed by those of calicheamicin $\gamma_1$, esperamicin C, dynemicin A, kedarcidin and C-1027 (Figure 8.26a). A common feature of these compounds is a highly unsaturated medium-sized ring, which contains a 1,5-diyne-3-ene arrangement of multiple bonds, —C≡C—CH=CH—C≡C—. They have become known as the *enediyne antibiotics* and have taken a place at the forefront of research in biology,

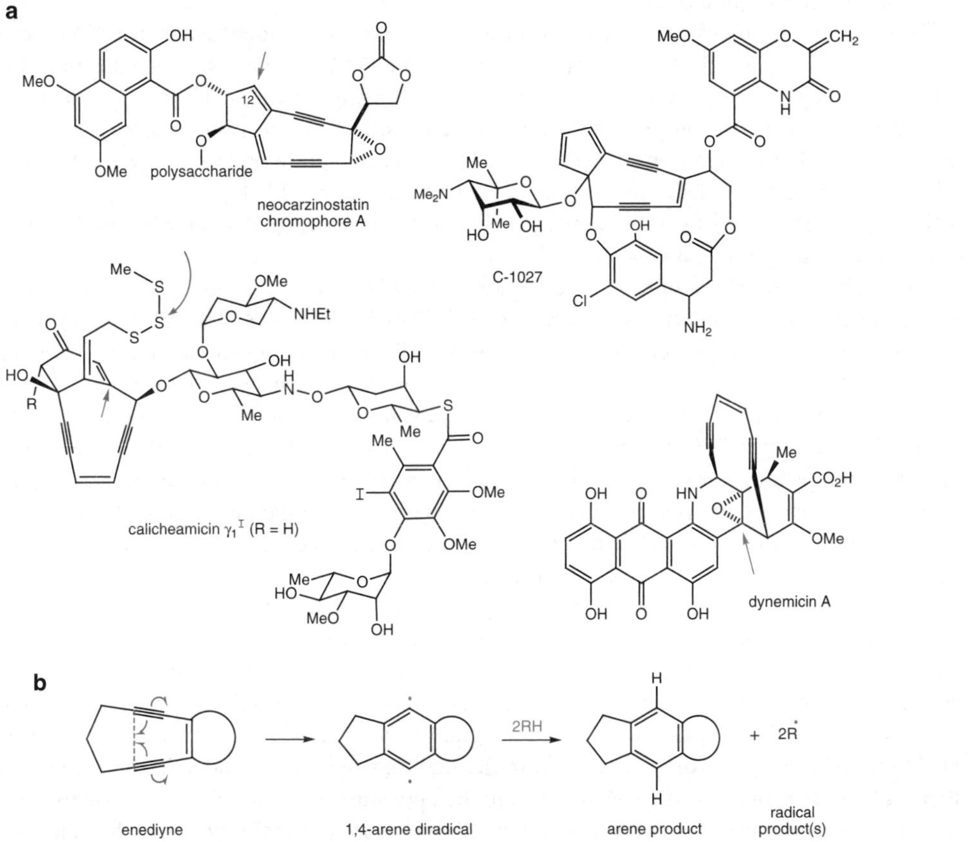


**Figure 8.26** *(a) Structures of the enediyne antibiotics: neocarzinostatin (chromophore), calicheamicin $\gamma_1^I$, esperamicin A (chromophore), dynemicin A and antibiotic* C-1027. *The site of thiol attack is indicated (→) for the first three antibiotics. (b) The Bergman enediyne cyclization gives a 1,4-benzenoid diradical (not quite the same in the case of neocarzinostatin), which then abstracts two hydrogen atoms to give the stable arene product*

chemistry and medicine since this group of compounds contains some of the most potent anti-tumour antibiotics known. They are about a thousand times more toxic than the clinically used **adriamycin** and anthracycline antibiotics and generally induce **apoptosis** through a caspase-mediated mitochondrial amplification loop.[54] In particular, the high cytotoxicity of calicheamicin has been harnessed by means of antibody directed drug delivery[55] to target human myeloid leukaemia tumour cells in a Food and Drug Administration (FDA) approved product, **Mylotarg™**.

The mode of action of the enediyne antibiotics involves single or double strand scission of DNA and depends on the formation of carbon radicals followed by hydrogen atom abstraction from specific nucleotides. The key enediyne reaction is an electrocyclisation, as reported by Bergman in 1972, which generates a 1,4-benzenoid diradical (Figure 8.26b). In the case of the very unstable antibiotic C-1027, this process takes place merely on warming the antibiotic to 50°C in solution in ethanol. For the other compounds, a distortion of the enediyne ring inhibits the **Bergman cyclisation**.[56] The rearrangement is 'triggered' by a chemical process that relaxes the ring and so facilitates a change in conformation of the enediyne to allow the electrocyclisation to take place spontaneously. In the cases of NCS, calicheamicin and esperamicin, this trigger is the attack of a thiol, possibly glutathione, at the site indicated (arrows in Figure 8.26a). In the case of dynemicin, opening of the epoxide follows initial biological reduction to a quinol by nucleophilic attack at the position indicated.

While the enediyne provides the warhead for strand breaking, sequence selectivity and orientation of the enediyne to the DNA target is delivered by minor groove binding and/or intercalation of peripheral components of these antibiotics.

The *neocarzinostatin* chromophore acts primarily by single-strand cleavage of DNA, and this requires oxygen. NCS first intercalates its naphthoate residue into the DNA duplex, which positions the remainder of the molecule in the minor groove (Section 8.7.1). Following activation of the molecule by thiol addition at C-12, Bergman cyclisation generates a **benzenoid diradical**, which abstracts a 5′-hydrogen atom from a residue in the DNA recognition sequence.[57] Such action takes place preferentially at dA and dT sites with at least 80% of the DNA cleavage resulting in the formation of 5′-aldehydes of A and T residues. Less than 20% of strand breaks result from pathways initiated by a second **hydrogen abstraction** in the alternate strand from a deoxyribose at C-4′ or C-1′.

The NMR structure in solution of a complex formed between *calicheamicin* and a DNA hairpin containing the preferred recognition sequence d(T4-C5-C6-T7)·(A17-G18-G19-A20) has been determined by Patel (Figure 8.27).[58] Sequence-specific binding of **calicheamicin** $\gamma_1^I$ to the (T-C-C-T) containing DNA hairpin duplex is favoured by the complementarity of fit through hydrophobic and hydrogen-bonding interactions between the antibiotic and the floor and walls of the minor groove of a minimally perturbed DNA helix (Section 9.7). Calicheamicin $\gamma_1^I$ binds with its arene-tetrasaccharide segment in an extended conformation spanning the (T-C-C-T)·(A-G-G-A) segment of the duplex minor groove. Its thiol-sugar and thiobenzoate rings are inserted in an edgewise manner deep into the minor groove with their faces sandwiched between its walls, where hydrophobic and hydrogen-bonding interactions account for the (TCCT) sequence recognition in the complex (Figure 8.27a).

This positioning of the arene-tetrasaccharide moiety orientates the enediyne ring deep in the minor groove, spanning both strands, such that its pro-radical carbon centers, C-3 and C-6, are proximal to the anticipated H-5′ and H-4′ sites for hydrogen atom abstraction. When a thiolate nucleophile adds in Michael fashion to the proximate α,β-unsaturated ketone, the resulting change in geometry of the enediyne triggers the Bergman cyclisation to generate a benzenoid 3,6-diradical. This is suitably orientated to abstract one 5′-hydrogen atom from the first deoxycytidine residue in the recognition sequence d(TCCT) and a second 4′-hydrogen atom from the opposite strand and this leads oxidatively to a **double strand cleavage** process (Figure 8.28). It is worth noting that the affinity of calicheamicin for DNA has been increased 1000-fold through synthesis of head-to-head and head-to-tail dimers with significant benefit to sequence-selectivity for TCCT and ACCT sequences.[59]

Esperamicin $A_1$ works in a similar fashion but with low sequence selectivity and favours cleavage at T > C > A > G. Its binding to DNA involves a combination of upstream intercalation of an

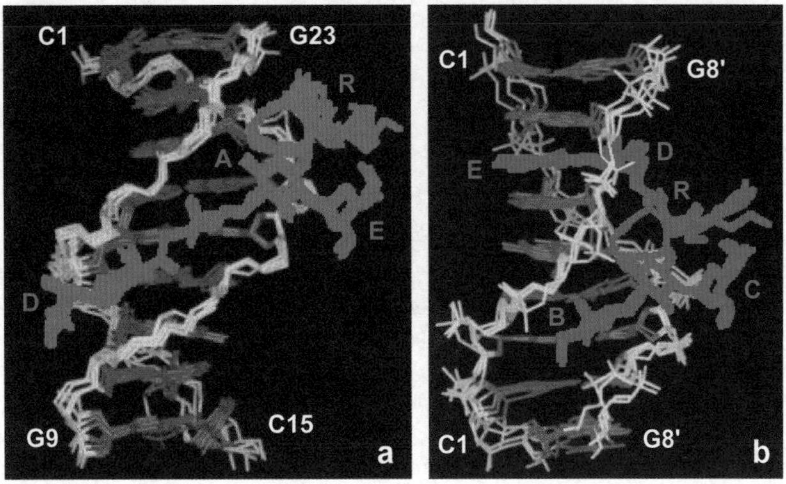


**Figure 8.27** *A superimposed set of NMR structures of the enediyne antibiotics (red) binding to DNA duplexes. In each case the orientation of the macrocyclic enediyne is shown spanning the groove and saccharide moieties buried deep in the minor groove. (a) Calicheamicin $\gamma_1^{I}$. (b) Esperamicin A showing intercalation of the anthranilate moiety*
(Adapted from Refs 58 and 60. © (1997), with permission from Elsevier)

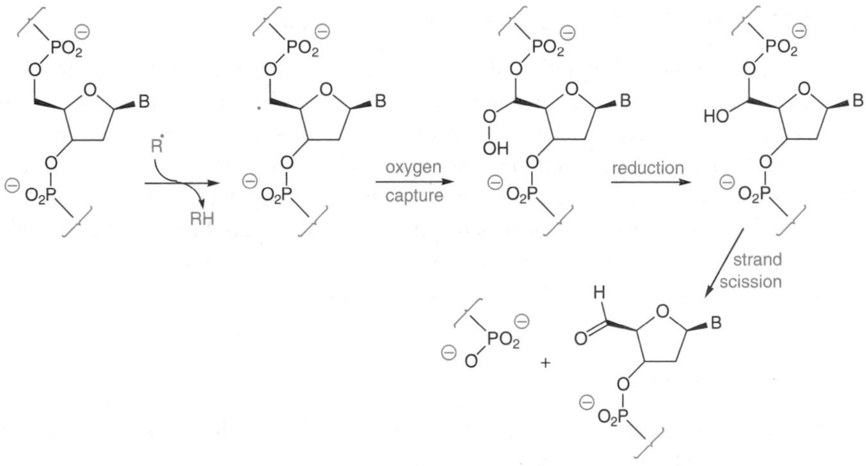


**Figure 8.28** *DNA single-strand cleavage by 5′-hydrogen abstraction by an aryl radical, followed by oxygenation and biological reduction*

ethoxyacrylyl-anthranilate moiety with downstream minor groove binding of a trisaccharide unit (Figure 8.27b).[60] Hydrogen atom abstraction follows the pattern H-5′ > H-1′, and this results in rather more single than double strand cuts. Intercalative DNA binding has also been identified for the antibiotic C-1027 through a combination of hydrodynamic and spectroscopic studies.[61] Lastly, it seems likely that dynemicin A also binds to DNA by a combination of intercalation and groove binding. It is activated by thiols (bioreduction) or by light and also causes both single and double strand DNA cleavage.

The general mechanism of strand cleavage by removal of a 5′-hydrogen is common to all these antibiotics, as shown in Figure 8.28. Processes involving hydrogen atom abstraction from C-4′ or C-1′ are illustrated later (Section 8.9.1, Figure 8.35).

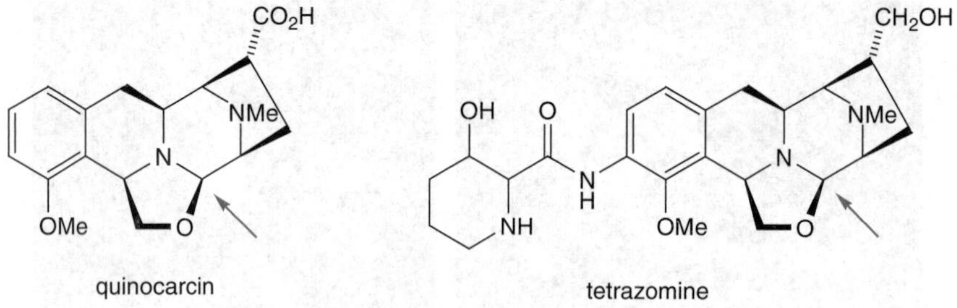


**Figure 8.29** *Antibiotics that interact with DNA by generation of superoxide leading to strand cleavage, probably through ring opening and formation of a peroxide radical at the positions indicated (→)*

### 8.7.4 Antibiotics Generating Superoxide

*Tetrazomine* is a secondary metabolite that is a member of the **quinocarcin**/saframycin class of anti-tumour agents. It has antibacterial activity as well as promising *in vivo* activity against leukaemia in mice. Quinocarcin has been used in clinical trials for a range of solid tumours (Figure 8.29). These compounds undergo a spontaneous reaction involving a stereospecific self-disproportionation of the oxazolidine ring to generate the **superoxide radical**, $HO_2^{\bullet}$. That leads on to a radical-initiated cleavage of DNA whose details are still under examination.[62]

## 8.8 PHOTOCHEMICAL MODIFICATION OF NUCLEIC ACIDS

The very serious concern about the depletion of the global ozone barrier is directly related to the action of UV light on nucleic acids. The UV effect is mutagenic at low doses, cytotoxic at high doses, and is linked to skin cancer in many cases where there is chronic, excessive exposure to sunlight, among whites, albino blacks, or for people with deficiencies in their repair genes. As a rule, a 10% reduction in the ozone layer causes *ca.* 20% increase in UV-radiation and a 40% increase in skin cancers.[63] The photolesions in DNA caused by direct excitation or triplet photosensitisation are largely confined to the pyrimidine bases, thymine and cytosine, while guanine is the main target for photo-oxidation.[64,65] In contrast, adenine is largely resistant to photomodification under all irradiation conditions.

### 8.8.1 Pyrimidine Photoproducts

Light of 240–280 nm excites the pyrimidine bases, C, T and U, to a higher **singlet state** ($^1S_1$) which has a lifetime of only a few picoseconds before it gives photohydrates (in which water has added to either face of the 5,6-double bond), decays, or passes into the triplet state. Uridine photohydrate (U*) dehydrates slowly to uridine in acidic or alkaline solution and is moderately stable at neutral pH ($t_{1/2}$ 9 h at 50°C). The cytidine photohydrate is some tenfold less stable ($t_{1/2}$ 6 h at 20°C) and either reverts to cytidine (90%) or is deaminated to give U* (10%) (Figure 8.30a).[66] This process effects a net conversion of C into U (Section 8.11.2). The formation of **photohydrates** of thymine has a very low quantum yield and its biological consequences are not significant.

All the major pyrimidines form **cyclobutane photodimers** on direct irradiation at 260–300 nm. The reaction is a [2 + 2] cycloaddition, mainly involving the triplet state. Of the four possible isomers for thymine dimer, T<>T, the *cis-syn* isomer is formed by irradiation of thymine in an ice matrix and is known to be the major product (>95%) formed by UV irradiation of native DNA. The *trans-syn* isomer is one of the four isomeric products produced by the **photosensitised irradiation** of thymidine in solution and accounts for some 2% of the native DNA T<>T. A larger proportion of this **thymine photodimer** is formed in denatured DNA (Figure 8.30b), where the *trans-syn*, *cis-syn* T<>T and T<>U dimers account for 1.6, 11.0 and 4.2% of total thymine.

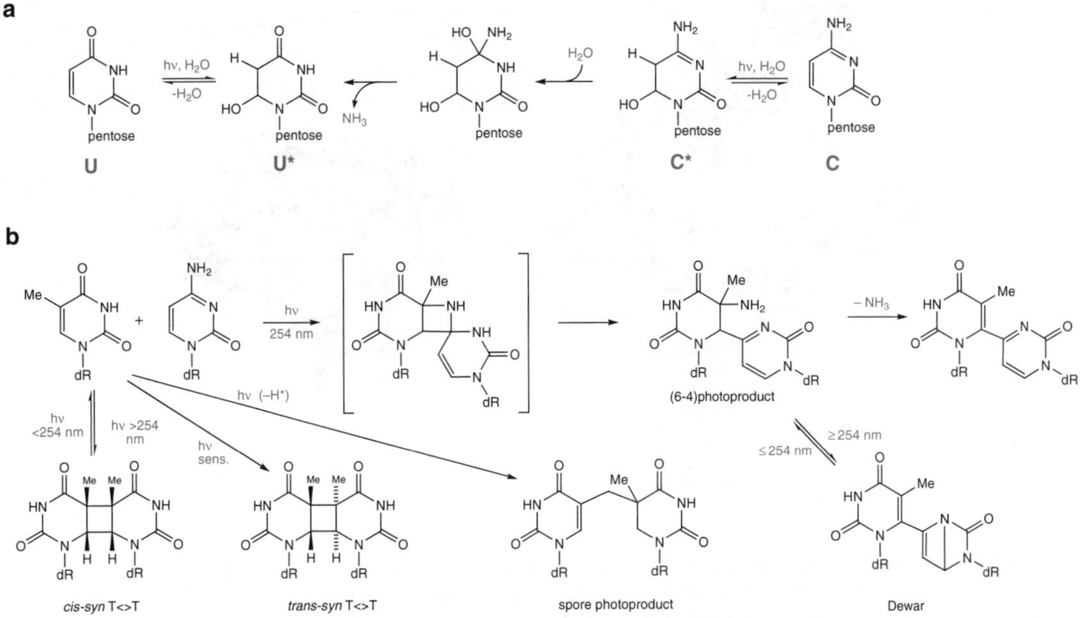


**Figure 8.30** *Pyrimidine photochemistry. (a) Photohydration of ribo- and deoxyribo-uridine and cytidine with deamination of cytidine photohydrate leading to uridine. (b) Pyrimidine [2 + 2]photodimers and other photoproducts formed from DNA*

The stereochemistry of these [2 + 2] cyclobutane dimers requires that Py<>Py formation in native DNA is predominantly an intrastrand process with photoaddition involving adjacent pyrimidines.[67] The *trans-syn* isomer may perhaps result from regions of Z-DNA, and NMR analysis of an oligomer containing this lesion has indicated that in this structure it causes a larger distortion of helical structure. The structure of a duplex DNA dodecamer containing a *cis-syn*-thymine photodimer d(GCACGAAT<>TAAG)·d(CTTAATTCGTGC) has been determined (Figure 8.31a).[68] It shows that the overall helical axis bends approximately 30° toward the major groove and unwinds approximately 9°. This structure is consistent with the ability of T<>T drastically to slow DNA synthesis opposite the dimer site, yet still directing the incorporation of A residues opposite to the dimer during **bypass** (Section 8.11).

T<>T photodimers revert to monomers on irradiation at wavelengths shorter than 254 nm, while dimers containing cytosine residues are easily deaminated by mild hydrolysis so that subsequent **photoreversion** provides yet another source of the C→T base transition by the following reaction sequence:

$$\mathrm{C} + \mathrm{T} \xrightarrow{h\nu} \mathrm{C} <> \mathrm{T} \xrightarrow{\mathrm{H_2O}} \mathrm{U} <> \mathrm{T} \xrightarrow{h\nu} \mathrm{U} + \mathrm{T}$$

Many other types of pyrimidine photoproduct have been isolated from irradiated DNA. The most noteworthy are the **pyrimidine(6-4)pyrimidone photoadducts** and the 'spore photoproduct'. The latter is formed from a radical generated by loss of a hydrogen atom from the methyl group of thymine (Figure 8.30b) and is the major UV-induced photoproduct in the dehydrated state of DNA found in bacterial spores. Pyrimidine(6-4)pyrimidone photoproducts are formed by UVC with an incidence of about 25% that of the cyclobutane photodimers, with T(6-4)C being twice as abundant as T(6-4)T. With UVB, such (6-4) products undergo isomerization to give a **'Dewar-benzene'** structure that on UVC irradiation reverts to its precursor (Figure 8.30b).

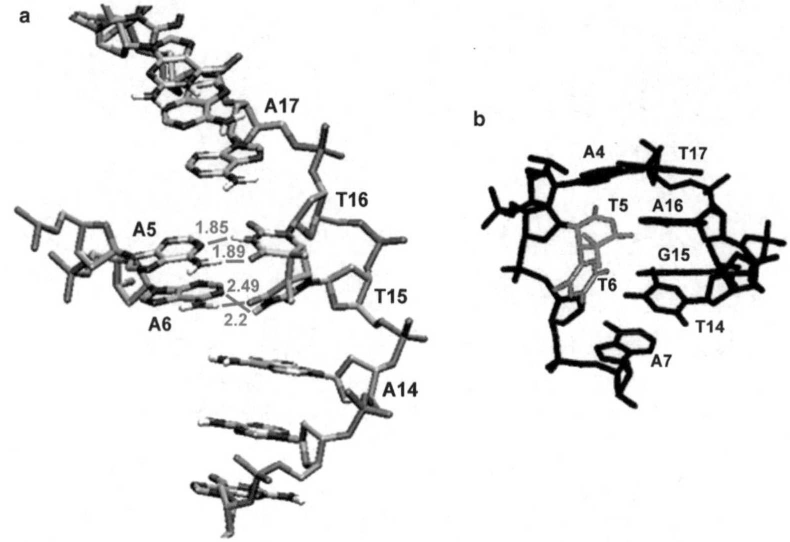


**Figure 8.31** *(a) X-ray crystal structure of a DNA T<>T photoproduct within a duplex dodecamer*
(Adapted from Ref. 68. © (2002), with permission from Elsevier)
*(b) NMR structure of a DNA (4-6)Py photoproduct*
(Adapted from Ref. 69. © (1995), with permission from Blackwell Publishing)

The structure of a DNA decamer containing a (6-4) adduct has been determined by NMR analysis.[69] Because (6-4) adducts are highly mutagenic and most often lead to a 3′-T$\rightarrow$C transition (85% replicating error frequency), a DNA duplex was built having a mismatched base-pair for the 3′-T residue with an opposed G residue. The resulting NMR structure shows that normal Watson–Crick-type hydrogen bonding is retained at the 5′-T of the lesion site while the O-2-carbonyl of the 3′-T residue forms hydrogen bonds with the imino and amino protons of the opposed G residue. Hydrogen bonding thus stabilizes the overall helix and restores the highly distorted conformation of the (6-4) adduct to the typical B-form-like DNA structure (Figure 8.31b). This structural feature may thereby explain the marked preference for the insertion of an A residue opposite the 5′-T and a G residue opposite the 3′-T of the (6-4) lesion during **trans-lesion synthesis**, and so lead to the predominant 3′-T $\rightarrow$ C transition.

Thomas Carell has undertaken a systematic analysis of how DNA sequence and duplex conformation can influence the relative formation of these photochemical lesions.[70] It appears that the production of *cis-syn* cyclobutane photodimers from adjacent pyrimidines resulting from UVC irradiation is dramatically inhibited in A-DNA and RNA conformations. The formation of pyrimidine photoproducts in B-DNA conformations is inhibited by the presence of flanking guanine residues.

The cyclobutane photodimers and the (6-4) photoproducts products can cause cell death, as based on four criteria.

- Dimers (80%) and (6-4) products (20%) are the major photoproducts at low doses.
- Action spectra for the formation of both types of photoproduct correlate well with that for cell death.
- Enzymatic excision repair of both lesions enhances cell survival.
- Cells deficient in excision repair are hypersensitive to the lethal effects of UV radiation.

The formation of these types of lesion is responsible for the development of a large proportion of non-melanoma skin cancers,[71,72] while the damage is limited by the action of many DNA repair enzymes (Section 8.11). The mutagenic nature of cyclobutane dimers, (6-4) products and photohydrates is complex.[73]

It results from (i) a combination of the relative yields for their photochemical formation (approx. 100:25:1), (ii) the capability of DNA polymerase to read through the different lesions, (iii) the errors that may result and (iv) the ability of enzymes to repair such lesions. It appears that bacterial replication past cyclobutane dimers is rather accurate whereas bypass of T(6-4)T products is highly mutagenic and leads to T $\rightarrow$ C transitions (see above). In yeast, Py<>Py cyclobutane dimers are the most mutagenic lesion. In mammalian systems, such as hamster cells, GC $\rightarrow$ AT transitions predominate (50%) with large amounts of transversions (23%) and tandem and non-tandem double mutations (20%). Cytosine-to-thymine transition mutations are generally formed through deamination of C<>T or T<>C dimers to U<>T or T<>U dimers, followed by monomerization to U, and not by deamination of a cytosine photohydrate.

In general, it now appears that:

- the (6-4) photoproduct is more mutagenic than Py<>Py while T<>T is poorly mutagenic;
- in bacteria the (6-4) product is likely to be the major premutagenic lesion;
- in mammalian cells, the (6-4) product is repaired more quickly than the Py<>Py making the cyclobutane dimers the major premutagenic lesion;
- in repair-deficient strains, the (6-4) product may be the dominant mutational lesion even though its photochemical yield is lower than for Py<>Py.

### 8.8.2 Psoralen–DNA Photoproducts

**Psoralens** are furocoumarins that have been widely used in the phototherapy of psoriasis and other skin disorders. They act as photochemical cross-linking agents for DNA.[74] This involves psoralen intercalation followed by two successive photochemical [2 + 2] cycloadditions, which create two cyclobutane linkages. Thymines are the preferred target so that cross-linking occurs mainly at d(TpA) sites. The products are predominantly the *cis-syn* stereoisomers and have an overall S-shape as a result of one thymine being above and the other below the plane of the psoralen (Figure 8.32).

Psoralen cross-linking of a duplex has been developed for the examination of triple-strand helix formation in homopurine–homopyrimidine DNA sequences (Section 2.3.6). Psoralen is attached by its C-5 position to a 5′-thiophosphate on a homopyrimidine undecamer nucleotide. On incubation with DNA which has the complementary sequence followed by UV irradiation, the two parent strands of the DNA become cross-linked at a TpA step present at the junction between the duplex and the triplex. The presence of a neighbouring triplex structure interferes with different stages of psoralen interstrand cross-link (ICL) processing: (i) the ICL-induced DNA repair synthesis in HeLa cell extracts is inhibited by the triplex structure; (ii) in HeLa cells, the ICL removal *via* a nucleotide excision repair (NER) pathway (Section 8.11.4) is delayed in the presence of a neighbouring triplex; and (iii) the binding to ICL of recombinant *Xeroderma pigmentosum* A protein, which is involved in pre-incision recruitment of NER factors, is

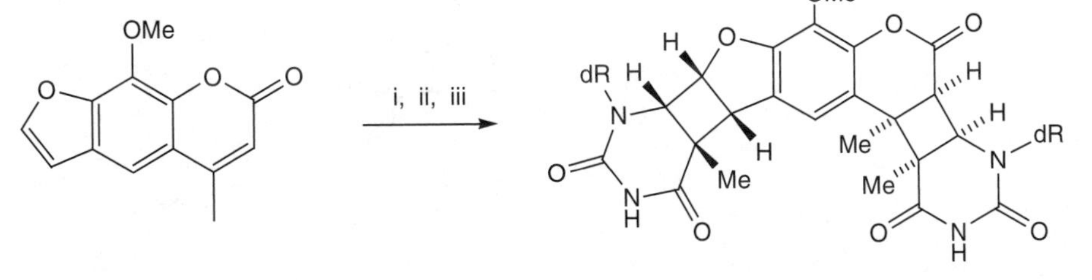


**Figure 8.32** *Photochemical binding of 4-methyl-8-methoxypsoralen to DNA and isolation of dithymidine photoproducts. Procedures: (i) DNA; (ii) hν 320–400 nm; and (iii)* $H^+$

impaired by the presence of the third DNA strand.[74] Such psoralen-oligonucleotide conjugates are probes for sequence-specific helix formation and may have application in site-directed mutagenesis or control of gene expression.

### 8.8.3 Purine Photoproducts

Adenine and guanine are intrinsically more photostable than the pyrimidines and tend to transfer photochemical excitation energy to neighbouring pyrimidines in DNA duplexes. However, UVB can lead to the formation of **adenine photoproducts** with an adjacent adenine or thymine base (Figure 8.33a).[75] While the quantum yields for these processes are low and product formation is easily quenched by base-pairing, the A–T adduct has been shown to be mutagenic.

Of greater significance is the photo-oxidation of cellular DNA resulting from UVA irradiation involving an unidentified endogenous photosensitiser. The reactions result from one-electron oxidation or hydrogen abstraction (type I) or from singlet oxygen (type II) processes. The four bases are the preferred substrates for type I reactions and reactivity depends on (low) ionisation potential in the order $G > A \sim T > C$. Guanine products predominate because of 'hole-transfer' in dsDNA, by which radical cation migration targets guanine, acting as a sink. Hydration of the initial guanine radical cation leads to **8-oxoguanine** in an oxidative process or to **FapyGua** by reduction. Alternatively, deprotonation of the guanine cation radical leads to imidazolone and oxazolone products (Figure 8.33b). Lipscomb and Rich have determined the X-ray structure of a DNA fragment containing 7,8-dihydro-8-oxoguanine (GO). The structure of the duplex form of d(CCAGOCGCTGG) has been determined to 1.6Å resolution. The 8-oxoGua is in the *anti*-conformation and forms Watson–Crick base-pairs with the opposite C. There is no steric clash of the 8-oxygen of G with backbone atoms.[76]

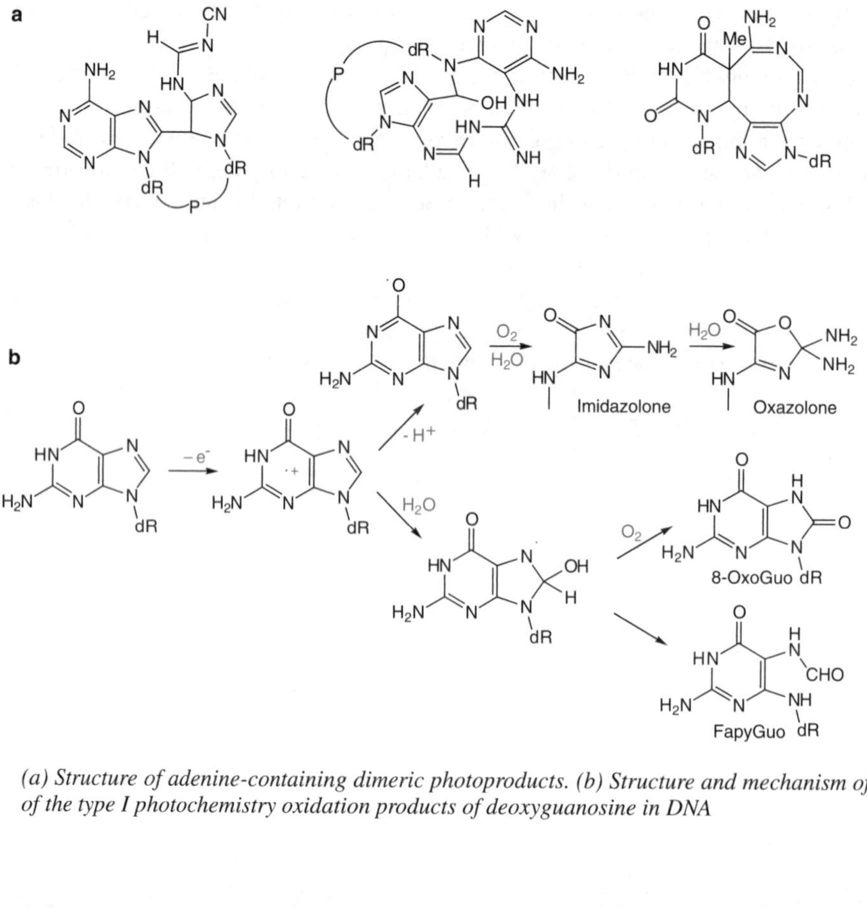


**Figure 8.33** *(a) Structure of adenine-containing dimeric photoproducts. (b) Structure and mechanism of formation of the type I photochemistry oxidation products of deoxyguanosine in DNA*

### 8.8.4 DNA and the Ozone Barrier

Life is shielded from DNA-damaging solar radiation by the ozone layer.[77,78] The amount of ozone involved is very small indeed. It is formed photochemically in the stratosphere by UVC radiation and is distributed throughout the atmosphere. Its maximum density is at some 20–25 km above the earth's surface, where its pressure is about 130 nbar. Were all the ozone to be concentrated at sea level it would form a layer only 3 mm thick!

The absorption spectrum of ozone is very similar to that of DNA (Figure 8.34), so it generally serves to prevent short-wavelength UVC radiation from reaching the earth's surface and causing DNA damage. To put that in perspective, it has been estimated that a 1-h exposure of DNA to ozone-filtered sunlight at sea level would generate about seven *cis-syn* thymine dimers per 1000 reactive sites. Were the ozone barrier level to be halved, the time would be reduced to 10 min. In the complete absence of ozone, the same DNA damage would occur in 10 s. Other calculations have estimated that for every 1% decrease in the ozone column there could be a 4% increase in the incidence of skin cancer.

One of the major causes for concern is the depletion of ozone resulting from the photochemical behaviour of **chlorofluorocarbons (CFCs)** in the upper atmosphere.[58,78] This interferes with the formation of ozone through the capture of oxygen atoms by chlorine atoms, as shown by the following equations.

Ozone formation $O_2 + h\nu \rightarrow O^{\bullet} + O^{\bullet}$

$O^{\bullet} + O_2 + N_2 \rightarrow O_3 + N_2$

Ozone depletion $Cl^{\bullet} + O_2 \rightarrow ClO$

$ClO + O_3 \rightarrow Cl^{\bullet} + O_2 + O_2$

A more cautious view is that the two dominant variables that determine the UV incidence at ground level integrated over 24 h are (i) the elevation of the sun above the horizon and (ii) the duration of daylight. These two factors, which ultimately relate to the earth's solar orbit, lead to variations that far exceed the changes predicted to occur at any middle latitude location as a consequence of any decline in the column of ozone since 1970. The situation over the Antarctic is, however, a special case. The large depletion in ozone over that area in springtime has resulted in UVB irradiances that are substantially larger than existed in that part of the world before the 1980s.

In 2003, the Director General of WHO Lee-Jong Wook said: "As ozone depletion becomes more marked and as people around the world engage more in sun-seeking behaviour, the risk of health complications from overexposure to ultraviolet radiation is becoming a substantial health concern."[79] The International

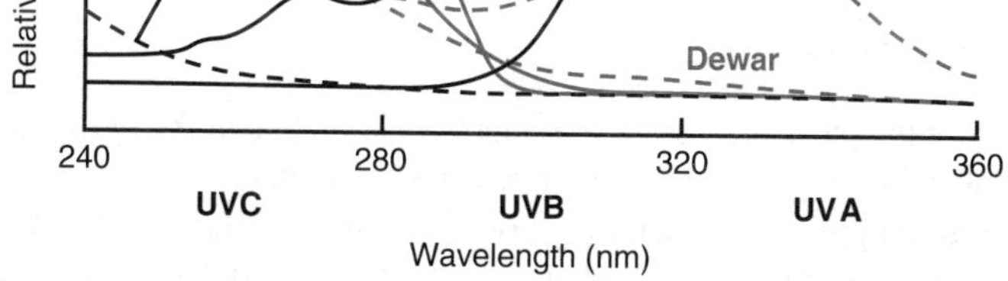

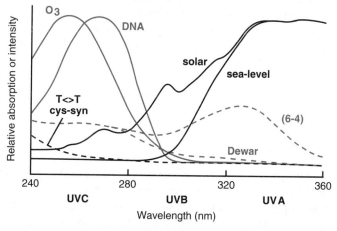


**Figure 8.34** *Absorption spectra for DNA and ozone (red) and the various photoproducts (broken lines) compared to the incidence of sunlight at sea level and above the atmosphere*

Agency for Research on Cancer has estimated that every year, between two and three million new cases of non-malignant melanomas and more than 130,000 new cases of melanoma skin cancer are seen worldwide. Furthermore, as reported by the WHO, melanomas and other skin cancers claimed the lives of 66,000 people in the world (2002).

International response to this threat has been strong. Some 165 countries have agreed to the Montreal Protocol of 1989, revised in London in 1990, to accelerate the phasing-out of **ozone-depleting substances** (ODS): initially chlorofluorocarbons (CFCs), halons and carbon tetrachloride by 2000 and methyl chloroform by 2005.[80] As of 2000, industrial production of chlorofluorocarbons has been arrested and atmospheric levels appear to show signs of coming under control.

## 8.9 EFFECTS OF IONIZING RADIATION ON NUCLEIC ACIDS

**X-rays, γ-radiation** and **high-energy electrons** all interact indirectly with nucleic acids in solution as a result of the formation of **hydroxyl radicals**, of solvated electrons, or hydrogen atoms from water. In aerobic conditions, the most important processes result from $HO^•$ radicals, which abstract a hydrogen atom forming an organic radical which then captures $O_2$. Measurements of the efficiency of these processes indicate that for every 1000 eV of energy absorbed, about 27 $HO^•$ radicals are formed of which 6 react with the pentoses and 21 react almost randomly with the four bases.[81]

### 8.9.1 Deoxyribose Products in Aerobic Solution

The hydroxyl radical is able to abstract a hydrogen atom from C-4′ or from any of the other four carbons in the sugar. The resultant pentose radicals capture oxygen to give a **hydroperoxide radical** at C-4′ or C-5′ and this leads directly to cleavage of a phosphate ester bond. Similar reactions at C-1′or C-2′ lead to alkali-labile phosphate esters that are cleaved on incubation with 0.1 M sodium hydroxide at room temperature in 10 min. In both cases, the bases are released intact (Figure 8.35). The use of Fenton's reagent to generate hydroxyl radicals is a key component of a DNA footprinting method (Section 5.5.7).

### 8.9.2 Pyrimidine Base Products in Solution

At least 24 different products have been isolated from γ-irradiation of thymidine in dilute, aerated solution and many more are formed in anoxic conditions or in the solid state. Cytosine and the purines show a similar diversity. The situation is simplified by limiting the study to oxygenated solutions when the principal site for reaction with the pyrimidines is the 5,6-double bond.[82] Under these conditions, **hydroxyhydroperoxides** are formed that are semi-stable for thymidine, but break down rapidly for deoxycytidine to give a range of products, of which the major ones are illustrated (Figure 8.36). About 10% of thymine modification occurs at the methyl group with the formation of 5-hydroxymethyldeoxyuridine. **Thymine-glycol** is also a significant lesion.

In anaerobic solution, γ-radiolysis gives 5,6-dihydrothymidine as the major product with a preference for formation of the 5(*R*)-stereoisomer.

### 8.9.3 Purine Base Products

The radiation chemistry of the purines is less well understood than that of the pyrimidines. Deoxyadenosine can add the $HO^•$ radical at C-8 to give either 8-hydroxydeoxyadenosine or, with cleavage of the imidazole ring, a 5-formamido-4-aminopyrimidine derivative of deoxyribose (Figure 8.37). Guanine has been even less well studied. However, this situation may change as a result of reports of the formation of **8-hydroxyguanine** from anaerobic irradiation associated, since there is considerable interest in this modified base as a strongly mutagenic lesion (Section 8.11.2).

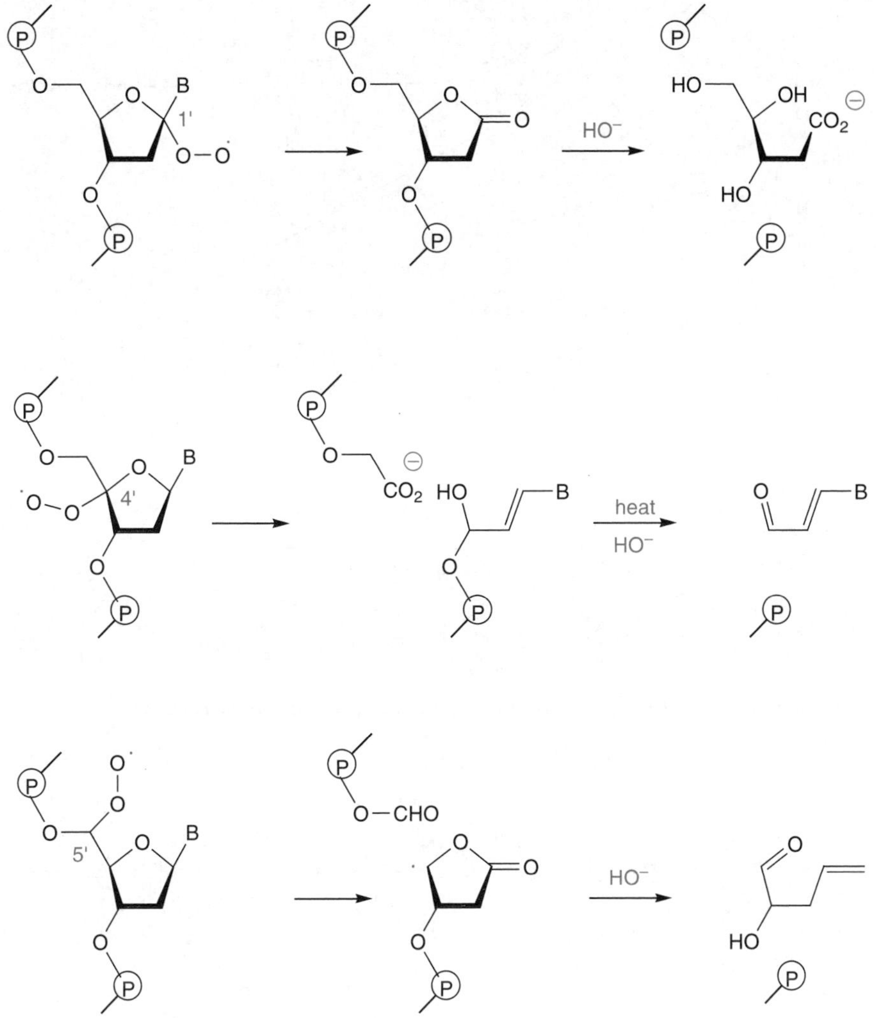


**Figure 8.35** *Breaks in DNA resulting from hydrogen atom abstraction and peroxide radical formation at C-1′ (upper), C-4′ (centre) and C-5′ (lower) in deoxynucleotides followed by mild alkaline treatment (0.1 M NaOH, 10 min, 20°C)*

## 8.10 BIOLOGICAL CONSEQUENCES OF DNA ALKYLATION

DNA damage is unavoidable and can be divided broadly into two categories. Some forms of damage, such as mismatches, deamination of bases, base-loss and oxidative damage, arise spontaneously. Others, such as ionizing radiation, UV radiation and chemical agents including ubiquitous alkylating agents (Section 8.5.3) arise through environmental influences. Many of these changes can result in errors during the steps of the DNA replication, recombination and repair, and so lead to modification of the structure of genetic material. Happily, many of these modifications can be removed successfully by DNA repair machinery.[83,84]

### 8.10.1 *N*-Alkylated Bases

**7-Alkylguanine** is the predominant product of **DNA alkylation**. However, methylation at this site appears not to change the base-pairing of G with C and so is an apparently innocuous lesion. Following N-7 alkylation,

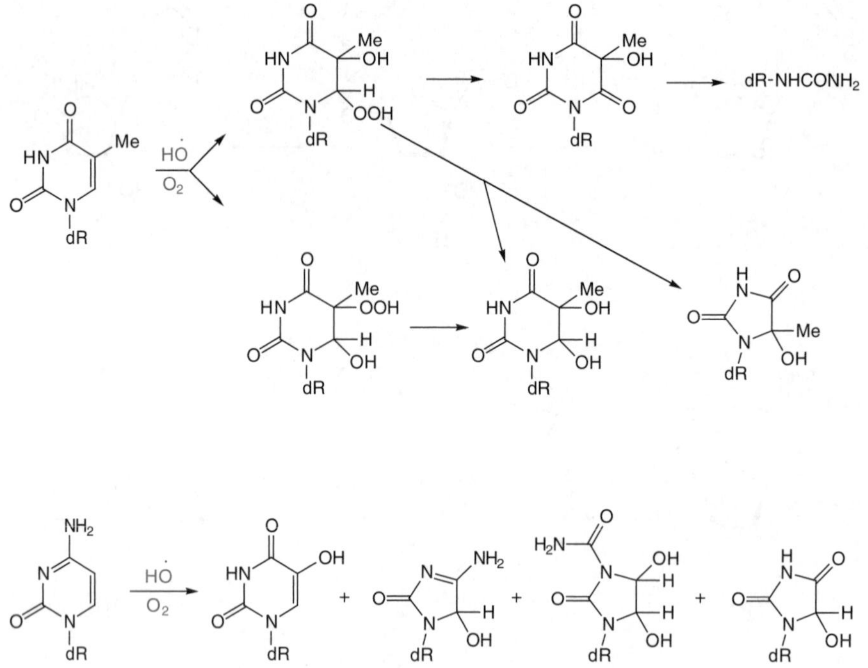


**Figure 8.36** *Products resulting from γ-radiolysis of deoxynucleosides in aerobic conditions*

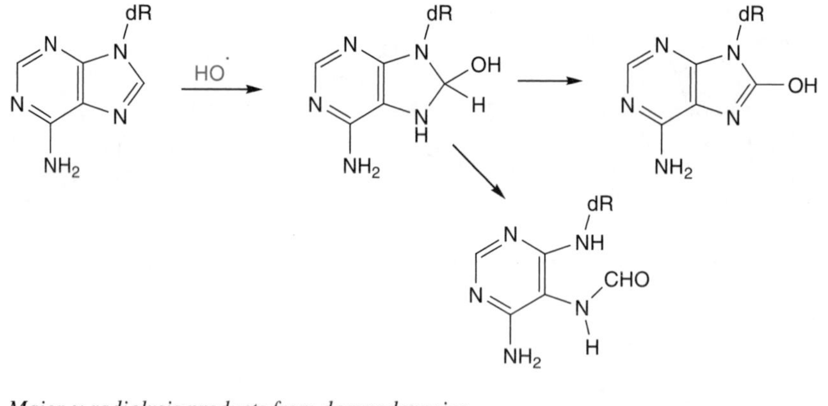


**Figure 8.37** *Major γ-radiolysis products from deoxyadenosine*

the glycosylic bond of a 7-alkylguanine residue undergoes slow and spontaneous hydrolysis, and so creates an **apurinic site**. This is a target for accurate repair (Section 8.11.2). In contrast, the cross-linking of two neighbouring guanines at N-7 by a nitrogen mustard or equivalent bifunctional alkylating agent (*e.g.* cis-platin, Section 8.5.4) is an important cell-killing event, and is equally significant for interstrand and intrastrand cross-links.

**3-Alkyladenines** are the major toxic lesion resulting from monofunctional alkylation of DNA. The alkyl group lies in the minor groove of the DNA double helix where it can block the progress of DNA polymerases (Section 8.11.3). **3-Alkylguanines** have a similar physiological effect but are much less prevalent, being formed ten times less frequently than 3-alkyladenines (Figure 8.11b). **1-Methyladenine** and **3-methylcytosine** also block DNA replication.[85]

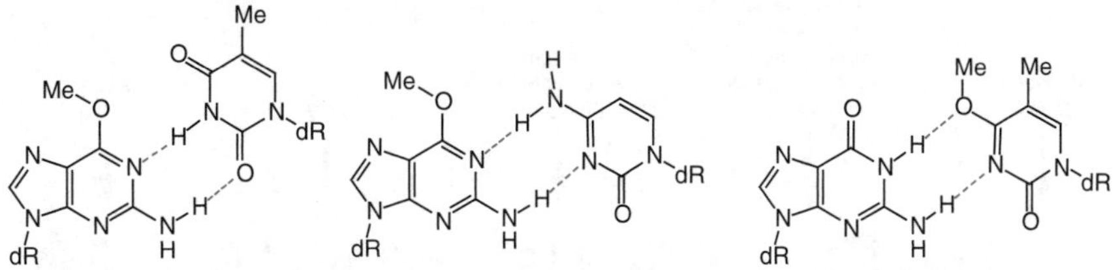


**Figure 8.38** *An $O^6$-MeG·T base-mispair (left), an $O^6$-MeG·C base-pair (centre), and a G·$O^4$-MeT base-mispair (right)*

## 8.10.2 *O*-Alkylated Lesions

$O^6$-Alkylguanines are locked in the **enol-tautomeric** form while guanine in DNA is normally in the *keto-form* (Chapter 2.1.2). Base-mismatch and melting studies have suggested that $O^6$-methylguanine forms an $O^6$-MeG·C base-pair that is *more* stable in a DNA duplex than an $O^6$-MeG·T **base-mispair**. However, in replication with various DNA polymerases, $O^6$-MeG residues do not block the polymerase and have been shown to direct preferential incorporation of thymine in place of cytosine. This result has led to a suggestion that the less stable $O^6$–MeG·T base-mispair may have a more Watson–Crick-like geometry and so better satisfy the demands of the DNA polymerase while the $O^6$-MeG·C is a **wobble base-pair** (Figure 8.38). NMR and X-ray structures have identified both these pairings while calculations suggest that the *anti*-conformation for the $O^6$-methyl group (shown) is less stable than the *syn*-conformation but only by about 1 kcal $mol^{-1}$.[86]

The net biological result is a G→A transition and this type of mutation is common in cells exposed to hard alkylating agents. It is a lethal lesion in human cells and, in particular, it has been identified with a single base transition for activation of the Ha-ras-1 proto-oncogene in the process of initiation of mammary tumours in rats with **N-methylnitrosourea** (MNU, Section 8.5.3), that is a result of the specific conversion of $G^{35}$ into $O^6$-MeG.[35,87]

$O^4$-Alkylthymines exist in the enol-tautomeric form and therefore they can base-pair with guanine. In model studies, both $O^4$-methylthymine and $O^4$-ethylthymine form base-mispairs with guanine (Figure 8.38) that do not block the replication of a defined DNA sequence *in vitro*. However, alkylpyrimidines are very minor products of DNA alkylation and their biological effects appear to be of low significance.

Lastly, the *O*-alkylation of the phosphate diesters in DNA gives **phosphate triesters**, but these are repairable (Section 8.11.1) and do not seem to be important either as cell-killing agents or mutagenic lesions. The vital processes for the biological repair of these and other types of DNA damage will now be described.

## 8.11 DNA REPAIR

DNA repair is essential for life on earth for the maintenance of genomic integrity. Estimates of the extent of DNA damage in a human cell range from $10^4$ to $10^6$ events per day, which calls for some $10^{16}$ to $10^{18}$ repair events per day in an adult person. All living cells possess a range of **DNA repair enzymes** in order to correct damage resulting from spontaneous chemical change, radiation and external chemical agents. Humans appear to be more effective than rodents in repairing DNA and are also better able to resist mutagenic agents. Moreover, a striking similarity has emerged between repair systems found in many species from bacteria to humans, although much of our knowledge comes from studies on bacteria or yeasts.

DNA damage has four possible biological consequences each of which is directly linked to DNA repair (Figure 8.39). First, the cell cycle can be arrested. This allows time for DNA repair prior to replication and cell division.[88] Where the extent of DNA damage is too large for effective repair, the cell may go into **apoptosis** (programmed cell death).[89] Finally, any surviving DNA damage can lead to **mutations** and cancer

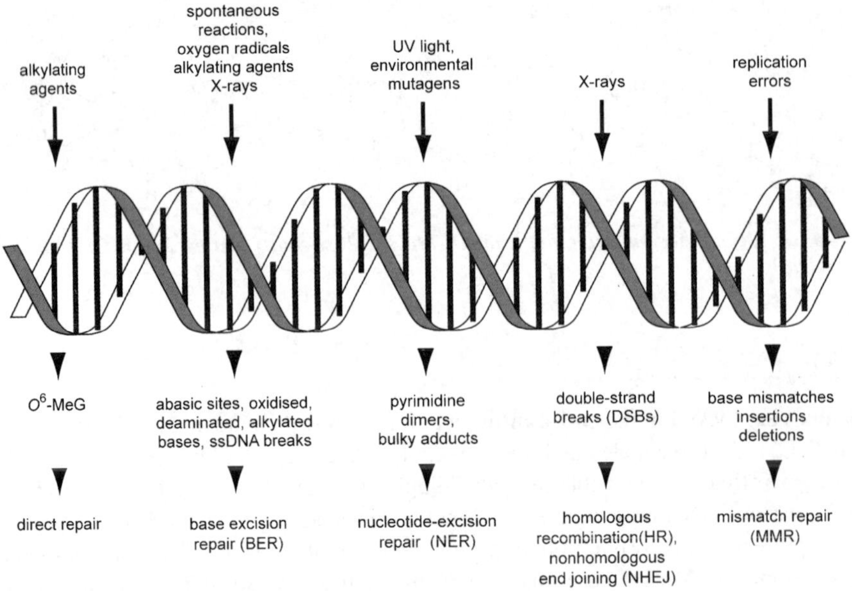


**Figure 8.39** *Most common DNA-damaging agents, lesions and repair pathways*

in mammals.[90] The various types of DNA repair will be considered in order of increasing molecular complexity, with primary emphasis on the chemistry of the repair processes (Sections 8.11.1 and 8.11.2). The repair of double strand breaks and other biological features of DNA repair are presented elsewhere (Section 6.7.2).

## 8.11.1 Direct Reversal of Damage

**Photoreactivation** is catalysed in bacteria by a **photolyase** (Section 10.5.4) that separates cross-linked pyrimidines by transfer of an electron from an excited FADH radical to the **pyrimidine dimer**. The pyrimidine dimer radical formed is unstable and decays to cleave the cyclobutane ring and transfer the electron back to the co-factor in the enzyme. The energy for the excitation of the FADH co-factor comes from a second chromophoric co-factor (5,10-methylenetetrahydrofolate or 7,8-didemethyl-8-hydroxy-5-deazariboflavin, depending on the source of the enzyme), which harvests blue-light photons. Interestingly, there is no evidence for DNA photolyase activity in humans. Thomas Carell has obtained an X-ray structure for a photodimer bound to the photolyase repair enzyme.[91]

**Demethylation** is effected by two distinct types of repair system.[92] The first is carried out by an ***O*$^6$-methylguanine-DNA methyltransferase** (Table 8.1)[93]. This sacrificial demethylation enzyme uses the SH group of a Pro-Cys-His sequence near its C-terminus as a methyl acceptor (Figure 8.40a). The same enzyme can also dealkylate other *O*$^6$-alkyl guanines including ethyl, 2-hydroxyethyl and 2-chloroethyl species. The *S*-alkylcysteine protein formed is inactive and cannot be reactivated. It thus follows that one molecule of the enzyme can only repair one *O*-methylated base. Human lymphoid cells that are resistant to alkyl-nitrosoureas have 10,000–25,000 copies of this enzyme per cell while mutants deficient in the repair of *O*$^6$-MeG have no enzyme. It is easy to see that the threshold of tolerance for alkylating agents may vary greatly between different species of cells and this contributes to the selective cytotoxicity of some anticancer alkylating agents, *e.g.* **Temozolomide** (Section 8.7). It has also led to the development of analogues of *O*$^6$-methylG, especially *O*$^6$-benzylG, as specific inhibitors of this methyltransferase with the aim of enhancing the biological activity of anticancer alkylating agents.

**Table 8.1** *Human repair enzymes*

| *Gene name* | *Enzyme; major altered base released* | *Chromosome location* |
|---|---|---|
| Direct reversal of damage | | |
| *MGMT* | $O^6$-MeG alkyltransferase | *10q26* |
| Base excision repair (BER) | | |
| *UNG* | U | *12q23-q24.1* |
| *SMUG1* | U | *12q13.11-q13.3* |
| *MBD4* | U or T opposite G at CpG sequences | *3q21-q22* |
| TDG | U, T or ethenoC opposite G | 12q24.1 |
| *OGG1* | 8-oxoG opposite C | *3p26.2* |
| *MYH* | A opposite 8-oxoG | *1p34.3-p32.1* |
| *NTHL1 (NTH1)* | Ring-saturated or fragmented pyrimidines | *16p13.3* |
| *MPG* | 3-meA, σA, hypoxanthine | *16p13.3* |
| *NEIL1* | Removes thymine glycol | 15q22-q24 |
| *NEIL2* | Removes oxidative products of C, U | 8p23 |
| *NEIL2* | Removes oxidative products of C, U | 8p23 |
| *NEIL3* | Removes fragmented/oxidized pyrimidines | 4q34.2 |
| Nucleotide excision (NER) (XP = xeroderma pigmentosum) | | |
| *XPC, RAD23B, CETN2* | Binds damaged DNA as complex | 3p25, 3p25.1, Xq28 |
| Mismatch excision repair (MMR) | | |
| *MSH2* | Mismatch and loop recognition MSH2, MSH3, MSH6 | 2p22-p21 |

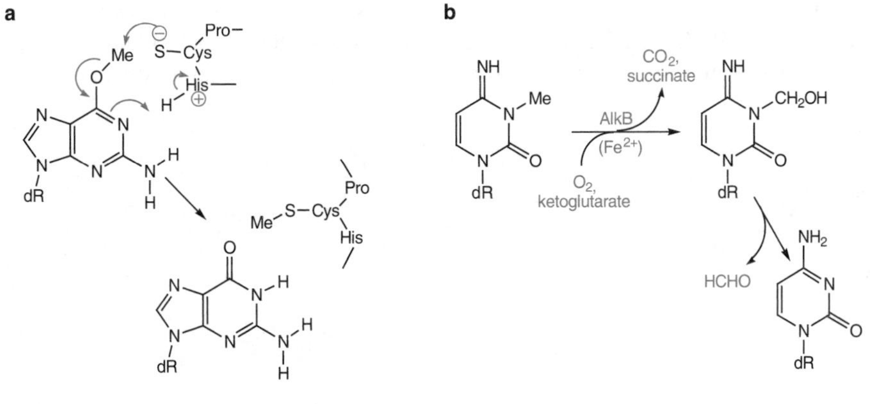


**Figure 8.40** *Repair of methylated DNA. (a) Ada protein demethylates $O^6$-methylguanosine through use of a Pro-Cys-His sequence. (b) Oxidative demethylation of 3-MeC by AlkB*

*Escherichia coli* has two methyltransferases, an inducible Ada protein that demethylates mainly $O^6$-MeG while the constitutive Ogt protein is more efficient for $O^4$-MeT and larger alkyl lesions.[94] Ada also demethylates DNA phosphate triesters (but only of the (*S*)-configuration) at a second cysteine site, Cys38. Methyl phosphate triesters are relatively innocuous lesions, not repaired in eukaryotes. Their repair in bacteria functions solely to upregulate the **adaptive Ada response** (involving the genes *ada-alkB, alkA* and *aidB*), which appears to have evolved to give protection against exogenous $S_N1$ alkylating agents.

AlkB is a bacterial repair enzyme that uses an active iron-oxo intermediate to hydroxylate 1-MeA and 3-MeC, both being products of $S_N2$ methylation of ssDNA. This $Fe^{2+}$-dependent dioxygenase gives hydroxymethylated derivatives that decompose to formaldehyde and the parent nucleotide (Figure 8.40b). Two human proteins, ABH2 and ABH3, also demethylate 1-MeA and 3-MeC but have only weak activity on 1-ethyladenine.[84]

### 8.11.2 Base Excision Repair of Altered Residues

In mammalian cells, **excision repair** is the most important mechanism and involves enzyme recognition of the modified base, usually resulting form oxidation, deamination, or alkylation (Table 8.1). There are two distinct types of excision repair process and both appear to be error-free: **base excision** and nucleotide excision.

Repair by base excision, BER, employs a **DNA glycosylase** that cleaves the glycosylic bonds of modified purine or pyrimidine nucleosides to generate an abasic site.[95] The next step is incision of the phosphate backbone by an **AP endonuclease** (Section 10.5.2), which operates close to apurinic and apyrimidinic sites. Subsequent exonuclease action then excises nucleotides to create a gap at least 30 residues long. Multiple human glycosylases have been identified, each with its own base-specificity. Four of these enzymes operate on 2′-deoxyuridine, which arises through misincorporation of dUMP (giving a U·A base-pair) or from deamination of dC (giving a U·G base-pair). It has been estimated that humans repair some 500 uracil residues per cell per day. **Uracil DNA glycohydrolase**, UDG, is highly conserved from archaebacteria to humans and is the product of the *UNG* gene. It is the most proficient of the uracil repair enzymes, having a reaction acceleration close to $10^{12}$, while its efficiency of discrimination against thymine and cytosine in favour of uracil exceeds $1{:}10^7$.

3-Methyladenine is the principal target for two 3-methyladenine-DNA glycosylases from *E. coli*, one of which also acts on 3-MeG, $O^2$-MeC and $O^2$-MeT residues. The human enzyme 3-methyladenine DNA glycosylase, the *MPG* gene product, removes a remarkably diverse group of damaged bases from DNA, including cytotoxic and mutagenic alkylation adducts of purines such as 3-MeA, 7-MeG and hypoxanthine. The structure of the human 3-methyladenine DNA glycosylase complexed to a mechanism-based pyrrolidine inhibitor (Figure 8.41a) shows the enzyme intercalated into the minor groove of DNA, causing the abasic pyrrolidine nucleotide to flip into the enzyme active site, where a bound water is poised for nucleophilic attack.

8-Oxoguanine is the major mutagenic base lesion in DNA caused by exposure to reactive oxygen species. It is excised by OGG1, a DNA glycosylase with an associated lyase activity for chain cleavage. It releases free **8-hydroxyguanine** specifically from 8-OH-G·C pairs and also introduces a chain break in a

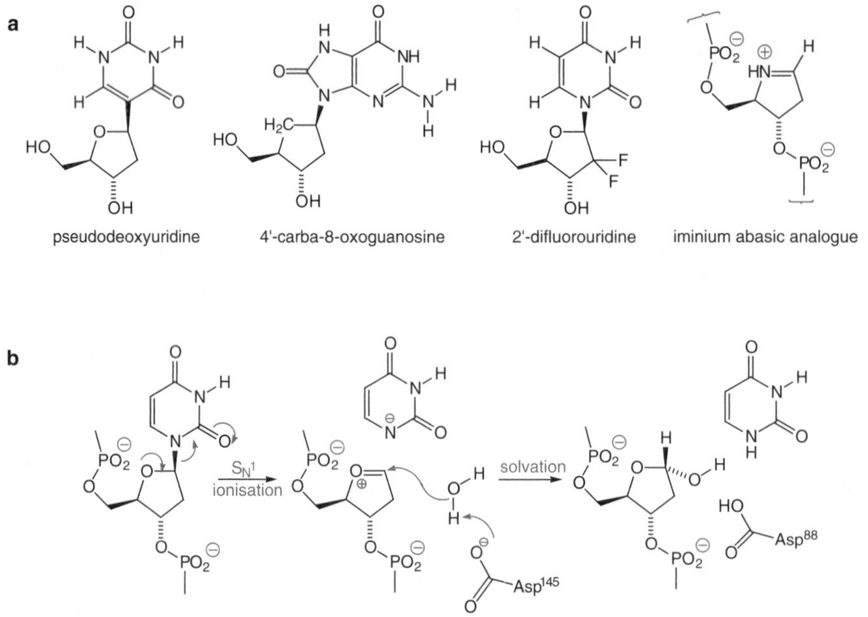


**Figure 8.41** *(a) Structures of some nucleoside analogues used to inhibit glycosylic bond cleavage. (b) 2-Step $S_N1$ mechanism of glycosylic bond cleavage by UDG*

double-stranded oligonucleotide (Table 8.1). The human enzyme MUTYH appears to be homologous to the bacterial repair glycosylase MutY, and removes adenines mis-paired with **8-oxoguanine**. The third component of the resistance to 8-oxoG is an **8-oxoGTP hydrolase**, which removes it from the nucleotide pool to circumvent misincorporation of 8-OH-G opposite A residues.

Recent studies have confirmed the role of reactive oxygen species in the pathogenesis of Alzheimer's disease (AD) and the accumulation of 8-oxo-2′-deoxyguanosine in AD brain has been discussed. It seems that reduced expression of **8-oxoguanine DNA glycosylase** (hOGG1-2a), an enzyme that repairs 8-oxo-2′-deoxyguanosine, may be involved in the pathomechanism of AD.[96]

### 8.11.3 Mechanisms and Inhibitors of DNA Glycohydrolases

All DNA glycosylases employ a **nucleotide-flipping** mechanism. This is a general strategy for enzymes operating on natural or modified DNA bases to gain access to target loci that are buried in the DNA stack or inaccessible from a groove. Such glycosylases fall into two distinct mechanistic classes: (i) **monofunctional glycosylases** and (ii) **bifunctional glycosylase/AP lyases**.

Considerable detail has been achieved in understanding the mechanism of monofunctional BER glycosylases through the use of inhibitory nucleotide analogues resistant to glycosylic bond cleavage.[95] Three general strategies have been employed (Figure 8.41a).

- C-Glycosides as in the case of pseudodeoxyuridine for UDG[97] and 2′-deoxyformycin A for MutY,
- Carbocyclic sugar nucleosides, as in the case of 8-oxocarbadG for OGG1[98] and
- 2-Fluoropentoses, as in the case of 2′-fluoroadenosine for MutY[99] and 2′-difluorodU for human TDG glycosylase.[100]

UDG is an example of an enzyme that 'flips-out' the aberrant nucleotide, as seen in an elegant sequence of X-ray structures by John Tainer, one including a transition state analogue.[97] Both QM-MM computation and kinetic isotope studies have identified the glycosylic bond cleavage as a dissociative process, much of the energy for which derives from stabilisation of an oxocarbenium ion at C-1 of the deoxyribose by the negative charges on the proximate phosphate residues in the same strand. The water that is captured by the oxocarbenium ion to give the deoxyribose AP-product is activated by an essential aspartic acid in an $S_N1$ mechanism (Figure 8.41b). The uracil anion is an adequate leaving group not needing general acid activation. Thus, the UDG enzyme appears to provide no chemical catalysis but serves simply to orientate the DNA substrate to maximise uracil recognition, stereoelectronic effects and the enabling contribution of the phosphate negative charges to C–N bond ionisation. In general, repair enzymes that involve **oxocarbenium ion intermediates** are strongly inhibited by iminium abasic sugar analogues (Figure 8.41a), which have proven to constitute a general strategy for studying glycohydrolases.[101,102]

The enzymes that catalyse depurination show greatly reduced base-specificities and also appear to use general acid catalysis at the purine to effect glycosyl bond cleavage. Less mechanistic detail is yet available to establish whether they employ $S_N1$ or $S_N2$ reaction pathways.

The bifunctional BER enzymes that cause chain cleavage as well as base-excision generally operate *via* a sugar-enzyme intermediate. These include OGG1, NEI1 and NTH1 (Table 8.1). Their mechanism is well exemplified for the bacteriophage **T4 Endonuclease V**, which cleaves the glycosylic bond of the 5′-residue in a thymine photodimer.[103] The N-terminal amino group of T4 Endo V attacks C-1′, displaces the base and forms an aldimine. This is followed by β-elimination of the 3′-phosphate to give the single-strand break (Figure 8.42).

### 8.11.4 Nucleotide Excision Repair

**Nucleotide excision**, NER, is the repair pathway that deals with bulky lesions, including those formed by UV radiation, various environmental mutagens, and some chemotherapeutic agents, and shows broad substrate acceptability.[104] There seems to be a general correlation between the efficiency of repair and the

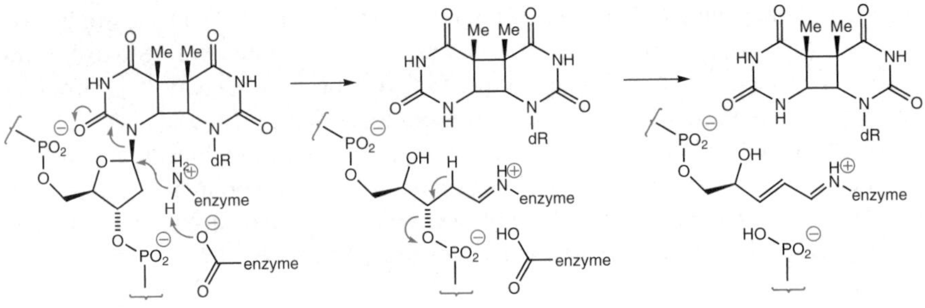

**Figure 8.42** *Mechanism of glycosylic bond cleavage and AP strand scission for T4 Endonuclease V*

extent of helical distortion resulting from the lesion. Nucleotide excision repair appears to be the dominant type of repair process in bacteria and possibly also in humans.[105]

One of the best understood repair systems of this sort is that from *E. coli*, whose *uvrABC* genes have been cloned and their protein products purified.[106] An endonuclease initiates the process by making a single-strand incision close to the damaged nucleotide. The uvrA protein (114 kDa) is an ATP-dependent DNA-binding protein that recognises and binds to the DNA photolesion and also to many other types of bulky base-modification. The uvrB (84 kDa) and uvrC (70 kDa) proteins can now bind and initiate the repair process by nicks in the damaged strand which are on either side of the lesion and some 12 bases apart. The excised nucleotide containing the lesion is now released by uvrD (DNA helicase II) while it is still bound to the protein complex. That leaves a single-stranded gap, which is filled by DNA polymerase I, which binds and fills the gap from the 3′-end until it approaches the 5′-terminus. Finally, ligation completes this **short-patch repair** (Figure 8.43a).

In humans, over 20 proteins contribute to this NER process to combat the damage resulting from UV radiation,[107] and deficiencies in this repair are responsible for the usually fatal condition of **Xeroderma pigmentosum**.

The DNA repair processes of eukaryotes are coming into focus at a molecular level. The fact that their DNA is organised into higher-order structures (Section 2.6.2) presumably has implications for their DNA repair processes, as has already been established for their transcription (Section 6.6) and replication (Section 6.6.4). What is clear is that a large number of loci are involved in the excision repair process for eukaryotes as compared to prokaryotes, yet the mammalian system is remarkably similar to the bacterial system.

### 8.11.5 Crosslink Repair

We have already seen that DNA cross-links can result from the alkylating activity of various natural, synthetic and photochemical agents (Sections 8.5, 8.7 and 8.8). Most DNA repair mechanisms rely on the redundant information inherent in the duplex to remove damaged nucleotides and replace them with normal ones, using the complementary strand as a template. **Interstrand cross-links** (ICLs) present a special case of NER and pose a unique challenge to the DNA repair machinery because both strands are damaged. The repair of ICLs by mammalian cells appears to involve dual incisions, both 5′ to the cross-link in one of the two strands. The net result is the generation of a 22- to 28-nucleotide-long gap immediately 5′ to the cross-link. This gap may act as a recombinogenic signal to initiate cross-link removal.[108] Several human genes involved in ICL repair have been identified and studies on their mode of action are advancing.

### 8.11.6 Base Mismatch Repair

The replication of DNA proceeds with a net error rate of around 1 in $10^{10}$. As polymerases have a nucleotide incorporation error rate of around 1 in $10^5$, and that is improved by their proofreading activity to $1{:}10^7$, a further error-reducing process of 1 in $10^3$ must exist. It is provided by **base-mismatch repair**,

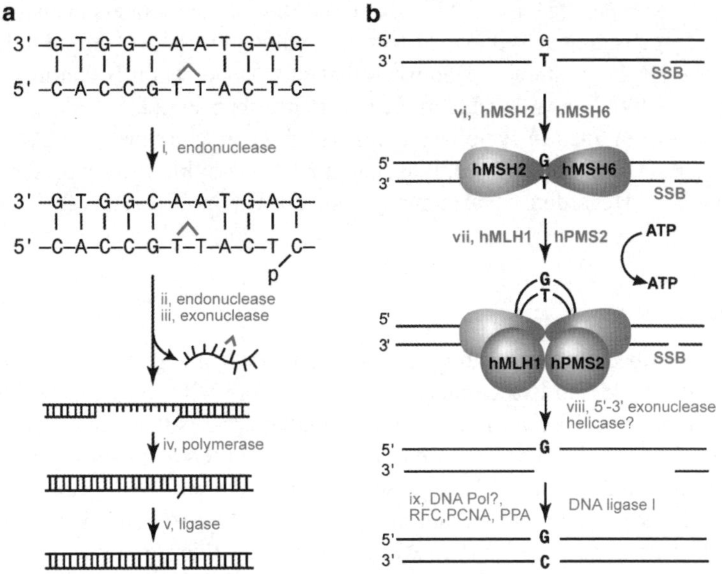


**Figure 8.43** *Mechanisms of DNA repair. (a) Bacterial base excision repair (**BER**) illustrated for the cis-syn-thymine photodimer. (i) An endonuclease makes an incision on 5′-side of the dimer; (ii) incision of 3′-side and removal of a 12-base oligomer containing the lesion; (iii) gap enlarged by exonuclease; (iv) polymerase resynthesis; and (v) ligation to complete patch (red line). (b) Human* mismatch repair *(**MMR**) illustrated for a G·T mismatch in a duplex having a nick to identify the daughter strand. (vi) The mismatch is recognised by the hMutα heterodimer; (vii) ATP drives the threading of loops through hMutSα and recruits hMutLα; (viii) exonuclease and helicase action degrade the error-containing strand (identified by a single strand break); and (ix) replication proteins fill the gap which is sealed by DNA ligase I*

MMR. This repair has to deal with mismatches between the four regular bases, by modified bases, and especially by nucleotide insertions and deletions arising from erroneous polymerase activity (often in repetitive mono- or di-nucleotide regions of DNA) and it restores normal Watson–Crick base-pairing. This subject has become increasingly significant through the discovery that defective MMR is associated with familial and sporadic gastrointestinal and endometrial cancer and is also common in some acute myeloid leukaemias that can follow successful chemotherapy for a primary malignancy.

MMR is primarily a DNA excision repair pathway. Its main targets in man are G·T, G·G, A·C and C·C and also $O^6$-MeG·T and $O^6$-MeG·C mismatches. Four key human proteins are involved in MMR: hMSH2, hMSH6, hMLH1 and hPMS2. First, a base-mismatch is recognised by a MutSα complex (a heterodimer of hMSH6 and hMSH2), which, in binding, recognises errors in the newly-synthesised daughter strand (Figure 8.43b). Some features of this complex have been revealed in two X-ray structures of bacterial MutS complexes. ATP hydrolysis then is linked to conformational changes that are associated with binding of MutLα (a dimer of hMLH1 and hPMS2) and formation of a loop structure. One or more 5′→3′exonucleases and helicase then degrade and elongate the daughter strand loop and the gap is repaired by Polδ and DNA ligation to seal the nick. Such MMR repair tracts are 10 to 100 times longer than in NER or BER, respectively.

## 8.11.7 Preferential Repair of Transcriptionally Active DNA

**Preferential repair** simply means that the more important parts get fixed first. It has been known for over 30 years that DNA repair is a heterogeneous process. In 1985, work by Phil Hanawalt established that a transcriptionally active gene is repaired preferentially to the genome overall. He introduced the term preferential repair to describe the phenomenon that is caused by interactions between a stalled pol II transcription complex and the excision repair enzymes.[109] For example, 70% of photodimers in an active *DHFR* gene in Chinese hamster ovary cells are excised in a 24-h period while only 10–20% of photodimers are removed from the genome overall. Later studies established that this phenomenon is confined to the transcribed DNA strand while virtually no repair occurs in the non-transcribed strand.

Selective repair is not confined to cyclobutane dimers; (6-4) photoproducts are also removed preferentially from active genes. On the other hand, it appears that *N*-methylpurine repair is not coupled to transcription while the situation regarding covalent modification by carcinogens, such as N-AAF (Section 8.6.1), remains to be clarified.

### 8.11.8 Post-Replication Repair

The repair systems described in the preceding sections maintain the integrity of the genetic message in the cell as long as repair is rapid and is accurately completed before DNA polymerase attempts to copy the damaged DNA. What happens if repair is slow or deficient? It appears that when DNA lesions block progression of the replication fork, a major response is the activation of **'error-prone repair'**, without removing the lesion. Xeroderma pigmentosum is likely caused by a defect in post-replication repair.

**SOS repair** is the name given to such enhanced repair (ER), which is induced by a wide variety of types of DNA damage.[110] This is how it seems to work in bacteria. When the advancing DNA polymerase reaches a lesion that blocks its further progress, for example T<>T or 3-MeA, a very sizeable gap is left opposite the lesion before replication starts again. This gap can either be made good by recombination processes, known as **recombination-repair** (Section 6.7.2), or by the induction of SOS repair.

The SOS process has been most thoroughly studied in *E. coli* and is one of the functions of the *recA* gene protein.[111] Several **recA proteins** identify and bind to any long, single-stranded stretch of DNA, which has been formed by the recBCD proteins. RecBCD produces ssDNA using its helicase/nuclease activities. Through interaction with ATP, recA now becomes activated as a protease towards the **lexA protein**, which is known to be a multifunctional repressor in control of several DNA repair enzymes. Among the many consequences of hydrolysis of lexA are the de-repression of *ruvAB* and *recA* genes and a reduction in the $3'\rightarrow5'$ exonuclease proofreading activity of DNA polymerase III. The net result is that pol III operates with decreased fidelity, not only in its ability to read through regions of damage such as T<>T but also elsewhere in the cell, and this situation persists until the activation of recA as a protease ends. At this point SOS repair is rapidly switched off.

The mechanism of repairing these large gaps, which can be over 1000 bases long, is highly variable. In particular, the specificity of elongation varies with several factors: the type of polymerase, the activity of the $3'$–$5'$ editing nuclease, and the particular type of DNA lesion. For example, in cases where the base modification behaves as a **non-instructional site**, there is a strong tendency for the incorporation of an adenine residue.[112] This explains why transversion mutations are seen when guanines are modified by **aflatoxin** or by **acetylaminofluorene** derivatives. Even then, further elongation in the gap is dependent on the base-sequence in the template strand on the $5'$-side of the lesion. The overall result is that 'long-patch' repair seems likely to be idiosyncratic for each type of lesion in each mutable site and, above all, is error-prone.

### 8.11.9 Bypass Mutagenesis

A single mutation that arises from damage to DNA can be described by a two-step mechanism. **Misincorporation** is the inclusion in the daughter strand of a nucleotide different from the Watson–Crick complement to the original residue(s) in the primer strand. **Bypass** is the process of continued chain elongation beyond the mis-inserted base at the site of the lesion.[113] This process has been carefully examined in *E. coli* through the insertion of synthetic cyclobutane photodimers into oligomers and their use as templates for pol I in primer extension reactions *in vitro* and in phage and bacterial replication.

When pol I arrives at a lesion in the template, chain extension of the primer strand can either be terminated or retarded. Three types of result have now been observed for **bypass replication** beyond a cyclobutane

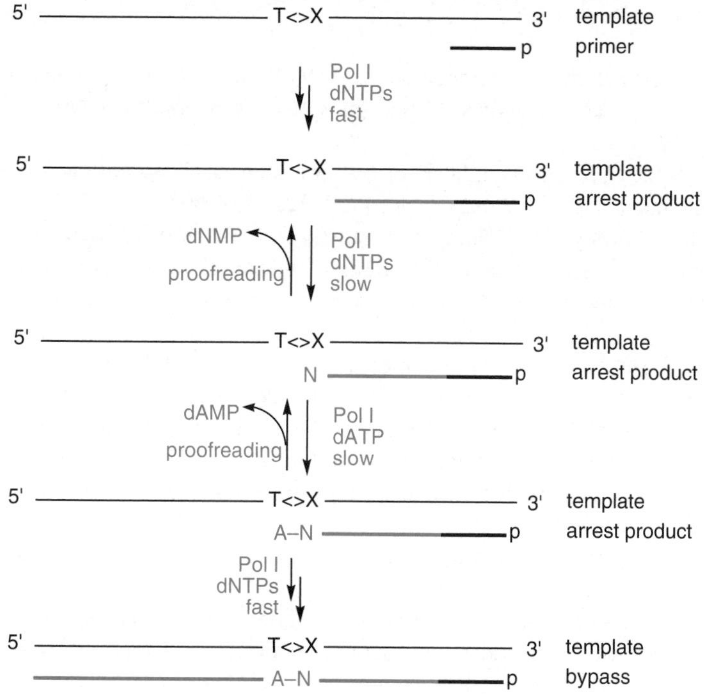


**Figure 8.44** *Bypass mutagenesis illustrated by the scheme employed to investigate the kinetics and mutagenic consequences of the bypass of DNA cyclobutane photoproducts by polymerase I in vitro*

dimer lesion. First, in the case of the lesion T<>T, pol I is able to bypass the dimer and incorporates adenines opposite the lesion with less than 95% efficiency (Figure 8.44). This explains why T<>T appears to be inherently non-mutagenic. Second, adenines are also incorporated when the dimer is T<>U, and this strongly supports the idea that T<>C and C<>T photodimers cause G $\rightarrow$ A transition mutations. This can be a result either of deamination of the cytidine residue to give a deoxyuridine photoproduct prior to replication, **deamination bypass**, or of mispairing of the T<>C dimer with the 5,6-saturated cytosine moiety in its imino tautomeric form, **tautomer bypass** (Section 2.1.2). Third, when T<>T dimers are incorporated into oligo(dl) tracts, specifically at positions -2 and -3 from the 5′-end of the $T_n$ tract, both -1 and -2 deletions are observed.

Studies of this nature can be expected to uncover the role of bipyrimidine (6-4) photoproducts and other DNA lesions in due course.[114]

The main DNA repair processes in humans are: direct repair (DR), base excision repair (BER), nucleotide excision repair (NER), mismatch repair (MMR), along with homologous recombination (HR) and non-homologous end-joining (NHEJ). Studies on human repair systems are advancing rapidly, especially into UV damage. Recent studies have shown that human DNA polymerase $\eta$ (Pol $\eta$) modulates susceptibility to skin cancer by promoting DNA synthesis past sunlight-induced cyclobutane pyrimidine dimers that escape nucleotide excision repair (NER). Pol $\eta$ bypasses a dimer with low fidelity and with higher error rates at the 3′-thymine than at the 5′-thymine and UV-induced mutagenesis is higher at the 3′-base of dipyrimidine sequences. Thus, in normal people and particularly in individuals with NER-defective *Xeroderma pigmentosum* who accumulate dimers, errors made by Pol $\eta$ during dimer bypass may contribute to mutagenesis and skin cancer.[115]

These human repair systems are key players in maintaining our genomic integrity while defects in many of these pathways have been linked to particular human hereditary syndromes. Our growing understanding of these processes is a prime area for the interplay of the chemistry and biology of nucleic acids and will continue to enhance our ability to identify and resist environmental hazards.

## REFERENCES


1. W. Szybalski, Effects of elevated temperatures on DNA and on some polynucleotides. Denaturation, renaturation, and cleavage of glycosidic and phosphate ester bonds. in *Thermobiology*, A.H. Rose (ed). 1967, 73–122.
2. M. Green and S.S. Cohen, Biosynthesis of bacterial and viral pyrimidines. II. Dihydrouracil and dihydrothymine nucleosides. *J. Biol. Chem.*, 1957, **225**, 397–407.
3. P. Cerutti, Sodium borohydride reduction of transfer RNA. *Methods Enzymol.*, 1971, **20**, 135–143.
4. D.H.R. Barton and R. Subramanian, Reactions of relevance to the chemistry of aminoglycoside antibiotics. Part 7. Conversion of thiocarbonates into deoxy-sugars. *J. Chem. Soc. Perkin Trans. 1*, 1977, 1718–1723.
5. C.A. Krauth, A.T. Shortnacy, J.A. Montgomery and J.A. Secrist III, Synthesis and antiviral evaluation of adenosine-N1-oxide and 1-(benzyloxy)adenosines. *Nucleos. Nucleot.*, 1989, **8**, 915–917.
6. Y. Vaishnav, E. Holwitt, C. Swenberg, H.C. Lee and L.S. Kan, Synthesis and characterization of stereoisomers of 5,6-dihydro-5,6-dihydroxythymidine. *J. Biomol. Struct. Dyn.*, 1991, **8**, 935–951.
7. S. Nampalli and S. Kumar, Efficient synthesis of 8-Oxo-dGTP: A mutagenic nucleotide. *Bioorg. Med. Chem. Letts.*, 2000, **10**, 1677–1679.
8. J.W. Trauger and P.B. Dervan, Footprinting methods for analysis of pyrrole-imidazole polyamide/DNA complexes. *Methods Enzymol.*, 2001, **340**, 450–466.
9. B.S. Ermolinsky and S.N. Mikhailov, Periodate oxidation in chemistry of nucleic acids: Dialdehyde derivatives of nucleosides, nucleotides, and oligonucleotides. *Russ. J. Bioorg. Chem.*, 2000, **26**, 429–449.
10. E. Pichersky, DNA sequencing by the chemical method. *Methods Mol. Biol.*, 1993, **23**, 255–259.
11. J.H. Phillips, J.H. and D.M. Brown, Mutagenic action of hydroxylamine. *Prog. Nucleic Acid Res. Mol. Biol.*, 1967, **7**, 349–368.
12. R. Shapiro, V. DeFate and M. Welcher, Deamination cytosine derivatives by bisulfite. Mechanism of the reaction. *J. Am. Chem. Soc.*, 1974, **96**, 906–912.
13. C.W. Carreras and D.V. Santi, The catalytic mechanism and structure of thymidylate synthase. *Ann. Rev. Biochem.*, 1995, **64**, 721–762.
14. X. Cheng and R.J. Roberts, AdoMet-dependent methylation, DNA methyltransferases and base flipping. *Nucleic Acids Res.*, 2001, **29**, 3784–3795.
15. J.A. Houghton and P.J. Houghton, 5-Halogenated pyrimidines and their nucleosides. *Handbook Exp. Pharmacol.*, 1984, **72**, 515–549.
16. M.J. Robins and S.R. Naik, Nucleic acid related compounds. III. Facile synthesis of 5-fluorouracil bases and nucleosides by direct fluorination. *J. Am. Chem. Soc.*, 1971, **93**, 5277–5278.
17. R. Shapiro and H. Yamaguchi, Nucleic acid reactivity and conformation. I. Deamination of cytosine by nitrous acid. *Biochim. Biophys. Acta*, 1972, **281**, 501–506.
18. G. Cheng, Y. Shi, S.J. Sturla, J.R. Jalas, E.J. McIntee, P.W. Villalta, M. Wang and S.S. Hecht, Reactions of formaldehyde plus acetaldehyde with deoxyguanosine and DNA: Formation of cyclic deoxyguanosine adducts and formaldehyde cross-links. *Chem. Res. Toxicol.*, 2003, **16**, 145–152.
19. R.F. Newbold, W. Warren, A.S.C. Medcalf and J. Amos, Mutagenicity of carcinogenic methylating agents is associated with a specific DNA modification. *Nature*, 1980, **283**, 596–599.
20. P. Blans and J.C. Fishbein, Determinants of selectivity in alkylation of nucleosides and DNA by secondary diazonium ions: Evidence for, and consequences of, a preassociation mechanism. *Chem. Res. Toxicol.*, 2004, **17**, 1531–1539.
21. R.L. Wurdeman, K.M. Church and B. Gold, DNA methylation by *N*-methyl-*N*-nitrosourea, *N*-methyl-*N*′-nitro-*N*-nitrosoguanidine, *N*-nitroso(1-acetoxyethyl)methylamine, and diazomethane. The mechanism for the formation of N7-methylguanine in sequence-characterized 5′-[32P]-end-labeled DNA. *J. Am. Chem. Soc.*, 1989, **111**, 6408–6412.
22. A. Fidder, G.W.H. Moes, A.G. Scheffer, G.P. van der Schans, R.A. Baan, L.P.A. de Jong and H.P. Benschop, Synthesis, characterization, and quantitation of the major adducts formed

between sulfur mustard and DNA of calf thymus and human blood. *Chem. Res. Toxicol.*, 1994, **7**, 199–204.

23. F.P. Guengerich, S. Langouet, A.N. Mican, S. Akasaka and M.P. Muller, Formation of etheno adducts and their effects on DNA polymerases. *IARC Sci. Publ. (Exocyclic DNA Adducts in Mutagenesis and Carcinogenesis)*, 1999, **150**, 137–145.
24. B. Singer and D. Grunberger, *Molecular Biology of Mutagens and Carcinogens*, Plenum Press, New York, 1983, 45–96.
25. C.S. Madsen, S.C. Ghivizzani, C.V. Ammini, M.R. Nelen and W.W. Hauswirth, Genomic footprinting of mitochondrial DNA: I. *In organello* analysis of protein-mitochondrial DNA interactions in bovine mitochondria. *Methods Enzymol.*, 1996, **264**, 12–22.
26. D.E. Bergstrom and X. Lin, Recent advances in palladium-mediated reactions of nucleosides. *Nucleos. Nucleot.*, 1991, **10**, 689–691.
27. P.R. Langer, A.A. Waldrop and D.C. Ward, Enzymic synthesis of biotin-labeled polynucleotides: Novel nucleic acid affinity probes. *Proc. Natl. Acad. Sci. USA*, 1981, **78**, 6633–6637.
28. J.M. Prober, G.L. Trainor, R.J. Dam, F.W. Hobbs, C.W. Robertson, R.J. Zagursky, A.J. Cocuzza, M.A. Jensen and K. Baumeister, A system for rapid DNA sequencing with fluorescent chain-terminating dideoxynucleotides. *Science*, 1987, **238**, 336–341.
29. E.R. Jamieson and S.J. Lippard, Structure, recognition, and processing of cisplatin-DNA adducts. *Chem. Rev.*, 1999, **99**, 2467–2498.
30. A.P. Silverman, W. Bu, S.M. Cohen and S.J. Lippard, Crystal structure of the asymmetric platinum complex $\{Pt(ammine)(cyclohexylamine)\}^{2+}$ bound to a dodecamer DNA duplex. *J. Biol. Chem.*, 2002, **277**, 49743–49749.
31. B. Spingler, D.A. Whittington and S.J. Lippard, 2.4 angstrom crystal structure of an oxaliplatin 1,2-d(GpG) intrastrand cross-link in a DNA dodecamer duplex. *Inorg. Chem.*, 2001, **40**, 5596–5602.
32. G.M. Blackburn and B. Kellard, Chemical carcinogens. *Chem. Ind. (Lond.)*, 1986, 607–613, 687–695, 770–779.
33. J.H. Weisburger, A perspective on the history and significance of carcinogenic and mutagenic N-substituted aryl compounds in human health. *Mutat. Res.*, 1997, **376**, 261–266.
34. B. Mao, A. Gorin, Z. Gu, B.E. Hingerty, S. Broyde and D.J. Patel, Solution structure of the aminofluorene-intercalated conformer of the syn [AF]-C8-dG adduct opposite α-2 deletion site in the NarI hot spot sequence context. *Biochemistry*, 1997, **36**, 14479–14490.
35. T.D. Bradshaw and A.D. Westwell, The development of the antitumour benzothiazole prodrug, phortress, as a clinical candidate. *Curr. Med. Chem.*, 2004, **11**, 1009–1021.
36. H. Ohgaki, S. Takayama and T. Sugimura, Carcinogenicities of heterocyclic amines in cooked food. *Mutat. Res.*, 1991, **259**, 399–410.
37. K. Kohda, K. Y, Y. Minoura and M. Tada, Separation and identification of N4-(guanosin-7-yl)-4-aminoquinoline 1-oxide, a novel nucleic acid adduct of carcinogen 4-nitroquinoline 1-oxide. *Carcinogenesis*, 1991, **12**, 1523–1525.
38. R.G. Harvey and N.E. Geacintov, Intercalation and binding of carcinogenic hydrocarbon metabolites to nucleic acids. *Acc. Chem. Res.*, 1988, **21**, 66–73.
39. W.M. Baird, L.A. Hooven and B. Mahadevan, Carcinogenic polycyclic aromatic hydrocarbon-DNA adducts and mechanism of action. *Environ. Mol. Mutagen.*, 2005, **45**, 106–114.
40. H. Ling, J.M. Sayer, B.S. Plosky, H. Yagi, F. Boudsocq, R. Woodgate, D.M. Jerina and W. Yang, Crystal structure of a benzo[a]pyrene diol epoxide adduct in a ternary complex with a DNA polymerase. *Proc. Natl. Acad. Sci. USA*, 2004, **101**, 2265–2269.
41. R.S. Iyer, B.F. Coles, K.D. Raney, R. Thier, F.P. Guengerich and T.M. Harris, DNA adduction by the potent carcinogen aflatoxin B j: Mechanistic studies. *J. Am. Chem. Soc.*, 1994, **116**, 1603–1609.
42. R.S. Iyer, M.W. Voehler and T.M. Harris, Adenine adduct of aflatoxin B1 epoxide. *J. Am. Chem. Soc.*, 1994, **116**, 8863–8869.

43. E.S. Newlands, M.F.G. Stevens, S.R. Wedge, R.T. Wheelhouse and C. Brock, Temozolomide: a review of its discovery, chemical properties, pre-clinical development and clinical trials. *Cancer Treat. Rev.*, 1997, **23**, 35–61.
44. J. Arrowsmith, S.A. Jennings, A.S. Clark and M.F.G. Stevens, Antitumor imidazotetrazines. 41. Conjugation of the antitumor agents mitozolomide and temozolomide to peptides and lexitropsins bearing DNA major and minor groove-binding structural motifs. *J. Med. Chem.*, 2002, **45**, 5458–5470.
45. M. Tomasz, Mitomycin C: small, fast and deadly (but very selective). *Chem. Biol.*, 1995, **2**, 575–599.
46. L.H. Hurley and D.R. Needham-Vandevanter, Covalent binding of antitumour antibiotics in the minor groove of DNA, mechanism of action of CC1065 and the pyrrolo[1,4]benzodiazepines. *Acc. Chem. Res.*, 1986, **19**, 230–237.
47. G. Subramaniam, M.M. Paz, G.S. Kumar, A. Das, Y. Palom, C.C. Clement, D.J. Patel and M. Tomasz, Solution structure of a guanine-N7-linked complex of the mitomycin C metabolite 2,7-diaminomitosene and DNA. Basis of sequence selectivity. *Biochemistry*, 2001, **40**, 10473–10484.
48. R.W. Armstrong, M.E. Salvati and M. Nguyen, Novel interstrand crosslinks induced by the antitumour antibiotic carzinophilin/azinomycin B. *J. Am. Chem. Soc.*, 1992, **114**, 3144–3145.
49. R.S. Coleman, R.J. Perez, C.H. Burk and A. Navarro, Studies on the mechanism of action of azinomycin B: definition of regioselectivity and sequence selectivity of DNA cross-link formation and clarification of the role of the naphthoate. *J. Am. Chem. Soc.*, 2002, **124**, 13008–13017.
50. L.H. Hurley, T. Reck, D.E. Thurston, D.R. Langley, K.G. Holden, R.P. Hertzberg, J.R.E. Hoover, G. Gallagher, L.F. Faucette, S.M. Mong and R.K. Johnson, Pyrrolo[1,4]benzodiazepine antitumor antibiotics: relationship of DNA alkylation and sequence specificity to the biological activity of natural and synthetic compounds. *Chem. Res. Toxicol.*, 1988, **1**, 258–268.
51. W.A. Remers, M.D. Barkley and L. Hurley, Pyrrolo(1,4)benzodiazepines. Unraveling the complexity of the structures of the tomaymycin-DNA adducts in various sequences using fluorescence, proton-NMR, and molecular modeling. in *Nucleic Acid Targeted Drug Design*, C.L. Propst and T.J. Perun (eds) QC Marcel Dekker, New York, 1992, 375–422.
52. M.L. Kopka, D.S. Goodsell, I. Baikalov, K. Grzeskowiak, D. Cascio and R.E. Dickerson, Crystal structure of a covalent DNA-drug adduct: anthramycin bound to C-C-A-A-C-G-T-T-G-G and a molecular explanation of specificity. *Biochemistry*, 1994, **33**, 13593–13610.
53. U. Galm, M.H. Hager, S.G. Van Lanen, J. Ju, J.S. Thorson and B. Shen, Antitumor antibiotics: bleomycin, enediynes, and mitomycin. *Chem. Rev.*, 2005, **105**, 739–758.
54. A. Prokop, W. Wrasidlo, H. Lode, R. Herold, F. Lang, G. Henze, B. Doerken, T. Wieder and P.T. Daniel, Induction of apoptosis by enediyne antibiotic calicheamicin theta II proceeds through a caspase-mediated mitochondrial amplification loop in an entirely Bax-dependent manner. *Oncogene*, 2003, **22**, 9107–9120.
55. D.L. Meyer and P.D. Senter, Recent advances in antibody drug conjugates for cancer therapy. *Ann. Rep. Med. Chem.*, 2003, **38**, 229–237.
56. M. Yus and F. Foubelo, Enediynes. Recent developments. *Recent Res. Dev. Org. Chem.*, 2002, **6**, 205–280.
57. I.H. Goldberg, Mechanism of neocarzinostatin action: role of DNA microstructure in determination of chemistry of bistranded oxidative damage. *Acc. Chem. Res.*, 1991, **24**, 191–198.
58. R.A. Kumar, N. Ikemoto and D.J. Patel, Solution structure of the calicheamicin gamma-1I-DNA complex. *J. Mol. Biol.*, 1997, **265**, 187–201.
59. K.C. Nicolaou, B.M. Smith, K. Ajito, H. Komatsu, L. Gomez-Paloma and Y. Tor, DNA-carbohydrate interactions. Specific binding of head-to-head and head-to-tail dimers of the calicheamicin oligosaccharide to duplex DNA. *J. Am. Chem. Soc.*, 1996, **118**, 2303–2304.
60. R.A. Kumar, N. Ikemoto and D.J. Patel, Solution structure of the esperamicin-$A_1$-DNA complex. *J. Mol. Biol.*, 1997, **265**, 173–186.
61. L. Yu, S. Mah, T. Otani and P. Dedon, The benzoxazolinate of C-1027 confers intercalative DNA binding. *J. Am. Chem. Soc.*, 1995, **117**, 8877–8878.

62. J.D. Scott and R.M. Williams, Chemistry and biology of the tetrahydroisoquinoline antitumor antibiotics. *Chem. Rev.*, 2002, **102**, 1669–1730.
63. A. Oikarinen and A. Raitio, Melanoma and other skin cancers in circumpolar areas. *Int. J. Circumpolar Health*, 2000, **59**, 52–56.
64. J. Cadet, S. Courdavault, J.-L. Ravanat and T. Douki, UVB and UVA radiation-mediated damage to isolated and cellular DNA. *Pure Appl. Chem.*, 2005, **77**, 947–961.
65. J. Cadet and R. Vigny, The photochemistry of nucleic acids, in *Bioorganic Photochemistry*, H. Morrison (ed). Wiley, New York, 1990, 1–272.
66. G.J. Fisher and H.E. Johns, Pyrimidine photohydrates. *Photochem. Photobiol. Nucleic Acids*, 1976, **1**, 169–224.
67. G.M. Blackburn and R.J.H. Davies, Photochemistry of nucleic acids. III. Structure of DNA-derived thymine photodimer. *J. Am. Chem. Soc.*, 1967, **89**, 5941–5945.
68. H. Park, K. Zhang, Y. Ren, S. Nadji, Sinha, J.-S. Taylor and C. Kang, Crystal structure of a DNA decamer containing a *cis-syn* thymine dimer. *Proc. Natl. Acad. Sci. USA*, 2002, **99**, 15965–15970.
69. J.-K. Kim and B.-S. Choi, The solution structure of DNA duplex-decamer containing the (6-4) photoproduct of thymidylyl(3′-5′)thymidine by NMR and relaxation matrix refinement. *Eur. J. Biochem.*, 1995, **228**, 849–854.
70. L.M. Kundu, U. Linne, M. Marahiel and T. Carell, RNA is more UV resistant than DNA: the formation of UV-induced DNA lesions is strongly sequence and conformation dependent. *Chem. Eur. J.*, 2004, **10**, 5697–5705.
71. J.S. Taylor, Unraveling the molecular pathway from sunlight to skin cancer. *Acc. Chem. Res.*, 1994, **27**, 76–82.
72. Y.-H. You, P.E. Szabó and G.P. Pfeifer, Cyclobutane pyrimidine dimers form preferentially at the major *p53* mutational hotspot in UVB-induced mouse skin tumors. *Carcinogenesis*, 2000, **21**, 2113–2118.
73. T. Douki and J. Cadet, Individual determination of the yield of the main UV-induced dimeric pyrimidine photoproducts in DNA suggest a high mutagenicity of CC photolesions. *Biochemistry*, 2001, **40**, 2495–2501.
74. F.-X. Barre, L.L. Pritchard and A. Harel-Bellan, Psoralen-coupled oligonucleotides: *in vivo* binding and repair, in *Triple Helix Forming Oligonucleotides*, C. Malvy, A. Harel-Bellan and L.L. Pritchard (eds). Kluwer, Boston, MA, 1999, 181–192.
75. N.J. Duker and P.E. Gallagher, Purine photoproducts. *Photochem. Photobiol.*, 1988, **48**, 35–39.
76. L.A. Lipscomb, M.E. Peek, M.L. Morningstar, S.M. Verghis, E.M. Miller, A. Rich, J.M. Essigmann and L.D. Williams, X-Ray structure of a DNA decamer containing 7,8-dihydro-8-oxoguanine. *Proc. Natl. Acad. Sci. USA*, 1995, **92**, 71923.
77. J.E. Frederick, Ultraviolet sunlight reaching the earth's surface. *Photochem. Photobiol.*, 1993, **57**, 1758.
78. M. McFarland and J. Kaye, Chlorofluorocarbons and ozone. *Photochem. Photobiol. Nucleic Acids*, 1992, **55**, 911–929.
79. J.-W. Lee, Children suffer most from the effects of ozone depletion, 2003, available at: www.who.int/mediacentre/news/releases/2003/pr66/en/.
80. I.H. Rowlands, The fourth meeting of the parties to the Montreal Protocol: Report and reflection. *Environ. Mol. Mutagen.*, 1993, **35**, 25–34.
81. C. von Sonntag, *The Chemical Basis of Radiation Biology*. Taylor & Francis, London, 1986.
82. R.J. Boorstein, J. Cadet, T. Hilbert, M. Lustig, R. O'Donnell, S. Zuo and G. Teebor, Formation and stabilities of pyrimidine modification in DNA. *Journal de Chimie Physique et de Physico-Chimie Biologique*, 1993, **90**, 837–852.
83. M.D. Evans and M.S. Cooke, Factors contributing to the outcome of oxidative damage to nucleic acids. *BioEssays*, 2004, **26**, 533–542.
84. A.B. Guliaev and B. Singer, DNA damage: alkylation. *Encyclopedia Biol. Chem.*, 2004, **1**, 609–613.
85. J.C. Delaney and J.M. Essigmann, Mutagenesis, genotoxicity, and repair of 1-methyladenine, 3-alkylcytosines, 1-methylguanine, and 3-methylthymine in alkB *Escherichia coli*. *Proc. Natl. Acad. Sci. USA*, 2004, **101**, 14051–14056.

86. T.E. Spratt and D.E. Levy, Structure of the hydrogen bonding complex of O6-methylguanine with cytosine and thymine during DNA replication. *Nucleic Acids Res.*, 1997, **25**, 3354–3361.
87. G.P. Margison, K.M.F. Santibanez and A.C. Povey, Mechanisms of carcinogenicity/chemotherapy by O6-methylguanine. *Mutagenesis*, 2002, **17**, 483–487.
88. Y. Luo and J.D. Leverson, New opportunities in chemosensitization and radiosensitization: modulating the DNA-damage response. *Expert Rev. Anticancer Ther.*, 2005, **5**, 333–342.
89. B. Kaina, DNA damage-triggered apoptosis: critical role of DNA repair, double-strand breaks, cell proliferation and signaling. *Biochem. Pharmacol.*, 2003, **66**, 1547–1554.
90. J. Rouse and S.P. Jackson, Interfaces between the detection, signaling, and repair of DNA damage. *Science*, 2002, **297**, 547–551.
91. A. Mees, T. Klar, P. Gnau, U. Hennecke, A.P.M. Eker, T. Carell and L.-O. Essen, Crystal structure of a photolyase bound to a CPD-like DNA lesion after in situ repair. *Science*, 2004, **306**, 1789–1793.
92. L.D. Samson, The repair of DNA alkylation damage by methyltransferases and glycosylases. *Essays Biochem.*, 1992, **27**, 69–78.
93. R.D. Wood, M. Mitchell, J. Sgouros and T. Lindahl, Human DNA repair genes. *Science*, 2001, **291**, 1284–1289.
94. M.H. Moore, J.M. Gulbis, E.J. Dodson, B. Demple and P.C. Moody, Crystal structure of a suicidal DNA repair protein: the Ada $O^6$-methylguanine-DNA methyltransferase from *E. coli*. *EMBO J.*, 1994, **13**, 1495–1501.
95. J.C. Fromme and G.L. Verdine, Base excision repair. *Adv. Protein Chem.*, 2004, **69**, 1–41.
96. M.A. Lovell, C. Xie and W.R. Markesbery, Decreased base excision repair and increased helicase activity in Alzheimer's disease brain. *Brain Res.*, 2000, **855**, 116–123.
97. S.S. Parikh, G. Walcher, G.D. Jones, G. Slupphaug, H.E. Krokan, G.M. Blackburn and J.A. Tainer, Uracil-DNA glycosylase-DNA substrate and product structures: conformational strain promotes catalytic efficiency by coupled stereoelectronic effects. *Proc. Natl. Acad. Sci. USA*, 2000, **97**, 5083–5088.
98. D.O. Zharkov, T.A. Rosenquist, S.E. Gerchman and A.P. Grollman, Substrate specificity and reaction mechanism of murine 8-oxoguanine-DNA glycosylase. *J. Biol. Chem.*, 2000, **275**, 28607–28617.
99. C.L. Chepanoske, S.L. Porello, T. Fujiwara, H. Sugiyama and S.S. David, Substrate recognition by *Escherichia coli* MutY using substrate analogs. *Nucleic Acids Res.*, 1999, **27**, 3197–3204.
100. O.D. Schärer, T. Kawate, P. Gallinari, J. Jiricny and G.L. Verdine, Investigation of the mechanisms of DNA binding of the human G/T glycosylase using designed inhibitors. *Proc. Natl. Acad. Sci. USA*, 1997, **94**, 4878–4883.
101. J.C. Fromme, A. Banerjee and G.L. Verdine, DNA glycosylase recognition and catalysis. *Curr. Opin. Struct. Biol.*, 2004, **14**, 43–49.
102. J.T. Stivers and Y.L. Jiang, A mechanistic perspective on the chemistry of DNA repair glycosylases. *Chem. Rev.*, 2003, **103**, 2729–2759.
103. K. Morikawa and M. Shirakawa, Three-dimensional structural views of damaged-DNA recognition: T4 endonuclease V, *E. coli* Vsr protein, and human nucleotide excision repair factor XPA. *Mutat. Res.*, 2000, **460**, 257–275.
104. J.E. Cleaver, Mending human genes: a job for a lifetime. *DNA Repair*, 2005, **4**, 635–638.
105. A. Sancar and J.T. Reardon, Nucleotide excision repair in *E. coli* and man. *Adv. Protein Chem.*, 2004, **69**, 43–71.
106. B. Van Houten and L. Grossman, Nucleotide excision repair, bacterial: the UvrABCD system. *Encyclopedia Biol. Chem.*, 2004, **3**, 134–142.
107. R. Dip, U. Camenisch and H. Naegeli, Mechanisms of DNA damage recognition and strand discrimination in human nucleotide excision repair. *DNA Repair*, 2004, **3**, 1409–1423.
108. G.-M. Li and S.R. Presnell, DNA mismatch repair and the DNA damage response. in *Encyclopedia of Biological Chemistry*, W.J. Lennarz and M.D. Lane (eds). 2004, 671–674.
109. P.C. Hanawalt, Genomic instability: Environmental invasion and the enemies within. *Mutat. Res.*, 1998, **400**, 117–125.

110. G.C. Walker, SOS-regulated proteins in translesion DNA synthesis and mutagenesis. *Trends Biochem. Sci.*, 1995, **20**, 416–420.
111. M.F. Goodman, Coping with replication 'train wrecks' in *Escherichia coli* using Pol V, Pol II and RecA proteins. *Trends Biochem. Sci.*, 2000, **25**, 189–195.
112. J.-S. Taylor, New structural and mechanistic insight into the A-rule and the instructional and non-instructional behavior of DNA photoproducts and other lesions. *Mutat. Res.*, 2002, **510**, 55–70.
113. Z. Livneh, UmuC, D lesion bypass DNA polymerase V. *Encyclopedia of Biological Chemistry*, W.J. Lennarz and M.D. Lane (eds). 2004, 308–312.
114. S.G. Kozmin, Y.I. Pavlov, T.A. Kunkel and E. Sage, Roles of *Saccharomyces cerevisiae* DNA polymerases Pol eta and Pol zeta in response to irradiation by simulated sunlight. *Nucleic Acids Res.*, 2003, **31**, 4541–4552.
115. S.D. McCulloch, R.J. Kokoska, C. Masutani, S. Iwai, F. Hanaoka and T.A. Kunkel, Preferential cis-syn thymine dimer bypass by DNA polymerase eta occurs with biased fidelity. *Nature*, 2004, **428**, 97–100.

CHAPTER 9

# Reversible Small Molecule–Nucleic Acid Interactions

## CONTENTS

## 9.1 INTRODUCTION

Most biological processes are reliant upon molecular recognition and the reversible interactions of one set of molecules with another. This is particularly the case where nucleic acids are concerned, since chromosome packaging and structural integrity as well as gene expression and DNA replication depend on numerous protein–DNA interactions as well as, in a more subtle way, interactions with ions and water. Because nucleic acids play a central role in critical cellular processes, such as cell division and protein expression, they are very attractive targets for small molecule therapeutics. A desirable goal is that malfunctions in cell replication or gene expression might be overcome by modulating nucleic acid activity through use of sequence- or structure-specific drugs. Such compounds would have a direct and beneficial role in the treatment of major diseases such as cancer.

Molecular recognition is a fundamental and underlying concept in chemistry and short, well-defined nucleic acid sequences that bind to small ligands are ideal model systems for gaining understanding of the basic principles of recognition. Nucleic acids, especially DNA, are less extensively folded and structurally less complex than proteins, and this makes them attractive target molecules for general biophysical studies.

The equilibrium binding of small molecules to nucleic acids has been an active area of research for over 40 years. This Chapter aims to focus on basic concepts and major issues by use of selected examples. Many of the biophysical and structural methods used in studying ligand binding to nucleic acids are outlined in Chapter 11.

In general there are more biophysical studies of interactions of drugs with DNA than with RNA. This is because DNA is a more attractive target for anticancer drugs and DNA has been much easier to synthesize chemically to obtain model systems. Therefore, this chapter focuses primarily on small molecule–DNA binding and reference should be made to specialized literature for discussions on ligand–RNA interactions.

High-resolution structures are necessary to reveal the overall three-dimensional shape of a complex, the conformations adopted by the two reacting species, as well as some of the molecular interactions in the final complex. But resolving the structure of a complex is insufficient to understand its formation. Various small molecule–nucleic acid complexes can have very similar structures, yet may have radically different underlying driving forces. To gain a fuller understanding, it is also necessary to have thermodynamic and kinetic information. In this chapter, the structures of various complexes are summarised and then rationalized in terms of binding thermodynamics or kinetics, where such information is available. This will support a better appreciation of the exquisite nature of small molecule–nucleic acid interactions.

## 9.2 BINDING MODES AND SITES OF INTERACTION

There are three principal ways in which low molecular weight ligands can interact with double-stranded DNA (Figure 9.1).

**Outside-edge binding**: This involves ligand binding (*e.g.* $Na^+$, $Mg^{2+}$ or polyamines, Section 9.4) to the outside of the helix through non-specific, primarily electrostatic interactions with the sugar–phosphate backbone.

**Intercalation**: A planar (or near planar) aromatic ring system (*e.g.* daunomycin, Section 9.6) inserts in between two adjacent base pairs, perpendicular to the helical axis.

**Groove binding**: A bound ligand (*e.g.* netropsin, Section 9.7) makes direct molecular contacts with functional groups on the edges of the bases that protrude into either the major or minor grooves.

Compounds that have the potential to be clinically useful are normally either intercalators or groove-binders. However, outside-edge, electrostatic interactions are also important, not least because the association of positively charged counterions with the DNA polyanion has a large effect on DNA conformation and stability.

The structure-specific recognition of higher-order nucleic acids, such as three-stranded triplexes or four-stranded quadruplex DNA molecules (Section 9.10.2) is a developing field. Drug binding to these structures can also be classified according to the above three groups. In addition, RNA–DNA heteroduplexes that usually adopt the A-conformation (Section 2.2) are also attractive drug targets. Single-stranded RNA can

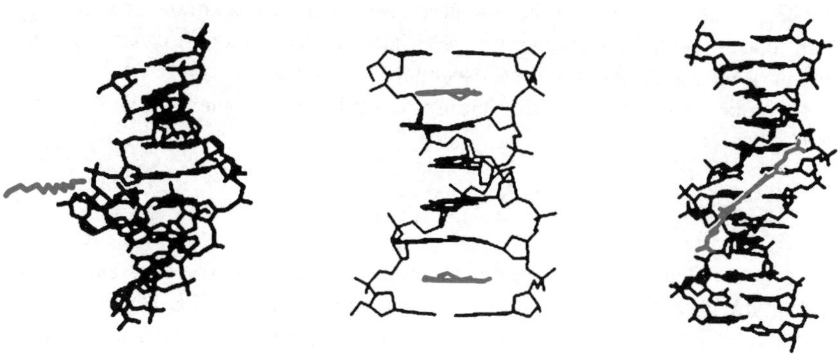


**Figure 9.1** *Schematic representations of the three primary binding modes for ligand–duplex DNA binding. The ligand is shown in red and the DNA chains are in black: (a) Outside-edge binding. (b) Intercalation. (c) Groove binding*

form intramolecular folds with regions of duplex, internal loops, bulges and hairpins, for example, in ribosomes, tRNA, or genomic viral RNA (Chapter 7). Such non-standard duplex folds in RNA offer a varied array of molecular architectures that can be exploited as drug targets, for example, those found in pathogenic RNA viruses such as HIV-1.

## 9.3 COUNTER-ION CONDENSATION AND POLYELECTROLYTE THEORY

Nucleic acids are highly charged polyanions, and thus polyelectrolyte effects have a major impact on their biological behaviour. For example, the polyanionic nature of nucleic acids directly affects the binding of charged drugs and proteins and the denaturation and folding of nucleic acids as well as the condensation and packaging of DNA and RNA in cells.

A nucleoside monophosphate has two ionizable hydroxyl groups of $pK_a < 1$ and *ca.* 6.8 (Figure 2.5). These $pK_a$ values are only slightly dependent on the location of the phosphate on the nucleoside or which of the bases is involved. At pH 7.0 and above, a terminal phosphate group has two negative charges while an internucleoside phosphate diester has one.

Monovalent cations bind to nucleic acids in a somewhat unusual manner that is unique to highly charged linear polyelectrolytes. In the early 1970s Gerald Manning described such binding as **counter-ion condensation**:[1–3] territorial binding of ions that are constrained to stay within a few angstroms of the DNA surface, yet are free to move along the helix. Hence cations associate with nucleic acids largely as a function of polymer charge density. To be stable, DNA must associate with counter-ions from solution. Additional counter-ions are associated with remaining charges on the polyanion through **Debye–Hückel interactions**.

The initial interaction of counter-ions with nucleic acids is referred to as condensation because the cations generally form a cloud around the charge density of the nucleic acid and are not bound at specific sites. The ions retain their inner sphere water of hydration and they move up and down the phosphate backbone. Secondary hydration layers of both the ion and the nucleic acid are affected by this interaction. With B-type duplex DNA, counter-ion condensation theory predicts an average of 0.76 monovalent counter-ions condensed per phosphate group, rising to a total of 0.88 counter-ions per phosphate group including Debye–Hückel type interactions.

The effect of counter-ion condensation is to reduce the effective charge on the nucleic acid. This in turn has a profound effect on solution properties, binding interactions, and stability of nucleic acids. A proportion of the overall free energy for binding intercalators and groove-binders can be derived from the **entropically favourable release of condensed counter-ions** upon binding. At constant temperature, the entropic effect resulting from counter-ion release can also lead to nucleic acid denaturation as well as to an apparent increase in binding affinity for cationic ligands as bulk salt concentration is decreased. Hence both the melting temperature (**$T_m$**, Section 2.5.1) and the observed equilibrium association constant (**$K_{obs}$**) are strongly dependent on salt concentration. When values for these thermodynamic parameters are measured, it is important to take account of the solution conditions, especially salt concentration.

In Manning's original formulation, condensation is described by a dimensionless structural parameter $\xi$

$$\xi = \frac{e^2}{\varepsilon kTb} \tag{9.1}$$

where $e$ is the electronic charge magnitude, $\varepsilon$ is the bulk solution dielectric constant, $k$ Boltzmann's constant, $T$ the temperature in K, and $b$ the average phosphate group spacing. If B-DNA is modelled as a cylinder with two negatively charged phosphates spaced 3.4 Å apart, then $b = 1.7$ Å and $\xi = 4.2$. The Manning theory states that the average fraction of monovalent cation associated with each DNA phosphate as a result of condensation ($\psi_c$) is defined as

$$\psi_c = (1 - \xi^{-1}) \tag{9.2}$$

For B-DNA $\psi_c = 0.76$. However, as noted by Thomas Record, cations not only associate *via* condensation but can also bind through Debye–Hückel interactions. This fraction of associated counterions is defined as $\psi_s$. Therefore, the overall fraction of monovalent cations associated with each phosphate can be defined as $\psi$ where

$$\psi = \psi_c + \psi_s = [1 - (2\xi)^{-1}] \tag{9.3}$$

For B-DNA the value of $\psi$ is 0.88, *i.e.* the double helix retains a net charge equal to 12% of the total number of phosphates. Record has developed a particularly useful formulation that describes the effect of salt concentration on nucleic acid equilibria.[4] He applied Wyman's concept of linkage to quantify **salt effects on ligand binding**. The magnitude of the apparent ligand binding constant $K_{obs}$ for the interaction of a charged ligand with DNA is found to be strongly dependent on monovalent salt concentration [MX], as embodied in the expression

$$\frac{\delta \ln K_{obs}}{\delta \ln[MX]} = -Z\psi \equiv SK \tag{9.4}$$

where $Z$ is the charge on the ligand and $SK$ is the slope of a plot of $\delta \ln K_{obs}$ versus $\delta \ln[MX]$. The negative sign of $Z\psi$ in Equation 9.4 indicates that $K_{obs}$ decreases with increasing concentrations of MX. The magnitude of the decrease depends on $Z$, the charge on the ligand. In this analysis, any possible contribution from changes in anion binding or hydration is omitted. For a cationic ligand $L$ binding to a nucleic acid site $D$ to form a ligand–nucleic acid complex $C$ in the presence of monovalent cation $M^+$, the thermodynamic equilibrium can be written as

$$D + L \rightleftharpoons C + Z\psi[M^+] \tag{9.5}$$

This equilibrium describes the release of bound counter-ions concomitant with the binding of a cationic ligand. The equilibrium constant for the above reaction is

$$K = \frac{[C] \cdot [M^+]^{Z\psi}}{[D] \cdot [L]} \tag{9.6}$$

Hence for the association of a positively charged ligand with DNA in the presence of excess monovalent cation,

$$\ln K_{obs} = \ln K_T - Z\psi \ln[MX] \quad (9.7)$$

and

$$\Delta G_{obs} = \Delta G_T + \Delta G_{pe} \quad (9.8)$$

where $\ln K_T$ and the corresponding free energy term $\Delta G_T$ refer to a standard state of $[MX]$ of 1 M . **$\Delta G_{pe}$ is the polyelectrolyte contribution to the overall free energy**

$$\Delta G_{pe} = Z\psi RT \ln[MX] \quad (9.9)$$

$\Delta G_{pe}$ is a favourable free energy contribution that arises from the entropically favoured release of counter-ions into bulk solvent on binding of a cationic ligand. By experimental determination of the value of $\delta \ln K_{obs}/\delta \ln[MX]$, $\Delta G_{pe}$ can be calculated at any given salt concentration through use of Equation 9.9. From experiments and simulations, it is clear that $\Delta G_{pe}$ can be quite large, even at modest salt concentrations, and it can contribute several kJ to the overall binding free energy. The difference between $\Delta G_{pe}$ and the experimental $\Delta G_{obs}$ (determined at the same salt concentration) is equal to $\Delta G_T$, which is **the nonpolyelectrolyte contribution to free energy**. $\Delta G_T$ is independent of salt concentration and represents the free energy arising from all types of interactions other than coupled polyelectrolyte effects, such as van der Waals interactions, hydrogen bonding, and hydrophobic effects. The use of polyelectrolyte theory to probe the energetics of drug–DNA interactions has been discussed in a useful review article.[5]

### 9.3.1 Intercalation and Polyelectrolyte Theory

To accommodate an intercalating ligand, the nucleic acid base pairs at the intercalation site must be separated by an additional 3.4 Å (Section 9.6). This lengthens the helix and **increases the inter-phosphate spacing**, parameter $b$ in Equation 9.1. It then leads to a decrease in $\xi$ and a decrease in the fraction of monovalent cations associated per phosphate group. For intercalation, there are two distinct factors that contribute to **the polyelectrolyte effect**: (1) release of condensed counter-ions following binding of a positively charged ligand (as discussed above), and (2) a contribution that arises from increased phosphate spacing that results from intercalation-induced conformational changes in the DNA. For a neutral intercalator, the observed binding constant is still dependent on salt concentration, since the intercalator induces a structural transition upon binding.

David Wilson has described a modification of Record's theory that accounts for ion release arising from a DNA structural transition.[6] He predicted values for $\delta \ln K_{obs}/\delta \ln[MX]$ of $-1.06$ and $-1.89$ for the binding of mono- and dicationic intercalators, respectively. Later Friedman and Manning used the same general concept but with a different theoretical model to predict values for $\delta \ln K_{obs}/\delta \ln[MX]$ of $-0.24$ and $-1.24$ for the binding of uncharged and monocationic intercalators, respectively.[7]

## 9.4 NON-SPECIFIC OUTSIDE-EDGE INTERACTIONS

There are several types of molecule that can interact non-specifically with the nucleic acid phosphate backbone through mainly electrostatic interactions. The archetypal **outside-edge binding drugs** are the polyamines, **spermine** and **spermidine** (Figure 9.2).

The nonspecific nature of the interaction between DNA and **polyamines** means that it is difficult to obtain high-resolution NMR or crystallographic structural data on these complexes. In fact these ligands are often used to reduce charge–charge effects in crystallographic studies, since they can rarely be seen at high-density sites in a resolved structure.

Polyamines such as spermine are ubiquitous in eukaryotic cells and these ligands are thought to play multiple roles in cellular function. One important role is in **DNA packaging** into chromatin, but the exact

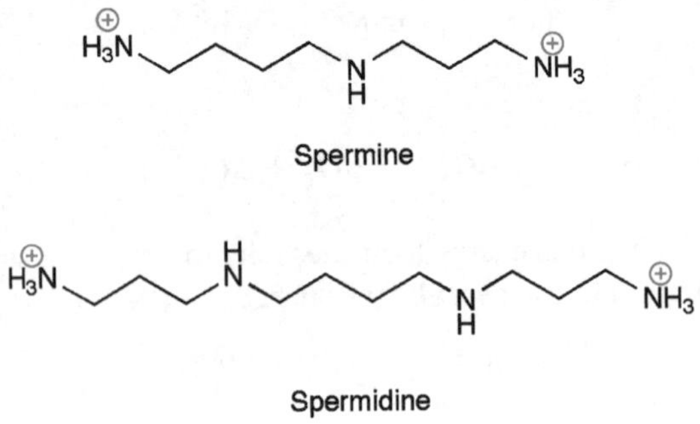


**Figure 9.2** *Structures of two outside-edge electrostatic DNA binding ligands, spermine and spermidine*

mechanism of action is unknown. However, it is likely that polyamines play an analogous role to histones in that they may neutralise some of the negative charges on the DNA backbone and hence promote DNA packaging. Direct binding of polyamines to DNA and the resulting modulation of protein–DNA interactions appears to play an important role in cell proliferation.[9]

## 9.5 HYDRATION EFFECTS AND WATER–DNA INTERACTIONS

**Hydration of DNA** is essential for it to maintain its folded conformation.[9] Indeed the hydration state and interactions with ions can exert large effects on nucleic acid conformation and hence influence the nature of interactions with other ligands such as proteins (Section 10.4.2). Early fibre diffraction studies (Section 1.4) showed that relative humidity has an effect on DNA conformation, such that B-type DNA is found at high humidity values (above 85%) whereas A-form DNA occurs with lower humidity (between 75–80%) and DNA becomes disordered if the relative humidity is lowered to between 55 and 75%. In addition, base-composition and salt concentration affect the values of relative humidity under which one or other conformation is favoured.

In practice, it is very challenging to study **water–DNA interactions** experimentally. Useful data have been obtained from gravimetric studies and infrared (IR) spectra of DNA films as a function of relative humidity. Visualization of specific water molecules from X-ray diffraction analysis of single DNA crystals has also proved difficult in the past. This is because a crystal is derived typically from aqueous solution that also contains cations such as $Mg^{2+}$ or spermine, and alcohols such as 2-methyl-2,4-pentanediol (MPD7). Thus at resolutions of 2 Å or above, it is difficult to distinguish whether a particular electron density corresponds to water, a cation, or a spermine molecule.

Much of our present understanding of DNA hydration (as well as other biophysical characteristics of DNA) has come from studies of the **Dickerson–Drew dodecamer** (Section 2.2.4, Figure 2.17). This oligonucleotide duplex, with the sequence $d(CGCGAATTCGCG)_2$, provided the first published B-DNA X-ray crystal structure. It is the best studied DNA sequence with over 68 isomorphous members of the $d(CGCXAATTYGCG)_2$ (X = G or A; Y = C or T) dodecamer family, and numerous biophysical and binding studies have been carried out with it in solution. In the early 1980s, Dickerson and co-workers obtained several structures of this sequence, for example, at room temperature, at 16 K or at 7°C for the 9-bromo derivative in MPD7.[10] Hydration was not readily visible in the room temperature structure due to high *B* values. However the 16 K and MPD7 structures were extensively hydrated with 65 solvent peaks and an average of three waters per phosphate. In the major groove, 19 solvent molecules formed a first hydration layer around nitrogen and oxygen atoms of the bases in that groove. A second hydration layer contained a further 36 solvent molecules that formed clusters spanning the major groove. In the minor groove, a zigzag pattern of water molecules formed a "**spine of hydration**" in all the structures (Section 2.2.4, Figure 2.18). This concept of a spine of hydration has become pervasive in DNA chemistry and is often invoked to explain a multitude of effects related to DNA stability and ligand binding.

Crystallographic studies have also shown that **minor groove width** is instrumental in dictating hydration patterns. The duplex d(CCAACGTTGG)$_2$ has a narrow minor groove that is about 4.2 Å wide and it has a spine of hydration similar to that originally observed in the Dickerson–Drew sequence. The helix d(CCAAGATTGG)$_2$ sequence has a wider, 7.2 Å minor groove within which two side-by-side ribbons of water molecules are found. Interestingly, the sequence d(CCAGGCCTGG)$_2$ has a minor groove of intermediate width, which is occupied only very sparsely by ordered water molecules and which does not possess either of the two other hydration patterns. It has been suggested that the A·T minor groove spine of hydration is important in stabilization of B-DNA against the B→A conformational transition. High A·T content DNA is associated with narrow minor grooves and resistance to the B→A transition, which has led to the assertion that the spine of hydration is more effective at stabilization of B-DNA than the two parallel ribbons of waters (Section 2.2.4).

### 9.5.1 Cation Binding in the Minor Groove

The seminal work of Dickerson and Drew has been central to the current view that DNA is conformationally polymorphic. While many DNA structures have been solved to a resolution of 2.0–3.0 Å, Loren Williams has achieved a resolution of 1.4 Å, thus allowing a more accurate analysis of the dodecamer structure and revealing some new insights into DNA hydration.[11–13] His principal conclusion is that **the primary spine of hydration is composed partly of sodium ions**. In other words, there is partial occupancy of primary sites of solvation in the minor groove by both water and sodium cations. There is also a **secondary, regular and highly ordered spine of water molecules on top of the primary spine** that does not interact directly with the DNA. This DNA structure is seen to be more conformationally ordered and not as heterogeneous as thought earlier.

However, in practice, partial occupancy (known as a sodium–water hybrid) is difficult to distinguish from a water molecule alone, especially if sodium occupancy is less than 50%. This is because sodium ions and water molecules have the same number of electrons and therefore yield electron density peaks of similar size. In addition, both water and sodium ions have a wide variation in their coordination geometries. To confirm **the hybrid–solvent model** proposed for the high-resolution Dickerson–Drew dodecamer, Williams carried out potassium substitution experiments[12] to take advantage of the fact that a potassium ion can readily be distinguished from water because of its larger number of electrons. These studies have confirmed that monovalent cations penetrate the primary layer of the spine of hydration and are also present in the secondary hydration layer. The spine of hydration forms only the bottom two layers of a four-layer arrangement of solvent and cations in the A·T minor groove, which are organised to form a repeating motif of fused hexagons (Figure 9.3).

This work has implications for binding of other small molecules to DNA minor grooves. A substantial favourable entropic contribution to binding free energy may arise from positioning positively charged, minor groove-binding ligands or cationic ancillary functional groups of intercalators (*e.g.* the amino sugar of daunomycin) into A·T minor grooves that contain sodium or potassium ions bound in specific sites. Favourable entropic effects will arise when monovalent metal cations are expelled from the minor groove by cationic ligands.

## 9.6 DNA INTERCALATION

### 9.6.1 The Classical Model

Leonard Lerman first proposed **the intercalation model** for ligand binding to DNA in the early 1960s.[14,15] Working with aminoacridines and proflavin, he postulated that DNA unwinds slightly to open a space into which the **flat polycyclic aromatic chromophores** can be inserted between adjacent base pairs. This mode of binding has now been established for a large number of planar heteroaromatic or polycyclic aromatic ring structures (Figure 9.4).

A common structural feature of all **intercalators** is that they possess an extended, electron-deficient, planar aromatic ring system that is often referred to as a chromophore. Such compounds typically have

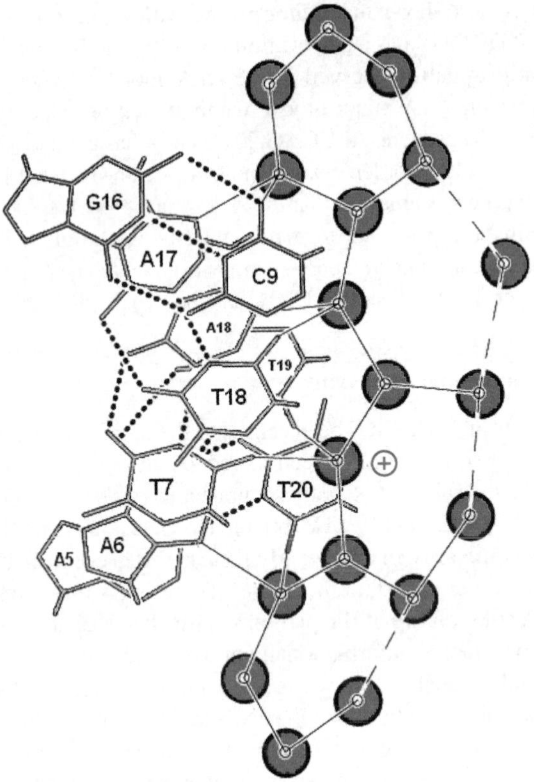


**Figure 9.3** *The hexagonal hydration motif observed in A-tract DNA at high resolution. This view is across the minor groove, roughly along the normal of the central hexagons. Solvent sites are indicated by red spheres, with solid lines designating molecular contacts. Distances greater than 3.4 Å are specified by dashed lines. Watson–Crick hydrogen bonds are indicated by dotted lines. The site of confirmed cation occupancy is indicated by a plus sign*
(Figure prepared and kindly provided by Loren D. Williams)

two or more six-membered rings that form a platform of approximately the same size as a DNA base pair. Intercalators are an important type of DNA binding agent, not least because some of these compounds have been used as drugs.

Simple intercalators, such as **ethidium, propidium** and **proflavin**, contain only the intercalating chromophore and they are often cationic at neutral pH. The archetypal **simple DNA intercalator** is **ethidium bromide**, which binds to two adjacent G·C base pairs (Figure 9.5). However, many ligands that bind by this mode are not "simple" intercalators but rather their intercalating moiety is decorated by a variety of chemical substituents, sugar rings or peptide groups. These additional, often non-planar, functional groups have a major role in the exact sequence specificity, thermodynamic stability and structural orientation of the bound ligand. This is because the ancillary groups occupy the DNA grooves, where they form favourable noncovalent interactions, such as van der Waals, hydrophobic and hydrogen bonding to the bases that protrude into the grooves. Whilst intercalation itself leads to only a slight preference for 5′-pyrimidine-purine-3′ dinucleotides, because of stacking requirements, more complex intercalators with various functional groups attached to the intercalating chromophore can give rise to direct sequence readout and hence sequence selectivity. Some intercalating molecules bind by nonclassical mechanisms that involve threading through the DNA base stack prior to intercalation (Section 9.6.6).

Over the past 40 years, numerous studies in intercalator–DNA systems have greatly increased our understanding of the intercalation process. When a ligand intercalates into the DNA stack, **the bases must be**

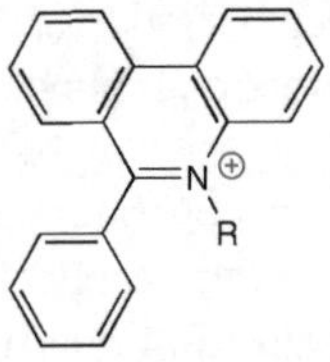

Ethidium, R = $CH_2CH_3$
Propidium, R = $(CH_2)_3N^{+}(CH_2CH_3)_2CH_3$

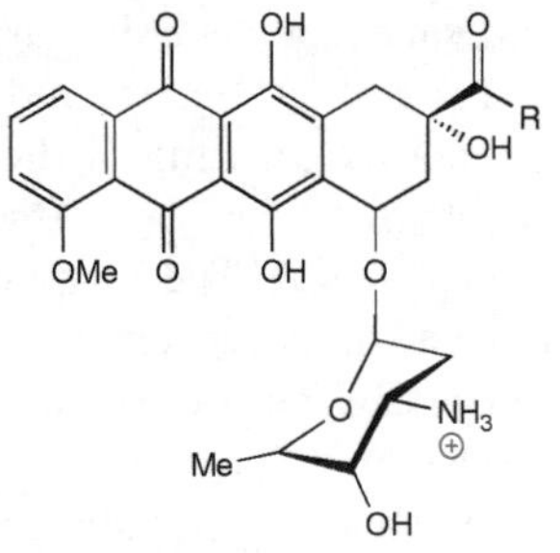

Daunomycin, R = H
Adriamycin, R = OH

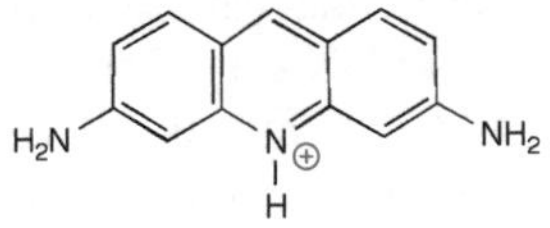

Proflavin

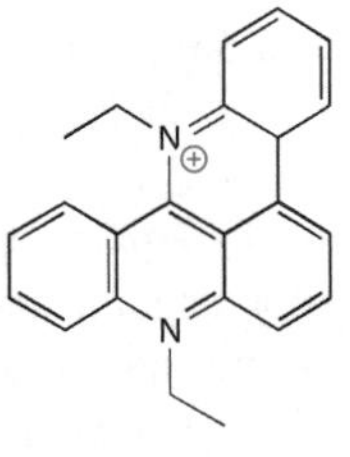

Acridines

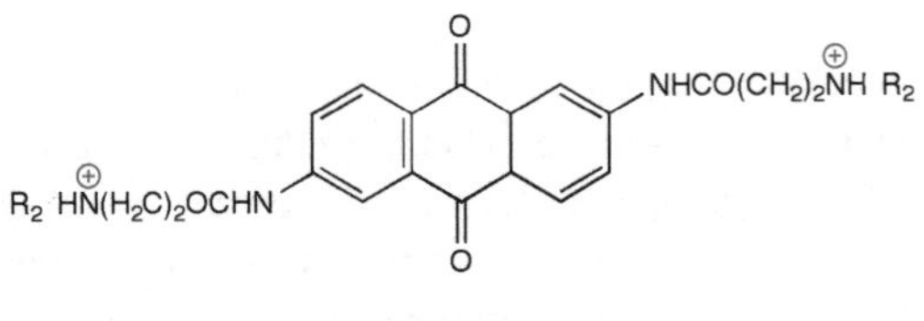

Anthracene-9,10-diones

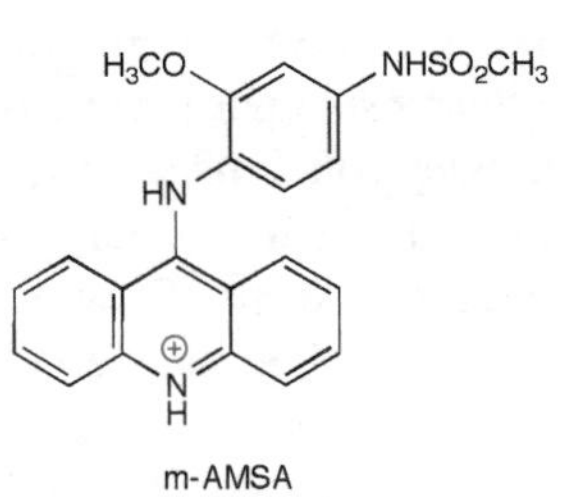

m-AMSA

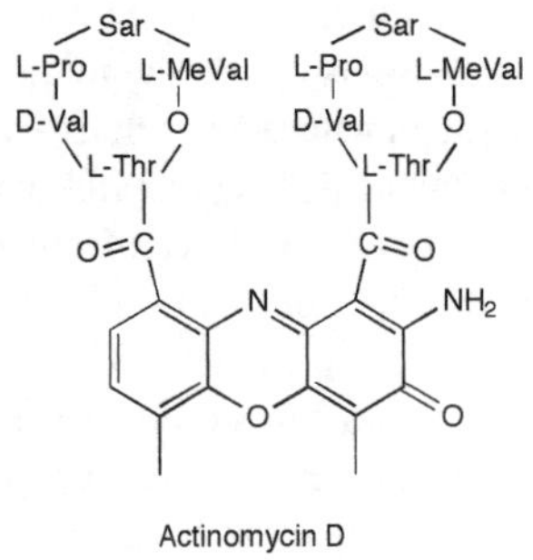

Actinomycin D

**Figure 9.4** *Structures of some DNA-intercalating compounds*

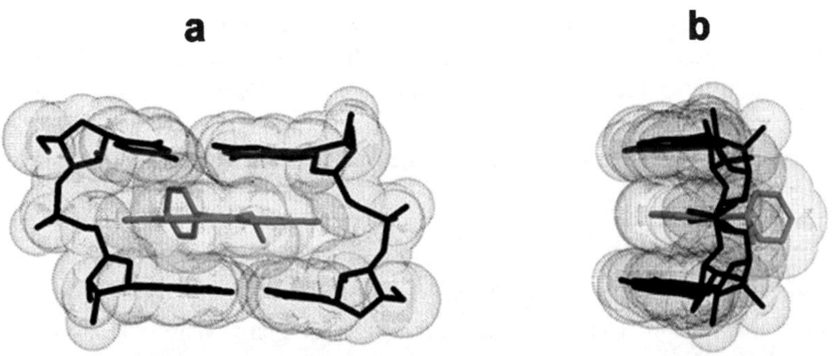


**Figure 9.5** *Two orthogonal views of ethidium (red) bound in between two adjacent G·C base pairs*

**separated by approximately 3.4 Å** (the van der Waals thickness of a phenyl ring) to accommodate the ligand. This base pair separation can only occur if there is rotation about torsional bonds in the phosphodiester backbone, such that unwinding around the helical axis occurs.

Intercalation of multiple ligands into a stretch of short rod-like DNA and concomitant **helix unwinding** leads to an overall **lengthening of the duplex**. This change in DNA length is reflected as a change in hydrodynamic character, and hence the **intrinsic viscosity** and **sedimentation coefficient** of the DNA change upon intercalation. The extent of helix unwinding produced by intercalation is dependent on the nature of the bound drug, especially its geometry and physico–chemical properties. The amount of helix unwinding can be determined experimentally by use of **covalently closed circular DNA**, with reference to a standard value of 26° per bound ethidium bromide drug. For canonical B-DNA, the normal backbone twist angle is 36°. Binding of ethidium reduces this twist angle to 10° and hence the DNA unwinds by 26°. Binding of proflavin or other related acridines causes unwinding of DNA by about 17°, whereas binding of anthracycline antibiotics results in only 11° unwinding of DNA. The effects of localised unwinding at the intercalation site are propagated along the DNA helix and result in structural perturbations over long distances from the bound drug. The helical twist angle for the two base pairs immediately surrounding the bound drug may be little changed from 36°. In such cases, there is usually a significant helical unwinding at adjacent base pairs. Hence the unwinding angle is the sum of cumulative changes in helical twist across all affected base pairs.

The process of **base-pair separation** that occurs during intercalation also produces a number of other changes in backbone conformation and base-pair and base-step geometry. In the X-ray-derived molecular structure of ethidium intercalated between a G·C dinucleotide (Figure 9.5), ethidium is stacked with its long axis parallel to the long axis of the adjacent base pair. The exocyclic amino groups point towards the diester oxygen atoms of the DNA phosphate groups and provide additional electrostatic and hydrogen-bond stabilization of the complex. The base pairs are slightly kinked, since the out-of-plane phenyl group limits full intercalation of the cationic phenanthridinium ring (Figure 9.5a). The phenyl and ethyl substituents of ethidium lie in the minor groove of the complex.

Various molecular events contribute to the overall free energy observed for intercalation reactions. Some of these are enthalpic in nature and others entropic. One of the most important energetic driving forces for drug binding is **the favourable hydrophobic transfer of the drug from bulk solvent into the DNA binding site**. The non-polar ligand is removed from the aqueous environment and this disrupts the hydration layer around the ligand, resulting in an entropically favourable release of site-specific water molecules. The DNA binding site must undergo an energetically unfavourable conformational transition to form a cavity into which the drug intercalates. Whilst this deformation is unfavourable from a point-of-view of configuration, it has additional features that give rise to a favourable entropy change. Localized unwinding of the helix at the intercalation site results in an increase in the distance between adjacent phosphates on the backbone. This gives rise to a reduction of localized charge density and hence the release of condensed counter-ions (Section 9.3.1). Additional counter-ions are released during the process of intercalation of the cationic ligand owing to the polyelectrolyte effect (Section 9.3).

Ligand intercalation is also associated with favourable enthalpic contributions to free energy arising from the formation of noncovalent interactions between the drug and base pairs. These noncovalent interactions involve several different forces, such as the hydrophobic effect, reduction of coulombic repulsion, van der Waals interactions, **π-stacking**, and hydrogen bonding. Quantitative apportionment of the overall binding free energy for intercalators and groove binders[16–18] is discussed later (Sections 9.6.4 and 9.7.5).

### 9.6.2 The Anthracycline Antibiotic Daunomycin

The **anthracycline antibiotics, daunomycin (daunorubicin)** and **adriamycin (doxorubicin)** as well as some derivatives have been frontline anticancer drugs for many years. Daunomycin is perhaps the best-studied intercalating drug to date with over 20 high-resolution structures known and numerous studies on the kinetics and thermodynamics of its interaction with DNA.[19]

*9.6.2.1 Structure.* The first X-ray crystal structure of a monointercalator–DNA complex was obtained by Wang and Rich in 1987 for the complex between **daunomycin and the hexamer d(CGTACG)$_2$** (Figure 9.6). Unlike other X-ray structures of simple intercalators bound to dinucleotides, daunomycin binds to DNA with its long axis almost perpendicular to the long axis of adjacent base pairs. The daunomycin amino sugar, which is attached to ring A of the anthracycline ring system (Figure 9.4), lies in the minor groove, while ring D, which contains a methoxy group, protrudes into the major groove. The other two anthracycline rings (B and C) lie between base pairs. Two bound daunomycin molecules intercalate, one at each of the C·G binding sites at the ends of the duplex. The C·G intercalation sites are separated by 3.4 Å and the cationic amino–sugar substituents, as well as ring A, largely fill the minor groove. In so doing, bound water and ions are expelled giving rise to a favourable entropic driving force for intercalation. The hydroxyl group on ring A donates a hydrogen bond to N-3 of guanine and it is a hydrogen-bond acceptor for the amino group of the same guanine. This gives rise to a favourable enthalpic contribution to binding as well as base pair and orientation specificity for the intercalated drug. The conformations of ring A and of the amino sugar are altered relative to the unbound drug to facilitate a tight fit of daunomycin into the right-handed minor groove.

The conformation of the drug-bound duplex DNA is also significantly altered relative to free DNA. In addition to the normal increased separation of base pairs at the intercalation site, the G·C base pairs are also shifted laterally towards the major groove, so that the helix axis changes position. There is no unwinding at the intercalation site itself and the base pairs here maintain a helical twist of 36°, but the base pairs at adjacent sites are unwound by 8°. Such relatively long-range induced conformational changes provide a graphic illustration of the **flexibility of DNA** and demonstrate the significant variations that can occur in duplex conformation in order to accommodate large and complex intercalators.

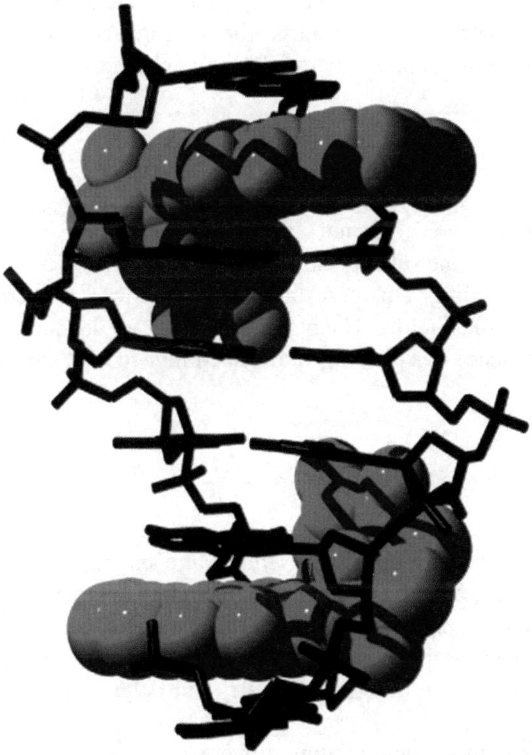

**Figure 9.6** *The structure of the daunomycin–DNA complex, (NDB: DDF001, PDB: 1D11). The two intercalated drug molecules are shown in space fill*

*9.6.2.2 Thermodynamics of Daunomycin: DNA Interactions.* The details of the energetics of anthracycline–DNA interactions have been studied by several groups.[19] In most thermodynamic characterizations, the initial step involves determination of the binding free energy. This is readily achieved by measurement of the equilibrium association (or binding) constant and use of the **standard Gibbs relationship**,

$$\Delta G = -RT \ln K_{\text{obs}} \tag{9.10}$$

where $K_{\text{obs}}$ is the observed equilibrium constant, $R$ the gas constant and $T$ the temperature. Brad Chaires has conducted equilibrium binding experiments and constructed **binding isotherms** that span a 100-fold range in free ligand concentration. Such essentially complete binding curves allow the free energy to be estimated in a model-free manner by use of **Wyman's concept of median ligand activity**. The free energy of ligation ($\Delta G_X$) is defined as the free energy to go from a state where no ligand is bound to a degree of saturation $\bar{X}$. For the binding of daunomycin to calf-thymus DNA, the value of $\Delta G_X$ was found to be $-32.6\,\text{kJ mol}^{-1}$ for the full ligation of a daunomycin binding site.

The daunomycin–DNA binding affinity is dependent on bulk salt concentration, as is explained by the polyelectrolyte theories of Record and Manning (Section 9.3). Daunomycin carries one positive charge at neutral pH, which arises from protonation of the amine ($pK_a = 8.2$) on the sugar moiety. Binding of cationic ligands such as daunomycin to DNA results in the entropically favourable release of condensed counter-ions from the DNA sugar–phosphate backbone. One practical consequence of polyelectrolyte theory is that it allows the observed binding free energy to be partitioned into two components, $\Delta G_{\text{pe}}$ and $\Delta G_T$, the polyelectrolyte and nonpolyelectrolyte contributions, respectively. The former is the free energy arising from counter-ion release. For daunomycin binding to DNA, $\Delta G_{\text{pe}}$ is $-4.2\,\text{kJ mol}^{-1}$ at a reference salt concentration of 0.2 M NaCl. This is about 13% of the total binding free energy.

Once the free energy for a binding interaction has been determined, the next step in a detailed thermodynamic characterization is to determine the **enthalpic and entropic components of the free energy**. These contributions are defined using the standard thermodynamic relationship,

$$\Delta G = \Delta H - T\Delta S \tag{9.11}$$

where $\Delta H$ and $\Delta S$ are the enthalpy and entropy changes accompanying binding. At 20°C, $\Delta H$ for the daunomycin–DNA interaction is $-37.6\,\text{kJ mol}^{-1}$. By use of Equation 9.11, $T\Delta S$ is then calculated as $-5.0\,\text{kJ mol}^{-1}$, and hence at 20°C the value of $\Delta S$ is $-17.1\,\text{J mol}^{-1}\,\text{K}^{-1}$. The negative sign for $\Delta S$ indicates that entropy decreases upon binding. This unfavourable entropy term is overcome by a large favourable (negative) enthalpy that drives the binding. The binding enthalpy for daunomycin is strongly dependent on temperature. This is a consequence of a **binding-induced change in heat capacity** ($\Delta C_p$) that accompanies the interaction:

$$\Delta C_p = \frac{\delta \Delta H}{\delta T} \tag{9.12}$$

By measurement of the enthalpy as a function of temperature and calculation of the slope of the resulting linear graph, the $\Delta C_p$ can be determined. For daunomycin binding to DNA, this value is $-669\,\text{J mol}^{-1}\,\text{K}^{-1}$. Determination of heat capacity change is useful, since this parameter can yield information on changes in hydrophobic interactions and removal of nonpolar solvent accessible surface area. It is an important, experimentally measurable quantity that is now used to carry out detailed partitioning of free energy terms (Section 9.6.4).

Studies of daunomycin and adriamycin binding to DNA have revealed important differences between the two drugs and served to illustrate the value of thermodynamic characterisation in the context of rational drug design. The structures of the two ligands bound to DNA are virtually superimposable and molecular interactions are equivalent, with only subtle differences in hydration and ion binding. However, adriamycin has a binding affinity for DNA that is 10-fold greater than that for daunomycin. In addition, binding of adriamycin to DNA is accompanied by a large favourable entropy term, which is the exact opposite of the situation for daunomycin. As a consequence, the binding enthalpy of adriamycin is significantly less than that measured for daunomycin, but the free energy is greater for adriamycin. These clear and unambiguous differences in the binding thermodynamics of two very similar drugs are not predicted from examination of the structures of the two drug–DNA complexes. Thus isostructural does *not* mean isoenergetic.

### 9.6.3 The Neighbour Exclusion Principle

Simple intercalators might be expected to occupy every potential intercalation site, *i.e.* in between every base pair in a DNA helix. Therefore, at saturation this would give rise to a binding stoichiometry of one base pair per bound intercalator. Molecular modelling studies where a drug molecule is bound at every possible intercalation site have led to energy-minimized structures with reasonable backbone torsion angles. However, solution studies have generated a different picture. In 1962 Cairns was the first to observe that even simple intercalators, such as proflavin, show saturation with DNA at a stoichiometry of one drug molecule per 2 base pairs. More complex intercalators such as daunomycin occupy 1 in 3 sites. Such empirical observations are a reflection of the **neighbour exclusion principle**, which states that intercalators can, at most, only bind at alternate possible base pair sites on DNA. Hence there is only ever a maximum of one intercalator between every second potential binding site. In the first instance, prior to addition of any ligand, all inter-base pair spaces have an equal potential to be intercalation sites for a non-specific compound (ignoring sequence preference). However, when an intercalator binds to a certain site, the neighbour exclusion principle states that binding of additional ligand molecules at sites immediately adjacent is inhibited.

Many attempts have been made to account for the molecular basis of neighbour exclusion. One possibility is that intercalator binding induces conformational changes at adjacent sites in the DNA and the new conformation is structurally or sterically unable to accommodate another mono-intercalator. Friedman and Manning have advanced an explanation based on counter-ion release and electrostatic effects. Briefly, binding of cationic intercalators has the effect of neutralization of some of the charge on the DNA backbone. In other words, binding of a cationic intercalator lengthens the DNA and hence increases the average charge spacing, and therefore **decreases the linear charge density**. In consequence, the binding-induced, energetically favourable release of condensed counter-ions is reduced and thus the observed equilibrium binding constant is also diminished. This effect leads to curvature in equilibrium binding isotherms that is similar to that predicted by neighbour exclusion theory. Since the local release of ions at the intercalation site would be greater than ion release at some distance from the binding site, then the local release of counter-ions could also lead to a lower free energy of binding at sites next to an intercalation site. In practice, it is likely that both conformational and electrostatic factors play a role in the neighbour exclusion phenomena. In addition, other factors may also be responsible for neighbour exclusion, for example, electrostatic repulsion between proximally bound drugs.

Various mathematical solutions have been developed that allow drug binding data to be described by the neighbour exclusion model.[20,21] In one of these approaches, the DNA is assumed to be a lattice sufficiently long to allow end effects to be ignored while all potential binding sites are assumed to be equivalent. McGhee and Von Hippel first wrote the **neighbour exclusion model** in this form in 1974:

$$\frac{r}{C_\mathrm{f}} = K(1 - nr)\left[\frac{1 - nr}{1 - (n - 1)r}\right]^{n-1} \tag{9.13}$$

where $r$ is the binding ratio (defined as the ratio of the concentration of bound drug ($C_\mathrm{b}$) over the total concentration of DNA binding sites ($S$), $C_\mathrm{f}$ the concentration of free drug, $K$ the equilibrium binding constant for all sites, *i.e.* any sequence dependence for drug binding is ignored, and $n$ the binding site size, *i.e.* the number of base pairs occupied by a bound drug. The neighbour exclusion model predicts that binding isotherms plotted in the form $r/C_f$ *versus* $r$ should be initially linear but should curve asymptotically as values of $r/C_f$ approach zero. This is exactly what is observed in experiments, and the neighbour exclusion model can describe accurately the binding of intercalators to DNA.[22]

However, it is common for non-integral numbers for the parameter $n$ to emerge from this analysis, which implies that the neighbour exclusion model is not adequate to describe the experimental data. Obvious reasons for this are that certain assumptions are not appropriate for the particular system under study. In such cases, more complex versions of the model that account for sequence specificity or the influence of co-operativity are required (Section 9.9).

### 9.6.4 Apportioning the Free Energy for DNA Intercalation Reactions

The likelihood of success in rational DNA-binding drug design programmes would be greatly enhanced if, in addition to structural studies, thermodynamic analyses were to be carried out for a wide range of ligand–DNA complexes. Determination of the free energy change for binding is useful but insufficient. As a minimum, the overall observed free energy should be examined in terms of component enthalpic and entropic terms. Modern advances in titration microcalorimetry have given experimentalists the ability to undertake novel and detailed **thermodynamic investigations into ligand–DNA interactions**. These tools allow the **drug–DNA binding free energy** to be apportioned in great detail.

One approach involves evaluation of a minimum set of free energy terms, which are assumed to be additive that can account for the experimentally determined value of $\Delta G_{obs}$.[16,18] Based on current concepts of ligand–DNA interactions, it is reasonable to describe $\Delta G_{obs}$ as being composed of at least five component free energy terms. Thus,

$$\Delta G_{obs} = \Delta G_{conf} + \Delta G_{r+t} + \Delta G_{hyd} + \Delta G_{pe} + \Delta G_{mol} \tag{9.14}$$

where $\Delta G_{conf}$ is the unfavourable free energy term arising from conformational changes in the DNA and ligand, $\Delta G_{r+t}$ is an unfavourable term that results from losses in rotational and translational degrees of freedom upon complex formation, $\Delta G_{hyd}$ is the free energy for the **hydrophobic transfer of ligand from bulk solvent into the DNA binding site**, $\Delta G_{pe}$ (Section 9.3) is the electrostatic (polyelectrolyte) contribution to the observed free energy that arises from coupled polyelectrolyte effects such as counter-ion release (entropically favourable) and $\Delta G_{mol}$ is the free energy term arising from weak non-covalent interactions, such as van der Waals, hydrogen bonding, dipole–dipole interactions, *etc.*

Each of the terms in Equation 9.14 can be estimated using semi-empirical or theoretical methods. At least two of the terms, $\Delta G_{conf}$ and $\Delta G_{r+t}$, make unfavourable contributions to binding free energy. Kinetic studies have revealed that the conformational transition in DNA that accompanies binding of a simple intercalator such as ethidium has a free energy cost of $+16.7\,\text{kJ mol}^{-1}$. Therefore, this value can be set as the energetic cost of structurally perturbing DNA to form an intercalation site. When a bimolecular complex is formed, there is an associated loss of rotational and translational degrees of freedom. This gives rise to a large entropic cost for binding, which is reflected in the free energy term $\Delta G_{r+t}$. There is some debate as to the precise value for this parameter.[16] However, a value of $+62.8\,\text{kJ mol}^{-1}$ for $\Delta G_{r+t}$ for a rigid body interaction is not unreasonable. Hence the two unfavourable free energy terms are set as constants for all intercalator binding reactions.

For a stable ligand–DNA complex to form, the remaining free energy terms must be large enough and favourable to overcome these two unfavourable contributions. It has been experimentally demonstrated for intercalators (and groove binders) that the major favourable free energy term that drives binding is the hydrophobic transfer of drug from the aqueous environment to the DNA binding site. It is possible to quantify

$\Delta G_{hyd}$ by using algorithms derived from semi-empirical relationships based on heats of solvent transfer of small molecules and protein folding–unfolding equilibria. Record has shown that

$$\Delta G_{hyd} = (80 \pm 10)\Delta C_p \tag{9.15}$$

From this equation, it is clear that $\Delta G_{hyd}$ can be estimated in a simple and direct manner by measurement of the heat capacity change associated with a binding interaction. A near-complete thermodynamic analysis, including a detailed partitioning of free energy, has been presented recently for a range of DNA intercalators[23] and reveals some interesting thermodynamic features of DNA intercalation by small molecules. The binding of most intercalators is enthalpically driven. For ethidium and daunomycin this is opposed by an unfavourable entropy term while for propidium and adriamycin, there is a small favourable entropy term. These differences in $T\Delta S$ are likely to be due to differences in hydration of the ligands involved. Heat capacity changes for ethidium, propidium, daunomycin, and adriamycin fall within a narrow range of about $-630\,\text{J mol}^{-1}\,\text{K}^{-1}$.

In general, heat capacity changes are smaller for intercalators than they are for groove binders. In addition, the heat capacity change for actinomycin D binding is much larger. Both these observations make sense when one considers that heat capacity change correlates with **changes in nonpolar and polar solvent accessible surface area**. For drug–DNA interactions, the following empirical relationship has been established:

$$\Delta C_p = 0.382(\pm 0.026)\Delta A_{np} - 0.121(\pm 0.077)\Delta A_p \tag{9.16}$$

where $\Delta A_{np}$ and $\Delta A_p$ are binding-induced changes in nonpolar and polar surface areas, respectively. The change in nonpolar surface area for ethidium is 407 Å$^2$, whereas for actinomycin D it is 1046 Å$^2$. Intercalation of positively charged ligands gives rise to a small favourable polyelectrolyte term. For actinomycin D, which is uncharged, the value of $\Delta G_{pe}$ is negligible.

For intercalation, the overwhelming energetic driving forces are hydrophobic interactions and other weak, non-covalent molecular interactions, such as van der Waals and hydrogen bonding. These free energy terms are large enough to overcome the $+79.5\,\text{kJ mol}^{-1}$ energetic cost that arises from conformational changes and losses in degrees of rotational and translational freedom that accompanies binding. In the case of actinomycin D, molecular interactions make a small unfavourable contribution to binding free energy. This is also the case for the groove binder, Hoechst 33258 (Section 9.7.5). Actinomycin D contains cyclic peptide moieties that, like Hoechst 33258, occupy the DNA minor groove. Binding of a ligand in the minor groove has the effect of expulsion of site-specifically bound water or cation molecules. Hence water-base pair hydrogen bonds are broken to form ligand-base pair hydrogen bonds. Therefore, from a thermodynamic point-of-view, a favourable energetic contribution arising from ligand hydrogen bonding may be limited.

### 9.6.5 Bisintercalation

**Bisintercalators** are bifunctional molecules that possess two planar intercalating aromatic ring systems covalently linked by chains of varying length. It is also possible to link three or more ring systems together using linkers. A good reason for designing and synthesising a bisintercalator is that it should have a significantly higher affinity and much slower dissociation kinetics than the monointercalator equivalents. The binding constant for a bisintercalator should be approximately the *square* of the monomer binding constant. Since biological activity is often closely correlated with binding affinity, bisintercalators should also have enhanced medicinal application.

Bisintercalators have a larger site size than their monointercalator counterparts, which can lead to increased sequence selectivity. Many simple monointercalators have a binding site size of 1 in 3 base pairs. Therefore, to avoid violation of the neighbour exclusion principle, a bisintercalator made from such monomers can be expected to span a site of at least 6 base pairs, which is the same size as the recognition sites of many sequence-specific restriction endonuclease enzymes (Section 5.3.1).

Bisintercalators are also ideal model systems for probing and evaluating neighbour exclusion effects, because they can be prepared with variable linker lengths. Ligands with short linkers may be induced to

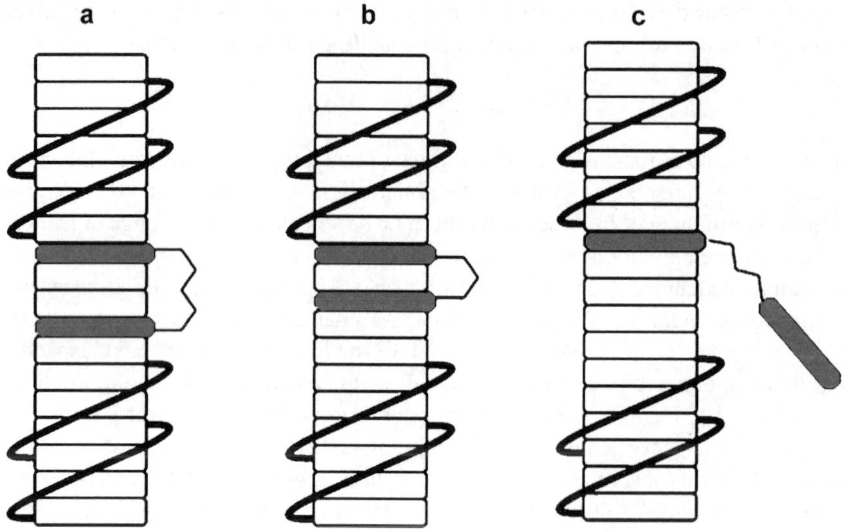


**Figure 9.7** *The effect of linker length on possible bisintercalator binding modes*

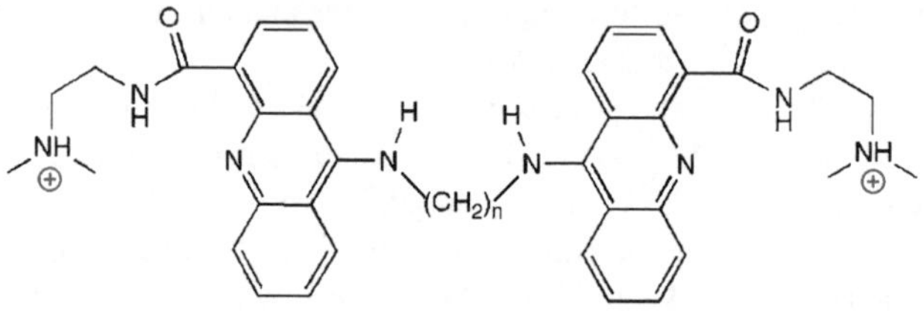


**Figure 9.8** *A flexibly linked bisintercalator formed from two dimethylaminoethylacridine-4-carboxamide moieties*

bind at adjacent base pair sites in violation of neighbour exclusion, whereas molecules with longer linkers would bind with at least one base pair separation as dictated by neighbour exclusion (Figure 9.7).

For example, **bisacridine** was developed as a transcription inhibitor (Figure 9.8). Structures of this type can bisintercalate into DNA forming a two base pair sandwich, with threading of the carboxamide side groups through the helix so that the protonated dimethylammonium groups hydrogen bond with O-6 and N-7 atoms of guanine. This compound has also been found to bind symmetrically to a DNA quadruplex that has junction-like properties.[24]

Simply by altering the number of methylene groups in the bisacridine, *i.e.* by varying the number $n$ in Figure 9.8, it is possible to **modulate linker length**. Hydrodynamic studies using viscosity measurements have shown that for $n < 4$ the molecules only monointercalate (Figure 9.7c). When $n = 6$ the molecules bisintercalate (Figure 9.7b), *i.e.* in violation of the neighbour exclusion principle. However, when $n > 8$, the molecule bisintercalates without violating neighbour exclusion principles (Figure 9.7a). For molecules with a linker length of $n = 6$, viscosity and NMR data are not in accord. Analysis of imino proton chemical shifts upon titration of bisacridine to the self-complementary oligonucleotides d(AT)$_5$ showed that bisintercalation occurred for $7 \leqslant n \leqslant 10$ but only monointercalation was observed when $4 \leqslant n \leqslant 6$. The aliphatic linker in these molecules is highly flexible and therefore it is possible that the $n = 6$ molecules might bind without violating neighbour exclusion if local distortions in the DNA could be induced. If such a distortion is not possible in the sequence used for NMR under those specific solution conditions, then the disparity between NMR and viscosity studies might be explained.

*9.6.5.1 Bisintercalating Anthracyclines: Bisdaunomycin.* In 1997 Brad Chaires and co-workers reported the structure-based design of a new **bisintercalating anthracycline antibiotic** based on two covalently linked daunomycin molecules.[25,26] The starting point for the design of this molecule (code named WP631) was the high-resolution X-ray structures of daunomycin monomers bound to DNA hexanucleotides (Figure 9.6). In this structure, two ligand molecules have their amino sugar groups in the minor groove and pointing towards one another with a separation of <7 Å. An obvious way forward was to join covalently the two $NH_2$ groups (one from each drug) to form a **bisanthracycline** that would have the potential to bisintercalate into DNA. A linker of about 7 Å in length was designed by Waldemar Priebe so as to fit into the DNA minor groove without any steric constraints to give the structure of WP631 (Figure 9.9).

Viscosity studies confirmed that WP631 bisintercalates. The ultra-tight binding observed for WP631 ($K_{obs} = 2.7 \times 10^{11}\,M^{-1}$) approaches that predicted for a bisintercalator comprised of two daunomycin moieties and is close to the affinity observed for many specific protein–DNA interactions. The overall thermodynamic profile shows that binding of WP631 is driven by a large favourable enthalpy change ($-126.4\,kJ\,mol^{-1}$). Binding is opposed by a substantial unfavourable entropy term, which at 20°C $T\Delta S$ is $-62.3\,kJ\,mol^{-1}$. This unfavourable entropic term is comparatively larger for WP631 than for daunomycin,

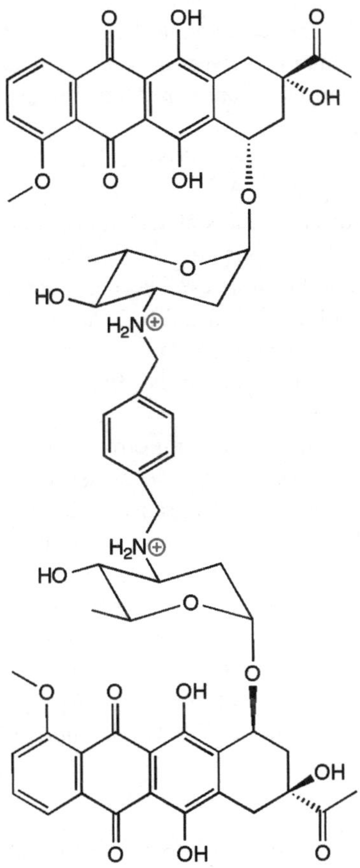


**Figure 9.9** *Structure of the bisanthracycline WP631. Two daunomycin molecules are covalently attached to each other through their 3′-$NH_2$ groups using a p-xylene linker*

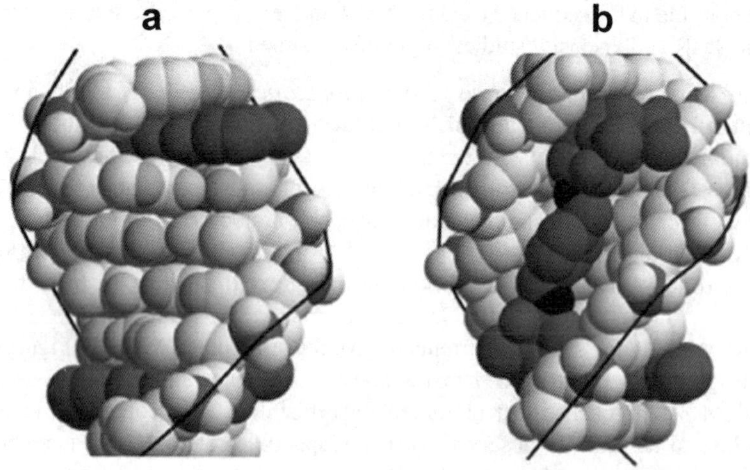


**Figure 9.10** *X-ray crystal structure of the daunomycin bisintercalator WP631-d(CGATCG)$_2$ complex, (NDB: DDF072; PDB: 1AGL). Two views are shown: left, looking into the major groove and right, looking into the minor groove. A solid black line traces the DNA backbone*
(Figure prepared and kindly provided by Loren D. Williams)

because the bisintercalator imposes a greater amount of rigidity in the DNA helix compared to daunomycin. There is considerable conformational freedom in the unbound WP631 with free rotation about many of the bonds in the linker. The process of bisintercalation results in restriction in these bonds, thus giving rise to unfavourable entropic effects.

Subsequent to the initial characterisation of WP631, a 2.2 Å X-ray crystal structure of the WP631-d(CGATCG)$_2$ complex was solved (Figure 9.10).[26]

This structure confirmed unambiguously that bisintercalation is the binding mode. Many of the structural features observed in crystals of the 2:1 daunomycin–DNA structures are conserved in the bisintercalator complex. For example, intercalation of the anthracycline rings at both terminal d(CpG) steps, and hydrogen bond interactions between OH-9 of the ligand and N-2 and N-3 of guanine. However, bisintercalation alters the DNA conformation relative to monointercalation, primarily in the centre of the oligonucleotide. Helix unwinding and other distortions propagate to the centre of the WP631 complex much more efficiently than in the daunomycin complex.

While the sugar constituents still occupy the minor groove, their exact position and interactions are different in the bisintercalator from daunomycin alone. The presence of the linker results in the amino sugar groups being lifted away from the floor of the minor groove. The net result is that hydrogen bonds between the 3′-amino groups of the sugar that are present in the daunomycin complex are absent in WP631. In daunomycin–DNA complexes, there are two water molecules in the minor groove that each forms two hydrogen bonds with the floor of the groove and one hydrogen bond to 3′-amino group of the daunomycin sugar. The linker of WP631 displaces these water molecules resulting in a forced contact between the hydrophobic linker and polar functional groups occupying the groove floor. The reduced hydrogen bonding, the greater separation of the amino sugars from the groove floor, and the desolvation of the minor groove may explain why the binding constant for WP631 falls short of being the *square* of the daunomycin binding constant.

### 9.6.6 Nonclassical Intercalation: The Threading Mode

One group of ligands that bind to DNA by intercalation contain substituent groups on opposite edges of the planar aromatic intercalating ring system. For the planar part of the molecule to intercalate, one of the substituent groups must be threaded through the base pair stack to the opposite side of the helix. This more complicated binding mechanism is reflected in the kinetics of association and dissociation. The substituents that are threaded through the base pair at the intercalation site are often bulky and/or polar and hence this may represent a kinetically unfavourable step in the binding mechanism. Examples of intercalators that bind to DNA *via* a **threading mechanism** include **naphthalene–bisamides, cationic porphyrins**, and the anticancer antibiotic **nogalamycin** (Figure 9.11).

The binding mechanism for simple intercalators such as ethidium and proflavin involves just two steps. First, the cationic intercalator forms a complex with the negatively charged sugar–phosphate backbone of DNA through electrostatic interactions. Then the intercalator diffuses up and down the helix in the anionic potential along the surface of the DNA until it comes across gaps in between adjacent base pairs that have formed due to the normal thermal or **"breathing" motions of the DNA**. This results in separated base pairs that form a cavity into which the simple intercalator can bind.

By contrast, threading intercalators, with their large and polar distal-substituted side chains or rings, necessarily involve much larger openings in the DNA base pair stack (*i.e.* a bubble-type structure), and these may only form after significant distortions and/or breaking of inter-base hydrogen bonds. These larger and more drastic dynamic motions in the helix occur with low frequency and hence the association kinetics for threading intercalators is much slower than for simple intercalators. Thus both association and dissociation rate constants should decrease relative to classical intercalation. In addition, threading of cationic side chains through a DNA duplex requires sequential formation and breakage of electrostatic interactions in the threading complex. Hence the effect of salt concentration on observed rate constants can be used to probe for the threading mode.

Although threading intercalators have more complex binding mechanisms with potentially unfavourable steps, they can still bind to DNA with high affinity. Once the side chain manages to traverse the intercalation site, the DNA can adopt a conformation that gives a large and favourable free energy for complex formation. The side chains present a kinetic barrier to binding but favourable interactions, both in the intercalation site and in the grooves, give rise to a stable final complex. The nature and extent of the kinetic barrier depends upon the size of the side chain, its orientation and polarity.

The nogalamycin chromophore is intercalated, with the uncharged nogalose sugar (the aglycone group) in the minor groove and the positively charged amino sugar in the major groove (Figure 9.11). In the drug–DNA complex, hydrogen bonds are observed between the drug and N-2 and O-6 atoms of a guanine. Hence there is simultaneous direct readout of sequence information in both the major and minor grooves. Nogalamycin has a binding preference for N·G or C·N sequences, where N can be any base, but the drug intercalates preferentially at the 5′-side of guanine or at the 3′-side of cytosine. The two sugars point in the

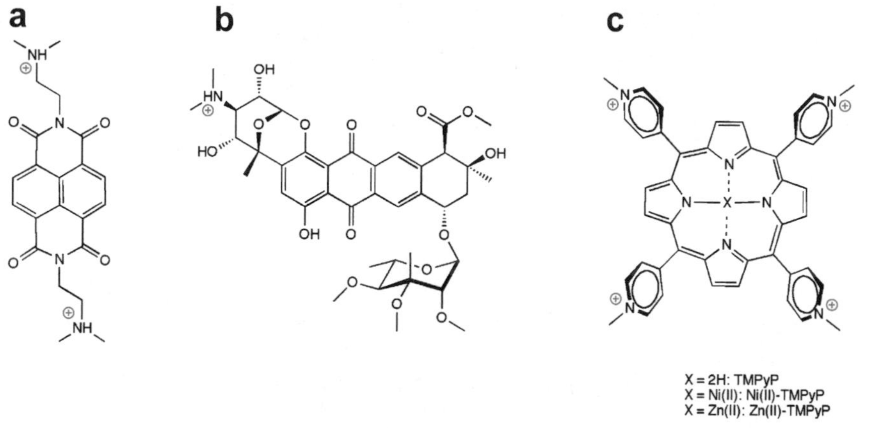


**Figure 9.11** *Structures of three intercalators that bind to DNA by a threading mode. (a) Naphthalene–bisimide. (b) Nogalamycin. (c) Cationic porphyrin*

same direction in the intercalation complex and they overlap the G·C base pair in the binding site. The nogalamycin–DNA contacts that give rise to specificity occur mainly in the major groove. This is in contrast to the daunomycin–DNA complex where interactions are predominantly in the minor groove. The uncharged nogalamycin sugar that occupies the minor groove appears to interact only very weakly with the DNA. In a way similar to daunomycin, the long axis of nogalamycin is perpendicular to the long axis of the base pairs. This arrangement leads to a slight buckling of base pairs that constitute the intercalation site, such that they wrap around the bound drug. This results in improved van der Waals contacts. As expected for a threading intercalator, the mechanism of nogalamycin binding to duplex DNA is not simple. For intercalation to occur, one or other of the attached sugar moieties has to thread through the intercalation site. These structural constraints lead to slow association and dissociation kinetics.

The general features of the threading mode have been illustrated by **stopped-flow kinetics** measurements conducted by Wilson and co-workers on two groups of **disubstituted anthracene-9,10-diones (anthraquinones)** synthesised in the laboratories of Jenkins and Neidle (Figure 9.12).[27]

Several such anthraquinone compounds have been synthesized and they have been shown to bind to duplex, triplex, and quadruplex DNA by intercalation. The 1,4-*bis*(amino) functionalized compound, mitoxantrone, was in clinical use for the treatment of breast cancer. Wilson, Jenkins and Neidle have systematically examined the effects of alteration of substituents on the anthraquinone ring as well as alteration of substituent positions. From molecular modelling studies, it was predicted that anthraquinones substituted at either the 1,4- or 1,8-positions with cationic side chains should intercalate into DNA with both side chains binding in the same groove through classical intercalation. However, drugs substituted at either the 1,5- or 2,6-positions on the anthraquinone ring should intercalate through a threading mode with one side chain occupying each groove.

DNA binding measurements, conducted as a function of salt concentration and temperature, concur with the predictions of molecular modelling. 1,4- or 1,8-difunctionalized anthraquinones have association rate constants of about $\geqslant 2 \times 10^6\,M^{-1}\,s^{-1}$, while for 1,5- or 2,6-compounds these values are $\leqslant 2 \times 10^5\,M^{-1}\,s^{-1}$. Hence the compounds that bind by threading have about 10-fold slower association kinetics than do classical intercalators. The dissociation rate constants are approximately $10\,s^{-1}$ or greater for classical intercalators and $\leqslant 5 - 6\,s^{-1}$ for threading intercalators. Interestingly, compensatory effects in association and dissociation rate constants give rise to equilibrium binding constants that are closer than kinetic constants for classical *versus* threading intercalation. The slower observed kinetics for the threading mode arise because

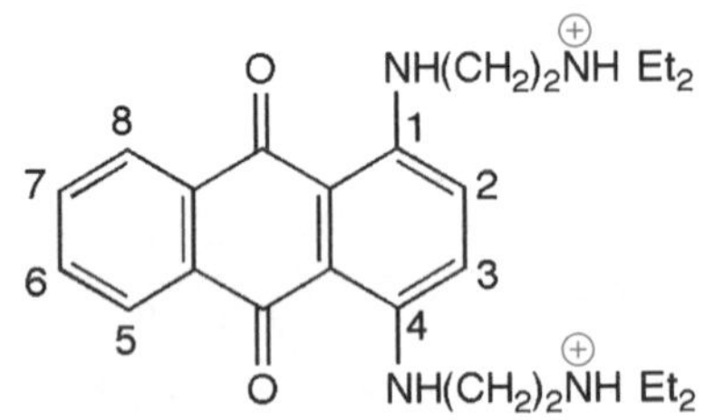


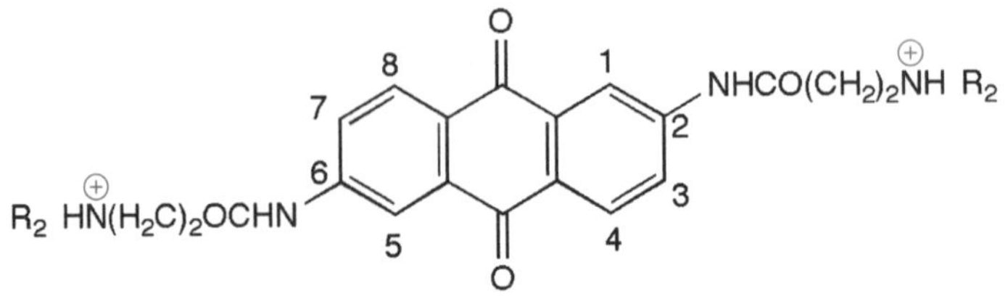


**Figure 9.12** *Functionalised anthracene-9,10-diones showing substituents at the 1,4- and 2,6-ring positions*

(1) the polar side chain must go from aqueous solvent through the hydrophobic region in between two adjacent base pairs, (2) the size and rigidity of the side chains may impede this passage, and (3) the rate of DNA thermal motion will result in the formation of a bubble-like structure that can accommodate the side chain.

## 9.7 INTERACTIONS IN THE MINOR GROOVE

### 9.7.1 General Characteristics of Groove Binding

A larger number of functional groups of the DNA bases are accessible in the wide major groove compared to the narrow minor groove of the B-helix. Therefore, most proteins have evolved to make sequence-specific interactions with DNA in the major groove. However, some proteins and many **small molecules interact with DNA in the minor groove**. The major and minor grooves differ in electrostatic potential, hydrogen-bonding characteristics, steric effects, and hydration. Furthermore, DNA structure and conformation is highly dependent on sequence context. For example, runs of successive adenosine residues give rise to a very narrow minor groove, where the base pairs have a higher propeller twist and the DNA becomes slightly bent from the helical axis. In general, small groove-binding molecules exhibit a preference for the minor groove, not least because this site of interaction provides better van der Waals contacts. In addition, many minor groove-binding ligands prefer A·T sites (compared to intercalators, which generally exhibit a G·C preference).

Minor groove-binding ligands often contain several **simple aromatic rings**, such as **pyrrole, furan, benzene** or **imidazole**. These are connected by **bonds with torsional freedom** to enable such ligands to twist and become **isohelical with the DNA minor groove**. They thus provide optimal shape complementarity with the DNA receptor and in most cases a typical crescent shape is adopted by these ligands (Figure 9.13).

The DNA minor groove is narrower in A·T-rich sequences compared to G·C-rich sequences and thus correctly twisted aromatic rings in minor groove-binding ligands fit better at into the A·T minor groove to make optimal van der Waals contacts with the helical chains that define the walls of the groove. Additional specificity (and to some extent stability) is derived from molecular contacts between the bound ligand and the edges of the base pairs on the floor of the groove. Hydrogen bonds can be accepted by A·T base pairs from the bound molecule to the **C-2 carbonyl oxygen of thymine** or the **N-3 nitrogen adenine**. Even though similar functional groups are available in G·C base pairs, the amino group of guanine presents a steric block to hydrogen bond formation at N-3 of guanine and at the O-2 carbonyl of cytosine. The inter-base hydrogen bond between the guanine amino group and the cytosine carbonyl oxygen lies in the minor groove and this interaction sterically inhibits penetration of small molecules into the minor groove at G·C sites. Hence the aromatic rings of many minor groove-binding ligands form close contacts with A-H-2 protons in the minor groove, but there is no room for the added steric bulk of the guanine amino group in G·C base pairs. Bernard Pullman has shown that **negative electrostatic potential is greater in A·T minor grooves** than it is in G·C minor grooves. This provides an additional important source for the observed A·T specificity of many minor groove-binding ligands, which often are positively charged. Synthesis of ligands capable of accepting hydrogen bonds from the guanine amino group is one way of enhancing minor groove-binding at G·C sites.

Unlike most intercalators, groove-binding molecules can extend to span many base pairs along the groove and hence they can exhibit very high levels of DNA sequence-specific recognition. A long established goal has been to produce molecules that can specifically recognise DNA sequences long enough to be unique in a biological context, for example, a eukaryotic regulatory sequence. Minor groove-binding drugs[28–30] provide some hope of achieving this goal (Section 9.7.4) and some archetypal minor groove ligands are described below.

### 9.7.2 Netropsin and Distamycin

**Netropsin** was first isolated from *Streptomyces netropsis* by Finlay in 1951 while Arcamone discovered **distamycin A** as a by-product of *Streptomyces distallicus* fermentation in 1958. Both drugs are

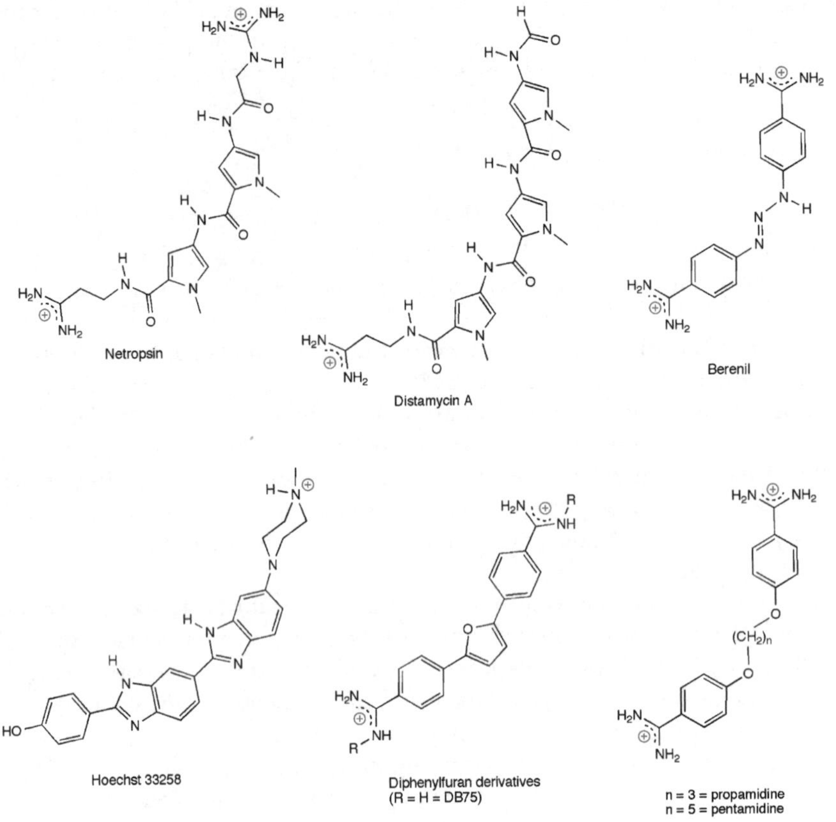


**Figure 9.13** *Structures of some DNA minor groove binding ligands*

**pyrrole–amidine antibiotics** (Figure 9.13) that have antibacterial and antiviral activity, presumably as a result of their interaction with DNA. Both compounds have been the focus of much study as DNA interactive ligands. These drugs have a binding preference for **A·T-rich minor grooves** over G·C-containing sequences. The preferred binding site is 5′-AATT and to a lesser extent 5′-ATAT. The rather similar sites 5′-TTAA and 5′-TATA are disfavoured sites of interaction. These differences in binding site preference have been interpreted in terms of **sequence-dependent variations in minor groove width** as well as due to the disruption induced by the unusually flexible TpA base pair step. In general, these drugs bind in a narrowed minor groove and the complex is stabilised by a combination of electrostatic interactions between the DNA and the cationic end groups, as well as by hydrogen bonds to N-3 of adenines and O-2 of thymines. In addition, non-bonded hydrophobic interactions are made with the sugar–phosphate backbone that defines the walls of the groove. The binding site size of netropsin is about four consecutive A·T base pairs. Since distamycin is a longer molecule it requires at least five or six contiguous A·T base pairs.

*9.7.2.1 Structures of Netropsin– and Distamycin–DNA Complexes.* Dickerson and co-workers solved the first X-ray crystal structure of a netropsin–DNA complex in 1985.[31] The DNA target was the Dickerson–Drew dodecamer: d(CGCGAATTCGCG)$_2$, referred to here as A2T2. As expected, netropsin binds in the central AATT of the duplex and it displaces several water molecules that are present in the drug-free structure. The three-ligand amide groups point inwards and form bifurcated or three-centred hydrogen bonds with N-3 of adenine and O-2 of thymine. The **netropsin fits tightly into the narrow minor**

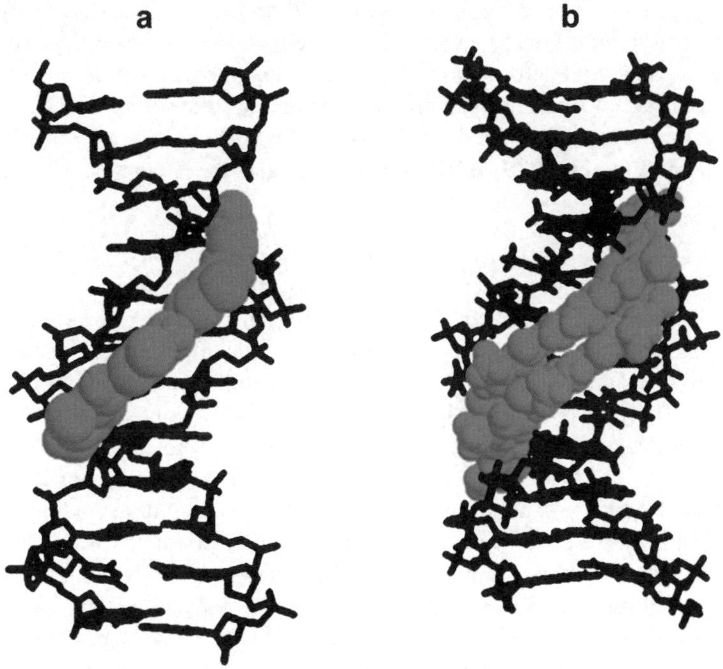


**Figure 9.14** *(a) X-ray structure of the distamycin-d(CGCAAATTTGCG)$_2$ complex (NDB: GDL003; PDB: 2DND). (b) Exactly the same complex solved using NMR, Note the different groove widths in the two structures* (Coordinates supplied by David E. Wemmer in a personal communication to I. Haq and T. C. Jenkins)

**groove** and forms van der Waals contacts with the atoms of DNA that constitute the sugar–phosphate walls of the groove. These contacts hold the ligand pyrrole rings approximately parallel to the walls of the groove and since there is helical twist in the groove, the two pyrrole rings are twisted by about 33° with respect to one another. The two cationic termini of netropsin are also in the central minor groove and they are associated with N-3 atoms of the outer adenine bases of the central four A·T base pairs. Steric interactions between pyrrole methyl groups and the DNA bases prevent deep penetration of the drug into the groove. The result is that some of the hydrogen bonds between the drug and DNA are quite long (3.3–3.8 Å) compared to standard values (less than 3 Å). The binding of netropsin causes a slight widening of the minor groove and a bending of the helical axis away from the binding site.

The binding of distamycin to DNA has many features that are similar to that of netropsin. A 2.2 Å resolution X-ray crystal structure of the 1:1 distamycin/A3T3 complex[32] (Figure 9.14**a**) shows a snug fit of the crescent-shaped ligand into the A·T minor groove.

NMR studies of the same distamycin–DNA interaction from David Wemmer reveal an interesting additional binding mode for distamycin[33] (Figure 9.14**b**). Sequences that contain more than four successive A·T base pairs in the minor groove, such as **A3T3, can accommodate two distamycin molecules** in the central AT-tract. Sequences that contain only four A·T base pairs favour the 1:1 complex. Hence with a run of six A·T base pairs, the 2:1 complex is favoured. The NMR data suggest that the distamycin molecules are stacked on each other with their charged groups arranged in opposite (antiparallel) directions. The fact that netropsin binds as a single molecule per binding site is probably because it is a dication and thus side-by-side 2:1 binding is inhibited due to drug–drug electrostatic repulsion.

Both distamycins in the dimer complex lie deep within the minor groove and form hydrogen bonds with adenine N-3 and thymine O-2, in the same way as in the 1:1 complex. An interesting feature of the dimer complex is that it requires the minor groove to widen two-fold compared to drug-free DNA or to the 1:1 complex (Figure 9.14). This structural rearrangement gives rise to a positive free energy contribution to binding, in a similar way to structural deformations that accompany intercalation. However, this unfavourable free energy is more than compensated by additional favourable free energy that arises from the greater hydrophobic interactions derived from binding of two drugs within one minor groove. This shows another example of the **conformational mobility of DNA** that is crucial to the interaction of proteins and ultimately to biological function.

*9.7.2.2 Energetics of Netropsin– and Distamycin–DNA Interactions.* Recently, there has been an upsurge in detailed thermodynamic studies of minor groove binding that go beyond simple determinations of binding constants. For netropsin, distamycin, and a number of other minor-groove specific ligands, it is now possible to apportion the overall binding free energy into **enthalpic and entropic terms**. In addition, recent developments in thermodynamic analysis allows a much more detailed parsing of free energy terms (Section 9.6.4 and Equation 9.14), which provides insight into the energetics of drug–DNA interactions.

The relative thermodynamic contributions of two distamycin molecules that bind to the duplex d(GCCAAATTGGC)·d(GCCAATTTGGC), referred to as A3T2, have been examined in the context of the thermodynamics of the 1:1 netropsin interaction with A3T2.[34] In all cases, drug binding is associated with an exothermic binding enthalpy, which for netropsin is $-31.4\,\mathrm{kJ\,mol^{-1}}$ and for distamycin first and second binding events are $-51.5$ and $-78.7\,\mathrm{kJ\,mol^{-1}}$, respectively. The drugs bind to DNA moderately tightly with $K_{obs}$ of $4.3 \times 10^7\,\mathrm{M^{-1}}$ for netropsin and $3.1 \times 10^7\,\mathrm{M^{-1}}$ for the first distamycin binding. The second distamycin molecule binds with a lower affinity of $3.3 \times 10^6\,\mathrm{M^{-1}}$ (20°C, pH 7.0 and $Na^+$ concentration of about 20 mM).

Thus netropsin and the first distamycin have very similar overall binding affinities and both interactions are enthalpically driven, with the first distamycin having a larger favourable enthalpy. One possible interpretation for these observations is that hydrogen bonding contributions are similar for the two complexes. The smaller number of favourable electrostatic interactions in the monocationic distamycin, compared to dicationic netropsin, is compensated by an increase in hydrophobic interactions in distamycin arising from its additional *N*-methylpyrrole ring. The magnitudes of the negative enthalpy values are dependent upon relative contributions from hydrogen bonding, van der Waals interactions, and overall hydration changes. The entropy contributions are of opposite sign (favourable for netropsin and unfavourable for distamycin), probably due to differences in release of counter-ions and uptake or release of water molecules. The second distamycin (in the 2:1 complex) binds with an eight-fold reduction in affinity and a $27.2\,\mathrm{kJ\,mol^{-1}}$ more favourable enthalpy. This finding is consistent with observations from the NMR structure, since there is an increase in van der Waals contacts due to side-by-side interactions in the 2:1 distamycin–DNA complex.

Variations in binding affinity with salt concentration are also revealing. The slopes of $\ln K_{obs}$ *versus* $\ln[Na^+]$ plots were found to be $-0.8$ for netropsin–A3T2 binding and $-0.56$ for distamycin–A3T3 binding. These values are lower than previous estimates for 1:1 binding of netropsin and distamycin to A2T2 and also lower than the theoretical value of $-1.76$ (for the binding of one dicationic ligand or two monocationic ligands) predicted from the polyelectrolyte theories of Record and Manning. Structural data provides a possible explanation for these results. The minor groove becomes widened upon the binding of the second distamycin, which results in a lower local charge density. Also, structured water around exposed hydrophobic groups of the bound drug may give rise to a different type of dielectric screening. For both ligands, calorimetrically measured binding-induced changes in heat capacity $\Delta C_p$ are significant in magnitude and negative in sign, consistent with theoretical estimates based on changes in solvent accessible surface area upon binding.[35]

### 9.7.3 Lexitropsins

**Lexitropsins** are a group of minor groove-specific ligands that are comprised of a range of different aromatic rings connected by peptide bonds. Netropsin and distamycin are examples of lexitropsins, where the sequence

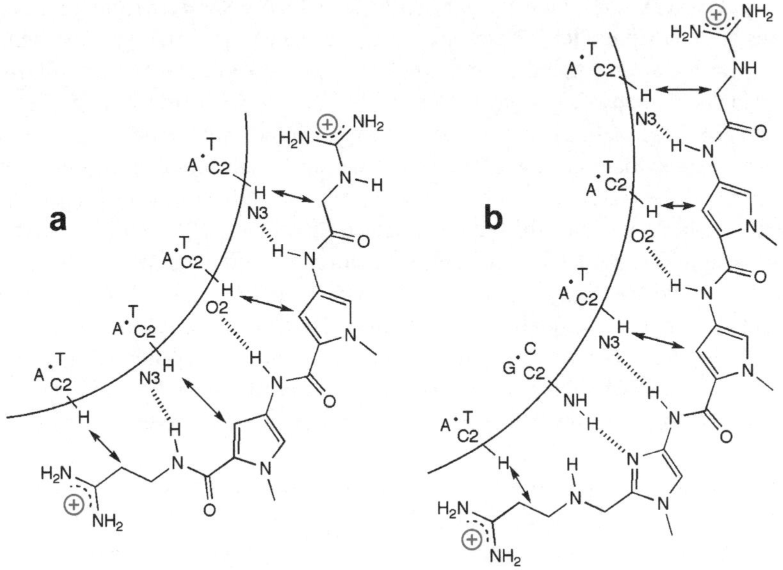


Netropsin (PyPy) ImPyPy

**Figure 9.15** *Schematic representation of the specific molecular interactions that occur between (a) netropsin and (b) a lexitropsin ImPyPy. Dotted lines indicate hydrogen bonds. Double-headed arrows represent nonbonded van der Waals contacts. The curved line is the floor of the minor groove*

of aromatic rings is pyrrole–pyrrole (PyPy) for netropsin and pyrrole–pyrrole–pyrrole (PyPyPy) for distamycin. Dickerson and Lown proposed that replacement of Py moieties in netropsin or distamycin by other heterocycles, such as 1-methylimidazole (Im), could yield compounds capable of recognising G·C base pairs as well as A·T sequences. In netropsin, pyrrole methyl groups are orientated towards the minor groove such that they make close contact with A·T base pairs. In lexitropsins, one or more of the pyrrole rings is replaced by, for example, an imidazole (Im) ring, which is capable of accepting a hydrogen bond. This alleviates the steric clash of a pyrrole with a guanine amino group in the minor groove and instead allows a specific hydrogen bond to form between the guanine amino group and the imidazole nitrogen atom, which facilitates increased minor groove interactions between the lexitropsin and G·C base pairs. DNA interactions with netropsin (PyPy) as well as an imidazole-containing **lexitropsin (ImPyPy)** are illustrated in Figure 9.15.

NMR and footprinting studies have confirmed the enhanced interactions between suitably modified lexitropsins and G·C-containing minor grooves, although the presence of an A·T base pair does not interfere with binding. Other examples of ligands in this group include ***bis*-furan** lexitropsins, **thiazole** compounds, and analogues of **Hoechst 33258**, where benzoxazole or pyridoimidazole groups have replaced the benzimidazole rings in Hoechst 33258. Binding studies indicate that whilst most of these lexitropsins lose their A·T specificity, they fail to become properly G·C-specific.

## 9.7.4 Hairpin Polyamides

**Hairpin polyamide** ligands provide important opportunities for development of high affinity, sequence-specific DNA recognition agents for the control of gene function and as potential anticancer therapeutics.[36–43] In the early 1990s, Dervan synthesised a polyamide containing an imidazole ring followed by two pyrrole moieties (**ImPyPy**) with the expectation that it would form a 1:1 complex with sequences of the type 5′-XGXXX, in which X can be either A or T. Unexpectedly, ImPyPy was found to bind to a five base pair sequence of the type 5′-XGXCX, (again X can be either A or T). An NMR structure of the complex showed that ImPyPy binds as an **antiparallel 2:1 dimer** in a similar manner to distamycin binding to A·T-tracts in the minor groove. In this arrangement, an imidazole ring is stacked against a pyrrole ring and thus the antiparallel pair can distinguish a G·C base pair from a C·G, *i.e.* the pair Im/Py targets G·C and Py/Im recognizes C·G. Each G·C base pair has an imidazole on the G strand and a pyrrole on the C strand. A hydrogen bond is formed between the N-3 atom of the imidazole ring and the amino group of guanine as well as a series of hydrogen bonds between polyamide NH groups and the edge of the G·C base pair.

Thermodynamic studies reveal that the sequence selectivity of the Im/Py pair is mostly driven by a large favourable enthalpy term. Imidazole nitrogen atoms do not hydrogen bond to the cytosine side of the base pair due to an unfavourable bond angle for this interaction. In addition, a Py/Py pair is degenerate and can recognize A·T and T·A base pairs equally well in preference to G·C or C·G base pairs.

These discoveries suggested a new design strategy: unsymmetrical polyamide ring pairs can be used for specific recognition of the DNA minor groove where one pair of rings in the ligand targets one base pair of DNA. An Im is used for a guanine and a Py for a cytosine in G·C or C·G base pairs, whereas two Py rings are used to target either A·T or T·A.

To test these pairing rules, Dervan changed an Im in one of the polyamides of a homodimer to Py to give the heterodimer ImPyPy/PyPyPy. As predicted, this pair of ligands binds specifically to the sequence 5′-XGXXX. Further, ImPyImPy binds to the minor groove of the sequence 5′-GCGC. By analogy with distamycin, binding of an alkyl chain is always to an A·T site, so that the exact recognition sequence is XGCGCX. The specificity is extremely high, such that even a single mismatch with a recognition motif, *e.g.* G·C in place of C·G, leads to a 10-fold drop in binding affinity.

Since **polyamide dimers** do not distinguish A·T from T·A base pairs in the minor groove, a third type of monomer was designed as a thymine-selective moiety, *N*-methyl-3-hydroxypyrrole (Hp). An Hp/Py pair recognizes a T·A base pair whereas a Py/Hp combination recognizes A·T base pairs. Both orientations disfavour binding to G·C and C·G base pairs. X-ray structures of two Hp-containing **polyamide–DNA complexes** confirm that the hydroxyl group of Hp fits snugly into the asymmetric cleft between adenine C-2 and thymine O-2 atoms. In addition, there is a specific hydrogen bond between the hydroxyl group of Hp and thymine O-2. Hp-containing polyamides bind with lower affinity than the equivalent Py-containing ones. The reason for this is not clear but may be related to factors such as partial melting of the T·A base pair recognized by Hp/Py or a lengthening of the amide–DNA hydrogen bond on the C-terminal of the Hp residue. Nevertheless, with the addition of this third type of monomer, it has become possible to create polyamides containing different combinations of Im, Py and Hp rings. Using the **Dervan Pairing Rules**, one can distinguish all four types of Watson–Crick base pairs and thus design polyamide ligands to target any specific DNA sequence (Figure 9.16).

**Linked polyamides** are a further synthetic development where two separate molecules are covalently joined together to reduce unfavourable entropy loss upon binding. It also results in an increase in specificity by ensuring that the two molecules bind at the same site. An additional problem is that once the number of rings in the polyamide ligand increases above six, affinity is decreased. This is because the ligand loses synchronization in either length or curvature with respect to the DNA groove. The problem is addressed by introduction of linkers with torsional flexibility, such as a β-alanine subunit, which allows the natural curvature of the polyamide to be relaxed. Molecules have been designed that can bind as overlapping homodimers to recognize up to 11 base pairs with subnanomolar dissociation constants. The eventual aim is to target a sequence of 16–18 base pairs, since these would be unique in the human genome.

### 9.7.5 Hoechst 33258

**Hoechst 33258** (Figure 9.13) is a fluorescent dye often used as a chromosome stain. Like other minor groove binders, Hoechst 33258 binds to A·T rich sequences preferentially. In fact it has a 50-fold greater affinity for AATT than for TATA and higher binding to AATT than to AAAA, which is in contrast to other A·T-specific minor groove drugs such as distamycin, netropsin and berenil.[44]

**a**

| Pair | G·C | C·G | T·A | A·T |
|---|---|---|---|---|
| Im/Py | ✓ | ✗ | ✗ | ✗ |
| Py/Im | ✗ | ✓ | ✗ | ✗ |
| Hp/Py | ✗ | ✗ | ✓ | ✗ |
| Py/Hp | ✗ | ✗ | ✗ | ✓ |

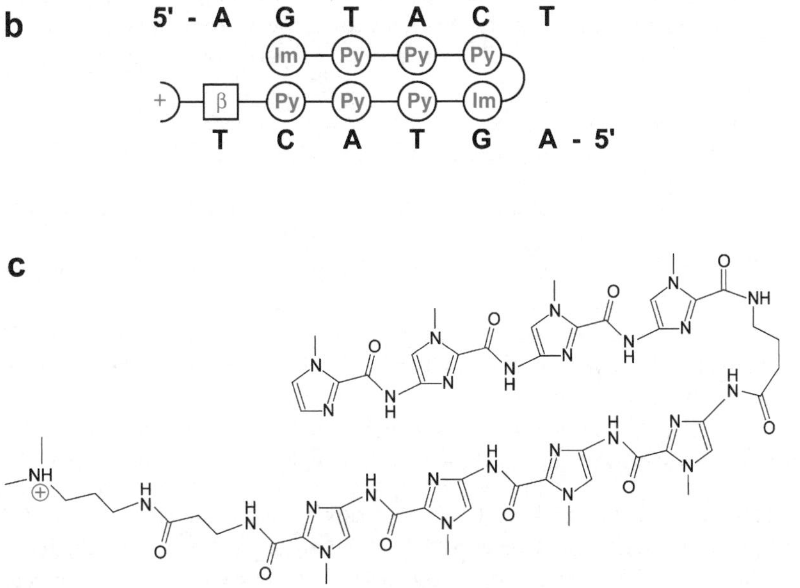


**Figure 9.16** *(a) The Dervan Pairing Rules, which describe how different combinations of imidazole (Im), pyrrole (Py) and hydroxypyrrole (Hp) rings placed opposite each other can be used to distinguish all four types of base pairs. (b) and (c) These pairing rules can be applied to target region of DNA with the sequence 5′-AGTACT (+ = dimethylaminopropylamine residue)*

*9.7.5.1 Structures of Hoechst 33258–DNA Complexes.* Hoechst 33258–DNA structures are extremely well studied. NMR analysis has revealed a number of locations and conformations of the drug in 5′-AATT sites. The phenolic ring is able to undergo rapid ring-flips, which involve 180° rotations about the phenol–benzimidazole bond, even though the phenolic ring is tightly accommodated between the walls of the minor groove. Dynamic breathing motions must occur in the DNA that results in transient widening of the minor groove to allow the ring to flip. The piperazine ring also adopts several different conformations. These two rings at the termini of the molecule are generally more mobile than the *bis*-benzimidazole core.

In one Hoechst 33258–A2T2 X-ray crystal structure, the ligand is located in the central 5′-AATT tract in a similar manner to netropsin. Hydrogen bonds are observed from the benzimidazole amide groups to O-2 atoms on thymine and N-3 atoms on adenine. Hoechst 33258 is isohelical with the curvature of the DNA minor groove and numerous molecular contacts are made with the walls of the groove. The phenolic ring of the ligand is twisted 8° with respect to the adjacent benzimidazole ring and the two benzimidazole rings are twisted 32° with respect to each other. The piperazine ring is only slightly puckered and is almost planar to the benzimidazole ring to which it is attached. The O-4′ atoms of the deoxyribose sugar of the DNA backbone are in the correct position to make favourable interactions with the π-electron system of the ligand as it sits within van der Waals contact distance of the groove walls. The differences between observed structures in NMR and X-ray experiments arise probably because of the different low-energy conformations that this ligand can adopt when bound to various DNA sequences.

*9.7.5.2 Thermodynamics of Hoechst 33258–DNA Interactions.* The fluorescent properties of Hoechst 33258 can be exploited to study the DNA binding character of this dye. **Equilibrium experiments** show that Hoechst 33258 binds tightly to calf-thymus DNA, poly(dA-dT)·poly(dA-dT), and to the oligonucleotides A2T2 and $d(CCGGAATTCCGG)_2$ with $K_{obs}$ values in the range 1–3 × $10^9\,M^{-1}$.[45,46] However, this tight binding occurs only once every 100 base pairs of calf-thymus DNA. In fact Hoechst 33258 binds to calf-thymus DNA with a spectrum of different binding modes each characterized by a different affinity, with some values of $K_{obs}$ in the ~$10^6\,M^{-1}$ range. Similarly with poly(dA-dT)·poly(dA-dT), Job Plots reveal a multiplicity of binding stoichiometries with between 1 and 6 Hoechst 33258 molecules bound per 5 A·T base pairs as well as stoichiometries that suggest 1 or 2 ligand molecules are bound per DNA phosphate.

Thus Hoechst 33258 binding to DNA is complex and depends strongly on sequence and DNA length. At least four distinct types of interaction between Hoechst 33258 and DNA are seen:

- *Sequence-mediated*: specific, high affinity, binding observed in X-ray structures and footprinting studies;
- *Charge-mediated*: non-specific interaction that involves the electrostatic binding of monocationic Hoechst 33258 with the DNA phosphate backbone;
- *Ligand-mediated*: non-specific type of binding that occurs when a molecule of free Hoechst 33258 in solution binds to other dye molecules that are already associated with the DNA helix, either in the minor groove or the backbone; and
- *Structure-mediated*: either specific or non-specific binding – a mode that confines Hoechst 33258 to a specific structural region, such as the minor groove, because of some geometric requirement.

The differences between these four modes of binding are subtle. They are not totally separate entities because more than one type of interaction contributes to the overall affinity of each bound ligand.

Sequence mediated (specific) binding of Hoechst 33258 to DNA has been examined in some detail using $d(CGCAAATTTGCG)_2$ (**A3T3**) as the receptor.[47–49] A detailed partitioning of the overall observed free energy for the Hoechst 33258–A3T3 interaction (Section 9.6.4) has been carried out by measuring, as directly as possible, a near complete thermodynamic profile for the interaction. This used a combination of ITC to measure $\Delta H$ and $\Delta C_p$ and fluorescence titrations to measure $K_{obs}$ and hence $\Delta G_{obs}$. From these measurements $T\Delta S$ and $\Delta G_{pe}$ were calculated using standard thermodynamic relationships. The measured binding constant of 3.2 ± 0.6 × $10^8$ M(duplex)$^{-1}$ at salt concentration of *ca.* 12 mM is similar to the Hoechst 33258 binding affinity ($K_{obs}$ = 3.15 × $10^8$ M(duplex)$^{-1}$) determined by Clegg under similar conditions for the less extended A·T-tract dodecamer $d(CGCGAATTCGCG)_2$. Quantification of the salt dependence yielded a value for *SK* of 0.99, which is consistent with the monocationic charge of Hoechst 33258. This allows the overall observed binding free energy ($\Delta G_{obs}$) to be broken down into its component polyelectrolyte and nonpolyelectrolyte terms (Section 9.3). The analysis showed that from a $\Delta G_{obs}$ of −48.9 kJ mol$^{-1}$, some −7.5 kJ mol$^{-1}$ (about 15%) is $\Delta G_{pe}$, *i.e.* from coupled polyelectrolyte effects such as the release of condensed counter-ions from the DNA upon ligand binding.

X-ray crystal structures of this complex show a network of hydrogen bonds and other minor groove interactions involving van der Waals contacts and hydrophobic interactions. The binding free energy should thus reflect a large nonpolyelectrolyte component. This is borne out by experimental data showing that out of a total free energy of −48.9 kJ mol$^{-1}$, some −41.4 kJ mol$^{-1}$ is due to non-polyelectrolyte effects. At each temperature studies, binding enthalpies for the interaction of A3T3 with Hoechst 33258 are positive and their magnitudes decrease with increasing temperature ($\Delta H$ = +42.7 kJ mol$^{-1}$ at 9.3°C and $\Delta H = +17.6\,\text{kJ mol}^{-1}$ at 30.1°C). The binding-induced change in heat capacity ($\Delta C_p$) is therefore estimated as $-1.38\,\text{kJ mol}^{-1}\,\text{K}^{-1}$. This gives an overall thermodynamic profile of positive enthalpy and entropy together with a negative heat capacity change, which is generally characteristic of hydrophobic interactions and correlates well with X-ray crystal structures of this complex. The binding reaction is entropically driven and the favourable entropy term arises from the release of structured water and cations from the minor groove and/or the ligand into bulk solvent upon binding. Another origin for the positive entropy is the release of condensed counter-ions from the DNA helix upon binding of the cationic ligand.

The heat capacity change can be related to changes in the surface area exposure (Section 9.6.4 and Equation 9.16). Binding of Hoechst 33258 to A3T3 results in a ~20% loss of solvent-accessible surface relative to the individual components. The majority of surface removed from exposure to solvent upon binding is nonpolar rather than polar. Application of Equation 9.16 to the Hoechst 33258–A3T3 system gives a calculated $\Delta C_p$ of $-1.37\,\text{kJ mol}^{-1}\,\text{K}^{-1}$ which is in good agreement with the calorimetrically determined heat capacity change of $-1.38\,\text{kJ mol}^{-1}\,\text{K}^{-1}$.

The empirical relationship used to calculate $\Delta C_p$ was originally based on the heats of solvent transfer of small model compounds such as amino acids, amides, and hydrocarbons. Later it was extended and applied to protein folding–unfolding equilibria and to protein–DNA interactions.[50] The Hoechst 33258–A3T3 calorimetric study was the first demonstration that the same relationship is valid for low molecular weight DNA binding drugs. The values used in Equation 9.16 are derived only on data from drug–DNA interactions as it is now established that change in surface area exposed to solvent on ligand binding provides a tentative link between structural and thermodynamic data.

There have been several reports of calorimetrically measured heat capacity changes for both groove binders and intercalators. In most cases there is good agreement with theoretical predictions of heat capacity change derived from predictive algorithms of the type shown in Equation 9.16. Figure 9.17 shows a good correlation of calculated *versus* measured heat capacity change for a variety of systems.

A detailed partitioning of $\Delta G_{obs}$ into its component parts gives considerable insight into the forces responsible for the binding process (Section 9.6.4 and Equation 9.14). In the Hoechst 33258–A3T3 system, the contribution of each of the five terms in Equation 9.14 to $\Delta G_{obs}$ has been estimated. The variation of $K_{obs}$ with salt concentration and application of polyelectrolyte theory revealed that $\Delta G_{pe}$ is $-7.36\,\text{kJ mol}^{-1}$.

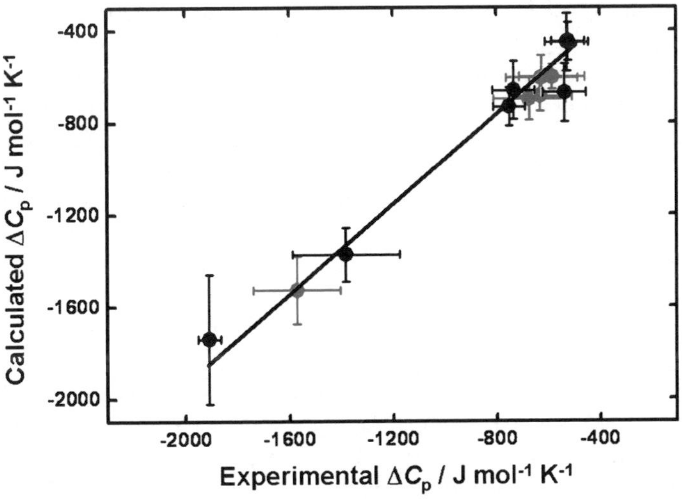


**Figure 9.17** *The relationship between calorimetrically measured heat capacity change and the heat capacity change calculated from changes in solvent accessible surface area for twelve different drug–DNA systems (red: intercalators, black: groove binders). The line is a linear fit to the data (r = 0.97)*

From evaluation of $\Delta C_p$ one can calculate the possible contribution to binding free energy that arises from the hydrophobic transfer of the drug from bulk solvent into the DNA minor groove ($\Delta G_{hyd}$) as $-110.4\,kJ\,mol^{-1}$. These data, as well as crystallographic studies, indicate that binding of dye to A3T3 induces negligible conformational changes in either the DNA or the drug. In effect it is a rigid body interaction. Therefore the value of $\Delta G_{conf}$ can be set to zero. The value of $\Delta G_{r+t}$ can be set at $+62.8\,kJ\,mol^{-1}$ by use of arguments discussed by Spolar and Record.[50] Thus the calculated free energy change for the Hoechst 33258–A3T3 interaction is $-55.0\,kJ\,mol^{-1}$, very close to the experimental value of $-48.9\,kJ\,mol^{-1}$ for $\Delta G_{obs}$. Therefore, by difference it is possible to estimate $\Delta G_{mol}$ (the contribution to free energy from weak noncovalent interactions between the drug and DNA), which is $+6.1\,kJ\,mol^{-1}$.

This analysis suggests that for this particular ligand–DNA complex, noncovalent molecular interactions such as hydrogen bonds make a net *unfavourable* contribution to binding free energy. Although this result is apparently at odds with NMR and crystallographic studies, the two positions can be reconciled when one considers the hydration state of the minor groove. Hydrogen-bond formation between the ligand and DNA is at the expense of hydrogen-bond breakage between the DNA and site-specific waters located in the minor groove. The net result is an iso-energetic exchange reaction involving Hoechst 33258 and water, such that a near zero free energy change results. Instead, the major driving force for Hoechst 33258 binding to A3T3 is the hydrophobic transfer of the drug from bulk solvent into the duplex minor groove binding site. This is reflected in the observed negative $\Delta C_p$ and the positive binding enthalpy. The large energetic cost resulting from losses in rotational and translational freedom that occurs upon complex formation is more than compensated for by the favourable contributions from hydrophobic transfer and polyelectrolyte effects. Thus affinity might be best modulated by changes in drug hydrophobicity. However, hydrogen bond formation and other noncovalent interactions are key modulators of specificity in drug–DNA interactions and they are also important in fine-tuning the free energy of binding in response to the interactions available in a given DNA binding site.

## 9.8 INTERCALATION *VERSUS* MINOR GROOVE BINDING

There are some interesting compounds that possess structural features that could, in principle, involve **binding by either intercalation or groove binding**. In reality, one is favoured over the other. What are the factors that dictate choice of binding mode? Specific examples are the ligand SN 6999 and the related compound chloroquine (Figure 9.18).

Both molecules contain a fused bicyclic **quinoline ring system**, yet biophysical and structural studies have shown that while chloroquine can intercalate, SN 6999 binds in the minor groove. SN 6999 has a curved shape that fits in the minor groove and it forms hydrogen bonds in a similar manner to other groove binding drugs. Modelling studies have shown that SN 6999 could form favourable contacts within a base pair stack as well as possible electrostatic interactions with the phosphate backbone in an intercalation complex. However, if such an intercalation complex formed instead of the groove-bound complex, it would be at the expense of favourable free energy from hydrogen bonding, non-bonded contacts with the groove walls, and solvent release from the minor groove. Hence for SN 6999, binding in the minor groove is much more favourable energetically than intercalation, even though the latter is possible. Chloroquine however does not have the correct hydrogen bond donor/acceptor groups or any optimal structure for appropriate minor groove interactions. Thus it binds to DNA by intercalation but only with low affinity.

SN 6999 is an example of a minor groove binder having a benzene-fused heterocycle. Conversely, there are examples of unfused aromatic cations that bind by intercalation rather than groove binding. Netropsin, distamycin, and Hoechst 33258 are all examples of ligands that have torsional freedom and this property allows these compounds to adopt an optimal twist for minor groove binding. Strekowski and Wilson have designed and synthesised a new group of **diphenylpyridine** compounds with unfused aromatic rings possessing torsional freedom. So one might expect them to be good minor groove-binding ligands (Figure 9.18). Molecular modelling and X-ray crystallography indicate that the molecule has a twist of between 20° and 25° between the phenyl and pyrimidine planes. The existence of twisted rings along with cationic end

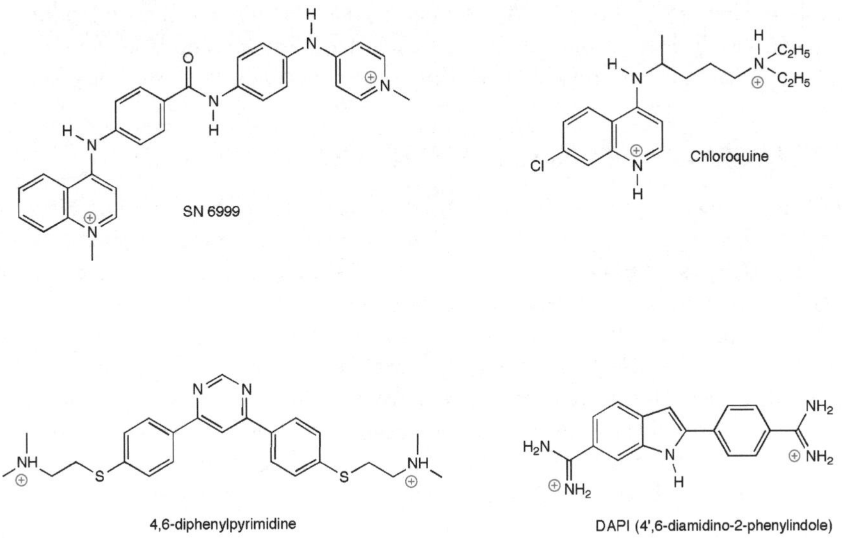


**Figure 9.18** *Structures of DNA binding ligands SN 6999, chloroquine, DAPI and 4,6-diphenylpyrimidine*

groups suggests that these compounds should bind in the minor groove. However, solution studies, (NMR, LD and DNA unwinding) have established unequivocally that these ligands are intercalators.

It is true to say that DNA base pairs can have significant intrinsic propeller twist, so there is no reason why unfused aromatic ring compounds should not intercalate and enhance the base pair interactions responsible for propeller twist. Yet, why should 4,6-diphenylpyridine (Figure 9.18) intercalate whilst similar compounds such as Hoechst 33258 bind in the minor groove?

For ligands of this nature, the only reasonable potential binding modes are intercalation or minor groove binding at AT-rich sites. This is because the major groove is too wide to form sufficient favourable contacts while G·C minor grooves present a steric block *via* the guanine amino group. The choice of whether the ligand intercalates or binds at A·T minor grooves is probably in large part determined by the presence or absence of functional groups capable of participating in hydrogen bonds, notably ligand amides. We have already seen that appropriately placed NH groups on the ligand can donate hydrogen bonds to N-3 atoms of adenine and O-2 atoms of thymine in forming strong minor groove complexes. Examination of the structure of 4,6-diphenylpyridine shows that it has no reasonable hydrogen bond donors, either in the rings or in the connecting bonds, and so there is no strong minor groove complex. However, the aromatic rings in the ligand can form good stacking interactions with the base pairs and hence it forms a strong intercalation complex.

The ligand **4′,6-diamidino-2-phenylindole (DAPI**, Figure 9.18) is an unfused aromatic diamidine that can select between intercalation and groove binding. Many biophysical studies have shown that DAPI is a minor groove binder at A·T sequences and DNA footprinting has identified a binding site size of about three base pairs, in accord with its molecular length. DAPI also has some structural similarity to the unfused intercalators discussed above and solution studies provided the surprising result that it can bind to G·C sequences by intercalation.

These findings can be rationalized in terms of DNA structure. Both A·T and G·C sequences form good intercalation sites; however, only A·T base pairs are favourable minor groove binding sites for unfused aromatic cations like DAPI, netropsin, Hoechst 33258, *etc*. Therefore DAPI selects the appropriate binding mode depending on sequence, it intercalates at G·C sites but binds in the minor groove at A·T sites. Intercalation of DAPI at G·C sites has a binding constant similar to other intercalators such as proflavin. When it binds at A·T minor grooves, it spans more base pairs and makes more specific contacts and hence it binds with higher affinity in the minor groove.

The conclusion is that for all drug–DNA interactions, intercalation and groove binding represent two potential low-energy wells in a continuous free energy surface. The binding mode with lower energy will depend on DNA sequence and conformation as well as the molecular features of the bound ligand.

## 9.9 CO-OPERATIVITY IN LIGAND–DNA INTERACTIONS

Genomic DNA is somewhat different from model duplexes used in structural and biophysical studies, since it possesses a continuous array of potential binding sites that are incorporated into a large nucleic acid–protein complex. Thus the biological activity of a DNA binding drug will be affected by co-operativity.

**Co-operativity** is an important and pervasive concept in biology. The phenomenon of allostery is well known in protein–ligand interactions. A type of negative co-operativity in drug–DNA interactions is neighbour exclusion (Section 9.6.3). Here each bound drug excludes one or more adjacent base pairs from being potential drug binding sites. This is often referred to as site-exclusion co-operativity. However, it is also possible for there to be co-operative effects between adjacent binding sites (composed of more than one base pair) and in such cases the data can be described by a modified version of the **McGhee and von Hippel relationship**.[20] (Section 9.6.3 and Equation 9.13). The modified form of this equation is

$$\frac{r}{C_f} = K(1 - nr)\left[\frac{(2\omega - 1)(1 - nr) + r - R}{2(\omega - 1)(1 - nr)}\right]^{n-1}\left[\frac{1 - (n + 1)r + R}{2(1 - nr)}\right]^2 \qquad (9.17)$$

$$R = \sqrt{\{[1 - (n + 1)r]^2 + 4\omega r(1 - nr)\}}$$

where $r$ is the ratio of bound drug to total DNA sites (in base pairs), $C_f$ the concentration of free drug, $K$ the binding constant for the interaction of a drug with an isolated site, and $n$ the neighbour exclusion parameter. The parameter $\omega$ is the co-operativity function; which in effect is the equilibrium constant for moving a bound drug molecule from a totally isolated binding site to a site contiguous to another bound ligand. If $\omega$ is greater than 1, this indicates positive co-operativity. Conversely, if $\omega$ is less than 1, there is negative co-operativity. When $\omega = 1$ there is no co-operativity and then Equation 9.17 can be simplified to Equation 9.13. This method of analysis is very useful for interactions involving long polynucleotides or natural DNA samples. Equation 9.17 can be incorporated into nonlinear least squares fitting routines to fit equilibrium binding isotherms and the parameter $\omega$ can thus be evaluated readily. However, binding data should primarily be fitted to the simplest model, therefore Equation 9.17 should only be used if it yields a significantly better statistical fit to the data compared to Equation 9.13.

## 9.10 SMALL MOLECULE INTERACTIONS WITH HIGHER-ORDER DNA

The vast majority of genomic DNA exists as a double-stranded, antiparallel structure in the B-conformation. However in the past fifteen years, it has become clear that DNA can adopt a number of **higher-order conformations**. These involve the self-assembly of three or four DNA strands or the intramolecular folding of a single strand such that it adopts a **triplex** or **quadruplex (tetraplex)** conformation (Section 2.3.6 and 2.3.7). Further, it has become apparent that, far from being simple biophysical curiosities, these multi-stranded DNA structures have considerable biological significance. This has led to triplex and quadruplex DNA becoming major drug targets. Below are summarised the key features of these higher-order structures and a discussion of how small molecules have been used to target them and effect their stabilization.

### 9.10.1 Triplex DNA and its Interactions with Small Molecules

**Triplex DNA** is formed when a single-stranded oligonucleotide specifically recognizes the major groove of duplex DNA and forms hydrogen bonds with the Watson–Crick base pairs (Section 2.3.6). Due to the inherent specificity of base–base interactions, triplex formation is a versatile and powerful tool for sequence-specific recognition of nucleic acids.

Studies on intermolecular triplex-forming oligonucleotides have been stimulated by their possible use as therapeutic agents. This is known as the **antigene concept of therapy**, as described by Claude Hélène and others.[51] A **triplex-forming oligonucleotide** (TFO) might be delivered to the nucleus of a cell to bind to a unique sequence in the genome and so form a triplex and inhibit the transcription of a single gene or cleave a specific site within a chromosome. For example, the ability to suppress the transcription of key genes may trigger apoptosis in a cancer cell.

A major practical difficulty is the relatively low stability of triplex DNA compared to duplex structures under physiological conditions. The presence of the third strand increases substantially the negative charge near to the environment of the anionic duplex backbone. Therefore, relatively high cation (*e.g.* $Na^+$ or $K^+$) concentrations are required to stabilise the triple helix. Several approaches have been investigated to overcome this problem, including the use of an intercalating ligand that can preferentially stabilize triplex DNA.

Duplexes and single-stranded DNAs can form triple helixes that have a number of different conformations.[52–54] In one particular conformation, a third strand comprised of either cytosines or thymidines orientates parallel to the polypurine strand of the duplex and binds to it through formation of Hoogsteen hydrogen bonds (Figures 2.33 and 9.19). The resulting base triplets are T·A–T and $C^+$·G–C. Alternatively, the third strand can be composed of purines, *i.e.* guanines or adenines that recognise and bind to the purine strand of a target duplex to form T–A·A or C–G·G base triplets. The third strand binds in an antiparellel orientation with respect to the polypurine sequence and interacts with the duplex through reverse Hoogsteen associations.

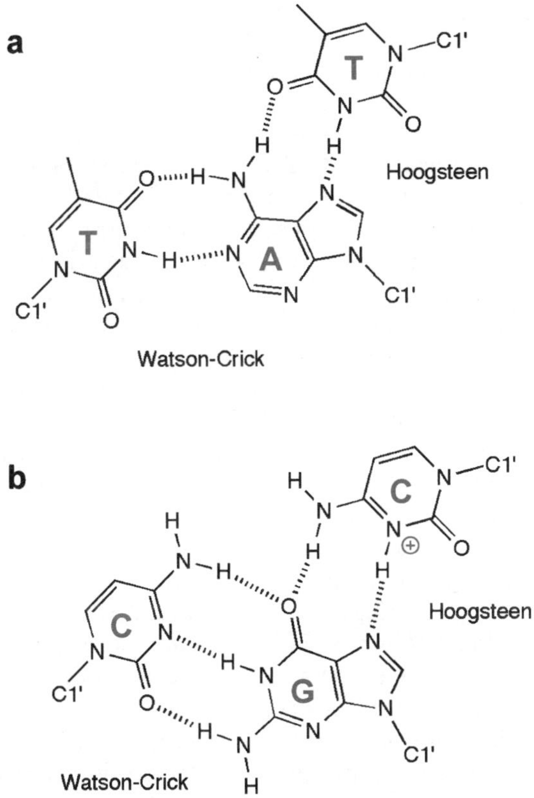


**Figure 9.19** *Hydrogen-bonding arrangement for two possible base triplets*

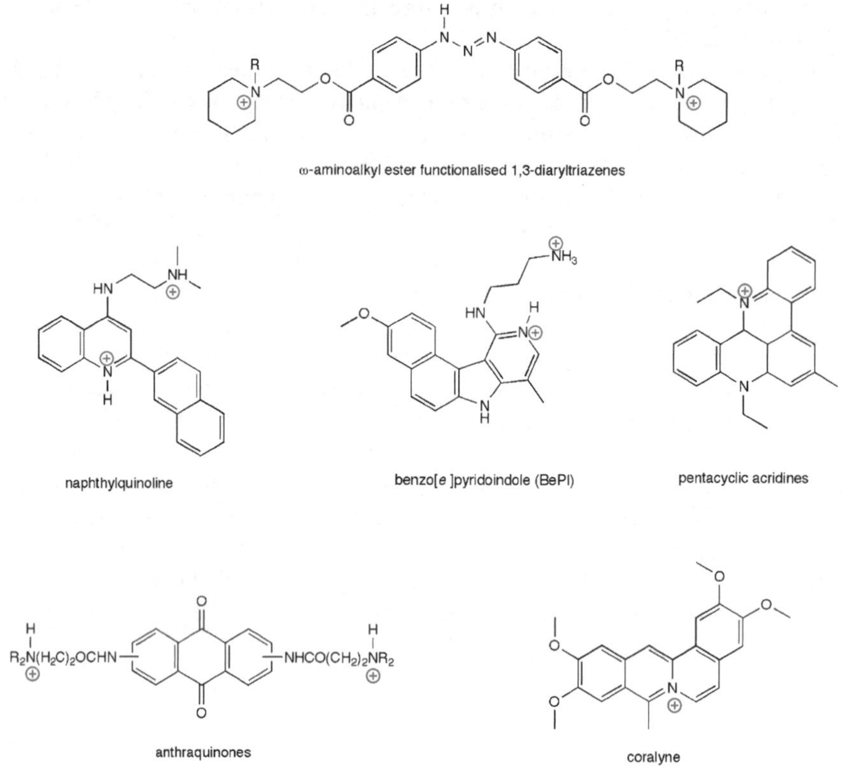


**Figure 9.20** *The structures of a range of triplex-selective ligands*

In recent years a number of **triple helix-specific ligands** have been characterized (Figure 9.20). The function of such ligands is to overcome the inherent instability of triplex DNA and effect triplex stabilization.

The addition of a third strand into the major groove of a DNA duplex leads to the creation of two new grooves on either side of the third strand in what was previously the major groove, and the minor groove is left largely unchanged. One of the new grooves on either side of the third strand is essentially featureless. However, the other groove resembles the minor groove, except that its walls have hydrophobic residues from pyrimidine C-5 methyl groups while the polar C-4 carbonyl/amino groups are buried.

This unique groove in triplex DNA with its bifurcated hydrogen-bonded centres on the floor, as well as the hydrophobic nature of the groove walls, has led to the development of a new group of ligands designed to target this previously unexploited binding site in triplex DNA.[55] Thus, Terry Jenkins has synthesized a series of *para-* and *meta-*carbonyl substituted **1,3-diaryltriazenes** (Figure 9.20) intended to target one of the two new grooves created during triplex formation. The concave inner surface of the ligand formed by triazene and phenyl rings has no protrusions and so in principle the ligand can penetrate deeply into the narrow triplex groove. The ligand possesses hydrogen bond acceptor groups, which can interact with hydrogen bond donors on the groove floor. In addition, the $\pi$-conjugated core residues make close van der Waals contacts with the hydrophobic groove walls by alignment of phenyl rings in both ligand and DNA. Solution studies confirm that these ligands bind preferentially to triplex DNA and stabilize this structure more than they stabilize a corresponding duplex. For example, sub-saturating amounts of the ligand produce a $\Delta T_m$ of ~2°C in duplex DNA compared to triplex melting, where the presence of ligand shifts $T_m$ 20–25°C higher relative to ligand-free DNA triplex. Binding of the triazene compounds is enthalpically driven, which is consistent with hydrogen bonding in the groove. It appears that these ligands bind through good surface matching, and the complex is stabilized by favourable groove wall–ligand and hydrogen bond base–ligand interactions.

Most strategies aimed at triplex stabilization have centred on intercalators, which can also bind to duplex DNA as well as to triplexes. For example, the prototypical duplex-binding ligand ethidium intercalates into triplexes. Indeed, it binds more strongly to triplexes composed of T·A–T base triplets than it does to the corresponding duplex. Conversely, ethidium binds more strongly to G·C duplexes than it does to $C^+$·G–C triplex, probably because of the positive charge on the cytosine. Other intercalators have been developed that increase triplex stability substantially.

One of the original triplex specific ligands is **benzo[*e*]pyridoindole (BePI**, Figure 9.20) developed by Claude Hélène. Subsequently, many structurally related analogues of BePI were also made, (for example benzoindoloquinoline, benzopyridoquinoxaline, benzoquinoquinoxaline).[56] Triplex-selective intercalators that contain fused or unfused heterocyles have also been synthesized by a number of other groups, such as **naphthyquinoline** (Wilson), **anthraquinones** (Jenkins) and **acridine derivatives** (Stevens) (Figure 9.20). Most ligands in this class have an extended planar aromatic ring system of optimal size to fit the extended molecular surface area presented by a base triplet (compared to a base pair). Therefore, there are maximal $\pi$–$\pi$ stacking interactions between the ligand and the base triplets that constitute the binding site. Most of the ligands have at least one cationic charge on the chromophore. Hence there is the possibility of favourable electrostatic interactions with increased negative charge density in the triplex DNA. Many of these ligands also possess pendant side chains that terminate in a protonatable amino group. This provides extra stability through electrostatic interactions with the sugar–phosphate backbone of the DNA.

### 9.10.2 Quadruplex DNA and its Interactions with Small Molecules

It is well known that four guanine bases can self-associate to form cyclic, planar tetramers of Hoogsteen hydrogen-bonded structures known as **G-tetrads** (Section 2.3.7). When two or more successive guanine residues are incorporated into a nucleic acid sequence, the DNA can fold into either an inter- or intramolecular, **four-stranded quadruplex** – or **tetraplex** – structure in which successive G-tetrads are stacked on top of each other (Figure 9.21).

The stability of quadruplex DNA is highly dependent on the type and concentration of counter-ion present. Most quadruplexes are stable in solutions containing between 100 and 200 mM $Na^+$ or $K^+$. It is believed that these monovalent cations coordinate to the eight carbonyl oxygen atoms from two stacked G-tetrads that protrude into the central core of the folded structure. A number of different quadruplex conformations can form *in vitro* depending on strand molecularity and orientation. When a sequence of DNA has a single run of successive guanine bases, four separate strands can combine to form a **parallel tetramolecular quadruplex structure**. If a sequence of DNA has two separate runs of guanines separated by three or four T or A bases, the single strand can fold back to form a **guanine-hairpin**. Two such hairpins can then self-associate to form a stable quadruplex. Finally, single-stranded sequences containing four separate runs of guanines, again separated by intervening T or A residues, can fold intramolecularly to form an **antiparallel quadruplex** (Figure 9.21d). Several review articles are available that discuss G-quadruplexes and their potential as drug targets.[57–61]

Much interest in quadruplex DNA stems from its potential role in telomere biology. **Telomeres** are specialized protein–DNA complexes at the 3′-termini of linear eukaryotic chromosomes (Section 6.4.5). A particularly remarkable feature of telomeric DNA is that its sequence is conserved across a large range of eukaryotes. The overall consensus sequence is $d(T_{1-3}(T/A)\mathbf{G_{3-4}})$. These sequence motifs are tandemly repeated for up to many thousands of base pairs. The G-rich strand extends beyond the duplex region to form a single-stranded overhang that is a few hundred nucleotides long. This widely conserved G-rich sequence in a wide range of species implies some conserved biological role.[62]

Telomeres have key functional importance in the regulation of cellular longevity and apoptosis (Section 6.6.5).[63] Human cells typically start out with telomeres that are about 15 kb in length and they are composed of repeating units of 5′-TTAGGG. Because DNA polymerase does not fully replicate the lagging strand, ~100 base pairs of DNA are lost from chromosome termini after each round of cell replication. This process limits

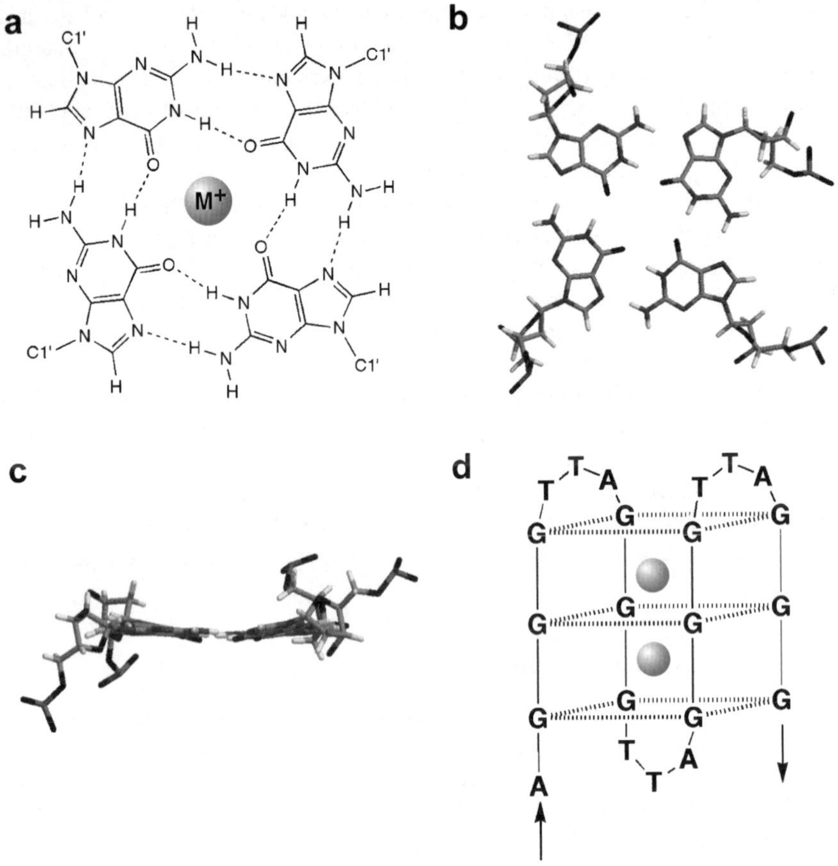


**Figure 9.21** *(a) Hydrogen-bonding arrangement of an individual guanine–tetrad. This type of guanine–guanine interaction leads to the formation of a central pore with negative electrostatic potential due to the carbonyl oxygen atoms. Quadruplex DNA folds only in the presence of monovalent cations, which coordinate in the central pore as shown. (b) NMR-derived view of the actual structure of a G-tetrad. (c) Planar nature of a G-tetrad. (d) Schematic of the folding topology adopted by a 22-nucleotide sequence containing the human telomere repeat unit TTAGGG in the presence of sodium ions*

the replicative life span of most somatic cells to about 50 rounds of mitosis. Germline cells, proliferative stem cells, and tumour cells need to circumvent this limit on cell replication to achieve longer life times, and overcome telomere shortening by expressing the enzyme telomerase. This enzyme, discovered by Elizabeth Blackburn in 1985, is a multisubunit ribonucleoprotein complex that has reverse transcriptase activity.

Several studies have indicated that 80–90% of human tumours exhibit elevated levels of telomerase activity. One approach to telomerase inhibition was initiated by Thomas Cech and David Prescott, who realised that when telomeric DNA is folded into a quadruplex secondary structure, it cannot act as a template for telomerase.[64] This suggested that small molecules might be used to induce quadruplex formation in telomeres and prevent telomerase from carrying out its function.

There have been several recent advances in development of ligands that are capable of quadruplex-specific recognition and telomerase inhibition,[59,61,65] as well as a number of NMR and X-ray crystallographic studies, of inter- and intramolecular quadruplexes.[66–72] Nearly all quadruplex architectures are built upon the guanine quartet (Figure 9.21). However, there is a large degree of structural heterogeneity in the arrangement of the four strands and the loops that connect them.

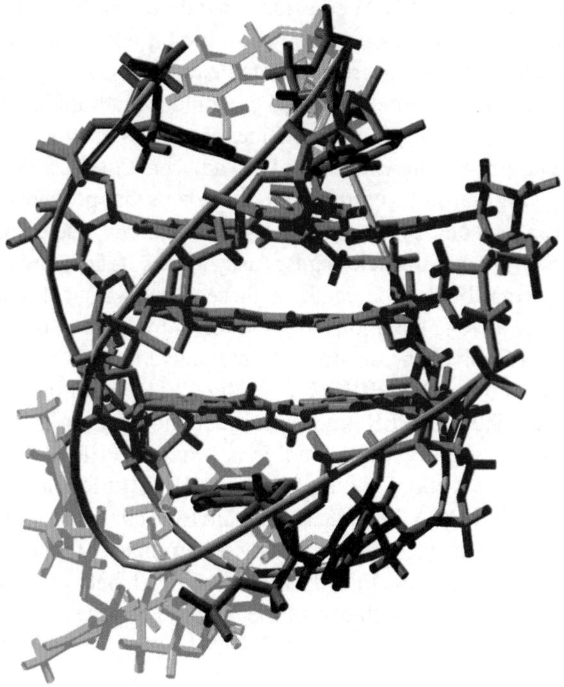

**Figure 9.22** *NMR solution structure of the human telomeric sequence 5′-AGGGTTAGGGTTAGGGTTAGGG in the presence of $Na^+$ (PDB: 143D). Guanine residues are in red, thymines in grey, adenines in black, and a grey ribbon highlights the DNA backbone*

One of the most important factors that influences folding topology is the type of monovalent cation present during folding. For example, in the fold of the human telomeric sequence (Figure 9.21), which is based on an NMR study in the presence of $Na^+$ (Figure 9.22)[67] the TTA loops are arranged in a lateral and diagonal conformation, and the strands alternate between parallel and antiparallel. By contrast, a recent X-ray structure in the presence of $K^+$ ions shows that all four strands are parallel and the TTA loops are extended out from the sides of the G-quartets.

It has been suggested that the antiparallel fold observed in an NMR study is solely a consequence of the presence of sodium ions, and the presence of potassium ions induces a transition to the parallel arrangement. The different conformations adopted by quadruplex structures in different cationic environments have a direct impact on the binding of drug molecules. In fact recent biophysical studies reveal considerable complexity in quadruplex unfolding and show that multiple species coexist in equilibrium across the temperature range. Thus there is no single static structure for the folded quadruplex and hence targeting quadruplex DNA with small molecules is not straightforward.

Despite these difficulties, several different classes of ligand have shown promise as structure-specific, quadruplex-binding agents and as telomerase inhibitors (Figure 9.23). Most of these compounds have **telomerase inhibitory effects** with $IC_{50}$ values in the low micromolar range but specific binding to quadruplex DNA over duplex is, at best, only modest.

Laurence Hurley has developed some cationic porphyrins for targeting quadruplex DNA and to effect telomerase inhibition. The structure of an archetypal porphyrin, [tetra-(*N*-methyl-4-pyridyl)-porphine] TMPyP4 (Figure 9.23) illustrates the planar arrangement of the aromatic rings, which give rise to a ligand that has similar molecular dimensions to a G-tetrad intercalation site. This has led to the proposal that compounds of this type will bind to quadruplex DNA through intercalative stacking. UV spectroscopy, circular dichroism,

NMR, and ITC have all shown that TMPyP4 does indeed stabilize both parallel and antiparallel quadruplex DNA. In addition, this compound has been shown to inhibit telomerase activity, but the exact mode of interaction of TMPyP4 with DNA is uncertain. Photocleavage assays suggest that the porphyrin "end-pastes" by stacking externally to the G-tetrad at the end of a model quadruplex based on the human telomere sequence. By contrast, calorimetric and spectroscopic measurements of stoichiometry as well as molecular modelling indicate that the porphyrin intercalates between adjacent G-tetrads. Very recent ITC experiments show that there is an initial, high-affinity, entropic binding event that probably is end-pasting. Following the initial tight binding event, secondary binding probably occurs by intercalation. Some 150 porphyrin analogues have been screened in a cell-free telomerase assay as well as an assay that monitors arrest of DNA synthesis. The results show that binding strength of these porphyrin analogues to quadruplex DNA is dependent on overall charge, side chain length and presence of hydrogen bonding groups.[57]

Another promising group of polycyclic quadruplex-interactive ligands is based on a **perylene moiety**, for example, the dicationic compound **PIPER** (*N,N′-bis*[2-(1-piperidino)-ethyl]-3,4,9,10-perylenetetracarboxylic diimide) (Figure 9.23). PIPER binds specifically to quadruplex DNA and there is only low-level binding to duplex DNA. In slightly alkali conditions, where PIPER aggregates, the selectivity of the ligand for quadruplex over duplex is about $10^3$. However at pH 7, PIPER is less selective and binds 10-fold better to quadruplex than to duplex DNA. The NMR structure for a complex of PIPER and quadruplex DNA shows that PIPER end-pastes, *i.e.* it stacks externally on to terminal G-tetrads. By doing so it can form a 2:1 quadruplex/PIPER complex, where the ligand forms a sandwich between two quadruplex molecules.[73] The ligand only has minimal π-overlap with two of the guanine residues in the terminal G-tetrad but the

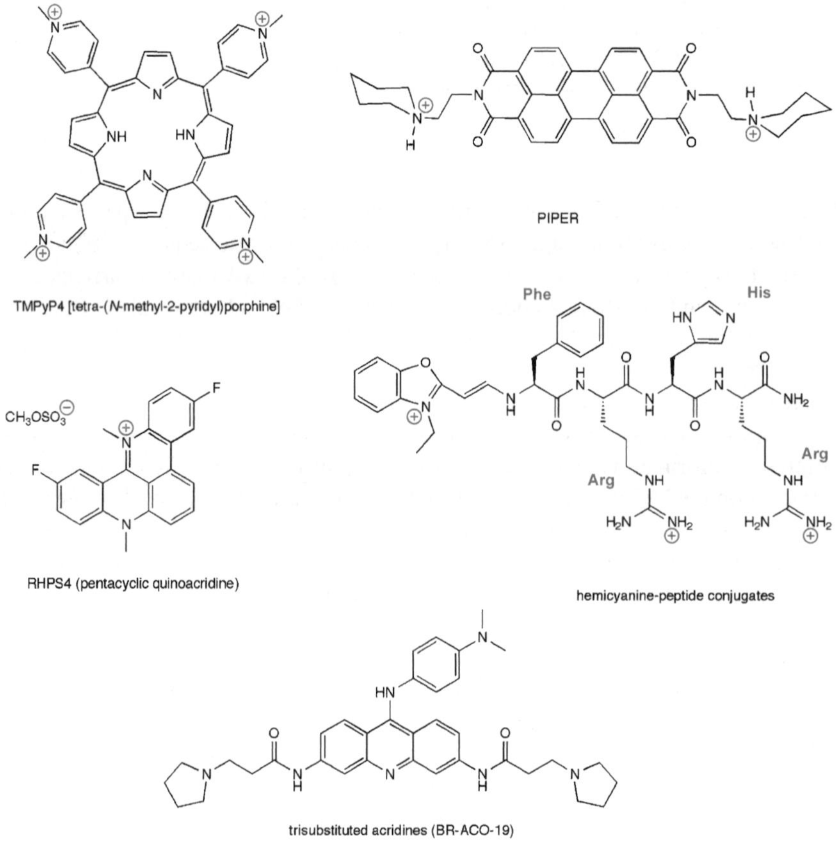


**Figure 9.23** *Chemical structures of quadruplex-binding small molecules*

presence of two positive charges on the ligand undoubtedly enhances quadruplex binding. PIPER also exhibits significant antitelomerase activity and behaves as a chaperone-like molecule, since it enhances the folding of quadruplexes. The ligand alters the kinetics of the association reaction for dimeric quadruplex formation from second- to first-order and it increases the initial rate for quadruplex formation by about 100-fold. Thus as well as binding to preformed quadruplex structures, ligands such as PIPER may also be able to induce quadruplex formation *in vivo* at appropriate sequences.

**Trisubstituted acridines** of the type shown in Figure 9.23 have been synthesized and exhibit very promising quadruplex affinity, selectivity and high levels of telomerase inhibition.[74] Based on an NMR structure of the quadruplex formed from a sequence model of the human telomere, the expectation was that these molecules would end-paste in a similar manner to PIPER and 2,6-disubstituted amidoanthraquinones. Disubstituted compounds have similar affinity for duplex and quadruplex DNA ($\sim 1 \times 10^6\,M^{-1}$) in that there is no appreciable structure selectivity. By contrast, trisubstituted analogues bind quadruplex DNA 30 to 40 times more strongly than to duplex DNA and telomerase inhibition correlates with quadruplex binding strength. Thus, activity for disubstituted compounds is either very modest or nonexistent, whereas trisubstituted compounds are about 100-fold more potent against telomerase activity, but without increased cellular toxicity.

Quadruplex formation may play many important biological roles. For example, promoter regions of certain genes such as β-globin genes, rat preproinsulin II gene, adenovirus serotype 2, retinoblastoma susceptibility genes, the *c-Myc* gene, and various other oncogenes, all contain guanine-rich regions that have the potential to form quadruplex structures. Thus **G-quadruplexes may play a role in regulating gene expression**, for example, inhibition of transcription. Hurley has shown that ligand-induced quadruplex formation in *c-Myc* promoter region leads to transcriptional down-regulation of that gene.[75] Conversely a G to A mutant in the quadruplex forming region leads to ~3-fold increase in the basal transcriptional activity of the *c-Myc* promoter. Thus, genomic sequences that have the propensity to form quadruplex DNA will become major drug targets in ongoing efforts to develop structure/sequence-specific agents for the treatment of human diseases.

Improvements in DNA and RNA synthetic techniques as well as design and synthesis of additional DNA-interactive drugs and model compounds continue to make available systems that are providing new information on the structure and energetics of nucleic acid complexes with small molecules and are dramatically advancing the field of reversible nucleic acid interactions. In addition, it is clear that both DNA and RNA are prime targets as receptors for drug development. While DNA has been the prime target for such studies, it can be expected that RNA will play an increasing role in the next period of time.

## REFERENCES


1. G.S. Manning, Polyelectrolytes. *Ann. Rev. Phys. Chem.*, 1972, **23**, 117–140.
2. G.S. Manning, The molecular theory of polyelectrolyte solutions with applications to the electrostatic properties of polynucleotides. *Quart. Rev. Biophys.*, 1978, **11**, 179–246.
3. G.S. Manning, Counterion binding in polyelectrolyte theory. *Acc. Chem. Res.*, 1979, **12**, 443–449.
4. M.T. Record Jr., C.F. Anderson and T.M. Lohman, Thermodynamic analysis of ion effects on the binding and conformational equilibria of proteins and nucleic acids: the roles of ion association or release, screening and ion effects on water activity. *Quart. Rev. Biophys.*, 1978, **11**, 103–178.
5. J.B. Chaires, Dissecting the free energy of drug binding to DNA. *Anti Cancer Drug Design.*, 1996, **11**, 569–580.
6. W.D. Wilson and I.G. Lopp, Analysis of cooperativity and ion effects in the interaction of quinacrine with DNA. *Biopolymers*, 1979, **18**, 3025–3041.
7. R.A.G. Friedman and G.S. Manning, Polyelectrolyte effects on site binding equilibria with application to the intercalation of drugs into DNA. *Biopolymers*, 1984, **23**, 2671–2714.
8. T. Thomas and T.J. Thomas, Polyamines in cell growth and cell death: molecular mechanisms and therapeutic applications. *Cell. Mol. Life Sci.*, 2001, **58**, 244–258.
9. E. Westhof and D.L. Beveridge, Hydration of nucleic acids. *Water Sci. Rev.*, 1990, **5**, 24–136.

10. R.E. Dickerson, H.R. Drew, B.N. Conner, R.M. Wing, A.V. Fratini and M.L. Kopka, The anatomy of A-, B- and Z-DNA. *Science*, 1982, **216**, 475–485.
11. X. Shui, L. McFail-Isom, G.G. Hu and L.D. Williams, The B-DNA dodecamer at high resolution reveals a spine of water on sodium. *Biochemistry*, 1998, **37**, 8341–8355.
12. X. Shui, C.C. Sines, L. McFail-Isom, D. VanDerver and L.D. Williams, Structure of the potassium form of the CGCGAATTCGCG: DNA deformation by electrostatic collapse around inorganic cations. *Biochemistry*, 1998, **37**, 16877–16887.
13. L. McFail-Isom, C.C. Sines and L.D. Williams, DNA structure: cations in charge? *Curr. Opin. Struct. Biol.*, 1999, **9**, 298–304.
14. L.S. Lerman, Structural considerations in the interaction of DNA and acridines. *J. Mol. Biol.*, 1961, **3**, 18–30.
15. L.S. Lerman, The structure of the DNA–acridine complex. *Proc. Natl. Acad. Sci. USA*, 1963, **49**, 94–102.
16. J.B. Chaires, Energetics of DNA–drug interactions. *Biopolymers*, 1998, **44**, 201–215.
17. J.B. Chaires, Drug–DNA interactions. *Curr. Opin. Struct. Biol.*, 1998, **8**, 314–320.
18. I. Haq, Thermodynamics of drug–DNA interactions. *Arch. Biochem. Biophys.*, 2002, **403**, 1–15.
19. J.B. Chaires, Energetics of Anthracycline–DNA interactions, in *Small Molecule DNA and RNA Binders: From Synthesis to Nucleic Acid Complexes*, M. Demeunynck, C. Bailly and W.D. Wilson (eds). Wiley–VCH Verlag GmbH, Weinheim, 2003, 461–481.
20. J.D. McGhee and P.H. von Hippel, Theoretical aspects of DNA–protein interactions: cooperative and noncooperativity binding of large ligands to a one-dimensional homogeneous lattice. *J. Mol. Biol.*, 1974, **86**, 469–489.
21. D.M. Crothers, Calculation of binding isotherms for heterogeneous polymers. *Biopolymers*, 1968, **6**, 575–584.
22. J.B. Chaires, Analysis and interpretation of ligand–DNA binding isotherms. *Methods Enzymol.*, 2001, **340**, 1–22.
23. J. Ren, T.C. Jenkins and J.B. Chaires, Energetics of DNA intercalation reactions. *Biochemistry*, 2000, **39**, 8439–8447.
24. S.C.M. Teixeira, J.H. Thorpe, A.K. Todd, H.R. Powell, A. Adams, L.P.G. Wakelin, W.A. Denny and C.J. Cardin, Structural characterisation of Bisintercalation in higher-order DNA at a junction like quadruplex. *J. Mol. Biol.*, 2002, **323**, 167–171.
25. J.B. Chaires, F. Leng, T. Przewloka, I. Fokt, Y.-H. Ling, R. Perez–Solar and W. Priebe, Structure-based design of a new bisintercalating anthracycline antibiotic. *J. Med. Chem.*, 1997, **40**, 261–266.
26. G.G. Hu, X. Shui, F. Leng, W. Priebe, J.B. Chaires and L.D. Williams, Structure of a DNA–bisdaunomycin complex. *Biochemistry*, 1997, **36**, 5940–5946.
27. F.A. Tanious, T.C. Jenkins, S. Neidle and W.D. Wilson, Substituent position dictates the intercalative DNA-binding mode for anthracene-9,10-dione antitumor drugs. *Biochemistry*, 1992, **31**, 11632–11640.
28. C. Zimmer and U. Wähnert, Nonintercalating DNA-binding ligands: specificity of their interaction and their use a stools in biophysical, biochemical and biological investigations of the genetic material. *Prog. Biophys. Molec. Biol.*, 1986, **47**, 31–112.
29. B.H. Geierstanger and D.E. Wemmer, Complexes of the minor groove of DNA. *Ann. Rev. Biophys. Biomol. Struct.*, 1995, **24**, 463–493.
30. P.R. Turner and W.A. Denny, The genome as a drug target: sequence specific minor groove binding ligands. *Curr. Drug Targets*, 2000, **1**, 1–14.
31. M.L. Kopka, C. Yoon, D. Goodsell, P. Pjura and R.E. Dickerson, The molecular origin of drug–DNA specificity in netropsin and distamycin. *Proc. Natl. Acad. Sci. USA*, 1985, **82**, 1376–1380.
32. M. Coll, C.A. Frederick, A.H.-J. Wang and A. Rich, A bifurcated hydrogen bonded conformation in the d(AT) base pairs of the DNA dodecamer d(CGCAAATTTGCG) and its complex with distamycin. *Proc. Natl. Acad. Sci. USA*, 1987, **84**, 8385–8389.
33. J.G. Pelton and D.E. Wemmer, Binding modes of distamycin A with $d(CGCAAATTTGCG)_2$ determined by two-dimensional NMR. *J. Am. Chem. Soc.*, 1990, **112**, 1393–1399.

34. D. Rentzeperis and L.A. Marky, Interaction of minor groove ligands to an AAATT/AATTT site: correlation of thermodynamic characterisation and solution structure. *Biochemistry*, 1995, **34**, 2937–2945.
35. A.N. Lane and T.C. Jenkins, Thermodynamics of nucleic acids and their interactions with ligands. *Quart. Rev. Biophys.*, 2000, **33**, 255–306.
36. P.B. Dervan, Design of sequence specific DNA-binding molecules. *Science*, 1986, **232**, 464–471.
37. J.W. Trauger, E.E. Bairde and P.B. Dervan, Recognition of DNA by designed ligands at subnanomolar concentrations. *Nature*, 1996, **382**, 559–561.
38. J.M. Gottesfeld, L. Neely, J.W. Trauger, E.E. Baird and P.B. Dervan, Regulation of gene expression by small molecules. *Nature*, 1997, **387**, 202–205.
39. D.E. Wemmer and P.B. Dervan, Targeting the minor groove of DNA. *Curr. Opin. Struct. Biol.*, 1997, **7**, 355–361.
40. S. White, J.W. Szewczyk, J.M. Turner, E.E. Baird and P.B. Dervan, Recognition of the four Watson–Crick base pairs in the DNA minor groove by synthetic ligands. *Nature*, 1998, **391**, 468–470.
41. D.S. Pilch, N. Poklar, E.E. Baird, P.B. Dervan and K.J. Breslauer, The thermodynamics of polyamide–DNA recognition: hairpin polyamide binding in the minor groove of duplex DNA. *Biochemistry*, 1999, **38**, 2143–2151.
42. P.B. Dervan and R.W. Bürli, Sequence-specific DNA recognition by polyamides. *Curr. Opin. Chem. Biol.*, 1999, **3**, 688–693.
43. P.B. Dervan and B.S. Edelson, Recognition of the DNA minor groove by pyrrole–imidazole polyamides. *Curr. Opin. Struct. Biol.*, 2003, **13**, 284–299.
44. A. Abu-Daya, P.M. Brown and K.R. Fox, DNA sequence preference of several AT-selective minor groove binding ligands. *Nucleic Acids Res.*, 1995, **23**, 3385–3392.
45. F.G. Loontiens, P. Regenfuss, A. Zechel, L. Dumortier and R.M. Clegg, Binding characteristics of Hoechst 33258 with calf-thymus DNA, poly[d(A-T)] and d(CCGGAATTCCGG): multiple stoichiometries and determination of tight binding with a wide spectrum of site affinities. *Biochemistry*, 1990, **29**, 9029–9039.
46. F.G. Loontiens, L.W. McLaughlin, S. Diekmann and R.M. Clegg, Binding of Hoechst 33258 and 4′,6-diamidino-2-phenylindole to self-complementary decadeoxynucleotides with modified exocyclic base substituents. *Biochemistry*, 1991, **30**, 182–189.
47. M.C. Vega, I. García–Sáez, J. Aymamí, T. Eritja, G.A. van der Marel, J.H. van Boom, A. Rich and M. Coll, Three-dimensional crystal structure of the A-tract DNA dodecamer $d(CGCAAATTTGCG)_2$ complexed with the minor groove binding drug Hoechst 33258. *Eur. J. Biochem.*, 1994, **222**, 721–726.
48. N. Spink, D.G. Brown, J.V. Skelly and S. Neidle, Sequence-dependent effects in drug–DNA interaction: the crystal structure of the Hoechst 33258 bound to the $d(CGCAAATTTGCG)_2$ duplex. *Nucleic Acids Res.*, 1994, **22**, 1607–1612.
49. I. Haq, J.E. Ladbury, B.Z. Chowdhry, T.C. Jenkins and J.B. Chaires, Specific binding of Hoechst 33258 to the $d(CGCAAATTTGCG)_2$ duplex: calorimetric and spectroscopic studies. *J. Mol. Biol.*, 1997, **271**, 244–257.
50. R.S. Spolar and M.T. Record Jr., Coupling of local folding to site-specific binding of proteins to DNA. *Science*, 1994, **263**, 777–784.
51. D. Praseuth, A.L. Guieysse and C. Hélène, Triple helix formation and the antigene strategy for the sequence-specific control of gene expression. *Biochim. Biophys. Acta*, 1999, **1489**, 181–206.
52. J.-S. Sun and C. Hélène, Oligonucleotide-directed triple-helix formation. *Curr. Opin. Struct. Biol.*, 1993, **3**, 345–356.
53. M.D. Frank-Kamenetskii and S.M. Mirkin, Triplex DNA structures. *Annu. Rev. Biochem.*, 1995, **64**, 65–95.
54. K.R. Fox, Targeting DNA with triplexes. *Curr. Med. Chem.*, 2000, **7**, 17–37.
55. T.C. Jenkins, Targeting multistranded DNA structures. *Curr. Med. Chem.*, 2000, **7**, 99–115.
56. J.L. Mergny, G. Duval–Valentin, C.H. Nguyen, L. Perrouault, B. Faucon, M. Rougée,T. Montenay-Garestier, E. Bisagni and C. Hélène, Triple helix-specific ligands. *Science*, 1992,**256**, 1681–1684.

57. H. Han and L.H. Hurley, G-quadruplex DNA: a potential target for anticancer drug design. *Trends Pharm. Sci.*, 2000, **21**, 136–142.
58. J.R. Williamson, G-quartet structures in telomeric DNA. *Annu. Rev. Biophys. Biomol. Struct.*, 1994, **23**, 703–730.
59. J.L. Mergny and C. Hélène, G-quadruplex DNA: a target for drug design. *Nature Med.*, 1998, **4**, 1366–1367.
60. D. Sen and W. Gilbert, Guanine quartet structures. *Methods Enzymol.*, 1992, **211**, 191–199.
61. P.J. Perry and T.C. Jenkins, Recent advances in the development of telomerase inhibitors for the treatment of cancer. *Exp. Opin. Invest. Drugs*, 1999, **8**, 1981–2008.
62. D. Rhodes and R. Giraldo, Telomere structure and function. *Curr. Opin. Struct. Biol.*, 1995, **5**, 311–322.
63. E.H. Blackburn, Structure and function of telomeres. *Nature*, 1991, **350**, 569–573.
64. A.M. Zahler, J.R. Williamson, T.R. Cech and D.M. Prescott, Inhibition of telomerase by G-quartet DNA structures. *Nature*, 1991, **350**, 718–720.
65. S. Neidle and G.N. Parkinson, Telomere maintenance as a target for anticancer drug discovery. *Nature Rev. Drug Disc.*, 2002, **1**, 383–393.
66. C. Kang, X. Zhang, R. Ratliff, R. Moyzis and A. Rich, Crystal structure of four-stranded *Oxytricha* telomeric DNA. *Nature*, 1992, **356**, 126–131.
67. Y. Wang and D.J. Patel, Solution structure of the human telomeric repeat $d[AG_3(T_2AG_3)_3]$ G-tetraplex. *Structure*, 1993, **1**, 263–282.
68. K. Phillips, Z. Dauter, A.I.H. Murchie, D.M. Lilley and B. Luisi, The crystal structure of a parallel-stranded quanine tetraplex at 0.95 Å resolution. *J. Mol. Biol.*, 1997, **273**, 171–182.
69. A. Kettani, S. Bouaziz, A. Gorin, H. Zhao, R.A. Jones and D.J. Patel, Solution structure of a $Na^+$ cation stabilized DNA quadruplex containing G·G·G·G and G·C·G·C tetrads formed by G-G-G-C repeats observed in adeno-associated viral DNA. *J. Mol. Biol.*, 1998, **282**, 619–636.
70. S. Haider, G.N. Parkinson and S. Neidle, Crystal structure of the potassium form of an *Oxytricha nova* G-quadruplex. *J. Mol. Biol.*, 2002, **320**, 189–200.
71. G.N. Parkinson, M.P.H. Lee and S. Neidel, Crystal structure of parallel quadruplexes from human telomeric DNA. *Nature*, 2002, **417**, 876–880.
72. S. Neidle and G.N. Parkinson, The structure of telomeric DNA. *Curr. Opin. Struct. Biol.*, 2003, **13**, 275–283.
73. O.Y. Fedoroff, M. Salazar, H. Han, V.V. Chemeris, S.M. Kerwin and L.H. Hurley, NMR-based model of a telomerase-inhibiting compound bound to G-quadruplex DNA. *Biochemistry*, 1998, **37**, 12367–12374.
74. M. Read, R.J. Harrison, B. Romagnoli, F.A. Tanious, S.H. Gowan, A.P. Reszka, W.D. Wilson, L.R. Kelland and S. Neidle, Structure-based design of selective and potent G-quadruplex-mediated telomerase inhibitors. *Proc. Natl. Acad. Sci. USA*, 2001, **98**, 4844–4849.
75. A. Siddiqui-Jain, C.L. Grand, D.J. Bearss and L.H. Hurley, Direct evidence for a G-quadruplex in a promoter region and its targeting with a small molecule to repress *c-MYC* transcription. *Proc. Natl. Acad. Sci. USA*, 2002, **99**, 11593–115.

CHAPTER 10

# Protein–Nucleic Acid Interactions

## CONTENTS

In this chapter, we will review the known principles of protein–nucleic acid interactions and explain how these molecules recognize and distinguish specific binding sites from all other potential binding sites in the cell. We will also discuss how nucleic acids are manipulated by macromolecular machines when they are packaged, copied, transcribed, translated, modified chemically, or transformed topologically.

## 10.1 FEATURES OF DNA RECOGNIZED BY PROTEINS

DNA structures may be classified (Section 2.2) according to idealized representations. For instance, in the B-form of duplex DNA (Figure 10.1 centre, see also Figure 2.17), the two polynucleotide strands wind around each other in anti-parallel directions, with the bases pairing within planes that lie at roughly right angles to the helical axis and the sugar–phosphate units forming repetitive linkages. To a first approximation, the striking helical character of the duplex results from the propensity of the bases to stack one on top of the other while accommodating the inextensible but flexible sugar–phosphate backbone.[1] Both the propensity for base stacking and Watson–Crick hydrogen bonding guide the successive pairing of bases in a highly co-operative manner, so that DNA strands that are several thousand base pairs in length can self-assemble spontaneously into stable duplexes. However, most other aspects of DNA assembly and recognition in living systems require the orchestrated actions of proteins.

The most conspicuous features of DNA that can be recognized by proteins and other ligands, such as drugs (see Chapter 9), are the **major and minor grooves** (Figure 10.1). These grooves arise from the helical geometry of the strands. In the B-form, the major groove is wide and accessible, whilst the minor groove is comparatively narrow. The grooves are hydrated canyons that can accommodate secondary structural elements of proteins. For instance, the major groove of the B-form is sufficiently wide (11.7 Å) and deep (8.8 Å) to accommodate an **α-helix** or two strands of a **β-ribbon**. In contrast, the minor groove of relaxed, undistorted B-form DNA is narrower (5.7 Å in width and 7.5 Å in depth on average) and is thus less accessible to secondary protein structures, although it is well suited to the insertion of a single peptide chain. The proteins that recognize DNA bury parts of their surfaces within these exposed grooves, sometimes through forces that deform the DNA shape, so as to optimise the complementarity of the molecular shapes.

Accessible within the major and minor grooves of the DNA are the **hydrogen-bond donors** and **acceptors** on the edges of the base pairs. A protein can contact these features directly or indirectly through one or more water molecules. It is also apparent that the negatively charged phosphate backbone may be a target for **electrostatic recognition**, and in stable complexes it is often matched by a positively charged surface of the protein. Associated with these phosphate groups is a highly mobile layer of counter-ions that

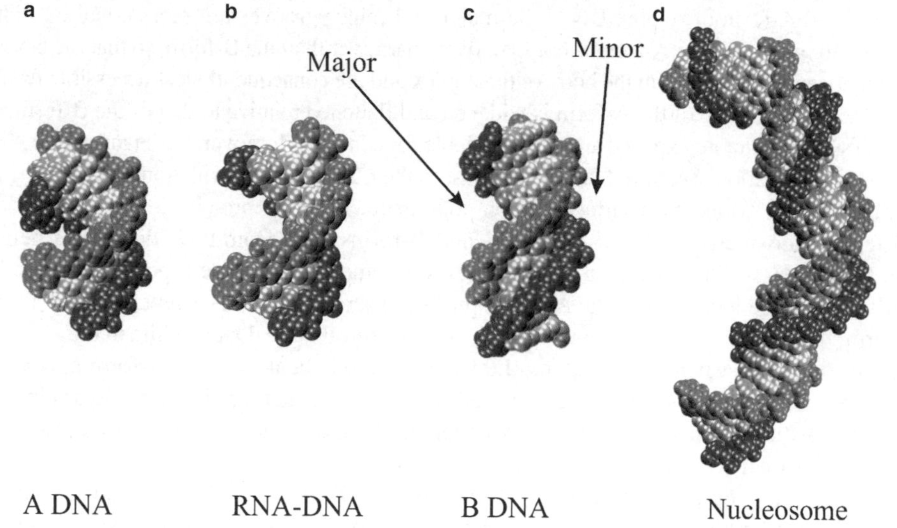


**Figure 10.1** *Space-filling representations of duplex nucleic acids. (a) The idealized A-form of DNA (PDB: 348D). (b) An RNA–DNA heteroduplex from the crystal structure of yeast RNA polymerase II (see Section 10.7.2) (PDB: 1I6H). (c) The idealized B-form DNA (PDB: 1ILC). (d) A segment of the DNA from the experimental structure of the nucleosome particle, illustrating its in-plane curvature (PDB: 1AOI). The sugar–phosphate backbone of each duplex is coloured red and blue, respectively, and the corresponding bases are pink and cyan. In the RNA–DNA duplex (b), the RNA backbone is red and the DNA backbone is blue. The arrows indicate the principal grooves*

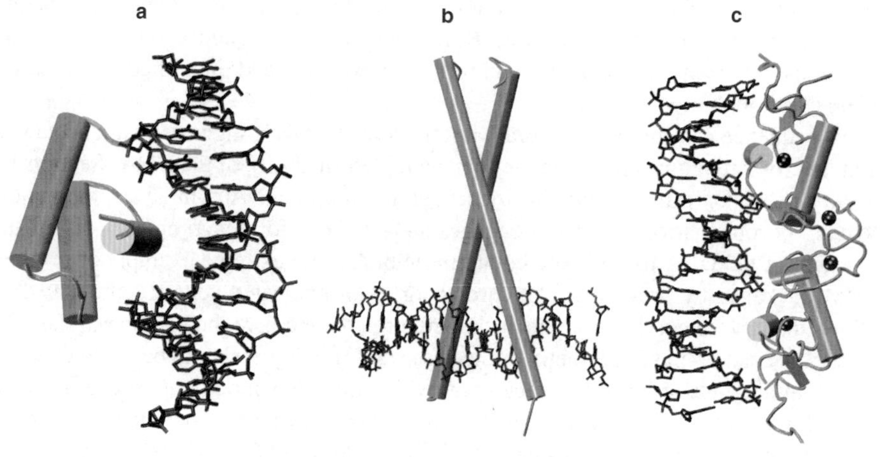


**Figure 10.2** *Representative helical motifs and their docking into the grooves of DNA. (a) The helix-turn-helix motif of the Drosophila engrailed protein (PDB: 1YSA). (b) Basic helix of the leucine zipper motif (PDB: 1YSA). (c) Glucocorticoid hormone receptor (PDB: 1GLU)*

partially neutralize the DNA charges and therefore contribute to duplex stability. These counter-ions are displaced when the complex forms and so they affect the free energy of binding.

Classic examples of protein/DNA complexes (Figure 10.2) illustrate how a dimeric protein engages with its DNA target. Both the protein and the DNA have **twofold symmetry**, often referred to as a **palindrome** on the DNA, and each monomer inserts an α-helix into the DNA major groove. Side chains from these helices (not shown here) make equivalent hydrogen-bonding interactions with the bases in the two symmetry-related DNA sites.

In the idealized A-form of duplex DNA, the major and minor grooves are more equal in width (Figure 10.1a). The major groove of the A-form is comparatively narrower than the B-form, so that the bases of the A-form are more deeply buried within the body of the duplex and are consequently less accessible from that side. In contrast, the minor groove of the A-form is widened and flattened relative to that of the B-form, so that the bases and the sugars are more exposed and thus available for contact. Moreover, the sugars in the A-form have the C3′-*endo* conformation (Section 2.2.2), in contrast to the C2′-*endo* conformation of the B-form, and they consequently have greater exposed surface for non-polar interaction with proteins in the minor groove.[2]

The different groove characteristics of the A- and B-forms arise from their distinctive helical geometries: the helical axis of the idealized B-form is straight and the structure repeats every ten base pairs (Figure 10.1) while the local axis of the A-form spirals in space and the helix repeats every 11 base pairs. In the B-form, the centres of the base pairs lie close to the central helical axis, while in the A-form they are displaced from it. Thus, when viewed along the length of the helical axis, the A-form appears to have a central channel, while the B-form is densely packed. The A-form is favoured under dehydrating conditions and by G:C rich sequences, and the transition between the A- and B-forms involves a sliding movement of the bases and adjustments in the backbone torsion angles.[3,4]

The A- and B-forms represent idealized conformations on the basis of fibre diffraction studies of DNA that have provided cylindrically averaged structural parameters (Section 2.3). More detailed stereochemical information has been obtained from high-resolution X-ray crystal structures of short segments of synthetic DNA. These structures show, for instance, that both grooves are well hydrated and contain intricately organized networks of hydrogen-bonding water molecules. Such bound water molecules are mostly displaced when a specific protein complex forms, but a small proportion may be retained in the interface to fill gaps because of imperfections in matching the surfaces (Section 10.4). An important finding from X-ray crystal structure analyses is that there are significant structural variations in duplex DNA, such that local structures often hardly seem to resemble either of the idealized A- or B-forms, but instead something intermediate between these forms (Section 2.3). The structural variations are related to the underlying DNA sequence itself because of the stacking preferences of the bases. Rules relating DNA sequence to conformation have been worked out qualitatively from mechanical principles,[1] and extended and quantified on the basis of chemical principles[5] (Section 2.3).

The fact that sequence can impart structural variations of the DNA suggests that a sequence could be recognized indirectly through its effects on the shape and deformability of the DNA (Section 10.4). As an example of the type of conformational adjustment seen in protein–DNA complexes, the major groove is often observed to become narrower when it engages an $\alpha$-helix to form a complex. This creates a gentle in-plane curvature of the DNA towards the bound protein. A more extreme example of in-plane bending occurs in DNA packaged by proteins in the chromosomes of nucleated cells (Figure 10.1d). The DNA segment follows the surface curvature of the packaging protein, and here the bending of the DNA is associated with periodic changes in the geometry of the base steps. Finally, it should be noted that a DNA-binding protein seldom amounts to a rigid surface upon which the DNA moulds its shape: in the process of forming a complex with DNA, portions of the protein surface may also adjust their conformation. Thus, recognition often involves a mutually **induced fit** of both protein and DNA (Section 10.4). Similar effects also occur in the formation of most RNA–protein complexes (Section 10.9).

In cells, RNA is often found in single-stranded form, and may fold into compact structures, such as an RNA hairpin (Figure 10.3a), especially if it contains regions of self-complementarity. The segment of RNA that separates the complementary sequences forms a loop in which the bases are exposed. Such a loop often acts as a target for protein recognition, since the unpaired bases are more readily accessible for contacts with amino acid residues than those in paired duplex regions: see for example the interaction of the protein U1 with an RNA hairpin (Figure 10.3a). RNA bases may also become exposed by internal bulges, which are formed wherever extra bases are accommodated on one strand within a duplex RNA region.

Duplex regions of RNA resemble the A-form of DNA, and consequently these regions have a deep major groove in which the bases are not readily accessible. Heteroduplexes of RNA and DNA, such as those formed transiently when genes are transcribed by RNA polymerase, also have a structure similar to

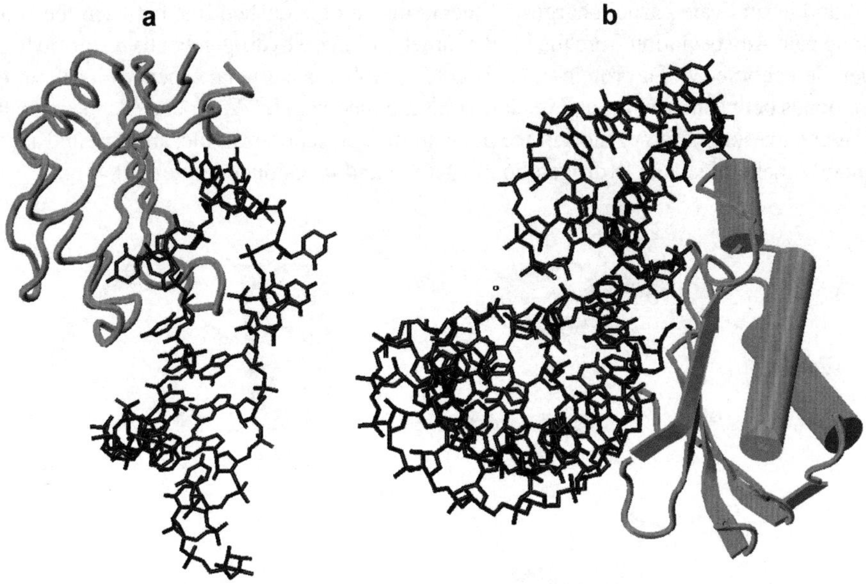


**Figure 10.3** *Representative complexes of protein with RNA. (a) An RNA hairpin, in complex with the RNP protein U1 from the eukaryotic intron-splicing machinery (PDB: 1URN). The bases of the single-stranded loop are recognized by contacts with the protein side chains. (b) A view of a section of the ribosome, showing an even more complicated RNA/protein interaction (PDB: 1IBM). A β-strand lies in the groove of the duplex RNA and the bases in a single-stranded region are contacted by an α-helical region*

A-form DNA (Figure 10.1a). These duplexes may be recognized more by shape and groove width than by specific hydrogen-bonding contacts. An example from the bacterial ribosome shows how a β-strand lies in the groove of a duplex RNA, while the bases in a single-stranded region are contacted by an α-helical region of the protein (Figure 10.3b). The intricate three-dimensional folds of RNA add to its ability to make an induced fit in recognition of protein, for example, during the charging of transfer RNA with amino acids by tRNA synthetases (Sections 7.3.2 and 10.9.3).

## 10.2 THE PHYSICAL CHEMISTRY OF PROTEIN–NUCLEIC ACID INTERACTIONS

Since the 1990s, many hundreds of crystal and NMR structures have been solved for protein–DNA and protein–RNA complexes. These structures illustrate the diversity of nucleic acid–protein recognition patterns. To understand such complicated molecular architectures, we need to explore the forces that hold the components together.

### 10.2.1 Hydrogen Bonding Interactions

Although energetically rather weak, **hydrogen bonding interactions** are a key component of protein–nucleic acid recognition, since collectively they confer sequence-specificity in the majority of cases. In solution, all exposed polar groups form hydrogen bonds with water molecules. When the solvent is removed from these groups, as happens in the formation of a protein/nucleic acid complex, they must make compensating interactions with other polar groups. The energy of the hydrogen-bonding interaction is optimal when the atoms of the **donor** and **acceptor** groups (X, H and Y) and the lone electron pair are all arranged linearly, as indicated in the scheme

$$X^{\delta-}—H^{\delta+}—Y^{\delta-}—R^{\delta+}$$

Here, the $\delta^-$ and $\delta^+$ indicate partial charges on the atoms and the dashed line between the H and Y is the axis of the lone pair. Any deviation from the ideal geometry of these hydrogen bonds results in higher energy. Thus, the detailed geometry of a protein–nucleic acid complex is a crucial aspect of its stability.

Hydrogen bonds between proteins and nucleic acids are mediated by donor and acceptor groups from the bases of DNA and the side chains of most of the polar amino acids. For example, arginine and asparagine side chains frequently make bidentate hydrogen bonds with G and A residues (Figure 10.4a). Such interactions

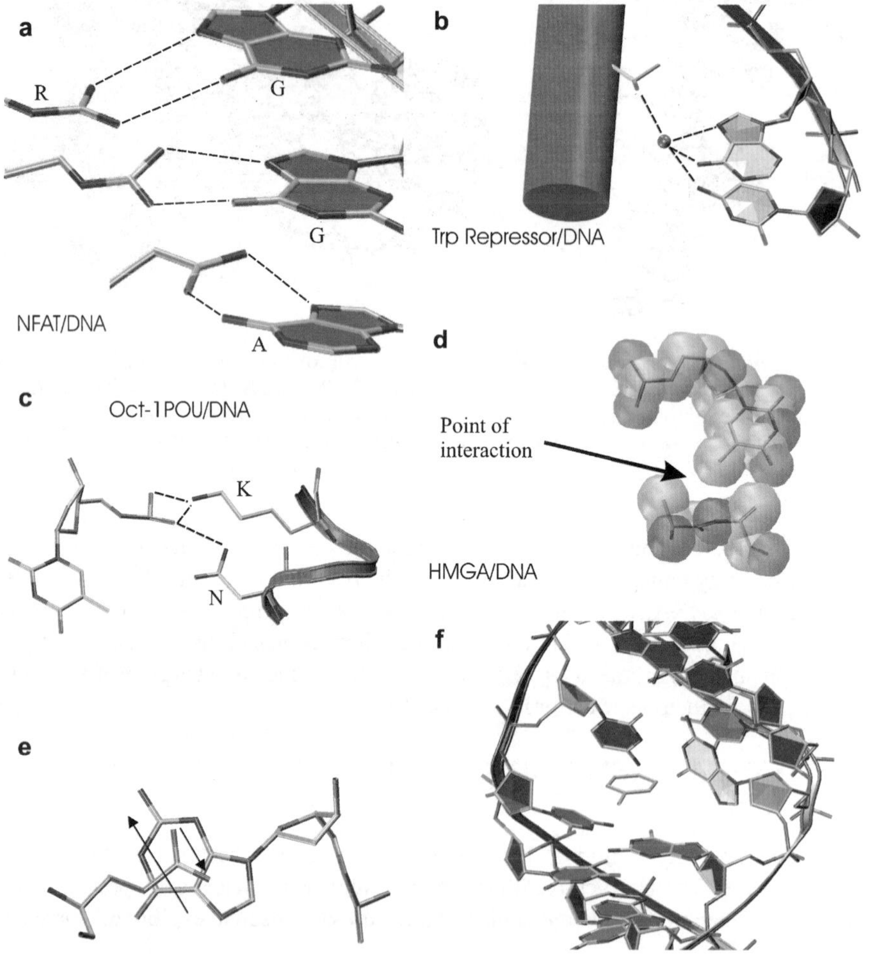


**Figure 10.4** *Representative interactions between side chains, water molecules and nucleic acids. (a) The arginine and asparagine (or glutamine) side chains can make bidentate hydrogen-bonding interactions with G and A, respectively (PDB: 1AO2). Hydrogen-bonding interactions are indicated by the dashed black lines. (b) Water molecules provide a non-covalent extension of the surface of the DNA-binding protein; here water (red) bridges between the hydroxyl group of a threonine side chain and the major-groove amino groups of a C/A base step in the trp repressor/operator complex (PDB: 1TRO). (c) Salt bridge and hydrogen-bonding interaction with the non-esterified phosphate oxygen atoms (PDB: 1OCT). (d) Recognition from contacts of non-polar atoms. Two alanine residues contact the 5-methyl carbon of a T in the HincII/DNA complex (PDB: 1CKT). The atoms are in space-filling representation, where carbon atoms are cyan, nitrogen atoms blue, oxygen red and phosphorus yellow. (e) Another intercalation interaction, where a glutamine side chain stacks onto a G base in the HincII restriction enzyme/DNA complex (PDB: 1KC6). The interaction may align favourably the permanent dipoles, indicated by the black arrows. (f) The intercalation of an aromatic residue between unstacked bases in the A domain of HMGB/DNA structure (PDB: 1KC6). Only the phenylalanine side chain of the protein is shown for clarity*

may be partially conjugated and can contribute to stability by 'stacking' like plates, akin to the stacking seen for the bases in the nucleic acid. Much less frequently, weaker hydrogen bonds can form between the DNA bases and the aromatic rings of Phe, Tyr and Trp and these are called **'π-hydrogen bonds'** (not shown). Water molecules can extend the surface of the DNA and bridge to an amino acid, for example the mediation of interaction between a Thr hydroxyl group and the major groove exocyclic amino groups at a C/A base step (Figure 10.4b). In addition, the amide backbone of a protein may participate in hydrogen-bond interactions with the nucleic acid, as either a donor (N—H) or acceptor (C═O) (not shown).

From the nucleic acid bases, the carbonyl oxygen atoms of G, C and T (U in RNA) are commonly acceptors, as are also the heterocyclic N-7 and N-3 atoms of purines. Potential donors are the exocyclic amino groups of A, G and C and the amido N—H functions of T (U in RNA), G and C. In the case of unpaired bases, the N-1 of purines and the N-3 of pyrimidines often form hydrogen bonds with protein. It might seem surprising that the most common site of hydrogen-bonding interaction in the recognition of a nucleic acid is not with the bases, but with the non-bridging oxygen atoms of phosphates. An example of this frequently occurring interaction is shown in Figure 10.4c. The bridging oxygen atoms of the phosphate backbone and the oxygen atom within the ribose sugar ring may act as acceptors of hydrogen bonds, but these are comparatively weaker interactions.

### 10.2.2 Salt Bridges

**Salt bridges** are electrostatic interactions between groups of opposite charge. In protein–nucleic acid complexes, salt bridges may be formed between the ionized, non-bridging oxygen atoms of the phosphate backbone and the protonated, positively charged guanidinium moiety of Arg, the imidazole ring of His, the ε-amino group of Lys, or the terminal α-amino group in a protein (*e.g.* Figure 10.4c). The importance of salt bridges in protein stability has been inferred by their frequent occurrence in the heat-stable proteins isolated from extreme thermophilic organisms, where the salt bridges are found in co-operative networks. Isolated salt bridges are entropically disfavoured because they place constraints on the side-chain conformations and cause desolvation of the interacting side chains. However, networks of mutually supporting salt bridges are favourable, since the co-operativity of their interaction compensates for the unfavourable entropy change. This principle may also be true for salt bridges formed between amino acid side chains and nucleic acid phosphate groups, which are often organized as intricate networks of mutually supporting interactions. Another type of salt bridge that is also observed in protein–nucleic acid complexes is an electrostatic interaction mediated by bound counter-ions, such as chloride ions.

Unlike hydrogen bonds, which require alignment of the interacting groups for optimal strength, salt bridges and counter-ion interactions are not directional. Salt bridges vary in strength in inverse proportion to the square of the distance between the individual charges and to the dielectric constant, so that they are greatly weakened as ionic strength increases and strengthened as the dielectric constant decreases. If salt bridges or bound counter-ions become buried and isolated from bulk solvent, their interaction energy is enhanced, especially if they are sequestered into a hydrophobic environment. Thus, the strength of salt bridges is highly context-dependent, but a typical value in proteins corresponds to about 40 kJ mol$^{-1}$. Salt-bridging interactions play an important role in making a general electrostatic match of the polyanionic surface of the DNA or RNA with the polycationic surface of the recognizing protein. Thus, a charged surface of the protein may be used as a rough gauge of the groove width or to discriminate single-stranded nucleic acids from double-stranded. However, electrostatic complementarity is unlikely to provide specificity by itself. Indeed, specificity must originate from other effects, such as the pattern of hydrogen bonds described above and complementary surface shapes (Sections 10.2.3 and 10.2.4).

### 10.2.3 The Hydrophobic Effect

The greatest contribution to the Gibbs free energy of complex formation is that of the **hydrophobic effect**. This short-range effect is sometimes referred to as a force. However, it is not a real force in the classical sense, with an associated scalar field, such as found for the electric forces in which two charged particles attract or repel each other. Instead, the hydrophobic effect is a consequence of the behaviour of bulk water at a non-polar surface.[6] In its optimal orientation, a water molecule can form four hydrogen bonds in a tetrahedral geometry: two through its lone pairs of electrons (the acceptors) and two through its protons (the donors). In bulk solvent at physiological temperatures and pressures, the water molecules are in continual exchange of hydrogen-bonding partners. When the water molecules encounter a non-polar molecule that cannot accept or donate, these hydrogen bonding interactions are broken. If two such non-polar molecules are in proximity, they will associate so as to minimise the disruption to the dynamic water interactions. In general, macromolecules such as proteins and nucleic acids tend to associate in such a way as to expose a minimum number of non-polar features on the external surface to the aqueous solvent. The non-polar parts of nucleic acid include the 5-methyl group of thymine, the heterocyclic carbon atoms within the purine and pyrimidine rings, and the ribose carbon atoms. These atoms are often contacted by the aliphatic portions of amino acid side chains in protein–nucleic acid complexes (*e.g.* Figure 10.4d).

The association of hydrophobic molecules in aqueous solvent is favoured **entropically**. This might seem surprising given that a hydrogen-bonding pattern is being affected and that bond breaking is associated mostly with **enthalpic** change. However, it is thought that water molecules become immobilized on non-polar surfaces because of restricted interactions with partner water molecules. Thus, by burying the exposed non-polar surface, water molecules are liberated and their translational and rotational entropy increases, with a concomitant lowering of the overall free energy. In this way, shape complementarity can help to bury non-polar surfaces optimally and to contribute to the affinity of a macromolecular interaction. Shape complementarity also favours optimal van der Waals interactions (Section 10.2.4).

### 10.2.4 How Dispersions Attract: van der Waals Interactions and Base Stacking

**Van der Waals interactions** arise from the **dispersive** forces originating from transient dipoles inherent in the electronic structure of atoms and molecules, and even molecules without a net permanent dipole can form these interactions. Dispersive forces are proportional to the inverse sixth power of the distance between the dipoles. Although the interactions are very weak and sensitive to structural fluctuations, their cumulative contribution becomes significant over an extensive intermolecular interface. In a specific protein–DNA complex, van der Waals contacts account for roughly two-thirds of the interfacial surface area, whilst the remaining one-third involves direct or water-mediated hydrogen-bonding interactions.[7] Thus, the contacting protein and nucleic acid surfaces in specific complexes are not only characterized by the stereochemical match of donors and acceptors, but also by the shape complementarity that produces a snug fit so as to benefit from the $r^{-6}$ law.

Dispersive effects also contribute to the stability of **base-stacking interactions** and thus are an important aspect of the sequence-dependent conformational effects mentioned earlier (Section 2.3.1). As a consequence of stacking and base-pair hydrogen bonding, the packing density of atoms within duplex DNA is greater than for equivalent atoms in related small molecules packing in crystalline lattices.[8] As a consequence of the hydrophobic effect, the bases tend to **propeller-twist** in the horizontal plane so as to optimize the buried surface area between neighbouring base steps. However, this feature is not sufficient in itself to account fully for the observed sequence-dependent conformational variations in duplex DNA. Aside from the general effect of hydrophobicity on the general shape of the molecule, electrostatic effects are important components of DNA conformation, and they tune the detailed geometry of the base stacking. Each of the aromatic atoms of the bases can be treated as sandwiches of charges, from which the conformational propensities of the heterocyclic bases can be calculated based on the optimized electrostatic interactions (Section 2.3.1). When combined with a simulated 'backbone link', these calculations can account for many of the experimentally observed conformations of different base steps.[5]

**Dipolar effects** also contribute to DNA conformation and protein–DNA interactions, but the importance of this effect is difficult to quantify since it depends greatly on the organization of fixed charges and on the dielectric constant. For example, a permanent dipole is associated with the G bases because of the concentration of electronegative atoms on the major groove side and electropositive atoms on the minor groove side. The dipole of an asparagine or glutamine side chain can match the local dipole of the nucleotide (Figure 10.4e). Dipoles also affect the stacking of the bases in G/C rich sequences. To a first approximation, the G·C base pair has a large dipole along its long axis resulting from the cluster of partial negative charge on the G and partial positive charge on the C. Dipoles align to maximize attraction and minimize repulsion, and it can be envisaged how the sandwiched G·C dipoles in a sequence like GGGG/CCCC favour high slide, whilst low slide is preferred for an alternating sequence such as GCGC/CGCG. In contrast, the A·T base pair contains only small patches of isolated positive and negative electrical charge along its long axis, which are relatively dispersed over the entire pair, and therefore do not amount to a substantial dipole.

Stacking interactions can also occur between aromatic residues in a protein and unstacked bases, for example in the intercalation of a Phe side chain between two unstacked nucleotides (Figure 10.4f).

## 10.3 REPRESENTATIVE DNA RECOGNITION MOTIFS

### 10.3.1 The Tree of Life and its Fruitful Proteins

Protein folds that are used in recognition of DNA (and RNA) may be grouped according to their structural and evolutionary relationships. But before grouping such folds into classes, it is useful to be aware of how organisms are grouped. All cellular life on Earth can be organized into three domains: the eukaryotes, the bacteria and the archaea. Eukaryotes and archaebacteria may have shared a common ancestor during a period of evolutionary history following the divergence from bacteria.[9] During the subsequent evolution of the archaea, there was a split into the sub-domains known as the *crenarchaeota* and the *euryarchaeota*. While eukaryotes and bacteria have many distinguishing features, the archaea share certain molecular similarities with both these groups. For instance, the DNA of eukaryotes is packed tightly into cellular nuclei, while bacteria and archaea both lack nuclei. On the other hand, the archaea control the expression of their genes by mechanisms that are more similar to those of the eukaryotes, and they use similar proteins in the initiation of transcription. In all three taxonomic domains, the core of the RNA polymerase is conserved in structure and function (Section 10.7.2). The packaging of DNA is important for all cells (Sections 6.4 and 10.6), and the *euryarchae* contain histone proteins that are similar to those that compact DNA in the nuclei of eukaryotes. In contrast, the *crenarchae* lack histones but have two DNA-binding proteins that compact relaxed or supercoiled DNA.[10]

In all organisms, gene expression is highly coordinated and regulated. In bacteria and archaea, related genes are clustered together in the genome, and a single DNA-binding protein is often used to control the transcription of individual clusters (Section 6.1). Representative examples in bacteria are the lac, met and trp repressors. By contrast, in the eukaryotes several 'transcription factors' may act together in different combinations, so that specificity for individual genes is achieved through protein–protein interactions. Another distinguishing aspect of eukaryotic genes is that their activities are generally repressed by chromatin structure. Thus, covalent modification of histones by acetylation and methylation plays an important role in gene expression by affecting the access of the chromatin to transcription factors.

DNA-binding domains in proteins from all of these taxonomic domains of life can be classified into the order of dozens of distinct groups according to recurring structural motifs. The structural element employed most frequently is the **α-helix**. Less frequently found are **β-ribbons**, **β-sheets** and **loops**. The major groove of DNA is most often the target for recognition. But in some cases recognition is mediated by binding in the minor groove. An example of this is the DNA complex of the TATA-binding protein (TBP) (Section 6.6.3). Some of the motifs are very ancient and can be found in archaea, bacteria and eukaryotes alike. Others are clearly more recent, for instance the homeodomain transcription factor which occurs only in higher eukaryotes (Figure 10.2a). Here, we describe a representative selection of these various folds. Interactive websites provide more details of many of the known motifs (http://www.biochem.ucl.ac.uk/bsm/prot_dna/prot_ dna_cover.html).

### 10.3.2 The Structural Economy of α-Helical Motifs

*10.3.2.1 The Helix-Turn-Helix Motif.* The **helix-turn-helix** motif was first characterized structurally in regulatory proteins of bacteria and bacteriophage and appears to be both ancient and widely occurring. Remarkably, many proteins that control transcription in eukaryotes also use a similar helix-turn-helix reading head to that found in bacteria, and this is likely to have arisen by convergent evolution. A classical example is the eukaryotic homeodomain, which forms a widely occurring family of factors whose members are often involved in controlling the expression of genes of development, such as the engrailed homeodomain from the fruit fly *Drosophila melanogaster* (Figure 10.2a).

Another important transcription factor that contains a helix-turn-helix motif is the eukaryotic transcription factor TFIIB, which is a component of a complex that binds roughly 30 bases ahead of the start site for transcription. This highly conserved protein is also found in archaea (where it is known as TFB). TFIIB forms a complex with the TBP (TATA Binding Protein) in which the DNA is highly distorted (Figure 10.5). The helix-turn-helix motif recognizes a short element known as the 5′-BRE that precedes the TATA element (Section 6.6.3), where the pseudo-symmetric TBP binds, and thus defines the direction of transcription.

There are many variations on the theme of the helix-turn-helix fold that are found in both eukaryotes and prokaryotes. It seems clear from the representative cases that the helix-turn-helix motif has evolved many times within completely different protein frameworks, and, as noted by Janet Thornton, the motif is likely to represent the convergent evolution of a stable solution to the problem of DNA recognition.

*10.3.2.2 Leucine Zippers.* The zipper proteins are dimeric transcription regulators that insert α-helices into the major groove of duplex DNA. One example is the protein domain called bZIP (Figure 10.2b). The bZIP domain, like many other transcription factors, binds as a **homodimer** to a palindromic DNA sequence. bZIP has two protruding α-helical DNA 'reading heads', and the motif from each monomer is buried in the DNA major groove. The two reading heads make symmetry-equivalent interactions with base pairs in two successive major grooves (**half-sites**). The dimer interface fixes the relative orientation of the two reading heads so that each motif reads the same short DNA sequence on each of the two strands of the palindromic site.

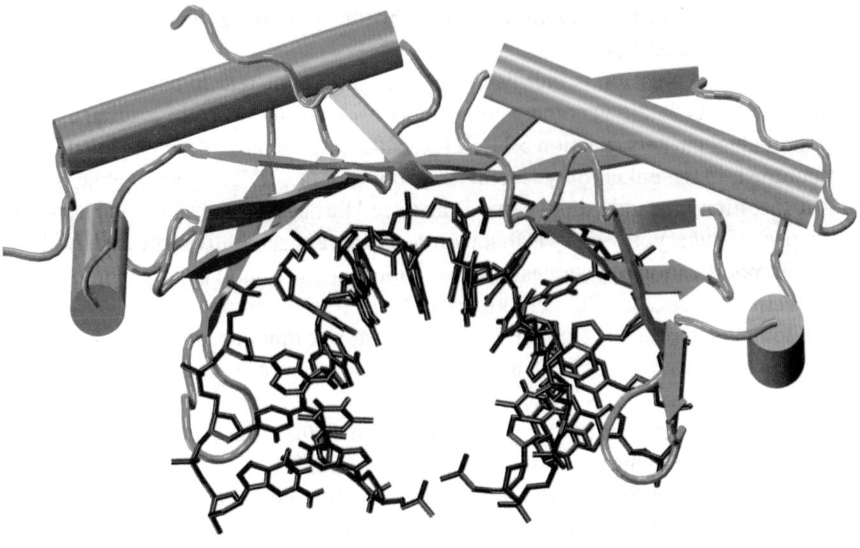

**Figure 10.5** *The human TBP-protein binding to the TATA element (PDB: 1IFH). The protein has pseudo-dimeric symmetry because of gene duplication of an ancestral precursor. The structure of the archaeal TBP is very similar*

The bZIP domain has a long pair of identical helices, which coil gently around one another in a left-handed sense. This creates a dimer interface known as a **leucine zipper** because leucines from each of the α-helices intermesh along a line of contact that is co-axial with the long axis of the helical pair. These amino acids are like protruding knobs on each α-helix that fit into complementary hydrophobic pockets on the partner helix. Such favourable 'knobs-into-holes' hydrophobic interactions hold the two parts together like a zipper. Because the α-helix has 3.6 amino acids in a helical turn, the leucines form a left-handed spiral on the surface of the α-helix. Consequently, the hydrophobic regions of those two helices match and maintain perfect register if the helices coil around one another in a gentle left-handed supercoil. This packing was deduced by Francis Crick from geometric principles more than 50 years ago.

There are many structural variations of the zipper. For instance, the zipper architecture is elaborated in the basic helix-loop-helix proteins, as represented by the developmental controlling factor Max. The dimerization interface of the zipper proteins can support homodimers and heterodimers and thus tune their binding affinity and specificity, and hence their repertoire for genetic regulation. Zipper transcription factors appear to occur exclusively in eukaryotes, and they are widely distributed within this taxonomic domain and are found in yeast, plants, insects and vertebrates. In multi-cellular organisms they often play roles in tissue-specific gene regulation.

### 10.3.3 Zinc-Bearing Motifs

Collectively, the zinc-bearing transcription factors form the largest group of DNA-binding proteins in the eukaryotes. Their abundance may be due to their simple, structurally economic folds, which are supported by metal binding. Further, their modularity may have allowed them to be shuffled and combined readily in the course of evolution to generate new proteins with different specificity and function.

*10.3.3.1 The β–β–α Zinc Fingers.* The first of the zinc-bearing DNA-binding motifs to be discovered were the '**zinc fingers**'. Transcription factors containing the zinc finger motif are found in a wide range of eukaryotic organisms, where they play a number of different genetic regulatory roles. In the human genome, roughly 3% of the genes that encode proteins specify zinc fingers.

The canonical structure of the zinc finger is composed of a peptide loop containing a two-stranded β-hairpin and an α-helix. The fold is stabilized by a hydrophobic core and coordination of $Zn^{2+}$ *via* the sulfur atoms of two cysteines and the imidazole nitrogen atoms of two histidines.[11] The inclination of the helices of the fingers in the transcription factor known as 'Zif268' allows stacking of side chains of the helix and loop regions, which are aligned to interact with the bases in the DNA major groove.[12] The modular finger structure permits the protein to track along the groove – mainly to one side of it – and to follow the helical trajectory.

*10.3.3.2 The Steroid and Nuclear Receptors.* Organisms as diverse as insects and vertebrates have transcription factors that are activated by small, hydrophobic signalling ligands, such as hormones. These proteins, known as the **steroid and nuclear receptors**, use one structural domain to recognize the signalling ligand and a second domain to recognize specific DNA sequences in the vicinity of the target gene. Representative examples are the receptors for estrogens, mineralocorticoids, progesterone, glucocorticoids and vitamin D (Figures 10.2c). Also in this class are the 'orphan' receptors which have no known ligand.

In these receptor proteins, the DNA-binding domain contains eight cysteines that form two peptide loops that each cap amphipathic α-helices. There is an imperfect structural repeat of the zinc-loop-α-helix, but these repeating units do not form independent finger-like structures. Instead, they associate through hydrophobic interaction of the two amphipathic α-helices to form a globular domain. The α-helix exposed by the amino-terminal module rests in the major groove of the DNA target, for example in the glucocorticoid receptor, where it directs base contacts. These receptors often bind to DNA as homo- or hetero-dimers and recognise the spacing between adjacent sites through protein–protein interactions (Section 10.4.6).

### 10.3.4 The Orientations of α-Helices in the DNA Major Groove

In Sections 10.3.2 and 10.3.3, a handful of representative examples of protein–DNA complexes were described in which an α-helix is docked snugly into the major groove. While it might seem that a snug fit implies restricted orientation, there is in fact a wide range of orientations observed for the α-helix within the major groove. Indeed, there may be great variation in the orientation of the recognition α-helix in the DNA even within a domain class whose members have the same fold and very similar amino acid sequences. Pabo and Nekludova[13] found a way to explain this observed variation by grouping complexes into two principal families according to how the residues along a '**ridge**' of the α-helix align with respect to the spiralling trajectory of the DNA. In one family, every fourth residue is aligned (*i.e.* the ridge formed by residues along the $i$, $i+4$ line) to maintain register with the contacting bases, whilst a second family uses the line connecting every third residue ($i$, $i+3$). For instance, the zinc fingers tend to use the $i$, $i+3$ geometry. Because the helical repeat of the relaxed α-helix is 3.6 residues in one complete turn, the $i$, $i+3$ line will form a left-handed spiral on the helix surface, while the $i$, $i+4$ line will form a right-handed spiral. Thus, the two families use different geometries to match the spiralling pattern of amino acids with the curving path of the major groove. These matches are only local, and typically correspond to contact with three to five bases.

There are interesting outliers to this grouping, such as the trp repressor and the GAL4 protein, where the α-helices tend to enter the major groove at a steep angle because of additional contacts made to the DNA by other structural elements of the protein. While these family groupings indicate the importance of matching the geometry of the α-helix with the surface curvature of the DNA, they do not provide detailed rules for the interaction, since the precise helix orientation is context-dependent.

### 10.3.5 Minor Groove Recognition *via* α-Helices

The docking of an α-helix into the major groove of DNA causes the groove to become slightly compressed, on account of the tendency of the phosphate backbones of the DNA to clamp down on the α-helix. The resulting compression of the major groove tends to bend the DNA towards the protein. By contrast, the accommodation of an α-helix into the narrow minor groove of B-form DNA tends to expand the groove tremendously, with the result that the DNA bends away from the body of the protein. Representative examples of minor groove recognition are provided by the widely occurring and well-conserved high-mobility group (HMG) proteins, which play structural roles in the dynamic organization of chromatin (Section 10.6.1).

### 10.3.6 β-Motifs

The specific recognition of DNA by the *Escherichia coli* met repressor protein resembles closely α-helix-DNA recognition, except that the amino acids, which directly contact the base pairs are from a **β-sheet** (which may be made up from a number of **β-strands**) that make a snug fit into the major groove (Figure 10.6a). With this mode of interaction, the met repressor protein recognises a conserved eight-base-pair sequence of DNA (5′-A**GACG**TCT), although only four of the eight individual bases are contacted (indicated in bold). The adjacent G and A bases form hydrogen bonds with lysine and threonine residues from different strands of the β-sheet, while the outer two bases A and C in AGAC are not contacted directly by the protein. Several phosphates on the DNA interact with various amino acids from the protein.

The met repressor protein has twofold rotational symmetry and it recognizes a palindromic DNA sequence, just as in the DNA complexes of the 434 repressor and bZIP proteins (Figure 10.2b). But in addition, the met-repressor dimers can bind to adjacent sites on the DNA, and the neighbouring dimers interact favourably through protein–protein interactions. This gives a beneficial co-operativity of binding (Section 10.4.6).

Another protein that uses a β-ribbon to contact the DNA is the integration host factor (IHF), which is a 20-kDa heterodimeric protein from *E. coli* that binds DNA with sequence-specificity and induces a sharp bend in the DNA, of greater than 160°.[14,15] IHF functions as an architectural protein that assists assembly of replication complexes and affects long-distance transcriptional regulation. The subunits of the heterodimer

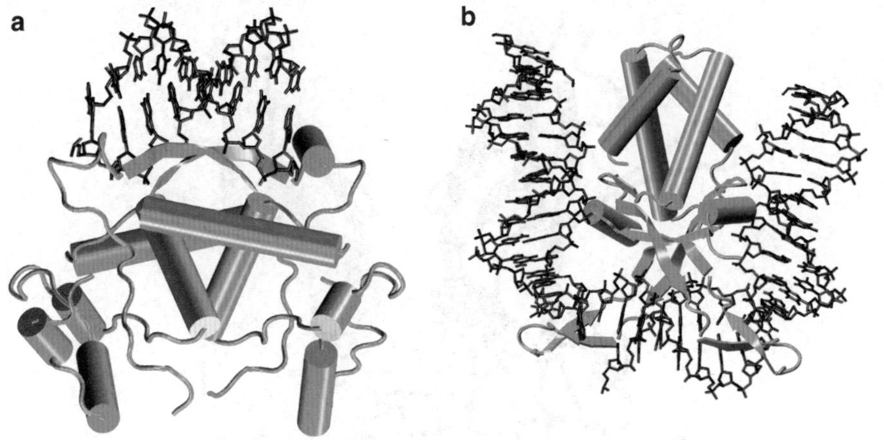

**Figure 10.6** *Representative β-motifs and their docking into the grooves of DNA. (a) The E. coli met repressor (PDB: 1TGH). Only a single dimer complex is shown in complex with a 'met box', but the repressor binds as a co-operative oligomer of dimers to adjacent met boxes (PDB: 1CMA). (b) The integration host factor (IHF) from bacteriophage. Here, two β-ribbons engage the minor groove*

are intertwined in a compact body from which two β-ribbons extend and interact with the DNA (Figure 10.6b). Unlike those of the met repressor, the β-ribbons of IHF interact with the DNA in the minor groove. Conserved proline residues from the tips of the β-ribbon intercalate between base steps to support the DNA trajectory. The protein binds a 35 base pair target, and although there is little apparent consensus between the sequences of the binding sites, IHF prefers to intercalate at pyrimidine–purine steps. IHF binds preferentially to these sites $10^3$–$10^4$ times more strongly than to random DNA sites. Pyrimidine–purine steps have the greatest flexibility of the dinucleotide steps (Section 10.4.5), and it seems that IHF may recognize these sites indirectly through their ability to become conformationally distorted. The IHF protein interacts extensively with the phosphate backbone and the minor groove of the DNA, where only three amino acid side chains form hydrogen bonds with the DNA bases. However, these hydrogen-bonding interactions contacts can be made with an arbitrary sequence. Thus, IHF recognises DNA sites through sequence-dependent structural deformation but not through base-specific contacts.

The TBP binds to the TATA element that is 27–30 bases upstream of the start site of many eukaryotic and archaea genes. This protein also uses a β-sheet as the means of recognition of DNA, like the met repressor and IHF, except that the sheet is fitted into a greatly widened minor groove, rather than the major groove (Figure 10.9, Section 10.4.6). While IHF lies on the concave side of the bent DNA, TBP lies on the convex side of its deformed target. In binding its target site, TBP unwinds and sharply bends the TATA element. The conserved β-sheet of TBP forms a saddle-shaped surface that matches the curvature of the exposed bases in the deformed minor groove. A similar distortion is seen in the crystal structure of the HMG protein bound to a TATA-like element, although an α-helix is used in that complex.[16]

The TBP/DNA interaction is highly conserved, as can be seen in the similar structures of the TBP/DNA complexes from plants, animals and archaea. The TBP protein is used as a scaffold for the recruitment of other general transcription factors, such as TFIIB (Figure 10.9b, Section 10.4.6). It is curious that such extreme distortion of the DNA by TBP and its associated proteins occurs and is so highly maintained in evolution. The distortion appears to be required to orient the DNA so that it enters the active site of the RNA polymerase, and it might facilitate unwinding of the DNA at the transcription start site (Section 10.7.2).

### 10.3.7 Loops and Others Elements

A polypeptide does not necessarily need a defined secondary structure to interact with DNA. For example, loops from certain DNA-binding proteins are used in the recognition of defined DNA sequences. Often

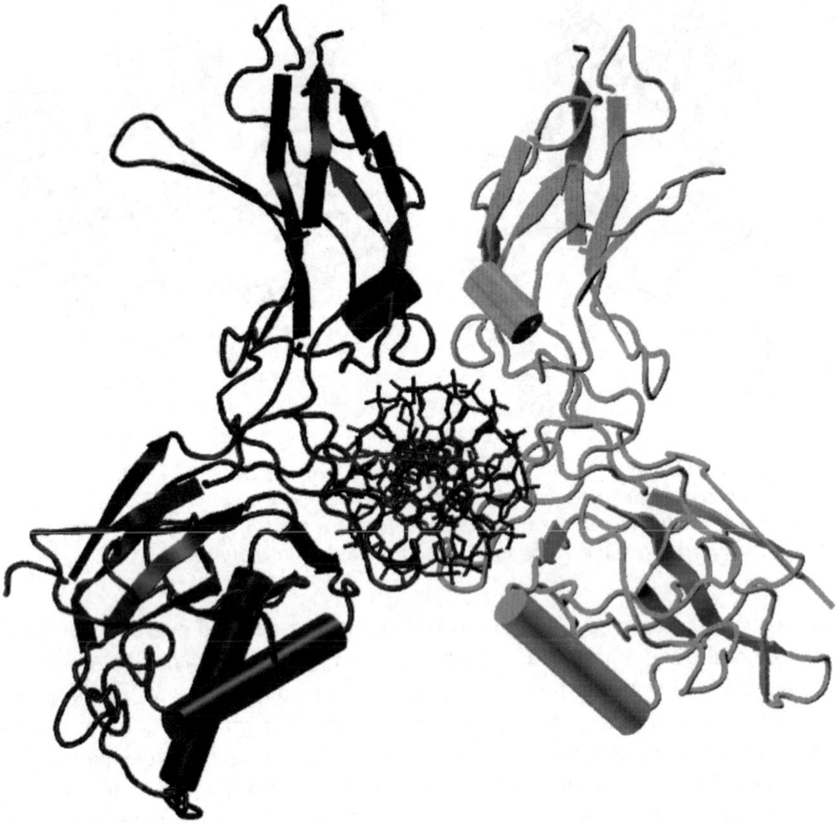

**Figure 10.7** *Loops and miscellaneous folds. Representative complexes from the Rel and Stat protein fold families. The Rel proteins, such as p50, can form heterodimers on different DNA elements. Here, the p50 (black)/p65 (red) heterodimer is shown (PDB: 1LE5)*

these loops connect strands of a β-sheet in a globular domain of the DNA-binding protein, and they penetrate deeply into major and/or minor grooves, where the side chains make hydrogen-bonding and van der Waals contacts with the bases, sugars and phosphates. This type of interaction is also very effective for achieving shape complementarity and for the optimization of buried surface area in the complex. Examples of this type of interaction come from the extensive family of proteins with a **β-layer structure**, as represented by the Rel and Stat proteins (Figure 10.7). The effectiveness of the mode of interaction of these proteins is shown in their specificity and very high binding constants (dissociation constants typically of $10^{-9}$ M).

Loops are also used by many restriction endonucleases, in conjunction with other structural elements, to embrace the target DNA and to activate hydrolytic attack of the phosphate backbone (Section 10.5).

### 10.3.8 Single-Stranded DNA Recognition

Single-stranded DNA does not fold into the same compact structure as duplex DNA, and it is consequently much more flexible. The hydrophobic bases are more readily exposed and these are targets for interaction with the protein. Consequently, most single-stranded DNA-binding proteins have surface patches that are positively charged, to match the phosphate backbone, as well as patches, which are hydrophobic, so as to match the bases.

The 'protection of telomere' protein (Pot1) provides an example of single-stranded DNA recognition (Figure 10.8a). This protein binds to the termini of eukaryotic chromosomes, known as telomeres, which have single-stranded 3′-overhangs of a G/T rich, repetitive sequence. The structural motif used by Pot1 is the widely

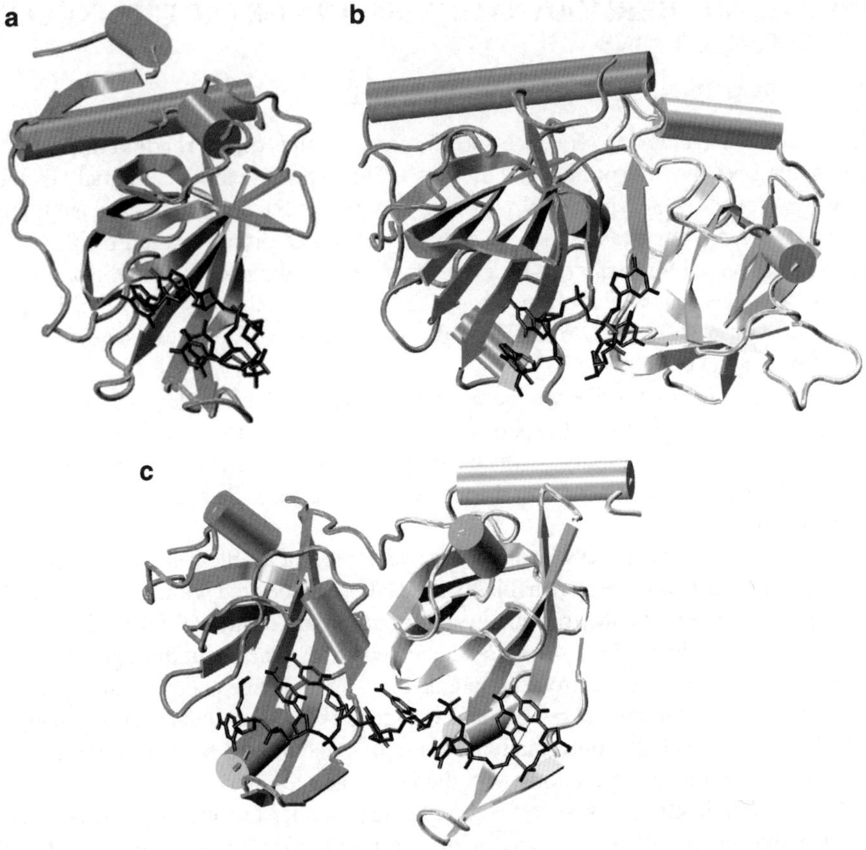


**Figure 10.8** *The OB fold and single-stranded DNA recognition. (a) The 'protection of telomere' protein (Pot1) from the fission yeast, Schizosaccharomyces pombe (PDB: 1QZG). (b) The Oxytrichia nova telomere end binding protein (TEBP) bound with specific DNA (PDB: 1K8G). The Pot1 and TEBP proteins bind single-stranded 3′ overhang at the termini of chromosomes. (c) The human Rep protein (PDB: 1JMC). The common OB core for all three proteins is shown in red and pink for comparison. The single-stranded DNA is shown in black*

occurring **oligonucleotide/oligosaccaride-binding fold** (the **OB fold**). This structure has an anti-parallel **β-barrel**, and it has an overall shape like the capstan of a ship, with part of the concave surface used to engage the single-stranded DNA. Comparison of the OB folds for several different proteins (Pot1, the telomere end-binding protein (TEBP) from the ciliated protozoan *Oxytricha nova*, and the human replication protein A) illustrates how the common core of all three proteins is used to form a scaffold for the binding of single-stranded DNA (Figure 10.8b and c). Interestingly, the OB fold also occurs in many RNA-binding proteins, such as the transfer RNA synthetases (Section 10.9.3) and key ribonucleases that process and turnover targeted RNA.

What is especially remarkable about the structure of the yeast Pot1 protein/DNA complex is that the single-stranded DNA itself participates in the recognition, by folding back to make intra-molecular G·T base pairs and base-to-phosphate hydrogen bonds. Disruption of the base pairing, for example by substituting a guanosine base for inosine (and so replacing an exocyclic amino group by a keto group), decreases the binding affinity by the equivalent of roughly 12 kJ mol$^{-1}$. The fold of the DNA presents some of the bases for stacking on aromatic and aliphatic side chains of the protein. The mode of interaction may be conserved in Pot1 homologues from other organisms, including the human protein.

## 10.4 KINETIC AND THERMODYNAMIC ASPECTS OF PROTEIN–NUCLEIC ACID INTERACTIONS

### 10.4.1 The Delicate Balance of Sequence-Specificity

The most common way of achieving sequence-specific recognition is **direct readout** of the DNA sequence by hydrogen bonding between the protein and the DNA. This type of readout commonly includes the participation of water molecules. A typical DNA–protein complex in the cell might have a binding energy of a few kJ $mol^{-1}$, so it seems that the binding energy and sequence discrimination of most protein–DNA complexes can be accounted for easily just by a few hydrogen-bonding contacts to the bases and the phosphate backbone. However, bases that are not contacted by the protein may also significantly affect the stability of certain complexes. This results in **indirect readout** whereby the sequence is recognized through its conformational effects.

The energy of indirect readout is likely to be around 5–20 kJ $mol^{-1}$ and is thus important for fine-tuning specificity. This has been demonstrated experimentally in studies of DNA-binding by the 434 repressor, where it is found that the binding affinity is affected by base substitutions in the central portion of the palindromic site, where there are no contacts with the repressor.[17] The bases there must influence the moulding of the DNA to the protein.

The formation of specific protein–DNA complexes may represent a delicate balance between favourable interactions on the one hand and energy penalties, such as those associated with distorting the DNA, on the other hand. In specific complexes, the free energy contributions are clearly all favourable and derive from base contacts, phosphate contacts, bulk solvent and counter-ion displacement through the optimal burying of surface area, and by stabilizing electrostatic interactions. The net free energy change ($\Delta G$) is offset by the entropy penalties of conformational restriction and induced-folding and the energy penalty for deforming the DNA, which involves both conformational restriction and altered base stacking. In a non-specific complex, there are fewer direct interactions with the bases, and the buried surface area is often smaller (Section 10.4.3). Consequently, a non-specific complex has smaller favourable contributions from these effects, while the entropic penalties associated with conformational effects are also smaller. The relative energy contributions are illustrated for specific and non-specific complexes of the restriction enzyme *Eco*RI with DNA (Figure 10.9). In this example, specific binding (left) is a fine balance of favourable effects (downward arrows) against unfavourable ones (upward arrows), to leave a small energy gain. In the energetic decomposition of the non-specific binding event (right), the free energy change is comparatively smaller. This decomposition is only illustrative because the changes are highly context-dependent.

Complexes in which the DNA is highly distorted tend to be entropy-driven, whilst those with a more 'relaxed' DNA conformation tend to be enthalpy-driven (Figure 10.9, right). There are several reasons for this, but the principal cause is the entropy cost associated with protein folding upon binding of the DNA in the relaxed case and the enthalpy cost of bending the DNA in the 'distorted' DNA case.

### 10.4.2 The Role of Water

There are broadly two effects to be considered for the role of water in protein–nucleic acid interactions: (1) the thermodynamics of assembly, and (2) the detailed optimization of surface interactions. In specific complexes, buried surface area is optimised, and the matching surfaces fit together well. The mobilization of water from the hydrated surfaces of the protein and of the oligonucleotide grooves and phosphate backbones of the nucleic acid provides an entropically favourable contribution to complex formation (Figure 10.9, right). These bulk solvent effects account in part for the noted changes in heat capacity associated with the formation of complexes.[18]

Crystallographic structures of nucleic acid–protein complexes frequently reveal ordered water molecules that lie at the molecular interfaces (where their diffraction limit is better than about 3 Å). These water molecules mediate the macromolecular interactions and also fill the occasional gap arising from interfacial imperfections. Water molecules participate in hydrogen-bonding networks that link side- and main-chain

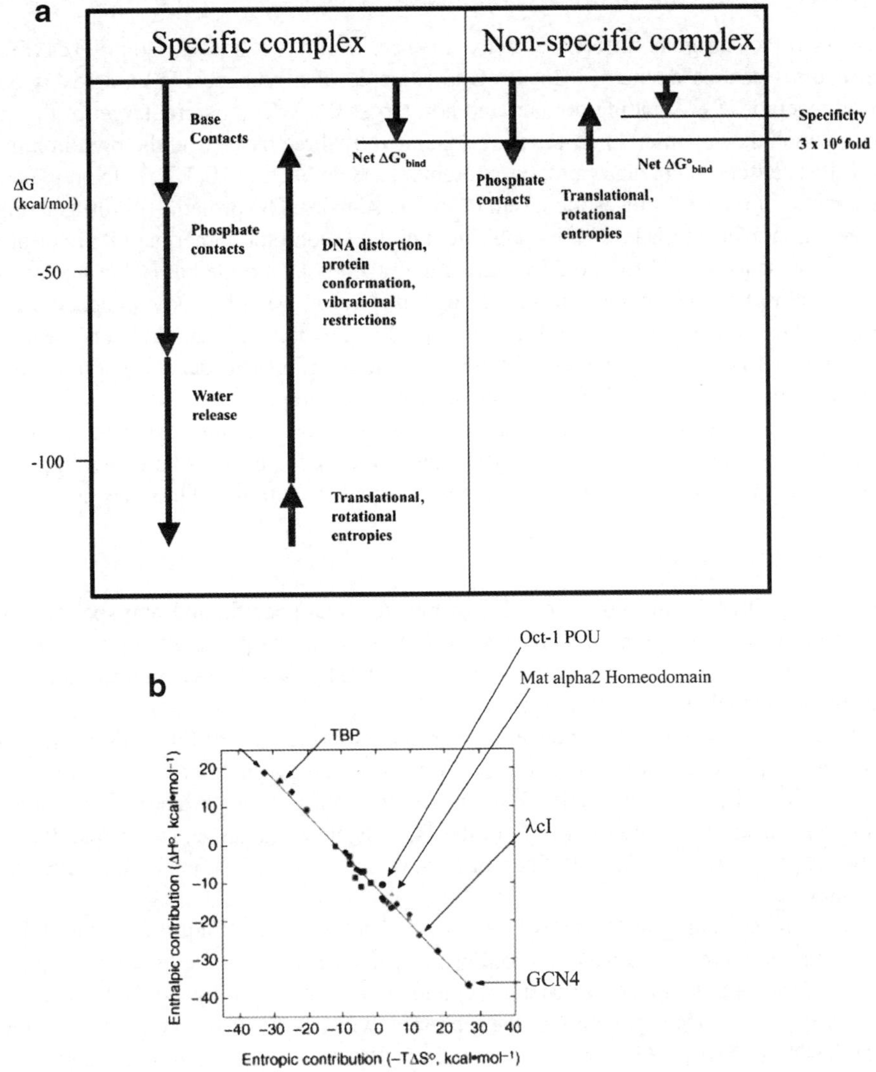


**Figure 10.9** *The energy decomposition of a representative specific protein–DNA complex. (a) The black arrows on the left are estimated contributions of hydrogen bonding and the release of bulk solvent and counter-ions. The energy penalty for deforming the DNA is shown by the downward arrow. On the right side the energetic decomposition of the non-specific binding event.* (Adapted from L. Jen-Jacobson *Biopolymers*, 1997, **44**, 153–180. © (1997), with permission from John Wiley and Sons, Inc.) *(b) An example of the thermodynamic correlations of protein–DNA complex formation with DNA distortion.* (Adapted from L. Jen-Jacobson, L.E. Engle and L.A. Jacobson, *Struct. Fold Design*, 2000, **8**, 1015–1023. © (2000), with permission from Elsevier)

atoms, the substituents of the bases, and the non-bridging oxygen atoms of the phosphate backbone. Such buried water molecules may be isolated from the bulk solvent and become in effect non-covalent extensions of the protein or DNA. In general, protein–DNA interfaces are significantly more hydrated than protein interiors.[8] In some cases, water-mediated hydrogen-bonding patterns account for base recognition (*e.g.* the trp repressor–operator complex).[19]

### 10.4.3 Specific versus Non-Specific Complexes

We have discussed the *energetic* differences between specific and non-specific complexes (Figure 10.9), but what *structural* features distinguish them? One example of a non-specific complex is between the glucocorticoid receptor DNA-binding domain and non-target DNA. The natural target of this protein is a palindrome in which the hexameric half-sites are separated by three bases, and the protein binds this as a homodimer with a self-complementary protein–protein interface (Section 10.3.3). In the non-specific complex, the separation of the half-sites is increased to four base pairs. The protein still forms a dimer on this DNA, and each monomer interacts with its half-site, but the second monomer is pulled out of alignment with its recognition sequence. There are fewer direct hydrogen bonds to the bases, but just as many interactions with the phosphate backbone. The surface match is not so good for the non-specific complex, so that the buried surface area is only about half that the specific interaction. In another non-specific complex, a layer of bound water molecules at the protein–nucleic acid interface is observed that must contribute an entropic penalty to the binding and disfavours the non-specific complex.[20]

The non-specific complex of the *Bam*HI restriction enzyme illustrates the role of both surface electrostatic and shape complementarity (Figure 10.10). Thus, target recognition is here associated with structural changes in the protein that alter both the surface charge distribution and the shape.

### 10.4.4 Electrostatic Effects

For most RNA- and DNA-binding proteins, the affinity for both specific and non-specific nucleic acid decreases sharply as the ionic strength increases. This arises from the polyelectrolyte character of DNA and RNA, which affect the binding of polycationic proteins and ligands. The salt concentration also affects the ability of proteins to slide along the longer target.

One curious exception to this salt-dependence rule occurs in the case of the transcription factor TBP from the hyperthermophilic archaea *Pyrococcus*,[21] where the binding affinity *increases* with ionic strength. This may be explained by a putative cation-binding site that bridges the surfaces of the protein and the DNA. The protein surface that is in proximity to the DNA is electronegative, rather than the more usual electropositive (*e.g.* the surface of *Bam*HI; Section 10.4.3), and the counter-ion binding helps to overcome the charge repulsion.

Subtle electrostatic switching may also explain how the met repressor (Figure 10.6a) is activated by its ligand, *S*-adenosylmethionine, to gain a 1000-fold increase in its affinity for the 'met box' target DNA. The met repressor undergoes little apparent structural change upon binding of ligand. But because the ligand is charged, it changes the surface potential of the repressor, which now makes a more favourable match to the surface of the DNA. Perhaps this effectively neutralizes the charges on the DNA, but does so in an asymmetrical fashion, to favour its conformational adjustment, not only to the bound protein, but also to its neighbour at an adjacent met box. Attractive protein–protein interactions also favour the binding of adjacent repressors, with the net result that they bind co-operatively to the DNA element (Section 10.4.6).

### 10.4.5 DNA Conformability

Duplex DNA can be treated globally as a moderately flexible rod, with two principal degrees of freedom (Section 2.3.1). One mode involves *twisting* along the local helical axis, but structural restraints to the twisting impart a torsional stiffness to the molecule. The **energy of twisting** is proportional to the square of the change of twist angle per unit length. The other mode is *bending*, and stereochemical restraints to this mode are associated with axial stiffness. The **energy for bending** is roughly proportional to the square of the curvature, where curvature is a bend-angle per unit length, which is the same as the reciprocal of the local radius of curvature. Bending is usually achieved by changing the roll angle at the base steps. The energy required for bending may be anisotropic, with a preferred azimuthal angle, according to the ability of the base steps to roll, and the projection of the roll with respect to the plane of curvature of the bend. When the stacking is maintained, the bending is smooth, as shown for the nucleosomal DNA (Figure 10.1d), but the DNA becomes kinked if the stacking is disrupted (*e.g.* the base step in Figure 10.4f).

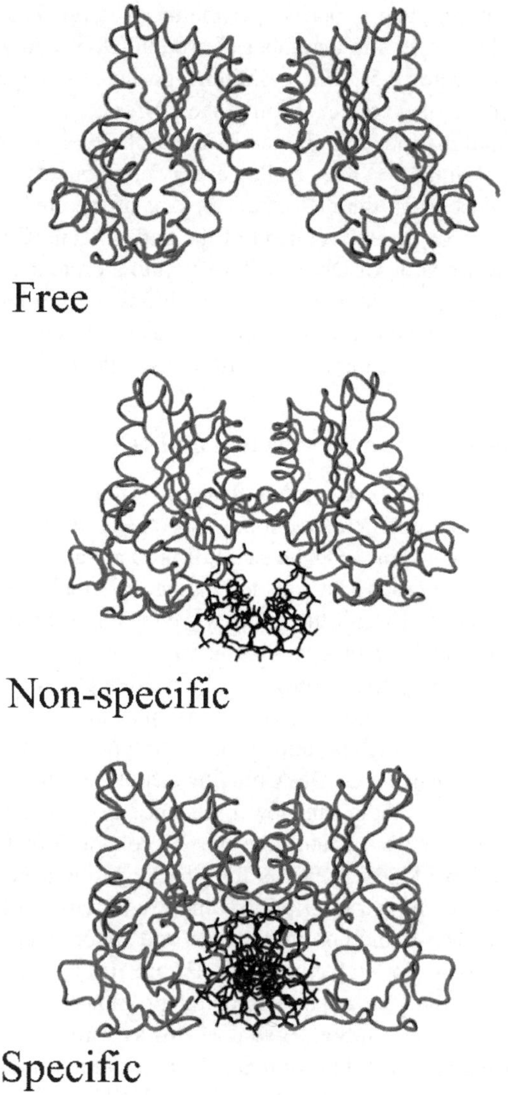


**Figure 10.10** *A representative example of a non-specific protein–DNA complex. The BamHI restriction endonuclease, showing the free state, bound to non-specific DNA and bound to specific DNA*

The energy of changing a DNA conformation is affected by many factors. For instance, the DNA can be curved by selectively neutralizing charges on one surface, which happens if one surface of the DNA interacts with a basic surface of a recognition protein. An important contribution to DNA conformation comes from the energy required to change the base stacking, since stacking and hydrogen bonding of the bases are key determinants of DNA structure.

The energy of DNA deformation contributes to indirect readout and, for a typical DNA–protein complex interacting over five or six bases, is likely to be in the order of 5–20 kJ $mol^{-1}$. In the case of the nucleosome, the calculated energy for DNA bending is 126 kJ $mol^{-1}$ for 124 base pairs.[22] Indirect readout is thus important for fine-tuning specificity. The different base steps have different stacking energies, with TpA having the least stacking energy and GpC the greatest of the ten possible base steps. Correspondingly, the most easily distorted base step is the TpA step, and this step commonly occurs in examples of indirect read-out. Conformational variability occurs more frequently for steps in the order pyrimidine–purine $>$

purine–purine = pyrimidine–pyrimidine > purine–pyrimidine. The propensity for DNA deformations to occur at pyrimidine–purine steps is seen in a number of protein–DNA complexes, including those of the IHF (Figure 10.6b) and TBP (Figure 10.5). The pyrimidine–purine step may adopt large roll or high twist angles in protein–DNA complexes. Severe distortion of the DNA occurs in restriction enzyme/DNA complexes, and here indirect readout may account for nearly half of the energy associated with cleavage specificity, corresponding to 25–30 kJ $mol^{-1}$.[23] The detailed energy components of base stacking in the context of protein–DNA complexes depend on the dielectric constant of the environment and the need to bury protein–DNA interfaces away from the environment of the bulk solvent. Currently, it is hard to quantify the energy associated with deformation of DNA or RNA within a protein–nucleic acid complex because in such a complex the environment of the nucleic acid is different from bulk solvent. For instance, the nucleic acid experiences a large change in dielectric constant as its protein partner surrounds it. This has a large effect on the energies of any conformational adjustments in the nucleic acid.

### 10.4.6 Co-operativity through Protein–Protein and DNA–Protein Interactions

Many of the proteins, already described, bind to DNA elements as homo- or heterooligomers, and thus make **homotypic** or **heterotypic** protein–protein interactions, respectively. In such complexes, each subunit may present a reading head that interacts with the DNA target. If properly arranged, the organization of binding sites on the DNA can provide **co-operative** benefits to binding, by simply helping to pre-organize the successive binding of the other subunits. Even if the proteins pre-exist in solution, there is still an implicit co-operative benefit because the matching of one subunit to its site pre-organizes the second site. This decreases the entropic cost of binding, and is known as the **chelate effect**.[24]

A typical control region for a eukaryotic gene is very complex and contains a myriad of binding sites for different proteins. In many cases, such proteins form homotypic or heterotypic protein–protein interactions, so that they mutually influence their DNA-binding affinity. Such co-operative effects might compensate for the comparatively weaker binding and lower sequence discrimination of eukaryotic transcription factors compared with prokaryotic repressors and activators. A complex organization that involves multiple interacting proteins may also have kinetic benefits for gene regulation. For example, multi-component assembly is involved in the pre-initiation transcription complexes of eukaryotes and archaea. Pre-initiation complexes assist RNA polymerases to bind promoter sites, and are often composed of the TATA-element binding protein (TBP) and the transcription factor TFIIB (Figure 10.11). Other components may be assembled onto the exposed surface of the TBP, depending on the promoter. Because the complex is asymmetric, it can specify the direction of transcription. How does this asymmetry arise?

The TBP protein has approximate twofold symmetry, so there are in principle two possible orientations for TBP to bind with respect to the start site of transcription. TBP has a saddle-like shape that engages a widened DNA minor groove, but clearly, TBP binding alone cannot specify the direction of transcription. Transcription factor TFIIB binds asymmetrically to the TBP by recognition of one exposed surface and makes a contact in the major groove on the 5′ side of the TATA element (at the 'BRE' recognition site) through a helix-turn-helix motif (Figure 10.11). Thus, it is the combination of protein–protein interactions between the TFIIB and the TBP and the protein–DNA interactions of the TFIIB that breaks the symmetry of the complex and defines the direction of transcription.

Another example of protein–protein interactions is provided by the hormone and nuclear receptors (Figure 10.2c). The DNA elements of steroid receptors usually comprise two hexameric elements that are arranged as an inverted repeat with a three-nucleotide separation of half-sites (*e.g.* the idealized glucocorticoid response element is 5′-AGAACAxxxTGTTCT-3′). The pseudo-dyad symmetry of the half-sites aligns the proteins, which interact in a 'head-to-head' orientation. The DNA-binding domains of these proteins bind through favourable protein–protein interactions, which give rise to co-operativity. The two monomers interact with the DNA through a dimerization interface, the proper alignment of which requires exact spacing between the two monomer-binding sites. The two monomers would clash sterically if they were to bind to a DNA sequence in which the half-site separation is either two or four bases.

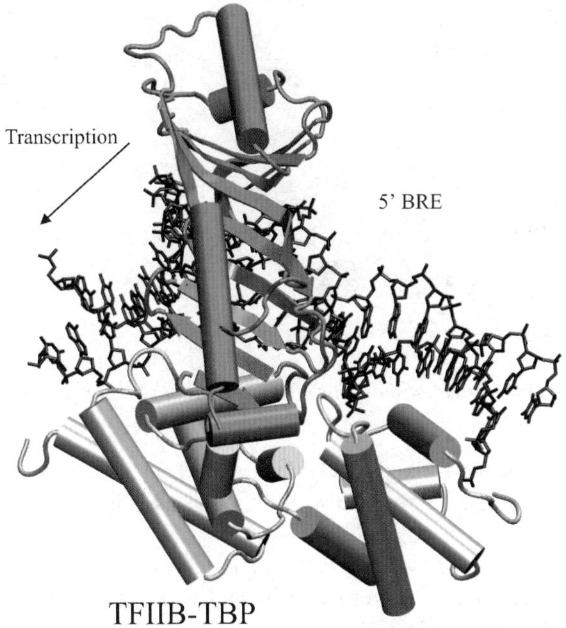


**Figure 10.11** *A representative example of a ternary complex with a specific DNA target. The TBP–TFIIB–DNA complex (PDB: 1AKH). The TBP (red) docks into the distorted minor groove. This perspective shows the recognition helix of TFIIB (pink) docked in the major groove at the recognition site. This binding helps to define the polarity of transcription initiation. The perspective is into the distorted minor groove engaged by the TBP (and is perpendicular to the view seen in Figure 10.5 in two axes)*

Several different nuclear receptor monomers recognize the same consensus DNA sequence: ACTGGA. However, they bind to different DNA sequences, which are distinguished by the spacing between the half-sites. These elements are bound by a pair of receptors, the relative orientation of which changes with the half-site separation. This brings different surfaces into register and thus explains how the proteins discriminate DNA targets.

Some proteins can undergo secondary, tertiary or quaternary conformational changes upon binding to DNA, and such changes can result in co-operativity. It has been proposed that the DNA element itself can act as an **allosteric ligand** that induces structural changes in the bound protein, and thereby influences the composition or conformation of multi-component regulatory complexes.[25] In this way, subtle changes in sequence can increase the combinatorial repertoire of a limited number of genetic regulatory proteins.

### 10.4.7 Kinetic and Non-Equilibrium Aspects of DNA Recognition

The high degree of specificity in protein–DNA recognition does not require high-affinity binding. For instance, most specific protein–DNA complexes are stabilized by the energy corresponding to only a few hydrogen bonds or the liberation of just a handful of bound water molecules – interactions that could be lacking in their non-specific complexes. Such weak binding may provide kinetic benefits by permitting protein–DNA complexes to associate and dissociate on timescales that match other processes in the cell. In many cases, the association rates may be close to maximal for facilitated diffusion.

Some dimeric DNA-binding proteins, such as the leucine zipper proteins (Section 10.3.2), dissociate into monomers at a significant rate. Dissociable dimeric and oligomeric proteins might have special kinetic properties for DNA association. Consider, for example, two alternative pathways that may describe the assembly of a complex. In one pathway, the dimerization occurs prior to binding the DNA, whilst in the second, the dimer forms on the DNA only after sequential binding of the subunits. Since in both cases there is a

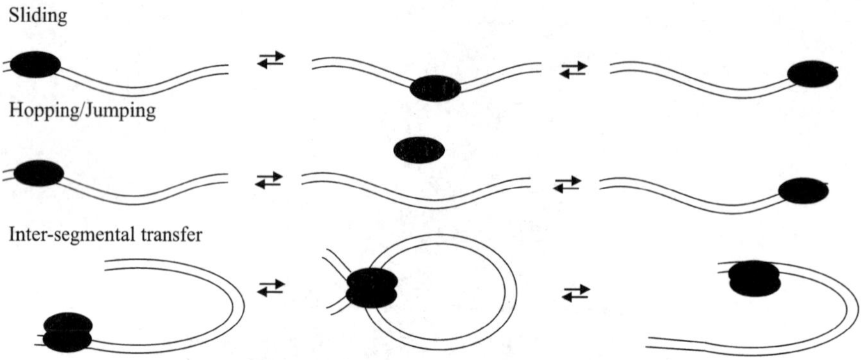


**Figure 10.12** *Possible modes for search mechanisms of DNA-binding proteins for their target sites*

closed thermodynamic cycle, the particular pathway chosen does not affect the equilibrium affinity. However, the second (monomer) pathway has been noted to have faster association rates for certain leucine-zipper proteins.[26]

Some proteins find their DNA target sites at rates approaching $10^3$-fold greater than those expected if site location were diffusion-controlled.[27] Three experimentally determined mechanisms account in part for this apparent acceleration (Figure 10.12).

1. **sliding** – involves diffusion in one dimension,
2. **hopping/jumping** – involves limited sliding diffusion, followed by dissociation and re-association at a different point on the DNA chain,[28] and
3. **inter-segmental transfer** – allows the direct transfer from one segment of DNA to another so as to slide or hop to different regions.[29]

In all cases, the rate enhancement is expected to be a factor of only sixfold over diffusion-controlled rates, since the diffusion rates are proportional to twice the dimensions of the search space. Most of the apparent rate acceleration may be due to electrostatic attraction, because of the polyelectrolyte character of the DNA.

In the cell, DNA is hardly ever in a free form. Thus, sliding-type diffusion is extremely limited, for example because of obstacles such as nucleosomes (Section 10.6.1). However, three-dimensional diffusion mechanisms such as hopping provide a means to maintain the search process even with such obstacles present.[30]

## 10.5 THE SPECIFICITY OF DNA ENZYMES

### 10.5.1 Restriction Enzymes: Recognition through the Transition State

Restriction enzymes are bacterial proteins that provide defence against foreign DNA. They recognize and cleave specific DNA sequences on both strands. Many, but not all, have twofold symmetry and so recognize palindromic sites. The enzymes recognize typically a sequence of four to eight base pairs and, not surprisingly, have extremely high specificity for well-defined target sites on DNA. Any relaxed specificity would have dire consequences for a bacterium, since its own DNA would be at risk of cleavage at sites not protected by its own methylase enzyme (Section 5.3.1).

The mechanism of restriction enzyme cleavage of DNA involves nucleophilic attack of a hydroxide ion on the scissile phosphorus atom to generate a 5′-phosphate group (*e.g.* for the enzyme *Hinc*II). The hydroxyl ion attack causes an in-line displacement of the 3′-hydroxyl group *via* a penta-coordinate transition state. Restriction enzymes require divalent cations for activity, which are used to activate the water molecule for attack and to stabilize the negative charge that develops transiently on the transition state.[31]

It might seem paradoxical that many restriction enzymes bind target and non-target DNA with similar affinities. Therefore, specificity must arise at the catalytic step, probably because of the distortions induced in the transition state. Such distortions are seen in the crystal structure of the *Eco*RI restriction endonuclease–cognate DNA, which was the first complex to indicate the potential role of **induced fit** of the DNA in catalysis.[32–34] The enzyme is dimeric and engages a palindromic DNA target sequence 5′-GAATTC-3′. At the dyad centre of symmetry, there is a 'kink' deformation in which the DNA is unwound by roughly 25° and the central ApT base step becomes partially unstacked. More modest bends occur at other steps, and the cumulative effect of the bends and kink is to **open the major groove** and improve its accessibility to direct protein–base contacts. The kink appears to be required to orient the phosphate backbone for hydrolytic attack.

By use of base analogues, Bernard Connolly and Linda Jen-Jacobson[34] found that removal of the N-6 amino group of the A residue at the central ApT step increased enzyme affinity by 4.0 kJ $mol^{-1}$, despite the loss of a hydrogen bond between the protein and the base. This removal results in relaxation of any steric hindrance in the kinked conformation, and may also cause a re-distribution of partial charges on the purine ring. How does the enzyme achieve its specificity for this particular sequence, despite the DNAs being energetically costly to deform? The explanation may lie with the kink at the centre, since the deformation of alternative base steps may have even greater energy penalties. Consequently, other DNA sequences make comparatively poorer substrates.

The principle that recognition occurs through DNA distortion is also borne out by the structure of the *Eco*RV endonuclease in complex with target DNA. Like *Eco*RI, *Eco*RV is dimeric, and it binds to a similar palindromic target site 5′-GATATC-3′. Also like *Eco*RI, the *Eco*RV–DNA complex reveals some striking deviations from uniform helical structure. The bases at the central TpA step of the target are unstacked and the major groove is compressed. This deformation brings the phosphate group near the dyad into the immediate vicinity of the catalytic residues. The bases at the TpA step are recognized *via* hydrophobic contacts to the thymine methyl groups; but their mechanical properties are also an important aspect of recognition. The TpA step is easily distorted into a high-roll conformation and the specificity in *Eco*RV relies on this distinctive property. These deviations produce an extensive surface-area contact between the protein and DNA, and this complex has one of the largest water-filled cavities of comparable protein–DNA complexes.[8]

While *Eco*RI binds to both target and non-target DNA with nearly equal affinity, *Eco*RV discriminates against non-target DNA at both the binding and the catalytic events.[23] Nevertheless, both binding and catalysis involve indirect readout. Studies with base analogues show that several functional groups in the DNA that are not contacted in the crystal structure nonetheless reduce enzymatic activity when they are deleted, principally by decreasing the catalytic cleavage rate ($k_{cat}$).[35] Other studies with base analogues indicate that direct and indirect readouts contribute roughly equally in the discrimination of specific *versus* non-specific targets, which corresponds to about 59 kJ $mol^{-1}$ binding energy.[23]

A comparison of the structure of *Eco*RV in the free form and in complex with DNA shows that the protein also undergoes some structural remodelling upon specific complex formation. Thus, both protein and DNA interactively change structure to form the specific complex.

In contrast to *Eco*RI and *Eco*RV, the related restriction enzyme *Hinc*II has apparently less stringent specificity, since it recognizes the sequences 5′-GTPyPuAC-3′, where Py (=pyrimidine) can be either C or T and Pu (=purine) is either G or A. As for the other enzymes, recognition involves indirect readout. Here, the DNA conformation is to some extent 'self-recognized' *via* interactions of the central purines, which stack even though they are on opposite strands.[20] This self-recognition *via* cross-strand purine–purine stacking probably explains the ability of the enzyme to distinguish Py–Pu from the other possible base step combinations at the central position of the recognition sequence.

### 10.5.2 DNA-Repair Endonucleases

The DNA of all organisms is modified continually by the spontaneous deamination of cytosine residues, or by the consequences of photoactivation or by the actions of harmful chemicals (Chapter 8). In every human cell, roughly $10^4$ DNA bases are repaired every day.

The major DNA-repair endonuclease APE1 cleaves DNA at sites where bases are missing. The crystal structure defines its active site and has helped to identify the way missing bases are recognized (Figure 10.13). Just as in the case of the bacterial restriction enzymes, the mechanism of action involves an activated hydroxyl group, which is directed to attack the scissile phosphate. Most enzymes have poor affinity

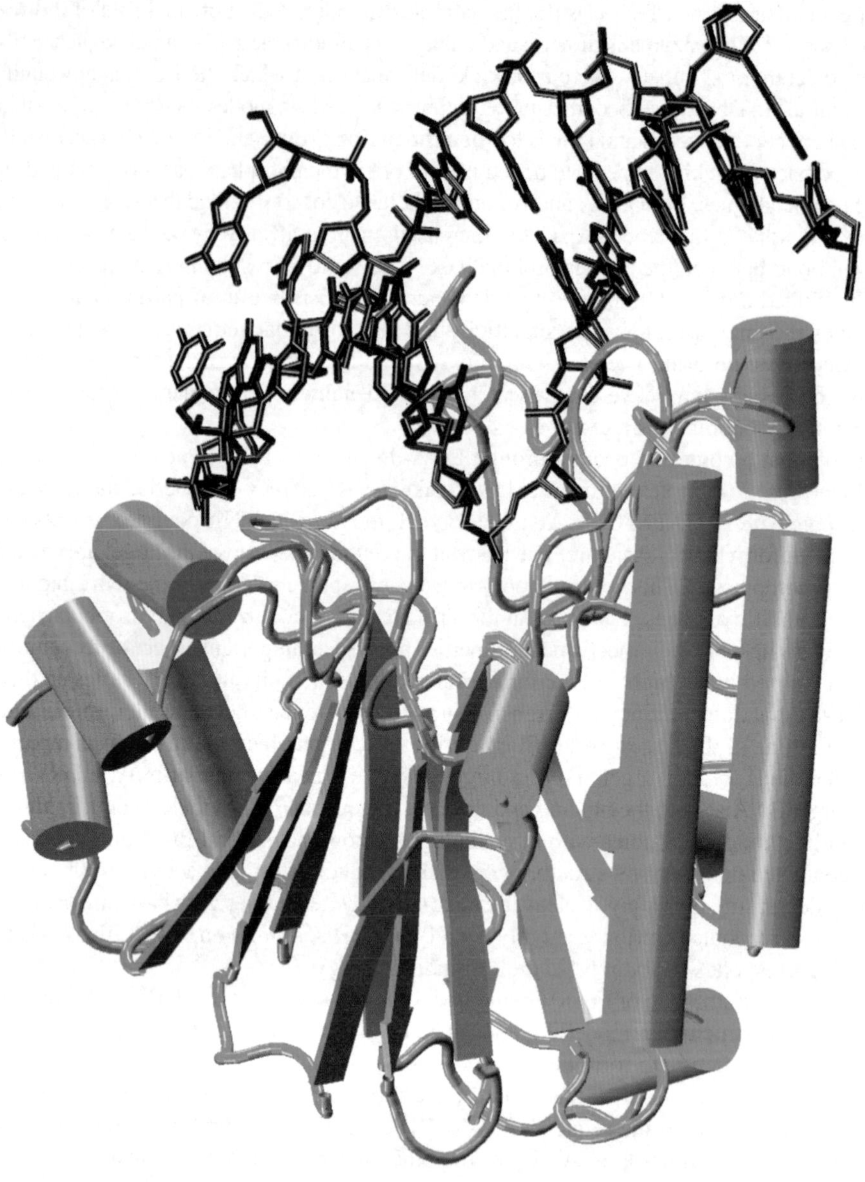

**Figure 10.13** *The structure of a representative repair enzyme, the human APE1 protein bound to DNA (PDB: 1DE8)*

for their products. However, APE1 appears to form very stable complexes with its product.[36] Thus, the enzyme holds onto the product after cleavage, since severed DNA poses great risk to the cell, and then delivers it to the next component of the repair machinery.

Certain repair enzymes cause the damaged nucleotide to be **flipped out** from the DNA double helix. In such enzymes, the nucleotide is accommodated in a specific binding pocket within the enzyme.

### 10.5.3 DNA Glycosylases

**Glycosylases** are enzymes that excise damaged DNA, as the first step towards repair. The enzyme uracil–DNA glycosylase functions to remove the base wherever uracil bases occur in DNA as a result of either misincorporation or deamination of cytosine. The structure of the human uracil DNA glycoslyase shows that the glycosylase completely encompasses the substrate. Another example is the enzyme 3-methyladenine DNA glycosylase, which excises that base from damaged DNA by intercalation into the minor groove using two β-strands, and by flipping the damaged base into the active site for base excision.[37]

### 10.5.4 Photolyases

Light is the primary source of energy for most life on earth, but ultraviolet light induces spontaneous damage to DNA in the way of covalent photo-adducts. One such example is the dimer formed from pyrimidines, cyclobutadipyrimidine. Organisms in all three domains of life have evolved enzymes that repair the light-induced damage by using the energy of light itself. These enzymes are carbon–carbon **photolyases**. Photolyases share a common fold and use a catalytic mechanism that involves conjugated co-factors for photo-induced electron transfer. One co-factor that is used directly in the catalytic process is the reduced form of flavin-adenine dinucleotide (FAD-$H_2$). The second chromophore acts as an antenna to transfer light energy to the FAD-$H_2$, which in turn supplies the electron to rupture the carbon–carbon bond. In most enzymes the antenna co-factor is 5,10-methenyl-tetrahydrofolate, but certain photolyases instead use 5-deazariboflavin. The active site engulfs the pyrimidine substrate, which is in van der Waals contact with the FAD. The FAD is sandwiched by aromatic stacking with side chains of two conserved tryptophan residues, which are thought to become radicals transiently in the transfer process.[38] There is little room to accommodate a duplex DNA. So these enzymes probably flip out the pyrimidine dimer from the duplex, in loose analogy with the flipping mechanism used by other DNA-repair enzymes (Sections 10.5.2 and 10.5.3).

### 10.5.5 Structure-Selective Nucleases

In DNA recombination, an intermediate is formed in which four strands join together (Figure 10.14a and Section 2.3.7) These structures are resolved back into two duplex DNA molecules by the activity of structure-specific endonucleases, which are found in all three domains of life. Site-specific recombinases (e.g. FLP protein, Figure 10.14b) use the hydroxyl group of either serine or tyrosine as the nucleophile for cutting and rejoining DNA. Topoisomerases also use a tyrosine hydroxyl group for a similar purpose (Section 10.8.3).

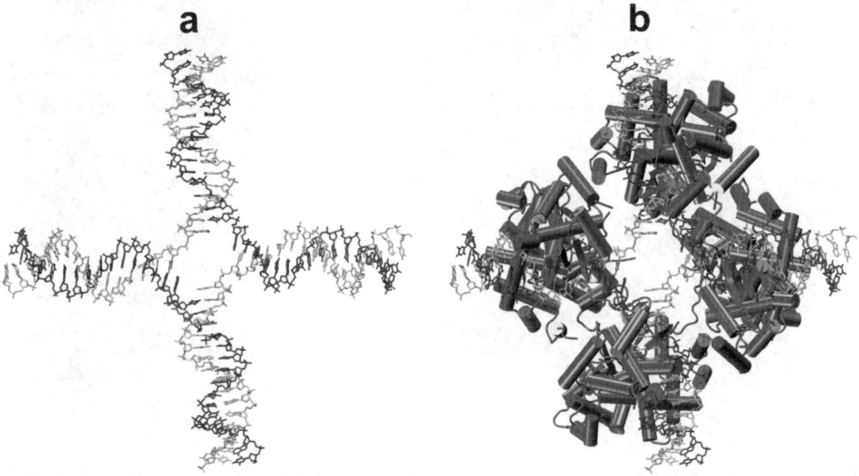


**Figure 10.14** *(a) The structure of a four-way DNA junction. (b) Its complex with the recombinase FLP protein (PDB: 1P4E)*

## 10.6 DNA PACKAGING

### 10.6.1 Nucleosomes and Chromatin of the Eukaryotes

A typical eukaryotic cell contains roughly 2 metres of DNA, which must be compacted by a factor of $10^6$ so as to fit within the nucleus. The DNA is assembled with proteins into complexes known as chromatin (Section 6.4). The fundamental packaging unit of the eukaryotic DNA is the **nucleosome**, an octameric complex of histone proteins and DNA (Figure 10.15). The histone octamer engages roughly one- and three-quarter turns of duplex DNA in a left-handed toroidal superhelix. It has been proposed that the efficiency of the histone-based system of packaging DNA was a key development in the expansion of the eukaryotic genome, and the divergence of this branch of life.[39]

The accessibility of DNA is determined primarily by the arrays of nucleosomes, which affect transcription activity, mostly through repression. The cell has elaborate machinery to 're-model' those arrays during gene expression. Such machines transduce the free energy of ATP binding and hydrolysis to displace nucleosomes. The machines are transiently fixed with respect to a target nucleosome, and it has been proposed that they displace the DNA so that it twists up to form a loop that can move around the nucleosome, causing the nucleosome to slide.[40] Modifications of the histones by acetylation and methylation provide signals for recruitment of the remodelling machinery. Recently, it has also been discovered that variant histones are used to mark boundaries for the repair of damaged DNA by recombination.

The crystal structure of the nucleosome has been solved at high resolution, from which one can observe in detail the protein–protein and protein–DNA contacts that are made for a particular sequence of DNA (Figure 10.15a). The histone octamer of the nucleosome is composed of two copies each of the four histone proteins H2A, H2B, H3 and H4. These proteins have similar sequences, and have undoubtedly evolved by gene duplication. Histones H3 and H4 are among the most highly conserved proteins in multi-cellular eukaryotes, while H2A and H2B are comparatively more variable, especially at their termini. Histones H3 and H4 form a heterotypic dimer that self-pairs into a tetramer through a four-helix bundle, in which the hydrophobic interfaces between helices intermesh in a 'knobs-into-holes' manner, just like the two-stranded coiled-coil described for leucine zippers (Section 10.3.2).

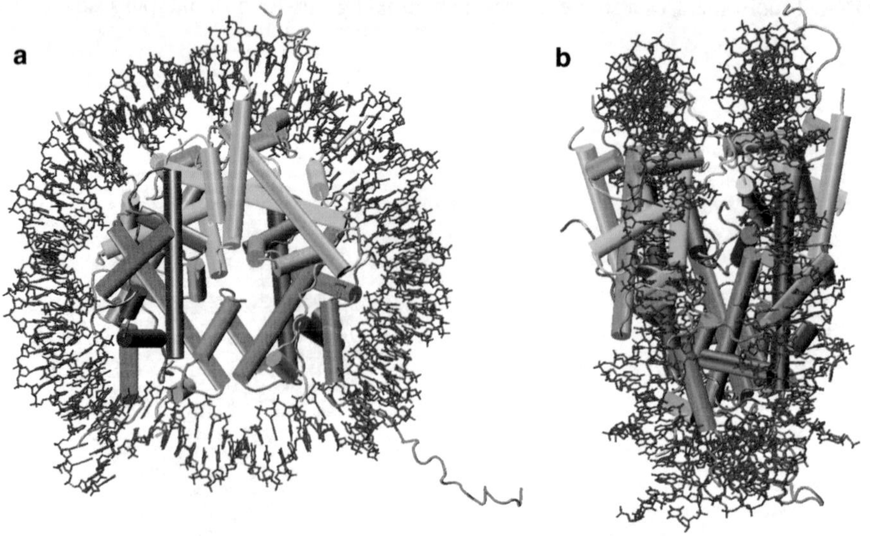


**Figure 10.15** *The nucleosome core particle (PDB: 1AOI). The histone amino termini protrude from the core particle; these are sites of covalent modification that affect chromatin structure and gene regulation. The histone dimers have been colour coded. (a) The dyad axis is vertical. (b) A side view: the dyad is now in the plane of the page*

The nucleosomal histones contact the DNA through α-helices and loops. Some of the long α-helices are oriented so that their helix dipoles form favourable interactions with the electronegative phosphate backbone. Water molecules and counter-ions are also involved in fitting together the protein and DNA surfaces. The DNA follows a highly curved trajectory on the surface of the protein (Figure 10.15b). This requires in-plane bending and a small accommodating change in twist. The bending is achieved by compression of the minor groove where it contacts the surface of the protein, and by widening of the minor groove on the outer surface of the curved DNA. The DNA on the nucleosome is overwound at about 10.2 base pairs/turn (in comparison with 10.6 base pairs/turn for relaxed DNA in solution). This overwinding aligns the minor grooves of the DNA, so that there are little gaps through which the terminal tails of the histone proteins protrude. Although the nucleosome has no sequence preference *per se*, there are notable preferences for periodic composition patterns to occupy the regions where the grooves compress or expand. There is a periodic variation in twist and roll at base steps, and these angles on average are larger in magnitude than for free DNA.

A string of nucleosomes can assemble into a helical structure known as the **solenoid**, in which there are six nucleosomes in a superhelical turn. The termini of the histones, which protrude from the surface as mentioned above, are thought to be important for mediating the contacts within the solenoid, and hence modifications such as methylation or acetylation of lysine amino groups on the histones can affect the assembly or the recruitment of other regulatory proteins. The solenoid assembly is consolidated by a fifth class of histone protein, known as the H1 (or H5) histones, which are not structurally related to the octamer core histones H2A, H2B, H3 and H4. These H1 and H5 histones have a DNA-binding motif that is remarkably similar to that found in the bacterial CAP gene activator protein and the SAP-1 protein. The H1/H5 proteins are probably located near the dyad of the nucleosome.

Binding sites within nucleosomes are occasionally exposed, but at sufficiently rapid rates to allow site access under physiological conditions.[41,42] The sequential binding of a number of factors independently to DNA sites exerts co-operativity through their mutual effects on the chromatin. The binding of one protein at its recognition site on the DNA may help to alter the nucleosome position and thus expose the binding site of a second protein, favouring its binding.

### 10.6.2 Packaging and Architectural Proteins in Archaebacteria and Eubacteria

Archaea in the *euryarchaeota* sub-branch also have histone-core-like proteins. There is a remarkable structural similarity of the archaeal histone dimer made by the HMfB and HMfb proteins with the eukaryotic H3–H4 dimer (Figure 10.16a). This similarity suggests that the H3–H4 tetramer may represent an earlier organization of chromatin during the early stages of the evolution of the eukaryotic cell.[10]

The *crenarchaeota* sub-branch appears not to have histones, but instead a variety of abundant non-sequence-specific DNA-binding proteins. For instance, the Sul7d protein binds non-specifically in the DNA minor groove by means of a three-stranded β-sheet, with intercalation of hydrophobic side chains between the bases (Figure 10.16b). Another protein used in packaging the DNA is the homodimeric Alba, which promotes negative supercoiling in DNA. The protein has two highly conserved loops that may contact the DNA.

The genomes of eubacterial species are also compacted by proteins. These organisms use small, basic proteins such as HU and FIS. The HU protein can supercoil DNA to form beaded structures.

## 10.7 POLYMERASES

### 10.7.1 DNA-Directed DNA Polymerases

**DNA-directed DNA polymerases** (E.C. 2.7.7.7) catalyse the sequential, DNA template-directed extension of the 3′-end of a DNA strand by addition of single nucleotides. The enzymes replicate genomic DNA with high fidelity and processivity. Crystal structures of DNA polymerases reveal the role of two metal ions in the active site. A mechanism has been proposed to be common for polynucleotide polymerases.[43] This involves a metal ion that activates the deprotonation of the 3′-hydroxyl group at the terminus of the growing chain,

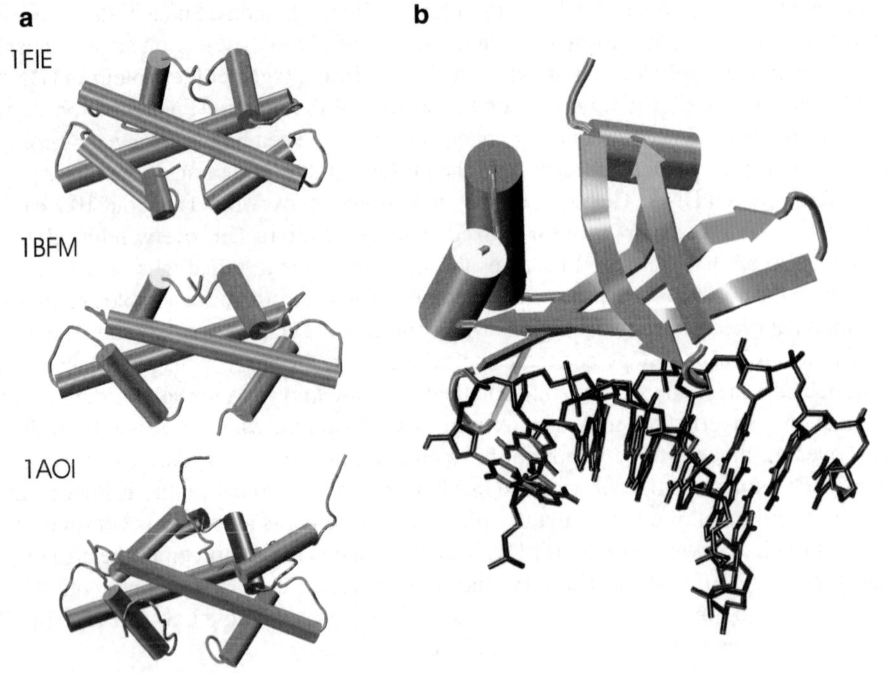


**Figure 10.16** *(a) A composite of DNA packaging proteins from the three domains of life. Comparison of the archaeal and eukaryotic histones (PDB: 1F1E, 1BFM and 1AOI). (b) Interaction of the SUL7d protein in the minor groove of DNA*

resulting in attack on the α-phosphate of the incoming **nucleoside triphosphate**. A second metal ion helps to orient the α-phosphate of the nucleotide triphosphate, so as to stabilize the penta-covalent transition state, and to assist the departure of the pyrophosphate product from the catalytic site. The proposed two-metal ion mechanism is based on both the DNA polymerase and the 3′–5′ exonuclease domains of DNA polymerase I, the latter of which serves a proofreading function to ensure fidelity in the replication of its DNA.

The DNA polymerases must have no specificity for particular DNA sequences, but they must be specific for correct Watson–Crick base pairing, for example as seen in the interaction of T7 DNA polymerase with DNA (Figure 10.17a). Studies of replication of DNA with nucleotides having chemically modified bases suggest that the fidelity is due largely to steric exclusion and shape-matching.[44] Mismatches are also discriminated by contacts to the minor groove of the DNA. Conformational changes are necessary to adapt the protein to the substrate during the polymerisation, and the polymerases have a striking modular organization in which different portions undergo movement, much like a **right hand** that grasps and manipulates substrates and translocates products (Figure 10.17b). Accordingly, these domains have been labelled 'thumb', 'palm' and 'fingers'.

## 10.7.2 DNA-Directed RNA Polymerases

The **DNA-directed RNA polymerases** (E.C. 2.7.7.6) contain a 400-kDa catalytic core that has a conserved structure and function in all three domains of life. RNA polymerases do not recognize or bind avidly to the promoter DNA themselves. Instead, they require the help of auxiliary proteins that recognize both the target and recruit the enzyme. Nearly a dozen basal transcription factors are involved in the initiation of transcription in eukaryotes; whereas in bacteria, things are much simpler, and initiation involves only one 'σ-factor' protein, of which there are many different types.

The RNA polymerases have an initiation and elongation phase in transcription. During the initiation phase, the polymerase binds to a promoter sequence, melts the DNA to form a transcription '**bubble**' and

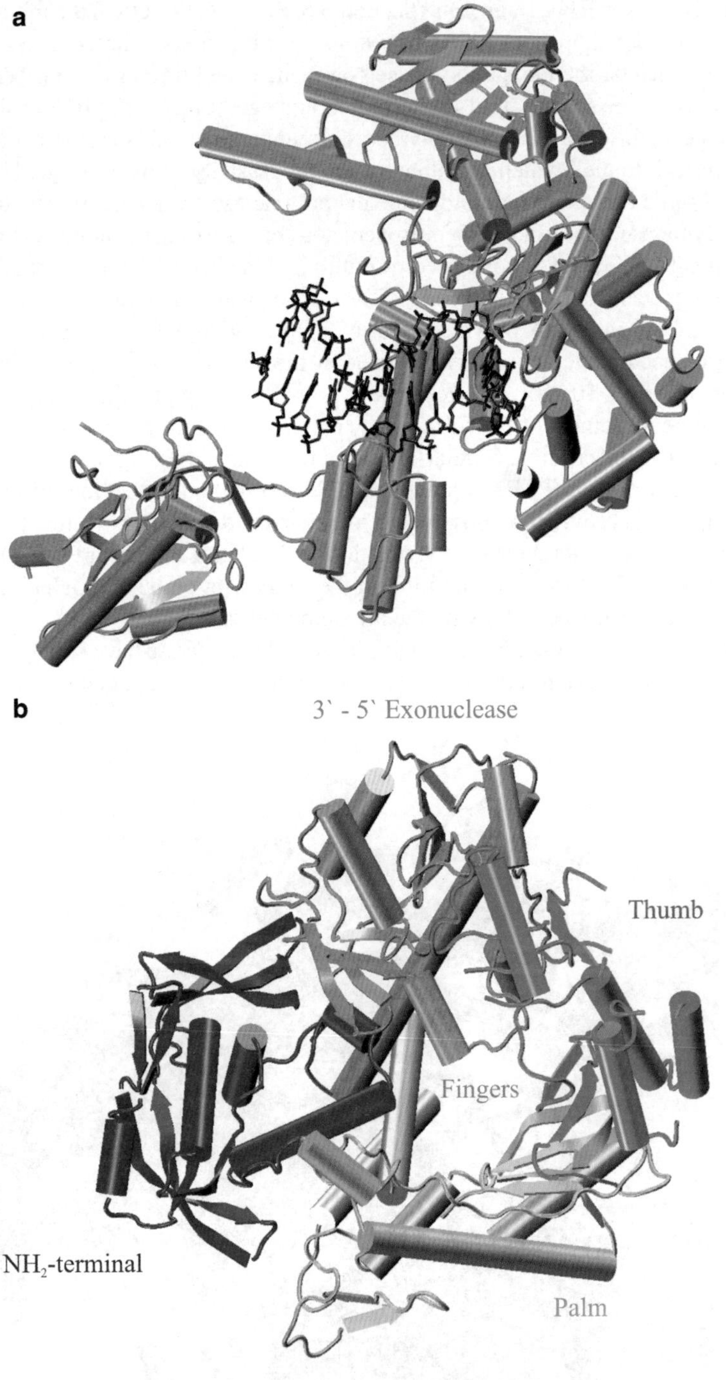


**Figure 10.17** *Representative bacterial DNA polymerases. (a) T7 bacteriophage DNA polymerase with duplex DNA (black) (PDB: 1T7P). Conformational changes are necessary to adapt the protein to the DNA substrate. (b) Thermophilic polymerase (PDB: 1QHT). Here, the polymerase adopts a more compact structure in the absence of DNA. Note the domain organization, with segments labeled 'thumb', 'palm' and 'fingers' according to polymerase nomenclature. The 3′–5′ exonuclease domain is coloured red*

initiates the synthesis of short RNA fragments (Section 6.6, Figure 6.17). Often the initiation is an abortive 'futile' cycle in which only short nucleotides of two to six bases are synthesized. With the synthesis of 10–12-nucleotide fragments, the elongation phase commences and transcription proceeds processively. The polymerase does not dissociate until the termination stage is reached. The transition to the longer RNA fragments appears to force the RNA polymerase to undergo an allosteric transition to a shape that creates an engulfing exit tunnel, which accounts for the processivity of the enzyme. Hence, a number of seemingly wasteful futile cycles are required while the polymerase awaits the proper structural switch. In the prokaryotic polymerase, the σ-factor is jettisoned, and this causes an analogous allosteric switch that ensures processivity. The σ-factor is composed of multiple domains, which individually might associate weakly, but when compacted, collectively bind tightly. Consequently, they can be dissociated sequentially.

The structure of a bacterial RNA polymerase in a complex with σ-factor and with a DNA promoter shows that sequence-specific interactions are mediated by the σ subunit; and that explains how the enzyme complex can recognize the −10 and −35 elements of promoters. The protein undergoes large conformational changes during engagement of the substrate, which may be required during the elongation so as to maintain processivity of transcription. Similar allosteric transitions are observed for the eukaryotic polymerase from yeast and the phage T7 RNA polymerase.[45,46] Formation of the transcription bubble is a key step to the initiation of transcription. A conserved tryptophan residue in the σ-factor stacks on the exposed downstream edge of the transcription bubble at position −12 relative to the start site. One of the domains of the σ-factor is essential for DNA melting to form the 'transcription bubble'. Here, the σ-factor forms single strand, sequence-specific contacts with the non-template strand.

The yeast RNA polymerase II (Figure 10.18) and eubacterial polymerase have two metal ions at the active site, which are probably both magnesium ions. One metal ion is persistent and the other may possibly exit

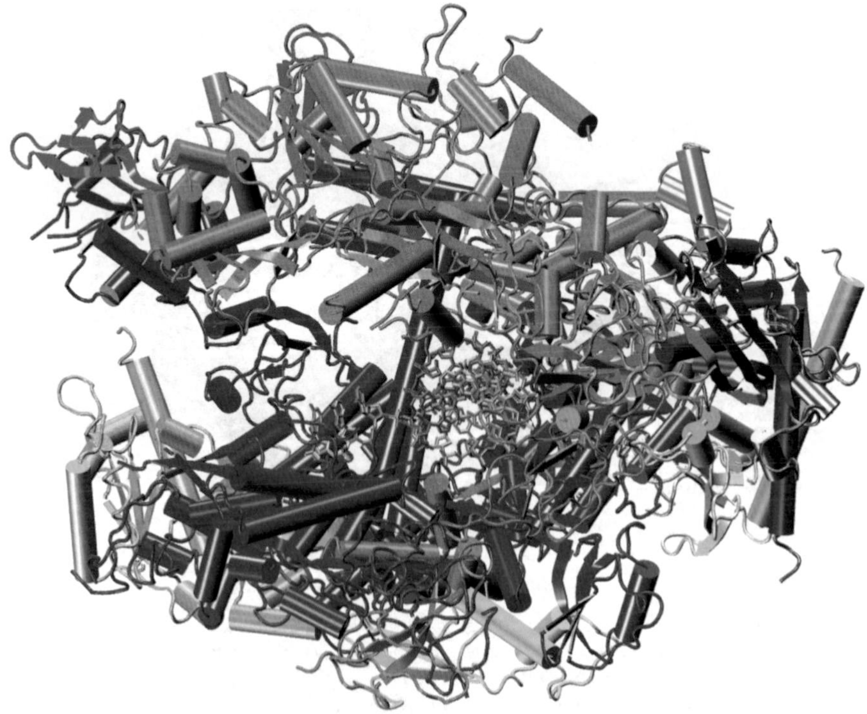

**Figure 10.18** *Structure of the elongation complex of yeast RNA polymerase II (PDB: 1I6H). Each protein subunit is colour coded differently. The view is along the axis of the central RNA–DNA heteroduplex (green and purple)*

with pyrophosphate that is released during the extension reaction. Thus, the RNA polymerase may share a common two-metal mechanism with the DNA polymerases.

## 10.8 MACHINES THAT MANIPULATE DUPLEX DNA

### 10.8.1 Helicases

Helicases unwind DNA and RNA duplexes, and they participate in nearly every cellular process that involves nucleic acids. In *E. coli*, more than a dozen different helicases have been identified, and these carry out a variety of tasks, ranging from strand separation during DNA replication to more complex processes in DNA repair and recombination and, in the case of RNA helicases, the degradation of messenger RNA.

Whereas certain repair enzymes might use the energy of DNA binding to flip a single base outside the duplex, a helicase must not only locally melt the DNA target, but must also move itself along the DNA to prevent re-annealing of the strands. This requires mechanical work, which is provided by the free energy of ATP binding and hydrolysis. Since helicases must displace proteins from the DNA, they are capable of generating tremendous force. For instance, helicases can produce sufficient force to displace the protein streptavidin from 5′-biotinylated DNA,[47] which binds with a dissociation constant on the order of $10^{-14}$ M. There are several families of DNA helicase in archaea, prokaryotes and eukaryotes, but mechanistically these enzymes can be divided into two classes according to their direction of translocation. One class moves in the 5′–3′ direction along single-stranded DNA and the second class moves in the opposite direction, from 3′–5′.

The first crystal structure of a 3′–5′ helicase was that of PcrA from *Bacillis stearothermophilus.*[48] The enzyme consists of several domains, including two that are structurally similar to the ATPase domain of the recombination protein RecA. Subsequent structures of PcrA helicase in complexes with a DNA substrate have provided details of the unwinding mechanism. There are two separable steps: (1) duplex destabilization, and (2) DNA translocation.[39] Duplex DNA is engaged on the surface of the protein, and one of the displaced single strands binds to a surface channel. The operation involves allosteric transitions in which protein sub-domains move relative to each other, so that there are changes in both the surface and the channel that engage the duplex DNA.

An example of a 5′–3′ helicase is the bacteriophage T7 gene 4 helicase, which is involved in strand separation during phage DNA replication (Figure 10.19). The protein forms a hexameric ring but, surprisingly, this ring has twofold rotational symmetry rather than the expected sixfold. ATP hydrolysis around the ring is associated with a rotation of each subunit about axes perpendicular to the ring. The nucleotide-binding sites (at the interfaces between the subunits) are therefore non-equivalent, and have differing affinities for the NTPs. As nucleotide hydrolysis occurs, the states of the three sites may interconvert from NTP-bound to NDP + phosphate binding to empty. This would allow a 'wave' of NTP hydrolysis to run around the ring.

This model shows some similarities with the 'binding-change' model associated with the F1-subunit of the mitochondrial ATPase, although the asymmetries arise in different manners (Figure 10.19b).

### 10.8.2 DNA Pumps

The duplex DNA of bacteriophages is packaged within protein-rich capsid shells at densities approaching those of crystals. Electrostatic repulsion of the compacted nucleic acid generates pressure, and this force is used to discharge the DNA into the host cell during the infection process. But the packaging of nucleic acid to such an extent in the first place requires tremendous work which is provided by motors that are fuelled by ATP binding and hydrolysis. The force generated by DNA pumps of bacteriophages during the initial stages of translocation is estimated to be in the order of 100 pico-Newtons, and the force grows as packaging proceeds.[49]

The crystal structure of the central component of the motor from *B. subtilis* bacteriophage $\phi$29 illustrates the principle of operation of the DNA pumps (Figure 10.20a). In essence, the translocation machinery is a rotary motor comprising a system of two rings of mismatched symmetry that interact with the DNA.[50,51] The central component contains 12 copies of a 36-kDa subunit. This forms a central channel with a diameter

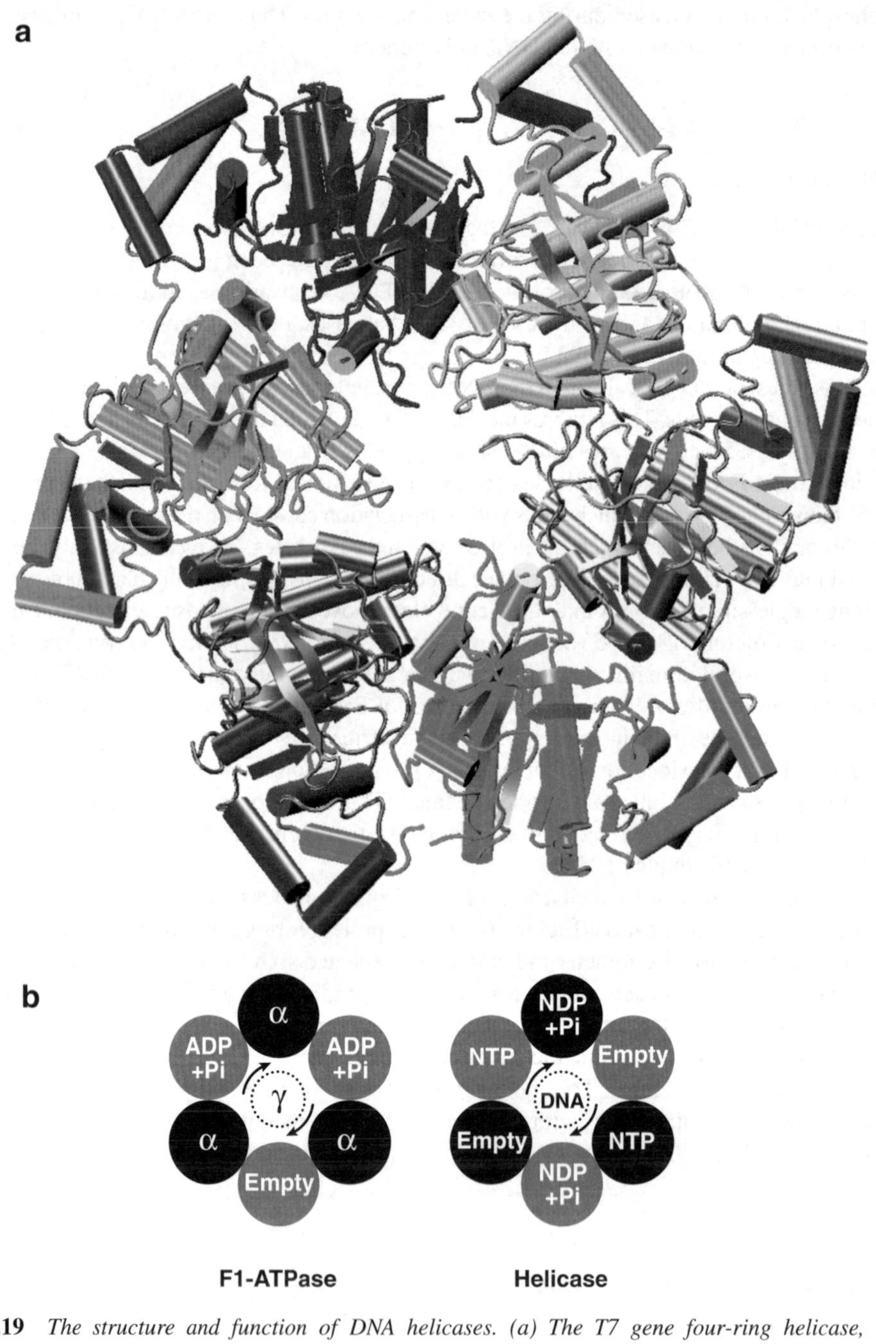


**Figure 10.19** *The structure and function of DNA helicases. (a) The T7 gene four-ring helicase, which is a homohexamer with three non-equivalent states (PDB: 1EOK). (b) The winding mechanism involves the DNA visiting different sites around the ring (right), which changes affinity for the nucleic acid as it binds and hydrolyses nucleotide triphosphate then releases the nucleoside diphosphate and inorganic phosphate. Because of the helical symmetry of the DNA, it translates during the cycle. The mechanism is analogous to the rotatory process of the F1-ATPase (left)*

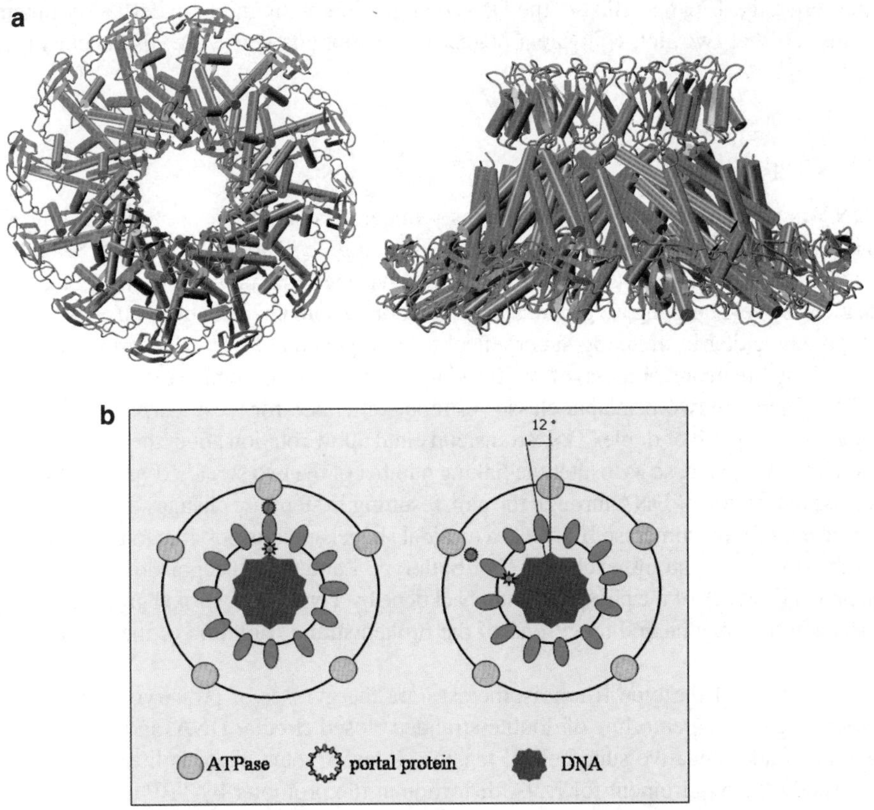


**Figure 10.20** *(a) The crystal structure of the central component of the motor from Bacillus subtilis bacteriophage $\phi$29 illustrates the principle of operation of DNA pumps (PDB: 1H5W). The panel on the left is a view down the molecular 12-fold axis and on the right is the side view. (b) A schematic of the mismatched symmetries of the five helicase subunits, the 12 portal proteins, and the duplex DNA. The asterisk indicates the point of contact of the DNA with the portal protein, which changes during rotation of the portal ring*
(Figure kindly provided by F. Anson, University of York)

at its narrowest point of 36 Å, which is large enough to accommodate the duplex DNA. The negative charge of the ring might electrostatically impede the exit of the compacted DNA. A second ring is composed of five subunits, and is therefore a symmetry mismatch with the 12-membered ring. The subunits are RNA–protein complexes that undergo conformational change upon binding and hydrolysis of ATP, and it is possible that this conformational switch may lead to asymmetry around the ring, just as in the case of the T7 helicase and F1 ATPase mentioned above (Figure 10.20b). The dodecameric connector may rotate in response to ATP binding and hydrolysis. To preserve its interaction with the connector, the DNA must translate along its helical axis by 1/5th its pitch, corresponding to two base pair steps for B-form DNA (Figure 10.20b). Thus, the pump is cleverly designed to convert the mismatch of molecular symmetries between the two rings and the DNA into a translational displacement of the DNA.

An analogous pumping mechanism may occur in the export of DNA by bacteria during transfer of plasmid DNA ('conjugation') *via* a hexameric ATPase. Here, the DNA must move across a lipid-bilayer membrane, which is a very strong permeability barrier to the nucleic acid. The pump acts as an energy-inducible, selective channel that transports the DNA vectorially.[52,53] The protomer of the hexameric pump has structural similarity with the RecA ATPase that forms helical filaments on single-stranded DNA.[54] It seems likely that, during the translocation process, each protomer may interact with the DNA in a similar fashion to

RecA. Perhaps, like the T7 ring helicase, the DNA conjugation pump might undergo asymmetric changes within the ring, so that two non-equivalent states of the subunits cause the displacement of the DNA duplex.

### 10.8.3 DNA Topoisomerases

Genomic DNA is packed densely and continuously undergoes superhelical strain during transcription, replication and (occasionally) recombination. For example, in *E. coli*, replication involves a rate of unwinding of 160 turns of duplex DNA per second: it can be envisaged that unwinding must be compensated at comparable rates to avoid tangling. Organisms in all taxonomic domains have evolved enzymes that manipulate DNA topology, either by relaxing superhelical tension generated from transcription, replication and recombination (a spontaneous process) or by introducing supercoiling (which requires an input of energy) (Section 2.3.5). There are two principal classes of topoisomerases for these purposes: type I and type II. The former break one strand of duplex DNA transiently and allow rotation about the unbroken strand before the broken strand is re-ligated, so as to alter the linking number of the two strands. Type II enzymes break both strands and pass one region of DNA through the gap, resulting in stepwise changes in linking number of two units. The structure of topoisomerase II reveals a toroidal shape, and it is possible to imagine the DNA passing through the ring during strand exchange. For both type I and type II topoisomerases, strand cleavage involves nucleophilic attack of the phosphodiester backbone by a tyrosine hydroxyl group, resulting in a covalent bond between the enzyme and the 5′ end of the broken strand. Rotation of the 3′ end is followed by re-ligation.

A unique sub-group of the type II topoisomerases are the gyrases of prokaryotes, which catalyse the ATP-dependent negative supercoiling of double-stranded closed circular DNA, and they thus maintain the bacterial genome under negative superhelical tension. In eukaryotes, there is little superhelical tension, which may explain the requirement for DNA distortion at the promoter by TBP and the requirement for helicase activity in one of the transcription factors (Section 10.3.6).

## 10.9 RNA–PROTEIN INTERACTIONS AND RNA-MEDIATED ASSEMBLIES

Just as in the recognition of DNA, many different protein structural motifs are used to recognize RNA. RNA–protein complexes that have been characterized so far have tended to be more complicated than protein–DNA complexes, since RNA structures are themselves more complex than double-stranded DNA, and include duplex regions, pseudo-knots and single-stranded regions, such as bulges and hairpin loops (Section 7.1).[55] Loops can be sub-divided into nearly a dozen different known classes of internal loops that occur between adjoining duplex regions and external loops in hairpins.

RNA–protein complexes are divided roughly into two main classes, depending on the mode of RNA binding: the **groove-binding mode** and **β-sheet binding**, which uses the sheet to bind unpaired RNA bases.[56] For groove binding, the α-helix is a predominant secondary structural element for recognition. In many RNA–protein complexes, backbone and base contacts are used with roughly equal frequency, in contrast to protein–DNA interactions where backbone contacts are dominant. Because of the intricate tertiary structure of the RNA, the packing density of the RNA–protein interface is often not as great as in complexes of protein with single- and double-stranded DNA. However, the recognition of RNA generally involves more van der Waals contacts than hydrogen bonds.[57]

Sequence-specific complexes of RNA and protein achieve their specificity through tight packing of comparatively non-polar surfaces. Thus, aromatic residues play key roles, probably by stacking on the unpaired bases. The 2′-hydroxyl group often protrudes from these RNA targets; it is more solvent-exposed than the other sugar oxygen atoms,[57] and is therefore a target for hydrogen-bonding interactions. Induced fit occurs frequently in protein–RNA recognition. For instance, assembly of almost all ribosomal protein–RNA complexes involve induced fit for either the protein, or the RNA or for both.[58]

### 10.9.1 Single-Stranded RNA Recognition

Here, we describe only a few, representative RNA/protein complexes that illustrate how recognition of single-stranded RNA occurs.

Genes in the tryptophan biosynthetic pathway in *E. coli* are tightly controlled in response to cellular concentrations of the amino acid Trp, and the dimeric protein known as the trp repressor modulates transcription rates of the trp operon. In many species of bacteria, the expression of these genes is also controlled post-transcriptionally through the binding of the messenger RNA by the trp RNA-binding attenuation protein.[59] This protein forms a molecular ring with 11 subunits. Each subunit binds the triplet GAG on the periphery of the ring (Figure 10.21a). The RNA is recognized by its unstacked bases, which are interdigitated with aliphatic and aromatic amino acid side chains.

An analogous mode of recognition is used by proteins that contain an **RNP-binding motif**, also known as an **RNA recognition motif** (RRM), which is found widely in all life forms. An example is the U1A spliceosome protein from the RNA-splicing machinery, which binds part of the of U1 snRNA (Figure 10.3a). A related mode of recognition of an RNA hairpin is seen in the structure of the MS2 bacteriophage capsid bound to the initiation site of assembly.[60]

Other components of the RNA spliceosome are the Sm proteins. In eukaryotes, the Sm proteins form heteroheptameric rings that are the main components of the spliceosomal small nuclear ribonucleoproteins (snRNPs).[61] Surprisingly, bacteria and archaea also contain an Sm-like protein that forms a hexamer and may serve as a carrier of small regulatory RNA and other RNA molecules (Figure 10.21b).[62–65] Thus, the Sm fold represents a common modular binding unit for oligomeric RNA-binding proteins.

The S1 and KH folds are other classes of small RNA-binding motifs that occur widely in nature. The S1 domain was originally identified in the ribosomal protein S1, and it belongs to the wider OB fold such as in the Pot protein mentioned earlier (Figure 10.8a). This motif occurs in the principal ribonucleases of messenger RNA decay in *E. coli*. In these enzymes, the conserved ligand-binding cleft has a groove of positive charge that contains exposed aromatic residues.[66] The K homology (KH) domain, first identified in the hnRNP K protein, is found in all taxonomic domains of life and often has a role in sequence-specific binding of RNA. They often are found as tandem copies that may be linked flexibly.[67]

### 10.9.2 Duplex RNA Recognition

The modular zinc-finger motif plays an important role in DNA recognition (Section 10.3.3), but it has also a second function in the cell in the recognition of RNA. For example, transcription factor TFIIIA contains nine zinc fingers, three of which can interact with a segment of the ribosomal 5S RNA (Figure 10.22). The fold of the 5S RNA involves both duplex regions and two loops. The helices are oriented so that the positive end of the helix dipole points towards the phosphate backbone, in a similar way to that seen in the histone–DNA complex in the nucleosome (Figure 10.15).

A completely different mode of recognition of duplex RNA is used by the p19 protein from tombavirus (a plant virus), which has no apparent sequence-specificity, but is specific for an optimal length of duplex RNA. Here, a β-sheet serves as a platform for interactions with the RNA, which are mostly with the phosphate backbone and sugar 2′-hydroxyl groups.[68] The p19 protein suppresses the plant's anti-viral defence by sequestering 19–20 long duplex RNAs that guide destruction of the viral mRNA by cellular ribonucleases.

### 10.9.3 Transfer RNA Synthetases

To achieve high fidelity in the translation of genetic information, the cell must ensure that each and every tRNA is charged correctly with its corresponding amino acid. This essential task is carried out by the **aminoacyl-transfer RNA synthetases** (Section 7.3.2). All synthetases catalyse a two-step process. The first step involves the binding of the substrates – ATP and amino acid by the tRNA synthetase and the formation of a covalent link between them. In the second step, the amino acid is transferred onto the tRNA.

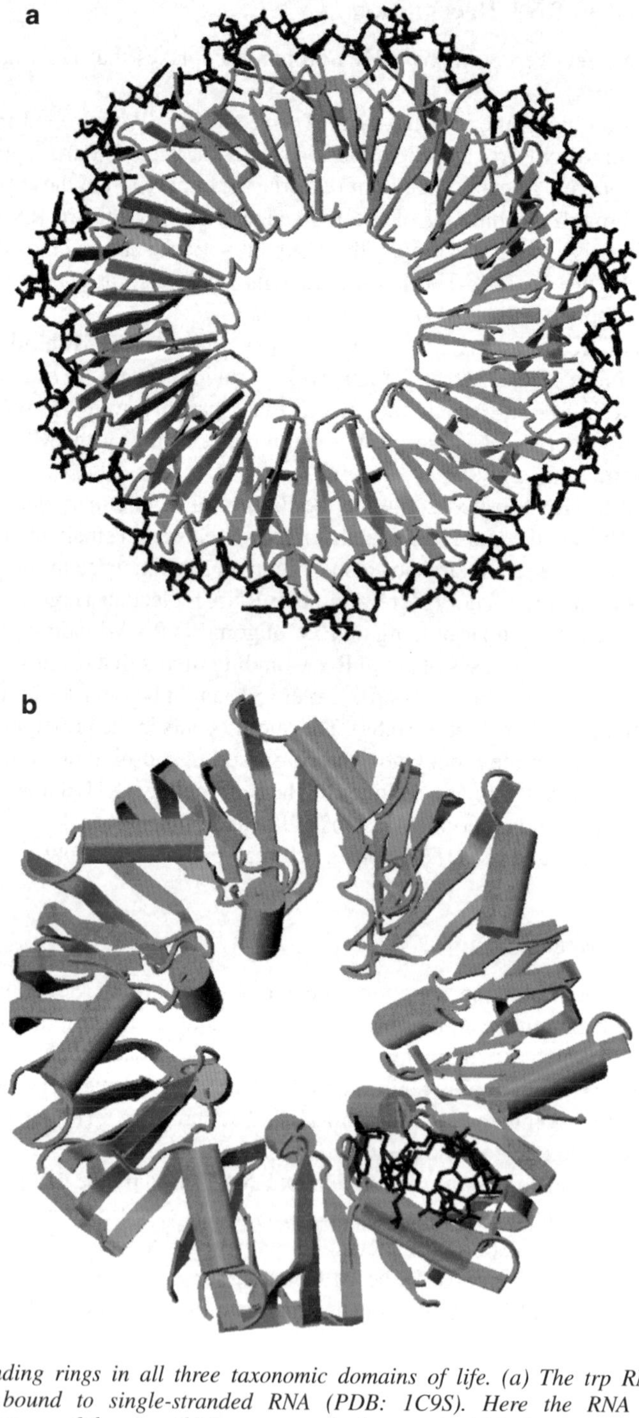

**Figure 10.21** *RNA-binding rings in all three taxonomic domains of life. (a) The trp RNA-binding attenuation protein bound to single-stranded RNA (PDB: 1C9S). Here the RNA is bound around the circumference of the ring. (b) In contrast the hexameric Af-Sm1 protein from the archaebacterium Archaeoglobus fulgidus forms a complex with short poly-U RNA in its central channel (PDB: 1I5L). A homologous protein is found in eukaryotes*

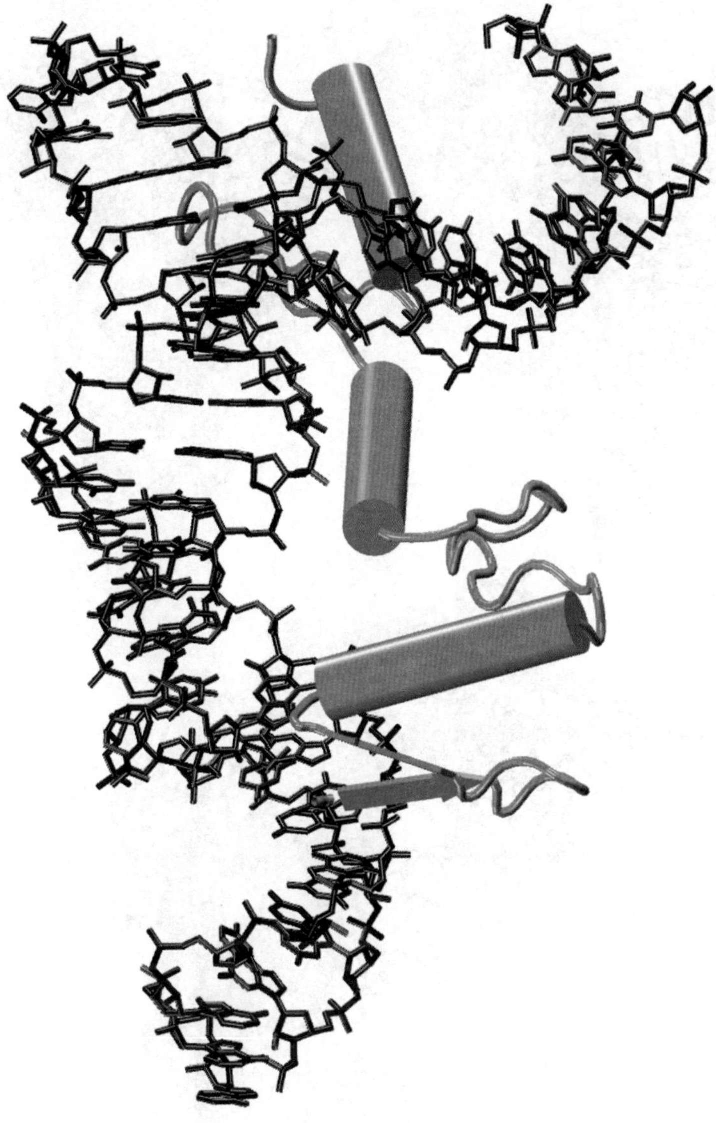

**Figure 10.22** *The complex of three of the nine zinc fingers for transcription factor TFIIIA with a portion of 5S RNA of the ribosome (PDB: 1VN6)*

All tRNAs have the same general shape, resembling the letter 'L', and all have a CCA sequence at the 3′ terminus that serves as amino-acid acceptor (Figure 7.6). Here, the ribose of the 3′-terminal A becomes linked to the amino acid *via* an ester linkage. The synthetases either link the amino acid onto the tRNA at the ribose 2′-hydroxyl group (Class I) or at the 3′-hydroxyl group (Class II).

The classes of synthetase have different structures and ways of recognizing their cognate tRNA. In all cases, a single domain binds the amino acid, ATP and the acceptor 3′-terminal A residue. The Class I group uses a variant of the '**Rossmann fold**' – a common dinucleotide-binding domain composed typically of six parallel β-strands. In the synthetases, only five parallel strands are found. Class II synthetases use instead an anti-parallel six-stranded β-sheet.

The first crystal structure of a tRNA-synthetase complex to be solved, which was also the first protein–RNA complex structure to be studied, was of *E. coli* glutaminyl-tRNA synthetase with its cognate tRNA

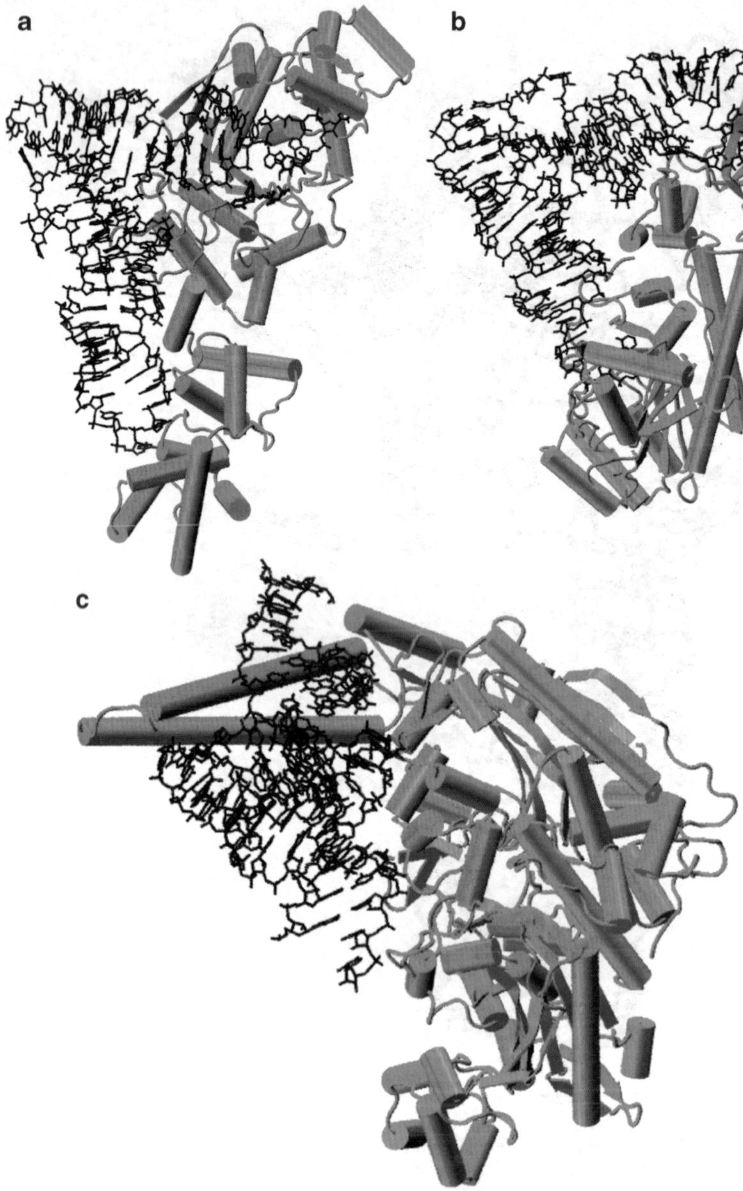


**Figure 10.23** *Representative amino acyl tRNA synthetases showing different modes of RNA–protein recognition. (a) Glutamyl-tRNA synthetase (PDB: 1N77). (b) Aspartyl-tRNA syntetase (PDB: 1JGO). (c) Seryl-tRNA synthetase (PDB: 1SER)*

(Figure 10.23a). This Class I synthetase binds on one side of the tRNA and contacts the sugar–phosphate backbone and the **minor groove** side of the acceptor stem. The acceptor stem is recognized by 'indirect readout' determined by the conformation of the CCA end. By contrast, the structure of the complex between the Class II yeast aspartyl-tRNA synthetase and its cognate tRNA shows that this enzyme binds tRNA from the **major groove** side (Figure 10.23b).

It may seem surprising that the determinants of recognition do not in all cases involve directly the anticodon triplet, but instead are often distributed throughout the tRNA. For example, the structure of seryl-tRNA synthetase complexed with tRNA (Figure 10.23c) shows a distinctive mode of recognition where an

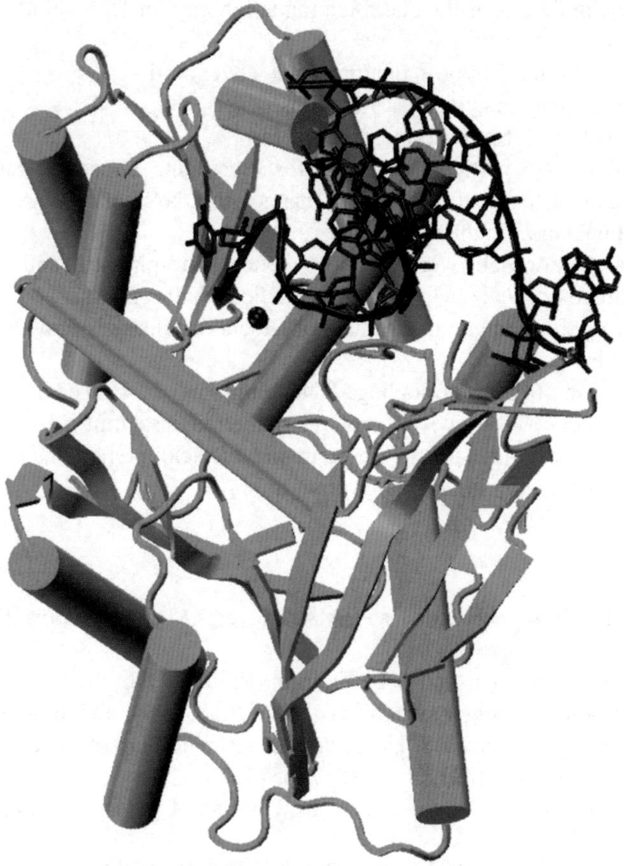

**Figure 10.24** *PIWI–siRNA complex. The domain is shown in red, the duplex composed of the siRNA guide strand and target mRNA strand is shown in black. The manganese ion is shown in black, situated in the binding pocket for the first residue of the guide strand*

extra loop of seryl-tRNA is recognized by the enzyme, but not the anticodon loop. A conserved N-terminal domain of the protein forms a coiled-coil region that is involved in RNA recognition.

### 10.9.4 Small Interfering RNA Recognition

Small interfering RNAs (siRNA) play an important role in gene regulation in eukaryotes (Section 5.7.2). These RNAs are bound by specific protein complexes and target the cleavage of complementary RNA transcripts. The crystal structure of a 16-mer siRNA duplex with the recognition domain (PIWI) has been reported. The structure shows how the RNA is recognized for length and without sequence preference. The guide siRNA strand (antisense strand) binds by use of a divalent metal ion to help place the complementary mRNA strand in a position to reach the putative catalytic site where degradation takes place (Figure 10.24).[69]

## WEB RESOURCES

There are many useful databases of structural and functional information of protein–nucleic acid complexes. Among these are the NDB: a database of nucleic acids and protein–nucleic acid complexes[70]. http://ndbserver.rutgers.edu

An interactive website with details of the classification taxonomy for DNA-binding motifs can be found at the address:

http://www.biochem.ucl.ac.uk/bsm/prot_dna/prot_dna_cover.html

Enzyme structural database with enzyme classifications:

http://www.biochem.ucl.ac.uk/bsm/enzymes/ index.html

Molecular graphics movies are particularly useful for illustrating allosteric transitions and proposed catalytic mechanisms. For instance, the allosteric changes in DNA helicase are shown in an idealized continuum of the structural states in the movie at:

http://sci.cancerresearchuk.org/labs/wigley/projects/helicase/morph1.html and
http://sci.cancerresearchuk.org/labs/wigley/projects/helicase/morph2.html

A graphics movie showing the catalytic mechanism of AP endonuclease is provided at:

http://www. scripps.edu/~jat/

A molecular graphics movie of the bacteriophage DNA pump may be viewed at:

http://bilbo.bio.purdue.edu/~viruswww/Rossmann_home/movies.shtml

ProNIT, a database for the thermodynamics of protein–nucleic acid interactions is available at:

http://gibk26.bse.kyutech.ac.jp/jouhou/pronit/pronit.html

## REFERENCES


1. C.R. Calladine, H.R. Drew, B.F. Luisi and A.A. Travers, *Understanding DNA*, 3rd edn, Elsevier, Amsterdam, 2004.
2. M.Y. Tolstorukov, R.L. Jernigan and V.B. Zhurkin, Protein–DNA hydrophobic recognition in the minor groove is facilitated by sugar switching. *J. Mol. Biol.*, 2004, **337**, 65–76.
3. J.M. Vargason, K. Henderson and P.S. Ho, A crystallographic map of the transition from B-DNA to A-DNA. *Proc. Natl. Acad. Sci. USA*, 2001, **98**, 7265–7270.
4. C.R. Calladine and H.R. Drew, A base-centred explanation of the B-to-A transition in DNA. *J. Mol. Biol.*, 1984, **178**, 773–782.
5. E.J. Gardiner, C.A. Hunter, M.J. Packer, D.S. Palmer and P. Willett, Sequence-dependent DNA structure: a database of octamer structural parameters. *J. Mol. Biol.*, 2003, **332**, 1025–1035.
6. A. Fersht *Enzyme Structure and Mechanism.* Freedman, New York, 1998.
7. N.M. Luscombe, R.A. Laskowski and J.M. Thornton, Amino acid–base interactions: a three-dimensional analysis of protein–DNA interactions at an atomic level. *Nucleic Acids Res.*, 2001, **29**, 2860–2874.
8. K. Nadassy, S.J. Wodak and J. Janin, Structural features of protein–nucleic acid recognition sites. *Biochemistry*, 1999, **38**, 1999–2017.
9. C.R. Woese, Interpreting the universal phylogenetic tree. *Proc. Natl. Acad. Sci. USA*, 2000, **97**, 8392–8396.
10. M.F. White and S.D. Bell, Holding it together: chromatin in the Archaea. *Trends Genet.*, 2002, **18**, 621–626.
11. J. Miller, A.D. McLachlan and A. Klug, Repetitive zinc-binding domains in the protein transcription factor IIIA from *Xenopus* oocytes. *EMBO J.*, 1985, **4**, 1609–1614.
12. Y. Choo and A. Klug, Physical basis of a protein–DNA recognition code. *Curr. Opin. Struct. Biol.*, 1997, **7**, 117–125.
13. C.O. Pabo and L. Nekludova, Geometric analysis and comparison of protein–DNA interfaces: Why is there no simple code for recognition? *J. Mol. Biol.*, 2000, **301**, 597–624.
14. P.A. Rice, S. Yang, K. Mizuuchi and H.A. Nash, Crystal structure of an IHF-DNA complex: a protein-induced DNA U-turn. *Cell*, 1996, **87**, 1295–1306.
15. T.W. Lynch, E.K. Read, A.N. Mattis, J.F. Gardner and P.A. Rice, Integration host factor: putting a twist on protein–DNA recognition. *J. Mol. Biol.*, 2003, **330**, 493–502.
16. F.V.T. Murphy, R.M. Sweet and M.E. Churchill, The structure of a chromosomal high mobility group protein–DNA complex reveals sequence-neutral mechanisms important for non-sequence-specific DNA recognition. *EMBO J.*, 1999, **18**, 6610–6618.

17. G.B. Koudelka, S.C. Harrison and M. Ptashne, Effect of non-contacted bases on the affinity of 434 operator for 434 repressor and Cro. *Nature*, 1987, **326**, 886–888.
18. S. Bergqvist, M.A. Williams, R. O'Brien and J.E. Ladbury, Heat capacity effects of water molecules and ions at a protein–DNA interface. *J. Mol. Biol.*, 2004, **336**, 829–842.
19. Z. Otwinowski, R.W. Schevitz, R.G. Zhang, C.L. Lawson, A. Joachimiak, R.Q. Marmorstein, B.F. Luisi and P.B. Sigler, Crystal structure of trp repressor/operator complex at atomic resolution. *Nature*, 1988, **335**, 321–329.
20. N.C. Horton, L.F. Dorner and J.J. Perona, Sequence selectivity and degeneracy of a restriction endonuclease mediated by DNA intercalation. *Nat. Struct. Biol.*, 2002, **9**, 42–47.
21. S. Bergqvist, M.A. Williams, R. O'Brien and J.E. Ladbury, Halophilic adaptation of protein–DNA interactions. *Biochem. Soc. Trans.*, 2003, **31**, 677–680.
22. A.V. Sivolob and S.N. Khrapunov, Translational positioning of nucleosomes on DNA: the role of sequence-dependent isotropic DNA bending stiffness. *J. Mol. Biol.*, 1995, **247**, 918–931.
23. A.M. Martin, M.D. Sam, N.O. Reich and J.J. Perona, Structural and energetic origins of indirect readout in site-specific DNA cleavage by a restriction endonuclease. *Nat. Struct. Biol.*, 1999, **6**, 269–277.
24. M.F. Perutz, *Mechanisms of Cooperativity and Allosteric Regulation in Proteins.* Cambridge University Press, Cambridge, 1990.
25. J.A. Lefstin and K.R. Yamamoto, Allosteric effects of DNA on transcriptional regulators. *Nature*, 1998, **392**, 885–888.
26. J.J. Kohler, S.J. Metallo, T.L. Schneider and A. Schepartz, DNA specificity enhanced by sequential binding of protein monomers. *Proc. Natl. Acad. Sci. USA*, 1999, **96**, 11735–11739.
27. A.D. Riggs, S. Bourgeois and M. Cohn, The lac repressor–operator interaction. 3. Kinetic studies. *J. Mol. Biol.*, 1970, **53**, 401–417.
28. S.E. Halford and J.F. Marko, How do site-specific DNA-binding proteins find their targets? *Nucleic Acids Res.*, 2004, **32**, 3040–3052.
29. B.A. Lieberman and S.K. Nordeen, DNA intersegment transfer, how steroid receptors search for a target site. *J. Biol. Chem.*, 1997, **272**, 1061–1068.
30. M. Kampmann, Obstacle bypass in protein motion along DNA by two-dimensional rather than one-dimensional sliding. *J. Biol. Chem.*, 2004, **279**, 38715–38720.
31. N.C. Horton, K.J. Newberry and J.J. Perona, Metal ion-mediated substrate-assisted catalysis in type II restriction endonucleases. *Proc. Natl. Acad. Sci. USA*, 1998, **95**, 13489–13494.
32. J.A. McClarin, C.A. Frederick, B.C. Wang, P. Greene, H.W. Boyer, J. Grable and J.M. Rosenberg, Structure of the DNA–Eco RI endonuclease recognition complex at 3 A resolution. *Science*, 1986, **234**, 1526–1541.
33. D.E. Draper, Protein–DNA complexes: the cost of recognition. *Proc. Natl. Acad. Sci. USA*, 1993, **90**, 7429–7430.
34. D.R. Lesser, M.R. Kurpiewski, T. Waters, B.A. Connolly and L. Jen-Jacobson, Facilitated distortion of the DNA site enhances EcoRI endonuclease-DNA recognition. *Proc. Natl. Acad. Sci. USA*, 1993, **90**, 7548–7552.
35. I.B. Vipond and S.E. Halford, Structure-function correlation for the EcoRV restriction enzyme: from non-specific binding to specific DNA cleavage. *Mol. Microbiol.*, 1993, **9**, 225–231.
36. C.D. Mol, T. Izumi, S. Mitra and J.A. Tainer, DNA-bound structures and mutants reveal abasic DNA binding by APE1 and DNA repair coordination [corrected]. *Nature*, 2000, **403**, 451–456.
37. A.Y. Lau, O.D. Scharer, L. Samson, G.L. Verdine and T. Ellenberger, Crystal structure of a human alkylbase-DNA repair enzyme complexed to DNA: mechanisms for nucleotide flipping and base excision. *Cell*, 1998, **95**, 249–258.
38. D.M. Popovic, A. Zmiric, S.D. Zaric and E.W. Knapp, Energetics of radical transfer in DNA photolyase. *J. Am. Chem. Soc.*, 2002, **124**, 3775–3782.
39. K. Sandman, S.L. Pereira and J.N. Reeve, Diversity of prokaryotic chromosomal proteins and the origin of the nucleosome. *Cell Mol. Life Sci.*, 1998, **54**, 1350–1364.

40. M.R. Singleton and D.B. Wigley, Multiple roles for ATP hydrolysis in nucleic acid modifying enzymes. *EMBO J.*, 2003, **22**, 4579–4583.
41. K.J. Polach and J. Widom, A model for the cooperative binding of eukaryotic regulatory proteins to nucleosomal target sites. *J. Mol. Biol.*, 1996, **258**, 800–812.
42. J.D. Anderson and J. Widom, Sequence and position-dependence of the equilibrium accessibility of nucleosomal DNA target sites. *J. Mol. Biol.*, 2000, **296**, 979–987.
43. T.A. Steitz, A mechanism for all polymerases. *Nature*, 1998, **391**, 231–232.
44. J.C. Delaney, P.T. Henderson, S.A. Helquist, J.C. Morales, J.M. Essigmann and E.T. Kool, High-fidelity *in vivo* replication of DNA base shape mimics without Watson–Crick hydrogen bonds. *Proc. Natl. Acad. Sci. USA*, 2003, **100**, 4469–4473.
45. T.H. Tahirov, D. Temiakov, M. Anikin, V. Patlan, W.T. McAllister, D.G. Vassylyev and S. Yokoyama, Structure of a T7 RNA polymerase elongation complex at 2.9 A resolution. *Nature*, 2002, **420**, 43–50.
46. Y.W. Yin and T.A. Steitz, Structural basis for the transition from initiation to elongation transcription in T7 RNA polymerase. *Science*, 2002, **298**, 1387–1395.
47. P.D. Morris and K.D. Raney, DNA helicases displace streptavidin from biotin-labeled oligonucleotides. *Biochemistry*, 1999, **38**, 5164–5171.
48. S.S. Velankar, P. Soultanas, M.S. Dillingham, H.S. Subramanya and D.B. Wigley, Crystal structures of complexes of PcrA DNA helicase with a DNA substrate indicate an inchworm mechanism. *Cell*, 1999, **97**, 75–84.
49. D.E. Smith, S.J. Tans, S.B. Smith, S. Grimes, D.L. Anderson and C. Bustamante, The bacteriophage straight phi29 portal motor can package DNA against a large internal force. *Nature*, 2001, **413**, 748–752.
50. A. Guasch, J. Pous, B. Ibarra, F.X. Gomis-Ruth, J.M. Valpuesta, N. Sousa, J.L. Carrascosa and M. Coll, Detailed architecture of a DNA translocating machine: the high-resolution structure of the bacteriophage phi29 connector particle. *J. Mol. Biol.*, 2002, **315**, 663–676.
51. A.A. Simpson, Y. Tao, P.G. Leiman, M.O. Badasso, Y. He, P.J. Jardine, N.H. Olson, M.C. Morais, S. Grimes, D.L. Anderson, T.S. Baker and M.G. Rossmann, Structure of the bacteriophage phi29 DNA packaging motor. *Nature*, 2000, **408**, 745–750.
52. S.N. Savvides, H.J. Yeo, M.R. Beck, F. Blaesing, R. Lurz, E. Lanka, R. Buhrdorf, W. Fischer, R. Haas and G. Waksman, VirB11 ATPases are dynamic hexameric assemblies: new insights into bacterial type IV secretion. *EMBO J.*, 2003, **22**, 1969–1980.
53. F.X. Gomis-Ruth, G. Moncalian, R. Perez-Luque, A. Gonzalez, E. Cabezon, F. de la Cruz and M. Coll, The bacterial conjugation protein TrwB resembles ring helicases and F1-ATPase. *Nature*, 2001, **409**, 637–641.
54. R.M. Story and T.A. Steitz, Structure of the recA protein–ADP complex. *Nature*, 1992, **355**, 374–376.
55. K. Nagai, RNA–protein complexes. *Curr. Opin. Struct. Biol.*, 1996, **6**, 53–61.
56. D.E. Draper, Themes in RNA–protein recognition. *J. Mol. Biol.*, 1999, **293**, 255–270.
57. S. Jones, D.T. Daley, N.M. Luscombe, H.M. Berman and J.M. Thornton, Protein–RNA interactions: a structural analysis. *Nucleic Acids Res.*, 2001, **29**, 943–954.
58. J.R. Williamson, Induced fit in RNA–protein recognition. *Nat. Struct. Biol.*, 2000, **7**, 834–837.
59. N.H. Hopcroft, A. Manfredo, A.L. Wendt, A.M. Brzozowski, P. Gollnick and A.A. Antson, The interaction of RNA with TRAP: the role of triplet repeats and separating spacer nucleotides. *J. Mol. Biol.*, 2004, **338**, 43–53.
60. K. Valegard, J.B. Murray, N.J. Stonehouse, S. van den Worm, P.G. Stockley and L. Liljas, The three-dimensional structures of two complexes between recombinant MS2 capsids and RNA operator fragments reveal sequence-specific protein–RNA interactions. *J. Mol. Biol.*, 1997, **270**, 724–738.
61. C. Kambach, S. Walke, R. Young, J.M. Avis, E. de la Fortelle, V.A. Raker, R. Luhrmann, J. Li and K. Nagai, Crystal structures of two Sm protein complexes and their implications for the assembly of the spliceosomal snRNPs. *Cell*, 1999, **96**, 375–387.

62. I. Toro, S. Thore, C. Mayer, J. Basquin, B. Seraphin and D. Suck, RNA binding in an Sm core domain: X-ray structure and functional analysis of an archaeal Sm protein complex. *EMBO J.*, 2001, **20**, 2293–2303.
63. C. Sauter, J. Basquin and D. Suck, Sm-like proteins in Eubacteria: the crystal structure of the Hfq protein from *Escherichia coli. Nucleic Acids Res.*, 2003, **31**, 4091–4098.
64. S. Thore, C. Mayer, C. Sauter, S. Weeks and D. Suck, Crystal structures of the *Pyrococcus abyssi* Sm core and its complex with RNA. Common features of RNA binding in archaea and eukarya. *J. Biol. Chem.*, 2003, **278**, 1239–1247.
65. M.A. Schumacher, R.F. Pearson, T. Moller, P. Valentin-Hansen and R.G. Brennan, Structures of the pleiotropic translational regulator Hfq and an Hfq-RNA complex: a bacterial Sm-like protein. *EMBO J.*, 2002, **21**, 3546–3556.
66. M. Schubert, R.E. Edge, P. Lario, M.A. Cook, N.C. Strynadka, G.A. Mackie and L.P. McIntosh, Structural characterization of the RNase E S1 domain and identification of its oligonucleotide-binding and dimerization interfaces. *J. Mol. Biol.*, 2004, **341**, 37–54.
67. H.A. Lewis, K. Musunuru, K.B. Jensen, C. Edo, H. Chen, R.B. Darnell and S.K. Burley, Sequence-specific RNA binding by a Nova KH domain: implications for paraneoplastic disease and the fragile X syndrome. *Cell*, 2000, **100**, 323–332.
68. K. Ye, L. Malinina and D.J. Patel, Recognition of small interfering RNA by a viral suppressor of RNA silencing. *Nature*, 2003, **426**, 874–878.
69. J.S. Parker, S.M. Roe and D. Barford, Structural insights into mRNA recognition from a PIWI domain-siRNA guide complex. *Nature*, 2005, **434**, 663–666.
70. H.M. Berman, W.K. Olson, D.L. Beveridge, J. Westbrook, A. Gelbin, T. Demeny, S.-H. Hsieh, A.R. Srinivasan and B. Schneider, The Nucleic Acid Database: a comprehensive relational database of three-dimensional structures of nucleic acids. *Biophys. J.*, 1992, **63**, 751–759.

CHAPTER 11

# Physical and Structural Techniques Applied to Nucleic Acids

## CONTENTS

## 11.1 SPECTROSCOPIC TECHNIQUES

### 11.1.1 Ultraviolet Absorption

The light absorption characteristics of nucleic acids[1] result from the combination of the strong ultraviolet (UV) absorption of the purine and pyrimidine bases in the 240–280 nm range modulated by the stereochemistry and conformational influences of a ribose-phosphate backbone that is essentially transparent to light of that wavelength (Section 2.1.3).

Oligonucleotides exhibit a strong UV absorption maximum $\lambda_{max}$ at approximately 260 nm and a molar extinction coefficient $\varepsilon$ of the order of $10^4\,dm^3\,mol^{-1}\,cm^{-1}$ (Table 2.2). This absorption arises almost entirely from complex electronic transitions in the purine and pyrimidine components. The intensity and exact position of $\lambda_{max}$ are functions not only of the base composition of the nucleic acid but also of the state of the base-pairing interactions present, the salt concentration of the solution and its pH. Most importantly, base–base stacking results in a decrease in $\varepsilon$ – a situation known as hypochromicity. This arises from dipole–dipole interactions that depend on the three-dimensional (3D) structure of an oligonucleotide and ranges in magnitude from 1–11% for deoxyribonucleoside phosphates to 30% for most helical polynucleotides. While some degree of stacking is apparent for all dimers, it appears that UpU, UpA, UpC, GpU and UpU are generally less stacked than other ribodinucleoside phosphates at pH 7. In practice, the effects of structure on the UV absorption of oligonucleotides are so complex that only the most basic interpretations can be made (Figure 11.1).

In practical terms it is possible to estimate the molar quantity of an oligonucleotide, which is for example chemically synthesised (Section 4.1), by measurement of the number of absorbance units at 260 nm ($A_{260}$, the optical density) of its solution in a UV spectrometer and by relating the value obtained to the sum of the extinction coefficients of the individual nucleotides (*e.g.* 8.8 for dT, 7.3 for dC, 11.7 for dG and 15.4 for dA for $A$ in $cm^2\,\mu mol^{-1}$). Such estimation makes no allowance for hypochromicity, which will depend on the particular sequence and whether it forms a secondary structure or not.

Ultraviolet absorption is a sensitive and convenient way to monitor the 'melting behaviour' of DNA and RNA. When the UV absorption of a nucleic acid sample is measured as a function of temperature, the resulting plot is known as a melting curve. The midpoint in the melting curve showing the increase in absorbance with increasing temperature is known as the melting temperature $T_m$. This is dependent on the base composition of the sample, the salt concentration of its solution and even the type of counter-ion present (Section 5.5.1). Such melting is a co-operative phenomenon and the observed melting curves become progressively

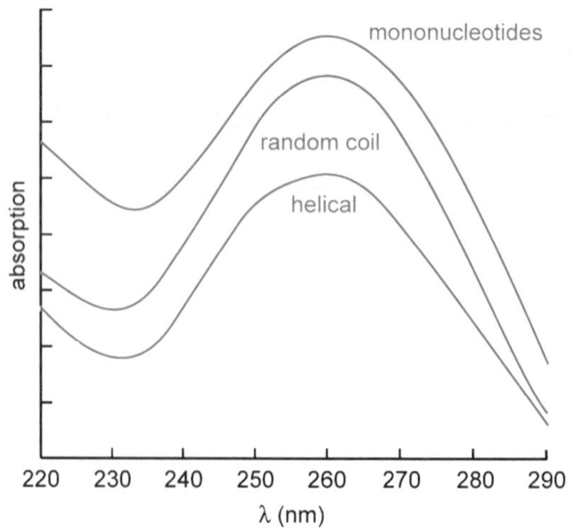


**Figure 11.1** *Typical UV absorption curves for equimolar base concentrations of mononucleotides, of single-stranded (random coil) oligonucleotide and of double-helical DNA*

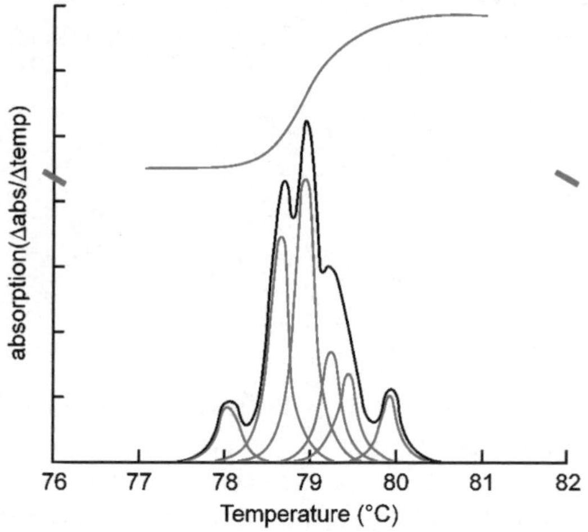


**Figure 11.2** *An oligonucleotide melting curve (red), the derivative of the curve (black) and the deconvolution of the derivative into its composite components (red)*

steeper with increasing length of the oligonucleotide. Simple sigmoid melting curves are observed for many DNA samples, but it is also possible to observe more complex, multi-phase melting in some cases (Figure 11.2). The deconvolution of such multi-phase melting curves makes it possible to examine the effects of base modification on the stability and nature of nucleic acid secondary structure in more detail.

### 11.1.2 Fluorescence

Fluorescence is defined as the emission of radiation as a molecule returns to its ground state from an excited electronic state.[2] In order to characterise the photoexcited emission from a molecule, it is necessary to determine the spectral distribution, photon yield, excited-state lifetime and polarisation of the emission, all as functions of excitation wavelength. The precision with which fluorescence intensity can be measured is very high, since photon-counting techniques eliminate several sources of uncertainty and error. Excitation and emission occur in timescales of $\sim 10^{-15}$ s, but the lifetime of an excited state has a duration of $\sim 10^{-9}$ s. Thus any physical process that takes place on a similar timescale as the fluorescence lifetime can be analysed by examining changes in the emission.

The fluorescence emission from nucleotides and dinucleoside phosphates is very weak at room temperature and can only be examined in frozen samples at ~80 K. Fluorescence spectroscopy is nevertheless invaluable for examining nucleic acid–ligand interactions. Many DNA-binding ligands have little or no fluorescence in aqueous solution. However, upon binding to a nucleic acid the ligand is in a hydrophobic environment and the solvent can no longer quench the intrinsic ligand fluorescence. Therefore fluorescence emission is a direct probe of the concentration of bound ligand.

**Fluorescence resonance energy transfer (FRET)** is a phenomenon in which the energy of an excited-state fluorescent donor molecules is transferred to an unexcited acceptor molecule *via* a dipole–dipole coupling. Importantly, the rate of energy transfer is dependent on the distance between donor and acceptor molecules, the spectral characteristics of the pair and the relative orientations of the donor and acceptor transition dipoles. FRET experiments have been used widely to determine proximity relationships in protein and nucleic acid systems since the 1940s when Förster derived the quantitative analysis for FRET. A key parameter in FRET is the efficiency of depopulation ($E_T$) and this is related to fluorescent lifetimes in the presence ($\tau_T$) and absence ($\tau$) of resonance energy transfer [Equation (11.1)]:

$$E_T = 1 - \frac{\tau_T}{\tau} \qquad (11.1)$$

The donor–acceptor distance ($R$) is related to $E_T$ by the system constant $R_o$, which is the distance at which there would be 50% of maximal energy transfer [Equations (11.2) and (11.3)]:

$$E_T = \frac{R_o^6}{R^6 + R_o^6} \tag{11.2}$$

$$R = R_o \left( \frac{1 - E_T}{E_T} \right)^{1/6} \tag{11.3}$$

From these relationships, FRET theory predicts that fluorescence transfer depends on the sixth power of the distance between partner molecules. In practice, the distances over which FRET can be measured vary between 40 and 100 Å (Figure 11.3). There are several factors that can influence the extent of fluorescence transfer, and hence precautions need to be taken in order to account for all possible energy transfer pathways and to properly quantify FRET measurements.

There are a few naturally occurring fluorophores found in nucleic acids, for example wyosine found in tRNA. More commonly synthetic fluorophores (*e.g.* Figure 8.13) can be introduced into oligonucleotides. Such modifications are extremely useful for studying nucleic acid structure and function. For FRET experiments, in general one or both of the donor and/or acceptor is either covalently or non-covalently attached to a nucleic acid, often *via* chemical synthesis.

Site-specific labelling of longer (thousands of base pairs) DNA samples remains problematic, although recent advances in the use of sequence-specific methyl transferases to add specific labels to defined DNA sites are promising. Normally DNA methyl transferases (MTases) transfer an activated methyl group to the N-6 position of an adenine or C-5/N-4 position of a cytosine within a specific cognate sequence. By use of a fluorophore and a flexible linker, sequence-specific labelled DNA can be obtained in a methyl transferase–catalysed reaction. This technique developed by Elmar Weinhold is known as **SMILing DNA**.

David Lilley and co-workers have used FRET to delineate the overall geometry of four-way DNA junctions[3] and the fold of the hammerhead ribozyme.[4] In addition, FRET is a useful method for distinguishing between intercalation and groove binding in small molecule–DNA interactions (Section 9.2), since in general FRET is observed during intercalation but not during groove binding.

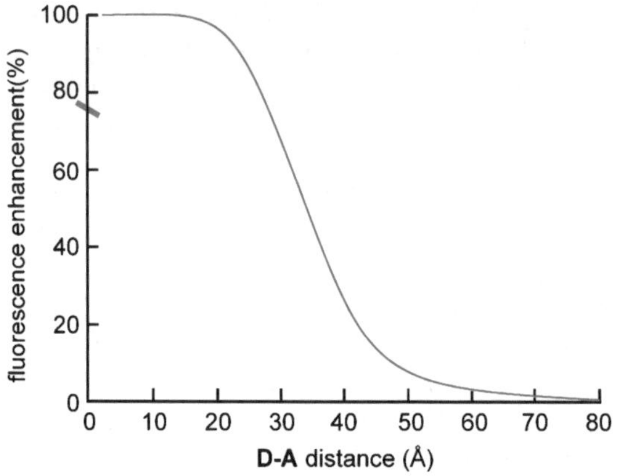


**Figure 11.3** *A typical relationship between fluorescence enhancement, resulting from FRET, and the distance between donor and acceptor fluorophores*

### 11.1.3 Circular and Linear Dichroism

**Circular dichroism** (CD) is a widely used form of chiral spectroscopy which has been applied to the study of nucleic acids.[5,6] A CD signal results from the differential absorption of left and right circularly polarised light. Circularly polarised light has chirality and therefore it will exhibit chiral discrimination. A chiral molecule will absorb left and right circularly polarised light to differing extents and this difference gives rise to the phenomenon of CD. Typically, data are presented as plots of $\Delta\varepsilon$ *versus* wavelength. Here $\Delta\varepsilon$ is the difference in molar extinction coefficient for left and right circularly polarised light ($\varepsilon_L - \varepsilon_R$).

Isolated purine and pyrimidine bases are planar, intrinsically optically inactive and hence do not exhibit a CD signal. However when incorporated into nucleosides and nucleotides, the glycosylic bond from the C-1′ atom of the sugar to either the N-9 of purines or N-1 of pyrimidines gives rise to a chiral perturbation of the UV absorption of the base. The CD signal of a nucleic acid increases with length due to the co-operativity of chiral interactions between contiguous bases. This effect occurs both as a result of sequence effects arising from nearest-neighbour interactions as well as from overall gross secondary structure.

The information derived from CD spectra is complementary to other types of optical spectroscopy, such as UV, infrared (IR) and linear dichroism (LD), and it provides a quick, convenient and accurate picture of the overall conformation and secondary structure of a particular nucleic acid solution. Reference CD spectra for a B-DNA duplex, single strands and a G3 quadruplex show that each type of secondary structure exhibits its own signature spectra (Figure 11.4). A- and Z-type DNA (Section 2.2) also have unique CD spectra, as do many nucleic acid–ligand complexes. CD spectroscopy also allows inter-conversions between different secondary structures to be monitored. For example titration of NaCl into a solution of poly(dG-dC) to salt concentrations above 4 M induces a structural transition in the polynucleotide from a standard right-handed B-helix to the left-handed Z-form. Increasing the temperature of a nucleic acid solution whilst measuring the CD signal allows DNA melting to be monitored directly and it is possible to observe the structural transition from folded duplex to random coil single strands.

Circular dichroism spectroscopy is also a useful tool for studying nucleic acid–ligand interactions. The CD spectrum of each component in solution is directly proportional to its concentration (**Beer's law**), and the total spectrum arises from the sum of all component spectra. If ligand binding induces extrinsic optical activity in the chromophores of the bound ligand, an *induced* CD signal is observed, which is directly proportional to the amount of nucleic acid–ligand complex formed, and hence it can be used to construct a binding isotherm. Alternatively, ligand binding may result in a conformational change in the nucleic acid, and the resultant change in the intrinsic CD signal of the macromolecule allows the binding to be quantified.

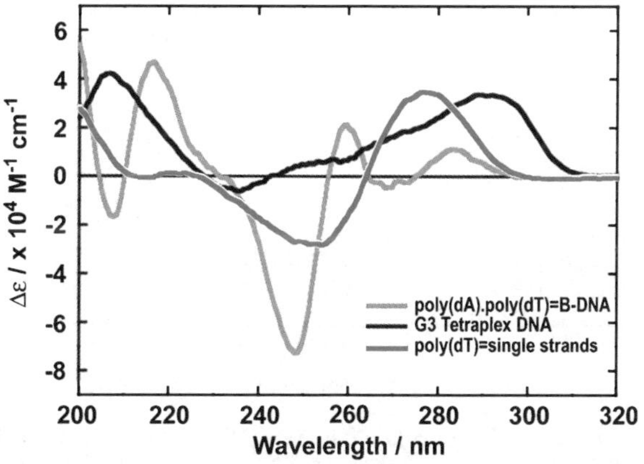


**Figure 11.4** *Circular dichroism spectra for three different conformations of DNA, a poly(dA)·poly(dT) B-type duplex, a guanine quadruplex (tetraplex) and a single-stranded random coil*

Modern CD spectrapolarimeters are often equipped with Peltier temperature control, automated titration and stopped-flow accessories which allows CD to be used to provide information on structure, thermodynamics (binding affinity, van't Hoff enthalpy, free energy) and kinetics (association and dissociation rate constants).

A closely related technique to CD is **linear dichroism** (LD). LD is a type of spectroscopy that yields useful information on DNA conformation in terms of base inclination and flexibility as well as the binding geometries of drug–DNA complexes. In this technique, the differential absorption of linearly or plane-polarised light is measured. A key parameter in LD is the **transition moment**, which is a vectoral property of light absorption related to a particular direction of the molecule. Light that is polarised parallel to the transition moment has a high probability of absorption in the region of spectral interest, whereas if light is polarised perpendicular to the transition moment, no absorption takes place. In practice, this means that intercalators that stack closely to base pairs have linear dichroism similar to the base pairs themselves. However, the dichroism of groove binders is frequently opposite to that of the base pairs, since they bind along the edges of the base pairs. Hence LD is a useful type of spectroscopy for assessing the binding mode of a drug to DNA.

### 11.1.4 Infrared and Raman Spectroscopy

Infrared and Raman spectroscopy are often regarded as closely related techniques in which the vibrational frequencies of localised parts of the sample molecule are observed.[7,8] Both techniques are largely non-destructive and can be used on microscopically small samples. A major advantage is that DNA can be analysed in crystals, gels or fibres, as well as in solution, and this has supported direct correlations between the observed vibrational frequencies and 3D structures derived by X-ray diffraction.

Infrared absorptions from nucleic acids are observed in the frequency range between 1800 and 700 $cm^{-1}$. The problem of strong IR absorption by water near 1600 $cm^{-1}$ and also below 1000 $cm^{-1}$ can be circumvented using $D_2O$ as solvent when the water signal at 1600 $cm^{-1}$ is shifted to about 1200 $cm^{-1}$ and the absorption at 1000 $cm^{-1}$ is shifted by an almost equivalent amount (Figure 11.5). So, measuring IR spectra in both solvents makes it possible to observe the full spectral range of interest. The use of $D_2O$ also causes small but significant shifts in nucleic acid absorptions resulting from deuterium exchange, and these can be used to monitor H–D exchange processes.

Fourier transform infrared absorption (FTIR) is so sensitive that it is possible to make measurements on very small crystals of nucleic acids and needs about 100 μg of material. IR spectra are largely unaffected by the external environment and so FTIR supports the identification of structurally significant bands by

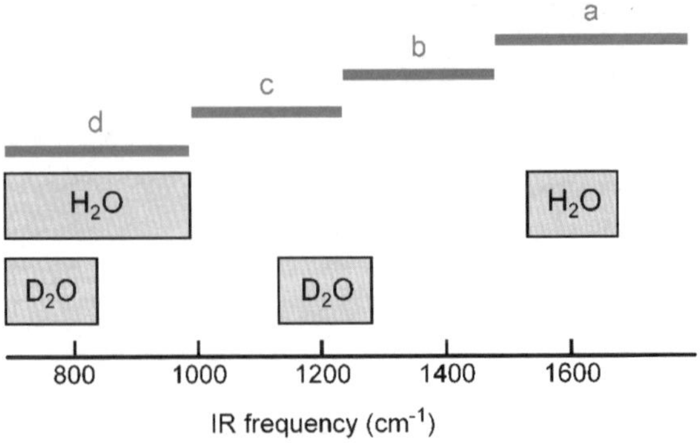


**Figure 11.5** *Schematic distribution of IR absorption bands of DNA and solvent. (a) 1800–1500 $cm^{-1}$ corresponding to stretching of C = X double bonds. (b) 1500–1250 $cm^{-1}$ corresponding to base-sugar entities (including glycosyl torsion angle effects). (c) 1250–1000 $cm^{-1}$ corresponding to phosphate and sugar absorptions. (d) Below 1000 $cm^{-1}$ associated with phosphodiester chain coupled with the sugar vibrations*

calibration with X-ray crystallography of nucleic acid samples. Such 'marker bands' are sensitive to helical type and can be used to show the existence of a conformational transition when different factors, such as temperature, hydration, concentration, or the nature and amount of cations are varied (B → A, B → C, B → Z-form, helix → coil, *etc.*). FTIR is also used to support studies on the recognition of DNA sequences by a wide variety of molecules, such as oligonucleotides (triple-stranded structures), drugs and proteins.

Raman spectroscopy also depends on the vibrational frequencies of groups within the molecule and, like IR, provides information concerning vibrational modes of nucleic acid components that are conformationally sensitive. The Raman technique has some useful advantages over IR technique. First, the incident radiation in Raman work is not strongly absorbed and so does little damage to the sample. Second, water has weak scattering properties and lacks absorption at the irradiation frequencies used for sample irradiation, so its presence in the sample is not a problem. Third, unlike IR, the intensity of the Raman bands is proportional to the concentration of the target species. As with IR, the accurate assignment of resonance lines can be simplified by calibration using samples of known X-ray structure.

Raman spectroscopy has been used to examine nucleic acids in a wide variety of situations including microcrystals and even within living cells.

## 11.2 NUCLEAR MAGNETIC RESONANCE

Nuclear magnetic resonance (NMR) spectroscopy is the method of choice for investigation of the conformation of short (up to about 25 base pairs) nucleic acid fragments in solution. Compared to other spectroscopic techniques routinely used, NMR is rather insensitive and requires millimolar sample concentrations. However, the structural information that may be gained from an NMR spectrum is much more detailed than that available from any other solution-phase technique and is complementary to that available from X-ray diffraction studies on solid samples. Indeed, interpretation of NMR data often requires an assumption of a structure, which may based on an X-ray diffraction structure of the specific or a related molecule. Solution-state NMR data can also reveal flexibility and dynamic behaviour.

The basis of NMR is that atomic nuclei are endowed with a property called **nuclear spin** and will align with an applied magnetic field ($B_o$).[9] The degree of this alignment is dependent not only on the strength of $B_o$ (magnets with field strengths up to 21.1 T are now commercially available) but also on the type of nucleus (different elements have different susceptibilities to magnetic fields) and its chemical environment (*e.g.* the attached bond and atom types, the 3D arrangement of those bonds, solvent, temperature). The magnet–nuclear spin alignment may be perturbed by the application of radiation from the **radio frequency** region of the **electromagnetic spectrum** (100–900 MHz currently utilised). For example a $^{1}H$ in a 14 T magnet would absorb in the region of 600 MHz, over a span of 12,000 Hz, and this may be considered as inducing a spin 'flip'. This excited state survives for a few seconds in some instances, which is why NMR spectral lines are sharp. The time taken to regain the ground state depends on the flexibility of the molecule and how easily it tumbles in solution. Thus, analysis of NMR data can confirm the bonding network in a molecule and the 3D arrangement of those bonds. It can establish the presence of flexibility and also identify and quantify molecular interactions.

In nucleic acids, the nuclei that are NMR active are $^{1}H$ and $^{31}P$, which have essentially 100% natural abundance. A proton has the most sensitive, non-radioactive, nuclear spin. Carbon and oxygen in their most abundant isotopes ($^{12}C$ and $^{16}O$) have zero spin and are therefore NMR silent. $^{15}N$ is NMR active but is very insensitive to magnetic fields and produces very broad line, unstructured spectra. It is possible to introduce chemically or enzymatically $^{13}C$ and $^{15}N$ isotopes, both of which give an NMR response. Fluorine ($^{19}F$) is another nucleus that may be introduced into chemically synthesised nucleic acids and is a very good probe of conformational flexibility. In a 14 T magnet, $^{31}P$ nuclei absorb energy around 243 MHz, $^{13}C$ around 151 MHz, $^{15}N$ around 61 MHz and $^{19}F$ around 564 MHz.

Whatever be the nuclear type observed, the spectrum has features common to all:[10] an *x*-axis that contains the frequency of the applied radiation and a *y*-axis that reports intensity, which relates to the concentration

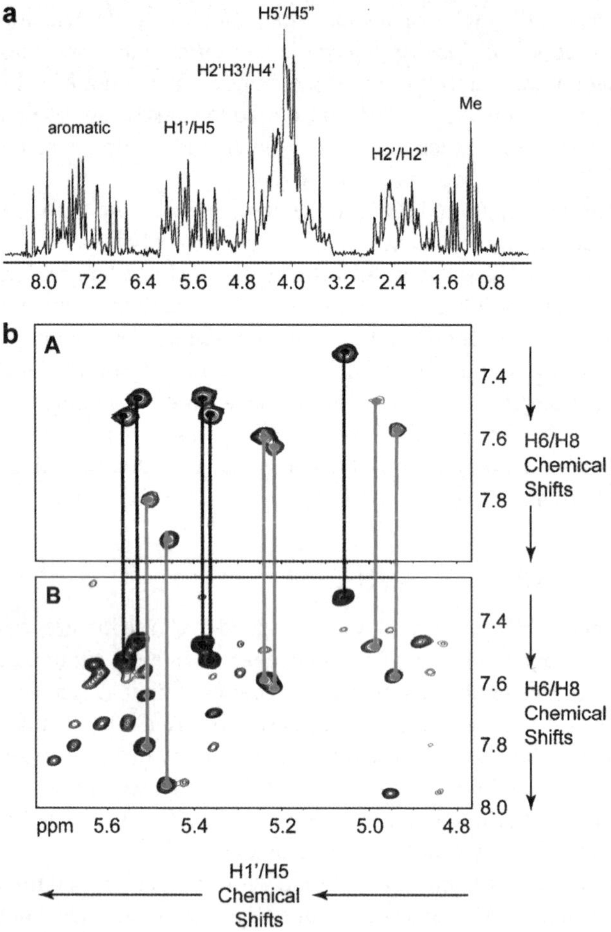


**Figure 11.6** *(a) A section of the 1D ¹H NMR spectrum for a 16 base-pair DNA/RNA hybrid 5′-d(ACTCGATTTCAT-AGCC)3′/5′-r(GGCUAUGAAAUCGAGA)3′. (b) The H5/H1′ to H6/H8 region of the (A) TOCSY and (B) NOESY spectrum. The black lines are for the DNA residues, the red lines for the RNA residues. This illustrates that nuclei that are close through bonds (shown in TOCSY spectrum) are generally also close through space (shown in NOESY spectrum)*

of the particular spin. The position or frequency of the spectral line is called the **chemical shift**, the fine structure observed on the lines is referred to as their **multiplicity**, and the separations within a multiplet are termed **scalar coupling**. A typical 1-(frequency) dimensional (1D) $^1$H spectrum for a DNA/RNA hybrid shows signals for over 300 different protons (Figure 11.6a), far too many to be readily distinguished and assigned a chemical shift. The overcrowding (signal overlap) problem is alleviated by introduction of further frequency dimensions.

A variety of experiments have been devised that enable scalar couplings (which are transmitted over covalent bonds), **nuclear Overhauser effects** (**nOes**, which are transmitted through space up to a distance of approximately 5 Å) and dynamic features to be detected in another frequency axis. For example a section of a 2D ($^1$H–$^1$H) **TOCSY** spectrum (revealing scalar coupling) and a **NOESY** spectrum (showing nOes) for the same DNA/RNA hybrid may be shown as a contour plot (Figure 11.6b). Each set of contours connects two protons (a cross-peak). Each cross-peak has a volume that relates to the size of the nOe (*i.e.* a large cross-peak volume means the protons are close in space) or the size of the coupling.

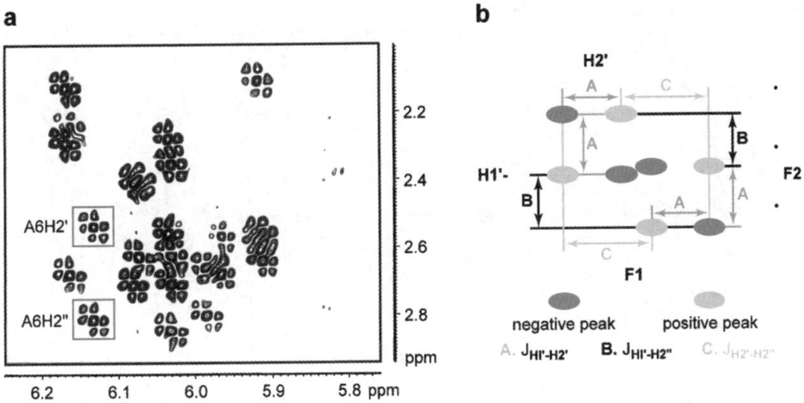


**Figure 11.7** *(a) A section of a PE-COSY spectrum for a DNA/RNA hybrid– cross-peaks connecting A6 H1′ to its H2′ and H2″ are boxed in red. (b) A sketch of one of the boxed regions showing how the coupling constants are measured*

Other 2D experiments permit accurate coupling constants to be measured and from these the conformation may be determined, for example of the ribose or dexoyribose sugar pucker. In a section of a **PE-COSY** (COSY stands for correlated spectroscopy) spectrum (Figure 11.7), note that the cross-peaks have a fine structure from which coupling constants (**J**) may be measured. The coupling between protons on adjacent carbon atoms (or any nuclear spin separated by three bonds, $^{3}$**J**) is related to the torsional angle (ϕ) between their bonds as described by the **Karplus Equation** (11.4):[9]

$$^{3}J_{HH} = 13.7\cos^2\phi - 0.73\cos\phi \quad (11.4)$$

With appropriately labelled samples, it is possible to measure $^{3}J$ for $^{1}H$–$^{1}H$, $^{1}H$–$^{13}C$, $^{1}H$–$^{31}P$ and $^{13}C$–$^{13}C$ nuclei and thereby describe not only the sugar pucker but also the geometry of the phosphate backbone. If the heterocyclic base is both $^{15}N$ and $^{13}C$ enriched, then the glycosylic torsion angle (see Section 2.1.4) may also be determined *via* measurement of the H1′ to C6 (or C8) coupling constant. This also connects a base to its sugar. Measurement of these heteronuclear couplings requires more complicated experiments that have acronyms such as **HMQC** (heteronuclear multiple quantum correlation) and **HMBC** (heteronuclear multiple bond correlation). These 2D experiments may be attached to COSY- or NOESY-type experiments to produce 3D and 4D spectra. If isotopically labelled material is not available to 'connect' a base to its own sugar, it is necessary to look for NOESY cross-peaks from the H1′ proton of the sugar to the H6 or H8 proton of the base. In this way, different nucleotide residues may be identified, but their sequence relationship is not yet established.

To assign a sequence position to each of the nucleoside residues, nOe connections have to be observed between adjacent residues.[11] The pattern of inter-nucleotide nOes is analysed with reference to a model: a B-type helix for DNA sequences, an A-type helix for RNA sequences. Analysis of the NOESY spectra involves a '**nOe-walk**', illustrated in practice in the DNA strand of a DNA/RNA hybrid (Figure 11.8). In the example shown, the $CH_3$ (Me) of the middle T in the $A_6T_7T_8$ stretch displays a NOESY cross-peak not only to its own ($T_7$) H6 but also to the H8 (or H6) of the residue 5′ to it ($A_6$ in this instance). $T_7$H6 on the other hand should share a cross-peak with the $CH_3$ of the 3′ residue ($T_8$ here) (highlighted in Figure 11.8) and so on.

From this analysis, the sequence ATT is confirmed. If there is only one of these sequences in the fragment, the residues may be numbered ($A_6T_7T_8$). The sequential assignment is further supported by observations of NOESY connections amongst sugar and aromatic protons. Qualitative analysis of some key cross-peak intensities can establish whether a sugar pucker is C3′-*endo* or C2′-*endo* (Section 2.1.4) and whether the

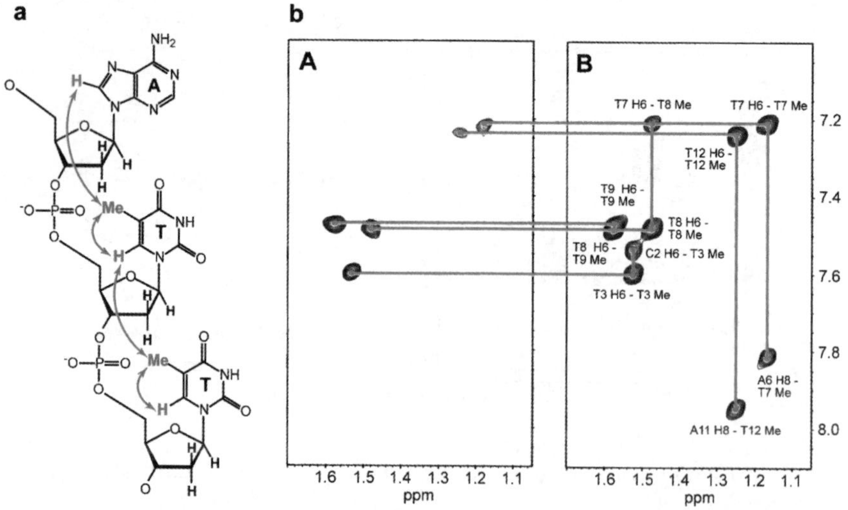


**Figure 11.8** *An example of a 'nOe-walk' for the sequential assignment of $^{1}H$ resonances. (a) The red arrows indicate key nOe connections. (b) The $CH_3$(Me) to H6/H8 region of the (A) TOCSY and (B) NOESY spectrum for the 16 base pair DNA/RNA hybrid (Figure 11.6). The 'walk' $A_6H8$ to $T_7Me$ to $T_7H6$ to $T_8Me$ to $T_8H6$ to $T_9Me$ is shown*

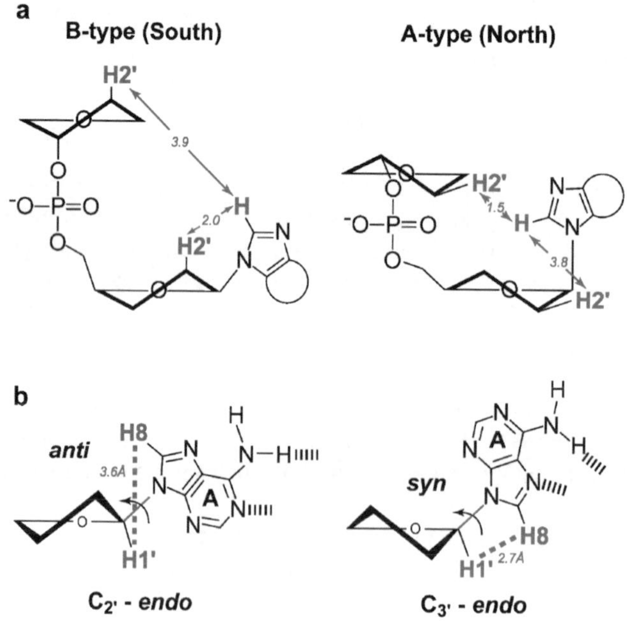


**Figure 11.9** *An illustration of the through-space connections sought to confirm: (a) the sugar pucker of the furanose ring or (b) the arrangement of the N-glycosylic bond. The numbers shown on the arrows are in Å. A short distance leads to a large NOESY cross-peak*

conformation about an *N*-glycosylic bond is *syn* or *anti* (Figure 11.9). This information lends support to the quantitative analysis of three-bond (or vicinal) scalar coupling constants, $^{3}J$.

Inter-strand hydrogen bonding may be detected indirectly by the observation of nOes among imino, amino and adenine H2 protons. If $^{15}N$-labelled bases are present, then the hydrogen bond can be detected directly through measurement of one-bond and two-bond inter-residue $^{1}H$–$^{15}N$ couplings.[12]

Once all or most of the $^1$H chemical shifts have been assigned, it simply remains to quantify NOESY cross-peaks. These, together with coupling constants from PE-COSY–type spectra, may be used in a variety of computational methods to generate 3D structures of nucleic acids. A simple method to quantify NOESY cross-peaks is to categorise them as 'small', 'medium' and 'large' by assignment of distance ranges, for example 3.7–5.8 Å, 1.7–3.8 Å and 0.7–2.8 Å, respectively. Through the use of approximately 500 nOe and coupling constant restraints, a structure for the DNA/RNA hybrid was obtained (Figure 11.10).

There is much less certainty about atom positions in NMR structures than in those usually achieved with X-ray crystallography. This is partly because in solution the molecule is flexible. But there may also be uncertainty due to inappropriate or an insufficient number of structural constraints. The generation of reliable

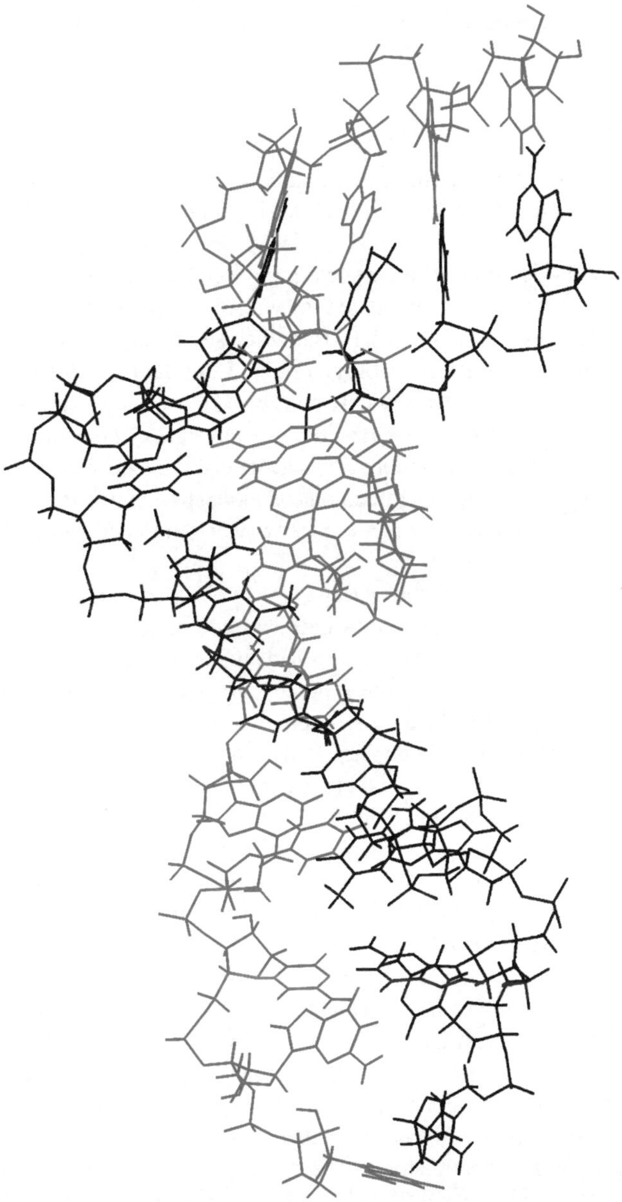

**Figure 11.10** *One of the structures generated from nOe and coupling constant measurements made for the DNA/RNA hybrid (Figure 11.6). The black strand is DNA and the red strand is RNA*

structures from NMR data has been an important issue until recently.[12,13] Approaches are now available to enable the measurement of **residual dipolar couplings**, which occur *via* a through-space process but over a longer range than nOes. To measure such couplings, it is necessary to partially align the nucleic acid molecule (not just the nuclear spin) with $B_o$, in a manner akin to a crystal lattice. To achieve molecular alignment, it is generally necessary for an agent to be introduced that has a preferred orientation, such as **phospholipid bicelles** or **filamentous phage Pf1**, to induce partial alignment of nucleic acid sequences. Such approaches are now aiding the conformational analysis of nucleic acids. NMR structures are now being generated of structural motifs, such as stem-loops, G-tetraplexes, I-motifs, pseudoknots and triplexes (Chapter 2),[14] as well as complexes of nucleic acids with drugs (Chapter 9) and proteins (Chapter 10).

Although the upper molecular mass limit for NMR structure determination for nucleic acids is around 25 kDa, a 100 kDa RNA has been studied by NMR.[15] Here only part of the RNA sequence was $^{13}C$ and $^{15}N$ enriched, thus simplifying the assignment procedure and allowing the RNA fragment to be studied within a larger molecule.

## 11.3 X-RAY CRYSTALLOGRAPHY

X-ray crystallography is the key method for determination of the 3D structure of biological macromolecules at atomic resolution, and therefore has had a major impact on our understanding of the structure and function of the nucleic acids.

In three decades of crystallographic structure determination, hundreds of structures of DNA, RNA and protein–nucleic acid complexes have accumulated, as witnessed by the 2334 structures deposited in the Nucleic Acid Database (NDB) as of March 16, 2004 (http://ndbserver.rutgers.edu),[16] and the 24,785 structures in the Research Collaboratory for Structural Biology/Protein Data Bank (RCSB/PDB) as of March 23, 2004 (http://www.rcsb.org/pdb).[17] Dramatic advances have been made in virtually all areas of X-ray crystallography, including crystallization (sparse-matrix crystallization screens) and crystal handling (cryo-protection), data collection and resolution (synchrotron radiation and CCD detectors), phasing (multi-wavelength anomalous dispersion, MAD), electron density map interpretation and model building (computer graphics and automatic chain tracing) and structure refinement (more computer power, simulated annealing and full-matrix least-squares refinement) (Figure 11.11a–e).[18–22]

The first step is the production of well-diffracting crystals. While oligonucleotides are relatively easy to crystallize, crystals with good diffraction qualities are less common. In order to obtain crystals with the necessary diffraction characteristics, it has proven desirable to consider the specific properties of a DNA and RNA molecule, to determine the physicochemical parameters that favour 3D assembly, and to analyse the inter-molecular contacts of the packing. In contrast to proteins, the packing environment of nucleic acids can stabilize or induce structural transitions. In general, it is necessary to vary the particular construct (lengths, sequence, blunt ends *vs.* dangling ends, *etc.*) and to test a large number of crystallization conditions.

Single-crystal diffraction data contain all the information needed to reconstruct the 3D structure of the unit cell and molecules in that structure except for the phase information, which can still present a formidable hurdle to solving some structures (Figure 11.11). The basic approaches to determining phase information are **multiple isomorphous replacement** (MIR), **multi- and single-wavelength anomalous dispersion** (MAD and SAD, respectively), and **molecular replacement** (MR). The first two techniques require the incorporation of 'heavy' atoms or heavy atom–containing derivatives into specific positions within the molecular lattice. Such incorporation can be achieved by either soaking the crystals in a solution containing the heavy atom, by crystallizing the compound in the presence of the heavy atom, or by incorporation of a heavy atom into the compound by chemical synthesis or modification, for example 5-bromo-2′-deoxyuridine. For MIR to be successful, it is an absolute requirement that crystals containing heavy atoms are of the same form as the original, so that their X-ray diffraction patterns can be used together with the diffraction pattern from the non-derivatised crystal to identify the positions of heavy atoms. The consequent phase and amplitude information can be used to calculate the phases of all other reflections. In many oligonucleotide structure determinations, the phase problem can be circumvented by use of MR, where a related structure or a model structure is used to

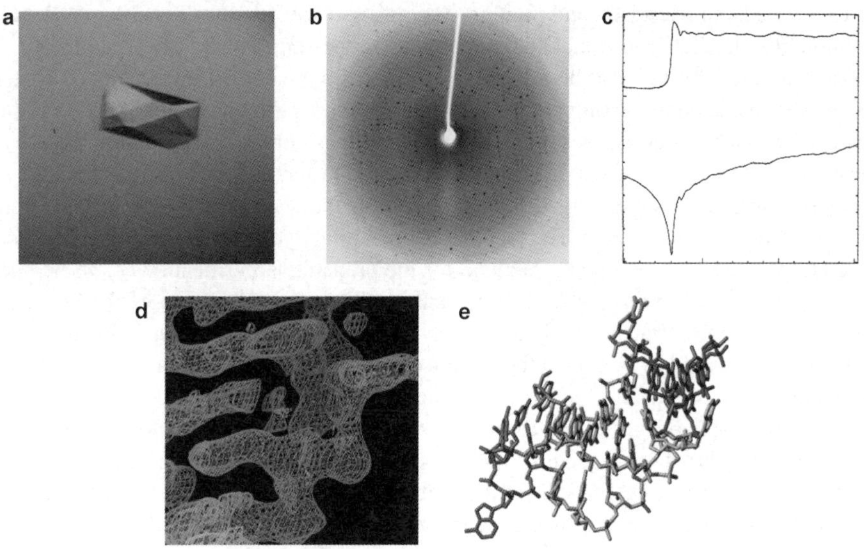


**Figure 11.11** *Steps in X-ray crystallographic structure solution. (a) A crystal. (b) An X-ray photograph. (c) Spot intensity measurement. (d) An electron density map. (e) A structural model*

search for possible solutions to the phase problem. By use of all the intensities and phase values from the diffraction pattern, it is then possible to calculate the electron density within all points in the unit cell by Fourier transformation. The quality and resolution of the electron density map depends on the number of reflections that can be included in the Fourier transformation and on the accuracy of their intensities and phases.

Oligonucleotides in crystals usually form close contacts with neighbouring molecules and their interactions can include base–base hydrogen bonding. Thus, the effect of crystal packing can be a more serious source of structural distortion. A difficulty in the interpretation of oligonucleotide crystal structures is the unknown extent to which crystallization conditions and crystal packing may have affected the structure. For example crystallization of nucleic acids usually requires a precipitant solvent, which often decreases the effective hydration of the molecule. For example dehydration of DNA can cause a structural change from a B-type to an A-type structure (Section 2.2). Occasionally oligonucleotides are crystallized in two different crystal forms. Provided the molecular structures are similar despite the different packing interactions, one may conclude that packing can be neglected in such cases as an important factor in the control of conformation. Crystal packing forces can sometimes be viewed as a blessing rather than a curse.[23] For example the GC/AT junction seems to be a flexible hinge, and packing forces can stabilize either straight or bent conformations, thus providing a source of information about DNA deformability.

If two or more crystal forms are available and/or the crystallographic asymmetric unit of a specific form contains several independent subunits, it may be possible to sample multiple conformations of the same molecule. Thus, alternative packing forces may trap particular conformational states of a molecule to yield dynamic information. Such an analysis has provided insights into the conformational basis for a bulge-mediated RNA self-cleavage reaction.[24]

## 11.4 HYDRODYNAMIC AND SEPARATION METHODS

### 11.4.1 Centrifugation

**Analytical ultracentrifugation** (**AUC**) is a powerful analytical tool that allows the hydrodynamic or thermodynamic solution state behaviour of proteins and nucleic acids to be characterised.[25,26] The AUC technique should not be confused with preparative procedures carried out by use of standard ultracentrifuges.

Classical experiments of Meselson and Stahl in 1957 utilised **equilibrium density gradient centrifugation** in caesium chloride (CsCl) to establish unambiguously the semi-conservative nature of DNA replication. A CsCl density gradient can also be established by diffusion under the influence of a centrifugal force to give a range of densities that encompass the buoyant densities of the molecular species to be separated. When centrifuged in such a gradient, a sample of DNA migrates until it reaches the point at which its own buoyant density equals that of the CsCl gradient. A number of different factors affect the buoyant density of DNA, for example the G + C content, the degree of hydration, the presence of other ions, pH or the presence of DNA binding or intercalating agents. The fact that intercalating dyes alter the buoyant density of DNA has been exploited to provide an experimental method for the preparative purification of DNA. The concentration of ethidium bromide bound to DNA is dependent on the superhelical state of the DNA. For example closed, circular plasmid DNA binds at a higher density than either nicked circular or linear DNA. Other cellular components are removed as a pellet at the bottom of the tube (*e.g.* RNA) or as a surface layer (proteins).

Modern, computer controlled, AUC instruments (such as the Beckman XL-I, which is equipped with both absorption and interference optics) allow characteristics such as molecular mass, stoichiometry of complexes, conformation and shape, diffusion and sedimentation, self-association and equilibrium constants to be studied quantitatively (Figure 11.12). The analytical ultracentrifuge spins the sample under a vacuum at a set spin speed and fixed temperature. At predefined time intervals, the instrument records the concentration distribution in the sample by measuring the absorbance. It is possible to attain speeds as high as 60,000 rpm ($\sim$250,000 $\times$ g). The sample itself is a solution in its native state and under biologically relevant conditions. It is held in specialised cells designed to withstand the high gravitational forces, which at the same time allow transmittance of light through the sample in the wavelength range 200–800 nm. The centrifuge cells accommodate both the sample and reference buffer, and the cells are assembled prior to each experiment from a cell housing, quartz or sapphire windows and one of several different types of centrepieces. Different types of rotors are available that can be used at different spin speeds or that contain either four or eight cell housings, one of which is used for the reference buffer in each case.

In general there are two different approaches to AUC: (1) **sedimentation velocity** and (2) **sedimentation equilibrium**. Sedimentation velocity is a hydrodynamic technique that is a probe for both mass and shape

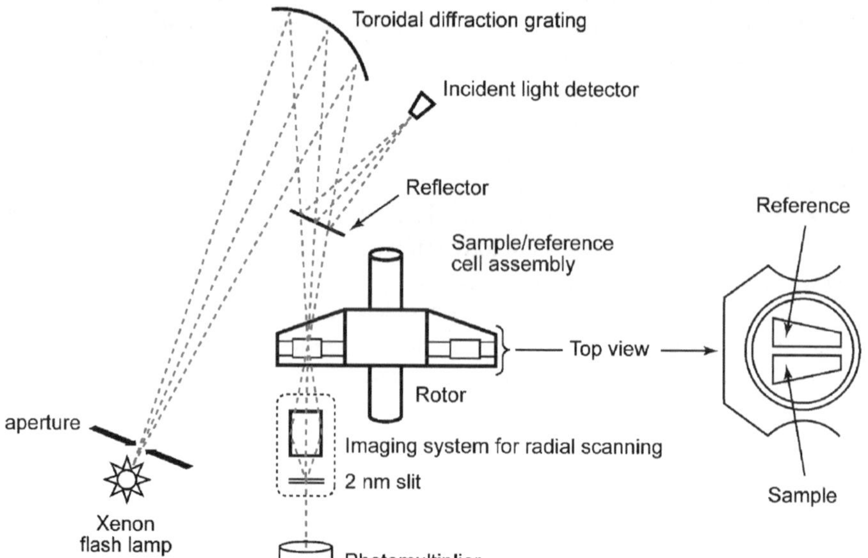


**Figure 11.12** *Schematic diagram of the optical arrangement of an analytical ultracentrifuge (this diagram is based on the Beckman Optima XL-A)*

of a macromolecule. A uniform solution of the sample is placed into the centrifuge cell, which is spun at high speed. This causes sedimentation of the macromolecule towards the bottom of the cell. This means that the macromolecule is depleted at the top of the cell (*i.e.* near the meniscus) and there is a sharp boundary, in terms of solute concentration, between the depleted region and the uniform concentration of the sedimenting molecule. A series of scans, where the sample concentration is measured as a function of radial distance, is taken at discrete time intervals and these measurements allow the rate of movement and broadening of the boundary to be monitored over time. From these data, the sedimentation coefficient ($s$) can be determined and this is directly proportional to the mass ($m$) of the solute and inversely proportional to the frictional coefficient ($f$), which is in effect a measure of size. These relationships are summarised in the Svedberg Equation (11.5).

$$s = \frac{v}{\omega^2 r} = \frac{m(1 - \bar{v}\rho_{solv})}{f} \qquad f = \frac{RT}{ND} \tag{11.5}$$

where $v$ is the velocity of the molecule, $\omega^2 r$ the strength of the centrifugal field ($\omega = 2\pi$.rpm/60 and $r$ is the radial distance from the centre of rotation), $\rho_{solv}$ the density of the solvent, $N$ Avogadro's number, $D$ the diffusional coefficient, $R$ the gas constant and $T$ the temperature.

Sedimentation equilibrium is a thermodynamic technique that is sensitive to the mass but not to the shape of the molecular species. The principal difference between sedimentation velocity and sedimentation equilibrium is that in the latter the initially uniform sample is spun at lower angular velocities than those employed in sedimentation velocity experiments. At the slower speeds used here, the macromolecule again starts to sediment towards the bottom of the cell and thus the concentration towards the bottom begins to increase. However, diffusion processes begin to oppose sedimentation, and after a suitable period of time the two opposing forces reach equilibrium. If several different species with different molecular weights exist, for example DNA that is free and bound to a ligand, or different association states of multi-subunit proteins, then each of the species will be distributed over the solution until it is at equilibrium. Higher molecular weight species will be nearer the bottom of the cell, whilst lower molecular weight species will be present at the top of the cell.

Sedimentation equilibrium is particularly useful for evaluation of the equilibrium association constant for reversible interactions such as ligand binding or protein self-association. The technique is sensitive to $K_{obs}$ values in the range 10–100 $M^{-1}$ but can also be used to measure affinities up to $10^7 M^{-1}$. The detailed mathematical theory that underlies both sedimentation velocity and equilibrium can be very complex, but there are several useful data analysis software packages that allow a non-expert to analyse and extract useful information.

Another type of ultracentrifugation is useful for examination of viscosity changes through use of sucrose density gradients. This method is sometimes used for studying various RNA species where separation depends largely on the size of the RNA molecule or for separating different classes of ribozymes. All these nucleic acid species have a higher buoyant density than the sucrose gradient and hence equilibrium is never reached. Thus separation depends on the different rates of migration of molecules through the sucrose gradient.

### 11.4.2 Light Scattering

Light scattering is another useful technique for determination of the molecular weight and size of biological macromolecules. Light scattering occurs when polarisable particles are exposed to an oscillating electric field present in a light beam. The varying field induces oscillating dipoles in the particles, which radiate light in all directions. The amount of light scattered is directly proportional to the product of the average molecular mass and the concentration of macromolecule. For globular structures such as proteins with molecular weight ~500 kDa or less, light scattering is uniform in all directions. Therefore, the amount of light scattered by a solution is measured at some angle relative to the incident laser beam. However for nucleic acids or other rod-like macromolecules, the scattering varies significantly with angle. Hence multiangle laser light scattering must be employed where light scattering is measured at a number of different angles. This allows the absolute molecular weight as well as the geometric size to be determined.

The fluctuation in **polarisability** observed in light scattering is caused by Brownian motion and is therefore related to the viscosity ($\eta$) of the solution and the diffusion coefficient ($D$) of the macromolecule. The diffusion coefficient is directly related to the hydrodynamic radius (**Stokes radius**) ($R_h$) as shown in Equation (11.6).

$$R_h = \frac{k_B T}{6\pi\eta D} \tag{11.6}$$

where $k_B$ is the Boltzmann constant and $T$ the temperature.

In solutions of large DNA molecules, one can observe the translational diffusion coefficients and also the rotational coefficients of the species. Also, because large DNA molecules exhibit a certain degree of flexibility, it is possible to determine the motions of small internal segments of these molecules.

**Dynamic light scattering** is a technique in which the time dependence of the light scattering from a small focussed region of the solution is measured over a timescale ranging from tenths of a millisecond to milliseconds. The time-dependent fluctuations in the intensity of scattered light are related to diffusion of molecules in and out of the region under study, which occur by Brownian motion. The data can be analysed to determine diffusion coefficients, or a distribution of diffusion coefficients if multiple species are present. In most cases, data is presented not in terms of the diffusion coefficient but rather in terms of particle size. The Stokes radius derived from this analysis gives the size of a spherical particle that would have a diffusion coefficient equal to the one observed for the protein or nucleic acid. Of course, most biological molecules, especially nucleic acids, are not spherical and their apparent hydrodynamic sizes are dependent on their shape, *i.e.* conformation, and molecular mass. Their diffusion is also affected by the hydration state. Hence the hydrodynamic size determined from dynamic light scattering could be significantly different from the true physical size observed in NMR or X-ray crystal structures.

### 11.4.3 Gel Electrophoresis

In free solution, the movement of DNA in an electric field is independent of shape and molecular mass and dependent only on charge. However when the DNA is exposed to an electric field in a gel matrix, the movement is dependent on size and shape as well as charge. Gels commonly used for nucleic acid electrophoresis are made of agarose and polyacrylamide. Both types of material consist of 3D networks of cross-linked polymer strands, which contain pores whose size varies according to the concentration of polymer used. The mobility of DNA in such gels is dependent mainly on size and shape, since the charge per unit length of DNA is effectively constant.

For linear DNA, there is an inverse relationship between size and rate of migration, such that mobility is approximately inversely proportional to $\log_{10}$ of the molecular mass. Horizontal **agarose gel electrophoresis** is useful for the separation of linear DNA molecules up to 2000 kbp. Staining with ethidium bromide can reveal the position of DNA in the gel or sometimes the DNA is labelled with $^{32}P$, in which case the DNA is detected by use of a phosphor imager or by autoradiography. **Polyacrylamide gel electrophoresis (PAGE)** is used for smaller DNA fragments. It can be used preparatively for species up to 1 kbp, and it forms the basis of rapid nucleic acid sequencing methods, when 8 M urea is added to denature the oligonucleotide.

One proposed explanation for the theoretical basis of size selection in gel electrophoresis is that the mobility is proportional to the volume fraction of the pores that can be entered. This theory predicts that mobility would decrease with increasing gel concentration and also with increasing molecular mass. This is consistent with the observation that very small nucleic acid fragments move independently of molecular mass.

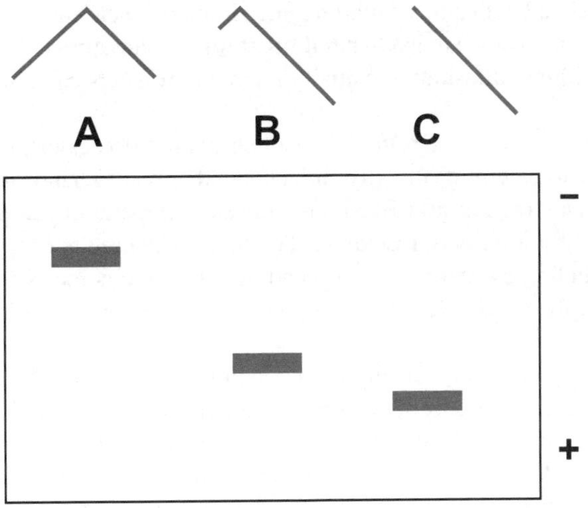


**Figure 11.13** *The electrophoretic mobility of curved DNA is dependent on the exact location of the bend in the DNA helix*

An alternative explanation assumes **reptation**, which is an end-to-end movement of the DNA through the pores of the gel. This theory explains the observed movement of DNA through gels more accurately, especially at high voltages where mobility becomes independent of molecular mass. In practice, the analysis of observed mobility in gels is made by comparison with and extrapolation from known standard samples.

The topological conformation of a DNA sample can have a marked effect on the rate of migration and the response of the sample to gel concentration and applied voltage. Relaxed circular DNA and supercoiled DNA migrate anomalously in agarose gel electrophoresis in comparison to linear DNA of similar molecular weight. The mobility of supercoiled DNA (Section 2.3.5) is also dependent on the linking number; as the linking number increases, so does the mobility of the sample. In addition, its mobility changes with the concentration of added ethidium bromide because increased intercalation of ethidium changes the number of superhelical turns.

The shape of linear DNA also has a strong effect on electrophoretic mobility. Bends or curvature in the DNA helix have the effect of slowing the migration relative to non-bent DNA. The extent of retardation depends on the end-to-end distance of the DNA and so is dependent on the location of the bend within the DNA helix (Figure 11.13). A qualitative theory, derived by Lumpkin and Zim, supports the general rule that mobility is related to the mean square end-to-end distance. In addition to the above examples, many elegant experiments in electrophoresis have been performed to examine the properties of poly(A) tracts, extra-base bulges, protein binding (gel-shift) and Holliday junctions.[27]

The fact that DNA of very high molecular mass does not separate well under normal electrophoretic conditions has led to the development of **pulsed field electrophoresis**. There are several variations of the technique, but in general the use of a voltage pulse with a resting period in between pulses changes the mobility of large DNA fragments and this can be used to separate DNA species of 1000 kbp or larger. The use of an alternating polarity field, with the positive pulse being longer than the negative pulse, enables the efficient resolution of very large DNA fragments. The application of inhomogeneous, perpendicular electric fields has been used to separate whole yeast chromosomes. In all of these applications, the size and duration of pulses can be tuned to obtain the specific resolution required.

### 11.4.4 Microcalorimetry

Calorimetry is a technique in which the heat of a reaction is measured. There are several types of calorimetry available that utilise a variety of different physical principles. Two direct methods are available for measuring the enthalpy change associated with either a binding interaction or a heat-induced conformational transition for nucleic acids and proteins. These are **isothermal titration calorimetry** (ITC) and **differential scanning calorimetry** (DSC).[28–33] Many ultra-high sensitivity microcalorimeters can now be obtained.

*11.4.4.1 Isothermal Titration Calorimetry.* Isothermal titration calorimetry is a powerful and flexible technique that is used to determine directly and independently the change in enthalpy ($\Delta H$) for almost any bimolecular binding interaction at a fixed and constant temperature. Depending on the magnitude of the binding constant and solubility of reactants, ITC can also be used to obtain binding isotherms, from which the equilibrium-binding constant ($K_b = K_a$) and binding stoichiometry ($n$) can be determined. This can therefore yield an almost complete thermodynamic profile for a binding interaction in a single experiment.

The calorimeter assembly consists typically of two identical cells (one reference and one sample) constructed of a metal alloy that has appropriate thermal properties, which is held within an adiabatic jacket. Detection of heat effects within the cells is achieved by placement of semiconductor thermopiles in between the cells and the jacket. Following injection of a ligand into the sample cell, the amount of thermal power that must be applied to actively balance the heat of reaction is measured. Thus the calorimeter monitors the heat effects in the sample cell and then applies appropriate power to maintain a zero temperature difference between sample and reference cells. The amount of heat applied to the cells is monitored continuously (**cell feedback** or CFB), and the amount of power applied is adjusted so as to drive the measured thermal power amplitude towards a stable baseline value. This is achieved by interfacing a computer to the system that continually monitors the output of a nanovoltmeter connected to record the output of the thermopiles. The experimentally measured quantity is the applied thermal power as a function of time required to return CFB to the stable baseline value subsequent to an injection of ligand (or system component), and is directly proportional to the reaction enthalpy. Modern ultrasensitive ITC instruments, such as the MicroCal VP-ITC, have baseline stability in the range $\pm 5$ ncal $s^{-1}$ and heat signals in the nanocalorie range can be detected.

In example data from an ITC experiment (Figure 11.14), Panel A shows primary data, where CFB is measured in $\mu$cal $sec^{-1}$ as a function of time. At equilibrium, CFB is held at a predefined value. Each peak represents the injection of a defined volume of ligand into a continually stirred sample cell. Addition of ligand leads to a binding interaction and concomitant heat effects; in this case an exothermic heat is observed. Each peak is integrated, with respect to time, to yield the binding isotherm in Panel B. After correction of background heats, due to dilution effects, and correction to a per mole basis the enthalpy is obtained directly and $K_{obs}$ and the stoichiometry ($n$) are obtained by fitting to an appropriate binding model, such that $n$ is the midpoint of the titration curve and $K_{obs}$ comes from the lineshape of the binding curve.

*11.4.4.2 Differential Scanning Calorimetry.* Differential scanning calorimetry involves the continuous measurement of the apparent specific heat of a system as a function of temperature. Hence, DSC can be used to examine physicochemical processes initiated by increases or decreases in temperature, for example phase transitions or conformational changes.

In general terms, a DSC instrument contains two cells, which are suspended in an adiabatic shield and connected *via* a multifunctional thermopile. In a typical experiment, the reference cell is filled with buffer and the sample cell is filled with buffer containing macromolecule (cell volume 0.5–1.5 mL). The temperature is increased to the 0.1–110°C range by use of electrical heaters that are in good thermal contact with the cells. During a heat-induced endothermic transition, the temperature of the sample cell falls behind that of the reference cell, since some of the energy required is used to induce the transition rather than to heat the solution. This lag is detected by the thermopile, so that extra electrical power is supplied to the sample cell to maintain it at a temperature identical to that of the reference cell. This additional energy is proportional to the energy associated with the thermally induced transition. Knowledge of the solute concentration enables a transformation of the observed electrical energy *versus* temperature profile into a curve that corresponds to an excess heat capacity *versus* temperature plot.

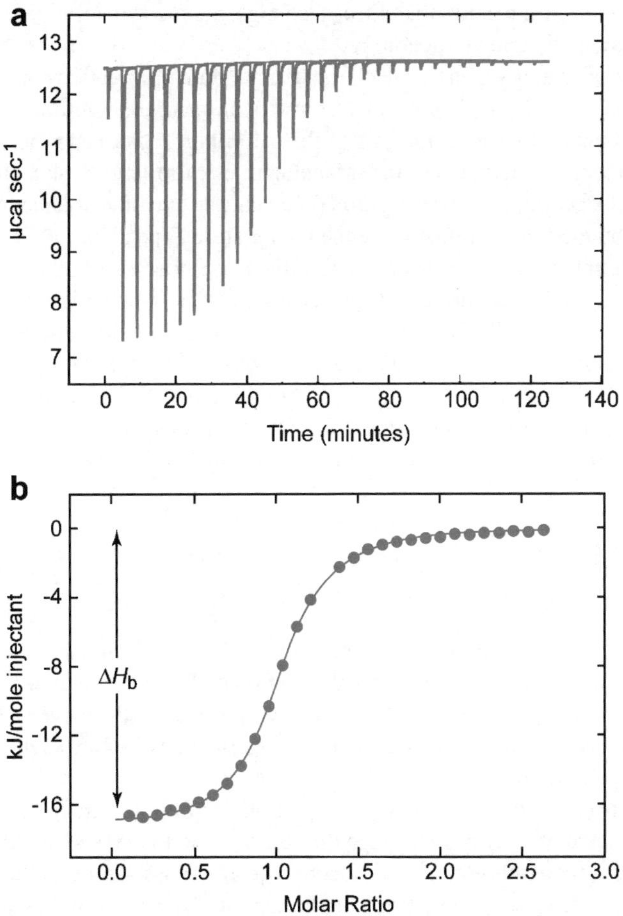


**Figure 11.14** *(a) Primary ITC data for ligand binding to DNA. Initially the baseline equilibrates at a preset value. After about 2 min an initial 3 μL preinjection is made and 5 min later the first experimental injection of 15 μL is made. The magnitude of the peaks is directly proportional to the power that is supplied to return cell feedback to the equilibrium value. (b) Integration of the peaks in Panel A with respect to time and correction to a per mole basis gives the binding isotherm*

A single DSC experiment can provide a large amount of thermodynamic information, much of which cannot be obtained by any other technique. The area under the experimental heat capacity curve can be used to determine the calorimetric transition enthalpy ($\Delta H_{\mathrm{cal}}$):

$$\Delta H_{\mathrm{cal}} = \int C_{\mathrm{p}} \cdot \mathrm{d}T \qquad (11.7)$$

This calorimetrically determined enthalpy is model-independent and is thus unrelated to the nature of the transition. The temperature at which excess heat capacity is at a maximum defines the **transition temperature** ($T_{\mathrm{m}}$). Differences in the initial and final baselines provide a measure of the heat capacity change that accompanies the transition. The experimental data can be converted into $C_{\mathrm{p,xs}}/T$ *versus* $T$ curves and, since (Equation (11.8)),

$$\Delta S = \int (C_{\mathrm{p,xs}}/T)\mathrm{d}T \qquad (11.8)$$

the area under such a curve represents the entropy change for the transition. From this information the value of $\Delta G$ can be evaluated at any temperature.

Differential scanning calorimetry data can also provide important information concerning the co-operativity of a thermal transition. This can be achieved by comparison of the magnitudes of the model-dependent van't Hoff enthalpy (obtained by shape analysis of the calorimetric data) and the calorimetric enthalpy. If $\Delta H_{vH} = \Delta H_{cal}$ and there are no chemical or instrumental kinetic limitations, such that $T_m$ is independent of both scan rate and concentration, then the transition proceeds in a two-state manner and meaningful thermodynamic data can be obtained by examination of the temperature dependence of an equilibrium property. If $\Delta H_{vH} < \Delta H_{cal}$, then the transition must involve a significant number of intermediate states. Conversely, if $\Delta H_{vH} > \Delta H_{cal}$, then aggregation is implicated. The $\Delta H_{vH}/\Delta H_{cal}$ ratio provides quantitative insight into the nature of the transition. Specifically, it provides a measure of the fraction of the structure that melts as a single thermodynamic entity (*i.e.* it defines the size of the co-operative unit). This is a unique advantage of DSC in the study of biological molecules and when combined with ITC, thermodynamic studies over a wide temperature range and under a variety of solution conditions can be carried out. These techniques offer a powerful tool to probe the energetics of macromolecule stability and ligand association reactions.

## 11.5 MICROSCOPY

### 11.5.1 Electron Microscopy

Electron microscopy (EM) is an invaluable tool for gaining structural information on systems that are too large for the atomic resolution methods of X-ray diffraction or NMR, and yet too small for optical microscopy.[34,35] Electron microscopy uses beams of electrons with wavelengths between 0.001 and 0.01 nm. The resolving power of EM can be below 2 nm but is largely restricted by factors such as radiation damage and the process of sample preparation.

The lenses in an electron microscope consist of shaped electric or magnetic fields. Electrons from a heated tungsten filament are accelerated across a voltage difference of up to 100 kV and the sample under study is examined in a vacuum. In **transmission EM** the electron beam is passed through the sample, then through a suitable lens system and is detected on a fluorescent plate or photographic film. In **scanning EM**, the electron beam is focused down to a point and scanned across the specimen. Secondary electrons are produced when electrons interact with the target specimen and these secondary electrons are detected and used to build up a raster image of the sample.

Both transmission and scanning EM require samples to be dehydrated and often coated by deposition of a film of either carbon or a metal such as tungsten. For nucleic acid analysis, one of the oldest means of sample preparation involves the combination of the nucleic acid with a thin film of denatured protein floating on water. The nucleic acid and protein film can be lifted off the surface onto a grid and then, after drying, coated with a fine layer of palladium or uranium. The process of coating, or shadow casting, involves depositing the metal from a heated filament that is oriented at an acute angle to the supporting grid. The effect of the metal deposition is to highlight regions that are raised above the grid. In order to achieve good contrast, either the samples can be shadowed from several angles or can be rotated during the coating procedure. The protein film has two benefits: one is that it supports and protects the DNA during preparation; the other is it causes thickening of the DNA strands. Nucleic acids will also bind directly to carbon supports that have been physically and chemically pre-treated in a suitable fashion. They can then be dried and shadowed in the same manner as in film deposition.

The technique of **cryogenic EM (cryo EM)**[36] uses samples that have been flash frozen into a thin film of vitreous water. The sample is prepared as an aqueous solution and applied to a carbon grid with holes of about 3 μm. Blotting of the grid followed by quick freezing in ethane at near its freezing temperature produces vitrified samples which can be directly imaged by EM.

While the contrast in cryo EM is relatively poor, it has the great advantage that a biological specimen is preserved in its solution conformation without many of the possible artefacts that can result from the dehydration and coating processes. Several images can then be recorded on the same sample, which is tilted between recordings, and used to reconstruct the full 3D shape of the sample (with resolution *ca.* 10 Å). Although the ultimate resolution of cryo EM is limited, crucial information about the interaction of nucleic acids with other macromolecules can be gained by the marriage of this tool with other structural methods. Cryo EM has also been used recently to investigate more complex nucleic acid incorporated processes, including those involved in transcription.

### 11.5.2 Scanning Probe Microscopy

The field of **scanning probe microscopy** (**SPM**) refers to a class of techniques, devised originally during the 1980s, which provide images of samples through use of a tiny probe (tip) to 'feel' the outermost properties of a flat surface. The members include the **scanning tunnelling microscope** (**STM**), the **atomic force microscope** (**AFM**), the scanning near-field optical microscope and the scanning capacitance microscope, but it is the first two of these which have contributed most to the structural analysis of nucleic acids.

*11.5.2.1 Scanning Tunnelling Microscopy.* Scanning tunnelling microscopy was developed in *ca.* 1982. This instrument exploits the 'electron tunnelling' effect: electrons are able to pass through a potential barrier when the distance of travel is small. The STM uses an atomically sharp **conducting probe** that scans closely (typically angstrom) across the sample immobilized on a **conducting substrate**. The movement of the probe or sample in *x*, *y* and *z* directions is realised by a piezoelectric crystal, most often in the form of a tube scanner. The scanner is capable of sub-angstrom resolution, with the *z*-axis conventionally perpendicular to the sample in all scanning probe miscroscopes.

By monitoring the tunnelling current as the probe is scanned over the sample, any spatial variation in the electronic topography of the surface is recorded. Due to the exponential dependence of the tunnelling current on the probe-sample separation, the recorded image contrast invariably arises from variations in the surface topography of the sample. Any variations in the electronic nature of the substrate under study, however, also contribute to the recorded signal. Although STM is capable of achieving atomic resolution, in analysis of biological molecules, image resolution is at best several nanometres.[37] Thus, whilst STM has been used for the imaging of some nucleic acid samples, the interpretation of the images has always been somewhat open to question.

Various sample preparation approaches have been used for STM analysis. The most straightforward involves deposition from a dilute solution, with the subsequent drying of solvent. The requisite conductive properties of the sample are normally acquired through deposition on conducting substrates (atomically flat gold and highly oriented pyrolytic graphite surfaces are frequently used), coating molecules deposited on mica with thin films of platinum–iridium–carbon or through variation in the relative humidity of the imaging environment. However, in reality the small number of suitable conducting surfaces, and the restrictions imposed by the working environment (in air or vacuum) have limited the analysis of nucleic acids by STM.

The limitations of STM subsequently led to the development of the **AFM** or **scanning force microscope** (**SFM**) (*ca.* 1986). The AFM utilises a similar instrumental setup to the STM, but replaces the conducting metal probe with a sharp probe (typically silicon or silicon nitride), which is situated at the apex of a cantilever spring of known stiffness (spring constant). As the probe is scanned over the surface of the sample, the cantilever bends or twists in response to forces acting between the probe and the sample. By reflection of a laser beam off the back surface of the cantilever onto a position sensitive photodiode, variation in the bending or twisting of the cantilever can be monitored to produce an image (Figure 11.15).

*11.5.2.2 Atomic Force Microscopy.* The AFM can be used in three different imaging formats, namely **contact, tapping** and **non-contact** mode, describing the way in which the probe interacts with the surface during imaging. For the imaging of biological samples, tapping mode is more commonly used. In this mode the probe intermittently comes into and out of contact with (*i.e.* lightly taps) the sample as is it scanned over the surface, avoiding the strong lateral forces associated with contact mode, which can denature or sweep away soft or poorly immobilized samples.

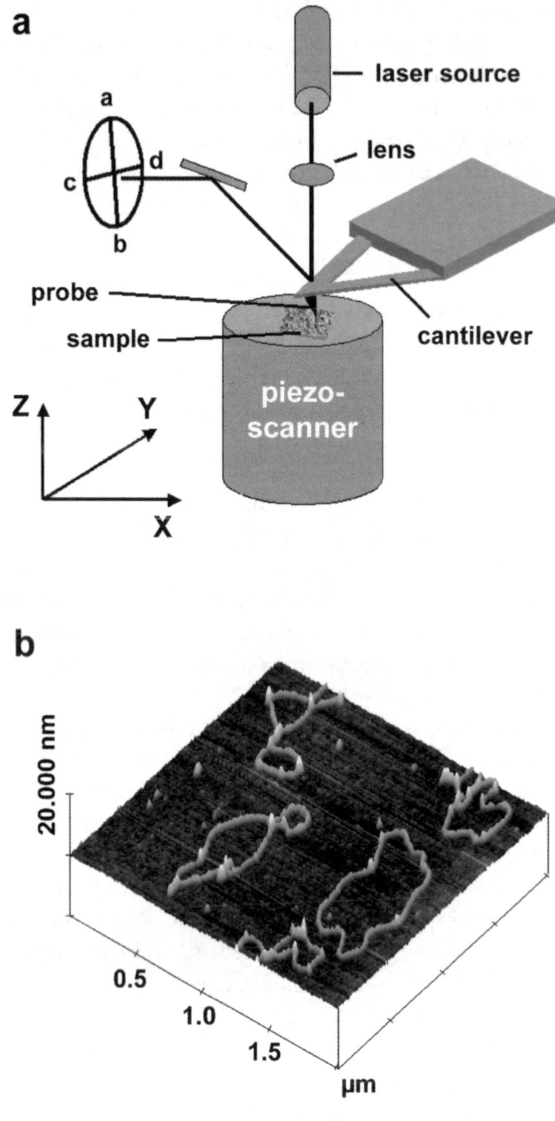


**Figure 11.15** *(a) Schematic of the principle features of an AFM instrument. The sizes of the AFM probe and cantilever chip are exaggerated for the purposes of illustration. (b) A representative 2 µm × 2 µm AFM image (obtained in air) displaying several pBR322 DNA plasmids deposited onto a mica substrate*

One of the big advantages of AFM for the structural analysis of biological samples is that it can image non-conducting surfaces in both air and liquid environments. Unsurprisingly, soon after its invention the AFM was applied to the analysis of many biological systems, including membrane proteins, DNA and RNA, protein–nucleic acid complexes and even more complex structures such as chromosomes.

To date, many investigations of nucleic acids have been performed using AFM.[38,39] By recording a series of images over a certain period of time, AFM is also able to provide dynamic information on processes involving nucleic acids, as well as time-lapse AFM movies of processes such as DNA condensation and the processing and/or degradation of DNA by enzymes such as the endonuclease *Eco*KI. The level of attainable image speed in conventional AFM imaging (normally 30–60 s per image) has been a major problem, however, for the imaging of such dynamic events. To a certain extent, this problem has been overcome recently by using purpose-developed small cantilevers, which are able to image faster (<2 s per image). Smaller cantilevers have high-resonant frequencies, which allow for high-scanning frequencies, but also low-spring constants, which enable any potential sample damage to be further minimised.

There are two main factors that can influence the level of resolution attainable in AFM images of nucleic acids. One is the sample preparation method employed for the immobilization of DNA to the underlying surface, and the other is the sharpness of the AFM probe. To this end, two immobilization methods have proved to be particularly useful for the analysis of DNA in aqueous environments. One method utilizes an unmodified mica surface with a buffer containing divalent metal ions (such as $Ni^{2+}$, $Co^{2+}$, $Zn^{2+}$, $Mn^{2+}$ or $Mg^{2+}$) or mica presoaked/pretreated with solutions of such ions (in this case a more concentrated divalent metal ion solution is needed). The other approach uses mica or alternative flat substrates (*e.g.* silicon) onto which positive charges have been introduced, for example through aminosilanation. This latter type of surface, however, tends to be more problematic and is more suitable for experiments in which divalent metal ions cannot be used without perturbing the DNA structure and/or the biomolecular functions under investigation.

One of the notable drawbacks for AFM is the probe convolution/broadening effect, which manifests as a size difference between the observed dimensions of the image feature and the actual molecular dimensions (normally the former is bigger). This effect arises when the radius of curvature of the imaging probe (typically 10–20 nm) is greater than the size of the feature to be imaged. Consequently, the apparent width of the double-stranded DNA helix in AFM images is typically 12–15 nm, rather than the 2 nm predicted from the crystal structure. Recent attempts to improve the sharpness of the probe (and thus the image resolution) have utilised single-walled carbon nanotubes attached to the apexes of AFM probes. Such probes have decreased the apparent DNA width to around 6–8 nm. Commercially available sharpened silicon oxide AFM probes (radius of curvature *ca.* 2 nm) are also available and are helpful in improving the level of attainable image resolution.

In addition to its imaging function, AFM can also be used to measure forces acting between the probe and surface. An approach termed **single molecule force spectroscopy** has been developed consequently to investigate the mechanical properties of single biopolymeric molecules, including DNA.[40] Here for example the opposite ends of a linear double-stranded DNA molecule are tethered between an opposing AFM probe and substrate, and then the molecule extended/stretched as the probe-substrate separation is increased. Pulling on single linear DNA molecules (with random sequence) produces force-extension curves (Figure 11.16). The plateau region at around 65 pN corresponds to the overstretching transition of DNA in which DNA is stretched from its B-DNA state to an overstretched state (around 1.7 times its B-DNA contour length). Upon further extension, a second transition occurs, during which the DNA can be melted into two single strands. Researchers have investigated the influence of different base sequences on this 'mechanical fingerprint', and the approach applied to the investigation of more complex molecular architectures including DNA/RNA hairpins. Recently, the effects of DNA-binding drugs on the mechanical properties of DNA have also been investigated (Figure 11.16). The observed changes were found to be dependent on the concentration of the applied drug and its mode of binding.

Although AFM is still a relatively new approach, it overcomes many of the limitations of its STM predecessor. While some improvements are still required before it will be utilised as a routine tool for the investigation of nucleic acid structure, its advantages over other microscopy approaches have meant that it is quickly becoming more accepted. The combination of AFM with other rapidly developing experimental approaches, such as single molecule fluorescence, also holds considerable promise for the future towards deepening our understanding of the basic processes in which nucleic acids are involved.

## 11.6 MASS SPECTROMETRY

The determination of intrinsic molecular masses of nucleic acids using mass spectrometry is widely accepted as one of the most accurate methods to detect nucleic acids.[41] The key technologies are the second-generation 'soft ionization' methods: **matrix-assisted laser desorption/ionization time-of-flight** (MALDI-TOF) and

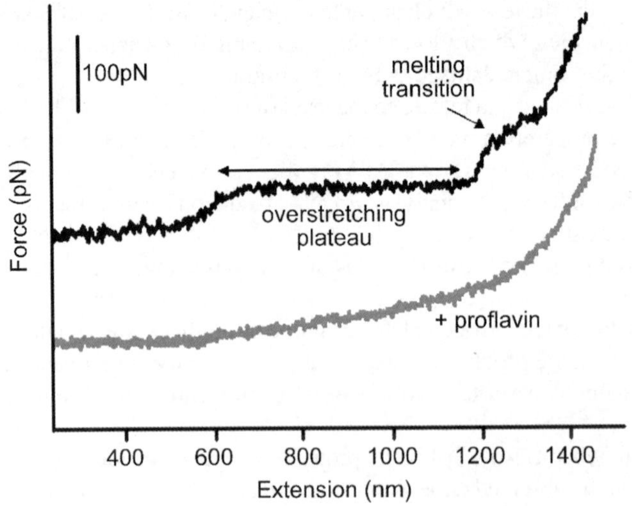


**Figure 11.16** *Representative force-extension curves of linear double-stranded DNA fragments recorded in 10 mM Tris buffer containing 1 mM EDTA and 150 mM NaCl (black) and the same buffer containing the intercalating drug proflavine (ca. 3 μg ml$^{-1}$) (grey)*

**electrospray ionization** (ESI) mass spectrometry (MS). Some of the current clinical applications of MS are (a) DNA sequencing,[42] (b) genotyping[43] and detection of genetic variations,[44] microsatellites,[45] short tandem repeats,[46] small insertions/deletions and (c) gene expression.

### 11.6.1 Matrix-Assisted Laser Desorption/Ionization Time-of-Flight Mass Spectrometry

Matrix-assisted laser desorption MS was introduced for proteins by Karas and Hillenkamp in 1988.[47] The nucleic acid sample is embedded in a crystalline matrix of a small molecule (*e.g.* 3-hydroxypicolinic acid). This target is excited by a pulse from an ultraviolet laser beam that, in high vacuum, results in intact molecules of the sample becoming desorbed into the gas phase and ionized by the UV radiation to give (mostly) singly charged ions. The resulting molecular ions are analysed by TOF MS, which sorts them by mass and thus removes any need for prior separation of mixtures. A resolution of 1 per 1000 and the detection of low femtomole quantities of DNA can be achieved routinely.[48] Oligonucleotides ranging from 2 to 2000 nucleotides are detected readily.[49] A typical MS measurement including acquisition and spectral interpretation takes <10 s. This is the preferred method for analysis of oligonucleotides up to mass 15,000 Da, *i.e.* approximately 50-mers. For example the MALDI-TOF spectrum of a crude oligonucleotide synthesised by solid-phase DNA synthesis (Section 4.1) is shown in Figure 11.17.

### 11.6.2 Electrospray Ionization Mass Spectrometry

In electrospray ionization MS, samples are ionized directly from micro-drops of solution and, as a result of their multiple charge, high mass nucleic acid molecules or protein-nucleic acid complexes can be observed at relatively low mass to charge (*m/z*) ratios. Several developments, especially time-of-flight analysers, nanoflow sampling and capillary gel electrophoresis interfacing, have greatly enhanced the performance of ESI MS. The advent of Fourier transform ion-cyclotron resonance (FT-ICR MS) has provided mass resolution in the range $10^4$–$10^6$. While ESI MS is slower than MALDI-TOF and is more sensitive to salt in the sample, it is the method of choice for studying species of mass up to several million Daltons. For example, the negative mode ESI mass spectrum of the oligodeoxyribonucleotide d(GCG TTC CCC CTT TGC G)

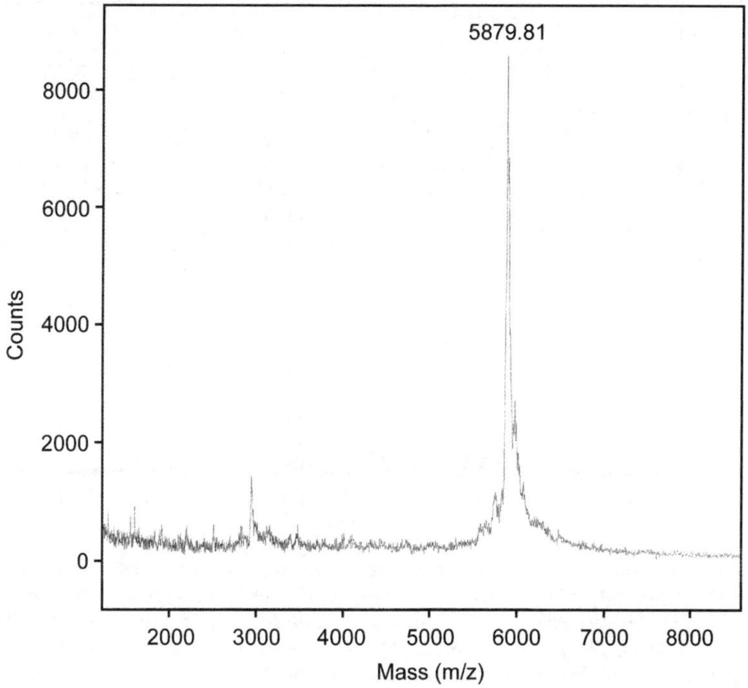


**Figure 11.17** *Matrix-assisted laser desorption/ionization time-of-flightmass spectrum of the crude products of a chemical synthesis of 19-mer oligodeoxynucleotide d(GGATTACAGGTATGAGCCA) showing a major component at 5879.8 (calculated 5876.8). This routine spectrum was obtained on an Applied Biosystems Voyager DE workstation (linear, negative ion mode) with a standard 3-hydroxypicolinic acid/ammonium acetate matrix. No internal mass calibration was used in this case, which would be required for more accurate mass measurement*

(Figure 11.18) appears as multiply charged peaks. The observed masses correspond to the oligonucleotide in which some of the phosphates are protonated. In some cases the replacement of a proton by sodium or potassium ions can be observed. The actual mass can be calculated from the accurate mass of any of the multiply charged species. For example $[M\text{-}4H]^{4-}$ (1198) has an accurate mass of 1198.4396, and since the observed masses correspond to *m/z*, the moleculer mass of the oligonucleotide is $4 \times 1198.4396 = 4793.7584 + 4H(4 \times 1.008) = 4797.7904$. The theoretically calculated mass is 4797.8091.

Electrospray ionization has emerged as a tool for studying non-covalent complexes and has enabled the measurement of spectra of intact viruses and whole ribosomes (Figure 11.19).[50,51] It is possible to measure directly the mass of a macromolecular complex as long as the binding interactions that support complex formation are maintained during its transit from solution micro-droplets into the gas phase. Nanoflow ESI[52] is a miniaturized version of ESI that facilitates this desolvation process because of the reduced size of droplets it generates compared with conventional ESI.

## 11.7 MOLECULAR MODELLING AND DYNAMICS

Despite the remarkable developments in experimental techniques to determine the molecular structure of nucleic acids, there remain many instances in which such information is not available. Certain nucleic acid structures (*e.g.* the triple helix) have proved remarkably difficult to crystallize, while the homogenous and repetitive nature of other structures makes NMR structure determination difficult or ambiguous. Molecular modelling can provide a method to predict the structure of such nucleic acids and also to study transient

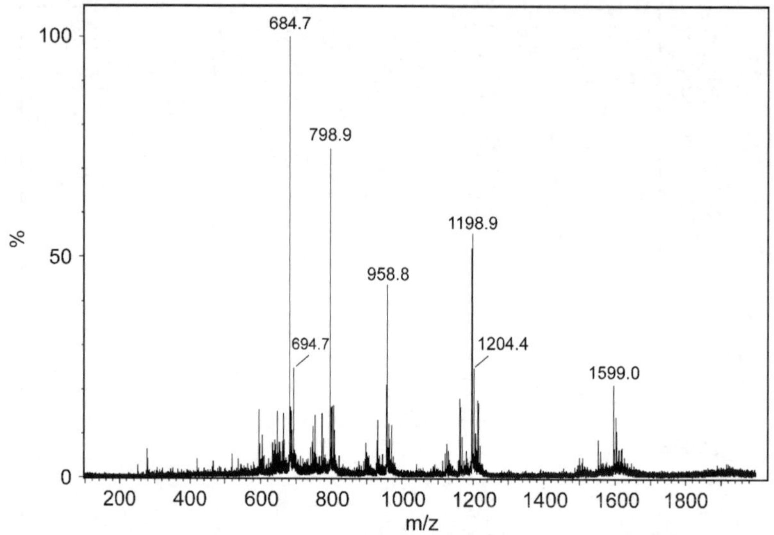


**Figure 11.18** *Electrospray ionization mass spectrum (negative mode) of oligodeoxyribonucleotide d(GCGTTCCCC-CTTTGCG) dissolved in methanol containing 0.5% diethylamine. The determined molecular mass is 4797.8091. The spectrum shows masses of the multiply charged species $[M\text{-}3H]^{3-}$ (1599), $[M\text{-}3H\text{-}Na]^{4-}$ (1204.4), $[M\text{-}4H]^{4-}$ (1198.9), $[M\text{-}5H]^{5-}$ (958.8), $[M\text{-}6H]^{6-}$ (798.9) and $[M\text{-}7H]^{7-}$ (684.7)* (Figure kindly provided by J. Brazier and R. Cosstick)

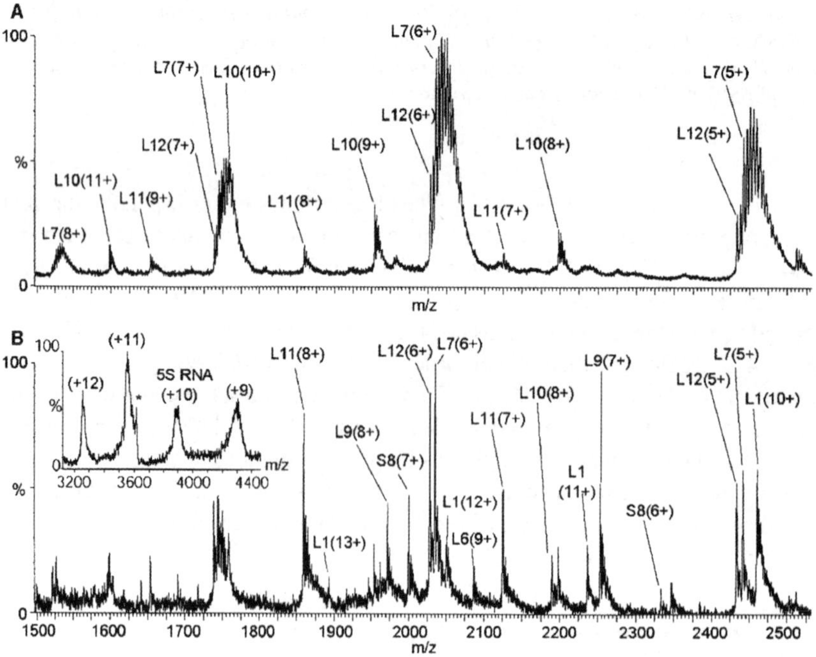


**Figure 11.19** *Mass spectra of E. coli ribosomes recorded from solution at pH 7.0 (a) and pH 4.5 (b). Peaks are labelled with the protein bound, and the numbers of positive charges are given in parenthesis. At pH 7.0, four proteins located in the mobile stalk region of the ribosome dominate the spectra. At pH 4.5, eight proteins are identified. Inset, MS/MS spectrum of the peak labelled *. Acceleration in the collision cell yields a mass consistent with that calculated for the 5S RNA. Individual charge states of the L7/L12 proteins reveal their metal-binding properties and post-translational modifications* (Reprinted from Ref. 50. © (2003), with permission from ASBMB)

or unstable structures. Molecular modelling also provides a method to relate the structural, dynamic and thermodynamic properties of nucleic acids to underlying theories of the physical processes that drive structure, bonding and recognition.

### 11.7.1 Molecular Mechanics and Energy Minimisation

Crick and Watson's model of the DNA duplex was constructed of wired-up physical shapes, but modern modelling is carried out almost exclusively through computational techniques. In theory, the most accurate predictions of molecular structure can come from quantum mechanical calculations, which treat the electronic structure of molecules in detail and from first principles. But despite the continuing rapid increase in computer power, the size of nucleic structures that are most often of interest precludes their day-to-day use. Instead, the most commonly used method is **molecular mechanics**.

Individual atoms are considered as spheres, and the bonds that connect them as springs. A set of equations, called a **force field**, determine how the energy of the system varies as bonds stretch or compress, bond angles vary or torsions rotate. Other terms in the equations describe the energetic consequences of nonbonded (van der Waals and electrostatic) interactions between atoms. A computer can then calculate the energy of the system as a function of the atomic coordinates. This gives rise to the idea of an **energy surface** (Figure 11.20a) and the simplest application of molecular mechanics, which is **energy minimisation**.

The 'true' structure of the molecule corresponds to a situation in which there is no net force acting on any atom, *i.e.* to a point on this surface with zero slope. Starting from an arbitrary initial guess for this true structure, the process of energy minimisation involves calculation of the forces and moving 'downhill' on this surface until a minimum is reached. However, the figure illustrates the important point that for most rather complex biomolecular structures, the energy surface has many minima. The real molecule is expected to adopt the global minimum energy conformation, but energy minimisation may only lead to a local minimum and thus predict an incorrect structure for the molecule. Because of the complexity of real energy surfaces, the process of searching for the true global energy minimum conformation can be very difficult and time-consuming, since there is no way of being absolutely sure that the global minimum has been found until the entire energy surface is examined.

### 11.7.2 Molecular Dynamics

One of the most widely used methods to try and overcome the problem of obtaining the global energy minimum of a structure is **molecular dynamics**. In this computer simulation method, masses and velocities, representative of a particular temperature, are assigned to each atom in some starting conformation. Using the solution of Newton's equations of motion, the conformation of the system at some later time-point is predicted. The forces acting on each atom, and thus the velocities, are then recalculated at this new conformation,

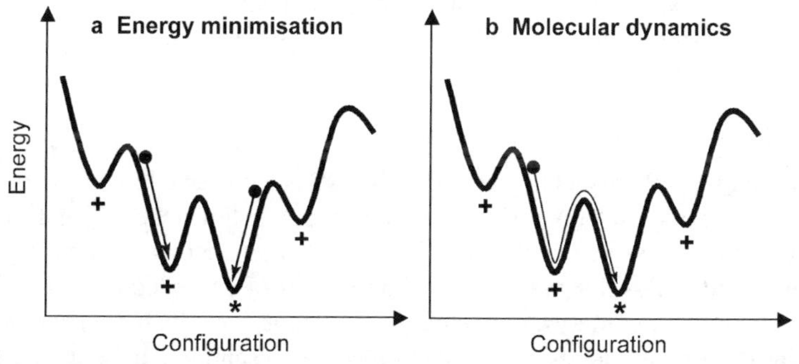


**Figure 11.20** *Each configuration of a molecule has an associated energy. In energy minimisation (a), different initial guesses (•) for the true structure may be optimised to local (+) rather than the global (*) energy minimum. Using molecular dynamics (b) the global energy minimum may be more reliably identified*

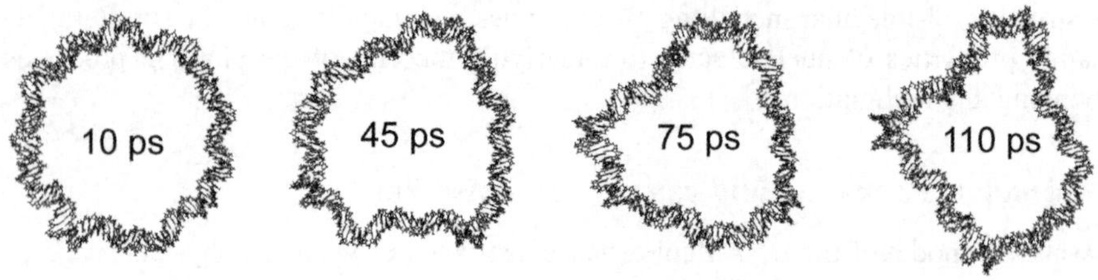


**Figure 11.21** *Snapshots taken from a short (120 ps) molecular dynamics simulation of a 126 base-pair DNA micro-circle, illustrating its conformational flexibility*

the motion over the next time step determined and so on. This process allows 'uphill' motion over the energy surface, and barriers between minima may be overcome, depending on their height relative to the simulation temperature (Figure 11.20b).

As well as a method for helping to locate the global energy minimum, molecular dynamics simulations can be regarded as a better representation of the physical state of the system – as a dynamic molecule that, at normal temperatures, does not have a single 'frozen' conformation. Rather the method samples a range of configurations, not just oscillating about the global minimum, but perhaps also significantly occupying other local energy minimum conformations. The flexibility of DNA can be considerable (Figure 11.21), and this is an important aspect of its function and recognition properties that often cannot be fully appreciated from crystallographic and (to a lesser extent) NMR data. The main limitation of molecular dynamics simulations is the timescale that is accessible. In order for the calculations predicting the trajectory of the molecule to be accurate, the time-step between which the forces acting on the system are recalculated must be very small – of the order of a femtosecond. Each calculation is computationally expensive and so current computer power typically limits such simulations to the order of a few tens of nanoseconds at best, which may take months to run. On this timescale, only quite modest motions of DNA can be observed directly: oscillations of small groups of atoms and localised bending and twisting motions. Many more biologically interesting conformational motions of DNA, for example winding and unwinding of the double helix and base-pair opening, only happen on the microsecond timescale or more slowly.

The basic output from a molecular dynamics simulation is a trajectory, *i.e.* the predicted behaviour of a single molecule as a function of time. However, the individual 'snapshots' from a molecular dynamics simulation can alternatively be regarded as a Boltzmann-weighted ensemble of structures from which thermodynamic quantities may be calculated. Modelling methods based on this approach are particularly valuable in that they provide a link between the microscopic behaviour of individual molecules – easy to simulate but difficult to observe experimentally – and the macroscopic properties of the system, which are much easier to measure, but can be difficult to interpret in terms of atomic-scale features. Molecular dynamics simulations of DNA have become increasingly sophisticated over the last few years[53] and are applied to a rapidly increasing volume of DNA structures and sequences.[54]

### 11.7.3 Mesoscopic Modelling

As indicated above, atomic-scale modelling, where each atom is represented by a sphere and each bond by a spring, does have some serious limitations. Current computer power limits the size of the molecules that may be studied this way to the order of a few thousand atoms (*e.g.* up to perhaps 200 base pairs) and, if one is interested in dynamics, to timescales of the order of a few tens of nanoseconds. To probe other sizes and timescales, modellers must adopt **mesoscopic modelling** methods, in which, for example a length of DNA is treated as a uniform elastic rod, or as a set of disks (each representing a base or base pair) connected to its neighbours by springs. These approaches are particularly used at present to study DNA supercoiling and packaging.

## REFERENCES


1. V.A. Bloomfield, D. Crothers and I. Tinoco, *Physical Chemistry of the Nucleic Acids*. Harper and Row, New York, 1974.
2. L. Brand and M.L. Johnson (eds), *Meth. Enzymol.*, 1997, **278**, 1–628.
3. D.M.J. Lilley, in *Nucleic Acids and Molecular Biology*, vol 4. D.M.J. Lilley and F. Eckstein (eds). Springer, Berlin, 1990, 55–77.
4. J.C. Penedo, T.J. Wilson, S.D. Jayasena, A. Khvorova and D.M.J. Lilley, Folding of the natural hammerhead ribozyme is enhanced by interaction of auxiliary elements. *RNA*, 2004, **10**, 880–888.
5. M. Eriksson and B. Nordén, Linear and circular dichroism of drug-nucleic acid complexes. *Meth. Enzymol.*, 2001, **340**, 68–99.
6. B. Nordén and T. Kurucsev, Analysing DNA complexes by circular and linear dichroism. *J. Mol. Recog.*, 1994, **7**, 141–156.
7. E. Taillandier and J. Liquier, Infrared spectroscopy of DNA. *Meth. Enzymol.*, 1992, **211**, 307–335.
8. G.J. Thomas and A.H.-J. Wang, in *Nucleic Acids and Molecular Biology*, vol 2. D.M.J. Lilley and F. Eckstein (eds), Springer, Berlin, 1988, 1–30.
9. H. Gunther, *NMR Spectroscopy – Basic Principles, Concepts, and Applications in Chemistry*. Wiley, Chichester, 1995.
10. T.L. James (ed), *Meth. Enzymol.*, Academic Press, San Deigo, 1995, **261**, 3–664.
11. K. Wuthrich, *NMR of Proteins and Nucleic Acids*. Wiley, New York, 1986.
12. L. Zidek, R. Stefl and V. Sklenar, NMR methodology for the study of nucleic acids. *Curr. Opin. Struct. Biol.*, **11**, 275–281.
13. E.T. Mollova and A. Pardi, NMR solution structure determination of RNAs. *Curr. Opin. Struct. Biol.*, 2000, **10**, 298–302.
14. D.E. Gilbert and J. Feigon, Multistrand DNA structures. *Curr. Opin. Struct. Biol.*, 1999, **9**, 305–314.
15. I. Kim, P.J. Lukavsky and J.D. Puglisi, NMR study of 100 kDa HCV IRES RNA using segmental isotope labeling. *J. Am. Chem. Soc.*, 2002, **124**, 9338–9339.
16. H.M. Berman, W.K. Olson, D.L. Beveridge, J. Westbrook, A. Gelbin, T. Demeny, S.-H. Hsieh, A.R. Srinivasan and B. Schneider, The nucleic acid database: a comprehensive relational database of three-dimensional structures of nucleic acids. *Biophys. J.*, 1992, **63**, 751–759.
17. H.M. Berman, J. Westbrook, Z. Feng, G. Gilliland, T.N. Bhat, H. Weissig, I.N. Shindyalov and B.P. E., The protein data bank. *Nucleic Acids Res.*, 2000, **28**, 235–242.
18. W.A. Hendrickson, Synchrotron crystallography. *Trends Biochem. Sci.*, 2000, **25**, 637–643.
19. C.W. Carter Jr. and R.M. Sweet (eds), *Meth. Enzymol. Macromolecular Crystallography, Part A*. Academic Press, New York, 1997.
20. C.W. Carter Jr. and R.M. Sweet (eds), *Meth. Enzymol. Macromolecular Crystallography, Part B*. Academic Press, New York, 1997.
21. C.W. Carter Jr. (ed), *Meth. Enzymol. Macromolecular Crystallography, Part C*. Academic Press, New York, 2003.
22. C.W. Carter Jr. (ed), *Meth. Enzymol. Macromolecular Crystallography, Part D*. Academic Press, New York, 2003.
23. R.E. Dickerson, D.S. Goodsell and S. Neidle, "… The tyranny of the lattice …" *Proc. Natl. Acad. Sci. USA*, 1994, **91**, 3579–3583.
24. V. Tereshko, S.T. Wallace, N. Usman, F. Wincott and M. Egli, X-ray crystallographic observation of "in-line" and "adjacent" conformations in a bulged self-cleaving RNA/DNA hybrid. *RNA*, 2001, **7**, 405–420.
25. P. Hensley, Defining the structure and stability of macromolecular assemblies in solution: the re-emergence of analytical ultracentrifugation as a practical tool. *Structure*, 1996, **4**, 367–373.
26. S.E. Harding and B.Z. Chowdhry (eds), *Protein Ligand Interactions: Hydrodynamics and Calorimetry*. Oxford University Press, Oxford, 2001.

27. V.A. Bloomfield, D.M. Crothers and J.I. Tinoco (eds), *Nucleic Acids Structures, Properties, and Functions*. University Science Books, Sausalito, CA, 2000.
28. I. Haq, C.B.Z., J.T.C. and C.J.B., Parsing the free energy of drug-DNA Interactions. *Meth. Enzymol.*, 2000, **323**, 373–405.
29. I. Haq, C.B.Z. and J.T.C., The interaction of drugs with higher order DNA structures. *Meth. Enzymol.*, 2001, **340**, 109–149.
30. I. Jelesarov and H.R. Bosshard, Isothermal titration calorimetry and differential scanning calorimetry as complementary tools to investigate the energetics of biomolecular recognition. *J. Mol. Recog.*, 1999, **12**, 3–18.
31. A. Cooper, Thermodynamic analysis of biomolecular interactions. *Curr. Opin. Chem. Biol.*, 1999, **3**, 557–563.
32. T. Wiseman, S. Williston, J.F. Brandts and L.-N. Lin, Rapid measurement of binding constants and heats of binding using a new titration calorimeter. *Anal. Biochem.*, 1989, **179**, 131–137.
33. J.M. Sturtevant, Biochemical applications of differential scanning calorimetry. *Ann. Rev. Phys. Chem.*, 1987, **38**, 463–488.
34. H.W. Fisher and R.C. Williams, Electron microscopic visualization of nucleic acids and of their complexes with proteins. *Ann. Rev. Biochem.*, 1979, **48**, 649–679.
35. R. Thresher and J. Griffith, Electron microscopic visualization of DNA and DNA-protein complexes as adjunct to biochemical studies. *Meth. Enzymol.*, 1992, **211**, 481–489.
36. J. Frank, Single particle imaging of macromolecules by cryo-electron microscopy. *Ann. Rev. Biophys. Biomol. Struct.*, 2002, **31**, 303–319.
37. P.G. Arscott and V.A. Blomfield, Scanning tunneling microscopy of nucleic acids. *Meth. Enzymol.*, 1992, **211**, 490–506.
38. H.G. Hansma, Surface biology of DNA by atomic force microscopy. *Ann. Rev. Phys. Chem.*, 2001, **52**, 71–92.
39. H.G. Abdelhady, S. Allen, M.C. Davies, C.J. Roberts, S.J.B. Tendler and W. P.M., Direct real-time molecular scale visualisation of the degradation of condensed DNA complexes exposed to DNase I. *Nucleic Acids Res.*, 2003, **31**, 4001–4005.
40. R. Krautbauer, L.H. Pope, T.E. Schrader, S. Allen and H.E. Gaub, Discriminating drug-DNA binding modes by single molecule force spectroscopy. *FEBS Lett.*, 2002, **510**, 154–158.
41. P.F. Crain and J.A. McCloskey, Applications of mass spectrometry to the characterization of oligonucleotides and nucleic acids. *Curr. Opin. Biotechnol.*, 1998, **9**, 25–34.
42. D.J. Fu, K. Tang, A. Braun, D. Reuter, B. Darnhofer-Demar, D.P. Little, M.J. O'Donnell, C.R. Cantor and H. Köster, Sequencing exons 5 to 8 of the p53 gene by MALDI-TOF mass spectrometry, *Nat. Biotech.*, 1998, **16**, 381–384.
43. P. Ross, L. Hall, I. Smirnov and L. Haff, High level multiplex genotyping by MALDI-TOF mass spectrometry. *Nat. Biotech.*, 1998, **16**, 1347–1351.
44. A. Braun, D.P. Little and H. Köster, Detecting CFTR gene mutations by using primer oligo base extension and mass spectrometry. *Clin. Chem.*, 1997, **43**, 1151–1158.
45. A. Braun, D.P. Little, D. Reuter, B. Muller-Mysok and H. Köster, Improved analysis of microsatellites using mass spectrometry. *Genomics*, 1997, **46**, 18–23.
46. P. Ross and P. Belgrader, Analysis of short tandem repeat polymorphisms in human DNA by matrix-assisted laser desorption/ionization mass spectrometry. *Anal. Chem.*, 1997, **69**, 3966–3972.
47. M. Karas and F. Hillenkamp, Laser desorption ionization of proteins with molecular masses exceeding 10000 Daltons. *Anal. Chem.*, 1988, **60**, 2299–2302.
48. D.J. Little, T.J. Cornish, M.J. O'Donnell, A. Braun, R.J. Cotter and H. Köster, MALDI on a chip: analysis of arrays of low-femtomole to subfemtomole quantities of synthetic oligonucleotides and DNA diagnostic products dispensed by a piezoelectric pipet. *Anal. Chem.*, 1997, **69**, 4540–4546.

49. S. Berkenkamp, F. Kirpekar and F. Hillenkamp, Infrared MALDI mass spectrometry of large nucleic acids. *Science*, 1998, **281**, 260–262.
50. C.L. Hanson, P. Fucini, L.L. Ilag, K.H. Nierhaus and C.V. Robinson, Evidence for conformational change in a ribosome elongation factor G complex. *J. Biol. Chem.*, 2003, **278**, 1259–1267.
51. C.L. Hanson and C.V. Robinson, Protein-nucleic acid interactions and the expanding role of mass spectrometry. *J. Biol. Chem.*, 2004, **279**, 24907–24910.
52. M. Wilm and M. Mann, Analytical properties of the nanoelectrospray ion source. *Anal. Chem.*, 1996, **68**, 1–8.
53. T.E. Cheatham and P.A. Kollman, Molecular dynamics simulations of nucleic acids. *Ann. Rev. Phys. Chem.*, 2000, **52**, 435–471.
54. D.L. Beveridge, G.K. Barreiro, S. Byun, D.A. Case, T.E. Cheatham, S.B. Dixit, E. Giudice, F. Lankas, R. Lavery, J.H. Maddocks *et al.*, Molecular dynamics simulations of the 136 unique tetranucleotide sequences of DNA oligonucleotides.1. Research design and results on d(C(p)G) steps. *Biophys. J.*, 2004, **87,** 3799–3813.

# Subject Index